The Condensed Protocols

From *Molecular Cloning: A Laboratory Manual*

Also from Cold Spring Harbor Laboratory Press

Related Laboratory Manuals

Molecular Cloning: A Laboratory Manual, Third Edition
PCR Primer: A Laboratory Manual, Second Edition
Protein–Protein Interactions: A Laboratory Manual, Second Edition
Proteins and Proteomics: A Laboratory Manual
Purifying Proteins: A Laboratory Manual
Using Antibodies: A Laboratory Manual

Other Related Titles

At the Bench: A Laboratory Navigator
Bioinformatics: Sequence and Genome Analysis, Second Edition
Lab Math: A Handbook of Measurements, Calculations, and Other Quantitative Skills for Use at the Bench
Lab Ref: A Handbook of Recipes, Reagents, and Other Reference Tools for Use at the Bench

The Condensed Protocols

From *Molecular Cloning: A Laboratory Manual*

Joseph Sambrook

Peter MacCallum Cancer Institute
East Melbourne, Australia

David W. Russell

University of Texas Southwestern Medical Center
Dallas, Texas

COLD SPRING HARBOR LABORATORY PRESS
Cold Spring Harbor, New York

The Condensed Protocols

From *Molecular Cloning: A Laboratory Manual*

Publisher	John Inglis
Acquisition Editors	John Inglis and Alexander Gann
Development Director	Jan Argentine
Project Manager	Judy Cuddihy
Developmental Editors	Judy Cuddihy and Kaaren Janssen
Project Coordinator	Maryliz Dickerson
Permissions Coordinator	Maria Fairchild
Production Editor	Rena Steuer
Desktop Editor	Susan Schaefer
Production Manager	Denise Weiss
Cover Design	Ed Atkeson (Berg Design) and Denise Weiss
Original and Redrawn Illustrations	Mark D. Curtis

Cover: The gene encoding green fluorescent protein was cloned from *Aequorea victoria*, a jellyfish found in abundance in Puget Sound, Washington State. This picture of a 50-mm medusa was taken on color film by flash photography and shows light reflected off various morphological features of the animal. The small, bright, roundish blobs in the photograph are symbiotic amphipods living on or in the medusa. The bright ragged area in the center is the jellyfish's mouth. Bioluminescence from *Aequorea* is emitted only around the margins of the medusae and cannot be seen in this image. As in most species of jellyfish, bioluminescence of *Aequorea* does not look like a soft overall glow, but occurs only at the rim of the bell and, given the right viewing conditions, appears as a string of nearly microscopic, fusiform green lights. The primary luminescence produced by *Aequorea* is actually bluish in color and is emitted by the protein aequorin. In a living jellyfish, light is emitted via the coupled green fluorescent protein, which causes the luminescence to appear green. The image and legend were kindly provided by Claudia Mills of the University of Washington, Friday Harbor. For further information, please see Mills C.E. 1999–2000. Bioluminescence of *Aequorea*, a hydromedusa. Electronic Internet document available at http://faculty. washington.edu/cemills/Aequorea.html. Published by the author; Web page established June 1999, but last updated 19 December 2005.

Library of Congress Cataloging-in-Publication Data

Sambrook, Joseph.
  The condensed protocols from molecular cloning : a laboratory manual /
Joseph Sambrook, David W. Russell.
    p. cm.
  Includes index.
  ISBN 0-87969-772-5 (hardcover : alk. paper) -- ISBN 0-87969-771-7
(pbk. : alk. paper)
    1. Molecular cloning--Laboratory manuals.   I. Russell, David W.
(David William), 1957- . II. Title.
    [DNLM: 1. Cloning, Molecular--Laboratory Manuals.   QH 440.5
S187c 2006]
QH442.2.S24 2006
572.8--dc22

                                            2005022077

10 9 8 7 6 5 4 3 2 1

Contents

Using Plasmid Vectors in Molecular Cloning

BACKGROUND INFORMATION

Background information found in *Molecular Cloning: A Laboratory Manual*, 3rd edition (hereafter MC3) unless otherwise indicated.

Choosing an appropriate strain of *E.coli*	MC3, pp. 1.14–1.16
Genetic markers and other properties of *E. coli* strains	MC3, pp. A3.6–A3.10
Properties of individual plasmids	MC3, pp. A3.2–A3.3
Types of plasmid vectors	MC3, pp. 1.11–1.14

Preparation of Plasmid DNA by Alkaline Lysis with SDS: Minipreparation

Plasmid DNA is isolated from small-scale (1–2 ml) bacterial cultures by treatment with alkali and SDS.

MATERIALS

CAUTION: Please see Appendix 4 for appropriate handling of materials marked with <!>.

Reagents and Solutions

Please see Appendix 1 for components of stock solutions, buffers, and reagents. Dilute stock solutions to the appropriate concentrations.
Alkaline lysis solutions I, II, and III
    Alkaline lysis solution II should be freshly prepared and used at room temperature.
Ethanol
Ethanol (70%)
Phenol:chloroform (1:1, v/v) <!> (optional; see Step 8)
TE (pH 8.0) containing 20 µg/ml RNase A

Vectors and Hosts

Colonies of transformed *E. coli* carrying plasmid of interest

Media and Antibiotics

LB, YT, or Terrific Broth containing appropriate antibiotic for plasmid selection

Centrifuges/Rotors/Tubes

Microfuge (4°C)

Additional Items

Vacuum aspirator (see Figure 1-1)

Additional Information

How alkaline lysis works	MC3, p. 1.31
Troubleshooting problems with alkaline lysis	MC3, p. 1.42

METHOD

1. Inoculate 2 ml of LB, YT, or Terrific Broth containing the appropriate antibiotic with a single colony of transformed bacteria. Incubate the culture overnight at 37°C with vigorous shaking.

2. Pour 1.5 ml of the culture into a microfuge tube. Centrifuge at maximum speed for 30 seconds at 4°C in a microfuge. Store the unused portion of the original culture at 4°C.

3. Remove the supernatant medium from the microfuge tube by aspiration, leaving the bacterial pellet as dry as possible (Figure 1-1).

4. Resuspend the bacterial pellet by vigorous vortexing in 100 µl of ice-cold alkaline lysis solution I.

5. Add 200 µl of freshly prepared alkaline lysis solution II to the bacterial suspension. Close the tube tightly and mix the contents by inverting the tube rapidly five times. *Do not vortex!* Store the tube on ice.

Adapted from Chapter 1, Protocol 1, p. 1.32 of MC3.

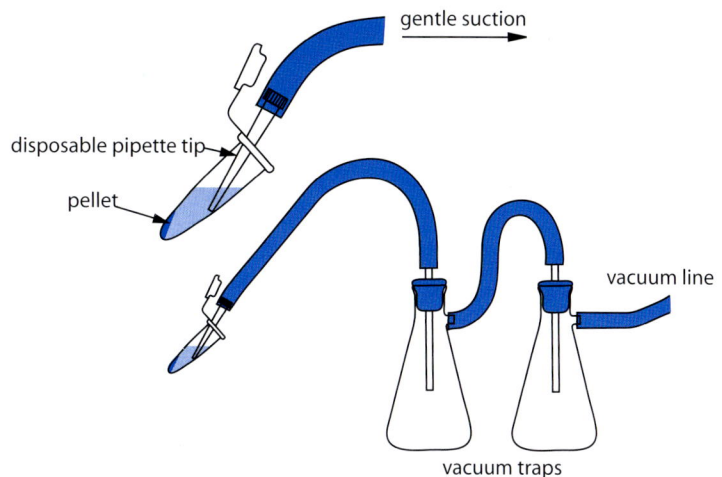

FIGURE 1-1 Aspiration of Supernatants

Hold the open microfuge tube at an angle, with the pellet on the upper side. Use a disposable pipette tip attached to a vacuum line to withdraw fluid from the tube. Insert the tip just beneath the meniscus on the lower side of the tube. Move the tip toward the base of the tube as the fluid is withdrawn. Use gentle suction to avoid drawing the pellet into the pipette tip. Keep the end of the tip away from the pellet. Finally, vacuum the walls of the tube to remove any adherent drops of fluid.

> Make sure that the entire surface of the tube comes into contact with alkaline lysis solution I.

6. Add 150 μl of ice-cold alkaline lysis solution III. Close the tube and disperse alkaline lysis solution III through the viscous bacterial lysate by inverting the tube several times. Store the tube on ice for 3–5 minutes.

7. Centrifuge the bacterial lysate at maximum speed for 5 minutes at 4°C in a microfuge. Transfer the supernatant to a fresh tube.

8. (*Optional*) Add an equal volume of phenol:chloroform. Mix the organic and aqueous phases by vortexing and then centrifuge the emulsion at maximum speed for 2 minutes at 4°C in a microfuge. Transfer the aqueous upper layer to a fresh tube.

9. Precipitate nucleic acids from the aqueous solution by adding 2 volumes of ethanol at room temperature. Mix the solution by vortexing and then allow the mixture to stand for 2 minutes at room temperature.

10. Collect the precipitated nucleic acids by centrifugation at maximum speed for 5 minutes at 4°C in a microfuge.

11. Remove the supernatant by gentle aspiration as described in Figure 1-1 above. Stand the tube in an inverted position on a paper towel to allow all of the fluid to drain away. Use a Kimwipe or disposable pipette tip to remove any drops of fluid adhering to the walls of the tube.

12. Add 1 ml of 70% ethanol to the pellet and invert the closed tube several times. Recover the DNA by centrifugation at maximum speed for 2 minutes at 4°C in a microfuge.

13. Remove all of the supernatant by gentle aspiration as described in Figure 1-1.

> Take care with this step, as the pellet sometimes does not adhere tightly to the tube.

14. Remove any beads of ethanol that form on the sides of the tube. Store the open tube at room temperature until the ethanol has evaporated and no fluid is visible in the tube (5–10 minutes).

15. Dissolve the nucleic acids in 50 μl of TE (pH 8.0) containing 20 μg/ml DNase-free RNase A (pancreatic RNase). Vortex the solution gently for a few seconds. Store the DNA solution at –20°C.

Preparation of Plasmid DNA by Alkaline Lysis with SDS: Midipreparation

This protocol describes how to isolate plasmid DNA from intermediate-scale (20–50 ml) bacterial cultures by treatment with alkali and SDS.

MATERIALS

CAUTION: Please see Appendix 4 for appropriate handling of materials marked with <!>.

Reagents and Solutions

Please see Appendix 1 for components of stock solutions, buffers, and reagents. Dilute stock solutions to the appropriate concentrations.

Alkaline lysis solutions I, II, and III

> For preparations of plasmid DNA that are to be further purified by chromatography (see Protocol 1.9), supplement sterile alkaline lysis solution I just before use with DNase-free RNase to a concentration of 100 µg/ml.
>
> Alkaline lysis solution II should be freshly prepared and used at room temperature.

Ethanol

Ethanol (70%)

Phenol:chloroform (1:1, v/v) <!>

STE (optional; see note to Step 3)

TE (pH 8.0) containing 20 µg/ml DNase-free RNase A

Vectors and Hosts

Colony of transformed *E. coli* carrying plasmid of interest

Media and Antibiotics

LB, YT, or Terrific Broth containing appropriate antibiotic for plasmid selection

Enzymes and Buffers

RNase A (pancreatic RNase), DNase-free

Centrifuges/Rotors/Tubes

Microfuge (4°C and room temperature)
Sorvall SS-34 rotor and 15-ml tubes (4°C)

Additional Items

Vacuum aspirator (see Figure 1-1)

Additional Information

How alkaline lysis works	MC3, p. 1.31
Troubleshooting problems with alkaline lysis	MC3, p. 1.42

Adapted from Chapter 1, Protocol 2, p. 1.35 of MC3.

METHOD

1. Inoculate 10 ml of LB, YT, or Terrific Broth containing the appropriate antibiotic with a single colony of transformed bacteria. Incubate the culture overnight at 37°C with vigorous shaking.

2. Transfer the culture into a 15-ml tube and recover the bacteria by centrifugation at 2000*g* (4000 rpm in a Sorvall SS-34 rotor) for 10 minutes at 4°C.

3. Remove the medium by gentle aspiration, leaving the bacterial pellet as dry as possible.

   (*Optional*) Wash bacterial pellet in STE. This step generally improves the efficiency of midiprep DNA with restriction enzymes.

4. Resuspend the bacterial pellet in 200 μl of ice-cold alkaline lysis solution I by vigorous vortexing and transfer the suspension to a microfuge tube.

5. Add 400 μl of freshly prepared alkaline lysis solution II to each bacterial suspension. Close the tube tightly and mix the contents by inverting the tube rapidly five times. *Do not vortex!* Store the tube on ice.

   Make sure that the entire surface of the tube comes into contact with alkaline lysis solution II.

6. Add 300 μl of ice-cold alkaline lysis solution III. Close the tube and disperse alkaline lysis solution III through the viscous bacterial lysate by inverting the tube several times. Store the tube on ice for 3–5 minutes.

7. Centrifuge the bacterial lysate at maximum speed for 5 minutes at 4°C in a microfuge. Transfer 600 μl of the supernatant to a fresh tube.

8. Add an equal volume of phenol:chloroform. Mix the organic and aqueous phases by vortexing and then centrifuge the emulsion at maximum speed for 2 minutes at 4°C in a microfuge. Transfer the aqueous upper layer to a fresh tube.

9. Precipitate nucleic acids from the supernatant by adding 600 μl of isopropanol at room temperature. Mix the solution by vortexing and then allow the mixture to stand for 2 minutes at room temperature.

10. Collect the precipitated nucleic acids by centrifugation at maximum speed for 5 minutes at room temperature in a microfuge.

11. Remove the supernatant by gentle aspiration as described in Step 3 above. Stand the tube in an inverted position on a paper towel to allow all of the fluid to drain away. Remove any drops of fluid adhering to the walls of the tube.

12. Add 1 ml of 70% ethanol to the pellet and recover the DNA by centrifugation at maximum speed for 2 minutes at room temperature in a microfuge.

13. Remove all of the supernatant by gentle aspiration as described in Step 3.

14. Remove any beads of ethanol that form on the sides of the tube. Store the open tube at room temperature until the ethanol has evaporated and no fluid is visible in the tube (2–5 minutes).

15. Dissolve the nucleic acids in 100 μl of TE (pH 8.0) containing 20 μg/ml DNase-free RNase A (pancreatic RNase). Vortex the solution gently for a few seconds and store at –20°C.

Plasmid DNA is isolated from large-scale (500 ml) bacterial cultures by treatment with alkali and SDS.

MATERIALS

CAUTION: Please see Appendix 4 for appropriate handling of materials marked with <!>.

Reagents and Solutions

Please see Appendix 1 for components of stock solutions, buffers, and reagents. Dilute stock solutions to the appropriate concentrations.

Alkaline lysis solutions I, II, and III

> For preparations of plasmid DNA that are to be further purified by chromatography (see Protocol 1.9), supplement sterile alkaline lysis solution I just before use with 20 mg/ml DNase-free RNase to a concentration of 100 µg/ml.

Chloramphenicol (34 mg/ml) <!>

> Required only for low- and moderate-copy-number plasmids

Ethanol (70%)

Isopropanol

Phenol:chloroform (1:1, v/v) <!>

STE

TE (pH 8.0) containing 20 µg/ml DNase-free RNase A

Vectors and Hosts

Colony or late-log-phase culture of transformed *E. coli* carrying plasmid of interest

Media and Antibiotics

LB, YT, or Terrific Broth containing appropriate antibiotic for plasmid selection

Enzymes and Buffers

DNase-free RNase A (pancreatic RNase)

Lysozyme (10 mg/ml)

Restriction enzymes

Gels/Loading Buffers

1% Agarose gel

Gel-loading buffer IV

Centrifuges/Rotors/Tubes

Microfuges (4°C and room temperature)

Sorvall SLC-1500 rotor and centrifuge bottles (4°C)

Additional Items

Step 18 of this protocol requires the reagents listed in Protocols 1.8–1.11.

Steps 18 and 19 of this protocol require the reagents listed in Protocol 5.1.

Vacuum aspirator (see Figure 1-1)

Additional Information	
Chloramphenicol	MC3, p. 1.143–1.144
Lysozymes	MC3, p. 1.153
Typical yields	MC3, p. 1.50
Troubleshooting	MC3, p. 1.42

METHOD

1. Inoculate 30 ml of LB, YT, or Terrific Broth containing the appropriate antibiotic with either a single colony of transformed bacteria or 0.1–1.0 ml of a small-scale liquid culture grown from a single colony.

2. Incubate the culture at the appropriate temperature with vigorous shaking until the bacterial culture enters the late log phase of growth (OD_{600} = ~0.6).

3. Warm 500 ml of LB, YT, or Terrific Broth containing the appropriate antibiotic to 37°C in a 2-liter flask. Inoculate the warmed medium with 25 ml of the late-log-phase bacterial culture and incubate for 2.5 hours at 37°C with vigorous shaking (250 cycles/minute on a rotary shaker).

4. For plasmids with low or moderate copy numbers, add 2.5 ml of 34 mg/ml chloramphenicol solution. The final concentration of chloramphenicol in the culture should be 170 µg/ml.

    For high-copy-number plasmids, do not add chloramphenicol.

5. Incubate the culture for an additional 12–16 hours at 37°C with vigorous shaking (250 cycles/minute on a rotary shaker).

6. Remove an aliquot (1–2 ml) of the bacterial culture to a fresh microfuge tube and store it at 4°C. Harvest the remainder of the bacterial cells from the 500-ml culture by centrifugation at 2700*g* (4100 rpm in a Sorvall SLC-1500 rotor) for 15 minutes at 4°C. Discard the supernatant. Stand the open centrifuge bottle in an inverted position.

7. Resuspend the bacterial pellet in 200 ml of ice-cold STE. Collect the bacterial cells by centrifugation as described in Step 6. Store the pellet of bacteria in the centrifuge bottle at –20°C.

8. Use one of the methods described in Protocol 1.1 or 1.4 to prepare plasmid DNA from the 1–2-ml aliquot of bacterial culture set aside in Step 6. Analyze the minipreparation plasmid DNA by digestion with the appropriate restriction enzyme(s) and agarose gel electrophoresis to ensure that the correct plasmid has been propagated in the large-scale culture.

9. Allow the frozen bacterial cell pellet from Step 7 to thaw at room temperature for 5–10 minutes. Resuspend the pellet in 18 ml (10 ml) of alkaline lysis solution I.

10. Add 2 ml (1 ml) of a freshly prepared solution of 10 mg/ml lysozyme.

11. Add 40 ml (20 ml) of freshly prepared alkaline lysis solution II. Close the top of the centrifuge bottle and mix the contents thoroughly by gently inverting the bottle several times. Incubate the bottle for 5–10 minutes at room temperature.

12. Add 20 ml (15 ml) of ice-cold alkaline lysis solution III. Close the top of the centrifuge bottle and mix the contents gently but well by swirling the bottle several times (there should no longer be two distinguishable liquid phases). Place the bottle on ice for 10 minutes.

13. Centrifuge the bacterial lysate at ≥20,000*g* (11,000 rpm in a Sorvall SLC-1500 rotor) for 30 minutes at 4°C in a medium-speed centrifuge. Allow the rotor to stop without braking. At the end of the centrifugation step, decant the clear supernatant into a graduated cylinder. Discard the pellet remaining in the centrifuge bottle.

14. Measure the volume of the supernatant. Transfer the supernatant together with 0.6 volume of isopropanol to a fresh centrifuge bottle. Mix the contents well and store the bottle for 10 minutes at room temperature.

15. Recover the precipitated nucleic acids by centrifugation at 12,000*g* (8000 rpm in a Sorvall SLC-1500 rotor) for 15 minutes at room temperature.

16. Decant the supernatant carefully and invert the open bottle on a paper towel to allow the last drops of supernatant to drain away. Rinse the pellet and the walls of the bottle with 70% ethanol at room temperature. Drain off the ethanol, and use a pasteur pipette attached to a vacuum line to remove any beads of liquid that adhere to the walls of the bottle. Place the inverted, open bottle on a pad of paper towels for a few minutes at room temperature.

17. Dissolve the damp pellet of nucleic acid in 3 ml of TE (pH 8.0).

18. Purify the crude plasmid DNA either by column chromatography (Protocol 1.9), precipitation with polyethylene glycol (Protocol 1.8), or equilibrium centrifugation in CsCl–ethidium bromide gradients (Protocols 1.10 and 1.11).

19. Check the structure of the plasmid by restriction enzyme digestion followed by gel electrophoresis.

Small-scale Preparation of Plasmid DNA by Boiling

Plasmid DNA is isolated by treating small-scale (1–2 ml) bacterial cultures with Triton X-100 and lysozyme, followed by heating. This method is not recommended for preparing plasmid DNA from strains of *E. coli* that express endonuclease A (*endA+* strains).

MATERIALS

Reagents and Solutions

Please see Appendix 1 for components of stock solutions, buffers, and reagents. Dilute stock solutions to the appropriate concentrations.
Ethanol (70% at 4°C)
Isopropanol
Sodium acetate (3.0 M, pH 5.2)
STET
TE (pH 8.0) containing 20 µg/ml RNase A

Vectors and Hosts

Colonies of transformed *E. coli* carrying plasmid of interest

Media and Antibiotics

LB, YT, or Terrific Broth containing appropriate antibiotic for plasmid selection

Enzymes and Buffers

Lysozyme (10 mg/ml), freshly prepared
RNase A (pancreatic RNase), DNase-free

Centrifuges/Rotors/Tubes

Microfuge (4°C)

Additional Items

Boiling-water bath
Vacuum aspirator (see Figure 1-1)

Additional Information

How boiling lysis works	MC3, p. 1.43
Lysozymes	MC3, p. 1.153

METHOD

1. Inoculate 2 ml of LB, YT, or Terrific Broth containing the appropriate antibiotic with a single colony of transformed bacteria. Incubate the culture overnight at 37°C with vigorous shaking.

2. Pour 1.5 ml of the culture into a microfuge tube. Centrifuge the tube at maximum speed for 30 seconds at 4°C in a microfuge. Store the unused portion of the culture at 4°C.

3. Remove the medium by gentle aspiration, leaving the bacterial pellet as dry as possible.

4. Resuspend the bacterial pellet in 350 µl of STET.

5. Add 25 µl of a freshly prepared solution of lysozyme. Close the top of the tube and mix the contents by gently vortexing for 3 seconds.

6. Place the tube in a boiling-water bath for *exactly* 40 seconds.

7. Centrifuge the bacterial lysate at maximum speed for 15 minutes at room temperature in a microfuge. Pour the supernatant into a fresh microfuge tube.

8. Precipitate the nucleic acids from the supernatant by adding 40 µl of 3.0 M sodium acetate (pH 5.2) and 420 µl of isopropanol. Mix the solution by vortexing and then allow the mixture to stand for 5 minutes at room temperature.

9. Recover the precipitated nucleic acids by centrifugation at maximum speed for 10 minutes at 4°C in a microfuge.

10. Remove the supernatant by gentle aspiration as described in Step 3 above. Stand the tube in an inverted position on a paper towel to allow all of the fluid to drain away. Use a Kimwipe or disposable pipette tip to remove any drops of fluid adhering to the walls of the tube.

11. Rinse the pellet of nucleic acid with 1 ml of 70% ethanol at 4°C. Remove all of the supernatant by gentle aspiration as described in Step 3.

12. Remove any beads of ethanol that form on the sides of the tube. Store the open tube at room temperature until the ethanol has evaporated and no fluid is visible in the tube (2–5 minutes).

13. Dissolve the nucleic acids in 50 µl of TE (pH 8.0) containing DNase-free RNase A (pancreatic RNase). Vortex the solution gently for a brief period. Store the DNA at –20°C.

Large-scale Preparation of Plasmid DNA by Boiling

Plasmid DNA is isolated by treating large-scale (500 ml) bacterial cultures with Triton X-100 and lysozyme, followed by heating. This method is not recommended for preparing plasmid DNA from strains of *E. coli* that express endonuclease A (*endA*+ strains).

MATERIALS

CAUTION: Please see Appendix 4 for appropriate handling of materials marked with <!>.

Reagents and Solutions

Please see Appendix 1 for components of stock solutions, buffers, and reagents. Dilute stock solutions to the appropriate concentrations.

Chloramphenicol (34 mg/ml) <!>
Ethanol (70% at 4°C)
Isopropanol
Sodium acetate (3.0 M, pH 5.2)
STE, ice cold
STET, ice cold
TE (pH 8.0)

Vectors and Hosts

Colony or late-log-phase culture of transformed *E. coli* carrying plasmid of interest

Media and Antibiotics

LB, YT, or Terrific Broth containing appropriate antibiotic for plasmid selection

Enzymes and Buffers

Lysozyme (10 mg/ml), freshly prepared
Restriction enzymes

Gels/Loading Buffers

1% Agarose gel
Gel-loading buffer IV

Centrifuges/Rotors/Tubes

Beckman SW 41 Ti rotor and tubes
Microfuge (4°C)
Sorvall SLC-1500 rotor and centrifuge bottles (4°C)
Sorvall SS-34 rotor and tubes (room temperature)

Additional Items

Boiling-water bath
Bunsen burner
Clamp to hold 50-ml Erlenmeyer flask

Gauze, 4-ply (optional; see Step 15)
Vacuum aspirator

Additional Information

Chloramphenicol	MC3, pp. 1.143–1.144
Expected yields of DNA	MC3, p. 1.50

METHOD

1. Inoculate 30 ml of LB, YT, or Terrific Broth containing the appropriate antibiotic either with a single colony of transformed bacteria or with 0.1–1.0 ml of a small-scale liquid culture grown from a single colony.

2. Incubate the culture at the appropriate temperature with vigorous shaking (250 cycles/ minute in a rotary shaker) until the bacterial culture reaches the late log phase of growth (OD_{600} = ~0.6).

3. Warm 500 ml of LB, YT, or Terrific Broth containing the appropriate antibiotic to 37°C in a 2-liter flask. Inoculate the warmed medium with 25 ml of the late-log-phase bacterial culture and incubate for 2.5 hours at 37°C with vigorous shaking (250 cycles/minute on a rotary shaker).

4. Add 2.5 ml of 34 mg/ml chloramphenicol. The final concentration of chloramphenicol in the culture should be 170 µg/ml. Incubate the culture for a further 12–16 hours at 37°C with vigorous shaking (250 cycles/minute on a rotary shaker).

5. Remove an aliquot (1–2 ml) of the bacterial culture to a fresh microfuge tube and store at 4°C. Harvest the remainder of the bacterial cells from the 500-ml culture by centrifugation at 2700*g* (4100 rpm in a Sorvall SLC-1500 rotor) for 15 minutes at 4°C. Discard the supernatant. Stand the open centrifuge bottle in an inverted position to allow all of the supernatant to drain away.

6. Resuspend the bacterial pellet in 200 ml of ice-cold STE. Collect the bacterial cells by centrifugation as described in Step 5. Store the pellet of bacteria in the centrifuge bottle at –20°C.

7. Prepare plasmid DNA from the 1–2-ml aliquot of bacteria set aside in Step 5 by the minipreparation protocol (either Protocol 1.1 or 1.4). Analyze the minipreparation DNA by digestion with the appropriate restriction enzyme(s) to ensure that the correct plasmid has been propagated in the large-scale culture.

8. Allow the frozen bacterial cell pellet from Step 6 to thaw for 5–10 minutes at room temperature. Resuspend the pellet in 10 ml of ice-cold STET. Transfer the suspension to a 50-ml Erlenmeyer flask.

9. Add 1 ml of a freshly prepared solution of 10 mg/ml lysozyme.

10. Use a clamp to hold the Erlenmeyer flask over the open flame of a Bunsen burner until the liquid *just* starts to boil. Shake the flask constantly during the heating procedure.

11. Immediately immerse the bottom half of the flask in a boiling-water bath. Hold the flask in the boiling water for exactly 40 seconds.

12. Cool the flask in ice-cold water for 5 minutes.

13. Transfer the viscous contents of the flask to an ultracentrifuge tube (Beckman SW 41 or its equivalent). Centrifuge the lysate at 150,000*g* (30,000 rpm in a Beckman SW 41 Ti rotor) for 30 minutes at 4°C.

14. Transfer as much of the supernatant as possible to a new tube. Discard the viscous liquid remaining in the centrifuge tube.

15. (*Optional*) If the supernatant contains visible strings of genomic chromatin or flocculent precipitate of proteins, filter it through 4-ply gauze before proceeding.

16. Measure the volume of the supernatant. Transfer the supernatant, together with 0.6 volume of isopropanol, to a fresh centrifuge tube(s). Store the tube(s) for 10 minutes at room temperature, after mixing the contents well.

17. Recover the precipitated nucleic acids by centrifugation at 12,000*g* (10,000 rpm in a Sorvall SS-34 rotor) for 15 minutes at *room temperature.*

    Salt may precipitate if centrifugation is carried out at 4°C.

18. Decant the supernatant carefully and invert the open tube(s) on a paper towel to allow the last drops of supernatant to drain away. Rinse the pellet and the walls of the tube(s) with 70% ethanol at room temperature. Drain off the ethanol and use a pasteur pipette attached to a vacuum line to remove any beads of liquid that adhere to the walls of the tube(s). Place the inverted, open tube(s) on a pad of paper towels for a few minutes at room temperature. The pellet should still be damp.

19. Dissolve the pellet of nucleic acid in 3 ml of TE (pH 8.0).

20. Purify the crude plasmid DNA either by chromatography on commercial resins (please see Protocol 1.9), precipitation with polyethylene glycol (Protocol 1.8), or equilibrium centrifugation in CsCl–ethidium bromide gradients (Protocols 1.10 and 1.11).

21. Check the structure of the plasmid by restriction enzyme digestion followed by gel electrophoresis.

Protocol 1.6 Preparation of Plasmid DNA: Toothpick Minipreparation

Plasmid DNA is prepared directly from bacterial colonies plucked from the surface of agar media with toothpicks.

MATERIALS

CAUTION: Please see Appendix 4 for appropriate handling of materials marked with <!>.

Reagents and Solutions

Please see Appendix 1 for components of stock solutions, buffers, and reagents. Dilute stock solutions to the appropriate concentrations.

Bromophenol blue solution (0.4% w/v) *or* cresol red solution (10 mM)

Cresol red is required if DNAs are to be analyzed by PCR.

KCl (4 M)

NSS solution, freshly made

0.2 N NaOH

0.5% SDS

20% sucrose

SYBR Gold <!> or other sensitive strain for DNA

Vectors and Hosts

Culture of transformed bacteria

Media and Antibiotics

LB, YT, or SOB (agar plates and liquid medium) containing the appropriate antibiotic for plasmid selection

Enzymes and Buffers

Restriction enzymes and buffers

Gels/Loading Buffers

0.7% Agarose gel *without* ethidium bromide, cast and run in TAE buffer

Gel-loading buffer IV

Additional Items

Step 12 of this protocol requires the reagents listed in Protocol 8.1.

Step 14 of this protocol requires the reagents listed in Protocol 1.1 or 1.4.

Sterile toothpicks

Water bath (70°C)

Additional Information

Cresol red	MC3, p. 1.53, note to Step 12
Electrophoresis buffers	MC3, p. 5.8 and note to Step 10 on p. 1.53
Ethidium bromide	MC3, pp. 5.14, 5.16, and A9.3–A9.4

   Adapted from Chapter 1, Protocol 6, p. 1.51 of MC3.

METHOD

1. Grow bacterial colonies, transformed with recombinant plasmid, on rich agar medium (LB, YT, or SOB) containing the appropriate antibiotic until they are approximately 2–3 mm in diameter (~18–24 hours at 37°C for most bacterial strains).

2. Use a sterile toothpick or disposable loop to transfer a small segment of a bacterial colony to a streak or patch on a master agar plate containing the appropriate antibiotic. Transfer the remainder of the colony to a numbered microfuge tube containing 50 µl of sterile 10 mM EDTA (pH 8.0).

3. Repeat Step 2 until the desired number of colonies has been harvested.

4. Incubate the master plate for several hours at 37°C and then store it at 4°C until the results of the gel electrophoresis (Step 11 of this protocol) are available. Colonies containing plasmids of the desired size can then be recovered from the master plate.

5. While the master plate is incubating, process the bacterial suspensions as follows: To each microfuge tube in turn, add 50 µl of a freshly made solution of NSS. Close the top of the tubes and then mix their contents by vortexing for 30 seconds.

6. Transfer the tubes to a 70°C water bath. Incubate the tubes for 5 minutes and then allow them to cool to room temperature.

7. To each tube, add 1.5 µl of a solution of 4 M KCl. Vortex the tubes for 30 seconds.

8. Incubate the tubes for 5 minutes on ice.

9. Remove bacterial debris by centrifugation at maximum speed for 3 minutes at 4°C in a microfuge.

10. Transfer each of the supernatants in turn to fresh microfuge tubes. Add to each tube 0.5 µl of a solution containing 0.4% bromophenol blue if the samples are to be analyzed only by agarose gel electrophoresis *or* 2 µl of 10 mM cresol red if the samples are to be analyzed both by PCR and by agarose gel electrophoresis. Load 50 µl of the supernatant into a slot (5 mm in length x 2.5 mm in width) cast in 0.7% agarose gel (5 mm thick).

11. After the bromophenol blue dye has migrated two-thirds to three-fourths the length of the gel, or the cresol red dye about one-half the length of the gel, stain the gel by soaking it in a DNA staining solution for 30–45 minutes at room temperature. Examine and photograph the gel under UV illumination.

12. If cresol red has been used at Step 10, analyze the supernatants by performing PCR as described in Protocol 8.1, using the remainder of each sample as a template.

13. Prepare small-scale cultures of the putative recombinant clones by inoculating 2 ml of liquid medium (LB, YT, or SOB) containing the appropriate antibiotic with bacteria growing on the master plate.

14. Use the small-scale bacterial cultures to generate minipreparations (please see Protocol 1.1 or 1.4) of the putative recombinant plasmids. Analyze the plasmid DNAs by digestion with restriction enzymes and agarose gel electrophoresis to confirm that they have the desired size and structure.

Large (>15 kb), closed circular plasmids are prepared (albeit inefficiently and in small yield) by lysing bacteria with SDS.

MATERIALS

CAUTION: Please see Appendix 4 for appropriate handling of materials marked with <!>.

Reagents and Solutions

Please see Appendix 1 for components of stock solutions, buffers, and reagents. Dilute stock solutions to the appropriate concentrations.

Chloramphenicol (34 mg/ml) <!>
  Required only for low- to moderate-copy-number plasmids
Chloroform <!>
EDTA (0.25 M, pH 8.0)
Ethanol
Ethanol (70%)
NaCl (5 M)
Phenol:chloroform (1:1, v/v) <!>
SDS (10% w/v)
STE, ice cold
TE (pH 8.0)
Tris-sucrose

Vectors and Hosts

Colony or late-log-phase culture of transformed *E. coli* carrying plasmid of interest

Media and Antibiotics

LB, YT, or Terrific Broth containing the appropriate antibiotic for plasmid selection

Enzymes and Buffers

Lysozyme (10 mg/ml)
Restriction endonucleases and buffers

Gels/Loading Buffers

0.7% Agarose gel containing ethidium bromide <!>

Centrifuges/Rotors/Tubes

Beckman Type 50 rotor (4°C) and 30-ml screw-capped Nalgene centrifuge tubes
Sorvall SLC-1500 or HS4 swing-out rotor and centrifuge bottles (4°C) *or* Sorvall HS4 rotor and tubes

Additional Items

Step 8 of this protocol requires the reagents listed in Protocols 1.1 and 1.4.
Step 21 of this protocol requires the reagents listed in either Protocol 1.9 or 1.10.

    Adapted from Chapter 1, Protocol 7, p. 1.55 of MC3.

Sturdy glass rod
Vacuum aspirator (see Figure 1-1)

Additional Information

Chloramphenicol   MC3, p. 1.143–1.144

METHOD

1. Inoculate 30 ml of LB, YT, or Terrific Broth containing the appropriate antibiotic with a single transformed bacterial colony or with 0.1–1.0 ml of a late-log-phase culture grown from a single transformed colony.

2. Incubate the culture with vigorous shaking until the bacteria enters the late log phase of growth (OD_{600} = ~0.6).

3. Warm 500 ml of LB, YT, or Terrific Broth containing the appropriate antibiotic to 37°C in a 2-liter flask. Inoculate the warmed medium with 25 ml of the late-log-phase bacterial culture and incubate for 2.5 hours at 37°C with vigorous shaking (250 cycles/minute on a rotary shaker).

4. For plasmids with low or moderate copy numbers, add 2.5 ml of 34 mg/ml chloramphenicol. The final concentration of chloramphenicol in the culture should be 170 µg/ml.

   For high-copy-number plasmids, do not add chloramphenicol.

5. Incubate the culture for an additional 12–16 hours at 37°C with vigorous shaking (250 cycles/minute on a rotary shaker)

6. Remove an aliquot (1–2 ml) of the bacterial culture to a fresh microfuge tube and store it at 4°C. Harvest the remainder of the bacterial cells from the 500-ml culture by centrifugation at 2700*g* (4100 rpm in a Sorvall SLC-1500 rotor) for 15 minutes at 4°C. Discard the supernatant. Stand the open centrifuge bottle in an inverted position.

7. Resuspend the bacterial pellet in 200 ml of ice-cold STE. Collect the bacterial cells by centrifugation as described in Step 6. Store the pellet of bacteria in the centrifuge bottle at –20°C.

8. Use one of the methods described in Protocol 1.1 or 1.4 to prepare plasmid DNA from the 1–2-ml aliquot of bacterial culture set aside in Step 6. Analyze the minipreparation plasmid DNA by digestion with the appropriate restriction enzyme(s) and agarose gel electrophoresis to ensure that the correct plasmid has been propagated in the large-scale culture.

9. Allow the frozen bacterial cell pellet from Step 7 to thaw for 5–10 minutes at room temperature. Resuspend the pellet in 10 ml of ice-cold Tris-sucrose solution. Transfer the suspension to a 30-ml plastic screw-cap tube.

10. Add 2 ml of a freshly prepared lysozyme solution (10 mg/ml) followed by 8 ml of 0.25 M EDTA (pH 8.0).

11. Mix the suspension by gently inverting the tube several times. Store the tube on ice for 10 minutes.

12. Add 4 ml of 10% SDS. Immediately mix the contents of the tube with a glass rod so as to disperse the solution of SDS evenly throughout the bacterial suspension. Be as gentle as possible to minimize shearing of the liberated chromosomal DNA.

13. As soon as mixing is completed, add 6 ml of 5 M NaCl (final concentration = 1 M). Use a glass rod to mix the contents of the tube gently but thoroughly. Place the tube on ice for at least 1 hour.

14. Remove high-molecular-weight DNA and bacterial debris by centrifugation at 71,000g (30,000 rpm in a Beckman Type 50 rotor) for 30 minutes at 4°C. Carefully transfer the supernatant to a 50-ml disposable plastic centrifuge tube. Discard the pellet.

15. Extract the supernatant once with phenol:chloroform and once with chloroform.

16. Transfer the aqueous phase to a 250-ml centrifuge bottle. Add 2 volumes (~60 ml) of ethanol at room temperature. Mix the solution well. Store the solution for 1–2 hours at room temperature.

17. Recover the nucleic acids by centrifugation at 5000g (5500 rpm in a Sorvall SLC-1500 rotor or 5100 rpm in a Sorvall HS4 swing-out rotor) for 20 minutes at 4°C.

18. Discard the supernatant. Wash the pellet and sides of the centrifuge tube with 70% ethanol at room temperature and then centrifuge as in Step 17.

19. Discard as much of the ethanol as possible and then invert the centrifuge bottle on a pad of paper towels to allow the last of the ethanol to drain away. Use a vacuum aspirator to remove droplets of ethanol from the walls of the centrifuge bottle. Stand the bottle in an inverted position until no trace of ethanol is visible. At this stage, the pellet should still be damp.

20. Dissolve the damp pellet of nucleic acid in 3 ml of TE (pH 8.0).

21. Purify the crude plasmid DNA either by chromatography on commercial resins (see Protocol 1.9) or by isopycnic centrifugation in CsCl–ethidium bromide gradients (see Protocols 1.10 and 1.11).

22. Check the structure of the plasmid by restriction enzyme digestion followed by gel electrophoresis.

Purification of Plasmid DNA by Precipitation with Polyethylene Glycol

Crude preparations of plasmid DNA are first treated with lithium chloride and RNase (to remove RNA). The plasmid DNA is then precipitated in a solution containing polyethylene glycol and $MgCl_2$.

MATERIALS

CAUTION: Please see Appendix 4 for appropriate handling of materials marked with <!>.

Reagents and Solutions

Please see Appendix 1 for components of stock solutions, buffers, and reagents. Dilute stock solutions to the appropriate concentrations.

> Chloroform <!>
> Ethanol
> Isopropanol
> LiCl (5 M)
> PEG-$MgCl_2$ solution <!>
> Phenol:chloroform (1:1, v/v) <!>
> Sodium acetate (3 M, pH 5.2)
> TE (pH 8.0)
> TE (pH 8.0) containing 20 µg/ml RNase A

Nucleic Acids/Oligonucleotides

> Crude plasmid preparation
> > Use material from either Protocol 1.3, Step 17 or Protocol 1.5, Step 19.

Centrifuges/Rotors/Tubes

> Corex glass tubes (15 and 30 ml)
> Microfuge (room temperature)
> Sorvall SS-34 rotor (4°C)

Additional Items

> Ice-water bath
> Vacuum aspirator

Additional Information

Absorption spectroscopy	MC3, pp. A8.20–A8.21
Lithium chloride	MC3, p. A8.12

METHOD

1. Transfer 3 ml of the crude large-scale plasmid preparation to a 15-ml Corex tube and chill the solution to 0°C in an ice bath.

2. Add 3 ml of an ice-cold solution of 5 M LiCl to the crude plasmid preparation, mix well, and centrifuge the solution at 12,000g (10,000 rpm in a Sorvall SS-34 rotor) for 10 minutes at 4°C.

3. Transfer the supernatant to a fresh 30-ml Corex tube. Add an equal volume of isopropanol. Mix well. Recover the precipitated nucleic acids by centrifugation at 12,000g (10,000 rpm in a Sorvall SS-34 rotor) for 10 minutes at room temperature.

4. Decant the supernatant carefully and invert the open tube to allow the last drops of supernatant to drain away. Rinse the pellet and the walls of the tube with 70% ethanol at room temperature. Carefully discard the bulk of the ethanol and then use a vacuum aspirator to remove any beads of liquid that adhere to the walls of the tube. Place the inverted, open tube on a pad of paper towels for a few minutes. The pellet should still be damp.

5. Dissolve the damp pellet of nucleic acid in 500 μl of TE (pH 8.0) containing 20 μg/ml RNase A. Transfer the solution to a microfuge tube and store it for 30 minutes at room temperature.

6. Extract the plasmid-RNase mixture once with phenol:chloroform and once with chloroform.

7. Recover the DNA by standard ethanol precipitation.

8. Dissolve the pellet of plasmid DNA in 1 ml of sterile H_2O and then add 0.5 ml of PEG-$MgCl_2$ solution.

9. Store the solution for ≥10 minutes at room temperature and then collect the precipitated plasmid DNA by centrifugation at maximum speed for 20 minutes at room temperature in a microfuge.

10. Remove traces of PEG by resuspending the pellet of nucleic acid in 0.5 ml of 70% ethanol. Collect the nucleic acid by centrifugation at maximum speed for 5 minutes in a microfuge.

11. Remove the ethanol by aspiration and repeat Step 10. Following the second rinse, store the open tube on the bench for 10–20 minutes to allow the ethanol to evaporate.

12. Dissolve the damp pellet in 500 μl of TE (pH 8.0). Measure the OD_{260} of a 1:100 dilution in TE (pH 8.0) of the solution and calculate the concentration of the plasmid DNA, assuming that 1 OD_{260} = 50 μg of plasmid DNA/ml.

13. Store the DNA in aliquots at –20°C.

Purification of Plasmid DNA by Chromatography

The following table summarizes the salient features of many of the commercial resins that are currently available for plasmid purification. Individual manufacturers supply detailed instructions, which should be followed to the letter.

TABLE 1-1 Commercially Available Resins and Their Uses

Resin	Manufacturer	Chemistry	Use	Notes
Qiagen	Qiagen www.qiagen.com	macroporous silica gel, anion-exchange (DEAE)	transfection of eukaryotic cells	some batch-to-batch variation; pH sensitive
QIAprep	Qiagen www.qiagen.com	silica gel	enzymatic manipulation	different columns available for purification of double- or single-stranded DNAs
Wizard	Promega www.promega.com	silica particle	additional ethanol precipitation required for transfection of eukaryotic cells	inexpensive, reproducible
FlexiPrep	GE Healthcare www.amershambio sciences.com	anion exchange	enzymatic manipulation	transfection requires further purification
GlassMAX	Invitrogen www.invitrogen.com	silica matrix	enzymatic manipulation	minipreps only
Perfectprep	Eppendorf 5 Prime www.eppendorf.com	silica matrix	transfection of eukaryotic cells	very fast; miniprep only
Strataprep Plasmid Miniprep Kit	Stratagene www.stratagene.com	silica resin, hydrophobic interaction	enzymatic manipulation	can be used for miniprep plasmid or purification of DNA fragments
Concert, rapid, and high-purity systems	Invitrogen www.invitrogen.com	silica gel	enzymatic manipulation and transfection of eukaryotic cells	mini- and maxipreps
RapidPURE Plasmid Mini Kit	Q-BIOgene www.qbiogene.com	silica-based (GENECLEAN) embedded membrane	enzymatic manipulation	minipreps
GENECLEAN II	Q-BIOgene www.qbiogene.com	silica-based (GENECLEAN) embedded membrane	enzymatic manipulation	for DNA fragments from 200 bp to 20 kb

Additional information may be found in MC3, pp. 1.62–1.64.

Purification of Closed Circular DNA by Equilibrium Centrifugation in CsCl–Ethidium Bromide Gradients: Continuous Gradients

Solutions containing plasmid DNA are adjusted to a density of 1.55 g/ml with solid CsCl. The intercalating dye, ethidium bromide, which binds differentially to closed circular and linear DNAs, is then added to a concentration of 200 µg/ml. During centrifugation to equilibrium, superhelical closed circular plasmid DNA and nonsuperhelical DNAs form bands at different buoyant densities.

MATERIALS

CAUTION: Please see Appendix 4 for appropriate handling of materials marked with <!>.

Reagents and Solutions

Please see Appendix 1 for components of stock solutions, buffers, and reagents. Dilute stock solutions to the appropriate concentrations.

CsCl rebanding solution (density = 1.55) (optional; see Steps 4 and 8)
    If rebanding solution is required, weigh 48 g of CsCl and add TE (pH 8.0) until the combined weight of the CsCl and TE is exactly 100 g.
CsCl (solid)
Ethanol
Ethidium bromide (10 mg/ml in H_2O) <!>
Light paraffin oil

Nucleic Acids/Oligonucleotides

Crude preparation of plasmid DNA
    Use material from Protocol 1.3, Step 16; Protocol 1.5, Step 18; or Protocol 1.7, Step 19.

Centrifuges/Rotors/Tubes

Corex glass tubes (15 ml) (preferred) or polypropylene tubes
Beckman vertical rotor (preferred) or angled rotor, and tubes
Sorvall SS-34 rotor

Additional Items

Disposable syringes (5 cc)
Hypodermic needles (21 and 18 gauge)
Refractometer
    Although not essential, a refractometer is extremely useful for estimating the density of CsCl solutions. For relationships among density, refractive index, and concentration of solutions of CsCl, see MC3, p. 1.156, Table 1-11.
Scotch tape or Time tape
Step 8 of this protocol requires the reagents listed in Protocol 1.12 or 1.13.
Top-loading balance

> **Additional Information**
>
> Cesium chloride–ethidium bromide density gradients
> > Radloff R., Bauer W., and Vinograd J. 1967. A dye-buoyant density method for the detection and isolation of closed circular duplex DNA: The closed circular DNA in HeLa cells. *Proc. Natl. Acad. Sci.* **57**: 1514–1521.
> Ethidium bromide   MC3, pp. 1.150–1.151, 5.14–5.15, and A9.3–A9.4

METHOD

1. Measure the mass of the crude plasmid DNA preparation. Measurement is best done by transferring the solution into a fresh tube that has been tared on a top-loading balance. Then, for every gram of plasmid DNA solution, add exactly 1.01 g of solid CsCl. Close the top of the tube to prevent evaporation and warm the solution to 30°C to facilitate the dissolution of the CsCl salt. Mix the solution gently until the salt is dissolved. The density (p) of the solution should be approximately 1.55 g/ml.

2. Add 100 µl of 10 mg/ml ethidium bromide for each 5 g of original DNA solution.

3. If Corex glass tubes are used, centrifuge the solution at 7700g (8000 rpm in a Sorvall SS-34 rotor) for 5 minutes at room temperature. If disposable polypropylene tubes are used, centrifuge at 1100g (3000 rpm in a Sorvall SS-34 rotor) for 10 minutes.

4. Use a pasteur pipette or a disposable syringe fitted with a large-gauge needle to transfer the clear, red solution under the scum and above the pellet to a tube suitable for centrifugation in an ultracentrifuge rotor. Top off the partially filled centrifuge tubes with light paraffin oil or a solution of CsCl in H_2O ($p = 1.55$) containing ethidium bromide at a concentration of 200 µg/ml. Make sure that the weights of tubes opposite each other in the rotor are equal. Seal the tubes according to manufacturer instructions.

5. Centrifuge the density gradients at 20°C as appropriate for the rotor:

Beckman NVT 65 rotor	366,000g (62,000 rpm)	for 6 hours
Beckman VTi 65 rotor	194,000g (45,000 rpm)	for 16 hours
Beckman Type 50 Ti rotor	180,000g (45,000 rpm)	for 48 hours
Beckman Type 65 Ti rotor	314,000g (60,000 rpm)	for 24 hours
Beckman Type 70.1 Ti rotor	331,000g (60,000 rpm)	for 24 hours

6. Gently remove the rotor from the centrifuge and place it on a flat surface. Carefully remove each tube and place it in a test tube rack covered with aluminum foil. In a dimly lit room (i.e., with the overhead fluorescent lights turned off), mount one tube in a clamp attached to a ring stand.

7. Collect the band of closed circular DNA (Figure 1-2).

   a. Use a 21-gauge hypodermic needle to make a small hole in the top of the tube to allow air to enter when fluid is withdrawn.

   b. Carefully wipe the outside of the tube with ethanol to remove any grease or oil and then attach a piece of Scotch tape or Time tape to the outside of the tube.

   c. Insert an 18-gauge hypodermic needle (beveled side up) into the tube through the tape so that the open, beveled side of the needle is positioned just below the lower DNA band (closed circular plasmid DNA).

   d. Collect the superhelical plasmid DNA in a glass centrifuge tube, taking care not to disturb the upper viscous band of chromosomal DNA.

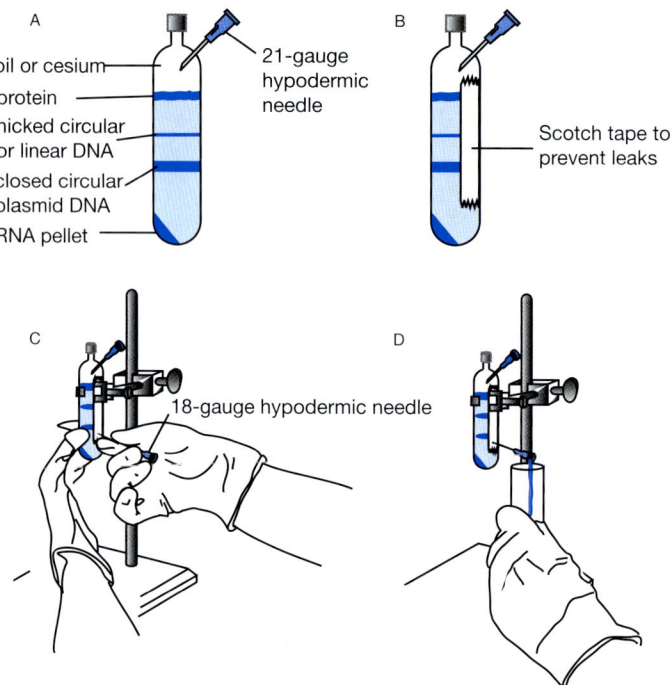

FIGURE 1-2 Collection of Superhelical Plasmid DNA from CsCl Gradients Containing Ethidium Bromide

Please see Step 7 for details.

8. Remove ethidium bromide from the DNA as described in one of the methods presented in Protocols 1.12 and 1.13.

Purification of Closed Circular DNA by Equilibrium Centrifugation in CsCl–Ethidium Bromide Gradients: Discontinuous Gradients

A solution containing plasmid DNA, saturating amounts of ethidium bromide, and CsCl (44% w/v) is layered between solutions of lesser (35% w/v CsCl) and greater (59% w/v CsCl) densities. During centrifugation to equilibrium, the superhelical closed circular plasmid DNA and nonsuperhelical DNAs form bands at different densities.

MATERIALS

CAUTION: Please see Appendix 4 for appropriate handling of materials marked with <!>.

Reagents and Solutions

Please see Appendix 1 for components of stock solutions, buffers, and reagents. Dilute stock solutions to the appropriate concentrations.

CsCl (solid)
CsCl solutions for three-step gradients
    See Table 1-2 on the next page.
Ethanol
Ethidium bromide (10 mg/ml in H_2O) <!>
TE (pH 8.0)

Nucleic Acids/Oligonucleotides

Crude preparation of plasmid DNA
    For each step gradient, use the crude plasmid DNA prepared from 50 ml of an overnight bacterial culture, as described in Protocol 1.3, Step 16; Protocol 1.5, Step 18; or Protocol 1.7, Step 19.

Centrifuges/Rotors/Tubes

Corex glass tubes (15 ml)
Beckman 5-ml polyallomer Quick-Seal tubes
Beckman Type 70.1 Ti rotor

Additional Items

Bone-marrow needles (18 gauge, 10 cm)
Hypodermic needles (18 and 21 gauge)
Hypodermic syringes (1, 3, and 5 cc)
Step 6 of this protocol requires the reagents listed in either Protocol 1.12 or 1.13.

Additional Information

Dorin M. and Bornecque C.A. 1995. Fast separations of plasmid DNA during discontinuous gradients in the preparative ultracentrifuge. *BioTechniques* **18:** 90–91.
Relationships among density, refractive index,
    and concentration of solutions of CsCl   MC3, p. 1.156, Table 1-11

TABLE 1-2 Three-step Discontinuous Gradient Layers

Layer	Molarity CsCl (w/w)	Refractive Index	Preparation
Top	2.806 (35%)	1.3670	Dissolve 4.720 g of CsCl in 8 ml of TE (pH 8.0). Adjust the volume to exactly 10 ml. Then add 120 µl of 10 mg/ml ethidium bromide.
Middle	3.870 (44%)	1.3792	Dissolve 0.8 g of CsCl in exactly 1 ml of the crude preparation of plasmid DNA. Then add 30 µl of 10 mg/ml ethidium bromide.
Bottom	6.180 (59%)	1.4052	Dissolve 10.4 g of CsCl in 7 ml of TE. Adjust the volume to exactly 10 ml. Then add 120 µl of 10 mg/ml ethidium bromide.

METHOD

1. Prepare CsCl layers for a three-step discontinuous gradient as described in Table 1-2. Use a 3-cc hypodermic syringe equipped with an 18-gauge bone-marrow (10 cm) needle to transfer 1.5 ml of the top layer (35%) CsCl solution to a 5-ml polyallomer ultracentrifuge tube (Beckman Quick-Seal or equivalent).

2. Use a 1-cc tuberculin syringe equipped with an 18-gauge bone-marrow (10 cm) needle to layer 0.5 ml of the middle layer (44%) CsCl solution, containing the plasmid DNA, into the bottom of the tube *under* the top layer solution.

   As a rule of thumb, the crude plasmid DNA prepared from no more than 50 ml of an overnight culture should be used per gradient. The crude plasmid preparation from a 100-ml culture should be reconstituted in approximately 0.9 ml of TE (pH 8.0), which is enough to form the middle layer of two discontinuous gradients.

3. Use a 5-cc hypodermic syringe equipped with an 18-gauge bone-marrow (10 cm) needle to fill the tube by layering the bottom layer (59%) CsCl solution *under* the middle layer CsCl solution.

4. Centrifuge the sealed tubes at 330,000*g* (60,000 rpm in a Beckman Type 70.1 Ti rotor) for 5 hours. Make sure that the weights of tubes opposite one another in the rotor are equal. Seal the tubes according to the manufacturer's instructions.

5. Collect the band of closed circular DNA (see Figure 1-2 in Protocol 1.10).

   a. Use a 21-gauge hypodermic needle to make a small hole in the top of the tube to allow air to enter when fluid is withdrawn.

   b. Carefully wipe the outside of the tube with ethanol to remove any grease or oil and then attach a piece of Scotch tape or Time tape to the outside of the tube.

   c. Insert a sterile 18-gauge hypodermic needle (beveled side up) into the tube through the tape so that the open, beveled side of the needle is positioned just below the lower DNA band (closed circular plasmid DNA).

   d. Collect the superhelical plasmid DNA in a glass centrifuge tube, taking care not to disturb the upper viscous band of chromosomal DNA.

6. Remove ethidium bromide from the DNA as described in one of the methods presented in Protocol 1.12 or 1.13.

Removal of Ethidium Bromide from DNA by Phase Extraction with Organic Solvents

This protocol describes how ethidium bromide is removed from DNA by phase extraction with organic solvents.

MATERIALS

CAUTION: Please see Appendix 4 for appropriate handling of materials marked with <!>.

Reagents and Solutions

Please see Appendix 1 for components of stock solutions, buffers, and reagents. Dilute stock solutions to the appropriate concentrations.

n-Butanol, saturated with H_2O <!>
Ethanol (optional; see Step 6)
Ethanol (70%) (optional; see Step 6)
Isoamyl alcohol, saturated with H_2O <!>
Phenol, equilibrated to pH 7.8 <!>
    May be needed at Step 13
Phenol:chloroform (1:1, v/v) <!>
    May be needed at Step 13
TE (pH 8.0) (optional; see Step 6)

Nucleic Acids/Oligonucleotides

Plasmid DNA, purified by CsCl–ethidium bromide centrifugation
    Use material from either Protocol 1.10, Step 7 or Protocol 1.11, Step 5.

Centrifuges/Rotors/Tubes

Sorvall H-400 rotor and 50-ml buckets (4°C)
Sorvall SS-34 rotor and tubes

Additional Items

Dialysis tubing (optional; see Step 6)
Microconcentrator (Millipore) (optional; see Step 6)

Additional Information

Ethidium bromide   MC3, pp. 1.150–1.151, 5.14–5.15, and A9.3–A9.4

METHOD

1. To a solution of DNA in a glass or polypropylene tube, add an equal volume of either water-saturated n-butanol or isoamyl alcohol. Close the cap of the tube tightly.

2. Mix the organic and aqueous phases by vortexing.

3. Centrifuge the mixture at $450g$ (1500 rpm in a Sorvall RT-6000 centrifuge with an H-400 rotor and 50-ml buckets) for 3 minutes at room temperature or stand the solution at room temperature until the organic and aqueous phases have separated.

4. Use a pasteur pipette to transfer the upper (organic) phase, which is now a beautiful deep pink color, to an appropriate waste container.

5. Repeat the extraction (Steps 1–4) four to six times until all of the pink color disappears from both the aqueous phase and organic phases.

6. Remove the CsCl from the DNA solution by ethanol precipitation (please follow Steps 7–12), by spin dialysis through a microconcentrator (Millipore), or by dialysis overnight (16 hours) against 2 liters of TE (pH 8.0) (change buffer frequently). If one of the latter two methods is used, proceed to Step 13.

7. To precipitate the DNA from the CsCl-DNA solution, measure the volume of the CsCl solution, add three volumes of H_2O, and mix the solution well.

8. Add 8 volumes of ethanol (1 volume is equal to that of the CsCl-DNA solution before dilution with H_2O in Step 7) to the DNA solution and mix well. Store the mixture for at least 15 minutes at 4°C.

9. Collect the precipitate of DNA by centrifugation at 20,000g (13,000 rpm in a Sorvall SS-34 rotor) for 15 minutes at 4°C.

10. Decant the supernatant to a fresh centrifuge tube. Add an equal volume of absolute ethanol to the supernatant. Store the mixture for at least 15 minutes at 4°C and then collect the precipitate of DNA by centrifugation at 20,000g (13,000 rpm in a Sorvall SS-34 rotor) for 15 minutes.

11. Wash the two DNA precipitates with 70% ethanol. Remove as much of the 70% ethanol as possible and allow any remaining fluid to evaporate at room temperature.

12. Dissolve the precipitated DNA in 2 ml of H_2O or TE (pH 8.0).

13. If the resuspended DNA contains significant quantities of ethidium bromide, as judged from its color or its emission of fluorescence when illuminated by UV light, extract the solution once with phenol and once with phenol:chloroform, and then again precipitate the DNA with ethanol.

14. Measure the OD_{260} of the final solution of DNA and calculate the concentration of DNA. Store the DNA in aliquots at –20°C.

Protocol 1.13 Removal of Ethidium Bromide from DNA by Ion-exchange Chromatography

This protocol describes how ethidium bromide is removed from DNA by chromatography through Dowex AG50, a cation-exchange resin.

METHOD

1. Before using, equilibrate the Dowex AG50 resin:

   a. Stir approximately 20 g of Dowex AG50 in approximately 100 ml of 1 M NaCl for 5 minutes. Allow the resin to settle and remove the supernatant by aspiration.

   b. Add approximately 100 ml of 1 N HCl and stir the slurry for an additional 5 minutes. Again allow the resin to settle and remove the supernatant by aspiration.

   c. Continue the process with two washes with H_2O (100 ml each), followed by one wash with 100 ml of TEN buffer.

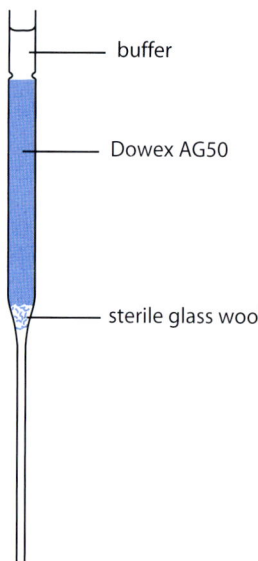

FIGURE 1-3 Removal of Ethidium Bromide from DNA by Chromatography through Dowex AG50

Please see Step 1 for details.

    d. Store the equilibrated resin at 4°C in TEN buffer containing 0.2% sodium azide.

2. Construct a 1-ml column of Dowex AG50 in a pasteur pipette (see Figure 1-3).

3. Remove the buffer above the resin and rinse the column with 2 column volumes of TE (pH 8.0). Apply the solution of DNA containing ethidium bromide and CsCl directly to the resin.

4. Immediately begin collecting the effluent from the column. After all of the DNA solution has entered the column, wash the resin with 1.2 column volumes of TE (pH 8.0) and continue to collect the eluate into a 30-ml Corex tube.

5. After the column has run dry, dilute the eluate with 2.5 column volumes of H_2O.

6. Precipitate the DNA by adding 8 volumes of ethanol followed by incubation for 15 minutes at 4°C. Collect the DNA by centrifugation at 17,000g (12,000 rpm in a Sorvall SS-34 rotor) for 15 minutes at 4°C.

7. Decant the supernatant to a fresh centrifuge tube. Add an equal volume of absolute ethanol to the supernatant. Store the mixture for at least 15 minutes at 4°C and then collect the precipitate of DNA by centrifugation at 20,000g (13,000 rpm in a Sorvall SS-34 rotor) for 15 minutes.

8. Wash the two DNA precipitates with 70% ethanol. Remove as much as possible of the 70% ethanol and allow any remaining fluid to evaporate at room temperature.

9. Dissolve the precipitated DNAs in a total of 2 ml of H_2O or TE (pH 8.0).

10. If the resuspended DNA contains significant quantities of ethidium bromide, as judged from its color or its emission of fluorescence when illuminated by UV light, extract the solution once with phenol and once with phenol:chloroform, and then again precipitate the DNA with ethanol.

11. Measure the OD_{260} of the final solution of DNA and calculate the concentration of DNA. Store the DNA in aliquots at −20°C.

Removal of Small Fragments of Nucleic Acid from Preparations of Plasmid DNA by Centrifugation through NaCl

Contamination of plasmid DNA by fragments of DNA and RNA is reduced to an acceptable level by centrifugation through 1 M sodium chloride.

MATERIALS

Reagents and Solutions

Please see Appendix 1 for components of stock solutions, buffers, and reagents. Dilute stock solutions to the appropriate concentrations.
Ethanol
NaCl (1 M), made from 5 M stock solution
Sodium acetate (3 M, pH 5.2)
TE (pH 8.0)

Enzymes and Buffers

RNase A (pancreatic RNase), DNase-free

Nucleic Acids/Oligonucleotides

Preparation of plasmid DNA
For example, use material from either Protocol 1.12, Step 14 or Protocol 1.13, Step 11.

Centrifuges/Rotors/Tubes

Beckman SW 50.1 rotor (20°C) and tubes
Microfuge (4°C)
Sorvall SS-34 rotor (4°C) and tubes

METHOD

1. Measure the volume of the plasmid preparation. Add 0.1 volume of 3 M sodium acetate (pH 5.2) and 2 volumes of ethanol. Store the mixture for 30 minutes at 4°C.

2. Recover the precipitate of nucleic acids by centrifugation at >10,000g (>9100 rpm in a Sorvall SS-34 rotor) for 15 minutes at 4°C. Decant as much of the supernatant as possible and then store the open tube on the bench for a few minutes to allow the ethanol to evaporate.

3. Dissolve the damp pellet of DNA in 0.5–1.0 ml of TE (pH 8.0) at a concentration of ≥100 μg/ml.

4. Add DNase-free RNase to a final concentration of 10 μg/ml. Incubate the mixture for 1 hour at room temperature.

5. Add 4 ml of 1 M NaCl in TE (pH 8.0) to a Beckman SW 50.1 centrifuge tube (or its equivalent). Use an automatic pipette with a disposable tip to layer up to 1 ml of the plasmid preparation on top of the 1 M NaCl solution. If necessary, fill the tube with TE (pH 8.0).

6. Centrifuge the solution at 150,000g (40,000 rpm in a Beckman SW 50.1 rotor) for 6 hours at 20°C. Carefully discard the supernatant.

7. Dissolve the pellet of plasmid DNA in 0.5 ml of TE (pH 8.0). Add 50 µl of 3 M sodium acetate (pH 5.2) and transfer the DNA solution to a microfuge tube.

8. Precipitate the DNA by addition of 2 volumes of ethanol and store the ethanolic solution for 10 minutes at 4°C. Recover the DNA by centrifugation at maximum speed for 15 minutes at 4°C in a microfuge. Decant as much of the supernatant as possible and then store the open tube on the bench for a few minutes to allow the ethanol to evaporate.

9. Dissolve the damp pellet of DNA in TE (pH 8.0).

Removal of Small Fragments of Nucleic Acid from Preparations of Plasmid DNA by Chromatography through Sephacryl S-1000

Contamination of plasmid DNA by small fragments of nucleic acid is reduced dramatically by size-exclusion chromatography through Sephacryl S-1000.

MATERIALS

CAUTION: Please see Appendix 4 for appropriate handling of materials marked with <!>.

Reagents and Solutions

Please see Appendix 1 for components of stock solutions, buffers, and reagents. Dilute stock solutions to the appropriate concentrations.

    Bromophenol blue sucrose solution, also known as gel-loading buffer IV
    Ethanol
    Phenol <!>
    Sephacryl equilibration buffer
        50 mM Tris-Cl (pH 8.0)
        5 mM EDTA
        0.5 M NaCl
    Sephacryl S-1000
    Sodium acetate (3 M, pH 5.2)
    TE (pH 8.0)
    TE (pH 8.0) containing 20 µg/ml DNase-free RNase A

Enzymes and Buffers

    RNase A (pancreatic RNase), DNase-free

Nucleic Acids/Oligonucleotides

    Preparation of plasmid DNA
        For example, use material from Protocol 1.12, Step 14 or Protocol 1.13, Step 11.

Gels/Loading Buffers

    0.7% Agarose gel
    Gel-loading buffer IV

Centrifuges/Rotors/Tubes

    Sorvall SS-34 rotor (4°C) and tubes

Additional Items

    Glass or plastic chromatography column (1 x 12 cm)

Additional Information

    Gómez-Márquez J., Freire M., and Segade F. 1987. A simple procedure for large-scale purification of plasmid DNA. *Gene* **54:** 255–259.

METHOD

1. Prepare a 1 x 10-cm column of Sephacryl S-1000, equilibrated in Sephacryl equilibration buffer.

2. Measure the volume of the plasmid preparation. Add 0.1 volume of 3 M sodium acetate (pH 5.2) and 2 volumes of ethanol. Store the mixture for 30 minutes at 4°C.

3. Recover the precipitate of nucleic acids by centrifugation at >10,000g (>9100 rpm in a Sorvall SS-34 rotor) for 15 minutes at 4°C. Drain off as much of the supernatant as possible and then store the open tube on the bench for a few minutes to allow the ethanol to evaporate.

4. Dissolve the damp pellet of nucleic acids in a small volume (<400 μl) of TE (pH 8.0) containing RNase A at a final DNA concentration of at least 100 μg/ml.

5. Incubate the mixture for 1 hour at room temperature.

6. Extract the solution once with an equal volume of phenol equilibrated in TE (pH 8.0).

7. Recover the aqueous layer and add 100 μl of gel-loading buffer IV. Layer the blue DNA solution on the column of Sephacryl S-1000.

8. Wash the DNA into the column and apply a reservoir of Sephacryl equilibration buffer. Immediately begin collecting 0.5-ml fractions.

9. When 15 fractions have been collected, clamp off the bottom of the column. At this stage, the blue dye should have traveled about half the length of the column.

10. Analyze 10 μl of each fraction by electrophoresis through 0.7% agarose gel to identify the fractions containing plasmid DNA.

11. Pool the fractions containing plasmid DNA and recover the DNA by precipitation with 2 volumes of ethanol for 10 minutes at 4°C and centrifugation at 10,000g (9200 rpm in a Sorvall SS-34 rotor) for 15 minutes at 4°C.

12. Decant as much of the supernatant as possible and then store the open tube on the bench for a few minutes to allow the ethanol to evaporate.

13. Dissolve the damp pellet in TE (pH 8.0).

Removal of Small Fragments of Nucleic Acid from Preparations of Plasmid DNA by Precipitation with Lithium Chloride

This protocol describes how high-molecular-weight RNA and proteins can be precipitated from preparations of plasmid DNA by high concentrations of LiCl and removed by low-speed centrifugation.

MATERIALS

Reagents and Solutions

Please see Appendix 1 for components of stock solutions, buffers, and reagents. Dilute stock solutions to the appropriate concentrations.

Ethanol
Ethanol (70% at 4°C)
Isopropanol
LiCl (4 M)
Sodium acetate (3 M, pH 5.2)
TE (pH 8.0)
TE (pH 8.0) containing 20 µg/ml DNase-free RNase A

Nucleic Acids/Oligonucleotides

Preparation of plasmid DNA
    For example, use material from either Protocol 1.12, Step 14 or Protocol 1.13, Step 11.

Centrifuges/Rotors/Tubes

Sorvall SS-34 rotor (4°C) and tubes

Additional Information

Kondo T., Mukai M., and Kondo Y. 1991. Rapid isolation of plasmid DNA by LiCl-ethidium bromide treatment and gel filtration. *Anal. Biochem.* **198**: 30–35.

METHOD

1. Measure the volume of the plasmid preparation. Add 0.1 volume of 3 M sodium acetate (pH 5.2) and 2 volumes of ethanol. Store the mixture for 30 minutes at 4°C.

2. Recover the precipitate of nucleic acids by centrifugation at >10,000g (>9100 rpm in a Sorvall SS-34 rotor) for 15 minutes at 4°C. Drain off as much of the supernatant as possible and then store the open tube on the bench for a few minutes to allow the ethanol to evaporate.

3. Dissolve the damp pellet in 1 ml of TE (pH 8.0) containing RNase A at a concentration of ≥100 µg/ml.

4. Add 3 ml of 4 M LiCl solution. Store the mixture on ice for 30 minutes.

5. Separate the plasmid DNA from the precipitated nucleic acids by centrifugation at 12,000g (10,000 rpm in a Sorvall SS-34 rotor) for 15 minutes at 4°C.

6. Transfer the supernatant to a fresh centrifuge tube and add 6 ml of isopropanol. Allow the plasmid DNA to precipitate for 30 minutes at room temperature.

7. Recover the precipitated plasmid DNA by centrifugation at 12,000g (10,000 rpm in a Sorvall SS-34 rotor) for 15 minutes at 4°C.

8. Carefully remove the supernatant and add 5–10 ml of 70% ethanol to the tube. Vortex the tube briefly and then recentrifuge at 12,000g for 10 minutes at 4°C.

9. Carefully remove the supernatant and store the open tube on the bench top for a few minutes until the ethanol has evaporated.

10. Dissolve the damp pellet of DNA in TE (pH 8.0).

Directional cloning requires that the plasmid vector be cleaved with two restriction enzymes that generate incompatible termini and that the fragment of DNA to be cloned carries termini that are compatible with those of the doubly cleaved vector.

MATERIALS

CAUTION: Please see Appendix 4 for appropriate handling of materials marked with <!>.

Reagents and Solutions

Please see Appendix 1 for components of stock solutions, buffers, and reagents. Dilute stock solutions to the appropriate concentrations.

ATP (10 mM)
> Not required if 10x ligation buffer contains ATP

Ethanol (4°C)

Ethanol (70% at 4°C)

Phenol:chloroform (1:1, v/v) <!>

Reagents required for transformation of *E. coli* (see Protocol 1.23, 1.24, 1.25, or 1.26)

Sodium acetate (3 M, pH 5.2)

TE (pH 8.0)

Vectors and Hosts

Transformation-competent *E. coli*

Vector DNA (superhelical plasmid) (10 µg)

Media and Antibiotics

SOB or LB agar plates containing the appropriate antibiotic

Enzymes and Buffers

Bacteriophage DNA ligase and 10x ligation buffer

Restriction enzymes and 10x buffers

Nucleic Acids/Oligonucleotides

Fragment of target DNA

Centrifuges/Rotors/Tubes

Microfuge tubes (0.5 ml)

Sorvall SS-34 rotor (4°C) and tubes

Additional Items

Spun columns

Water bath or heating/cooling block (16°C)

Additional Information

DNA ligases	MC3, pp. A4.30–A4.31
Spun-column chromatography	MC3, pp. A8.30–A8.31

METHOD

1. Digest the vector (10 μg) and foreign DNA with the two appropriate restriction enzymes.

2. Purify the digested foreign DNA by extraction with phenol:chloroform and standard ethanol precipitation.

3. Purify the vector DNA by spun-column chromatography followed by standard ethanol precipitation.

4. Reconstitute the precipitated DNAs separately in TE (pH 8.0) at a concentration of approximately 100 μg/ml. Calculate the concentration of the DNA (in pmole/ml), assuming that 1 bp has a mass of 660 daltons.

5. Transfer appropriate amounts of the DNAs to sterile 0.5-ml microfuge tubes as follows:

Tube	DNA
A and D	vector (30 fmoles [~100 ng])
B	insert (foreign) (30 fmoles [~ 10 ng])
C and E	vector (30 fmoles) plus insert (foreign) (30 fmoles)
F	superhelical vector (3 fmoles [~10 ng])

The molar ratio of plasmid vector to insert DNA fragment should be approximately 1:1 in the ligation reaction. The final DNA concentration should be approximately 10 ng/μl.

   a. To Tubes A–C, add:

10x ligation buffer	1.0 μl
bacteriophage T4 DNA ligase	0.1 Weiss unit
10 mM ATP	1.0 μl
H_2O	to 10 μl

   b. To Tubes D and E, add:

10x ligation buffer	1.0 μl
10 mM ATP	1.0 μl
H_2O	to 10 μl
no DNA ligase	

   Omit ATP from the ligation mixture if the 10x ligation buffer already contains ATP.

   The DNA fragments can be added to the tubes together with the H_2O and then warmed to 45°C for 5 minutes to melt any cohesive termini that have reannealed during fragment preparation. Chill the DNA solution to 0°C before the remainder of the ligation reagents is added.

6. Incubate the reaction mixtures overnight at 16°C or for 4 hours at 20°C.

7. Transform competent *E. coli* with dilutions of each of the ligation reactions as described in Protocol 1.23, 1.24, 1.25, or 1.26. As controls, include known amounts of a standard preparation of superhelical plasmid DNA to check the efficiency of transformation.

Tube	DNA	Ligase	Expected Number of Transformed Colonies
A	vector	+	~0 (~10^4 fewer than Tube F)[a]
B	insert	+	0
C	vector and insert	+	~10-fold more than Tube A or D
D	vector	–	~0 (~10^4 fewer than Tube F)
E	vector and insert	–	some, but fewer than Tube C
F	superhelical vector	–	>2 x 10^5

[a]Transformants arising from ligation of vector DNA alone are due either to failure of one or both restriction endonucleases to digest the DNA to completion or to ligation of the vector to residual amounts of the small fragment excised from the multiple cloning site.

Attaching Adapters to Protruding Termini

Adapters are short double-stranded synthetic oligonucleotides that carry an internal restriction endonuclease recognition site and single-stranded tails at one or both ends. Adapters are used to exchange restriction sites at the termini of linear DNA molecules. They may be purchased in phosphorylated and unphosphorylated forms.

MATERIALS

CAUTION: Please see Appendix 4 for appropriate handling of materials marked with <!>.

Reagents and Solutions

Please see Appendix 1 for components of stock solutions, buffers, and reagents. Dilute stock solutions to the appropriate concentrations.
ATP (10 mM)
Ethanol
Ethanol (70%)
Phenol:chloroform (1:1, v/v) <!>
Sodium acetate (3 M, pH 5.2)
TE (pH 8.0)

Vectors and Hosts

Plasmid vector DNA
Transformation-competent *E. coli*

Media and Antibiotics

SOB or LB agar plates containing the appropriate antibiotic

Enzymes and Buffers

Bacteriophage T4 DNA ligase and 10x ligation buffer
Bacteriophage T4 polynucleotide kinase and 10x linker kinase buffer
Enzymes and buffers for dephosphorylation of DNA (see Protocol 1.20)
Restriction enzymes and 10x buffers

Nucleic Acids/Oligonucleotides

Phosphorylated synthetic DNA adapter, dissolved in TE (pH 8.0) at a concentration of
approximately 400 µg/ml (~50 mM for a hexamer)
Plasmid vector
Target DNA fragment (100–200 ng)

Centrifuges/Rotors/Tubes

Microfuge (4°C) and 0.5-ml microfuge tubes

Additional Items

Spun columns
Step 9 of this protocol requires the reagents listed in Protocol 1.17.

Water bath (65°C)
Water bath or heating/cooling block (4°C)

Additional Information

Adapters	MC3, pp. 1.160–1.161
Spun-column chromatography	MC3, pp. A8.30–A8.31

METHOD

1. To phosphorylate the adapters, add to a sterile microfuge tube:

synthetic oligonucleotide or adapter	0.5–2.0 µg, dissolved in TE (pH 8.0)
10x linker kinase buffer	1.0 µl
10 mM ATP	1.0 µl
H_2O	to 10 µl
bacteriophage T4 polynucleotide kinase	1.0 unit

   Incubate the reaction for 1 hour at 37°C.

2. To ligate the phophorylated adapters to a DNA fragment with complementary protruding ends, set up a ligation reaction as follows:

DNA fragment	100–200 ng
phosphorylated adapter	10- to 20-fold molar excess
10x ligation buffer	1.0 µl
bacteriophage T4 DNA ligase	0.1 Weiss unit
10 mM ATP	1.0 µl
H_2O	to 10 µl

   Incubate the ligation mixture for 6–16 hours at 4°C.

   To achieve the maximum efficiency of ligation, set up the reactions in as small a volume as possible (5–10 µl). Omit ATP from the ligation mixture if the 10x ligation buffer already contains ATP.

3. Inactivate the DNA ligase by incubating the reaction mixture for 15 minutes at 65°C.

4. Dilute the ligation reaction with 10 µl of the appropriate 10x restriction enzyme buffer. Add sterile H_2O to a final volume of 100 µl followed by 50–100 units of restriction enzyme.

5. Incubate the reaction for 1–3 hours at 37°C.

6. Extract the restriction digest with phenol:chloroform and recover the DNA by standard ethanol precipitation.

7. Collect the precipitated DNA by centrifugation at maximum speed for 15 minutes at 4°C in a microfuge and resuspend the DNA in 50 µl of TE (pH 8.0).

8. Pass the resuspended DNA through a spun column (see Appendix 3, p. 748) to remove excess adapters and their cleavage products.

9. The modified DNA fragment can now be ligated to a plasmid vector with protruding ends that are complementary to those of the cleaved adapter (please see Protocol 1.17).

Blunt-ended Cloning into Plasmid Vectors

Target DNA is ligated to a blunt-ended plasmid DNA, and the products of the ligation reaction are used to transform competent *E. coli.* The maximum number of "correct" clones can generally be obtained from ligation reactions containing equimolar amounts of plasmid and target DNAs, with the total DNA concentration being <100 µg/ml. Blunt-ended ligation catalyzed by bacteriophage T4 DNA ligase is suppressed by high concentrations (5 mM) of ATP and polyamines such as spermidine.

MATERIALS

CAUTION: Please see Appendix 4 for appropriate handling of materials marked with <!>.

Reagents and Solutions

Please see Appendix 1 for components of stock solutions, buffers, and reagents. Dilute stock solutions to the appropriate concentrations.

ATP (10 mM)
> Not required if 10x ligation buffer contains ATP

Ethanol

Ethanol (70% at 4°C)

Phenol:chloroform (1:1, v/v) <!>

Polyethylene glycol 8000 (PEG 8000, 30% w/v) <!>

Sodium acetate (3 M, pH 5.2)

TE (pH 8.0)

Vectors and Hosts

Transformation-competent *E. coli*

Vector DNA

Media and Antibiotics

SOB or LB agar containing appropriate antibiotic

Enzymes and Buffers

Bacteriophage DNA ligase and 10x ligation buffer

Restriction enzymes and 10x buffers

Nucleic Acids/Oligonucleotides

Target DNA (blunt-ended fragment)

Vector DNA (plasmid)

Gels/Loading Buffers

Agarose or polyacrylamide gels (optional; see Steps 2 and 3)

Additional Items

Step 4 of this protocol requires the reagents listed in Protocol 1.20.

Step 7 of this protocol requires the reagents listed in Protocols 1.23–1.26.

Water bath or heating/cooling block (20°C)

> **Additional Information**
>
> Bercovich J., Grinstein S., and Zorzopoulos J. 1992. Promega enzyme resource guide: Cloning enzymes. *BioTechniques* **12:** 190–193.
>
> Blunt-ended ligation   MC3, pp. A4.31–A4.32

METHOD

1. In separate reactions, digest 1–10 μg of the plasmid DNA and foreign DNA with the appropriate restriction enzyme(s) that generate blunt ends.

2. Purify the digested foreign DNA and vector DNA by extraction with phenol:chloroform and standard ethanol precipitation.

3. Reconstitute the precipitated DNAs separately in TE (pH 8.0) at a concentration of approximately 100 μg/ml. Calculate the concentration of the DNAs (in pmole/ml) assuming that 1 bp has a mass of 660 daltons.

4. Dephosphorylate the plasmid vector DNA as described in Protocol 1.20.

5. Transfer appropriate amounts of the DNAs to sterile 0.5-ml microfuge tubes as follows:

Tube	DNA
A and E	vector[a] (60 fmoles [~100 ng])
B	foreign DNA[b] (60 fmoles [~10 ng])
C and F	vector[a] (60 fmoles) plus foreign DNA (60 fmoles)[c]
D	linearized vector (contains 5′-terminal phosphates) (60 fmoles)
G	superhelical vector (6 fmoles [~10 ng])

[a]Vector DNA is dephosphorylated as described in Protocol 1.20.
[b]Linkers may be ligated to foreign target DNA.
[c]The molar ratio of plasmid vector to insert DNA fragment should be approximately 1:1 in the ligation reaction. The total DNA concentration in the ligation reaction should be approximately 10 ng/μl.

   a. To Tubes A, B, and C, add:

10x ligation buffer	1.0 μl
bacteriophage T4 DNA ligase	0.5 Weiss unit
5 mM ATP	1.0 μl
H$_2$O	to 8.5 μl
30% PEG 8000	1–1.5 μl

   b. To Tubes D, E, and F, add:

10x ligation buffer	1.0 μl
5 mM ATP	1.0 μl
H$_2$O	to 8.5 μl
30% PEG 8000	1–1.5 μl
no DNA ligase	

   To achieve the maximum efficiency of ligation, set up the reactions in as small a volume as possible (5–10 μl).

   The DNA fragments can be added to the tubes together with the H$_2$O and then warmed to 45°C for 5 minutes to help dissociate any clumps of DNA that have formed during fragment preparation. Chill the DNA solution to 0°C before the remainder of the ligation reagents are added.

   Omit ATP from the ligation mixture if the 10x ligation buffer already contains ATP.

6. Incubate the reaction mixtures overnight at 16°C or for 4 hours at 20°C.

7. Transform competent *E. coli* with dilutions of each of the ligation reactions, using one of the methods described in Protocols 1.23–1.26. As controls, include known amounts of a standard preparation of superhelical plasmid DNA to check the efficiency of transformation.

Tube	DNA	Ligase	Expected Number of Transformants
A	vector[a]	+	~0[b]
B	insert	+	0
C	vector[a] and insert	+	~5-fold more than Tube F
D	vector[a]	–	~0
E	vector[c]	–	~50-fold more than Tube D
F	vector[a] and insert	–	~50-fold more than Tube D
G	superhelical vector	–	2×10^5

[a]Dephosphorylated.
[b]Transformants arising from ligation of dephosphorylated vector DNA alone are due to failure to remove 5′ residues during treatment with alkaline phosphatase.
[c]Not dephosphorylated.

Dephosphorylation of Plasmid DNA

During ligation in vitro, T4 DNA ligase catalyzes the formation of a phosphodiester bond between adjacent nucleotides only if one nucleotide carries a 5'-phosphate residue and the other carries a 3'-hydroxyl terminus. Recircularization of vector DNA can therefore be minimized by removing the 5'-phosphate residues from both termini of the linear, double-stranded plasmid DNA with alkaline phosphatase.

MATERIALS

CAUTION: Please see Appendix 4 for appropriate handling of materials marked with <!>.

Reagents and Solutions

Please see Appendix 1 for components of stock solutions, buffers, and reagents. Dilute stock solutions to the appropriate concentrations.

EDTA (0.5 M, pH 8.0)
   Required only if CIP is used for dephosphorylation
EGTA (0.5 M, pH 8.0) (optional; see Step 6)
Ethanol (4°C)
Ethanol (70% at 4°C)
Phenol <!>
Phenol:chloroform (1:1, v/v) <!>
SDS (10% w/v)
   Required only if CIP is used for dephosphorylation
Sodium acetate (3 M, pH 5.2)
TE (pH 8.0)
Tris-Cl (10 mM, pH 8.3)

Enzymes and Buffers

Calf intestinal alkaline phosphatase (CIP) *or* shrimp alkaline phosphatase (SAP) and buffers
Proteinase K (10 mg/ml)
   Required only if CIP is used for dephosphorylation
Restriction enzyme(s) and buffer(s)

Nucleic Acids/Oligonucleotides

Closed circular plasmid DNA (10 µg)

Gels/Loading Buffers

0.7% Agarose gel
Gel-loading buffer IV

Centrifuges/Rotors/Tubes

Microfuge (4°C)

Additional Items

Water bath 55°C (for CIP) or 65°C (for SAP)

Additional Information

Properties of alkaline phosphatases   MC3, pp. 9.92–9.93 and A4.37

METHOD

1. Digest a reasonable quantity (~10 μg) of closed circular plasmid DNA with a two- to three-fold excess of the desired restriction enzyme for 1 hour.

2. Remove an aliquot (0.1 μg) and analyze the extent of digestion by electrophoresis through a 0.7% agarose gel containing ethidium bromide, using undigested plasmid DNA as a marker. If digestion is incomplete, add more restriction enzyme and continue the incubation.

3. When digestion is complete, extract the sample once with phenol:chloroform and recover the DNA by standard precipitation with ethanol. Store the ethanolic solution on ice for 15 minutes.

4. Recover the DNA by centrifugation at maximum speed for 10 minutes at 4°C in a microfuge and dissolve the DNA in 110 μl of 10 mM Tris-Cl (pH 8.3).

5. To the remaining 90 μl of the linearized plasmid DNA, add 10 μl of 10x CIP or 10x SAP buffer and the appropriate amount of calf intestinal phosphatase (CIP) or shrimp alkaline phosphatase (SAP). Incubate as described in the table below.

TABLE 1-3 Conditions for Dephosphorylation of 5′-phosphate Residues from DNA

Type of Terminus	Enzyme/Amount per Mole DNA Ends	Incubation Temperature/Time
5′ Protruding	0.01 unit CIP[a]	37°C/30 min
	0.1 unit SAP	37°C/60 min
3′ Protruding	0.1–0.5 unit CIP[b]	37°C/15 min
		then
		55°C/45 min
	0.5 unit SAP	37°C/60 min
Blunt	0.1–0.5 unit CIP[b]	37°C/15 min
		then
		55°C/45 min
	0.2 unit SAP	37°C/60 min

[a]After the initial 30-minute incubation, add a second aliquot of CIP enzyme and continue incubation for another 30 minutes at 37°C.

[b]Add a second aliquot of CIP just before beginning the incubation at 55°C.

6. Inactivate the phosphatase activity.

To inactivate CIP at the end of the incubation period

Add SDS and EDTA (pH 8.0) to final concentrations of 0.5% and 5 mM, respectively. Mix well and add proteinase K to a final concentration of 100 μg/ml. Incubate for 30 minutes at 55°C.

Alternatively, CIP can be inactivated by heating to 65°C for 30 minutes (or 75°C for 10 minutes) in the presence of 5 mM EDTA or 10 mM EGTA (both at pH 8.0).

or

To inactivate SAP

Incubate the reaction mixture for 15 minutes at 65°C in the dephosphorylation buffer.

7. Cool the reaction mixture to room temperature and then extract it once with phenol and once with phenol:chloroform.

8. Recover the DNA by standard precipitation with ethanol. Mix the solution again and store it for 15 minutes at 0°C.

9. Recover the DNA by centrifugation at maximum speed for 10 minutes at 4°C in a microfuge. Wash the pellet with 70% ethanol at 4°C and centrifuge again.

10. Carefully remove the supernatant and leave the open tube on the bench to allow the ethanol to evaporate.

11. Dissolve the precipitated DNA in TE (pH 8.0) at a concentration of 100 µg/ml. Store the DNA in aliquots at –20°C.

Addition of Synthetic Linkers to Blunt-ended DNA

Linkers are small self-complementary pieces of synthetic DNA that are used to equip blunt-ended termini of DNA with restriction sites as an aid to cloning.

MATERIALS

CAUTION: Please see Appendix 4 for appropriate handling of materials marked with <!>.

Reagents and Solutions

Please see Appendix 1 for components of stock solutions, buffers, and reagents. Dilute stock solutions to the appropriate concentrations.

Ammonium acetate (2 M), diluted from 4 M stock (pH 4.8)

ATP (5 mM, Step 2 and 10 mM, Step 1)

> 5 mM ATP is not required if the 10x ligation buffer already contains ATP.

Ethanol

Ethanol (70%)

Phenol:chloroform (1:1, v/v) <!>

Sodium acetate (3 M, pH 5.2)

TE (pH 8.0)

Enzymes and Buffers

Bacteriophage T4 DNA ligase and 10x ligase buffer

Bacteriophage T4 polynucleotide kinase

10x Linker kinase buffer

> 60 mM Tris-Cl (pH 7.6)
> 100 mM $MgCl_2$
> 100 mM dithiothreitol
> 2 mg/ml bovine serum albumin

Restriction enzyme (to cleave linkers) and 10x buffer

Nucleic Acids/Oligonucleotides

Blunt-ended DNA

> Approximately 2 pmoles of blunt termini are required per reaction.

Synthetic linker

> Dissolved in TE (pH 8.0) at a concentration of approximately 50 μM

Centrifuges/Rotors/Tubes

Microfuge (4°C) and microfuge tubes (0.5 ml and standard size)

Additional Items

Spun column (see Appendix 3, p. 748)

Water bath (65°C)

Additional Information

Properties of linkers	MC3, pp. 1.98–1.100
Spun-column chromatography	MC3, pp. A8.30–A8.31

METHOD

1. Assemble the following reaction mixture in a sterile 0.5-ml microfuge tube:

10x linker kinase buffer	1.0 µl
10 mM ATP	1.0 µl
synthetic linker dissolved in TE (pH 8.0)	2.0 µg[1]
H$_2$O	to 9 µl
bacteriophage T4 polynucleotide kinase	10 units

   [1]Approximately 250 pmoles of a dodecamer.

   Incubate the reaction for 1 hour at 37°C.

2. Calculate the concentration of termini in the preparation of blunt-ended DNA and then assemble the following ligation mixture in the order given in a sterile 0.5-ml microfuge tube:

   50 µg of a 1-kb segment of double-stranded DNA = 78.7 nmoles or 157.4 nM of termini.

blunt-ended DNA	2 pmoles of termini
phosphorylated linkers	150–200 pmoles of termini
H$_2$O	to 7.5 µl
10x ligation buffer	1.0 µl
5 mM ATP (free acid)	1.0 µl
bacteriophage T4 DNA ligase	1.0 Weiss unit

   Incubate the reaction mixtures for 12–16 hours at 4°C.

   Omit ATP from the ligation mixture if the 10x ligation buffer already contains ATP.

   To achieve the maximum efficiency of ligation, set up the reactions in as small a volume as possible (5–10 µl).

3. Inactivate the bacteriophage T4 DNA ligase by heating the reaction mixture to 65°C for 15 minutes.

4. Cool the ligation mixture to room temperature and then add:

10x restriction enzyme buffer	10 µl
restriction enzyme	50 units
sterile H$_2$O	to a final volume of 100 µl

   Incubate the reaction for 1–3 hours at 37°C.

5. Purify the restricted DNA by extraction with phenol:chloroform. Precipitate the DNA with 2 volumes of ethanol in the presence of 2 M ammonium acetate.

6. Collect the precipitated DNA by centrifugation at maximum speed for 15 minutes at 4°C in a microfuge and dissolve the pellet in 50 µl of TE (pH 8.0).

7. Pass the resuspended DNA through a spun column (see Appendix 3, p. 748) to remove excess linkers.

8. Recover the DNA by standard ethanol precipitation and dissolve the precipitate in 10–20 µl of TE (pH 8.0).

The modified DNA fragment can now be ligated as described in Protocol 1.17 into a plasmid (or bacteriophage) vector with protruding ends that are complementary to those introduced by the linker.

Ligating Plasmid and Target DNAs in Low-melting-temperature Agarose

Ligation in low-melting-temperature agarose is much less efficient than ligation with purified DNA in free solution and requires a large amount of DNA ligase. The method is used chiefly for rapid subcloning of segments of DNA in dephosphorylated vectors and assembling recombinant constructs.

MATERIALS

CAUTION: Please see Appendix 4 for appropriate handling of materials marked with <!>.

Enzymes and Buffers

Bacteriophage T4 DNA ligase
2x Bacteriophage T4 DNA ligase mixture (freshly prepared)

1 M Tris-Cl (pH 7.6)	1.0 µl
100 mM MgCl$_2$	2.0 µl
200 mM dithiothreitol	1.0 µl
10 mM ATP	1.0 µl
H$_2$O	4.5 µl
Bacteriophage T4 DNA ligase	1 Weiss unit

10 µl is required for each reaction.
10x Restriction buffer
Restriction enzyme (to prepare target DNA)

Nucleic Acids/Oligonucleotides

Dephosphorylated plasmid DNA (~100 ng per ligation)
Target DNA (~250 ng)

Gels/Loading Buffers

Low-melting-temperature agarose gel containing ethidium bromide <!> (0.5 µg/ml)

Additional Items

Clean razor blade
Water bath or heating block (70°C)

Additional Information

Low-melting-temperature agarose   MC3, p. 5.7

METHOD

1. Use the appropriate restriction enzyme(s) to digest an amount of target DNA sufficient to yield approximately 250 ng of the desired fragment. Perform the digestion in a volume of 20 µl or less.

2. Separate the DNA fragments by electrophoresis through a low-melting/gelling-temperature agarose gel containing ethidium bromide (0.5 µg/ml).

3. Examine the agarose gel under long-wavelength UV illumination. From the relative fluorescent intensities of the desired bands, estimate the amounts of DNA that they contain. Use a clean razor blade to cut out the desired bands in the smallest possible volume of agarose (usually 40–50 μl). Leave a small amount of each band in the gel to mark the positions of the DNA fragments and then photograph the dissected gel.

4. Place the excised and trimmed slices of gel in separate, labeled microfuge tubes.

5. Melt the agarose by heating the tubes to 70°C for 15 minutes in a heating block. Estimate the volume of the melted agarose in the tube and calculate the volume that would contain approximately 200 ng of DNA.

6. In a sterile microfuge tube warmed to 37°C, combine the following:

dephosphorylated plasmid DNA	60 fmoles
foreign DNA fragment	120–240 fmoles (in a volume of 10 μl or less)

   Approximately 100 ng of dephosphorylated DNA is required for each ligation.

   Mix the contents of the tube quickly with a sterile disposable pipette tip before the agarose solidifies.

7. In separate tubes, set up two additional ligations as controls, one containing only the dephosphorylated plasmid vector and the other containing only the fragment of foreign DNA.

8. Incubate the three tubes for 5–10 minutes at 37°C and then add to each tube 10 μl of ice-cold 2x bacteriophage T4 DNA ligase mixture. Mix the contents of the tube quickly with a sterile disposable pipette tip before the agarose solidifies. Incubate the reactions for 12–16 hours at 16°C.

The Hanahan Method for Preparation and Transformation of Competent *E. coli:* High-efficiency Transformation

This procedure generates competent cultures of *E. coli* that can be transformed at high frequencies (5×10^8 transformed colonies/µg of superhelical plasmid DNA).

MATERIALS

CAUTION: Please see Appendix 4 for appropriate handling of materials marked with <!>.

Reagents and Solutions

Please see Appendix 1 for components of stock solutions, buffers, and reagents. Dilute stock solutions to the appropriate concentrations.

DMSO <!>
DnD solution
 1.53 g of dithiothreitol
 9 ml of DMSO
 100 µl of 1 M potassium acetate (pH 7.5)
 H_2O to 10 ml
 Sterilize by filtration through a Millex SR membrane unit (Millipore). Dispense 160-µl aliquots into sterile 0.5-ml microfuge tubes. Close the tubes tightly and store them at –20°C.
Transformation buffers

Vectors and Hosts

Plasmid DNA for transformation
Strain of *E. coli* suitable for transformation

Media and Antibiotics

SOB agar plates
SOB agar plates containing 20 mM $MgSO_4$ and the appropriate antibiotic
SOB medium containing 20 mM $MgSO_4$
SOC medium

Centrifuges/Rotors/Tubes

Sorvall SLC-1500 rotor (4°C) and centrifuge bottles

Additional Items

Circulating water bath (42°C)
Corning tissue-culture flasks, used to store transformation buffer
Freezer (–70°C)
Inoculating loop (platinum wire)
Liquid nitrogen bath
Nalgene filters (0.45-µm pore size)
Sterile 50-ml polypropylene centrifuge tubes
Sterile 17 x 100-mm polypropylene tubes
Vacuum aspirator

> **Additional Information**
>
> Factors affecting transformation   MC3, pp. 1.105–1.106
>
> Hanahan D. 1983. Studies on transformation of *Escherichia coli* with plasmids. *J. Mol. Biol.*
>   **136**: 557–580.
>
> Overview of transformation   MC3, pp. 1.24–1.26

METHOD

IMPORTANT: All steps in this protocol should be carried out aseptically.

1. Prepare transformation buffer.

   Standard transformation buffer (TFB) is used when preparing competent cells for immediate use. Frozen storage buffer (FSB) is used to prepare stocks of competent cells that are to be stored at –70°C.

To prepare standard transformation buffer

a. Prepare 1 M MES by dissolving 19.52 g of MES in 80 ml of pure H_2O (Milli-Q or its equivalent). Adjust the pH of the solution to 6.3 with 5 M KOH and add pure H_2O to bring the final volume to 100 ml. Sterilize the solution by filtration through a disposable prerinsed Nalgene filter (0.45-μm pore size). Divide into 10-ml aliquots and store at –20°C.

b. Prepare TFB by dissolving all the solutes listed below in approximately 500 ml of H_2O and then add 10 ml of 1 M MES buffer (pH 6.3). Adjust the volume of the TFB to 1 liter with pure H_2O.

Reagent	Amount per Liter	Final Concentration
1 M MES (pH 6.3)	10 ml	10 mM
$MnCl_2 \cdot 4H_2O$	8.91 g	45 mM
$CaCl_2 \cdot 2H_2O$	1.47 g	10 mM
KCl	7.46 g	100 mM
Hexamminecobalt chloride	0.80 g	3 mM
H_2O	to 1 liter	

c. Sterilize the TFB by filtration through a disposable prerinsed Nalgene filter (0.45-μm pore size). Divide the solution into 40-ml aliquots in tissue-culture flasks (e.g., Corning or its equivalent) and store them at 4°C.

To prepare frozen storage buffer

a. Prepare 1 M potassium acetate by dissolving 9.82 g of potassium acetate in 90 μl of pure H_2O (Milli-Q or its equivalent). Adjust the pH of the solution to 7.5 with 2 M acetic acid and add pure H_2O to bring the final volume to 100 ml. Divide the solution into aliquots and store at –20°C.

b. Prepare FSB by dissolving all of the solutes listed below in approximately 500 ml of pure H_2O. After the components are dissolved, adjust the pH of the solution to 6.4 with 0.1 N HCl. Too high a pH cannot be adjusted by adding base; instead, discard the solution and and begin again. Adjust the volume of the final solution to 1 liter with pure H_2O.

Reagent	Amount per Liter	Final Concentration
1 M potassium acetate (pH 7.5)	10 ml	10 mM
$MnCl_2 \cdot 4H_2O$	8.91 g	45 mM
$CaCl_2 \cdot 2H_2O$	1.47 g	10 mM
KCl	7.46 g	10 mM
Hexamminecobalt chloride	0.80 g	3 mM
Glycerol	100 ml	10% (v/v)
H_2O	to 1 liter	

   c. Sterilize the solution by filtration through a disposable prerinsed Nalgene filter (0.45-μm pore size). Dispense the solution into 40-ml aliquots and store the aliquots in tissue-culture flasks (e.g., Corning or its equivalent) at 4°C. During storage, the pH of the solution drifts down to a final value of 6.1–6.2 but then stabilizes.

2. Use an inoculating loop to streak *E. coli* of the desired strain directly from a frozen stock onto the surface of an SOB agar plate. Incubate the plate for 16 hours at 37°C.

3. Transfer four or five well-isolated colonies into 1 ml of SOB containing 20 mM $MgSO_4$. Disperse the bacteria by vortexing at moderate speed and then dilute the culture in 30–100 ml of SOB containing 20 mM $MgSO_4$ in a 1-liter flask.

4. Grow the cells for 2.5–3.0 hours at 37°C, monitoring the growth of the culture.

   It is essential that the density of the culture not exceed 10^8 cells/ml ($OD_{600} \leq 0.4$).

5. Transfer the cells to sterile, disposable, ice-cold 50-ml polypropylene tubes. Cool the cultures to 0°C by storing the tubes on ice for 10 minutes.

6. Recover the cells by centrifugation at 2700*g* (4100 rpm in a Sorvall SLC-1500 rotor) for 10 minutes at 4°C.

7. Decant the medium from the cell pellets. Stand the tubes in an inverted position for 1 minute to allow the last traces of medium to drain away.

8. Resuspend the pellets by swirling or gentle vortexing in approximately 20 ml (per 50-ml tube) of ice-cold TFB or FSB transformation buffer. Store the resuspended cells on ice for 10 minutes.

9. Recover the cells by centrifugation at 2700*g* (4100 rpm in a Sorvall SLC-1500 rotor) for 10 minutes at 4°C.

10. Decant the buffer from the cell pellets. Stand the tubes in an inverted position for 1 minute to allow the last traces of buffer to drain away.

11. Resuspend the pellets by swirling or gentle vortexing in 4 ml (per 50-ml tube) of ice-cold TFB or FSB. Proceed either with Step 12a if the competent cells are to be used immediately or with Step 12b if the competent cells are to be stored at –70°C and used at a later date.

12. Prepare competent cells for transformation.

   To prepare fresh competent cells

   a. Add 140 μl of DnD solution into the center of each cell suspension. Immediately mix by swirling gently and then store the suspension on ice for 15 minutes.

   b. Add an additional 140 μl of DnD solution to each suspension. Mix by swirling gently and then store the suspension on ice for an additional 15 minutes.

   c. Dispense aliquots of the suspensions into chilled, sterile, 17 × 100-mm polypropylene tubes. Store the tubes on ice.

   To prepare frozen stocks of competent cells

   a. Add 140 μl of DMSO per 4 ml of resuspended cells. Mix gently by swirling and store the suspension on ice for 15 minutes.

   b. Add an additional 140 μl of DMSO to each suspension. Mix gently by swirling and return the suspensions to an ice bath.

   c. Working quickly, dispense aliquots of the suspensions into chilled, sterile microfuge tubes or tissue-culture vials. Immediately snap-freeze the competent cells by immersing the tightly closed tubes in a bath of liquid nitrogen. Store the tubes at –70°C until needed.

d. When needed, remove a tube of competent cells from the –70°C freezer. Thaw the cells by holding the tube in the palm of the hand. Just as the cells thaw, transfer the tube to an ice bath. Store the cells on ice for 10 minutes.

e. Use a chilled, sterile pipette tip to transfer the competent cells to chilled, sterile 17 × 100-mm polypropylene tubes. Store the cells on ice.

Include all of the appropriate positive and negative controls.

13. Add the transforming DNA (up to 25 ng per 50 μl of competent cells) in a volume not exceeding 5% of that of the competent cells. Swirl the tubes gently several times to mix their contents. Set up at least two control tubes for each transformation experiment, including a tube of competent bacteria that receives a known amount of a standard preparation of superhelical plasmid DNA and a tube of cells that receives no plasmid DNA at all. Store the tubes on ice for 30 minutes.

14. Transfer the tubes to a rack placed in a preheated 42°C circulating water bath. Store the tubes in the rack for exactly 90 seconds. Do not shake the tubes.

15. Rapidly transfer the tubes to an ice bath. Allow the cells to cool for 1–2 minutes.

16. Add 800 μl of SOC medium to each tube. Warm the cultures to 37°C in a water bath and then transfer the tubes to a shaking incubator set at 37°C. Incubate the cultures for 45 minutes to allow the bacteria to recover and to express the antibiotic resistance marker encoded by the plasmid.

17. Transfer the appropriate volume (up to 200 μl per 90-mm plate) of transformed competent cells onto agar SOB medium containing 20 mM $MgSO_4$ and the appropriate antibiotic.

18. Store the plates at room temperature until the liquid has been absorbed.

19. Invert the plates and incubate them at 37°C. Transformed colonies should appear in 12–16 hours.

The Inoue Method for Preparation and Transformation of Competent *E. Coli:* "Ultra-competent" Cells

This method to prepare optimally competent *E. coli* differs from other procedures in that the bacterial culture is grown at 18°C rather than the conventional 37°C. Incubating large-scale cultures at 18°C can be a challenge. One solution is to move the bacterial shaker into a cold room. However, if this is impossible, the cultures may be grown at 20–23°C with little loss of transforming efficiency.

MATERIALS

CAUTION: Please see Appendix 4 for appropriate handling of materials marked with <!>.

Reagents and Solutions

Please see Appendix 1 for components of stock solutions, buffers, and reagents. Dilute stock solutions to the appropriate concentrations.

DMSO <!>
Inoue transformation buffer

Vectors and Hosts

Plasmid DNA for transformation
Strain of *E. coli* suitable for transformation

Media and Antibiotics

LB or SOB medium
SOB agar plates containing 20 mM $MgSO_4$ and the appropriate antibiotic
    Standard SOB contains 10 mM $MgSO_4$.
SOB or LB agar plates
SOC medium

Centrifuges/Rotors/Tubes

Sorvall SLC-1500 rotor (4°C) and centrifuge bottles

Additional Items

Circulating water bath (42°C)
Freezer (–70°C)
Ice-water bath
Liquid nitrogen bath
Nalgene filters (0.45-μm pore size)
Sterile microfuge tubes
Sterile 17 x 100-mm polypropylene tubes, chilled in ice
Vacuum aspirator

Additional Information

Inoue H., Nojima H., and Okayama H. 1990. High efficiency transformation of *Escherichia coli* with plasmids. *Gene* **96:** 23–28.

METHOD

1. Prepare Inoue transformation buffer (chilled to 0°C before use).

   a. Prepare 0.5 M PIPES (pH 6.7) (piperazine-1,2-bis[2-ethanesulfonic acid]) by dissolving 15.1 g of PIPES in 80 ml of pure H_2O (Milli-Q or its equivalent). Adjust the pH of the solution to 6.7 with 5 M KOH and then add pure H_2O to bring the final volume to 100 ml. Sterilize the solution by filtration through a disposable prerinsed Nalgene filter (0.45-μm pore size). Divide into aliquots and store frozen at –20°C.

   b. Prepare Inoue transformation buffer by dissolving all of the solutes listed below in 800 ml of pure H_2O and then add 20 ml of 0.5 M PIPES (pH 6.7). Adjust the volume of the Inoue transformation buffer to 1 liter with pure H_2O.

Reagent	Amount per Liter	Final Concentration
$MnCl_2 \cdot 4H_2O$	10.88 g	55 mM
$CaCl_2 \cdot 2H_2O$	2.20 g	15 mM
KCl	18.65 g	250 mM
PIPES (0.5 M, pH 6.7)	20 ml	10 mM
H_2O	to 1 liter	

   c. Sterilize Inoue transformation buffer by filtration through a prerinsed 0.45-μm Nalgene filter. Divide into aliquots and store at –20°C.

2. Pick a single bacterial colony (2–3 mm in diameter) from a plate that has been inoculated with a strain of *E. coli* suitable for transformation and incubated for 16–20 hours at 37°C. Transfer the colony into 25 ml of SOB medium (LB may be used instead) in a 250-ml flask. Incubate the culture for 6–8 hours at 37°C with vigorous shaking (250–300 rpm).

3. Use this starter culture to inoculate three 1-liter flasks, each containing 250 ml of SOB. The first flask receives 10 ml of starter culture, the second receives 4 ml, and the third receives 2 ml. Incubate all three flasks for 14 hours at 18–22°C with moderate shaking.

4. Read the OD_{600} of all three cultures. Continue to monitor the OD every 45 minutes.

5. When the OD_{600} of one of the cultures reaches exactly 0.55, transfer the culture vessel to an ice-water bath for 10 minutes. Discard the two other cultures.

6. Harvest the cells by centrifugation at 2500*g* (3900 rpm in a Sorvall SLC-1500 rotor) for 10 minutes at 4°C.

7. Pour off the medium and store the open centrifuge bottle on a stack of paper towels for 2 minutes. Use a vacuum aspirator to remove any drops of remaining medium adhering to the walls of the centrifuge bottle or trapped in its neck.

8. Resuspend the cells gently in 80 ml of ice-cold Inoue transformation buffer.

9. Harvest the cells by centrifugation at 2500*g* (3900 rpm in a Sorvall SLC-1500 rotor) for 10 minutes at 4°C.

10. Pour off the medium and store the open centrifuge tube on a stack of paper towels for 2 minutes. Use a vacuum aspirator to remove any drops of remaining medium adhering to the walls of the centrifuge tube or trapped in its neck.

11. Resuspend the cells gently in 20 ml of ice-cold Inoue transformation buffer.

12. Add 1.5 ml of DMSO. Mix the bacterial suspension by swirling and then store it in ice for 10 minutes.

13. Working quickly, dispense aliquots of the suspensions into chilled, sterile microfuge tubes. Immediately snap-freeze the competent cells by immersing the tightly closed tubes in a bath of liquid nitrogen. Store the tubes at –70°C until needed.

14. When needed, remove a tube of competent cells from the –70°C freezer. Thaw the cells by holding the tube in the palm of the hand. Just as the cells thaw, transfer the tube to an ice bath. Store the cells on ice for 10 minutes.

15. Use a chilled, sterile pipette tip to transfer the competent cells to chilled, sterile 17 x 100-mm polypropylene tubes in an ice bath. Store the cells on ice.

Include all of the appropriate positive and negative controls.

16. Add the transforming DNA (up to 25 ng per 50 μl of competent cells) in a volume not exceeding 5% of that of the competent cells. Swirl the tubes gently several times to mix their contents. Set up at least two control tubes for each transformation experiment, including a tube of competent bacteria that receives a known amount of a standard preparation of superhelical plasmid DNA and a tube of cells that receives no plasmid DNA at all. Store the tubes on ice for 30 minutes.

17. Transfer the tubes to a rack placed in a preheated 42°C circulating water bath. Store the tubes in the rack for exactly 90 seconds. Do not shake the tubes.

18. Rapidly transfer the tubes to an ice bath. Allow the cells to cool for 1–2 minutes.

19. Add 800 μl of SOC medium to each tube. Warm the cultures to 37°C in a water bath and then transfer the tubes to a shaking incubator set at 37°C. Incubate the cultures for 45 minutes to allow the bacteria to recover and to express the antibiotic resistance marker encoded by the plasmid.

20. Transfer the appropriate volume (up to 200 μl per 90-mm plate) of transformed competent cells onto agar SOB medium containing 20 mM $MgSO_4$ and the appropriate antibiotic.

21. Store the plates at room temperature until the liquid has been absorbed.

22. Invert the plates and incubate them at 37°C. Transformed colonies should appear in 12–16 hours.

Preparation and Transformation of Competent *E. coli* Using Calcium Chloride

This protocol, first developed during the early dawn of molecular cloning (1972), is used to prepare batches of competent bacteria that yield 5×10^6 to 2×10^7 transformed colonies/μg of supercoiled plasmid DNA.

MATERIALS

Reagents and Solutions

Please see Appendix 1 for components of stock solutions, buffers, and reagents. Dilute stock solutions to the appropriate concentrations.

$CaCl_2$ (1 M, sterile and ice cold)

$MgCl_2$-$CaCl_2$ solution (sterile and ice cold)
80 mM $MgCl_2$
20 mM $CaCl_2$
Standard transformation buffer STB (see Protocol 1.23, Step 1) may be substituted for the $MgCl_2$-$CaCl_2$ solution in Step 5 of this protocol.

Vectors and Hosts

Plasmid DNA for transformation
Strain of *E. coli* suitable for transformation

Media and Antibiotics

SOB agar plates containing 20 mM $MgSO_4$ and the appropriate antibiotic
Standard SOB contains 10 mM $MgSO_4$.
SOB medium
SOC medium

Centrifuges/Rotors/Tubes

Sorvall SS-34 rotor (4°C)
Sterile 17 x 100-mm polypropylene tubes
Sterile 50-ml polypropylene centrifuge tubes

Additional Items

Circulating water bath (42°C)

Additional Information

Cohen S.N., Chang A.C.Y., and Hsu L. 1972. Nonchromosomal antibiotic resistance in bacteria: Genetic transformation of *Escherichia coli* by R-factor DNA. *Proc. Natl. Acad. Sci.* **69:** 2110–2114.

METHOD

1. Pick a single bacterial colony (2–3 mm in diameter) from a plate that has been inoculated with a strain of *E. coli* suitable for transformation and incubated for 16–20 hours at 37°C. Transfer the colony into 100 ml of SOB medium (LB may be used) in a 1-liter flask. Incubate

Adapted from Chapter 1, Protocol 25, p. 1.116 of MC3.

the culture for 3 hours at 37°C with vigorous agitation, monitoring the growth of the culture.

It is essential that the density of the culture not exceed 10^8 cells/ml (1 $OD_{600} \leq 0.4$).

2. Transfer the bacterial cells to sterile, disposable, ice-cold 50-ml polypropylene tubes. Cool the cultures to 0°C by storing the tubes on ice for 10 minutes.

3. Recover the cells by centrifugation at 2700g (4100 rpm in a Sorvall SS-34 rotor) for 10 minutes at 4°C.

4. Decant the medium from the cell pellets. Stand the tubes in an inverted position on a pad of paper towels for 1 minute to allow the last traces of media to drain away.

5. Resuspend each pellet by swirling or gentle vortexing in 30 ml of sterile, ice-cold $MgCl_2$-$CaCl_2$ solution.

6. Recover the cells by centrifugation at 2700g (4100 rpm in a Sorvall SS-34 rotor) for 10 minutes at 4°C.

7. Decant the medium from the cell pellets. Stand the tubes in an inverted position on a pad of paper towels for 1 minute to allow the last traces of media to drain away.

8. Resuspend the pellet by swirling or gentle vortexing in 2 ml of sterile, ice-cold 0.1 M $CaCl_2$ (or TFB) for each 50 ml of original culture.

   When preparing competent cells, thaw a 10-ml aliquot of the $CaCl_2$ stock solution and dilute it to 100 ml with 90 ml of pure H_2O. Sterilize the solution by filtration through a prerinsed Nalgene filter (0.45-μm pore size) and then chill it to 0°C.

   For many strains of *E. coli*, standard TFB (Protocol 1.23) may be used instead of $CaCl_2$ with equivalent or better results.

9. At this point, either use the cells directly for transformation as described in Steps 10–16 below or dispense into aliquots and freeze at –70°C (please see Protocol 1.23, Step 12).

Include all of the appropriate positive and negative controls.

10. To transform the $CaCl_2$-treated cells directly, transfer 200 μl of each suspension of competent cells to a sterile, chilled 17 x 100-mm polypropylene tube using a chilled micropipette tip. Add DNA (no more than 50 ng in a volume of 10 μl or less) to each tube. Mix the contents of the tubes by swirling gently. Store the tubes on ice for 30 minutes.

11. Transfer the tubes to a rack placed in a preheated circulating water bath at 42°C. Store the tubes in the rack for exactly 90 seconds. Do not shake the tubes.

12. Rapidly transfer the tubes to an ice bath. Allow the cells to chill for 1–2 minutes.

13. Add 800 μl of SOC medium to each tube. Incubate the cultures for 45 minutes in a water bath set at 37°C to allow the bacteria to recover and to express the antibiotic resistance marker encoded by the plasmid.

14. Transfer the appropriate volume (up to 200 μl per 90-mm plate) of transformed competent cells onto agar SOB medium containing 20 mM $MgSO_4$ and the appropriate antibiotic.

15. Store the plates at room temperature until the liquid has been absorbed.

16. Invert the plates and incubate at 37°C. Transformed colonies should appear in 12–16 hours.

Transformation of *E. coli* by Electroporation

Electrocompetent bacteria are prepared by growing cultures to mid-log phase, washing the bacteria extensively at low temperature, and then resuspending them in a solution of low ionic strength containing glycerol. DNA is introduced during exposure of the bacteria to a short high-voltage electrical discharge.

MATERIALS

Reagents and Solutions

Please see Appendix 1 for components of stock solutions, buffers, and reagents. Dilute stock solutions to the appropriate concentrations.

Glycerol (10% v/v), molecular biology grade, ice cold

Pure H_2O

Milli-Q or its equivalent, sterilized by filtration through prerinsed 0.45-μm filters. Store at 4°C.

Vectors and Hosts

Plasmid DNA for transformation

Ideally, the concentration of the DNA to be electroporated should be 1–10 μg/ml in H_2O or TE (pH 8.0). When using the products of a DNA ligation reaction for electroporation, the reaction mixture should be diluted 10- to 20-fold in H_2O. Alternatively, the ligated DNA can be purified by spun-column chromatography. For the construction of cDNA libraries, where high efficiencies are required and cotransformants are undesirable, total DNA concentrations of <10 ng/ml are recommended.

Strain of *E. coli* suitable for transformation

Media and Antibiotics

GYT medium, ice cold

10% (v/v) glycerol

0.125% (w/v) yeast extract

0.25% (w/v) tryptone

LB medium, warmed to 37°C

SOB agar plates containing 20 mM $MgSO_4$ and the appropriate antibiotic

Standard SOB contains 10 mM $MgSO_4$.

SOC medium

Centrifuges/Rotors/Tubes

Sorvall SLC-1500 rotor (4°C) and centrifuge bottles (ice cold)

Additional Items

Electroporation device and cuvettes (0.2-cm gap)

Freezer (–70°C)

Ice-water bath

Sterile 0.5-ml microfuge tubes

Adapted from Chapter 1, Protocol 26, p. 1.119 of MC3.

METHOD

1. Inoculate a single colony of *E. coli* from a fresh agar plate into a flask containing 50 ml of LB medium. Incubate the culture overnight at 37°C with vigorous aeration (250 rpm in a rotary shaker).

2. Inoculate two aliquots of 500 ml of prewarmed LB medium in separate 2-liter flasks with 25 ml of the overnight bacterial culture. Incubate the flasks at 37°C with agitation (300 cycles/ minute in a rotary shaker). Measure the OD_{600} of the growing bacterial cultures every 20 minutes.

3. When the OD_{600} of the cultures reaches 0.4, rapidly transfer the flasks to an ice-water bath for 15–30 minutes. Swirl the culture occasionally to ensure that cooling occurs evenly. In preparation for the next step, place the centrifuge bottles in an ice-water bath.

4. Transfer the cultures to ice-cold centrifuge bottles. Harvest the cells by centrifugation at 1000*g* (2500 rpm in a Sorvall SLC-1500 rotor) for 15 minutes at 4°C. Decant the supernatant and resuspend the cell pellet in 500 ml of ice-cold pure H_2O.

5. Harvest the cells by centrifugation at 1000*g* (2500 rpm in a Sorvall SLC-1500 rotor) for 20 minutes at 4°C. Decant the supernatant and resuspend the cell pellet in 250 ml of ice-cold 10% glycerol.

6. Harvest the cells by centrifugation at 1000*g* (2500 rpm in a Sorvall SLC-1500 rotor) for 20 minutes at 4°C. Decant the supernatant and resuspend the pellet in 10 ml of ice-cold 10% glycerol.

7. Harvest cells by centrifugation at 1000*g* (2500 rpm in a Sorvall SLC-1500 rotor) for 20 minutes at 4°C. Carefully decant the supernatant and use a pasteur pipette attached to a vacuum line to remove any remaining drops of buffer. Resuspend the pellet by gentle swirling in 1 ml of ice-cold GYT medium.

8. Measure the OD_{600} of a 1:100 dilution of the cell suspension. Dilute the cell suspension to a concentration of 2×10^{10} to 3×10^{10} cells/ml ($1.0\ OD_{600} = {\sim}2.5 \times 10^8$ cells/ml) with ice-cold GYT medium.

9. Transfer 40 µl of the suspension to an ice-cold electroporation cuvette (0.2-cm gap) and test whether arcing occurs when an electrical discharge is applied (please see Step 16 below). If so, wash the remainder of the cell suspension once more with ice-cold GYT medium to ensure that the conductivity of the bacterial suspension is sufficiently low (< 5 mEq).

10. To use the electrocompetent cells immediately, proceed directly to Step 12. Otherwise, store the cells at –70°C until required. For storage, dispense 40-µl aliquots of the cell suspension into sterile, ice-cold, 0.5-ml microfuge tubes, drop into a bath of liquid nitrogen, and transfer to a –70°C freezer.

11. To use frozen electrocompetent cells, remove an appropriate number of aliquots of cells from the –70°C freezer. Store the tubes at room temperature until the bacterial suspensions are thawed and then transfer the tubes to an ice bath.

12. Pipette 40 µl of the freshly made (or thawed) electrocompetent cells into ice-cold, sterile 0.5-ml microfuge tubes. Place the cells on ice, together with an appropriate number of bacterial electroporation cuvettes.

13. Add 10 pg to 25 ng of the DNA to be electroporated in a volume of 1–2 µl to each microfuge tube and incubate the tube on ice for 30–60 seconds. Include all of the appropriate positive and negative controls.

14. Set the electroporation apparatus to deliver an electrical pulse of 25 µF capacitance, 2.5 kV, and 200-ohm resistance.

15. Pipette the DNA/cell mixture into a cold electroporation cuvette. Tap the solution to ensure that the suspension of bacteria and DNA sits at the bottom of the cuvette. Dry condensation and moisture from the outside of the cuvette. Place the cuvette in the electroporation device.

16. Deliver a pulse of electricity to the cells at the settings indicated in Step 14. A time constant of 4–5 msec with a field strength of 12.5 kV/cm should register on the machine.

17. As quickly as possible after the pulse, remove the electroporation cuvette and add 1 ml of SOC medium at room temperature.

18. Transfer the cells to a 17 x 100-mm or 17 x 150-mm polypropylene tube and incubate the cultures with gentle rotation for 1 hour at 37°C.

19. Plate different volumes (up to 200 µl per 90-mm plate) of the electroporated cells onto SOB agar medium containing 20 mM $MgSO_4$ and the appropriate antibiotic.

20. Store the plates at room temperature until the liquid has been absorbed.

21. Invert the plates and incubate them at 37°C. Transformed colonies should appear in 12–16 hours.

Screening Bacterial Colonies Using X-gal and IPTG: α-Complementation

α-Complementation occurs when two inactive fragments of *E. coli* β-galactosidase associate to form a functional enzyme. Many plasmid vectors carry a short segment of DNA containing the coding information for the first 146 amino acids of β-galactosidase. Vectors of this type are used in host cells that express the carboxy-terminal portion of the enzyme. Although neither the host nor the plasmid-encoded fragments of β-galactosidase are themselves active, they can associate to form an enzymatically active protein. Lac+ bacteria that result from α-complementation are easily recognized because they form blue colonies in the presence of the chromogenic substrate X-gal. However, insertion of a fragment of foreign DNA into the polycloning site of the plasmid almost invariably results in production of an amino-terminal fragment that is no longer capable of α-complementation. Bacteria carrying recombinant plasmids therefore form white colonies. The development of a simple blue-white color test greatly simplified the identification of recombinants constructed in plasmid vectors. This protocol describes how to use this color test to identify transformed colonies of *E. coli* whose ability to display α-complementation has been compromised. Such transformants are likely to contain recombinant plasmids of the desired structure.

MATERIALS

Reagents and Solutions

Please see Appendix 1 for components of stock solutions, buffers, and reagents. Dilute stock solutions to the appropriate concentrations.

IPTG (20% w/v)

> IPTG (isopropyl-β-D-thiogalactoside) is a nonfermentable analog of lactose that inactivates the *lacZ* repressor and therefore induces transcription of the *lac* operon. However, most strains of *E. coli* used for α-complementation do not synthesize significant quantities of *lac* repressor. Consequently, there is usually no need to induce synthesis of the host- and plasmid-encoded fragments of β-galactosidase for histochemical screening of transformed colonies. Only if the bacterial strain carries an I^Q allele of the *lac* repressor or if the plasmid carries a *lacI* gene should IPTG be used to induce synthesis of fragments of the enzyme.

X-gal (2% w/v)

Vectors and Hosts

Liquid culture of *E. coli* transformed by a recombinant construct

Media and Antibiotics

LB agar plates (90 or 150 mm) containing the appropriate antibiotic for selection of transformants
Molten top agar

Additional Items

Disposable and sterile polypropylene tubes (17 x 100-mm)
Heating block (45°C)

Additional Information

α-Complementation	MC3, pp. 1.149–1.150
Alternative protocol (Direct Application of X-gal and IPTG to Agar Plates)	MC3, p. 1.125

TABLE 1-4 Components for Top Agar

Size of Plate (mm)	Amount of Reagent		
	Molten Top Agar	X-gal	IPTG[a]
90	3 ml	40 µl	7 µl
150	7 ml	100 µl	20 µl

[a]May not be required; please see the entry on IPTG in the Materials list.

METHOD

1. Dispense aliquots of molten top agar into 17 × 100-mm tubes. Place the tubes in a 45°C heating block until they are needed.

2. Remove the first tube from the heating block. Working quickly, add 0.1 ml of a suspension containing <3000 (for a 90-mm plate) and <10,000 (for a 150-mm plate) viable transformants of *E. coli*. Close the top of the tube and invert it several times to disperse the bacteria through the molten agar.

3. Open the tube and add the appropriate amounts of X-gal and IPTG (if required) as shown in the table above. Close the top of the tube and gently invert it several times to mix the contents.

4. Quickly pour the molten top agar into the center of a hardened agar plate containing the appropriate antibiotic and distribute the solution by swirling.

5. Repeat Steps 2–4 until all of the samples have been plated.

6. Allow the soft agar to harden at room temperature, wipe any condensation from the lid of the plates, and incubate the plates in an inverted position for 12–16 hours at 37°C.

7. Remove the plates from the incubator and store them for several hours at 4°C to allow the blue color to develop.

8. Identify colonies carrying recombinant plasmids.

   • Colonies that carry wild-type plasmids contain active β-galactosidase. These colonies are pale blue in the center and dense blue at their periphery.

   • Colonies that carry recombinant plasmids do not contain active β-galactosidase. These colonies are creamy white or eggshell blue, sometimes with a faint blue spot in the center.

9. Select and culture colonies carrying recombinant plasmids.

Screening Bacterial Colonies by Hybridization: Small Numbers

This procedure is used to screen a small number of bacterial colonies (<200) that are dispersed over several agar plates and are to be screened by hybridization to the same radiolabeled probe. The colonies are gridded onto a master plate and onto a nitrocellulose or nylon filter laid on the surface of a second agar plate. After a period of growth, the colonies on the filer are lysed and processed for hybridization. The master plate is stored until the results of the screening procedure become available.

MATERIALS

Vectors and Hosts

Agar plate containing *E. coli* transformed by a nonrecombinant plasmid
Population of *E. coli* transformed by a recombinant plasmid construct

Media and Antibiotics

LB agar plates containing the appropriate antibiotic

Additional Items

Graph paper with 1-cm-square grid
Hypodermic needle (18 gauge) attached to 5-cc syringe
India ink
Nitrocellulose or nylon filters (sterile) to fit agar plates
Sterile toothpicks or inoculating loops

METHOD

1. Place a sterile nitrocellulose or nylon filter on an agar plate containing the selective antibiotic.

2. Draw a numbered grid on a piece of graph paper (1-cm-square grid). Number the base of each agar master plate and place the plate on the grid. Draw a mark on the side of the plate at the 6 o'clock position.

   Each 90-mm plate can accommodate approximately 100 bacterial streaks.

3. Use sterile toothpicks or inoculating loops to transfer bacterial colonies one by one onto the filter on the test plate and then onto the master agar plate that contains the selective antibiotic but no filter. Make small streaks 2–3 mm in length (or dots) arranged according to the grid pattern under the dish. Streak each colony in an identical position on both plates.

4. Finally, as a negative control, streak a colony containing a nonrecombinant plasmid onto the filter and the master plate.

5. Invert the plates and incubate them at 37°C until the bacterial streaks have grown to a width of 0.5–1.0 mm (typically 6–16 hours).

6. Mark the filter in three or more asymmetric locations by stabbing through it and into the agar of the test plate with an 18-gauge needle, attached to a syringe, dipped in waterproof black drawing ink (India Ink). Mark the master plate in approximately the same locations.

7. Seal the master plate with Parafilm and store it at 4°C in an inverted position until the results of the hybridization reaction are available.

8. Lyse the bacteria adhering to the filter and bind the liberated DNA to the nitrocellulose or nylon filter using the procedures described in Protocol 1.31. Proceed with hybridization as described in Protocol 1.32.

Screening Bacterial Colonies by Hybridization: Intermediate Numbers

Bacterial colonies growing on agar plates are transferred en masse to nitrocellulose filters. The spatial arrangement of colonies on the plates is preserved on the filters. After transfer, the filters are processed for hybridization to an appropriate radiolabeled probe, while the original (master) plate is incubated for a few hours to allow the bacterial colonies to regrow in their original positions. The procedure works best with 90-mm plates containing < 2500 colonies.

MATERIALS

CAUTION: Please see Appendix 4 for appropriate handling of materials marked with <!>.

Vectors and Hosts

Population of *E. coli* transformed by a recombinant plasmid construct
Use bacteria transformed by one of the methods described in Protocols 1.23–1.26.

Media and Antibiotics

LB or SOB agar plates containing chloramphenicol (170–200 µg/ml) <!> (optional; see Step 6)
SOB or LB agar plates (90 mm) containing the appropriate antibiotic

Additional Items

Blunt-ended forceps
Graph paper with 1-cm-square grid
Hypodermic needle (23 gauge) attached to 3-cc syringe
Nitrocellulose or nylon filters (sterile and detergent-free) to fit agar plates
Reagents listed in Protocols 1.31 and 1.32
Sterile toothpicks or inoculating loops
Waterproof black drawing ink (India ink)
Whatman 3MM circular filter papers

Additional Information

Alternative protocol (Rapid Lysis of Colonies and Binding
of DNA to Nylon Filters)   MC3, p. 1.131

METHOD

1. Plate the transformed *E. coli* culture onto 90-mm LB or SOB agar plates, at dilutions calculated to generate up to 2500 transformed colonies. When the colonies reach an average size of 1.5 mm, transfer the plates from the incubator to a cold room.

2. Number the nylon or nitrocellulose dry filters with a soft-lead pencil or a ball-point pen, wet them with water, and interleave them between dry Whatman 3MM filters. Wrap the stack of filters loosely in aluminum foil and sterilize them by autoclaving (15 psi [1.05 kg/cm²] for 10 minutes on liquid cycle).

3. Place a dry, sterile, detergent-free nitrocellulose or nylon filter, numbered side down, on the surface of the LB (or SOB) agar medium, in contact with the bacterial colonies (plated in Step 1), until it is completely wet.

4. Once the filter is in place, key the filter to the underlying medium by stabbing in three or more asymmetric locations through the filter with a 23-gauge needle attached to a syringe, dipped in waterproof black drawing ink.

5. Grip the edge of the filter with blunt-ended forceps and, in a single smooth movement, peel the filter from the surface of the agar.

6. Proceed with one of the following options as appropriate:

   • Lyse the bacteria adhering to the filter and bind the liberated DNA to the nitrocellulose or nylon filter using the procedures described in Protocol 1.31. Proceed with hybridization as described in Protocol 1.32.

   • Place the filter, colony side up, on the surface of a fresh LB (or SOB) agar plate containing the appropriate antibiotic. After incubation for a few hours, when the colonies have grown to a size of 2–3 mm, remove the filter and proceed with lysis and hybridization as described in Protocols 1.31 and 1.32.

   • Amplify the colonies on the filter by transferring the filter to an agar plate containing chloramphenicol (170–200 µg/ml) and incubating them for 12 hours at 37°C. Proceed with lysis and hybridization (Protocols 1.31 and 1.32).

     This amplification procedure, which results in a more intense signal when the transformants are screened by hybridization, is necessary only when the copy number of the plasmid is low or when highly degenerate oligonucleotides are used as probes.

   • Use the filter to prepare a second replica:

     a. Place the filter colony side up on the surface of a fresh LB (or SOB) agar plate containing the appropriate antibiotic.

     b. Lay a dry sterile nitrocellulose filter carefully on top of the first and key to it as described in Step 4 above.

     c. Incubate the "filter sandwich" for several hours at 37°C.

     d. Proceed with lysis and hybridization (Protocols 1.31 and 1.32), keeping the filters as a sandwich during the lysis and neutralization steps, but peeling them apart before the final wash.

7. Incubate the master plate for 5–7 hours at 37°C until the colonies have regrown. Seal the plate with Parafilm and store it at 4°C in an inverted position.

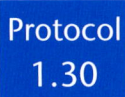

Protocol 1.30

Screening Colonies by Hybridization: Large Numbers

This procedure is used to plate, replicate, and subsequently screen large numbers of bacterial colonies (up to 2×10^4 colonies per 90-mm plate or 10^4 colonies per 90-mm plate).

MATERIALS

CAUTION: Please see Appendix 4 for appropriate handling of materials marked with <!>.

Vectors and Hosts

Liquid culture of *E. coli,* transformed with recombinant plasmids or a cDNA library
  Use bacteria transformed by one of the methods described in Protocols 1.23–1.26.

Media and Antibiotics

LB or SOB agar plates containing chloramphenicol (170–200 µg/ml) <!> (optional; see Step 12)
LB or SOB agar plates containing the appropriate antibiotic and 25% (v/v) glycerol
LB or SOB plates containing the appropriate antibiotic
  Plates that are 2–3 days old give the best results.

Additional Items

Bent glass rod to spread bacteria on filters
Needle (18 gauge) attached to 3-cc syringe
Nitrocellulose or nylon filters (sterile and detergent-free) to fit agar plates
Step 14 of this protocol requires the reagents listed in Protocols 1.23–1.26.
Whatman 3MM filters

Additional Information

Hanahan D. and Meselson M. 1983. Plasmid screening at high colony density. *Methods Enzymol.* **100:** 333–342.

METHOD

1. Number the dry nylon or nitrocellulose filters with a soft-lead pencil or a ball-point pen, wet them with water, and interleave them between dry Whatman 3MM filters. Wrap the stack of filters loosely in aluminum foil and sterilize them by autoclaving (15 psi [1.05 kg/cm²] for 10 minutes on liquid cycle).

2. Use sterile, blunt-ended forceps to lay a sterile nylon or nitrocellulose filter, numbered side down, on a 2–3-day-old LB (or SOB) agar plate containing the appropriate antibiotic. When the filter is thoroughly wet, peel it from the plate and replace it, numbered side up, on the surface of the agar.

3. Apply the bacteria, in a small volume of liquid (0.2 ml), to the center of the filter on the surface of the agar plate. Use a sterile glass spreader to disperse the fluid evenly, leaving a border 2–3 mm wide around the circumference of the filter free of bacteria.
     The filter will become a master from which replicas can be made.

4. Incubate the plate (noninverted) with the lid ajar for a few minutes in a laminar flow hood to allow the inoculum to evaporate. Then close the lid, invert the plate, and incubate at 37°C until small colonies (0.1–0.2-mm diameter) appear (~8–10 hours).

5. If desired, replica filters may be prepared at this stage (proceed with Step 6). Otherwise, prepare the bacterial colonies for storage at –20°C:

   a. Transfer the filter colony side up to a labeled LB (or SOB) agar plate containing the appropriate antibiotic and 25% glycerol.

   b. Incubate the plate for 2 hours at 37°C.

   c. Seal the plate well with Parafilm and store it in an inverted position in a sealed plastic bag at –20°C.

To prepare replica filters

6. Lay the master nitrocellulose or nylon filter colony side up on a sterile Whatman 3MM filter.

7. Number a damp, sterile nitrocellulose or nylon filter, and lay it on the master filter. Take care to prevent air bubbles from becoming trapped between the two filters.

8. Cover the filter sandwich with a second 3MM circle and place the bottom of a petri dish on top of the 3MM paper. Press down firmly on the petri dish with the palm of the hand to facilitate transfer of bacteria from the master filter to the replica.

9. Dismantle the petri dish bottom and top 3MM paper, and orient the two nitrocellulose or nylon filters by making a series of holes with an 18-gauge needle attached to a syringe.

10. Peel the filters apart. Lay the replica on a fresh LB (or SOB) agar plate containing the appropriate antibiotic.

    A second replica filter can be made at this stage by repeating Steps 6–9. Key the second replica to the existing holes in the master filter.

11. Place the second replica filter (if made) and the master filter on a fresh LB (or SOB) agar plate containing the appropriate antibiotic and incubate all plates at 37°C until colonies appear (4–6 hours).

12. At this stage, when the bacterial colonies are still growing rapidly, the filter may be transferred to an agar plate containing chloramphenicol and incubated for an additional 12 hours at 37°C.

    This amplification procedure, which results in a more intense signal when the transformants are screened by hybridization, is necessary only when the copy number of the plasmid is low or when highly degenerate oligonucleotides are used as probes.

13. Transfer the master nitrocellulose filter to a fresh LB (or SOB) agar plate containing the appropriate antibiotic and 25% glycerol. Then freeze it as described in Step 5.

14. Lyse the bacteria adhering to the replica filters and bind the liberated DNA to the nitrocellulose or nylon filter using the procedures described in Protocol 1.31. Proceed with hybridization as described in Protocol 1.32.

Protocol 1.31 — Lysing Colonies and Binding DNA to Filters

In this protocol, alkali is used to liberate DNA from bacterial colonies on nitrocellulose or nylon filters. The DNA is then fixed to the filter by UV cross-linking or baking under vacuum.

MATERIALS

Reagents and Solutions

Please see Appendix 1 for components of stock solutions, buffers, and reagents. Dilute stock solutions to the appropriate concentrations.
Denaturation solution
Neutralizing solution
SDS (10% w/v) (optional; see Steps 1 and 3)
2x SSPE

Vectors and Hosts

Nitrocellulose or nylon filters carrying colonies of *E. coli* transformants (see Protocols 1.28–1.30)

Additional Items

Blunt-ended forceps
Glass or plastic trays for processing batches of filters
Glass pipette (10 ml) or stout glass rod
Whatman 3MM paper sheets
Vacuum baking oven preset to 80°C *or* UV cross-linking device

---

METHOD

1. Cut four pieces of Whatman 3MM paper to an appropriate size and shape and fit them neatly onto the bottoms of four or five (see Step 5 below) glass or plastic trays. Saturate each of the pieces of 3MM paper with one of the following solutions:

   10% SDS (optional)
   denaturing solution
   neutralizing solution
   2x SSPE

2. Pour off any excess liquid and roll a 10-ml pipette along the sheet to smooth out any air bubbles that occur between the 3MM paper and the bottom of the container.

3. Use blunt-ended forceps to peel the nitrocellulose or nylon filters from their plates and place them colony side up on the SDS-impregnated 3MM paper for 3 minutes.

4. After the first filter has been exposed to the SDS solution for 3 minutes, transfer it to the second tray containing 3MM paper saturated with denaturing solution. Transfer the remainder of the filters in the same order in which they were removed from their agar plates. Expose each filter to the denaturing solution for 5 minutes.

5. Transfer the filters to the third sheet of 3MM paper, which has been saturated with neutralizing solution. Leave the filters for 5 minutes.

   (*Optional*) Transfer the filters to a second tray containing neutralizing solution.

6. Transfer the filters to the last sheet of 3MM paper, which has been saturated with 2x SSPE. Leave the filters for 5 minutes.

7. Dry the filters using one of the methods below.

If the DNA is to be fixed to the filters by baking

Lay the filters, colony side up, on a sheet of dry 3MM paper and allow them to dry at room temperature for at least 30 minutes.

If the DNA is to be fixed to the filters by cross-linking with UV light

Lay the filters on a sheet of 2x SSPE-impregnated 3MM paper or on dry 3MM paper, depending on the manufacturer's recommendation.

8. Fix the DNA to the filters using one of the methods below.

For baking

Sandwich the dried filters between two sheets of dry 3MM paper and fix the DNA to the filters by baking for 1–2 hours at 80°C in a vacuum oven.

For cross-linking with UV light

Follow the manufacturer's instructions for fixing DNA to filters using a commercial device.

9. Hybridize the DNA immobilized on the filters to a labeled probe as described in Protocol 1.32.

Hybridization of Bacterial DNA on Filters

This protocol describes procedures to hybridize DNA from transformed colonies immobilized on filters with radiolabeled probes and to recover from a master plate the corresponding colonies that hybridize specifically to the probe.

MATERIALS

CAUTION: Please see Appendix 4 for appropriate handling of materials marked with <!>.

Reagents and Solutions

Please see Appendix 1 for components of stock solutions, buffers, and reagents. Dilute stock solutions to the appropriate concentrations.

Formamide (optional) <!>
Prehybridization/hybridization solution (for hybridization in aqueous solvents)
6x SSC or 6x SSPE
5x Denhardt's reagent
0.5% (w/v) SDS <!>

or

Prehybridization/hybridization solution (for colony/plaque lifts)
50% formamide
6x SSC or 6x SSPE
0.05x BLOTTO (see Appendix 1, p. 704)
Prewashing solution
6x SSC or 6x SSPE
Wash solution 1
2x SSC
0.1% (w/v) SDS <!>
Wash solution 2
1x SSC
0.1% (w/v) SDS <!>
Wash solution 3
0.1x SSC
0.1% (w/v) SDS <!>

Media and Antibiotics

LB or YT medium containing the appropriate antibiotic

Nucleic Acids/Oligonucleotides

^{32}P-labeled double-stranded DNA (specific activity ~5 x 10^7 cpm/μg)
or radiolabeled synthetic oligonucleotide(s)

Additional Items

Adhesive dot labels
Boiling-water bath
Filters with immobilized DNA from transformed bacterial colonies
Use filters prepared as described in Protocol 1.31.

Glass baking dishes or plastic trays for processing batches of filters

Mylar film

Radioactive ink or chemiluminescent markers (Glogos [Stratagene])

Rotating platform in a incubator set at 50°C

Sealable container(s) for hybridization (heat-sealable plastic bags or plastic boxes with leak-proof lids)

Step 15 of this protocol requires the reagents listed in Protocol 1.1 or 1.4 as well as, possibly, Protocol 8.12.

Sterile wooden toothpicks *or* inoculating needle

Water-soluble glue stick
    UHU (Saunders)

Whatman 3MM paper sheets

X-ray film or phosphorimager

Additional information

Formamide and its uses in molecular cloning	MC3, pp. 6.59–6.60
Hybridization chambers	MC3, pp. 1.139 and 6.51
Prehybridization and hybridization solutions	MC3, p. 1.141

METHOD

1. Float the baked or cross-linked filters on the surface of a tray of 2x SSC until they have become thoroughly wetted from beneath. Submerge the filters for 5 minutes.

2. Transfer the filters to a glass baking dish containing at least 200 ml of prewashing solution. Stack the filters on top of one another in the solution. Cover the dish with Saran Wrap and transfer it to a rotating platform in an incubator. Incubate the filters for 30 minutes at 50°C.

3. Gently scrape the bacterial debris from the surfaces of the filters using Kimwipes soaked in prewashing solution. This scraping ensures removal of colony debris and does not affect the intensity or sharpness of positive hybridization signals.

4. Transfer the filters to 150 ml of prehybridization solution in a glass baking dish. Incubate the filters with agitation for 1–2 hours or more at the appropriate temperature (i.e., 68°C when hybridization is to be carried out in aqueous solution; 42°C when hybridization is to be carried out in 50% formamide).

5. Denature ^{32}P-labeled double-stranded DNA by heating to 100°C for 5 minutes. Chill the probe rapidly in ice water.

    Single-stranded probes need not be denatured.

6. Transfer the filters to a small sealable container. Add the probe to the prehybridization solution covering the filters. Incubate at the appropriate temperature until 1–3 $C_o t_{1/2}$ is achieved. During the hybridization, keep the containers holding the filters tightly closed to prevent the loss of fluid by evaporation.

    Use between 2×10^5 and 1×10^6 cpm of ^{32}P-labeled probe (specific activity ~5×10^7 cpm/μg) per milliliter of prehybridization solution. Using more probe will cause the background of non-specific hybridization to increase, whereas using less will reduce the rate of hybridization.

    Hybridization mixtures containing radiolabeled single-stranded probes may be stored at 4°C for several days and reused without further treatment. In some cases, hybridization probes prepared from double-stranded DNA templates can be reused after freezing the solution, thawing, and boiling for 5 minutes in a chemical fume hood.

7. When the hybridization is complete, remove the hybridization solution and immediately immerse the filters in a large volume (300–500 ml) of wash solution 1 at room temperature. Agitate the filters gently and turn them over at least once during washing. After 5 minutes, transfer the filters to a fresh batch of wash solution and continue to agitate them gently. Repeat the washing procedure twice more.

8. Wash the filters twice for 0.5–1.5 hours in 300–500 ml of wash solution 2 at 68°C.

9. Air dry the filters at room temperature on 3MM paper. Streak the underside of the filters with a water-soluble glue stick and arrange the filters (numbered side up) on a clean, dry, flat sheet of 3MM paper. Press the filters firmly against the 3MM paper to ensure sticking.

10. Apply adhesive dot labels marked with either radioactive ink or chemiluminescent markers to several asymmetric locations on the 3MM paper. Cover the filters and labels with Saran Wrap. Use tape to secure the wrap to the back of the 3MM paper and stretch the wrap over the paper to remove wrinkles.

11. Analyze the filters by phosphorimaging or exposing them to X-ray film (Kodak XAR-2, XAR-5, or their equivalents) for 12–16 hours at –70°C with an intensifying screen.

12. Develop the film and align it with the filters using the marks left by the radioactive ink. Use a nonradioactive fiber-tip pen in a color other than black to mark the film with the positions of the asymmetrically located dots on the numbered filters.

13. Tape a piece of clear Mylar or other firm transparent sheet to the film. Mark on the clear sheet the positions of positive hybridization signals. Also mark (in a different color) the positions of the asymmetrically located dots. Remove the clear sheet from the film. Identify the positive colonies by aligning the dots on the clear sheet with those on the agar plate.

14. Use a sterile toothpick or inoculating needle to transfer each positive bacterial colony into 1–2 ml of rich medium (e.g., LB, YT, or Terrific Broth) containing the appropriate antibiotic.

15. After a period of growth, plasmid DNA can be isolated from the culture by one of the minipreparation methods described in Protocols 1.1 and 1.4 and can be further analyzed by restriction endonuclease digestion or by PCR.

Bacteriophage λ and Its Vectors

BACKGROUND INFORMATION

Background information found in *Molecular Cloning: A Laboratory Manual,* 3rd edition (hereafter MC3) unless otherwise indicated.

Construction of bacteriophage λ vectors	MC3, pp. 2.18–2.24
E. coli strains used to propagate bacteriophage λ vectors	MC3, pp. 2.28–2.29
Lytic infection by bacteriophage λ	MC3, pp. 2.2–2.15
Physical and genetic map of wild-type bacteriophage λ	MC3, p. 2.5

Plating Bacteriophage λ

This protocol describes how to isolate individual plaques from a stock of bacteriophage λ. Each plaque derives from infection of a single bacterium by a single bacteriophage particle. Because each plaque contains the progeny of a single virus particle, the bacteriophages derived from a single plaque are essentially genetically identical to one another.

MATERIALS

Reagents and Solutions

Please see Appendix 1 for components of stock solutions, buffers, and reagents. Dilute stock solutions to the appropriate concentrations.

$MgSO_4$ (10 mM)

SM

SM plus gelatin (0.01% w/v)

All three of these solutions should be sterile.

Vectors and Hosts

Bacteriophage λ stocks

E. coli strain

Select a strain of *E. coli* equipped with the appropriate genetic markers to support the robust growth of the strain of bacteriophage λ. For a description of *E. coli* strains commonly used to propagate bacteriophage λ vectors and recombinants, see MC3, pp. 2.28 and 2.29, Table 2-3.

Media and Antibiotics

NZCYM or LB agar plates

Freshly poured plates are too wet for use in plaque assays. To prevent running and smearing of plaques, store the plates for 2 days at room temperature before use. They can then be transferred to plastic sleeves and stored at 4°C. Plates stored in this way should be placed at room temperature for 1–2 hours before use. Warming the plates reduces problems of condensation and allows the top agarose to spread across the entire surface before it sets.

Alternative methods for drying plates include incubating the plates for 2 hours with their lids ajar in a laminar flow hood or (less desirable) in a 37°C incubator. In an emergency, when there is no time to dry the plates, remove any droplets of water from the lids of the plates after the top agarose has set and insert a piece of sterile circular filter paper into each lid. During incubation of the inverted plates at 37°C, the filter absorbs much of the humidity and reduces streaking of plaques. For additional information, please see the note to Step 12.

Whenever possible, plate each dilution of bacteriophage λ in duplicate.

NZCYM or LB medium

NZCYM or LB top agarose (0.7%)

Agarose is preferred to agar in media used for plating bacteriophage λ to minimize contamination with polysaccharide inhibitors that can interfere with subsequent enzymatic analysis of bacteriophage λ DNA.

Centrifuges/Rotors/Tubes

Sorvall SS-34 rotor (room temperature)

Additional Items

Heating block or water bath (47°C)
Shaker incubator
Sterile plastic or glass tubes (13 x 75 or 17 x 100 mm)
Water bath (47°C)

Additional Information

Additional protocols (Plaque-Assay of Bacteriophages that Express
β-galactosidase and Macroplaques)   MC3, pp. 2.30–2.31

METHOD

1. Inoculate 50 ml of NZCYM or LB in a 250-ml conical flask with a single bacterial colony of the appropriate *E. coli* strain. Grow the culture overnight at 37°C with moderate agitation.

2. Centrifuge the cells at 4000*g* (5800 rpm in a Sorvall SS-34 rotor) for 10 minutes at room temperature.

3. Discard the supernatant and resuspend the cell pellet in 20 ml of sterile 10 mM $MgSO_4$. Measure the OD_{600} of a 1/100 dilution of the resuspended cells and dilute the cells with 10 mM $MgSO_4$ to a final concentration of 2.0 OD_{600}.

4. Store the suspension of plating bacteria at 4°C.

5. Melt top agarose by heating in a microwave oven for a short period of time. Store aliquots of the melted agarose (3 ml for 90-mm plates; 7 ml for 150-mm plates) on a heating block or in a water bath at 47°C to keep the solution molten.

6. Prepare tenfold serial dilutions of the bacteriophage stocks (in SM plus gelatin). Mix each dilution by *gentle* vortexing or by tapping on the side of the tube.

7. Dispense 0.1 ml of plating bacteria from Step 4 into a series of sterile tubes (13 x 75 or 17 x 100 mm).

8. Add 0.1 ml of each dilution of bacteriophage stock to a tube of plating bacteria. Mix the bacteria and bacteriophages by shaking or gentle vortexing.

9. Incubate the mixture for 20 minutes at 37°C to allow the bacteriophage particles to adsorb to the bacteria. Remove the tubes from the water bath and allow them to cool to room temperature.

10. Add an aliquot of molten agarose to the first tube. Mix the contents of the tube by gentle tapping or vortexing for a few seconds and, *without delay*, pour the entire contents of the tube onto the center of a labeled agar plate. Try to avoid creating air bubbles. Swirl the plate gently to ensure an even distribution of bacteria and top agarose. Repeat the procedure until the contents of all of the tubes have been transferred to separate labeled plates.

11. Replace the lids on the plates. Allow the top agarose to harden by standing the plates for 5 minutes at room temperature. Invert the closed plates and incubate them at 37°C.

    With some *E. coli* strains and bacteriophage vectors, better plaques are formed when the plates are incubated at temperatures other than 37°C. For example, when using the Stratagene *E. coli* strains SRBρ and SRB(P2)ρ as hosts, an incubation temperature of 39°C is recommended. In addition, λgt10 vectors produce better plaques on *E. coli* hfl⁻strains when incubated at 39°C.

12. Continue incubating the plates overnight and then count individual plaques and calculate the titer of the undiluted stock of bacteriophage λ (usually expressed in pfu/ml). Individual plaques may be picked for further analysis as described in Protocol 2.2.

    Plaques begin to appear after approximately 7 hours of incubation and should be counted or picked after 12–16 hours, when they are 2–3 mm in diameter. If the plates are too dry, the plaques wil grow more slowly and will not reach their full size. If the plates are too fresh and wet, droplets of moisture will sweat onto the surface of the top agarose during incubation at 37°C and cause the developing plaques to streak and run into one another.

Picking Bacteriophage λ Plaques

This protocol describes a general method to pick and store plaques.

MATERIALS

CAUTION: Please see Appendix 4 for appropriate handling of materials marked with <!>.

Reagents and Solutions

Please see Appendix 1 for components of stock solutions, buffers, and reagents. Dilute stock solutions to the appropriate concentrations.
Chloroform <!>
SM, sterile

Vectors and Hosts

Bacteriophage λ, grown as well-isolated plaques on bacterial lawn (see Protocol 2.1)

Additional Items

Borosilicate pasteur pipettes, fitted with a rubber pipette bulb
or
Automatic micropipette, fitted with yellow disposable tips
Polypropylene tubes (13 x 75 mm)

Additional Information

Long-term storage of bacteriophage λ stocks   MC3, p. 2.36

METHOD

1. Place 1 ml of SM in labeled sterile microfuge tubes or polypropylene test tubes. Add 1 drop (~50 μl) of chloroform to each tube.

2. Use a borosilicate pasteur pipette equipped with a rubber bulb, or a micropipette, to stab through the chosen plaque of bacteriophage λ into the hard agar beneath. Apply mild suction so that the plaque, together with the underlying agar, is drawn into the pipette.

3. Wash out the fragments of agar from the borosilicate pasteur pipette into the tubes containing SM/chloroform (prepared in Step 1). Let the capped tubes stand for 1–2 hours at room temperature to allow the bacteriophage particles to diffuse from the agar. To assist the elution of the virus, rock the tubes gently on a rocking platform. Store the bacteriophage suspension at 4°C.

   An average bacteriophage plaque yields approximately 10^6 infectious bacteriophage particles, which can be stored indefinitely at 4°C in SM/chloroform without loss of viability. The virus recovered from a plaque can be used as described in Protocols 2.3 and 2.4 to prepare larger stocks of bacteriophages by plate lysis or liquid culture methods.

Protocol 2.3

Preparing Stocks of Bacteriophage λ by Plate Lysis and Elution

A reliable method to prepare plate lysates and to recover infectious bacteriophages by elution from the top agar.

MATERIALS

CAUTION: Please see Appendix 4 for appropriate handling of materials marked with <!>.

Reagents and Solutions

Please see Appendix 1 for components of stock solutions, buffers, and reagents. Dilute stock solutions to the appropriate concentrations.
   Chloroform <!>
   SM

Vectors and Hosts

   Bacteriophage λ stock
      Prepare as described in Protocol 2.2.
   E. coli plating bacteria
      Prepare as described in Protocol 2.1.

Media and Antibiotics

   NZCYM or LB agar plates
      For best results, use freshly poured plates (90 or 150 mm) that have been equilibrated to room temperature. The older and dryer the plates, the lower the titer of the resulting plate stock.
   NZCYM top agarose, melted and dispensed into 3-ml (for 90-mm plates) or 7-ml (for 150-mm plates) aliquots and stored at 47°C
      Agarose is preferred to agar in media used for plating bacteriophage λ to minimize contamination with polysaccharide inhibitors that can interfere with subsequent enzymatic analysis of bacteriophage λ DNA.

Centrifuges/Rotors/Tubes

   Sorvall SS-34 rotor (4°C)

Additional Items

   Heating block or water bath (47°C)
   Pasteur pipettes
   Polypropylene tubes (13 x 75 mm)
   Step 9 of this protocol requires the reagents listed in Protocol 2.1.

Additional Information

   Alternative protocol (Preparing Stocks of Bacteriophage λ by
      Plate Lysis and Scraping)   MC3, p. 2.37
   Making and using λ macroplaques   MC3, p. 2.31
   Methods for long-term storage of bacteriophage λ stocks   MC3, p. 2.36

METHOD

1. Prepare infected cultures for plating.

   For a 90-mm diameter petri dish

   Mix 10^5 pfu of bacteriophage (usually ~1/10 of a resuspended individual plaque or 1/100 of a macroplaque with 0.1 ml of plating bacteria).

   For a 150-mm petri dish

   Mix 2×10^5 pfu with 0.2 ml of plating bacteria.

   Always set up at least one control tube containing uninfected cells. Incubate the infected and control cultures for 20 minutes at 37°C to allow the virus to attach to the cells.

   > When preparing stocks of bacteriophage λ that grow poorly, increase the inoculum to 10^6 pfu per 0.1 ml of plating bacteria.

2. Add 3 ml of molten top agarose (47°C) (90-mm plate) or 7.0 ml of molten top agarose (47°C) (150-mm plate) to the first tube of infected cells. Mix the contents of the tube by gentle tapping or vortexing for a few seconds and, *without delay*, pour the entire contents of the tube onto the center of a labeled agar plate. Try to avoid creating air bubbles. Swirl the plate gently to ensure an even distribution of bacteria and top agarose. Repeat this step until the contents of each of the tubes have been transferred onto separate plates.

3. Incubate the plates *without inversion* for approximately 12–16 hours at 37°C.

4. Remove the plates from the incubator and add SM (5 ml to each 90-mm plate or 10 ml to each 150-mm plate). Store the plates for several hours at 4°C on a gently shaking platform.

5. Use a separate pasteur pipette for each plate to transfer as much of the SM as possible into sterile screw- or snap-cap polypropylene tubes.

6. Add 1 ml of fresh SM to each plate, swirl the fluid gently, and store the plates for 15 minutes in a tilted position to allow all of the fluid to drain into one area. Remove the SM and combine it with the first harvest.

7. Add 0.1 ml of chloroform to each of the tubes containing SM, vortex the tubes briefly, and remove the bacterial debris by centrifugation at 4000g (5800 rpm in a Sorvall SS-34 rotor) for 10 minutes at 4°C.

8. Transfer the supernatants to fresh polypropylene tubes and add 1 drop of chloroform to each tube. Store the resulting bacteriophage plate stocks at 4°C.

9. Measure the concentration of infectious virus particles in each stock by plaque assay as described in Protocol 2.1.

   > The titer of plate stocks should be approximately 10^9 to 10^{10} pfu/ml and should remain stable as long as the stock is properly stored.

Preparing Stocks of Bacteriophage λ by Small-scale Liquid Culture

High-titer stocks of bacteriophage λ are easily prepared by infecting small-scale bacterial cultures.

MATERIALS

CAUTION: Please see Appendix 4 for appropriate handling of materials marked with <!>.

Reagents and Solutions

Please see Appendix 1 for components of stock solutions, buffers, and reagents. Dilute stock solutions to the appropriate concentrations.
   Chloroform <!>
   SM

Vectors and Hosts

   Bacteriophage λ stock
        Prepare as described in Protocol 2.2.
   Plate culture of an appropriate strain of *E. coli*

Media and Antibiotics

   NZCYM or LB medium

Additional Items

   Shaker incubator (37°C)
   Step 8 of this protocol requires the reagents listed in Protocol 2.1.
   Sterile polypropylene tubes (13 x 75 mm)

Additional Information

   Methods for long-term storage of bacteriophage λ stocks   MC3, p. 2.36

METHOD

1. Inoculate a single colony of an appropriate *E. coli* strain into 5 ml of NZCYM or LB medium in a sterile polypropylene culture tube. Incubate the culture overnight with vigorous shaking at 30°C.

2. Transfer 0.1 ml of the fresh overnight bacterial culture (prepared in Step 1) to a sterile 17 x 100-mm polypropylene culture tube with a loose-fitting cap. Infect the culture with approximately 10^6 pfu of bacteriophage λ in 50–100 μl of SM.

3. Incubate the infected culture for 20 minutes at 37°C to allow the bacteriophage particles to adsorb to the bacteria.

4. Add 4 ml of medium, prewarmed to 37°C, and incubate the culture with vigorous agitation until lysis occurs (usually 8–12 hours at 37°C).

5. After lysis has occurred, add 2 drops (~100 μl) of chloroform and continue incubation for 15 minutes at 37°C.

      Adapted from Chapter 2, Protocol 4, p. 2.38 of MC3.

6. Centrifuge the culture at 4000g (5800 rpm in a Sorvall SS-34 rotor) for 10 minutes at 4°C.

7. Recover the supernatant, add 1 drop (~50 μl) of chloroform, and store the virus stock at 4°C.

8. Measure the concentration of infectious particles in the stock by plaque assay as described in Protocol 2.1.

   The titer of the stock should be approximately 10^9 to 10^{10} pfu/ml and remain stable as long as the stock is properly stored.

Large-scale Growth of Bacteriophage λ: Infection at Low Multiplicity

After infection with bacteriophage λ at low multiplicity, a bacterial culture is transferred to a large volume of medium and incubated until complete lysis of the host cells occurs.

MATERIALS

CAUTION: Please see Appendix 4 for appropriate handling of materials marked with <!>.

Reagents and Solutions

Please see Appendix 1 for components of stock solutions, buffers, and reagents. Dilute stock solutions to the appropriate concentrations.
  Chloroform <!>
  SM

Vectors and Hosts

Bacteriophage λ stock
  Prepare as described in either Protocol 2.3 or 2.4.
Plate culture of an appropriate strain of *E. coli*

Media and Antibiotics

NZCYM or LB medium
  See note to Step 1.

Centrifuges/Rotors/Tubes

Sorvall SS-34 rotor (room temperature)

Additional Items

Glass tubes (12 x 75 mm)
Shaker incubator (37°C)
Step 8 of this protocol requires the reagents listed in Protocol 2.1.
Sterile polypropylene tubes (13 x 75 mm)

Additional Information

Alternative protocol (Large-scale Growth of Bacteriophage λ: Infection
  at High Multiplicity)   MC3, p. 2.42

METHOD

1. Inoculate 100 ml of NZCYM in a 500-ml conical flask with a single colony of an appropriate bacterial host. Incubate the culture overnight at 37°C with vigorous agitation.

   For a starter culture, prepare 100 ml of NZCYM in a 500-ml conical flask. For subsequent large-scale culture, prepare 4 x 500-ml aliquots of NZCYM in 2-liter flasks, prewarmed to 37°C. Four additional 500-ml aliquots may be needed for Step 9.

2. Measure the OD_{600} of the culture. Calculate the cell concentration assuming that $1\ OD_{600} = 1 \times 10^9$ cells/ml.

3. Withdraw four aliquots, each containing 10^{10} cells. Centrifuge each aliquot at $4000g$ (5800 rpm in a Sorvall SS-34 rotor) for 10 minutes at room temperature. Discard the supernatants.

4. Resuspend each bacterial pellet in 3 ml of SM.

5. Add the appropriate number of infectious bacteriophage particles and swirl the culture to ensure that the inoculum is dispersed rapidly throughout the culture.

   The number of bacteriophage particles used is crucial. For strains of bacteriophage λ that grow well (e.g., EMBL3 and 4 and λgt10), add 5×10^7 pfu to each suspension of 10^{10} cells; for bacteriophages that grow relatively poorly (e.g., the Charon series), it is better to increase the starting inoculum to 5×10^8 pfu.

6. Incubate the infected cultures for 20 minutes at 37°C with intermittent swirling.

7. Add each infected aliquot to 500 ml of NZCYM, prewarmed to 37°C in 2-liter flasks. Incubate the cultures at 37°C with vigorous shaking (300 cycles/minute in a shaker incubator).

8. Begin to monitor the cultures for lysis after 8 hours. Concomitant growth of bacteria and bacteriophages should occur, resulting in lysis of the culture after 8–12 hours. When lysis is observed, proceed to Step 10.

9. If lysis is not apparent after 12 hours, check a small sample of the cultures for evidence of bacteriophage growth.

   a. Transfer two aliquots (1 ml each) of the infected culture into glass tubes.

   b. Add 1 or 2 drops of chloroform (~50–100 μl) to one of the tubes and incubate both tubes for 5–10 minutes at 37°C with intermittent shaking.

   c. Compare the appearance of the two cultures by holding the tubes up to a light. If infection is near completion but the cells have not yet lysed, the chloroform will cause the cells to burst and the turbid culture will clear to the point at which it is translucent. In this case, proceed to Step 10.

10. Add 10 ml of chloroform to each flask and continue the incubation for an additional 10 minutes at 37°C with shaking.

11. Cool the cultures to room temperature and precipitate the bacteriophage particles as described in Protocol 2.6.

Precipitation of Bacteriophage λ Particles from Large-scale Lysates

This protocol describes how bacteriophage λ particles are recovered from bacterial lysates by precipitation with polyethylene glycol at high ionic strength.

MATERIALS

CAUTION: Please see Appendix 4 for appropriate handling of materials marked with <!>.

Reagents and Solutions

Please see Appendix 1 for components of stock solutions, buffers, and reagents. Dilute stock solutions to the appropriate concentrations.

Chloroform <!>
NaCl (solid)
PEG 8000 <!>
Use approximately 50 g for each 500-ml culture.
SM

Vectors and Hosts

Lysed infected cultures of *E. coli* (4 x 500 ml)
Prepare as described in Protocol 2.5.

Enzymes and Buffers

Pancreatic DNase I (1 mg/ml) in
10 mM Tris-Cl (pH 7.6)
150 mM NaCl
1 mM MgCl$_2$
RNase A (pancreatic RNase) (1 mg/ml) in TE (pH 7.6)

Centrifuges/Rotors/Tubes

Sorvall SLC-1500 and SS-34 rotors (4°C)

Additional Items

Magnetic stirrer and magnetic bar (in cold room at 4°C)
Wide-bore pipette (10 ml) equipped with a rubber bulb

METHOD

1. Cool the lysed cultures containing bacteriophage λ to room temperature. Add pancreatic DNase I and RNase, each to a final concentration of 1 µg/ml. Incubate the lysed cultures for 30 minutes at room temperature.

2. To each 500-ml culture, add 29.2 g of solid NaCl (final concentration ~1 M). Swirl the cultures until the salt has dissolved. Store the cultures for 1 hour on ice.

3. Remove debris by centrifugation at 11,000g (8300 rpm in a Sorvall SLC-1500 rotor) for 10 minutes at 4°C. Combine the supernatants from the four cultures.

4. Measure the volume of the pooled supernatants and then transfer the preparation to a clean 2-liter flask. Add solid PEG 8000 to a final concentration of 10% w/v (i.e., 50 g per 500 ml of supernatant). Dissolve the PEG by slow stirring on a magnetic stirrer at room temperature.

5. Transfer the solution to polypropylene centrifuge bottles, cool the bacteriophage/PEG solution in ice water, and store the centrifuge bottles for at least 1 hour on ice to allow the bacteriophage particles to precipitate.

6. Recover the precipitated bacteriophage particles by centrifugation at 11,000*g* (8300 rpm in a Sorvall SLC-1500 rotor) for 10 minutes at 4°C. Discard the supernatants and stand the inverted centrifuge bottles in a tilted position for 5 minutes to allow the remaining fluid to drain away from the pellet. Remove any residual fluid with a pipette.

7. Use a wide-bore pipette equipped with a rubber bulb to resuspend the bacteriophage pellet gently in SM without generating bubbles (8 ml for each 500 ml of supernatant from Step 3). Place the centrifuge bottles on their sides for 1 hour at room temperature so that the SM covers and soaks the pellets.

8. Extract the PEG and cell debris from the bacteriophage suspension by adding an equal volume of chloroform. Vortex the mixture gently for 30 seconds. Separate the organic and aqueous phases by centrifugation at 3000*g* (4300 rpm in a Sorvall SS-34 rotor) for 15 minutes at 4°C. Recover the aqueous phase, which contains the bacteriophage particles.

Assaying the DNA Content of Bacteriophage λ Stocks and Lysates by Gel Electrophoresis

The DNA content of bacteriophage λ stocks and crude lysates is easily and rapidly estimated by gel electrophoresis.

MATERIALS

CAUTION: Please see Appendix 4 for appropriate handling of materials marked with <!>.

Reagents and Solutions

Please see Appendix 1 for components of stock solutions, buffers, and reagents. Dilute stock solutions to the appropriate concentrations.
2.5x SDS-EDTA dye mix

Vectors and Hosts

Bacteriophage λ lysates or stocks
Prepare by using one of the methods described in Protocol 2.3, 2.4, or 2.5.

Enzymes and Buffers

Pancreatic DNase I (1 mg/ml) in DNase I dillution buffer
10 mM Tris-Cl (pH 7.6)
150 mM NaCl
1 mM MgCl$_2$

Nucleic Acids/Oligonucleotides

Bacteriophage λ DNA (for use as a control)

Gels/Loading Buffers

0.7% Agarose gel, cast and run in 0.5x TBE containing 0.5 µg/ml ethidium bromide <!>

Additional Items

Water baths (37°C and 65°C)

METHOD

1. Make a working solution of pancreatic DNase I (1 µg/ml) as follows: Dilute 1 µl of a stock solution of DNase I (1 mg/ml) with 1 ml of ice-cold DNase I dilution buffer.

2. Mix the solution by gently inverting the closed tube several times. Take care to avoid bubbles and foam. Store the solution in ice until needed. Discard the working solution after use.

3. Transfer 10 µl of crude bacteriophage lysate or stock to a microfuge tube. Add 1 µl of the working solution of pancreatic DNase and incubate the mixture for 30 minutes at 37°C.

4. Add 4 µl of 2.5x SDS-EDTA dye mixture and incubate the closed tube for 5 minutes at 65°C.

5. Load the sample onto an 0.7% agarose gel containing 0.5 µg/ml ethidium bromide.
   As controls, use samples containing 5, 25, and 100 ng of purified bacteriophage λ DNA.

Adapted from Chapter 2, Protocol 7, p. 2.45 of MC3.

6. Perform electrophoresis at <5 V/cm until the bromophenol blue has migrated 3–4 cm.

7. Examine the gel under UV illumination. Use the intensity of fluorescence of the DNA standards as a guide to estimate the amount of bacteriophage λ DNA in the test sample.

   Ten μl of a high-titer lysate should contain between 10 and 50 ng of bacteriophage DNA.

Purification of Bacteriophage λ Particles by Isopycnic Centrifugation through CsCl Gradients

Isopycnic centrifugation through CsCl gradients is used to prepare infectious bacteriophage λ particles of the highest purity that are essentially free of contaminating bacterial nucleic acids. DNA extracted from nonrecombinant bacteriophage λ vectors purified in this way can be used to prepare λ arms. DNA extracted from purified recombinant bacteriophages is a source of cloned DNA fragments.

MATERIALS

Reagents and Solutions

Please see Appendix 1 for components of stock solutions, buffers, and reagents. Dilute stock solutions to the appropriate concentrations.
   CsCl (solid, molecular biology grade)
   Ethanol
   SM

Vectors and Hosts

   Suspension of bacteriophage λ particles
      Prepare as described in Protocol 2.6.

Centrifuges/Rotors/Tubes

   Beckman SW 28 rotor and Ultra-Clear tubes (4°C)
   Beckman Ti 50 or SW 50.1 rotor and Ultra-Clear tubes (4°C)

Additional Items

   Hypodermic needles (21 gauge)
   Scotch tape or Time tape

Additional Information

   Alternative protocol (Purification of Bacteriophage λ Particles by
      Isopycnic Centrifugation through CsCl Equilibrium Gradients) MC3, p. 2.51
   Detailed explanation of CsCl step and equilibrium gradients
      used to purify bacteriophage λ   MC3, p. 2.49,
                                                             information panels
   Yamamoto K.R., Alberts B.M., Benzinger R., Lawhorne L., and Treiber G. 1970. Rapid bacterio-
      phage sedimentation in the presence of polyethylene glycol and its application to large-scale
      virus purification. *Virology* **40:** 734–744.

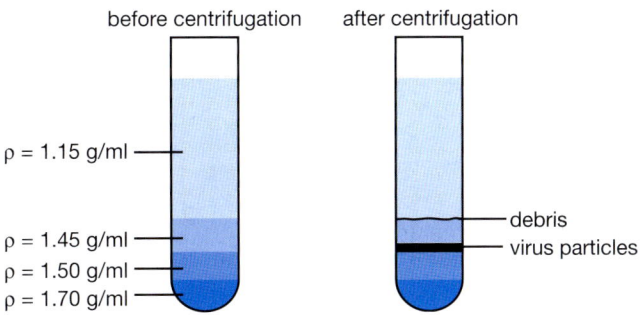

FIGURE 2-1 CsCl Gradients for Purifying Bacteriophage λ

After centrifugation, the bacteriophage particles form a visible band at the interface between the 1.45- and 1.50-g/ml CsCl layers.

METHOD

1. Prepare three solutions of different densities by adding solid CsCl to SM, as indicated below. Store the solutions at room temperature.

CsCl Solutions Prepared in SM (100 ml) for Step Gradients

Density ρ (g/ml)	CsCl (g)	SM (ml)	Refractive Index η
1.45	60	85	1.3768
1.50	67	82	1.3815
1.70	95	75	1.3990

Measure the volume of the bacteriophage suspension and add 0.5 g of solid CsCl per ml of bacteriophage suspension. Place the suspension on a rocking platform until the CsCl is completely dissolved.

2. Prepare enough CsCl step gradients in clear plastic centrifuge tubes (e.g., Beckman Ultra-Clear tubes) to fractionate the bacteriophage suspension. The gradients should occupy approximately 60% of the volume of the ultracentrifuge tubes. For example, in a Beckman SW 28 tube, which holds 38 ml, each step gradient consists of 7.6 ml of each of the three CsCl solutions. The step gradients are made either by carefully layering three solutions of CsCl of decreasing density on top of one another or by layering solutions of decreasing density under one another.

   Balance tubes (if required) should contain the same volumes of the three CsCl solutions.

3. Make a mark with a permanent felt-tipped marker on the outside of the tube opposite the position of the interface between the ρ = 1.50-g/ml layer and the ρ = 1.45-g/ml layer (Figure 2-1).

4. Measure the volume of the preparation of bacteriophage λ and adjust the density to 1.15 g/ml by adding 0.5 g of solid CsCl per bacteriophage suspension. Mix the solution gently until the CsCl is dissolved and then carefully layer the bacteriophage suspension over the step gradients. Centrifuge the gradients at 87,000*g* (22,000 rpm in a Beckman SW 28 rotor) for 2 hours at 4°C.

5. Collect the bacteriophage particles by puncturing the side of the tube as follows (Figure 2-2).

   a. Carefully wipe the outside of the tube with ethanol to remove any grease or oil and attach a piece of Scotch tape or Time tape to the outside of the tube, level with the band of bacteriophage particles.

   b. Use a 21-gauge hypodermic needle (no syringe-barrel required) to puncture the tube through the tape and collect the band of bacteriophage particles.

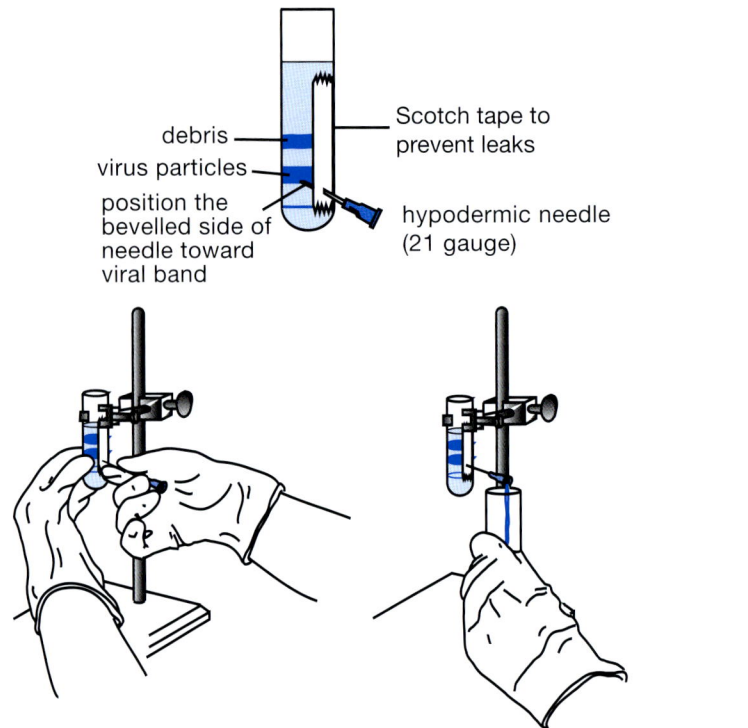

debris

virus particles

position the bevelled side of needle toward viral band

Scotch tape to prevent leaks

hypodermic needle (21 gauge)

FIGURE 2-2 Collection of Bacteriophage λ Particles by Side Puncture

6. Place the suspension of bacteriophage particles in an ultracentrifuge tube that fits a Beckman Ti 50 or SW 50.1 rotor (or equivalent) and fill the tube with CsCl solution (ρ = 1.5 g/ml in SM). Centrifuge at 150,000*g* (41,000 rpm in a Beckman Ti 50 rotor) for 24 hours at 4°C or at 160,000*g* (36,000 rpm in a Beckman SW 50.1 rotor) for 24 hours at 4°C.

7. Collect the band of bacteriophage particles as described in Step 5. Store the bacteriophage suspension at 4°C in the CsCl solution in a tightly capped tube.

   Bacteriophage λ is extremely sensitive to chelators such as EDTA. Mg^{++} must be present in all solutions used to prepare and store bacteriophage λ.

8. (*Optional*) If necessary, the bacteriophage particles can be further purified and concentrated by a second round of equilibrium centrifugation in CsCl. Transfer the bacteriophage suspension to one or more ultracentrifuge tubes that fit a Beckman SW 50.1 rotor (or its equivalent). Fill the tubes with a solution of CsCl in SM (ρ = 1.5) and centrifuge the tubes at 160,000*g* (36,000 rpm in a Beckman SW 50.1 rotor) for 24 hours at 4°C. When centrifugation is complete, collect the bacteriophage particles.

Purification of Bacteriophage λ Particles by Centrifugation through a Glycerol Step Gradient

This method can be used to purify bacteriophage λ particles from small volumes (5–30 ml) of lysed infected bacterial cultures (Protocol 2.6). DNA extracted from nonrecombinant bacteriophage λ vectors purified in this way can be used to prepare λ arms. DNA extracted from recombinant bacteriophages is a source of cloned DNA fragments.

MATERIALS

Reagents and Solutions

Please see Appendix 1 for components of stock solutions, buffers, and reagents. Dilute stock solutions to the appropriate concentrations.
EDTA (0.5 M, pH 8.0)
Glycerol solutions (5% and 40% in SM)
SM

Vectors and Hosts

Suspension of bacteriophage λ particles
Prepare as described in Protocol 2.6.

Enzymes and Buffers

Pancreatic DNase I (1 mg/ml) in
10 mM Tris-Cl (pH 7.6)
150 mM NaCl
1 mM MgCl$_2$
RNase A (pancreatic RNase) (1 mg/ml) in TE (pH 7.6)

Centrifuges/Rotors/Tubes

Beckman SW 28 or SW 41 rotor and Ultra-Clear tubes (4°C)

Additional Information

Vande Woude G.F., Oskarsson M., Enquist L.W., Nomura S., Sullivan M., and Fischinger P.J. 1979. Cloning of integrated Moloney sarcoma proviral DNA sequences in bacteriophage lambda. *Proc. Natl. Acad. Sci.* **76:** 4464–4468.

METHOD

1. Prepare a glycerol step gradient in a Beckman SW 41 polycarbonate tube (or its equivalent; one tube is needed for each 5 ml of bacteriophage suspension):

   a. Pipette 3 ml of a solution consisting of 40% glycerol in SM into the bottom of the tube.

   b. Carefully layer 4 ml of a solution consisting of 5% glycerol in SM on top of the 40% glycerol solution.

   c. Carefully layer the bacteriophage suspension on top of the 5% glycerol layer. Fill the tube with SM.

2. Centrifuge the step gradient at 151,000*g* (35,000 rpm in a Beckman SW 41 or SW 28 rotor) for 60 minutes at 4°C.

3. Discard the supernatant and resuspend the bacteriophage pellet in 1 ml of SM per liter of original culture.

4. Add pancreatic DNase I and RNase to final concentrations of 5 μg/ml and 1 μg/ml, respectively. Incubate the reaction mixture for 30 minutes at 37°C.

5. If proceeding to isolate DNA from the purified preparation of bacteriophage λ, add EDTA from a 0.5 M stock (pH 8.0) to a final concentration of 20 mM. Extract λ DNA as described in Protocol 2.11, beginning at Step 5.

Purification of Bacteriophage λ Particles by Pelleting/Centrifugation

This method can be used to recover bacteriophage λ particles from lysed infected bacterial cultures (Protocol 2.6). Although the resulting preparations are not as pure as those obtained by isopycnic centrifugation (Protocol 2.8), they yield high-quality DNA. This method is chiefly used to recover cloned DNA fragments from recombinant λ bacteriophages.

MATERIALS

Reagents and Solutions

Please see Appendix 1 for components of stock solutions, buffers, and reagents. Dilute stock solutions to the appropriate concentrations.
EDTA (0.5 M, pH 8.0)
SM

Vectors and Hosts

Suspension of bacteriophage λ particles
Prepare as described in either Protocol 2.9 or 2.10.

Centrifuges/Rotors/Tubes

Beckman SW 28 rotor and Ultra-Clear tubes (4°C)

METHOD

1. Transfer the bacteriophage suspension into a tube for use in a Beckman SW 28 rotor (or its equivalent).

2. Collect the bacteriophage particles by centrifugation at 110,000g (25,000 rpm in a Beckman SW 28 rotor) for 2 hours at 4°C.

3. Carefully pour off and discard the supernatant.

4. Add 1–2 ml of SM to the pellet and store it overnight at 4°C, preferably on a slowly rocking platform.

5. The following morning, pipette the solution gently up and down to ensure that all of the bacteriophage particles have been resuspended. If proceeding to isolate DNA from the purified preparation of bacteriophage λ, add EDTA from a 0.5 M stock solution (pH 8.0) to a final concentration of 20 mM. Extract λ DNA as described in Protocol 2.11, beginning at Step 5.

Extraction of Bacteriophage λ DNA from Large-scale Preparations Using Proteinase K and SDS

DNA is best isolated from bacteriophage λ particles by digesting the viral coat proteins with a powerful protease such as proteinase K, followed by extraction with phenol:chloroform. This procedure describes isolation of DNA from large-scale preparations of bacteriophage λ but can easily be adapted for use with smaller-scale preparations of the virus.

MATERIALS

CAUTION: Please see Appendix 4 for appropriate handling of materials marked with <!>.

Reagents and Solutions

Please see Appendix 1 for components of stock solutions, buffers, and reagents. Dilute stock solutions to the appropriate concentrations.

Chloroform <!>
EDTA (0.5 M, pH 8.0)
Ethanol
Ethanol (70%)
Phenol equilibrated to pH 7.8 <!>
Phenol:chloroform (1:1, w/v) <!>
SDS (10% w/v)
Sodium acetate (3 M, pH 5.2)
TE (pH 7.6 and pH 8.0)

Enzymes and Buffers

Proteinase K
Restriction enzymes and buffers

Gels/Loading Buffers

0.7% Agarose gel, cast and run in 0.5x TBE containing 0.5 μg/ml ethidium bromide <!>

Centrifuges/Rotors/Tubes

Sorvall SS-34 rotor and polypropylene centrifuge tubes (room temperature)

Additional Items

Dialysis tubing
Magnetic stirrer and bar (room temperature and 4°C)
Water bath (56°C)
Wide-bore glass pipette

Additional Information

Bacteriophage λ DNA is long enough (~50 kb) to be sensitive to shearing forces. To avoid breaking the DNA during extraction, see Minimizing Damage to Large DNA Molecules MC3, p. 2.110

METHOD

1. Transfer the bacteriophage suspension to a section of dialysis tubing sealed at one end with a knot or a plastic closure. Close the other end of the dialysis tube. Place the sealed tube in a flask containing a 1000-fold volume excess of dialysis buffer and a magnetic stir bar. Dialyze the bacteriophage suspension for 1 hour at room temperature with slow stirring.

2. Transfer the dialysis tube to a fresh flask of buffer and dialyze the bacteriophage suspension for 1 additional hour.

3. Transfer the bacteriophage suspension into a polypropylene centrifuge tube.

4. To the dialyzed bacteriophage suspension, add 0.5 M EDTA (pH 8.0) to a final concentration of 20 mM.

5. Add proteinase K to a final concentration of 50 μg/ml.

6. Add SDS to a final concentration of 0.5% and mix the solution by gently inverting the tube several times.

7. Incubate the digestion mixture for 1 hour at 56°C and then cool the mixture to room temperature.

8. Add an equal volume of equilibrated phenol to the digestion mixture. Mix the organic and aqueous phases by gently inverting the tube several times until an emulsion forms.

9. Separate the phases by centrifugation at 3000*g* (5000 rpm in a Sorvall SS-34 rotor) for 5 minutes at room temperature. Use a wide-bore pipette to transfer the aqueous phase to a clean tube.

10. Extract the aqueous phase once with a 1:1 mixture of equilibrated phenol and chloroform.

11. Recover the aqueous phase as described above (Step 9) and repeat the extraction with an equal volume of chloroform. For large-scale preparations, proceed to Step 12. For smaller-scale quantities (bacteriophage from 50- to 100-ml cultures),

    a. Recover the bacteriophage DNA by standard ethanol precipitation for 30 minutes at room temperature.

    b. Redissolve the DNA in an appropriate volume of TE (pH 7.6) and proceed to Step 14.

12. Transfer the aqueous phase to a dialysis sac.

13. Dialyze the preparation of bacteriophage DNA overnight at 4°C against three changes of a 1000-fold volume of TE (pH 8.0).

14. Measure the absorbance of the solution at 260 nm and calculate the concentration of the DNA.

    1 OD_{260} = 50 μg/ml of double-stranded DNA. A single particle of bacteriophage contains approximately 5×10^{-11} μg of DNA. The yield of bacteriophage DNA usually ranges from 500 μg to several mg per liter, depending on the titer of the bacteriophage in the lysed culture.

15. Check the integrity of the DNA by analyzing aliquots (0.5 μg) that are undigested or have been cleaved by appropriate restriction enzyme(s). Analyze the DNAs by electrophoresis through a 0.7% agarose gel, using markers of an appropriate size.

16. Store the stock of bacteriophage DNA at 4°C.

Extraction of Bacteriophage λ DNA from Large-scale Preparations Using Formamide

Formamide can be used instead of proteinase K to dissociate bacteriophage λ coat proteins from the viral λ. The procedure is rapid and works best with large-scale preparations of bacteriophage λ.

MATERIALS

CAUTION: Please see Appendix 4 for appropriate handling of materials marked with <!>.

Reagents and Solutions

Please see Appendix 1 for components of stock solutions, buffers, and reagents. Dilute stock solutions to the appropriate concentrations.
EDTA (0.5 M, pH 8.0)
Ethanol
Ethanol (70%)
Formamide, deionized <!>
NaCl (5 M)
TE (pH 8.0)
Tris-Cl (2.0 M, pH 8.5)

Vectors and Hosts

Suspension of bacteriophage λ particles
Prepare as described in either Protocol 2.9 or 2.10.

Enzymes and Buffers

Restriction enzymes and buffers

Gels/Loading Buffers

0.7% Agarose gel, cast and run in 0.5x TBE containing 0.5 μg/ml ethidium bromide <!>

Centrifuges/Rotors/Tubes

Beckman SW 28 rotor and Ultra-Clear tubes (4°C)
Microfuge (room temperature)

Additional Items

Borosilicate pasteur pipette (sealed) or Shepherd's crook
Step 1 of this protocol requires the reagents and equipment used in Protocol 2.11.
Water bath (37°C)

Additional Information

Bacteriophage λ DNA is long enough (~50 kb) to be sensitive to shearing forces. To avoid breaking the DNA during extraction, see Minimizing Damage to Large DNA Molecules

MC3, p. 2.110

METHOD

1. If necessary, remove CsCl from the preparation of bacteriophage particles as described in Steps 1–4 of Protocol 2.11.

2. Measure the volume of the preparation of bacteriophage particles.

3. Add 0.1 volume of 2 M Tris (pH 8.5), 0.05 volume of 0.5 M EDTA (pH 8.0), and 1 volume of deionized formamide. Incubate the solution for 30 minutes at 37°C.

4. Precipitate the bacteriophage λ DNA by adding 1 volume (equal to the final volume in Step 3) of H_2O and 6 volumes (each equal to the final volume in Step 3) of ethanol.

5. Hook the precipitate of bacteriophage λ DNA onto the end of a sealed borosilicate pasteur pipette or a Shepherd's crook and transfer it to a microfuge tube containing 70% ethanol.

6. Collect the DNA pellet by brief centrifugation (10 seconds) in a microfuge.

7. Discard the supernatant and store the open tube on the bench for a few minutes to allow the ethanol to evaporate. Redissolve the damp pellet of DNA in 300 μl of TE (pH 8.0) by tapping on the side of the tube. Try to avoid vortexing.

8. Reprecipitate the DNA by adding 6 μl of 5 M NaCl and 750 μl of ethanol. Collect the precipitated DNA and redissolve it as described in Steps 6 and 7.

9. Check the integrity of the DNA by analyzing aliquots (0.5 μg) that are undigested or have been cleaved by appropriate restriction enzyme(s). Analyze the DNAs by electrophoresis through a 0.7% agarose gel using markers of an appropriate size.

10. Store the stock of bacteriophage DNA at 4°C.

Preparation of Bacteriophage λ Cleaved with a Single Restriction Enzyme for Use as a Cloning Vector

Bacteriophage λ vectors that allow genetic selection of recombinant bacteriophages (e.g., the EMBL series, λ2001, λDASH, λZAP, and λgt10) can be prepared for cloning by simple digestion by one or more restriction enzymes.

MATERIALS

CAUTION: Please see Appendix 4 for appropriate handling of materials marked with <!>.

Reagents and Solutions

Please see Appendix 1 for components of stock solutions, buffers, and reagents. Dilute stock solutions to the appropriate concentrations.

ATP (10 mM)
   Not required if 10x ligase buffer already contains ATP
Chloroform <!>
EDTA (0.5 M, pH 8.0)
Ethanol
Ethanol (70%)
Phenol:chloroform (1:1, v/v) <!>
Sodium acetate (3 M, pH 7.0)
   Use in Step 6 in place of sodium acetate at pH 5.2.
TE (pH 7.6 and pH 8.0)

Vectors and Hosts

Suspension of bacteriophage λ particles
   Prepare as described in either Protocol 2.9 or 2.10.

Enzymes and Buffers

Bacteriophage T4 DNA ligase and 10x ligation buffer
Restriction enzymes and buffers

Nucleic Acids/Oligonucleotides

Bacteriophage λ DNA
   Prepare as described in either Protocol 2.11 or 2.12.
DNA markers for gel electrophoresis

Gels/Loading Buffers

0.7% Agarose gel, cast and run in 0.5x TBE containing 0.5 µg/ml ethidium bromide <!>
Gel-loading buffer IV

Additional Items

Bacteriophage λ packaging mixture (commercial)
Step 7e of this protocol requires the reagents listed in Protocol 2.1.
Water bath (68°C)

Additional Information

In vitro packaging   MC3, pp. 2.110–2.111

METHOD

1. Mix 25–50 μg of bacteriophage λ DNA with TE (pH 8.0) to give a final volume of 170 μl.

2. Add 20 μl of the appropriate 10x restriction enzyme buffer. Remove two aliquots, each containing 0.2 μg of undigested bacteriophage λ DNA. Store the aliquots of undigested DNA on ice.

3. Add a threefold excess (75–150 units) of the appropriate restriction enzyme and incubate the digestion mixture for 1 hour at the temperature recommended by the manufacturer.

4. Cool the reaction to 0°C on ice. Remove another aliquot (0.2 μg). Incubate this aliquot and one of the two aliquots of undigested DNA (Step 2 above) for 10 minutes at 68°C to disrupt the cohesive termini of the bacteriophage DNA. Add a small amount (10 μl) of gel-loading buffer IV and immediately load the samples onto a 0.7% agarose gel.

   This step is not as easy as it sounds. The left and right arms of bacteriophage λ DNA carry complementary termini 12 bases in length that can reanneal with one another. The resulting hydrogen-bonded DNA species can be easily confused with uncleaved bacteriophage λ DNA. For this reason, it is important to load and run the gel immediately after the DNA samples have been removed from the 68°C water bath.

   If digestion is incomplete, warm the reaction to the appropriate temperature, add more restriction enzyme (50–100 units), and continue the incubation at the optimal temperature recommended by the manufacturer.

5. When digestion is complete, add 0.5 M EDTA (pH 8.0) to a final concentration of 5 mM and extract the digestion mixture once with phenol:chloroform and once with chloroform.

6. Recover the DNA from the aqueous phase by standard ethanol precipitation in the presence of 0.3 M sodium acetate (pH 7.0). Redissolve the DNA in 100 μl of TE (pH 7.6). Determine the concentration by measuring absorbance at 260 nm.

   Do not use sodium acetate at pH 5.2 because the EDTA may precipitate from solution.

7. Remove an aliquot of DNA (0.5 μg) and test for its ability to be ligated as follows:

   a. Adjust the volume of the DNA solution to 17 μl with H_2O.

   b. Add 2 μl of 10x ligation buffer and, if necessary, 1 μl of 10 mM ATP.

   c. Remove 5 μl of the mixture prepared in Step b and store on ice.

   d. Add 0.2–0.5 Weiss unit of bacteriophage T4 DNA ligase to the remainder of the mixture (Step b) and incubate the reaction for 2 hours at 16°C.

   e. Use a commercially available bacteriophage λ packaging mixture to package 0.1 μg of the ligated and unligated samples and 0.1 μg of the undigested vector DNA from Step 2. Determine the titer (pfu/ml) of each packaged reaction as described in Protocol 2.1.

      The packaging efficiency of the digested vector should increase by nearly three orders of magnitude after ligation. The packaging efficiency of the ligated sample should be approximately 10% of that of undigested vector DNA.

Preparation of Bacteriophage λ DNA Cleaved with Two Restriction Enzymes for Use as a Cloning Vector

Many replacement vectors (e.g., the EMBL series, λ2001, and λDASH) contain a series of restriction sites, arranged in opposite orientations, at each end of the central stuffer fragment. Digestion of these vectors with two different restriction enzymes yields left and right arms, a stuffer fragment, and short segments of the polycloning sites. These can be easily removed from the arms by differential precipitation with isopropanol or spun-column chromatography.

MATERIALS

CAUTION: Please see Appendix 4 for appropriate handling of materials marked with <!>.

Reagents and Solutions

Please see Appendix 1 for components of stock solutions, buffers, and reagents. Dilute stock solutions to the appropriate concentrations.

ATP (10 mM)
> Not required if 10x ligase buffer already contains ATP

Chloroform <!>
Ethanol
Ethanol (70%)
Phenol:chloroform (1:1, v/v) <!>
Sodium acetate (3 M, pH 5.2)
Sodium acetate (3 M, pH 7.0)
> Use in Step 6 in place of sodium acetate at pH 5.2.

TE (pH 7.6 and pH 8.0)

Vectors and Hosts

Suspension of bacteriophage λ particles
> Prepare as described in either Protocol 2.9 or 2.10.

Enzymes and Buffers

Bacteriophage T4 DNA ligase and 10x ligation buffer
Restriction enzymes and buffers

Nucleic Acids/Oligonucleotides

Bacteriophage λ DNA
> Prepare as described in either Protocol 2.11 or 2.12.

DNA markers for gel electrophoresis

Gels/Loading Buffers

0.7% Agarose gel, cast and run in 0.5x TBE containing 0.5 µg/ml ethidium bromide <!>
Gel-loading buffer IV

Additional Items

Bacteriophage λ packaging mixture (commercial)
Step 10e of this protocol requires the reagents listed in Protocol 2.1.
Water bath for incubation of restriction reactions

Additional Information

In vitro packaging   MC3, pp. 2.110–2.111

METHOD

1. Mix 25–50 μg of bacteriophage λ DNA purified from a replacement vector with TE (pH 8.0) to give a final volume of 170 μl.

2. Add 20 μl of one of the two appropriate 10x restriction enzyme buffers. Remove two aliquots, each containing 0.2 μg of undigested bacteriophage λ DNA to serve as controls. Store the aliquots of undigested DNA on ice.

3. Add a fourfold excess (100–200 units) of one of the two restriction enzymes and incubate the digestion mixture for 4 hours at the temperature recommended by the manufacturer.

4. Cool the reaction to 0°C on ice. Remove two aliquots (0.2 μg). Incubate one of these aliquots (save the other for analysis in Step 10 below) and one of the two aliquots of undigested DNA (Step 2 above) for 10 minutes at 68°C to disrupt the cohesive termini of the bacteriophage DNA. Add a small amount (~10 μl) of gel-loading buffer IV and *immediately* electrophorese the samples through an 0.7% agarose gel.

   > If the restriction enzyme digestion is complete, no DNA will migrate at the position of the undigested control bands. Instead, three or more (depending on the number of cleavage sites in the vector) smaller DNA fragments will be seen. The number and yield of these smaller fragments should be examined carefully to ensure that no partial digestion products are present.

5. Purify the DNA by extracting twice with phenol:chloroform and once with chloroform.

6. Recover the DNA by standard ethanol precipitation.

7. Redissolve the DNA in TE (pH 8.0) at a concentration of 250 μg/ml. Add the appropriate 10x restriction buffer and digest the DNA with the second restriction enzyme. Use a fourfold excess of enzyme and incubate the reaction for 4 hours.

8. Purify the DNA by extracting twice with phenol:chloroform and once with chloroform. Recover the DNA by standard ethanol precipitation.

9. Redissolve the DNA in TE (pH 7.6) at a concentration of 300–500 μg/ml. Store an aliquot (0.2 μg) at –20°C.

10. To determine the effectiveness of the digestion procedure, set up trial ligation reactions using 0.2 μg of the vector digested with only the first enzyme (the aliquot set aside at Step 4 above) and 0.2 μg of the final preparation (Step 9). Package equivalent amounts of DNA (0.1 μg) from each ligation mixture and titrate the infectivity of the resulting bacteriophage particles.

    a. Adjust the volumes of the two DNA solutions to 17 μl with H₂O.

    b. Add to each sample 2 μl of 10x ligation buffer and, if necessary, 1 μl of 10 mM ATP.
       > Omit ATP if using a commercial ligase buffer that contains ATP.

    c. Remove 10-μl aliquots of each of the mixtures prepared in Step b and store the aliquots on ice.

    d. Add 0.2–0.5 Weiss unit of bacteriophage T4 DNA ligase to the remainder of the mixtures (Step b) and incubate the reactions for 2 hours at 16°C.

    e. Use a commercial bacteriophage λ packaging mixture to package 0.1 μg of the ligated and unligated samples and 0.1 μg of the undigested vector DNA from Step 2. Determine the titer (pfu/ml) of each packaged reaction as described in Protocol 2.1.
       > Digestion with a single restriction enzyme should reduce infectivity by a factor of about 10. After ligation, infectivity should be restored, essentially to the levels of undigested DNA. After digestion with the second restriction enzyme, infectivity should be further reduced by 30- to 50-fold and should not be appreciably restored after ligation.

Alkaline Phosphatase Treatment of Bacteriophage λ DNA

Removal of the 5′-phosphate groups from the internal termini of bacteriophage λ arms is used to prevent self-ligation and suppress the background of nonrecombinant bacteriophages during cloning.

MATERIALS

CAUTION: Please see Appendix 4 for appropriate handling of materials marked with <!>.

Reagents and Solutions

Please see Appendix 1 for components of stock solutions, buffers, and reagents. Dilute stock solutions to the appropriate concentrations.

ATP (10 mM)
> Not required if 10x ligase buffer already contains ATP

EDTA (0.5 M, pH 8.0)
Ethanol
Ethanol (70%)
λ Annealing buffer
> 100 mM Tris-Cl (pH 7.6)
> 10 mM MgCl$_2$

Phenol:chloroform (1:1, v/v) <!>
SDS (10% w/v)
Sodium acetate (3 M, pH 5.2 and pH 7.0)
TE (pH 7.6 and pH 8.0)

Enzymes and Buffers

Bacteriophage T4 DNA ligase and 10x ligation buffer
Calf intestinal phosphatase (CIP) and 10x dephosphorylation buffer
> 100 mM Tris-Cl (pH 8.3)
> 10 mM ZnCl$_2$
> 10 mM MgCl$_2$

Proteinase K

Nucleic Acids/Oligonucleotides

Bacteriophage λ DNA
> Prepare as described in either Protocol 2.11 or 2.12.

Gels/Loading Buffers

0.6% Agarose gel, cast and run in 0.5x TBE containing 0.5 µg/ml ethidium bromide <!>

Centrifuges/Rotors/Tubes

Microfuge (room temperature)

Additional Items

Dog toenail clippers
Steps 6 and 13 of this protocol require the reagents listed in Protocols 2.1, 2.13, and 2.14.
Water bath for restriction reactions
Water baths (16°C, 42°C, 56°C, and 68°C)

Additional Information

Minimizing damage to large DNA molecules	MC3, p. 2.110
In vitro packaging	MC3, pp. 2.110–2.111

METHOD

1. Dissolve 50–60 µg of DNA of the appropriate bacteriophage λ vector in a final volume of 150 µl of λ annealing buffer. Incubate the DNA for 1 hour at 42°C to allow the ends of the viral DNA containing the *cos* sites to anneal.

2. Add 20 µl of 10x ligase buffer, 20 µl of 10 mM ATP (if necessary), and 0.2–0.5 Weiss unit of bacteriophage T4 DNA ligase/µg of DNA. Incubate the reaction for 1–2 hours at 16°C.

   Omit ATP if using a commercial ligase buffer that contains ATP.

3. Extract the ligation reaction with phenol:chloroform.

   During ligation, the λ DNA will form closed circles and long concatemers and become sensitive to shearing. Handle the ligated DNA carefully! Do not vortex. Perform the phenol:chloroform extraction by gently inverting the tube to elicit emulsion formation.

4. Separate the organic and aqueous phases by centrifugation for 1 minute at room temperature in a microfuge. Remove the aqueous phase containing the viral DNA to a new tube using an automatic pipetting device equipped with a disposable tip that has been snipped with dog toenail clippers to increase the diameter of the hole.

5. Recover the DNA by standard ethanol precipitation. Remove the 70% ethanol supernatant and store the open tube on the bench to allow the ethanol to evaporate. Redissolve the damp pellet of DNA in 150 µl of TE (pH 8.0).

   Check that the ligation of *cos* termini has succeeded by heating an aliquot (0.2 µg) of the ligated DNA for 5 minutes at 68°C in TE. Chill the DNA in ice water and then electrophorese the DNA immediately through a 0.6% agarose gel. As controls, use (i) bacteriophage λ DNA that has been heated but not ligated and (ii) bacteriophage λ DNA that has been ligated but not heated.

   Ligation should convert the bacteriophage λ DNA to closed circular and concatenated forms that show no change in migration after heating. The unligated, heated, control DNA should migrate as a linear molecule, approximately 50 kb in length.

6. Digest the ligated DNA with one or more restriction enzymes as described in Protocol 2.13 or 2.14.

7. Repeat Steps 3 and 4 (above).

8. Add 0.1 volume of 3 M sodium acetate (pH 7.0) and 2 volumes of ethanol. Recover the precipitate of DNA by centrifugation for 10 minutes at 4°C in a microfuge. Rinse the pellet with 1 ml of 70% ethanol and recentrifuge for 2 minutes. Remove the 70% ethanol supernatant and store the open tube on the bench to allow the ethanol to evaporate.

9. Dissolve the digested and ethanol-precipitated DNA at a concentration of 100 µg/ml in 10 mM Tris-Cl (pH 8.3) and store an aliquot (0.2 µg) on ice. Treat the remainder of the DNA with an excess of CIP for 1 hour at 37°C as follows:

a. Add 0.1 volume of 10x dephosphorylation buffer and 0.01 unit of CIP for every 10 μg of bacteriophage λ DNA.

b. Mix; incubate the reaction for 30 minutes at 37°C. Add a second aliquot of CIP and continue incubation for an additional 30 minutes.

10. Add SDS and EDTA (pH 8.0) to final concentrations of 0.5% and 5 mM, respectively. Mix the solution by gentle vortexing and add proteinase K to a final concentration of 100 μg/ml. Incubate the mixture for 30 minutes at 56°C.

11. Cool the reaction mixture to room temperature and purify the bacteriophage λ DNA by extracting once with phenol:chloroform and once with chloroform. Recover the DNA by ethanol precipitation in the presence of 0.3 M sodium acetate (pH 7.0).

12. Dissolve the DNA in TE (pH 7.6) at a concentration of 300–500 μg/ml. Store the dephosphorylated DNA at –20°C in aliquots of 1–5 μg.

13. Measure the efficiency of dephosphorylation by ligating a portion (0.2 μg) of the digested vector before and after treatment with CIP (for ligation conditions, please see Protocol 2.13). Package the DNA into bacteriophage particles (for packaging conditions, please see Protocol 2.14) and titrate the infectivity.

Purification of Bacteriophage λ Arms: Centrifugation through Sucrose Density Gradients

This method may be used to prepare the arms of any bacteriophage λ vector.

MATERIALS

CAUTION: Please see Appendix 4 for appropriate handling of materials marked with <!>.

Reagents and Solutions

Please see Appendix 1 for components of stock solutions, buffers, and reagents. Dilute stock solutions to the appropriate concentrations.

ATP (10 mM)
> Not required if 10x ligation buffer already contains ATP

n-Butanol <!>
EDTA (0.5 M, pH 8.0)
Ethanol
Ethanol (70%)
$MgCl_2$ (1 M)
NaCl (5 M)
Phenol:chloroform (1:1, v/v) <!>
Sodium acetate (3 M, pH 5.2)
Solutions of 10% sucrose (w/v) and 40% sucrose (w/v) in 1 M NaCl, 20 mM Tris (pH 8.0), 5 mM EDTA
TE (pH 7.6)
Tris-Cl (1 M, pH 7.6 and pH 8.0)

Enzymes and Buffers

Bacteriophage T4 DNA ligase and 10x ligation buffer

Nucleic Acids/Oligonucleotides

Bacteriophage λ DNA (at least 60 µg)
> Prepare as described in either Protocol 2.11 or 2.12.

Gels/Loading Buffers

Gel-loading buffer IV
Thick (0.6 cm) agarose gel (0.5% and 0.7%) poured and run in 0.5x TBE containing 0.5 µg/ml ethidium bromide <!>

Centrifuges/Rotors/Tubes

Beckman SW 28 rotor and Ultra-Clear tubes (15°C)

Additional Items

Dialysis tubing
Gradient-making device (for 38-ml gradients)
Hypodermic needles (21 gauge)
Large-bore yellow tips (may be made with dog nail clippers)
Microfuge (room temperature)
Nitrocellulose filters (0.22 µm), used to sterilize sucrose solutions
Step 2 of this protocol requires the reagents listed in either Protocol 2.13 or 2.14.
Water bath or heating/cooling blocks (16°C, 37°C, 42°C, and 68°C)

METHOD

This method begins with an optional step that may be performed before cleaving the vector DNA with restriction enzymes. In the "Ligation First" technique, the cohesive termini of the vector DNA are ligated together to generate linear and closed circular concatemeric structures. These concatemers are then cleaved into left and right arms (that remain joined together) and the stuffer fragment (discarded during subsequent sucrose gradient centrifugation). The virtue of Ligation First is that it selects vector molecules with intact *cos* sites that are packaged with high efficiency.

i. Incubate the undigested bacteriophage λ DNA for 1 hour at 42°C in 150 μl of 0.1 M Tris-Cl (pH 7.6), 10 mM $MgCl_2$ to allow the cohesive termini to anneal.

ii. Add 20 μl of 10x ligation buffer (see Protocol 2.13), 20 μl of 10 mM ATP (if necessary), and 0.2–0.5 Weiss unit of bacteriophage T4 DNA ligase/μg of DNA. Incubate the reaction mixture for 1–2 hours at 16°C.

iii. Extract the ligated DNA once with phenol:chloroform.

> During ligation, the bacteriophage λ DNA forms closed circles and long concatemers and will be more sensitive to shearing. Handle the ligated DNA carefully! Do not vortex. Perform the phenol:chloroform extraction by gently inverting the tube to emulsify the two phases.

iv. Centrifuge the emulsion for 1 minute at room temperature in a microfuge to separate the organic and aqueous phases. Transfer the aqueous phase containing the viral DNA to a new tube using an automatic pipetting device equipped with a large-bore tip.

v. Recover the DNA by standard ethanol precipitation in the presence of 0.3 M sodium acetate (pH 5.2).

vi. Proceed with Step 1 of this protocol to digest the concatenated DNA with the appropriate restriction enzymes and fractionate the cleaved DNA by centrifugation through a sucrose gradient.

1. Prepare two sucrose solutions, one containing 10% (w/v) sucrose and another containing 40% (w/v) sucrose in a buffer of 1 M NaCl, 20 mM Tris-Cl (pH 8.0), and 5 mM EDTA (pH 8.0). Sterilize the two solutions by filtration through 0.22-μm nitrocellulose filters. Prepare one or more 38-ml (10–40% w/v) sucrose gradients in clear ultracentrifuge tubes. Store the gradients for 1–2 hours at 4°C in a quiet place until they are needed (Step 4).

> Continuous sucrose density gradients are best made in a gradient-making device such as those supplied by Bio-Rad or Techware. Each gradient should take 10–20 minutes to pour at room temperature using a gradient maker. Each gradient can accommodate 60–75 μg of digested bacteriophage λ DNA.

2. Digest and analyze approximately 60 μg of the bacteriophage λ vector DNA as described in Protocol 2.13 or 2.14. After standard ethanol precipitation, dissolve the DNA in TE (pH 7.6) at a concentration of 150 μg/ml. Set aside an aliquot (0.2 μg) for use as an electrophoretic control (Step 7).

3. Add $MgCl_2$ (1 M) to a final concentration of 10 mM and incubate the solution of bacteriophage DNA for 1 hour at 42°C to allow the cohesive termini of bacteriophage λ DNA to anneal. Analyze an aliquot (0.2 μg) by electrophoresis through a 0.7% agarose gel to determine whether annealing has occurred.

4. Load onto each gradient no more than 75 μg of annealed, digested, bacteriophage λ DNA in a volume of 500 μl or less.

> More DNA can cause the gradient to be overloaded and lead to poor separation of the stuffer fragments from the arms.

5. Centrifuge the gradients at 120,000*g* (26,000 rpm in a Beckman SW 28 rotor) for 24 hours at 15°C.

6. Collect 0.5-ml fractions through a 21-gauge needle inserted through the bottom of the centrifuge tube.

7. Take two 15-µl aliquots from every third fraction and dilute each with 35 µl of H_2O. Add 8 µl of gel-loading buffer IV, heat one aliquot from each fraction to 68°C for 5 minutes, and leave the second aliquot untreated. Analyze all of the samples by electrophoresis through a thick, 0.5% agarose gel. Use as markers intact bacteriophage λ DNA and the aliquot of digested DNA set aside in Step 2.

> Adjust the sucrose and salt concentrations of the markers to match those of the samples; otherwise, their electrophoretic mobilities will not be comparable.

8. After photographing the gel, locate and pool the fractions that contain the annealed arms (Figure 2-3).

> Be careful not to include fractions that are visibly contaminated with undigested bacteriophage λ DNA or fractions that contain significant quantities of unannealed left or right arms or stuffer fragment(s).

9. Dialyze the pooled fractions against a 1000-fold excess of TE (pH 8.0) for 12–16 hours at 4°C, with at least one change of buffer.

> Be sure to allow for a two- to threefold increase in volume during dialysis.

10. Extract the dialyzed sample several times with *n*-butanol to reduce its volume to less than 3 ml.

11. Recover the dialyzed DNA by standard ethanol precipitation.

12. Dissolve the DNA in TE (pH 7.6) at a concentration of 300–500 µg/ml.

13. Measure the concentration of the DNA spectrophotometrically (1 OD_{260} = ~50 µg/ml) and analyze an aliquot by electrophoresis through a 0.5% agarose gel to assess its purity. Store the DNA at –20°C in aliquots of 1–5 µg.

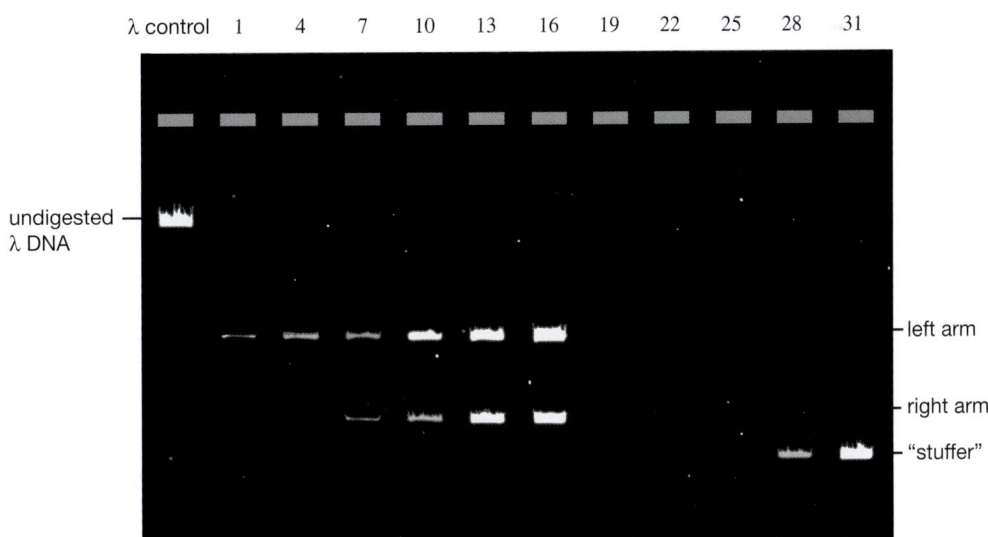

FIGURE 2-3 Preparation of the Arms of Bacteriophage λ DNA by Sucrose Gradient Centrifugation

In this experiment, the DNA of bacteriophage λ vector Charon 28 was digested with *Bam*HI and centrifuged through a 10–40% linear sucrose gradient. Aliquots from every third gradient fraction were heated to 68°C for 5 minutes, prepared for electrophoresis, and analyzed on a 0.5% agarose gel. The positions of the left arm (23.5 kb), right arm (9 kb), "stuffer" fragments (6.5 kb), and undigested λ DNA are indicated. Fractions 1–16 containing the arms were pooled.

Partial Digestion of Eukaryotic DNA for Use in Genomic Libraries: Pilot Reactions

This protocol is used to establish conditions for restriction enzyme digestion of eukaryotic genomic DNA that will generate fragments of a size appropriate for construction of genomic libraries. To construct a genomic library, the average length of the starting genomic DNA should be at least eight times the capacity of the vector. This size range ensures that the majority of DNA molecules created by partial digestion with restriction enzyme(s) are derived from internal segments of the high-molecular-weight DNA and therefore carry termini that are compatible with those of the vector arms.

MATERIALS

CAUTION: Please see Appendix 4 for appropriate handling of materials marked with <!>.

Reagents and Solutions

Please see Appendix 1 for components of stock solutions, buffers, and reagents. Dilute stock solutions to the appropriate concentrations.
Tris-Cl (10 mM, pH 8.0)

Enzymes and Buffers

Restriction enzyme and 10x buffer

Nucleic Acids/Oligonucleotides

DNA size markers (e.g., oligomers, bacteriophage λ DNA, or plasmids)
High-molecular-weight genomic DNA (as described in Protocol 6.4)

Gels/Loading Buffers

0.6% Agarose gel, cast and run in 0.5x TBE containing 0.5 μg/ml ethidium bromide <!>
or
Pulsed-field gel (see MC3, pp. 5.55–5.90)
Gel-loading buffer IV

Additional Items

Sealed glass capillary tube
Water baths or heating blocks (37°C and 70°C)
Wide-bore yellow tips (can be made using dog nail clippers)

METHOD

1. Set up pilot reactions using the same batch of genomic DNA that will be used to prepare fragments for cloning.

   a. Dilute 30 μg of high-molecular-weight eukaryotic DNA to 900 μl with 10 mM Tris-Cl (pH 8.0) and add 100 μl of the appropriate 10x restriction enzyme buffer.

      The best results are obtained if the same batch of 10x buffer is used in both the pilot reactions and the large-scale reaction.

 b. Use a sealed glass capillary to mix the solution gently. Mixing ensures that the high-molecular-weight DNA is distributed evenly throughout the restriction enzyme buffer.

 c. After mixing, store the diluted DNA for 1 hour at room temperature to allow any residual clumps of DNA to disperse (please also see the note to Step 8, below).

   If the concentration of the high-molecular-weight DNA is low, it is best to increase the volume of the pilot reactions and concentrate the DNA after digestion by standard ethanol precipitation. This approach minimizes the possibility of shearing the high-molecular-weight DNA. Each pilot reaction should contain at least 1 μg of DNA to allow the heterogeneous products of digestion to be detected by staining with ethidium bromide.

2. Label a series of microfuge tubes 1–10. Use a wide-bore glass capillary or disposable plastic pipette tip to transfer 60 μl of the DNA solution to a microfuge tube (Tube 1). Transfer 30 μl of the DNA solution to each of the nine additional labeled microfuge tubes. Incubate the tubes on ice.

3. Add 2 units of the appropriate restriction enzyme to Tube 1.

  Use a sealed glass capillary to mix the restriction enzyme with the DNA. Do not allow the temperature of the mixture to rise above 4°C.

4. Use a fresh pipette tip to transfer 30 μl of the reaction from Tube 1 to the next tube in the series. Mix as before, and continue transferring the reaction to successive tubes. Do not add anything to the tenth tube (the no-enzyme control), but discard 30 μl from the ninth tube.

5. Incubate the reactions for 1 hour at 37°C.

6. Inactivate the restriction enzyme by heating the reactions to 70°C for 15 minutes.

7. Cool the reactions to room temperature and add the appropriate amount of gel-loading buffer IV.

  Use a sealed glass capillary as a stirring rod to mix the solutions gently.

8. Use a wide-bore plastic pipette tip or a disposable wide-bore glass capillary to transfer the solutions to the wells of a 0.6% agarose gel or, even better, to the lanes of an agarose gel for pulsed-field electrophoresis (please see Chapter 5). Perform electrophoresis.

  When separating the partial digestion products by conventional agarose gel electrophoresis, it is essential to run the gel under conditions of maximum resolution. Use the same batch of buffer to cast the gel and to fill the gel tank before electrophoresis. The gel should be run slowly (<1 V/cm) at 4°C to prevent smearing of the fragments of DNA.

9. Compare the size of the digested eukaryotic DNA with that of DNA standards composed of oligomers of bacteriophage λ DNA and plasmids. Identify the partial digestion conditions that result in a majority of the genomic DNA migrating in the desired size range.

Partial Digestion of Eukaryotic DNA for Use in Genomic Libraries: Preparative Reactions

The results of the pilot experiment (Protocol 2.17) are used to establish conditions for partial digestion of eukaryotic DNA in the large-scale reactions described here.

MATERIALS

CAUTION: Please see Appendix 4 for appropriate handling of materials marked with <!>.

Reagents and Solutions

Please see Appendix 1 for components of stock solutions, buffers, and reagents. Dilute stock solutions to the appropriate concentrations.

Ammonium acetate (10 M)
n-Butanol <!>
Ethanol
Phenol:chloroform (1:1, v/v) <!>
Sodium acetate (3 M, pH 5.2)
Solutions of 10% sucrose (w/v) and 40% sucrose (w/v) in 1 M NaCl, 20 mM Tris-Cl
  (pH 8.0), 5 mM EDTA
TE (pH 7.6 and pH 8.0)
Tris-Cl (1 M, pH 8.0)

Enzymes and Buffers

Restriction enzyme and 10x buffer

Nucleic Acids/Oligonucleotides

DNA size markers (e.g., oligomers of bacteriophage λ DNA or plasmids)
High-molecular-weight genomic DNA, as described in Protocol 6.4

Gels/Loading Buffers

0.6% Agarose gel, cast and run in 0.5x TBE containing 0.5 µg/ml ethidium bromide <!>
or
Pulsed-field gel (see MC3, pp. 5.55–5.90)
Gel-loading buffer IV

Centrifuges/Rotors/Tubes

Beckman SW 28 rotor and Ultra-Clear tubes (20°C)

Additional Items

Dialysis tubing
Hypodermic needles (21 gauge)
Nitrocellulose filters (0.22 µm), used to sterilize sucrose solutions
Water baths or heating blocks (37°C and 68°C)

METHOD

1. Prepare two sucrose solutions, one containing 10% (w/v) sucrose and another containing 40% (w/v) sucrose in a buffer of 1 M NaCl, 20 mM Tris-Cl (pH 8.0), 5 mM EDTA (pH 8.0). Sterilize the two solutions by filtration through 0.22-μm nitrocellulose filters. Prepare one or more 38-ml (10–40% w/v) sucrose gradients in clear ultracentrifuge tubes. Store the gradients for 1–2 hours at 4°C in a quiet place until they are needed (Step 5).

2. Set up a series of digestions, each containing 100 μg of high-molecular-weight DNA.

   a. Use three different concentrations of restriction enzyme that straddle the optimal concentration determined in the pilot experiments (Protocol 2.17).

      The best results are obtained if the same batch of 10x buffer is used in both the pilot reactions and the large-scale reaction.

   b. Incubate the reactions for the appropriate time with the restriction enzyme.

   c. Analyze an aliquot of the partially digested DNA by gel electrophoresis to ensure that the digestion has worked according to prediction. Until the results are available, store the remainder of the sample at 0°C.

3. Gently extract the digested DNA twice with phenol:chloroform.

4. Recover the DNA by standard precipitation with ethanol in the presence of 0.3 M sodium acetate (pH 5.2) and dissolve it in 200 μl of TE (pH 8.0).

5. Heat the DNA sample (100 μg) for 10 minutes at 68°C. After cooling to 20°C, gently layer the sample on the top of the gradient. Centrifuge the gradients at 83,000g (25,000 rpm in a Beckman SW 28 rotor) for 22 hours at 20°C.

6. Use a 21-gauge needle or a gradient fractionation device to puncture the bottom of the tube and collect 350-μl fractions.

7. Mix 10 μl of every other fraction with 10 μl of H_2O and 5 μl of gel-loading buffer IV. Analyze the size of the DNA in each fraction by electrophoresis through a 0.6% agarose gel, using oligomers of plasmid DNA or other high-molecular-weight standards as markers. Adjust the sucrose and salt concentrations of the markers to correspond to those of the samples.

8. Following electrophoresis, pool the gradient fractions containing DNA fragments of the desired size (e.g., 35–45 kb for construction of libraries in cosmids and 20–25 kb for construction of libraries in bacteriophage λ vectors).

9. Dialyze the pooled fractions against 2 liters of TE (pH 8.0) for 12–16 hours at 4°C, with a change of buffer after 4–6 hours.

   Leave space in the dialysis sac for the sample to expand two- to threefold in volume.

10. Extract the dialyzed DNA several times with an equal volume of *n*-butanol until the volume is reduced to approximately 1 ml.

11. Precipitate the DNA with ethanol at room temperature in the presence of 2 M ammonium acetate (from a 10 M stock solution).

12. Recover the DNA by centrifugation and dissolve the DNA in TE (pH 8.0) at a concentration of 300–500 μg/ml. Analyze an aliquot of the DNA (0.5 μg) by electrophoresis through a conventional 0.6% agarose gel or by pulsed-field electrophoresis to check that the size distribution of the digestion products is correct. Store the DNA at 4°C.

13. To establish genomic DNA libraries, ligate the fractionated DNA to the arms of bacteriophage λ vectors as described in Protocol 2.19. For the preparation of cosmid libraries, please see Protocol 4.1.

Ligation of Bacteriophage λ Arms to Fragments of Foreign DNA

Pilot ligations and packaging reactions are used to establish the amounts of fragmented genomic DNA and bacteriophage λ arms that yield the maximum number of recombinants. Additional ligation and packaging reactions may then be set up to yield a comprehensive library of genomic DNA.

MATERIALS

CAUTION: Please see Appendix 4 for appropriate handling of materials marked with <!>.

Reagents and Solutions

Please see Appendix 1 for components of stock solutions, buffers, and reagents. Dilute stock solutions to the appropriate concentrations.

ATP (10 mM)
> Not required if 10x ligation buffer contains ATP

SM

SM plus gelatin

Vectors and Hosts

E. coli plating bacteria
> Prepare as described in Protocol 2.1.

Media and Antibiotics

NZCYM or LB agar medium
NZCYM or LB top agarose

Enzymes and Buffers

Bacteriophage T4 DNA ligase and 10x ligase buffer
Restriction enzyme(s) and 10x buffers

Nucleic Acids/Oligonucleotides

Bacteriophage λ arms
> Prepare as described in Protocols 2.11–2.15.

Genomic DNA, of an appropriate size for the vector
> Prepare as described in Protocol 2.18.

Gels/Loading Buffers

0.7% Agarose gel, cast and run in 0.5x TBE containing 0.5 µg/ml ethidium bromide <!>

Additional Items

Bacteriophage λ packaging mix (commercial)
Water bath or heating/cooling block (16°C)

Adapted from Chapter 2, Protocol 19, p. 2.84 of MC3.

METHOD

1. Use Table 2-1 as a guide to set up a series of ligation reactions that contain the following:

bacteriophage λ arms	0.5–1.0 µg
partially digested genomic DNA	6–1200 ng
10x ligation buffer	0.5–1.0 µl
10 mM ATP (if necessary)	0.5–1.0 µl
bacteriophage T4 DNA ligase	0.5–1.0 µl
H₂O	to 5 or 10 µl

   Set up two control reactions in which the vector and insert DNAs are each ligated in the absence of the other. Incubate the ligation reactions for 4–16 hours at 16°C.

   TABLE 2-1 Amounts of Insert DNA Used in Trial Ligations Containing 1 µg of Bacteriophage λ Arms

Size of Potential Insert DNA (kb)	Amount of Insert DNA (ng)
2–4	6–200
4–8	12–400
8–12	24–600
12–16	36–800
16–20	48–1000
20–24	60–1200

2. Package an aliquot (10–25%) of each of the ligation reactions into bacteriophage particles, following the instructions provided by the manufacturer of the packaging extract.

3. Make a series of tenfold dilutions (10^{-1} to 10^{-5}) of the packaging reactions, using as a diluent SM plus gelatin or an equivalent buffer recommended by the manufacturer of the packaging extract.

4. Assay the number of plaque-forming units in 1 and 10 µl of each dilution as described in Protocol 2.1.

5. From the ligation reaction yielding the largest number of infectious bacteriophage particles, pick 6–12 plaques and prepare a small amount of recombinant DNA from each as described in Protocol 2.23.

6. Check the size of the inserts of genomic DNA by digestion with the appropriate restriction enzymes, followed by electrophoresis through a 0.7% agarose gel, using appropriate size markers.

7. If the bacteriophages are recombinants and contain inserts of the desired size, establish a genomic DNA library by setting up multiple ligation and packaging reactions. The ratio of insert to vector DNA in these reactions should be that which generated the greatest number of recombinant plaques in the trial reactions.

8. Estimate the total number of recombinant plaques generated in the large-scale ligation and packaging reactions. Calculate the depth to which a library of this size would cover the target genome.

   To provide fivefold coverage of a mammalian genome (3×10^9 bp), a bacteriophage λ library containing inserts whose average size is 20 kb would contain 2×10^6 independent recombinants.

Amplification of Genomic Libraries

Libraries of recombinant bacteriophages may be amplified by growing plate stocks directly from the packaging mixtures generated in Protocol 2.18.

MATERIALS

CAUTION: Please see Appendix 4 for appropriate handling of materials marked with <!>.

Reagents and Solutions

Please see Appendix 1 for components of stock solutions, buffers, and reagents. Dilute stock solutions to the appropriate concentrations.
Chloroform <!>
SM

Vectors and Hosts

Library recombinant λ bacteriophages carrying segments of genomic DNA
Prepare as described in Protocol 2.18.
Plating bacteria (Protocol 2.1)

Media and Antibiotics

NZCYM agar plates (150 mm)
NZCYM top agarose
6.5 ml of top agarose is required per 150-mm plate.

Centrifuges/Rotors/Tubes

Sorvall SS-34 rotor (4°C)

Additional Items

Step 7 of this protocol requires the reagents and techniques listed in Protocol 2.1.
Sterile tubes (17 x 100 mm)
Water bath (37°C)
Water bath or heating block (47°C)

Additional Information

Alternative method of amplification of genomic libraries (panel on
in situ amplification)
MC3, p. 2.95

METHOD

1. To amplify a bacteriophage λ library, mix aliquots of the packaging mixture containing 10,000–20,000 recombinant bacteriophages in a volume of 50 µl or less with 0.2 ml of plating bacteria in a 17 x 100-mm tube. Incubate the infected culture for 20 minutes at 37°C.

   If enough incubator space is available, large glass baking dishes can be used instead of 150-mm plates.

2. Add 6.5 ml of melted top agar/agarose (47°C) to the first aliquot of infected bacteria. Mix the contents of the tube by tapping or gentle vortexing and spread the infected bacteria onto the surface of a freshly poured 150-mm plate of bottom agar. Repeat the procedure with the remaining infected cultures.

    As many as 450,000 bacteriophages may be mixed with 1.4 ml of bacteria and plated in 75 ml of top agar/agarose on 500 ml of bottom agar in a 23 x 33-cm glass baking dish.

3. Incubate the plates for a maximum of 8–10 hours at 37°C.

    Do not allow the plaques to grow so large that they touch one another.

4. Overlay the plates with 12 ml of SM (or 150 ml of SM if baking dishes are used). Store the plates overnight at 4°C on a level surface.

5. Harvest the SM from all of the plates into a single, sterile, polypropylene centrifuge tube or bottle. Wash each plate with an additional 4 ml of SM and combine the washings with the primary harvest. Add 0.2 ml of chloroform to the resulting amplified bacteriophage stock. Store the stock for 15 minutes at room temperature with occasional gentle shaking to allow time for the chloroform to lyse all of the infected cells.

6. Remove cell and agarose debris by centrifugation at 4000g (5800 rpm in a Sorvall SS-34 rotor) for 5 minutes at 4°C.

7. Transfer the supernatant to a sterile glass tube or bottle. Divide the amplified bacteriophage library into aliquots and store them at 4°C. Measure the titer of the library by plaque assay (Protocol 2.1).

Transfer of Bacteriophage DNA from Plaques to Filters

This protocol and Protocol 2.22 are used to identify and isolate recombinants containing DNA sequences of interest.

MATERIALS

Reagents and Solutions

Please see Appendix 1 for components of stock solutions, buffers, and reagents. Dilute stock solutions to the appropriate concentrations.

Denaturation solution
1.5 M NaCl
0.5 M NaOH

Neutralizing solution
0.5 M Tris-Cl (pH 7.4)

SM plus gelatin (0.01% w/v)

2x SSPE
0.3 M NaCl
20 mM NaHPO$_4$
2 mM EDTA

SDS (10% w/v) (optional; see Step 14)

Vectors and Hosts

Bacteriophage λ library
Prepare as described in Protocol 2.19.

E. coli plating bacteria
Prepare as described in Protocol 2.1.

Media and Antibiotics

NZCYM agar plates
Use well-dried, 2-day-old plates. If the plates are not well dried, the layer of top agarose will peel off when the nylon or nitrocellulose filters are removed.

NZCYM top agarose

Additional Items

Blunt-ended plastic forceps

Cross-linking device (e.g., Stratalinker [Stratagene] or microwave oven or vacuum oven set at 80°C)

Hypodermic syringe and 21-gauge needles

Nitrocellulose or nylon filters (e.g., Millipore HATF detergent-free nitrocellulose filters or Pall Biodyne™ A nylon filters)

Plastic cafeteria trays or Pyrex glass dishes, for processing batches of filters

Sterile tubes (13 x 75 or 17 x 100 mm)

Water baths (37°C, 47°C, and 65°C)

Waterproof black drawing ink (India ink)

Whatman 3MM, 85-mm filter papers and precut sheets

> **Additional information**
>
> | | |
> |---|---|
> | Alternative protocol (Rapid Transfer of Plaques to Filters) | MC3, p. 2.95 |
> | Discussion of the relative merits of nylon and nitrocellulose filters | MC3, pp. 2.91 and 6.37–6.38 |

METHOD

1. Prepare the nitrocellulose or nylon filters for transfer:

   a. Number the dry filters with a soft-lead pencil or a ball-point pen.

      Prepare enough filters to make one or two replicas from each agar plate. In the latter case, number two sets of filters 1A, 1B, 2A, 2B, etc.

   b. Soak the filters in water for 2 minutes.

   c. Arrange the filters in a stack with each filter separated from its neighbor by an 85-mm-diameter Whatman 3MM filter paper.

   d. Wrap the stack of filters in aluminum foil and sterilize them by autoclaving for 15 minutes at 15 psi (1.05 kg/cm^2) on liquid cycle.

2. Make a dilution (in SM plus gelatin) of the bacteriophage stock or library. Mix aliquots of the diluted bacteriophage with the appropriate amount of freshly prepared plating bacteria (please see Table 2-2). Incubate the infected bacterial cultures for 20 minutes at 37°C.

   The diluted stock (100 µl) should contain approximately 15,000 infectious bacteriophages when using 90-mm plates or 50,000 infectious particles when using 150-mm plates.

   When screening bacteriophage libraries, it is best to infect the cells as a single pool. For example, when using ten 150-mm plates, 1 ml of diluted bacteriophage stock containing 500,000 pfu would be added to 3 ml of freshly prepared plating bacteria. After 20 minutes at 37°C, equal volumes of the infected culture are then distributed into 13 x 75 or 17 x 100-mm tubes for plating. This procedure ensures that each plate contains approximately the same number of plaques.

   To find a bacteriophage that carries a particular genomic sequence, it may be necessary to screen one million or more recombinants in a genomic DNA library. Table 2-2 shows the maximum number of plaques that can be efficiently screened in culture dishes of different sizes.

   TABLE 2-2 Number of Plaques in Culture Dishes of Various Sizes

Size of petri dish	90 mm	150 mm
Total area of dish	63.9 cm^2	176.7 cm^2
Volume of bottom agar	30 ml	80 ml
Volume of plating bacteria	0.1 ml	0.3 ml
Volume of top agarose	2.5 ml	6.5–7.5 ml
Maximum number of plaques/dish	15,000	50,000

3. Add to each aliquot of infected cells 2.5 or 6.5 ml of molten (47°C) top agarose. Pour the contents of each tube onto separate, numbered, 90- or 150-mm agar plates.

4. Close the plates, allow the top agarose to harden, and incubate the plates at 37°C in an inverted position until plaques appear and just begin to make contact with one another (10–12 hours).

   To minimize plate-to-plate variation, it is crucial that each plate be heated to the same extent when placed in the incubator.

5. Chill the plates for at least 1 hour at 4°C to allow the top agarose to harden.

6. Remove the plates from the cold room or refrigerator. Make imprints of the plaques on each plate using the first set of labeled filters. Place a dry, labeled, circular nitrocellulose or nylon

filter neatly onto the surface of the top agarose so that it comes into direct contact with the plaques. Handle the filter with gloved hands.

7. Mark the filter in three or more asymmetric, peripheral locations by stabbing through it and into the agar beneath with a 21-gauge needle attached to a syringe containing waterproof black drawing ink.

8. After 1–2 minutes, use blunt-ended forceps (e.g., Millipore) to peel the filters from each plate in turn.

9. Transfer each filter, plaque side up, to a sheet of Whatman 3MM paper (or its equivalent) impregnated with denaturing solution in a plastic cafeteria tray or Pyrex dish for 1–5 minutes.

    Make sure that excess denaturing solution does not rinse over the sides of the nitrocellulose or nylon filters. When transferring the filters, use the edge of the cafeteria tray or dish to remove as much fluid as possible from the underside of the filters.

10. Transfer the filters, plaque side up, to a sheet of Whatman 3MM paper impregnated with neutralizing solution for five minutes.

    If nitrocellulose filters are used, repeat the neutralizing step using a fresh impregnated sheet of 3MM paper.

11. Prepare the filters for fixing.

    To fix the DNA to the filters by microwaving or baking

    If using a microwave oven, proceed directly to Step 13. If baking in a vacuum oven, transfer the filters, plaque side up, to a sheet of dry 3MM paper or a stack of paper towels. Allow filters to dry for at least 30 minutes at room temperature.

    To fix the DNA to nylon filters by cross-linking with UV light

    Place the filters on a sheet of Whatman 3MM paper impregnated with 2x SSPE and move the tray of 2x SSPE containing the filters to the vicinity of the UV-light cross-linker.

12. After the first set of filters has been processed, use the second set of filters to take another imprint of the plaques, if required. Make sure that both sets of filters are keyed to the plate at the same positions.

    Generally, the second set of filters is left in contact with the plaques for 3 minutes or until the filter is completely wet.

13. Fix the DNA from the plaques to the filter.

    To fix by treatment in a microwave oven

    Place the damp filters on a sheet of dry Whatman 3MM paper and irradiate them for 2–3 minutes at full power in a microwave oven.

    To fix by baking

    Arrange the dried filters (Step 10) in a stack with adjacent filters separated by a sheet of dry Whatman 3MM paper. Bake the stack of filters for 1–2 hours at 80°C in a vacuum oven.

    Check the oven after 30 minutes and wipe away any condensation from the door. The filters must be baked under vacuum rather than stewed.

    To fix by cross-linking with UV light

    Perform the procedure using a commercial device for this purpose and follow the manufacturer's instructions.

    Do not allow the filters to dry out before cross-linking.

14. After baking or cross-linking, loosely wrap the dry filters in aluminum foil and store them at room temperature. Alternatively, if hybridization is to be performed within a day or so, wash the filters for 30 minutes at 65°C in 0.1x SSC or SSPE, 0.5% SDS and store them wet in sealed plastic bags.

Hybridization of Bacteriophage DNA on Filters

Using hybridization with ^{32}P-labeled probes, it is possible to identify a single recombinant that carries the desired target sequence on a filter that carries the imprint of 15,000 or more plaques.

MATERIALS

CAUTION: Please see Appendix 4 for appropriate handling of materials marked with <!>.

Reagents and Solutions

Please see Appendix 1 for components of stock solutions, buffers, and reagents. Dilute stock solutions to the appropriate concentrations.

Chloroform <!>
Prehybridization solution <!>
    6x SSC or SSPE
    5x Denhardt's solution
    0.5% SDS
    100 µg/ml salmon sperm DNA
    1 µg/ml poly(A)
SM
Wash solution 1
    2x SSC or SSPE
    0.1% SDS
Wash solution 2
    1x SSC or SSPE
    0.1% SDS
Wash solution 3
    1x SSC
    0.1% SDS

Vectors and Hosts

Suspension of bacteriophage λ particles
    Prepare as described in either Protocol 2.9 or 2.10.

Nucleic Acids/Oligonucleotides

Filters containing immobilized bacteriophage λ DNA
    Prepare as described in Protocol 2.21.
Radiolabeled probe <!>
    Prepare as described in either Chapter 9 or 10.

Additional Items

Glue stick (UHU [Saunders])
Incubation chamber, present at the appropriate hybridization temperature
Radioactive ink <!> or chemiluminescent markers (e.g., Gloglos [Stratagene])
Rocking platform to hold incubation chamber, used as orientation markers during autoradiography
Step 12 of this protocol requires the reagents and techniques described in Protocol 2.1.
Water bath (100°C)

Whatman 3MM filter paper sheets
X-ray film (e.g., Kodak XAR-2)

Additional Information

Alternative hybridization chambers	MC3, p. 2.97
Prehybridization and hybridization solutions	MC3, p. 1.141
Preparation of radioactive ink	MC3, pp. 2.97–2.98
Using formamide as a solvent for hybridization	MC3, pp. 6.59–6.60

METHOD

1. If the filters are dry, float the baked or cross-linked filters on the surface of 2x SSPE until they have become thoroughly wetted from beneath. Submerge the filters for 5 minutes.

   Make sure that no air bubbles are trapped under the filters. The filters should change from white to a bluish color as the aqueous solvent penetrates the pores of the filter. Make sure that there are no white spots or patches remaining on the filters before proceeding to Step 2.

2. Transfer the filters to a Pyrex dish or other hybridization chamber containing prehybridization solution. Use 3 ml of prehybridization solution per 82-mm filter or 5 ml per 145-mm filter. Incubate the filters with gentle agitation on a rocking platform for 1–2 hours or more at the appropriate temperature (i.e., 68°C when hybridization is to be performed in aqueous solution; 42°C when hybridization is to be performed in 50% formamide).

   Whatever type of container is used, the important point is that the filters must be completely covered by the prehybridization solution.

3. Denature ^{32}P-labeled double-stranded probes by heating for 5 minutes at 100°C. Chill the probe rapidly in ice water. Single-stranded probes need not be denatured.

   Between 2×10^5 and 1×10^6 cpm of ^{32}P-labeled probe (specific activity 5×10^7 cpm/µg) should be used per milliliter of hybridization solution. Using more probe causes the background of non-specific hybridization to increase; using less reduces the rate of hybridization.

4. Add the denatured probe to the prehybridization solution covering the filters. Incubate the filters for 12–16 hours at the appropriate temperature (please see Protocol 6.10).

   Keep the containers holding the filters tightly closed to prevent the loss of fluid by evaporation.

   To maximize the rate of annealing of the probe with its target, hybridizations are usually carried out in solutions of high ionic strength (6x SSC or 6x SSPE) at a temperature that is 20–25°C below the melting temperature (please see Protocols 1.28–1.30 or 6.10). Both SSPE and SSC work equally well when hybridization is performed in aqueous solvents. However, when formamide is included in the hybridization buffer, 6x SSPE is preferred because of its greater buffering capacity.

5. When hybridization is complete, *quickly* remove filters from the hybridization solution and *immediately* immerse them in a large volume (300–500 ml) of wash solution 1 at room temperature. Agitate the filters gently, turning them over at least once during washing. After 5 minutes, transfer the filters to a fresh batch of wash solution and continue to agitate them gently. Repeat the washing procedure twice more.

   At no stage during the washing procedure should the filters be allowed to dry or stick together.

6. Wash the filters twice for 1–1.5 hours in 300–500 ml of wash solution 2 at 68°C.

   With experience, it is possible to use a handheld monitor to test whether washing is complete. If the background is still too high or if the experiment demands washing at high stringencies, immerse the filters for 60 minutes in 300–500 ml of wash solution 3 at 68°C.

7. Dry the filters in the air at room temperature on sheets of Whatman 3MM paper or stacks of paper towels. Streak the underside of the filters with a water-soluble glue stick and arrange the filters (numbered side up) on a clean, dry, flat sheet of 3MM paper. Firmly press the fil-

ters against the 3MM paper to ensure that they do not move. Apply adhesive labels marked with radioactive ink or chemiluminescent markers to several asymmetric locations on the 3MM paper. These markers serve to align the autoradiograph with the filters. Cover the filters and labels with Saran Wrap/cling film. Use tape to secure the wrap to the back of the 3MM paper and stretch the wrap over the paper to remove wrinkles.

8. Expose the filters to X-ray film (Kodak XAR-2, XAR-5, or their equivalent) for 12–16 hours at –70°C with an intensifying screen.

9. Develop the film and align it with the filters using the marks left by the radioactive ink or fluorescent marker. Use a nonradioactive, red, fiber-tip pen to mark the film with the positions of the asymmetrically located dots on the numbered filters.

10. Identify the positive plaques by aligning the orientation marks with those on the agar plate.

    When duplicate sets of filters are hybridized to the same probe, there is less chance of confusing a background smudge with a positive plaque. Pick only those plaques that yield convincing hybridization signals on both sets of filters for further analysis. When screening a genomic library for a single-copy gene, expect to find no more than one positive clone per 10^5 plaques screened. When screening cDNA libraries, the number of positives depends on the abundance of the mRNA of interest.

11. Pick each positive plaque as described in Protocol 2.2 and store in 1 ml of SM containing a drop (50 µl) of chloroform.

12. To purify a hybridization-positive plaque, plate an aliquot (usually 50 µl of a 10^{-2} dilution) of the bacteriophages that are recovered from the cored agar plug and proceed with subsequent rounds of screening by hybridization.

Rapid Analysis of Bacteriophage λ: Purification of λ DNA from Plate Lysates

This protocol is used to purify small amounts of bacteriophage λ DNA that are suitable for use as substrates for restriction enzymes and templates for DNA and RNA polymerases.

MATERIALS

CAUTION: Please see Appendix 4 for appropriate handling of materials marked with <!>.

Reagents and Solutions

Please see Appendix 1 for components of stock solutions, buffers, and reagents. Dilute stock solutions to the appropriate concentrations.

Chloroform <!>
Ethanol
Ethanol (70%)
Isopropanol
λ High-salt buffer
    20 mM Tris-Cl (pH 7.4)
    1.0 M NaCl
    1 mM EDTA (pH 8.0)
λ Low-salt buffer
    20 mM Tris-Cl (pH 7.4)
    0.2 M NaCl
    1 mM EDTA (pH 8.0)
Phenol:chloroform (1:1, v/v) <!>
SM
TE (pH 8.0)
TM

Vectors and Hosts

Bacteriophage λ recombinant, grown as well-isolated plaques on a bacterial lawn (Protocol 2.1)
E. coli plating bacteria (Protocol 2.1)

Media and Antibiotics

NZCYM or LB agarose plates
    Freshly poured plates (145-mm diameter) that have been equilibrated to room temperature give the best results. Agarose is preferred to agar to minimize contaminants that can interfere with enzymatic analysis of DNA.
NZCYM or LB top agarose

Centrifuges/Rotors/Tubes

Sorvall SS-34 rotor and tubes (4°C and room temperature)

Additional Items

Borosilicate pasteur pipettes
DE52, preswollen (Whatman), used as a 2:1 slurry of DE52 in TM
Elutip-d columns (Schleicher & Schuell)
Microfuge (4°C)

Adapted from Chapter 2, Protocol 23, p. 2.101 of MC3.

Rotating platform
Sterile tubes (17 x 100 mm)
Water bath or heating blocks (37°C and 47°C)

Additional Information

Additional protocol (Removing Polysaccharides by Precipitation with CTAB)	MC3, p. 2.105
Other methods of analysis	MC3, pp. 2.104 and 2.105

Xu S.-Y. 1986. A rapid method for preparing phage lambda DNA from agar plate lysates. *Gene Anal. Tech.* **3:** 90–91.

METHOD

1. Use a borosilicate pasteur pipette to pick 8–10 well-isolated bacteriophage plaques from a plate derived from a genetically pure, plaque-purified bacteriophage stock. Place the plaques in 1 ml of SM and 50 μl of chloroform. Store the suspension for 4–6 hours at 4°C to allow the bacteriophage particles to diffuse from the top agarose.

2. In a 17 x 100-mm sterile tube, mix 50–100 μl of the bacteriophage suspension (~10^5 pfu) with 150 μl of plating bacteria. Incubate the infected culture for 20 minutes at 37°C. Add 7.0 ml of molten (47°C) top agarose (0.7%) and spread the bacterial suspension on the surface of a freshly poured 150-mm plate containing NZCYM agarose.

3. Incubate the inverted plate at 37°C until the plaques cover almost the entire surface of the plate (7–9 hours).

   Do not incubate the plates for too long because confluent lysis will occur, which reduces the yield of bacteriophage DNA.

4. Add 7 ml of TM directly onto the surface of the top agarose. Allow the bacteriophage particles to elute during 4 hours of incubation at 4°C with constant, gentle shaking.

5. Transfer the bacteriophage λ eluate to a centrifuge tube and remove the bacterial debris by centrifugation at 4000g (58,000 rpm in a Sorvall SS-34 rotor) for 10 minutes at 4°C. A small aliquot of cleared lysate can be set aside at this step as a bacteriophage stock solution. Store the stock at 4°C over a small volume of chloroform.

6. Dispense 10 ml of a 2:1 slurry of DE52 resin into a centrifuge tube and sediment the resin by centrifugation at 500g (2000 rpm in a Sorvall SS-34 rotor) for 5 minutes at room temperature. Remove the supernatant from the resin pellet and place the centrifuge tube on ice.

7. Resuspend the DE52 in the cleared TM and allow the bacteriophage particles to absorb to the resin by rocking the centrifuge tube for 3 minutes at room temperature.

8. Centrifuge the TM/DE52 slurry at 4000g (5800 rpm in a Sorvall SS-34 rotor) for 5 minutes. Carefully transfer the supernatant to a fresh centrifuge tube and repeat the centrifugation step. Discard the pellet after each centrifugation.

9. Transfer the supernatant from the second centrifugation to a fresh centrifuge tube. Extract the supernatant, which contains the bacteriophage λ particles, once with phenol:chloroform.

10. Transfer the aqueous phase, which contains the bacteriophage λ DNA, to a fresh polypropylene tube and add an equal volume of isopropanol. Store the mixture for 10 minutes at –70°C.

11. Collect the precipitated bacteriophage DNA by centrifugation at 16,500g (12,000 rpm in a Sorvall SS-34 rotor) for 20 minutes at 4°C.

12. Drain the isopropanol from the centrifuge tube and allow the pellet of DNA to air dry.

13. Redissolve the DNA pellet in 2 ml of low-salt buffer.

14. Purify the bacteriophage DNA by chromatography on an Elutip-d column:

    a. Use a syringe to push 1–2 ml of high-salt buffer through the Elutip-d column.

    b. Push 5 ml of low-salt buffer through the column.

    c. Attach the 0.45-μm prefilter to the column and slowly push the DNA sample (Step 13) through the column.

    d. Rinse the column with 2–3 ml of low-salt buffer.

    e. Remove the prefilter and elute the DNA with 0.4 ml of high-salt buffer. Collect the eluate at this step in a 1.5-ml microfuge tube.

15. Add 1 ml of ethanol to the solution of eluted DNA, invert the tube several times, and incubate the mixture on ice for 20 minutes. Collect the precipitated DNA by centrifugation in a microfuge, discard the supernatant, and rinse the pellet of DNA with 0.5 ml of 70% ethanol. Discard the supernatant and allow the pellet of DNA to dry in the air. Dissolve the pellet of DNA in 50 μl of TE (pH 8.0).

    Resuspend the DNA by tapping on the side of the tube. Try to avoid vortexing. If the DNA proves difficult to dissolve, incubate the tube for 15 minutes at 50°C.

    If minipreparations are working well, expect to isolate approximately 5 μg of purified bacteriophage DNA from 5 x 10^{10} infectious particles.

Rapid Analysis of Bacteriophage λ: Purification of λ DNA from Liquid Cultures

This protocol is used to purify small amounts of recombinant DNAs cloned in robust strains of bacteriophage λ such as λgt10, λgt11, λZAP, or ZipLox. The DNAs are suitable for use as substrates for restriction enzymes and templates for DNA and RNA polymerases.

MATERIALS

CAUTION: Please see Appendix 4 for appropriate handling of materials marked with <!>.

Reagents and Solutions

Please see Appendix 1 for components of stock solutions, buffers, and reagents. Dilute stock solutions to the appropriate concentrations.

Chloroform <!>
Ethanol
Ethanol (70%)
Isopropanol
λ High-salt buffer
  20 mM Tris-Cl (pH 7.4)
  1.0 M NaCl
  1 mM EDTA (pH 8.0)
λ Low-salt buffer
  20 mM Tris-Cl (pH 7.4)
  0.2 M NaCl
  1 mM EDTA (pH 8.0)
Phenol:chloroform (1:1, v/v) <!>
SM
TE (pH 8.0)

Vectors and Hosts

Bacteriophage λ recombinant, grown as well-isolated plaques on a bacterial lawn
  (Protocol 2.1)
E. coli strain
  Inoculate a single colony of an appropriate strain of *E. coli* into 25 ml of NZCYM and incubate overnight at 30°C. Measure the OD_{600} of the overnight culture and calculate the number of cells/ml using the conversion factor 1 OD_{600} = 1 x 10^9 cells/ml.

Media and Antibiotics

NZCYM medium

Centrifuges/Rotors/Tubes

Sorvall SS-34 rotor and tubes (4°C and room temperature)

Additional Items

Bacterial shaking incubator (30°C)
Borosilicate pasteur pipettes
DE52, preswollen (Whatman), used as a 2:1 slurry of DE52 in TM
Elutip-d columns

Microfuge (4°C)
Rotating platform
Sterile tubes (17 × 100 mm)
Water bath or heating blocks (37°C and 47°C)

Additional Information

Other methods of analysis   MC3, pp. 2.104 and 2.105
   Xu S.-Y. 1986. A rapid method for preparing phage lambda DNA from agar plate lysates.
   Gene Anal. Tech. **3:** 90–91.

METHOD

1. Use a borosilicate pasteur pipette to pick a single well-isolated bacteriophage plaque to place into 1 ml of SM containing a drop of chloroform in a small sterile polypropylene tube. Store the suspension for 4–6 hours at 4°C to allow the bacteriophage particles to diffuse from the top agarose.

2. In a 17 × 100-mm tube, mix 0.5 ml of the bacteriophage suspension (~3 × 10^6 bacteriophages) with 0.1 ml of an overnight culture of bacteria. Incubate the culture for 15 minutes at 37°C.

3. Add 4 ml of NZCYM medium and incubate the culture for approximately 9 hours at 37°C with vigorous agitation.

   The culture should be clear, with very little debris evident.

4. Add 0.1 ml of chloroform to the culture and continue incubation for an additional 15 minutes at 37°C with vigorous agitation. Transfer the lysate to a 5-ml polypropylene centrifuge tube. Centrifuge at 800*g* (2600 rpm in a Sorvall SS-34 rotor) for 10 minutes at 4°C.

5. Transfer the supernatant to a fresh tube and remove the bacterial debris by centrifugation at 4000*g* (5800 rpm in a Sorvall SS-34 rotor) for 10 minutes at 4°C. A small aliquot of cleared lysate can be set aside at this step as a bacteriophage stock solution. Store the stock at 4°C over chloroform.

6. Dispense 10 ml of a 2:1 slurry of DE52 resin into a fresh centrifuge tube and sediment the resin by centrifugation at 500*g* (2000 rpm in a Sorvall SS-34 rotor) for 5 minutes at room temperature. Remove the supernatant from the resin pellet and place the centrifuge tube on ice.

7. Resuspend the DE52 in the cleared bacteriophage λ supernatant and allow the bacteriophage particles to absorb to the resin by rocking the centrifuge tube for 3 minutes at room temperature.

8. Centrifuge the bacteriophage λ supernatant/DE52 slurry at 4000*g* (5800 rpm in a Sorvall SS-34 rotor) for 5 minutes. Carefully transfer the supernatant to a fresh centrifuge tube and repeat the centrifugation step. Discard the pellet after each centrifugation.

9. Transfer the supernatant from the second centrifugation to a fresh centrifuge tube. Extract the supernatant, which contains the bacteriophage λ particles, once with phenol:chloroform.

10. Transfer the aqueous phase, which contains the bacteriophage λ DNA, to a fresh polypropylene tube and add an equal volume of isopropanol. Store the mixture for 10 minutes at –70°C.

11. Collect the precipitated bacteriophage DNA by centrifugation at 16,500*g* (12,000 rpm in a Sorvall SS-34 rotor) for 20 minutes at 4°C.

12. Drain the isopropanol from the centrifuge tube and allow the pellet of DNA to air dry.

13. Dissolve the DNA pellet in 2 ml of low-salt buffer.

14. Purify the bacteriophage DNA by chromatography on an Elutip-d column as described in Protocol 2.23.

15. Mix 1 ml of ethanol with the solution of eluted DNA and incubate the mixture on ice for 20 minutes. Collect the precipitated DNA by centrifugation in a microfuge, discard the supernatant, and rinse the pellet of DNA with 0.5 ml of 70% ethanol. Discard the supernatant and allow the ethanol to evaporate. Dissolve the damp pellet of DNA in 50 μl of TE (pH 8.0).

    Resuspend the DNA by tapping on the side of the tube. Try to avoid vortexing. If the DNA proves difficult to dissolve, incubate the tube for 15 minutes at 50°C.

    If minipreparations are working well, expect to isolate approximately 5 μg of purified bacteriophage DNA from 5×10^{10} infectious particles. It is possible to estimate the quantity of bacteriophage DNA present in a plate lysate or liquid culture lysate by direct agarose gel electrophoresis (please see Protocol 2.7).

    For restriction analysis and agarose gel electrophoresis, digest a 5–10-μl aliquot of the resuspended DNA.

Working with Bacteriophage M13 Vectors

BACKGROUND INFORMATION

Background information found in *Molecular Cloning: A Laboratory Manual*, 3rd edition (hereafter MC3) unless otherwise indicated.

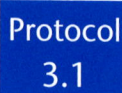

Plating Bacteriophage M13

Bacteriophage M13 forms turbid plaques on lawns of male strains of *E. coli*.

MATERIALS

CAUTION: Please see Appendix 4 for appropriate handling of materials marked with <!>.

Reagents and Solutions

Please see Appendix 1 for components of stock solutions, buffers, and reagents. Dilute stock solutions to the appropriate concentrations.
  IPTG solution (20% w/v)
  X-gal solution (2% w/v) <!>

Vectors and Hosts

E. coli F′ strain
Bacteriophage M13 stock in LB or YT medium
    LB or YT medium from a fully grown liquid culture of bacteria-infected bacteriophage M13
    contains between 10^{10} and 10^{12} pfu/ml. A bacteriophage M13 plaque suspended in 1 ml of
    LB or YT yields between 10^7 and 10^8 pfu.

Media and Antibiotics

LB agar plates containing tetracycline or kanamycin
    These media are needed only if a tetracycline-resistant strain of *E. coli* such as XL1-Blue, or a
    kanamycin-resistant strain such as XL1-Blue MRF′ Kan, is used to propagate the virus.
  or
M9 minimal agar plates, supplemented with 0.4% glucose, 5 mM $MgSO_4 \cdot 7H_2O$, and
  0.01% thiamine
    Minimal agar plates are needed when using *E. coli* strains that carry a deletion of the proline
    biosynthetic operon (Δ[*lac-proAB*]) in the bacterial chromosome and complementing *proAB*
    genes in the F′ plasmid.
LB or YT agar plates containing 5 mM $MgCl_2$
LB or YT medium
LB or YT top agarose containing 5 mM $MgCl_2$

Additional Items

Bacterial shaker (37°C)
Sterile tubes (13 x 75 or 17 x 100 mm)
Water bath or heating block (47°C)

Additional Information

Commonly used antibiotic solutions   MC3, p. A2.6, Table A2-1
Growth cycle of bacteriophage M13   MC3, pp. 3.2–3.7

METHOD

1. Streak a master culture of a bacterial strain carrying an F′ plasmid onto either a supplemented minimal (M9) agar plate or an LB plate containing tetracycline (XL1-Blue) or kanamycin (XL1-Blue MRF′ Kan). Incubate the plate for 24–36 hours at 37°C.

2. To prepare plating bacteria, inoculate 5 ml of LB or YT medium in a 20-ml sterile culture tube with a single, well-isolated colony picked from the agar plate prepared in Step 1. Agitate the culture for 6–8 hours at 37°C in a rotary shaker. Chill the culture in an ice bath for 20 minutes and then store it at 4°C. These plating bacteria can be stored for periods of up to 1 week at 4°C.

    Do not grow the cells to saturation, as this will increase the risk of losing the pili encoded by the F′ plasmid.

3. Prepare sterile tubes (13 x 75 or 17 x 100 mm) containing 3 ml of melted LB or YT medium top agar or agarose, supplemented with 5 mM $MgCl_2$. Allow the tubes to equilibrate to 47°C in a heating block or water bath.

4. Label a series of sterile tubes (13 x 75 or 17 x 100 mm) according to the dilution factor and amount of bacteriophage stock to be added (please see Step 5). Deliver 100 µl of plating bacteria from Step 2 into each of these tubes.

5. Prepare tenfold serial dilutions (10^{-6} to 10^{-9}) of the bacteriophage stock in LB or YT medium. Dispense 10 or 100 µl of each dilution to be assayed into a sterile tube containing plating bacteria from Step 4. Mix the bacteriophage particles with the bacterial culture by vortexing gently.

6. Add 40 µl of 2% X-gal solution and 4 µl of 20% IPTG solution to each of the tubes containing top agar. Immediately pour the contents of one of these tubes into one of the infected cultures. Mix the culture with the agar/agarose by gentle vortexing for 3 seconds and then pour the mixture onto a labeled plate containing LB or YT agar medium supplemented with 5 mM $MgCl_2$ and equilibrated to room temperature. Swirl the plate gently to ensure an even distribution of bacteria and top agar.

    Work quickly so that the top agar spreads over the entire surface of the agar before it sets.

7. Repeat the addition of top agar with X-gal and IPTG for each tube of infected culture prepared in Step 5.

8. Replace the lids on the plates and allow the top agar/agarose to harden for 5 minutes at room temperature. Wipe excess condensation off the lids with Kimwipes. Invert the plates and incubate them at 37°C.

    Pale blue plaques begin to appear after 4 hours. The color gradually intensifies as the plaques enlarge and is complete after 8–12 hours of incubation.

Growing Bacteriophage M13 in Liquid Culture

Most manipulations with M13, including preparations of viral stocks and isolation of single- and double-stranded DNAs, begin with small-scale liquid cultures that are infected with an M13 plaque, picked from an agar plate.

MATERIALS

Vectors and Hosts

Bacteriophage M13 plaques
    From an agar plate (see either Protocol 3.1 or 3.6)
E. coli F' strain
    Grown as well-isolated colonies on an agar plate

Media and Antibiotics

LB medium containing tetracycline (or kanamycin)
    Media containing antibiotics are needed only if a tetracycline-resistant strain of *E. coli* such as XL1-Blue, or a kanamycin-resistant strain such as XL1-Blue MRF' Kan, is used to propagate the virus.
or
M9 minimal agar plates, supplemented with 0.4% glucose, 5 mM $MgSO_4 \cdot 7H_2O$, and 0.01% thiamine
    Supplemented minimal agar plates are needed when using *E. coli* strains that carry a deletion of the proline biosynthetic operon ($\Delta[lac-proAB]$) in the bacterial chromosome and the complementing *proAB* genes in the F' plasmid.
2x YT medium containing 5 mM $MgCl_2$

Additional Items

Bacterial shaker (37°C)
Inoculating needle or sterile applicator sticks
Sterile tubes (13 x 75 or 17 x 100 mm)

Additional Information

Commonly used antibiotic solutions	MC3, p. A2.6, Table A2-1
Growth times of bacteriophage M13	MC3, p. 3.49
Picking bacteriophage M13 plaques	MC3, p. 3.22

METHOD

1. Inoculate 5 ml of supplemented M9 medium (or, for antibiotic-resistant strains, LB medium with the appropriate antibiotic) with a single freshly grown colony of *E. coli* carrying an F' plasmid. Incubate the culture for 12 hours at 37°C with moderate shaking.

2. Transfer 0.1 ml of the *E. coli* culture into 5 ml of 2x YT medium containing 5 mM $MgCl_2$. Incubate the culture for 2 hours at 37°C with vigorous shaking.

3. Dilute the 5-ml culture into 45 ml of 2x YT containing 5 mM $MgCl_2$ and dispense 1-ml aliquots into as many sterile tubes (13 x 75 or 17 x 100 mm) as there are plaques to be prop-

agated. Dispense two additional aliquots for use as positive and negative controls for bacteriophage growth. Set these cultures aside for use at Step 7.

4. Dispense 1 ml of YT or LB medium into sterile 13 x 75-mm tubes. Prepare as many tubes as there are plaques. Dispense two additional aliquots for use as positive and negative controls for bacteriophage growth.

5. Prepare a dilute suspension of bacteriophage M13 by touching the surface of a plaque with the end of a sterile inoculating needle and immersing the end of the needle into the YT or LB medium. Pick one blue M13 plaque as a positive control for bacteriophage growth. Also pick an area of the *E. coli* lawn from the plate that does not contain a plaque as a negative control.

6. Allow the suspension to stand for 1–2 hours at room temperature for the bacteriophage particles to diffuse from the agar.

7. Use 0.1 ml of the bacteriophage suspension (Step 6) as an inoculum to infect 1-ml cultures of *E. coli* (Step 3) for isolation of viral DNA. Incubate the inoculated tubes for 5 hours at 37°C with moderate shaking.

    Alternatively, transfer a plaque directly into the *E. coli* culture.

    To minimize the possibility of selecting deletion mutants, grow cultures infected with recombinant M13 bacteriophages for the shortest time that will produce a workable amount of single-stranded DNA (usually 5 hours).

8. Transfer the culture to a sterile microfuge tube and centrifuge at maximum speed for 5 minutes at room temperature. Transfer the supernatant to a fresh microfuge tube without disturbing the bacterial pellet.

9. Transfer 0.1 ml of the supernatant to a sterile microfuge tube.

    This high-titer stock (10^{10} to 10^{12} pfu/ml) can be stored indefinitely at 4°C or –20°C without loss of infectivity.

10. Use the remaining 1 ml of the culture supernatant to prepare single-stranded bacteriophage DNA (Protocol 3.4). Use the bacterial cell pellet to prepare double-stranded RF DNA (Protocol 3.3).

Preparation of Double-stranded (Replicative Form) Bacteriophage M13 DNA

The double-stranded replicative form (RF) of bacteriophage M13 is isolated from infected cells using methods similar to those used to purify plasmid DNA. Several micrograms of RF DNA can be isolated from a 1–2-ml culture of infected cells.

MATERIALS

CAUTION: Please see Appendix 4 for appropriate handling of materials marked with <!>.

Reagents and Solutions

Please see Appendix 1 for components of stock solutions, buffers, and reagents. Dilute stock solutions to the appropriate concentrations.
Alkaline lysis solution I
Alkaline lysis solution II, freshly prepared
Alkaline lysis solution III
Ammonium acetate (7.5 M) (optional; see Step 10)
Ethanol
Ethanol (70%)
Phenol:chloroform (1:1, v/v) <!>
TE (pH 8.0) containing 20 µg/ml RNase A

Vectors and Hosts

E. coli culture infected with bacteriophage M13
Prepare as described in Protocol 3.2.

Enzymes and Buffers

Restriction enzymes and 10x buffers

Gels/Loading Buffers

0.8% Agarose gel, cast and run in 0.5x TBE containing 0.5 mg/ml ethidium bromide <!>
Gel-loading buffer IV

Centrifuges/Rotors/Tubes

Microfuge (4°C and room temperature)

METHOD

1. Centrifuge 1 ml of the M13-infected cell culture at maximum speed for 5 minutes at room temperature in a microfuge to separate the infected cells from the medium. Transfer the supernatant to a fresh microfuge tube and store at 4°C. Keep the infected bacterial cell pellet on ice.

    The supernatant contains M13 bacteriophage housing single-stranded DNA. If desired, prepare M13 DNA from this supernatant at a later stage (Protocol 3.4).

 Adapted from Chapter 3, Protocol 3, p. 3.23 of MC3.

2. Centrifuge the bacterial cell pellet for 5 seconds at 4°C and remove residual medium with an automatic pipetting device.

3. Resuspend the cell pellet in 100 μl of ice-cold alkaline lysis solution I by vigorous vortexing.

    Make sure that the bacterial pellet is completely dispersed in alkaline lysis solution I.

4. Add 200 μl of freshly prepared alkaline lysis solution II to the tube. Close the tube tightly and mix by inverting the tube rapidly five times. *Do not vortex.* Store the tube on ice for 2 minutes after mixing.

5. Add 150 μl of ice-cold alkaline lysis solution III to the tube. Close the tube to disperse alkaline lysis solution III through the viscous bacterial lysate by inverting the tube several times. Store the tube on ice for 3–5 minutes.

6. Centrifuge the bacterial lysate at maximum speed for 5 minutes at 4°C in a microfuge. Transfer the supernatant to a fresh tube.

7. Add an equal volume of phenol:chloroform. Mix the organic and aqueous phases by vortexing and then centrifuge the tube at maximum speed for 2–5 minutes at room temperature. Transfer the aqueous (upper) phase to a fresh tube.

8. Precipitate the double-stranded DNA by adding 2 volumes of ethanol. Mix the contents of the tube by vortexing and then allow the mixture to stand for 2 minutes at room temperature.

9. Recover the DNA by centrifugation at maximum speed for 5 minutes at 4°C in a microfuge.

10. Remove the supernatant by gentle aspiration. Stand the tube in an inverted position on a paper towel to allow all of the fluid to drain away. Remove any drops of fluid adhering to the walls of the tube.

    An additional ethanol precipitation step here helps to ensure that the double-stranded DNA is efficiently cleaved by restriction enzymes.
    - Dissolve the pellet of RF DNA in 100 μl of TE (pH 8.0).
    - Add 50 μl of 7.5 M ammonium acetate, mix well, and add 300 μl of ice-cold ethanol.
    - Store the tube for 15 minutes at room temperature or overnight at –20°C and then collect the precipitated DNA by centrifugation at maximum speed for 5–10 minutes at 4°C in a microfuge. Remove the supernatant by gentle aspiration.
    - Rinse the pellet with 250 μl of ice-cold 70% ethanol, centrifuge again for 2–3 minutes, and discard the supernatant.
    - Allow the pellet of DNA to air dry for 10 minutes and then dissolve the DNA as described in Step 12.

11. Add 1 ml of 70% ethanol at 4°C and centrifuge again for 2 minutes. Remove the supernatant as described in Step 10 and allow the pellet of nucleic acid to air dry for 10 minutes.

12. To remove RNA, resuspend the pellet in 25 μl of TE (pH 8.0) with RNase. Vortex briefly.

13. Analyze the double-stranded RF DNA by digestion with appropriate restriction endonucleases followed by electrophoresis through an agarose gel.

    The yield of RF DNA expected from 1 ml of an infected bacterial culture is 2–5 μg, enough for 5–10 restriction digests.

Preparation of Single-stranded Bacteriophage M13 DNA

Bacteriophage M13 single-stranded DNA is prepared from virus particles secreted by infected cells into the surrounding medium. The filamentous particles are concentrated by precipitation from a high-ionic-strength buffer with polyethylene glycol. Subsequent extraction with phenol releases the single-stranded DNA, which is then collected by precipitation with ethanol. This protocol is generally used to prepare single-stranded DNA from a small number of M13 isolates. Typically, the yield of single-stranded DNA is 5–10 µg/ml infected culture.

MATERIALS

CAUTION: Please see Appendix 4 for appropriate handling of materials marked with <!>.

Reagents and Solutions

Please see Appendix 1 for components of stock solutions, buffers, and reagents. Dilute stock solutions to the appropriate concentrations.

Chloroform <!>
Ethanol
Ethanol (70%)
Phenol <!>
> Water saturated and equilibrated to pH 7.0 (see MC3, p. A1.23)

Polyethylene glycol 8000 (20% w/v PEG, 2.5 M NaCl) <!>
Sodium acetate (3 M, pH 5.2)
TE (pH 8.0)

Vectors and Hosts

E. coli culture infected with bacteriophage M13
> Prepare as described in Protocol 3.2.
> Uninfected culture of *E. coli*

Gels/Loading Buffers

1.2% Agarose gel, cast and run in 0.5x TBE containing 0.5 mg/ml ethidium bromide <!>
Gel-loading buffer IV

Centrifuges/Rotors/Tubes

Microfuge (4°C and room temperature)

Additional Item

Vacuum aspirator fitted with a disposable yellow tip

Additional Information

PEG 8000   MC3, p. 1.154

METHOD

1. Transfer 1 ml of the infected and uninfected cultures to separate microfuge tubes and centrifuge the tubes at maximum speed for 5 minutes at room temperature. Transfer each supernatant to a fresh microfuge tube at room temperature.

Adapted from Chapter 3, Protocol 4, p. 3.26 of MC3.

2. To each supernatant, add 200 μl of 20% PEG in 2.5 M NaCl. Mix the solution well by inverting the tube several times, followed by gentle vortexing. Allow the tube to stand for 15 minutes at room temperature.

3. Recover the precipitated bacteriophage particles by centrifugation at maximum speed for 5 minutes at 4°C in a microfuge.

4. Carefully remove all of the supernatants using a vacuum aspirator fitted with a disposable pipette tip. Centrifuge the tubes again for 30 seconds and remove any residual supernatant.

   A tiny, pinhead-sized pellet of precipitated bacteriophage particles should be visible at the bottom of the tubes. No pellet should be visible in the negative control tube in which a portion of the uninfected *E. coli* lawn was inoculated.

5. Allow the bacteriophage pellet to soak in 100 ml of TE (pH 8.0) for 5–10 minutes and then vortex the tube for 2x 20 seconds.

   It is important to resuspend the bacteriophage pellet completely to allow efficient extraction of the single-stranded DNA by phenol in the next step.

6. To the resuspended pellets, add 100 μl of equilibrated phenol. Mix well by vortexing for 30 seconds. Allow the sample to stand for 1 minute at room temperature and then vortex for another 30 seconds.

7. Centrifuge the samples at maximum speed for 3–5 minutes at room temperature in a microfuge. Transfer as much as is easily possible of the upper, aqueous phases to a fresh microfuge tube.

   Do not try to transfer all of the aqueous phase. Much cleaner preparations of single-stranded DNA are obtained when approximately 5 μl of the aqueous phase is left at the interface.

8. Recover the M13 DNA by standard precipitation with ethanol in the presence of 0.3 M sodium acetate. Vortex briefly to mix. Incubate the tubes for 15–30 minutes at room temperature or overnight at –20°C.

9. Recover the precipitated single-stranded bacteriophage DNA by centrifugation at maximum speed for 10 minutes at 4°C in a microfuge.

10. Remove the supernatants by gentle aspiration, being careful not to disturb the DNA pellets (often only visible as a haze on the side of the tube). Centrifuge the tubes again for 15 seconds and remove any residual supernatant.

11. Add 200 μl of cold 70% ethanol and centrifuge at maximum speed for 5–10 minutes at 4°C. Immediately remove the supernatant by gentle aspiration.

    At this stage, the pellets are not firmly attached to the wall of the tube. It is therefore important to work quickly and carefully to avoid losing the DNA.

12. Invert the open tubes on the bench for 10 minutes to allow any residual ethanol to drain and evaporate. Dissolve the pellets in 40 μl of TE (pH 8.0). Warm the solution to 37°C for 5 minutes to speed dissolution of the DNA. Store the DNA solutions at –20°C.

    The yield of single-stranded DNA is usually 5–10 μg/ml of the original infected culture.

13. Estimate the DNA concentration by mixing 2-μl aliquots of the DNA from Step 12, each with 1 μl of gel-loading buffer IV. Load the samples into the wells of a 1.2% agarose gel cast in 0.5x TBE and containing 0.5 μg/ml ethidium bromide. As controls, use varying amounts of M13 DNA preparations of known concentrations. Examine the gel after electrophoresis for 1 hour at 6 V/cm. Estimate the amount of DNA from the intensity of the fluorescence.

    Usually, 2–3 μl of a standard bacteriophage M13 DNA preparation is required for each set of four dideoxy cycle sequencing reactions using dye primers.

Large-scale Preparation of Single-stranded and Double-stranded Bacteriophage M13 DNA

This protocol, a scaled-up version of Protocols 3.3 and 3.4, is used chiefly to generate large stocks of double-stranded DNA of strains of M13 that are routinely used as cloning vectors. Large amounts of single-stranded DNA of an individual recombinant may occasionally be needed for specific purposes, e.g., to generate many preparations of a particular radiolabeled probe or to construct large numbers of site-directed mutants.

MATERIALS

CAUTION: Please see Appendix 4 for appropriate handling of materials marked with <!>.

Reagents and Solutions

Please see Appendix 1 for components of stock solutions, buffers, and reagents. Dilute stock solutions to the appropriate concentrations.
Ethanol
Ethanol (70%)
NaCl (solid)
Phenol <!>
    Water saturated and equilibrated to pH 7.0 (see MC3, p. A1.23)
Phenol:chloroform (1:1, v/v) <!>
Polyethylene glycol 8000 <!>
Sodium acetate (3 M, pH 5.2)
STE
TE (pH 8.0)
Tris-Cl (10 mM, pH 8.0)

Vectors and Hosts

Bacteriophage M13 stock
    Prepare as described in Protocol 3.2.
E. coli F' plating bacteria
    Prepare as described in Protocol 3.1.

Media and Antibiotics

LB or YT medium containing 5 mM $MgCl_2$, prewarmed to 37°C

Gels/Loading Buffers

0.8% Agarose gel, cast and run in 0.5x TBE containing 0.5 mg/ml ethidium bromide <!>

Centrifuges/Rotors/Tubes

Corex glass centrifuge tubes (30 ml)
Sorvall SLC-1500 and SS-34 rotors, tubes, and bottles (4°C)

Additional Items

Bacterial shaking incubator
Magnetic stirrer and magnetic bar

 Adapted from Chapter 3, Protocol 5, p. 3.30 of MC3.

Step 5 of this protocol requires the reagents listed in Protocols 1.3, 1.8, 1.9, and 1.10.
Sterile tubes (13 x 75 *or* 17 x 100 mm)

Additional Information

PEG 8000   MC3, p. 1.154

METHOD

1. Transfer 2.5 ml of plating bacteria to a sterile tube (13 x 75 or 17 x 100 mm). Add 0.5 ml of bacteriophage M13 stock (~5 x 10^{11} pfu) and mix by tapping the side of the tube. Incubate the infected cells for 5 minutes at room temperature.

2. Dilute the infected cells into 250 ml of fresh LB or YT medium containing 5 mM MgCl$_2$ pre-warmed to 37°C in a 2-liter flask. Incubate for 5 hours at 37°C with constant, vigorous shaking.

3. Harvest the infected cells by centrifugation at 4000*g* (5000 rpm in a Sorvall SLC-1500 rotor) for 15 minutes at 4°C. Recover the supernatant, which may be used for large-scale preparations of single-stranded bacteriophage M13 DNA, as described in Steps 7–17 below.

4. Resuspend the bacterial pellet in 100 ml of ice-cold STE. Recover the washed cells by centrifugation at 4000*g* (5000 rpm in a Sorvall SLC-1500 rotor) for 15 minutes at 4°C.

5. Isolate the bacteriophage M13 closed circular RF DNA by the alkaline lysis method (please see MC3, Protocol 1.3). Scale up the volumes of lysis solutions appropriately. Purify the DNA by precipitation with PEG, by column chromatography, or by equilibrium centrifugation in CsCl–ethidium bromide gradients.

6. Measure the concentration of the DNA spectrophotometrically and confirm its integrity by agarose gel electrophoresis. Store the closed circular DNA in small (1–5-μg) aliquots at –20°C.

   The expected yield of RF DNA from 250 ml of infected cells is approximately 200 μg.

7. To isolate single-stranded DNA from the bacteriophage particles in the infected cell medium, transfer the 250-ml supernatant from Step 3 to a 500-ml beaker containing a magnetic stirring bar.

8. Add 10 g of PEG and 7.5 g of NaCl to the supernatant. Stir the solution for 30–60 minutes at room temperature.

9. Collect the precipitate by centrifugation at 10,000*g* (7800 rpm in a Sorvall SLC-1500 rotor) for 20 minutes at 4°C. Invert the centrifuge bottle for 2–3 minutes to allow the supernatant to drain and then use Kimwipes to remove the last traces of supernatant from the walls and neck of the bottle.

   Avoid touching the thin whitish film of precipitated bacteriophage particles on the side and bottom of the centrifuge bottle.

10. Add 10 ml of 10 mM Tris-Cl (pH 8.0) to the bottle. Swirl the solution in the bottle and use a pasteur pipette to rinse the sides of the bottle thoroughly. When the bacteriophage pellet is dissolved, transfer the solution to a 30-ml Corex centrifuge tube.

11. To the bacteriophage suspension, add an equal volume of equilibrated phenol, seal the tube with a silicon rubber stopper, and mix the contents by vortexing vigorously for 2 minutes.

12. Centrifuge the solution at 3000*g* (5000 rpm in a Sorvall SS-34 rotor) for 5 minutes at room temperature. Transfer the upper aqueous phase to a fresh tube and repeat the extraction with 10 ml of phenol:chloroform.

If there is a visible interface between the organic and aqueous layers, extract the aqueous supernatant once more with chloroform.

13. Transfer equal amounts of the aqueous phase to each of two 30-ml Corex tubes. Add 0.5 ml of 3 M sodium acetate (pH 5.2) and 11 ml of ethanol to each tube. Mix the solutions well and then store them for 15 minutes at room temperature.

14. Recover the precipitate of single-stranded DNA by centrifugation at 12,000g (10,000 rpm in a Sorvall SS-34 rotor) for 20 minutes at 4°C. Carefully remove all of the supernatant.

    Most of the precipitated DNA will collect in a thin film along the walls of the centrifuge tubes.

15. Add 30 ml of 70% ethanol at 4°C to each tube and centrifuge at 12,000g (10,000 rpm in a Sorvall SS-34 rotor) for 10 minutes at 4°C. Carefully remove as much of the supernatant as possible, invert the tubes to allow the last traces of supernatant to drain away from the precipitate, and wipe the neck of the tubes with Kimwipes.

16. Allow the residual ethanol to evaporate at room temperature. Dissolve the pellets in 1 ml of TE (pH 8.0). Store the DNA at –20°C.

17. Measure the concentration of the DNA spectrophotometrically and confirm its integrity by agarose gel electrophoresis. Store the closed circular DNA in small (10–50 μg) aliquots at –20°C.

    The yield of single-stranded DNA expected from 250 ml of infected cuture is 250 μg to 1 mg.

Cloning into Bacteriophage M13 Vectors

This protocol describes three standard methods to construct bacteriophage M13 recombinants: (1) ligating insert DNA to a linearized vector, prepared by cleavage of M13 RF with a single restriction enzyme; (2) using alkaline phosphatase to suppress self-ligation of the linearized vector; and (3) using M13 RF cleaved with two restriction enzymes for directional cloning.

MATERIALS

CAUTION: Please see Appendix 4 for appropriate handling of materials marked with <!>.

Reagents and Solutions

Please see Appendix 1 for components of stock solutions, buffers, and reagents. Dilute stock solutions to the appropriate concentrations.

ATP (10 mM)
> Not required if 10x ligation buffer contains ATP

Ethanol

Ethanol (70%)

IPTG (20% w/v)

Phenol:chloroform (1:1, v/v) <!>

Sodium acetate (3 M, pH 5.2)

X-gal (2% w/v) <!>

Vectors and Hosts

Bacteriophage M13 RF DNA

E. coli-competent cells carrying an F′ plasmid
> Prepare as described in Protocol 1.25 or purchase from commercial sources.

E. coli F′ plating bacteria
> Prepare as described in Protocol 3.1 or purchase from commercial sources.

Media and Antibiotics

YT or LB agar plates

YT or LB medium

YT or LB top agarose

Enzymes and Buffers

Bacteriophage T4 DNA ligase and 10x buffer

Calf or shrimp alkaline phosphatase and 10x buffer (optional; see Step 4)

Restriction enzymes and 10x buffers

Nucleic Acids/Oligonucleotides

Foreign DNA

Test DNA

Gels/Loading Buffers

0.8% Agarose gel, cast and run in 0.5x TBE containing 0.5 µg/ml ethidium bromide <!>

Additional Items

Step 4 of this protocol requires the reagents listed in Protocol 1.20.
Sterile 5-ml culture tubes (e.g., Falcon 2054, Becton Dickinson)
Water baths or heating/cooling blocks (12–16°C, 42°C, and 47°C)

Additional Information

Table showing typical results obtained from this protocol   MC3, p. 3.38

METHOD

1. Digest 1–2 µg of the bacteriophage M13 vector RF DNA to completion with a three- to five-fold excess of the appropriate restriction enzyme(s). Set up a control reaction containing M13 RF DNA but no restriction enzyme(s).

2. At the end of the incubation period, remove a small sample of DNA (50 ng) from each of the reactions and analyze the extent of digestion by electrophoresis through an 0.8% agarose gel. If digestion is incomplete (i.e., if any closed circular DNA is visible), add more restriction enzyme(s) and continue the incubation.

3. When digestion is complete, purify the M13 DNA by extraction with phenol:chloroform followed by standard precipitation with ethanol in the presence of 0.3 M sodium acetate (pH 5.2). Dissolve the DNA in TE (pH 8.0) at a concentration of 50 µg/ml.

4. If required, dephosphorylate the linearized vector DNA by treatment with calf alkaline phosphatase or shrimp alkaline phosphatase. At the end of the dephosphorylation reaction, inactivate the alkaline phosphatase by heat and/or by digestion with proteinase K, followed by extraction with phenol:chloroform (for details, please see Protocol 1.20).

5. Recover the linearized M13 DNA as outlined in Step 3. Dissolve the dephosphorylated DNA in TE (pH 7.6) at a concentration of 50 µg/ml.

6. Generate individual restriction fragments of foreign DNA by cleavage with the appropriate restriction enzymes and purify them by agarose gel electrophoresis. Dissolve the final preparation of foreign DNA in TE (pH 7.6) at a concentration of 50 µg/ml.

When ligating DNAs with complementary cohesive termini, please follow Steps 7–9 below. For methods to set up blunt-ended ligation reactions, please see Protocol 1.19.

7. In a microfuge tube (Tube A), mix together approximately 50 ng of vector DNA and a one- to fivefold molar excess of the target (foreign) DNA fragment(s). The combined volume of the two DNAs should not exceed 8 µl. If necessary, add TE (pH 7.6) to adjust the volume to 7.5–8.0 µl. As controls, set up three ligation reactions containing the following:

Tube	DNA
B	the same amount of vector DNA, but no foreign DNA
C	the same amount of vector DNA and a one- to fivefold molar excess of the target DNA fragment(s)
D	the same amount of vector DNA together with an equal amount by weight of a test DNA that has been successfully cloned into bacteriophage M13 on previous occasions

As a test DNA, we routinely use a standard preparation of bacteriophage λ DNA cleaved with restriction enzymes that recognize tetranucleotide sequences and generate termini that are complementary to the M13 vectors to be used.

8. Add 1 µl of 10x ligation buffer and 1 µl of 10 mM ATP to all four reactions (Tubes A–D).

    Omit ATP if using a commercial buffer that contains ATP.

9. Add 0.5 Weiss unit of bacteriophage T4 DNA ligase to Tubes A, B, and D. Mix the components by gently tapping the side of each tube for several seconds. Incubate the ligation reactions for 4–16 hours at 12–16°C.

    At the end of the ligation reaction, analyze 1 µl of each ligation reaction by electrophoresis through an 0.8% agarose gel. Bands of circular recombinant molecules containing vector and fragment(s) of foreign DNA should be visible in the test reaction (Tube A) but not in the control (Tube C).

    After ligation, the reactions may be stored at –20°C until transformation.

10. Prepare and grow an overnight culture of plating bacteria (please see Protocol 3.1) in YT or LB medium at 37°C with constant shaking.

11. Thaw an aliquot of frozen competent cells of the desired strain carrying an F' plasmid. Allow the cells to thaw at room temperature and then place them on ice for 10 minutes.

12. Transfer 50–100 µl of the competent F' bacteria to each of 16 sterile 5-ml culture tubes (Falcon 2054, Becton Dickinson) that have been chilled to 0°C.

13. Immediately add 0.1-, 1.0-, and 5-µl aliquots of the ligation reactions and controls (Tubes A–D) to separate tubes of competent cells. Mix the DNAs with the bacteria by tapping the sides of the tubes gently for a few seconds. Store on ice for 30–40 minutes. Include two transformation controls, one containing 5 pg of bacteriophage M13 RF DNA and the other containing no added DNA.

14. While the ligated DNA is incubating with the competent cells, prepare a set of 16 sterile culture tubes containing 3 ml of melted YT or LB top agarose. Store the tubes at 47°C in a heating block or water bath until needed in Step 16.

15. Transfer the tubes containing the competent bacteria and DNA to a water bath equilibrated to 42°C. Incubate the tubes for exactly 90 seconds. Immediately return the tubes to an ice-water bath.

16. Add 40 µl of 2% X-gal, 4 µl of 20% IPTG, and 200 µl of the overnight culture of *E. coli* cells (Step 10) to the tubes containing the melted top agarose prepared in Step 14, and mix the contents of the tubes by gentle vortexing for a few seconds. Transfer each sample of the transformed bacteria to the tubes. Cap the tubes and mix the contents by gently inverting the tubes three times. Pour the contents of each tube in turn onto a separate labeled LB agar plate. Swirl the plate to ensure an even distribution of bacteria and top agarose.

17. Close the plates and allow the top agarose to harden for 5 minutes at room temperature. Use a Kimwipe to remove any condensation from the top of the plate, invert the plates, and incubate at 37°C.

    Plaques will be fully developed after 8–12 hours.

Analysis of Recombinant Bacteriophage M13 Clones

This protocol describes a rapid method to analyze the size of the single-stranded DNA of M13 recombinants.

MATERIALS

CAUTION: Please see Appendix 4 for appropriate handling of materials marked with <!>.

Reagents and Solutions

Please see Appendix 1 for components of stock solutions, buffers, and reagents. Dilute stock solutions to the appropriate concentrations.

SDS (2% w/v)

20x SSC

Vectors and Hosts

Bacteriophage M13 nonrecombinant plaques in top agarose
> Prepare as described in Protocol 3.1.

Bacteriophage M13 recombinant plaques in top agarose
> Prepare as described in Protocol 3.6.

E. coli F′ plating bacteria
> Prepare as described in Protocol 3.1 or purchase from commercial sources.

Nucleic Acids/Oligonucleotides

Single-stranded recombinant bacteriophage M13 DNAs
> Choose previously characterized recombinants that carry foreign sequences of known size to use as positive controls during gel electrophoresis.

Gels/Loading Buffers

0.7% Agarose gel, cast and run in 0.5x TBE containing 0.5 µg/ml ethidium bromide <!>

Gel-loading buffer IV

Additional Items

Step 1 of this protocol requires the reagents listed in Protocol 3.2.

Step 7 of this protocol may require the reagents listed in Protocols 2.21 and 2.22.

Water bath or heating/cooling block (65°C)

Additional Information

Alternative protocol (Screening Bacteriophage M13 Plaques
by Hybridization)   MC3, p. 3.41

METHOD

1. Prepare stocks of putative recombinant bacteriophages from single plaques, grown in an appropriate F′ host, as described in Protocol 3.2.

   > As controls, prepare stocks of several nonrecombinant bacteriophages (picked from well-isolated dark blue plaques).

2. Use a micropipettor with a sterile tip to transfer 20 µl of each of the supernatants into a fresh microfuge tube. Store the remainder of the supernatants at 4°C until needed.

3. To each 20-µl aliquot of supernatant, add 1 µl of 2% SDS. Tap the sides of the tubes to mix the contents and then incubate the tubes for 5 minutes at 65°C.

4. To each tube, add 5 µl of gel-loading buffer IV. Again mix the contents of the tubes by tapping and then analyze each sample by electrophoresis through an 0.7% agarose gel. Run the gel at 5 V/cm. As positive controls, use single-stranded DNA preparations of previously characterized M13 recombinants that carry foreign sequences of known size.

5. When the bromophenol blue has traveled the full length of the gel, photograph the DNA under UV illumination.

6. Compare the electrophoretic mobilities of the single-stranded DNAs liberated from the putative recombinants with those of the DNAs liberated from the control nonrecombinant bacteriophages.

   > The single-stranded DNAs of recombinants carrying sequences of foreign DNA longer than 200–300 nucleotides migrate slightly more slowly than empty vector through 0.7% agarose gels. Once recombinants of the desired size have been identified, single-stranded DNAs can be prepared from supernatants stored at 4°C (Step 2).

7. If necessary, confirm the presence of foreign DNA sequences by transferring single-stranded DNAs from the gel to a nitrocellulose or nylon membrane (please see Protocol 2.21) and hybridizing to an appropriate radiolabeled probe (Protocol 2.22). Soak the gel in 10 volumes of 20x SSC for 45 minutes and then transfer the DNA directly to the membrane.

Producing Single-stranded DNA with Phagemid Vectors

This protocol describes methods to superinfect bacteria carrying a recombinant phagemid with a high-titer stock of an appropriate helper virus and to assay the yield of filamentous virus particles that carry single-stranded copies of the phagemid DNA.

MATERIALS

CAUTION: Please see Appendix 4 for appropriate handling of materials marked with <!>.

Reagents and Solutions

Please see Appendix 1 for components of stock solutions, buffers, and reagents. Dilute stock solutions to the appropriate concentrations.
Kanamycin (10 mg/ml)
SDS (2% w/v)

Vectors and Hosts

Bacteriophage M13K07 (helper)
> M13K07 may be obtained commercially and propagated as described in Steps 1–3 of this protocol. Store stocks of helper virus at 4°C in 2x YT medium or at –20°C in 2x YT medium containing 50% (v/v) glycerol.

E. coli F' strain
> For a listing of strains suitable for propagation of bacteriophage M13, see MC3, Tables 3-2 (pp. 3.12–3.13) and 3-4 (p. 3.42).

E. coli strain DH11S
> This strain should be plated on supplemented minimal agar plates.

E. coli strain DH11S, transformed with bacteriophage M13 phagemid vector
> This strain of E. coli may be transformed with the phagemid vector as described in Protocol 3.6. The transformed strain may then be propagated as a culture as described in Protocol 3.2.

E. coli strain DH11S, transformed with bacteriophage M13 recombinant phagemid clone carrying foreign DNA
> This strain of E. coli may be transformed with the recombinant phagemid vector as described in Protocol 3.6. The transformed strain may then be propagated as a culture as described in Protocol 3.2.

Media and Antibiotics

Supplemented M9 minimal agar plates
YT agar plates containing 60 µg/ml ampicillin
2x YT containing 25 µg/ml kanamycin
2x YT containing 60 µg/ml ampicillin
2x YT medium

Gels/Loading Buffers

0.7% Agarose gel, cast and run in 0.5x TBE containing 0.5 µg/ml ethidium bromide <!>
Gel-loading buffer IV

Centrifuges/Rotors/Tubes

Microfuge (4°C) and sterile microfuge tubes

> **Additional Items**
>
> Bacterial shaker (37°C)
> Step 14 of this protocol requires the reagents listed in Protocol 3.4.
> Steps 2 and 5 of this protocol require the reagents listed in Protocol 3.1.
> Sterile tubes (17 x 100 mm)
> Water bath or heating block (65°C)
>
> **Additional Information**
>
> Properties, growth, and use of phagemids   MC3, pp. 3.42–3.44

METHOD

The key to success in using phagemids is to prepare a stock of helper virus whose titer is accurately known.

1. In 20 ml of 2x YT medium, establish a culture of *E. coli* strain DH11S from a single colony freshly picked from supplemented minimal agar plates. Incubate the culture at 37°C with moderate agitation until the OD_{600} reaches 0.8.

2. Prepare a series of tenfold dilutions of bacteriophage M13K07 in 2x YT medium and plate aliquots of the bacteriophage as described in Protocol 3.1 to obtain well-isolated plaques on a lawn of DH11S cells.

3. Pick well-separated, single plaques and place each plaque in 2–3 ml of 2x YT medium containing kanamycin (25 µg/ml) in a 15-ml culture tube. Incubate the infected cultures for 12–16 hours at 37°C with moderate agitation (250 cycles/minute).

   Kanamycin is used in this protocol to ensure that all bacterial cells containing a phagemid genome are infected by the helper M13K07 bacteriophage. During propagation of M13K07 (e.g., Steps 1–3), there is selection for bacteriophage genomes that have lost the p15A origin and the Tn*903* transposon. For this reason, it is essential to include kanamycin in the medium used to prepare the stock of helper virus in this step.

In the following steps, use stocks of M13K07 derived from single freshly picked plaques.

4. Transfer the infected cultures to 1.5-ml sterile microfuge tubes and centrifuge them at maximum speed for 2 minutes at 4°C in a microfuge. Transfer the supernatants to fresh tubes and store them at 4°C.

5. Measure the titer of each of the bacteriophage stocks by plaque formation (Protocol 3.1) on a strain of *E. coli* F′ (TG1, DH11S, NM522, or XL1-Blue) that supports the growth of bacteriophage M13.

   The titer of infectious bacteriophage particles in the stocks should be 10^{10} pfu/ml. Discard any stock with a lower titer.

6. Streak DH11S cells transformed by (i) the recombinant phagemid and (ii) the empty (parent) phagemid vector onto two separate YT agar plates containing 60 µg/ml ampicillin. Incubate the plates for 16 hours at 37°C.

7. Pick (i) several colonies transformed by the recombinant phagemid and (ii) one or two colonies transformed by the parent vector into sterile 17 x 100-mm tubes that contain 2–3 ml of 2x YT medium containing 60 µg/ml ampicillin.

8. To each culture, add M13K07 helper bacteriophage to achieve a final concentration of 2×10^7 pfu/ml. Incubate the cultures for 1.0–1.5 hours at 37°C with strong agitation (300 cycles/minute).

9. Add kanamycin to the cultures to a final concentration of 25 µg/ml. Continue incubation for an additional 14–18 hours at 37°C.

10. Transfer the cell suspensions to microfuge tubes and separate the bacterial cells from the growth medium by centrifugation at maximum speed for 5 minutes at room temperature in a microfuge. Transfer the supernatants to fresh tubes and store them at 4°C.

11. Combine 40 µl of each supernatant with 2 µl of 2% SDS in 0.5-ml microfuge tubes. Mix the contents of the tubes by tapping and then incubate the tubes for 5 minutes at 65°C.

12. Add 5 µl of gel-loading buffer IV to each sample of the phagemid DNA, mix the samples, and load them into separate wells of an 0.7% agarose gel.

13. Perform electrophoresis for several hours at 6 V/cm until the bromophenol blue has migrated approximately half the length of the gel. Examine and photograph the gel by UV light.

    Yields vary depending on the size and nature of foreign DNA in the phagemid, but are generally about 1 µg/ml of culture volume.

14. Isolate single-stranded phagemid DNA from the supernatants containing the largest amount of single-stranded DNA. Follow the steps outlined in Protocol 3.4, scaling up the volumes two- to threefold.

Working with High-capacity Vectors

BACKGROUND INFORMATION

Background information found in *Molecular Cloning: A Laboratory Manual*, 3rd edition (hereafter MC3) unless otherwise indicated.

Bacterial artificial chromosomes (BACs)	MC3, pp. 4.48–4.51
Choosing a vector	MC3, pp. 4.7–4.10
Commercially available libraries	MC3, p. 4.9
Constructing libraries of genomic DNA	MC3, pp. 4.6–4.7
Cosmids and cosmid libraries	MC3, pp. 4.11–4.17
P1 cloning systems	MC3, pp. 4.35–4.41
Properties of high-capacity vectors	MC3, pp. 4.1–4.5 and 4.7–4.10
Yeast artificial chromosomes	MC3, pp. 4.58–4.65

Construction of Genomic DNA Libraries in Cosmid Vectors

The protocol describes how to construct a library of 35–45-kb fragments of genomic DNA in the double *cos* site cosmid vector, SuperCos-1. The steps include:

- Linearization and dephosphorylation of SuperCos-1 DNA

- Partial digestion of high-molecular-weight genomic DNA with *Mbo*I

- Dephosphorylation of genomic DNA

- Ligation of cosmid arms to genomic DNA: Packaging and plating recombinants

- Isolation and analysis of recombinant cosmids: Validation of the library

MATERIALS

CAUTION: Please see Appendix 4 for appropriate handling of materials marked with <!>.

Reagents and Solutions

Please see Appendix 1 for components of stock solutions, buffers, and reagents. Dilute stock solutions to the appropriate concentrations.

Bacteriophage λ packaging mixture (commercially available)
Chloroform <!>
Ethanol
Ethanol (70%)
Phenol:chloroform (1:1, v/v) <!>
SM
Sodium acetate (3 M, pH 5.2)
TE (pH 8.0)

Vectors and Hosts

Bacteriophage λ stock
E. coli plating bacteria
> Used to measure the infectivity of cosmid DNA packaged into bacteriophage λ particles. Strains commonly used for this purpose include XL1-Blue, ED8767, NM554, and DH5αMCR. Choose a strain of plating bacteria appropriate for the cosmid vector.

Media and Antibiotics

TB
TB agar plates containing 25 µg/ml kanamycin
TB medium containing 25 µg/ml kanamycin

Enzymes and Buffers

Alkaline phosphatase (CIP or SAP) and 10x buffer
Bacteriophage T4 DNA ligase and 10x ligase buffer containing ATP (1 mM)
Restriction enzymes *Bam*HI, *Mbo*I, and *Xba*I, and 10x buffers
Restriction enzymes that cleave cosmid vector but are unlikely to cleave genomic insert
  (e.g., *Not*I, *Sal*I)

Nucleic Acids/Oligonucleotides

Bacteriophage λ DNA
Bacteriophage λ DNA digested with *Hind*III
High-molecular-weight genomic DNA
Marker DNA: linear bacteriophage λ DNA
Superhelical SuperCos-1 DNA

Gels/Loading Buffers

0.7% and 0.8% Agarose gels, cast and run in 0.5x TBE containing 0.5 µg/ml ethidium bromide <!>
Pulsed-field gels (or 0.5% agarose gels)

Additional Items

Bacterial shaker (37°C)
Step 10 of this protocol requires the reagents listed in MC3, Protocol 2.17.
Step 23 of this protocol requires the reagents listed in MC3, Protocol 1.1.
Water bath (65°C)
Water bath or heating block (37°C)

Additional Information

Alkaline phosphatases	MC3, pp. 9.92–9.93
In vitro packaging	MC3, p. 2.111
Minimizing damage to large DNAs	MC3, p. 2.10
Pulsed-field electrophoresis	MC3, pp. 5.55–5.60 and Protocols 5.14–5.20

METHOD

1. Combine 20 µg of superhelical SuperCos-1 DNA with 50 units of *Xba*I in a volume of 200 µl of 1x *Xba*I digestion buffer and incubate the reaction mixture for 2–3 hours at 37°C.

2. After 2 hours of incubation, transfer an aliquot (~1 µl) of the reaction mixture to a fresh microfuge tube. Analyze the aliquot of cosmid DNA by electrophoresis through an 0.8% agarose gel, using as controls (i) 50–100 ng of superhelical SuperCos-1 DNA and (ii) 50–100 ng of a bacteriophage λ DNA digested with *Hind*III.

3. Extract the digestion reaction once with phenol:chloroform and once with chloroform.

4. Transfer the aqueous phase to a fresh microfuge tube and recover the linearized cosmid DNA by standard precipitation with ethanol and subsequent washing in 70% ethanol. Store the open tube in an inverted position on a bed of paper towels to allow the ethanol to drain and evaporate. Dissolve the damp pellet of DNA in 180 µl of H_2O. Remove a 100-ng aliquot of the DNA for use as a control.

5. Add 20 µl of 10x dephosphorylation buffer to the remainder of the DNA solution. Add 0.1 unit of alkaline phosphatase (CIP or SAP) and incubate the reaction for 30 minutes at 37°C. Add a second aliquot (0.1 unit) of enzyme and continue digestion for an additional 30 minutes. Transfer the reaction to a water bath set at 65°C and incubate for 30 minutes to inactivate alkaline phosphatase. Remove two 100-ng aliquots of the DNA for use as controls.

6. Extract the reaction mixture once with phenol:chloroform and once with chloroform. Recover the linearized, dephosphorylated SuperCos-1 DNA by standard precipitation with ethanol and subsequent washing in 70% ethanol. Dissolve the damp pellet of DNA in 180 µl of H_2O.

7. Transfer an aliquot of the dephosphorylated DNA (50–100 ng) to a fresh microfuge tube and store it on ice. Add 20 μl of 10x *Bam*HI restriction buffer to the remainder of the desphosphor-ylated DNA. Add 40 units of *Bam*HI and incubate the reaction for 2–3 hours at 37°C.

8. After 2 hours of incubation, remove a second aliquot of DNA to a separate microfuge tube. Analyze both aliquots of DNA by agarose gel electrophoresis.

    After digestion with *Bam*HI, the linear 7.9-kb fragment of dephosphorylated SuperCos-1 DNA should be quantitatively cleaved into two DNA fragments of approximately 1.1 and 6.8 kb. If traces of the 7.9-kb DNA are still visible, add 10 more units of *Bam*HI to the digest and contin-ue incubation at 37°C until the reaction has gone to completion.

9. Extract the digestion reaction once with phenol:chloroform and once with chloroform. Recover the DNA by standard precipitation with ethanol followed by washing with 70% ethanol. Dissolve the damp pellet of DNA in 20 μl of H_2O and store the solution at 4°C until needed.

10. Establish the conditions for partial digestion of a 30-μg sample of high-molecular-weight genomic DNA with *Mbo*I. The aim is to establish conditions that produce the highest yield of DNA fragments with a modal size of 38–52 kb.

11. Using the conditions for partial digestion established in Step 10, set up three large-scale reac-tions each containing 100 μg of high-molecular-weight genomic DNA and amounts of *Mbo*I that bracket the optimal concentration, as determined in Step 10. At the end of the incuba-tion period, check the size of an aliquot of each partially digested DNA by agarose gel elec-trophoresis, as described in the note to Step 10.

12. Pool the samples of partially digested genomic DNA that contain the greatest amount of DNA in the 38–52-kb range. Extract the pooled DNAs once with phenol:chloroform and once with chloroform. Recover the DNA by standard precipitation with ethanol and subsequent wash-ing in 70% ethanol. Dissolve the damp pellet of DNA in 180 μl of TE (pH 8.0).

    Resuspension is best accomplished by allowing the DNA pellet to soak in TE overnight at 4°C. Do not vortex the DNA. Instead, mix the DNA by gently tapping the sides of the tube. If only small amounts of DNA are available, as is the case when constructing cosmid libraries from flow-sorted eukaryotic chromosomes or from purified YACs, then, instead of ethanol precipitation, purify the DNA by drop dialysis against TE (pH 8.0) with floating membranes, as described in Protocol 4.5.

13. To the resuspended partially digested genomic DNA, add 20 μl of 10x dephosphorylation buffer and 2 units of CIP or SAP. Immediately withdraw a sample of DNA for use as a control:

    a. Remove an aliquot of the reaction containing 0.1–1.0 μg of DNA to a small (0.5 ml) microfuge tube containing 0.3 μg of a linearized plasmid in 1x dephosphorylation buffer (e.g., pUC cleaved with *Bam*HI).

    b. Set up a second control that contains 0.3 μg of the same linearized plasmid in 10 μl of 1x dephosphorylation buffer.

       Do not add CIP or SAP to this control.

    c. Follow the instructions in Steps 14–16.

14. Incubate the large-scale dephosphorylation reaction and the two controls for 30 minutes at 37°C. Then transfer the three reactions to a water bath set at 65°C and incubate them for 30 minutes to inactivate CIP.

15. Cool the reactions to room temperature. Purify the DNAs by extracting once with phenol:chloroform and once with chloroform. Recover the DNA by standard precipitation with ethanol and subsequent washing in 70% ethanol.

16. Dissolve the two control DNAs in 10 μl of 1x ligation buffer. Add 0.1 Weiss unit of bacterio-phage T4 DNA ligase to each tube and incubate them for 3 hours at room temperature. Examine the ligated DNAs by agarose gel electrophoresis.

17. Allow the DNA in the large-scale reaction to dissolve overnight at 4°C in a small volume of H$_2$O. Aim for a final concentration of approximately 500 μg of DNA/ml. Estimate the concentration of DNA by agarose gel electrophoresis, or better, by measuring A$_{260}$.

18. Set up a series of ligation reactions (final volume 20 μl) containing:

cosmid arms DNA (Step 9)	2 μg
dephosphorylated genomic DNA (Step 17)	0.5, 1, or 2.5 μg
10x ligation buffer containing ATP (1 mM)	2 μl
bacteriophage T4 DNA ligase	2 Weiss units

    Incubate the ligation reactions for 12–16 hours at 16°C.

19. Package 5 μl of each of the ligation reactions into bacteriophage λ particles (equivalent to 0.5 μg of vector arms) using a commercial packaging kit and following the conditions recommended by the supplier. After packaging, add 500 μl of SM and 20 μl of chloroform to the reactions and then store the diluted reactions at 4°C.

20. Measure the titer of the packaged cosmids in each of the packaging reactions by transduction into an appropriate *E. coli* host. Mix 0.1 ml of a 10^{-2} dilution of an aliquot of each reaction with 0.1 ml of SM and 0.1 ml of fresh plating bacteria. Allow the bacteriophage particles containing the recombinant cosmids to adsorb by incubating the infected bacterial cultures for 20 minutes at 37°C. Add 1 ml of TB medium and continue the incubation for an additional 45 minutes at 37°C to allow expression of the kanamycin resistance gene in the SuperCos-1 vector.

    Store the remainder of the packaging mixtures at 4°C until Steps 21–24 have been completed (2–3 days).

21. Spread 0.5 and 0.1 ml of the bacterial culture onto separate TB agar plates containing kanamycin (25 μg/ml). After incubating the plates overnight at 37°C, count the number of bacterial colonies.

22. Pick 12 individual colonies and grow small-scale (2.5 ml) cultures in TB containing 25 μg/ml kanamycin for periods of no longer than 6–8 hours. Shake the cultures vigorously during incubation.

23. Isolate cosmid DNA from 1.5 ml of each of the 12 small-scale bacterial cultures using the alkaline lysis method, described in Protocol 1.1.

24. Digest 2–4 μl of each of the DNA preparations with restriction enzymes (e.g., *Not*I and *Sal*I) that cleave the cosmid vector but are unlikely to cleave the cloned insert of genomic DNA. Analyze the sizes of the resulting fragments by electrophoresis through a 0.7% agarose gel.

    Use as markers linear bacteriophage λ DNA and *Hind*III fragments of bacteriophage λ DNA, which can be prepared easily in the laboratory and are also available commercially.

25. Calculate the proportion of colonies that carry inserts.

26. Estimate the average size of the inserts by isolating a few dozen clones and measuring the size of inserts by pulsed-field gel electrophoresis.

27. Calculate the "depth" of the library, i.e., the number of genome equivalents it contains.

    If the library is satisfactory in size and quality, proceed to plate and screen the library by hybridization (Protocol 4.2) or, alternatively, to amplify and store the library (Protocols 4.3 and 4.4).

Protocol 4.2 — Screening an Unamplified Cosmid Library by Hybridization: Plating the Library on Filters

Unamplified cosmid libraries are plated at high density onto 150-mm nitrocellulose or nylon filters and screened by hybridization.

MATERIALS

Reagents and Solutions

Bacteriophage λ packaging mixture (commercially available)

Vectors and Hosts

E. coli plating bacteria
> Used to measure the infectivity of cosmid DNA packaged into bacteriophage λ particles. Strains commonly used for this purpose include XL1-Blue, ED8767, NM554, and DH5αMCR. Choose a strain of plating bacteria appropriate for the cosmid vector.

Media and Antibiotics

TB
TB agar plates (150-mm diameter) containing 25 µg/ml kanamycin

Additional Items

Blunt-ended forceps
Glass plates (two), thick and sterilized with ethanol
17-gauge Hypodermic needle
Nitrocellulose or nylon filters (137-mm diameter), sterile (see Protocol 1.29, Step 2) and detergent-free
Step 16 of this protocol requires the reagents listed in Protocols 1.31 and 1.32.
Sterile 13 x 75-mm tubes
Sterile bacterial spreader
Water bath or heating block (37°C)
Whatman No. 1 filters (137-mm diameter)

Additional Information

Additional protocol (Reducing Cross-hybridization)   MC3, p. 4.27

METHOD

1. Calculate the volume of the packaging reaction (Protocol 4.1, Steps 20 and 21) that will generate 30,000–50,000 transformed bacterial colonies.

2. Set up a series of sterile 13 x 75-mm tubes containing this volume of packaging reaction and 0.2 ml of plating bacteria.
   > The actual number of tubes to set up for plating the packaging reaction depends on the genome size, average size of fragments cloned, and the coverage of the genome required to find the particular clone of interest. In most cases, 15–20 tubes should be sufficient.

3. Incubate the tubes for 20 minutes at 37°C.

4. To each tube, add 0.5 ml of TB. Continue the incubation for an additional 45 minutes.

5. Place sterile, numbered, nitrocellulose or nylon filters onto a series (equal in number to the series of tubes in Step 2) of 150-mm TB agar plates containing kanamycin (25 µg/ml).

6. Use a sterile spreader to smear the contents of each tube over the surface of a filter on an agar plate. After the inoculum has been absorbed into each filter, transfer the plates to a 37°C incubator for several hours to overnight (12–15 hours).

   Try to avoid spreading the inoculum within 3 mm of the edge of the master filters.

7. Place a sterile, numbered nitrocellulose or nylon 137-mm filter on a fresh TB agar plate containing kanamycin (25 µg/ml).

8. Place a sterile Whatman No. 1 filter on a thick, sterile glass plate.

9. Use blunt-ended forceps to remove the replica filter from the fresh TB agar plate (Step 7) and place it on the Whatman No. 1 filter.

10. Again use forceps to remove a master filter now carrying transformed colonies from its TB agar plate (Step 6) and place it, colony side down, exactly on top of the numbered replica filter on the Whatman No. 1 filter. Cover the two filters with another sterile Whatman No. 1 filter.

11. Place a second sterile glass plate on top of the stack of filters. Press the plates together.

12. Remove the upper glass plate and the uppermost Whatman No. 1 filter. Use a 17-gauge hypodermic needle to key the two numbered nitrocellulose or nylon filters to each other by making a series of holes (~5 will do), placed asymmetrically around the edge of the filters.

13. Peel the two nitrocellulose or nylon filters apart, and working quickly, replace them on their TB agar plates containing kanamycin (25 µg/ml).

14. Incubate the master and replica filters for a few hours at 37°C, until the bacterial colonies are 0.5–1.0 mm in diameter.

15. Seal the master plates in Parafilm and store them at 4°C in an inverted position.

16. Lyse the colonies on the replica filter (Protocol 1.31) and process the filters for hybridization to radiolabeled probes (Protocol 1.32).

    Replica filters can be used to replicate the library again or may be stored frozen at –70°C.

Protocol 4.3 — Amplification and Storage of a Cosmid Library: Amplification in Liquid Culture

Amplification of cosmid libraries should be avoided wherever possible because it may result in distorted representation of cloned genomic sequences. If amplification should become necessary, it is best to expand the library by growth in liquid medium, as described here.

MATERIALS

Reagents and Solutions

Bacteriophage λ packaging mixture (commercially available)
Glycerol
> Sterilize by autoclaving for 20 minutes at 15 psi (1.05 kg/cm²) on liquid cycle.

Vectors and Hosts

E. coli plating bacteria
> Used to measure the infectivity of cosmid DNA packaged into bacteriophage λ particles. Strains commonly used for this purpose include XL1-Blue, ED8767, NM554, and DH5αMCR. Choose a strain of plating bacteria appropriate for the cosmid vector.

Media and Antibiotics

TB
TB agar plates (142-mm diameter) containing 25 μg/ml kanamycin
TB containing 25 μg/ml kanamycin

Centrifuges/Rotors/Tubes

Sorvall SLC-1500 and SS-34 rotors (4°C) (SS-34 rotor optional; see note to Step 3)

Additional Items

Step 9 of this protocol also requires the reagents listed in Protocols 1.31, 1.32, and 4.2.
Sterile 13 x 75-mm tubes
Sterile freezing vials, e.g., NUNC CryoTube vials
Water bath or heating block (37°C)

Additional Information

Alternative protocol (Amplification on Plates)	MC3, p. 4.324
Storage of bacterial cultures	MC3, p. A8.5

METHOD

1. Calculate the volume of the packaging reaction (Protocol 4.1, Steps 20 and 21) that will generate 30,000–50,000 transformed bacterial colonies.

2. Set up a series of sterile test tubes and into each tube deliver 0.2 ml of plating bacteria followed by the volume of packaging reaction determined in Step 1.

  Adapted from Chapter 4, Protocol 3, p. 4.28 of MC3.

The actual number of tubes to set up for plating the packaging reaction depends on the genome size, average size of fragments cloned, and the coverage of the genome required to find the particular clone of interest. In most cases, 15–20 tubes should be sufficient.

3. Incubate the tubes for 20 minutes at 37°C.

    Large amounts of packaging mixture can inhibit attachment of the bacteriophage particles to the plating bacteria. If the concentration of packaged bacteriophages is low ($<10^4$ transducing units/ml of packaging mixture), use more plating bacteria, e.g., 5 ml/ml of packaging reaction. After incubating the cells for 20 minutes at 37°C, recover the bacteria by centrifugation at 5000g (6500 rpm in a Sorvall SS-34 rotor) for 10 minutes at 4°C. Resuspend the cells in 0.5 ml of TB and proceed to Step 4.

4. Add 0.5 ml of TB to each tube. Continue incubation for an additional 45 minutes.

5. Inoculate 0.25-ml aliquots of each culture of infected cells into 100-ml volumes of TB medium containing 25 μg/ml kanamycin in 250-ml flasks.

6. Incubate the inoculated cultures with vigorous shaking at 37°C until the cells reach an optimal density of 0.5–1.0 OD_{600}.

7. Pool the cultures and recover the cells by centrifugation at 5000g (5500 rpm in a Sorvall SLC-1500 rotor) for 15 minutes at 4°C. Resuspend the cells in a volume of TB that is equal to 0.1× the volume of the original pooled cultures.

    Cosmid DNA can be isolated (please see Protocol 4.1, Step 23) from aliquots of cells taken before the addition of glycerol. This stock of DNA can be used as a template in PCR to determine whether a particular DNA sequence of interest is present in the library.

8. Add sterile glycerol to the cell suspension to a final concentration of 15% (v/v). Mix the suspension well by inverting the closed tube several times. Dispense aliquots (0.5–1.0 ml) of the bacterial suspension into sterile vials. Store the tightly closed vials at –70°C.

9. To screen the library, thaw an aliquot of frozen cells rapidly at 37°C and plate 30,000–50,000 bacteria onto each of a series of numbered nitrocellulose or nylon filters (137-mm diameter) as described in Protocol 4.2 beginning with Step 5. Proceed with lysing the colonies on the replica filters (Protocol 1.31) and processing the filters for hybridization to labeled probes (Protocol 1.32).

Amplification and Storage of a Cosmid Library: Amplification on Filters

Amplification of cosmid libraries may result in distorted representation of cloned genomic sequences and should be avoided wherever possible. In this method of amplification, distortion of the library is rarely a problem because at no stage are bacteria containing different recombinant cosmids grown in competition with one another. The major problem with this method is its tediousness.

MATERIALS

Reagents and Solutions

Bacteriophage λ packaging mixture (commercially available)

Vectors and Hosts

E. coli plating bacteria
> Used to measure the infectivity of cosmid DNA packaged into bacteriophage λ particles. Strains commonly used for this purpose include XL1-Blue, ED8767, NM554, and DH5αMCR. Choose a strain of plating bacteria appropriate for the cosmid vector.

Media and Antibiotics

TB
TB agar plates (150-mm diameter) containing 25 µg/ml kanamycin

Centrifuges/Rotors/Tubes

Sorvall SS-34 rotor (4°C)

Additional Items

Blunt-ended forceps
Nitrocellulose or nylon filters (137-mm diameter), sterile (see Protocol 1.29, Step 2) and detergent-free
Step 7 of this protocol requires the reagents listed in Protocol 4.2.
Step 8 of this protocol requires the reagents listed in Protocols 1.31 and 1.32.
Sterile 13 x 75-mm tubes
Sterile bacterial spreader
Water bath or heating block (37°C)

Additional Information

Alternative protocol (Amplification on Plates)   MC3, p. 4.34

METHOD

1. Calculate the volume of the packaging reaction (Protocol 4.1, Steps 20 and 21) that will generate 30,000–50,000 transformed bacterial colonies.

2. Set up a series of sterile test tubes and into each tube deliver 0.2 ml of plating bacteria followed by the volume of packaging reaction determined in Step 1.

The actual number of tubes to set up for plating the packaging reaction depends on the genome size, average size of fragments cloned, and the coverage of the genome required to find the particular clone of interest. In most cases, 15–20 tubes should be sufficient.

3. Incubate the tubes for 20 minutes at 37°C.

   Large amounts of packaging mixture can inhibit attachment of the bacteriophage particles to the plating bacteria. If the concentration of packaged bacteriophages is low (<10^4 transducing units/ml of packaging mixture), use more plating bacteria, e.g., 5 ml/ml of packaging reaction. After incubating the cells for 20 minutes at 37°C, recover the bacteria by centrifugation at 5000*g* (5500 rpm in a Sorvall SS-34 rotor) for 10 minutes at 4°C. Resuspend the cells in 0.5 ml of TB and proceed to Step 4.

4. Add 0.5 ml of TB to each tube. Continue incubation for an additional 45 minutes.

5. Place sterile, numbered filters onto a series (equal in number to the series of tubes in Step 2) of 150-mm TB agar plates containing kanamycin (25 µg/ml).

6. Use a sterile spreader to smear the contents of each tube over the surface of a filter on an agar plate. After the inoculum has absorbed into each filter, transfer the plates to a 37°C incubator for several hours to overnight (12–15 hours).

7. Make a replica of each of the master filters as described beginning with Protocol 4.2, Step 7. Store the filters at –70°C.

8. To screen the library, thaw the replica filters and proceed with lysing the colonies on the filters (Protocol 1.31) and processing the filters for hybridization to labeled probes (Protocol 1.32).

Working with Bacteriophage P1 and Its Cloning Systems

This protocol describes methods for recovery and purification of recombinant clones of bacterio-phage P1 or PAC DNAs from bacteria. Because of their large size, these DNAs are sensitive to shearing forces and must be handled carefully. This protocol generally yields P1 DNA that works well as a substrate or template in enzymatic reactions.

MATERIALS

CAUTION: Please see Appendix 4 for appropriate handling of materials marked with <!>.

Reagents and Solutions

Please see Appendix 1 for components of stock solutions, buffers, and reagents. Dilute stock solutions to the appropriate concentrations.

Alkaline lysis solution I
Alkaline lysis solution II, freshly prepared
Alkaline lysis solution III
Ammonium acetate (0.5 M) (from 10 M stock)
Ethanol
Ethanol (70%)
IPTG (1 mM) <!> (optional; see note to Step 1)
Isopropanol
$MgCl_2$ (1 M)
PEG 8000 (40% w/v) <!>
Phenol:chloroform (1:1, v/v) <!>
Sodium acetate (0.3 M, pH 5.2)
TE (pH 8.0)
TE (pH 8.0) containing 20 µg/ml RNase A

Vectors and Hosts

E. coli strain transformed with a nonrecombinant bacteriophage P1 or PAC vector
E. coli strain transformed with a recombinant bacteriophage P1 or PAC vector

Media and Antibiotics

LB containing 25 µg/ml kanamycin (optional; use in Step 1 with IPTG)
TB containing 25 µg/ml kanamycin

Enzymes and Buffers

Restriction enzymes and 10x buffers

Gels/Loading Buffers

Pulsed-field gels (or 0.5% agarose gels)
See Step 15 and information on pulsed-field electrophoresis in MC3, p. 4.18 and Chapter 5.

Centrifuges/Rotors/Tubes

Sorvall SS-34 rotor (4°C)

Adapted from Chapter 4, Protocol 5, p. 4.35 of MC3.

Additional Items

Bacterial shaker (37°C)
Centrifuge tubes (15 ml)
Corex glass centrifuge tubes (30 ml)
Falcon tubes (50 ml) or Erlenmayer flasks (sterile)
Microfuge (4°C)
Water bath (65°C)

Additional Information

Additional protocol (Purification of High-molecular-weight DNA by Drop Analysis)	MC3, p. 4.44
Alternative protocol (Purification of High-molecular-weight Circular DNA by Chromatography on Qiagen Resin)	MC3, p. 4.35
Cloning into P1 vectors	MC3, pp. 4.37–4.39
Design of P1 vectors	MC3, pp. 4.35–4.37
PAC vectors	MC3, pp. 4.40–4.41
Polyethylene glycol	MC3, p. 1.154
Pulsed-field gel electrophoresis	MC3, pp. 5.55–5.60 and Protocols 5.14–5.20
Screening of P1 libraries	MC3, pp. 4.39–4.40

METHOD

1. Transfer 10 ml of TB containing 25 µg/ml kanamycin into each of two 50-ml Falcon tubes or Erlenmeyer flasks. Inoculate one tube with a single colony of bacteria containing the recombinant P1 or PAC. Inoculate the other tube with a single colony of bacteria transformed by the vector alone. Grow the cultures to saturation, with vigorous shaking for 12–16 hours at 37°C.

   The addition of IPTG (1 mM) to cultures of cells carrying P1 recombinants inactivates the *lac* repressor and leads to induction of the P1 lytic replicon, which results in an increase in the copy number of the plasmid DNA from 1 to approximately 20 copies/cell.

   For induction with IPTG, cultures are grown in LB containing 25 µg/ml kanamycin until the OD_{600} reaches approximately 0.8, at which point IPTG is added to a final concentration of 0.5–1 mM. The cells are incubated for an additional 3 hours at 37°C and then harvested.

2. Transfer each culture to a 15-ml centrifuge tube. Harvest the cells by centrifugation at 3500*g* (5400 rpm in a Sorvall SS-34 rotor) for 5 minutes at 4°C. Resuspend each cell pellet in 3 ml of sterile H_2O and repeat the centrifugation step.

3. Resuspend each cell pellet in 2 ml of alkaline lysis solution I and place on ice.

4. Add 3 ml of alkaline lysis solution II and gently invert the tube several times to mix the solutions. Transfer the tube to an ice bath for 10 minutes.

   IMPORTANT: Because recombinant P1 and PAC clones are large enough to be sensitive to shearing, keep vortexing, pipetting, and shaking to a minimum during the isolation of DNA. Wherever possible, transfer by pouring the DNA from one tube to another. When pipetting cannot be avoided, use wide-bore pipette tips.

5. Add 3 ml of ice-cold alkaline lysis solution III to each cell suspension and mix the solution by gently inverting the tube several times. Store the tube on ice for 10 minutes.

6. Pellet the cellular debris by centrifugation at 12,000*g* (10,000 rpm in a Sorvall SS-34 rotor) for 10 minutes at 4°C.

7. Decant the supernatant (~7 ml) into a 30-ml Corex centrifuge tube and add an equal volume of isopropanol. Mix the solutions by gently inverting the tube several times and collect the

precipitated nucleic acids by centrifugation at 12,000g (10,000 rpm in a Sorvall SS-34 rotor) for 10 minutes at 4°C.

8. Remove the supernatant by gentle aspiration. Invert the tube on a Kimwipe tissue until the last drops of fluid have drained away. Use a pasteur pipette attached to a vacuum line to remove any drops remaining attached to the wall of the tube. Dissolve the pellet of nucleic acid in 0.4 ml of 0.3 M sodium acetate (pH 5.2). Heat the solution to 65°C briefly (for a few minutes) to assist in dissolving the nucleic acid.

9. Transfer the DNA solution to a microfuge tube. Extract the solution once with an equal volume of phenol:chloroform. Separate the aqueous and organic phases by centrifugation at maximum speed for 5 minutes at room temperature in a microfuge. Transfer the upper aqueous phase to a fresh microfuge tube.

10. Add 1 ml of ice-cold ethanol. Mix the solutions by inverting the tube several times. Collect the precipitated DNA by centrifugation at maximum speed for 10 minutes at 4°C. Rinse the DNA pellet with 0.5 ml of 70% ethanol and centrifuge again for 2 minutes.

11. Carefully remove the supernatant. Store the open inverted tube at room temperature until no traces of ethanol are visible. Add 0.4 ml of TE plus RNase to the pellet and keep the closed tube at 37°C. Periodically during the next 15 minutes, shake the tube gently to assist in dissolving the DNA. Continue the incubation for a total of 2 hours.

12. Add 4 μl of 1 M MgCl$_2$ and 200 μl of 40% PEG solution. Mix well and collect the precipitated DNA by centrifugation at maximum speed for 15 minutes at 4°C in a microfuge.

13. Remove the supernatant by aspiration and resuspend the pellet in 0.5 ml of 0.5 M ammonium acetate. Add 1 ml of ethanol, mix the solutions by inverting the tube several times, and collect the precipitate by centrifugation at maximum speed for 15 minutes at 4°C in a microfuge.

14. Decant the supernatant and rinse the pellet twice with 0.5 ml of ice-cold 70% ethanol. Store the open inverted tube at room temperature until no traces of ethanol are visible. Resuspend the damp pellet in 50 μl of TE (pH 8.0).

15. For restriction enzyme analysis, digest 5–15 μl (~1 μg) of resuspended DNA and analyze the products either by conventional 0.5% agarose gel electrophoresis or by pulsed-field gel electrophoresis.

Transferring P1 Clones between *E. coli* Hosts

Problems with low yield or poor quality can sometimes be overcome by transferring the P1 or PAC recombinant into a strain of *E. coli* that does not express Cre recombinase.

MATERIALS

CAUTION: Please see Appendix 4 for appropriate handling of materials marked with <!>.

Vectors and Hosts

Closed circular recombinant bacteriophage P1 DNA
> Prepare as described in Protocol 4.5.

E. coli (e.g., DH10B), electrocompetent
> Prepare as described in Protocol 1.26.

Media and Antibiotics

LB agar plates containing 25 µg/ml kanamycin
SOC medium, prewarmed to 37°C
TB containing 25 µg/ml kanamycin
TB containing 25 µg/ml kanamycin and 30% (v/v) glycerol

Enzymes and Buffers

Restriction enzymes and 10x buffers

Gels/Loading Buffers

0.6% Agarose gel, cast and poured in 0.5x TBE containing 0.5 µg/ml ethidium bromide <!>

Centrifuges/Rotors/Tubes

Sorvall SS-34 rotor (4°C)

Additional Items

Electroporation device and ice-cold cuvettes (0.1 cm)
Step 9 of this protocol requires the reagents listed in Protocol 4.5.
Sterile 13 x 75- or 17 x 100-mm culture tubes

METHOD

1. Dilute 2–3 µg of P1 plasmid DNA to a concentration of 60 ng/µl in sterile H_2O. Set up a control lacking P1 DNA (sterile H_2O only) and carry the control through the electroporation procedure in parallel with the DNA sample.

2. Thaw vials of electrocompetent cells on ice and prechill 0.1-cm electroporation cuvettes.

3. Combine 20 µl of cells and 1 µl of P1 DNA in the cold cuvette.

4. Set the electroporation device to 1.8 kV, 200 ohms, and 25 µF.

5. Shock the cells. The optimum time constant is usually about 5 msec.

6. Immediately add 0.5 ml of prewarmed (37°C) SOC medium to the cell suspension. Transfer the suspension to a culture tube and incubate the suspension for 1 hour at 37°C with moderate agitation.

7. Plate 100-μl aliquots of the cell suspensions on LB plates containing 25 μg/ml kanamycin. Incubate the plates overnight at 37°C.

   The control plates (no P1 DNA) should remain sterile; plates from the culture treated with P1 DNA should contain several hundred transformed colonies.

8. Transfer 10–12 colonies into separate 11-ml aliquots of TB containing 25 μg/ml kanamycin. Incubate the cultures overnight at 37°C with vigorous agitation.

9. Prepare P1 DNA as described in Protocol 4.5.

   The remainder of the overnight cultures can be stored at –80°C in TB/kanamycin containing 30% (v/v) glycerol.

10. Perform digestions with several different restriction endonucleases and compare the patterns of the newly isolated DNAs with that of the original recombinant by agarose gel electrophoresis.

Protocol 4.7

Working with Bacterial Artificial Chromosomes

This protocol describes how to transform *E. coli* with BAC DNA by electroporation.

MATERIALS

CAUTION: Please see Appendix 4 for appropriate handling of materials marked with <!>.

Reagents and Solutions

Please see Appendix 1 for components of stock solutions, buffers, and reagents. Dilute stock solutions to the appropriate concentrations.
Chloramphenicol (34 mg/ml) in ethanol <!>
Store at –20°C.

Vectors and Hosts

BAC DNA
or
Culture of *E. coli* strain transformed by a BAC isolate
E. coli (e.g., DH10B), electrocompetent
Prepare as described in Protocol 1.26. Required only if the BAC isolate is supplied as purified DNA

Media and Antibiotics

LB agar plates containing 12.5 µg/ml chloramphenicol <!>
LB freezing buffer
LB medium containing 12.5 µg/ml chloramphenicol

Enzymes and Buffers

Restriction enzymes and 10x buffers
Choose enzymes appropriate for measuring the size of the genomic DNA inserts.

Gels/Loading Buffers

Equipment for pulsed-field gel electrophoresis (PFGE), and PFGE markers (from New England BioLabs, Bio-Rad, etc.)
See also MC3, Protocol 5.16.

Additional Items

Electroporation device and cuvettes (0.1 cm), chilled to 0°C
Required only if the BAC isolate is supplied as purified DNA
Step 4 of this protocol requires the reagents listed in Protocol 5.8.
Step 5 of this protocol requires the reagents listed in either Protocol 6.10 or 8.12.

Additional Information

BAC cloning vectors	MC3, pp. 4.48–4.49
Pulsed-field gel electrophoresis	MC3, pp. 5.55–5.60 and Protocols 5.14–5.20
Screening BAC libraries	MC3, pp. 4.50–4.51

METHOD

1. Prepare fresh BAC transformants.

   ### If the BAC is supplied in the form of DNA

   a. Transform *E. coli* (strain DH10B) with BAC DNA by electroporation.

      Because the efficiency of transformation by large BACs decreases dramatically as a function of the voltage applied during electroporation, it is best to set up a series of electroporation reactions at voltages ranging from 13 to 25 kV/cm.

   b. Plate 2.5, 25, and 250 μl of each batch of electroporated bacteria onto LB agar plates containing 12.5 μg/ml chloramphenicol. Incubate the plates for 16–24 hours at 37°C.

   ### If the BAC is supplied as a transformed bacterial stock

   a. Streak the culture without delay onto LB agar plates containing 12.5 μg/ml chloramphenicol.

   b. Incubate the plates for 12–16 hours at 37°C.

2. Select 12 individual transformants and inoculate these into 5-ml aliquots of LB medium containing chloramphenicol (12.5 μg/ml). Grow the cultures overnight at 37°C with vigorous shaking.

3. Use a loopful of each 5-ml culture to set up cultures of the 12 transformants in LB freezing medium. When these cultures have grown, transfer them to a –20°C freezer for storage.

4. Purify the BAC DNA from 4.5 ml of each 5-ml culture from Step 2, as described in Protocol 4.8.

5. Analyze the BAC DNA.

   a. Confirm by Southern hybridization (please see Protocol 6.10) or by PCR that the BACs contain the chromosomal region of interest.

   b. Measure the size of the inserts by digestion with restriction enzymes and pulsed-field gel electrophoresis.

6. On the basis of the results, select one or more of the BACs for further analysis.

Isolation of BAC DNA from Small-scale Cultures

BAC DNAs are prepared from 5-ml cultures of BAC-transformed cells by a modification of the standard alkaline lysis method (Protocol 1.1). The yield typically varies between 0.1 and 0.4 μg of BAC DNA per culture.

MATERIALS

CAUTION: Please see Appendix 4 for appropriate handling of materials marked with <!>.

Reagents and Solutions

Please see Appendix 1 for components of stock solutions, buffers, and reagents. Dilute stock solutions to the appropriate concentrations.

Alkaline lysis solution I, ice cold
Alkaline lysis solution II, freshly prepared (room temperature)
Alkaline lysis solution III, ice cold
Ethanol
Ethanol (70%)
Isopropanol
Sodium acetate (3 M, pH 5.2)
STE, ice cold
TE (pH 8.0)

Vectors and Hosts

E. coli strain carrying a recombinant BAC
See Protocol 4.7.

Media and Antibiotics

LB containing 12.5 μg/ml chloramphenicol <!>

Enzymes and Buffers

Restriction enzymes and 10x buffers

Nucleic Acids/Oligonucleotides

DNA markers for pulsed-field gel electrophoresis (from New England BioLabs, Bio-Rad, etc.)
See also MC3, Protocol 5.16.

Additional Items

Bacterial shaker (37°C)
Chromatography resin, e.g., QIAprep Spin Miniprep Kit, Qiagen Inc.
Microfuge (4°C)
Sterile 17 x 100-mm tubes

Additional Information

Chromatography resins	MC3, pp. 1.62–1.64
Pulsed-field gel electrophoresis	MC3, pp. 5.55–5.60 and Protocols 5.14–5.20

METHOD

1. Prepare 5-ml cultures of BAC-transformed *E. coli* in LB medium containing 12.5 μg/ml chloramphenicol and grow the cultures overnight at 37°C with vigorous shaking.

2. Collect the bacterial cells by centrifugation at 2000*g* (4100 rpm in a Sorvall SS-34 rotor) for 5 minutes at 4°C. Decant the medium carefully and remove any residual drops by aspiration.

3. Add 5 ml of ice-cold STE to each tube and resuspend the bacterial pellet by pipetting. Recover the cells by centrifugation as in Step 2.

   Cleaner preparations of BAC DNA are obtained if the cells are washed briefly in ice-cold STE at this stage.

4. Resuspend the cells in 200 μl of ice-cold alkaline lysis solution I. Transfer the cells to an ice-cold microfuge tube. Place the tube on ice.

   The cell suspension may be gently vortexed to break up clumps of cells.

5. Add 400 μl of freshly prepared alkaline lysis solution II to the tube. Gently invert the closed tube several times. Place the tube on ice.

6. Add 300 μl of ice-cold alkaline lysis solution III to the tube. Gently invert the closed tube several times. Place the tube on ice for 5 minutes.

7. Remove the precipitated cell debris by centrifugation at maximum speed for 5 minutes at 4°C in a microfuge. Decant the supernatant into a fresh microfuge tube. Add 900 μl of isopropanol at room temperature and mix the contents of the tube by gentle inversion.

8. Immediately collect the precipitated nucleic acids by centrifugation at maximum speed for 5 minutes at room temperature in a microfuge. Discard the supernatant and carefully rinse the pellet with 1 ml of 70% ethanol. Centrifuge the tube for 2 minutes at room temperature and remove the ethanol by aspiration. Allow the pellet of nucleic acid to dry for 5–10 minutes in air. Dissolve the damp pellet in 50 μl of TE (pH 8.0).

9. Digest the BAC DNA with restriction endonucleases.

   If the DNA preparations are resistant to cleavage, further purification by column chromatography on Qiagen resin usually eliminates the problem.

10. Analyze the digested BAC DNA by pulsed-field gel electrophoresis, using DNA markers of an appropriate size.

Isolation of BAC DNA from Large-scale Cultures

The procedure for isolation of BAC DNA is scaled up to accommodate 500-ml cultures, which, on average, yield 20–25 µg of purified BAC DNA.

MATERIALS

CAUTION: Please see Appendix 4 for appropriate handling of materials marked with <!>.

Reagents and Solutions

Please see Appendix 1 for components of stock solutions, buffers, and reagents. Dilute stock solutions to the appropriate concentrations.

Alkaline lysis solution I, ice cold
Alkaline lysis solution II, freshly prepared (room temperature)
Alkaline lysis solution III, ice cold
Ethanol
Ethanol (70%)
Isopropanol
Phenol:chloroform (1:1, v/v) <!>
Sodium acetate (3 M, pH 5.2)
STE, ice cold
TE (pH 8.0)

Vectors and Hosts

E. coli strain carrying a recombinant BAC
See Protocol 4.7.

Media and Antibiotics

LB containing 12.5 µg/ml chloramphenicol <!>

Enzymes and Buffers

Lysozyme
Pancreative RNase (RNase A), free of DNase
Restriction enzymes and 10x buffers

Nucleic Acids/Oligonucleotides

DNA markers for pulsed-field gel electrophoresis (from New England BioLabs, Bio-Rad, etc.)

Gels/Loading Buffers

Pulsed-field gels (or 0.5% agarose gels)
See information on pulsed-field gel electrophoresis in MC3, p. 4.18 and Chapter 5.

Centrifuges/Rotors/Tubes

Sorvall SLC-1500 rotor and Oak Ridge tubes (4°C) (NALGENE)

Additional Items

Bacterial shaker (37°C)
Chromatography resin, e.g., QIAprep Spin Miniprep Kit, Qiagen Inc. (optional; see Step 13)
Microfuge (4°C)
Sterile 17 x 100-mm tubes

Additional Information

Chromatography resins	MC3, pp. 1.62–1.64
Lysozyme	MC3, p. 1.153
Pulsed-field gel electrophoresis	MC3, pp. 5.55–5.60 and Protocols 5.14–5.20

METHOD

1. Inoculate 500 ml of LB medium containing 12.5 µg/ml of chloramphenicol with 50 µl of a saturated overnight culture of BAC-transformed cells. Incubate the 500-ml culture for 12–16 hours at 37°C with vigorous agitation (300 cycles/minute) until the cells reach saturation.

2. Harvest the cells from the culture by centrifugation at 2500g (3900 rpm in a Sorvall SLC-1500 rotor) for 15 minutes at 4°C. Pour off the supernatant and invert the open centrifuge bottle to allow the last drops of the supernatant to drain away.

3. Resuspend the bacterial pellet in 100 ml of ice-cold STE. Collect the bacterial cells by centrifugation as described in Step 2.

4. Resuspend the bacterial pellet in 24 ml of alkaline lysis solution I containing DNase-free RNase (100 µg/ml). Add lysozyme to a final concentration of 1 mg/ml.

   Make sure that the cells are completely resuspended and that the suspension is free of clumps.

5. Add 24 ml of freshly prepared alkaline lysis solution II. Close the top of the centrifuge bottle and mix the contents thoroughly by gently inverting the bottle several times. Incubate the bottle for 5 minutes on ice.

6. Add 24 ml of ice-cold alkaline lysis solution III. Close the top of the centrifuge bottle and mix the contents gently but thoroughly by swirling the bottle until there are no longer two distinguishable liquid phases. Place the bottle on ice for 5 minutes.

7. Centrifuge the bacterial lysate at 15,000g (9600 rpm in a Sorvall SLC-1500 rotor) for 10 minutes at 4°C. At the end of the centrifugation step, decant the clear supernatant into a polypropylene centrifuge bottle. Discard the pellet remaining in the centrifuge bottle.

8. Add an equal volume of phenol:chloroform. Mix the aqueous and organic phases by gently inverting the tube several times. Separate the phases by centrifugation at 3000g (4300 rpm in a Sorvall SLC-1500 rotor) for 15 minutes at room temperature.

9. Use a wide-bore pipette to transfer the aqueous layer to a fresh centrifuge bottle and add an equal volume of isopropanol. Invert the bottle several times to mix well.

10. Mark the tube on one side and place it in a centrifuge rotor with the marked side facing away from the center of the rotor. Marking in this way will aid in the subsequent identification of the nucleic acid pellet. Recover the precipitated nucleic acids by centrifugation at 15,000g (9600 rpm in a Sorvall SLC-1500 rotor) for 15 minutes at *room temperature*.

11. Decant the supernatant carefully and invert the bottle on a paper towel to allow the last drops of supernatant to drain away. Rinse the pellet and the walls of the bottle with 20 ml of 70%

ethanol at room temperature. Drain off the ethanol and place the inverted tube on a pad of paper towels for a few minutes at room temperature to allow the ethanol to evaporate.

12. Gently dissolve the pellet of BAC DNA in 0.2 ml of TE (pH 8.0). Assist in the dissolution of the DNA by tapping the sides of the bottle rather than vortexing. Measure the concentration of DNA by absorption spectroscopy (please see MC3, Appendix 8).

13. Digest the BAC DNA with restriction endonucleases.

    If the DNA preparation is resistant to cleavage, further purification by column chromatography on Qiagen resin usually eliminates the problem.

14. Analyze the digested BAC DNA by pulsed-field gel electrophoresis, using DNA markers of an appropriate size.

Protocol 4.10 Working with Yeast Artificial Chromosomes

This protocol outlines a procedure to validate and store yeast strains carrying yeast artificial chromosomes (YACs).

MATERIALS

Reagents and Solutions

Please see Appendix 1 for components of stock solutions, buffers, and reagents. Dilute stock solutions to the appropriate concentrations.
   Glycerol (30% v/v) in YPD medium

Vectors and Hosts

   S. cerevisiae carrying recombinant YAC clone
      Often available commercially; see MC3, p. 4.86.

Media and Antibiotics

   Selective medium
      See Yeast Media panel in MC3, p. 4.65.
   YPD agar plates
   YPD medium

Gels/Loading Buffers

   Pulsed-field gels
      See information on pulsed-field gel electrophoresis in MC3, p. 4.18 and Chapter 5.

Additional Items

   Bacterial shaker (30°C)
   Culture tubes or Erlenmeyer flasks (25 ml)
   Screw-capped freezing vials (2 ml), e.g., NUNC CryoTube Vials

Additional Information

   Step 3 of this protocol requires the reagents listed in Protocol 6.7.
   Step 4 of this protocol requires the reagents listed in either Protocol 5.17 or 5.18.
   Step 5 of this protocol requires the reagents listed in either Protocol 6.10 or 8.12.

METHOD

1. Immediately on the arrival of clones in the laboratory, streak the cultures onto selective media and incubate for 48 hours at 30°C to obtain single colonies.

   Yeast colonies may be analyzed directly by PCR (please see Protocol 4.13).

2. Transfer each of 6–12 individual colonies into 10 ml of YPD medium. Incubate the cultures with vigorous agitation (300 cycles/minute) at 30°C overnight. The cells should reach saturation (OD_{600} = 2.0–3.0, ~3 x 10^7 cells/ml) during this time.

3. Extract yeast DNA from 9 ml of each of the cultures following the steps described in Protocol 6.7.
   Store the unused portions of the cultures at 4°C.

4. Analyze the size of the YAC in each of the DNA preparations by pulsed-field gel electrophoresis.

5. Confirm by either Southern hybridization or PCR that the target sequence is present in the YAC DNA.

6. If the results are satisfactory, i.e., if the cultures contain YACs of the same size, if there is no sign of instability or rearrangement, and if the target sequences are present, choose one or two of the cultures for long-term storage.

7. Prepare 2-ml vials containing 0.5 ml of sterile (30% v/v) glycerol in YPD medium.

8. Add 0.5 ml of the yeast culture and mix the contents of the tube by gentle vortexing.

9. Transfer vials to –70°C.
   Yeast can be recovered from storage by transferring a small frozen sample to a YPD agar plate.

Protocol 4.11

Growth of *S. cerevisiae* and Preparation of DNA

This protocol describes methods for isolation of DNA from a strain of *S. cerevisiae* carrying a recombinant YAC. Because the linear YAC DNAs are sensitive to shearing forces, pipettes with wide-bore tips should be used to transfer DNAs. The method is suitable for preparing DNA that will be used for agarose gel electrophoresis, Southern blotting, subcloning, genomic library construction, PCR, or other methods that do not require intact high-molecular-weight DNA. The expected yield from a 10-ml culture is 2–4 μg of yeast DNA.

MATERIALS

CAUTION: Please see Appendix 4 for appropriate handling of materials marked with <!>.

Reagents and Solutions

Please see Appendix 1 for components of stock solutions, buffers, and reagents. Dilute stock solutions to the appropriate concentrations.

Ammonium acetate (10 M)
Ethanol
Ethanol (70%)
Phenol:chloroform (1:1, v/v) <!>
TE (pH 8.0)
TE (pH 8.0) containing 20 μg/ml RNase A
Triton/SDS solution
    10 mM Tris (pH 8.0)
    2% (v/v) Triton X-100
    1% (w/v) SDS
    100 mM NaCl
    1 mM EDTA
    Sterilize by passage through a 0.22-μm filter.

Vectors and Hosts

S. cerevisiae carrying recombinant YAC clone
    Often available commercially; see MC3, p. 4.86.

Media and Antibiotics

YPD medium

Gels/Loading Buffers

Pulsed-field gels
    See information on pulsed-field gel electrophoresis in MC3, p. 4.18 and Chapter 5.

Centrifuges/Rotors/Tubes

Sorvall SS-34 rotor (4°C)

Additional Items

Bacterial shaker (30°C)
Culture tubes or Erlenmeyer flasks (25 ml)
Glass beads, acid washed (Sigma-Aldrich)

 Adapted from Chapter 4, Protocol 11, p. 4.67 of MC3.

METHOD

1. Inoculate a yeast colony containing the YAC clone of interest into 10 ml of YPD medium and incubate overnight with shaking at 30°C.

    The cells should reach saturation (OD_{600} = 2.0–3.0, ~3 x 10^7 cells/ml) during this time.

    If the DNA to be extracted will be used in pulsed-field gel electrophoresis, follow the steps in Protocol 5.14.

2. Collect the cells by centrifugation at 2000g (4100 rpm in a Sorvall SS-34 rotor) for 5 minutes at 4°C.

3. Remove the medium, replace with 1 ml of sterile H_2O, and resuspend the cells by gentle vortexing.

4. Collect the cells by centrifugation as in Step 2.

5. Remove the wash, resuspend cells in 0.5 ml of sterile H_2O, and transfer to a sterile 1.5-ml microfuge tube.

6. Collect the cells by centrifugation at maximum speed for 5 seconds at room temperature in a microfuge and remove the supernatant.

7. Add 0.2 ml of Triton/SDS solution to the cells and resuspend the cell pellet by tapping the side of the tube.

8. Add 0.2 ml of phenol:chloroform and 0.3 g of glass beads to the cells, and vortex the cell suspension for 2 minutes at room temperature. Add 0.2 ml of TE (pH 8.0) and mix the solution by vortexing briefly.

9. Separate the organic and aqueous phases by centrifugation at maximum speed for 5 minutes at room temperature in a microfuge. Transfer the aqueous upper layer to a fresh microfuge tube, taking care to avoid carrying over any of the material at the interface.

10. Add 1 ml of ethanol to the aqueous solution, cap the centrifuge tube, and gently mix the contents by inversion.

11. Collect the precipitated DNA by centrifugation at maximum speed for 2–5 minutes at 4°C in a microfuge. Remove the supernatant with a drawn-out pasteur pipette. Centrifuge the tube briefly (2 seconds) and remove the last traces of ethanol from the bottom of the tube.

12. Resuspend the nucleic acid pellet in 0.4 ml of TE (pH 8.0) with RNase A and incubate the solution for 5 minutes at 37°C.

13. Add to the solution an equal volume of phenol:chlorofom and extract the RNase-digested solution, mixing by inversion rather than vortexing. Separate the aqueous and organic phases by centrifugation at maximum speed for 5 minutes at room temperature in a microfuge and transfer the aqueous layer to a fresh microfuge tube.

14. Add 80 µl of 10 M ammonium acetate and 1 ml of ethanol to the aqueous layer. Mix the solution by gentle inversion and store the tube for 5 minutes at room temperature.

15. Collect the precipitated DNA by centrifugation for 5 minutes in a microfuge. Decant the supernatant and rinse the nucleic acid pellet with 0.5 ml of 70% ethanol. Centrifuge at maximum speed for 2 minutes and remove the ethanol rinse with a drawn-out pasteur pipette. Centrifuge the tube briefly (2 seconds) and remove the last traces of ethanol from the bottom of the tube. Allow the pellet of DNA to air dry for 5 minutes and then dissolve the pellet in 50 µl of TE (pH 8.0).

    The preparation should contain 2–4 µg of yeast DNA.

Protocol 4.12 Small-scale Preparations of Yeast DNA

Yeast DNA is prepared by digestion of the cell wall and lysis of the resulting spheroplasts with SDS. This method reproducibly yields several micrograms of yeast DNA that can be efficiently cleaved by restriction enzymes and used as a template in PCR.

MATERIALS

Reagents and Solutions

Please see Appendix 1 for components of stock solutions, buffers, and reagents. Dilute stock solutions to the appropriate concentrations.

Isopropanol
Potassium acetate (5 M)
SDS (10% w/v)
Sodium acetate (3 M, pH 7.0)
Sorbitol buffer
    1 M sorbitol
    0.1 M EDTA (pH 7.5)
TE (pH 7.4)
TE (pH 8.0) containing 20 µg/ml RNase A
Yeast resuspension buffer
    50 mM Tris (pH 7.4)
    20 mM EDTA (pH 7.5)

Vectors and Hosts

S. cerevisiae culture

Media and Antibiotics

YPD medium

Enzymes and Buffers

Zymolyase 100T

Gels/Loading Buffers

Pulsed-field gels
    See information on pulsed-field gel electrophoresis in MC3, p. 4.18 and Chapter 5.

Centrifuges/Rotors/Tubes

Sorvall SS-34 rotor (4°C)

Additional Items

Bacterial shaker (30°C)
17 x 100-mm Centrifuge tubes
Culture tubes or Erlenmeyer flasks (25 ml)
Water baths (37°C and 65°C)
Wide-bore yellow tips (use dog toenail clippers)

Adapted from Chapter 4, Protocol 12, p. 4.70 of MC3.

> **Additional Information**
>
> Alternative method for preparing yeast DNA   Protocol 6.7

METHOD

1. Set up 10-ml cultures of yeast in YPD medium. Incubate the cultures overnight at 30°C with moderate agitation.

2. Transfer 5 ml of the cells to a centrifuge tube. Collect the cells by centrifugation at 2000*g* (4100 rpm in a Sorvall SS-34 rotor) for 5 minutes at 4°C. Store the unused portion of the culture at 4°C.

3. Resuspend the cells in 0.5 ml of sorbitol buffer. Transfer the suspension to a microfuge tube.

4. Add 20 μl of a solution of Zymolyase 100T (2.5 mg/ml in sorbitol buffer) and incubate the cell suspension for 1 hour at 37°C.

5. Collect the cells by centrifugation in a microfuge for 1 minute. Remove the supernatant by aspiration.

6. Resuspend the cells in 0.5 ml of yeast resuspension buffer.

7. Add 50 μl of 10% SDS. Close the top of the tube and mix the contents by rapidly inverting the tube several times. Incubate the tube for 30 minutes at 65°C.

8. Add 0.2 ml of 5 M potassium acetate and store the tube for 1 hour on ice.

9. Pellet the cell debris by centrifugation at maximum speed for 5 minutes at 4°C in a microfuge.

10. Use a wide-bore pipette tip to transfer the supernatant to a fresh microfuge tube at room temperature.

11. Precipitate the nucleic acids by adding an equal volume of room-temperature isopropanol. Mix the contents of the tube and store it for 5 minutes at room temperature.

    Do not allow the precipitation reaction to proceed for >5 minutes.

12. Recover the precipitated nucleic acids by centrifugation at maximum speed for 10 *seconds* in a microfuge. Remove the supernatant by aspiration and allow the pellet to air dry for 10 minutes.

13. Dissolve the pellet in 300 μl of TE (pH 8.0) containing 20 μg/ml pancreatic RNase A. Incubate the digestion mixture for 30 minutes at 37°C.

14. Add 30 μl of 3 M sodium acetate (pH 7.0). Mix the solution and then add 0.2 ml of isopropanol. Mix once again and recover the precipitated DNA by centrifugation at maximum speed for 20 seconds in a microfuge.

15. Remove the supernatant by aspiration and allow the pellet to air dry for 10 minutes. Dissolve the DNA in 150 μl of TE (pH 7.4).

Analyzing Yeast Colonies by PCR

Yeast colonies are suspended in complete PCR buffer and transferred to a thermal cycler for 35 cycles of PCR. The products of the amplification reaction are analyzed by gel electrophoresis.

MATERIALS

CAUTION: Please see Appendix 4 for appropriate handling of materials marked with <!>.

Reagents and Solutions

Please see Appendix 1 for components of stock solutions, buffers, and reagents. Dilute stock solutions to the appropriate concentrations.

10x Colony PCR buffer
0.125 M Tris-Cl (pH 8.5)
0.56 M KCl

dNTP solution (containing all four dNTPs, each at a concentration of 10 mM) (pH 8.0, PCR grade)

$MgCl_2$ (25 mM)

Vectors and Hosts

Colonies of *S. cerevisiae* carrying recombinant YAC of interest

Media and Antibiotics

YPD medium

Enzymes and Buffers

Taq DNA polymerase

Nucleic Acids/Oligonucleotides

DNAs to be used as markers during gel electrophoresis
Oligonucleotide primers

The primers should be 20–24 nucleotides in length, specific for the target DNA sequences, free of potential secondary structures, and should contain no less than 10 and no more than 15 G and C residues. For advice on the design of oligonucleotide primers, see MC3, pp. 8.13–8.16.

Gels/Loading Buffers

Agarose or polyacrylamide gel <!>

Centrifuges/Rotors/Tubes

Microfuge and 0.5-ml thin-walled microfuge tubes
Sorvall SS-34 rotor (4°C)

Additional Items

Thermal cycler, equipped with a heated lid

Additional Information

Calculating melting temperatures of oligonucleotides	MC3, pp. 10.47–10.48
Taq DNA polymerase	MC3, pp. 8.108–8.109

METHOD

1. In a sterile 0.5-ml microfuge tube, mix in the following order:

10x colony PCR buffer	2 µl
25 mM MgCl$_2$	1.2 µl
10 mM dNTPs	0.4 µl
oligonucleotide primers	10 pmoles of each primer
Taq polymerase	5 units (0.2 µl)
H$_2$O	to 20 µl

2. Use a disposable yellow pipette tip to transfer a small amount of a yeast colony (0.10–0.25 µl) to the reaction mixture.

   It is important not to be too greedy when sampling the yeast colony because cell wall components inhibit the PCR.

3. Transfer the PCR tube to the thermocycler, programmed as follows, and start the program.

Cycle Number	Denaturation	Annealing	Polymerization
1	4 min at 95°C		
2–35	1 min at 95°C	1 min at 55°C	1 min at 72°C
Last			10 min at 72°C

   These times are suitable for 25–50-µl reactions assembled in thin-walled 0.5-ml tubes and incubated in thermal cyclers such as the Perkin-Elmer 9600 or 9700, Mastercycler (Eppendorf), and PTC-100 (MJ Research). Times and temperatures may need to be adapted to suit other types of equipment and reaction volumes.

4. Analyze the products of the PCR by electrophoresis through an agarose or polyacrylamide gel, using markers of suitable size.

   If amplification of the target sequence is weak or erratic, repeat the reactions using a polymerization temperature 2–3°C below the calculated melting temperature of the oligonucleotide primer that is richer in A+T. If the results are still unsatisfactory, convert the yeast cells to spheroplasts by removing the cell walls with Zymolyase 100T before beginning the protocol. This takes only 1 hour and almost always clears up any problems. Alternatively, grow 10-ml liquid (YPD) cultures of the colonies under test and make small-scale preparations of yeast DNA (please see Protocol 4.12).

Isolating the Ends of Genomic DNA Fragments Cloned in High-capacity Vectors: Vectorette Polymerase Chain Reactions

This protocol describes the use of vectorette PCR and single-site PCR to amplify the terminal sequences of genomic sequences cloned in high-capacity vectors such as PACs and YACs.

MATERIALS

CAUTION: Please see Appendix 4 for appropriate handling of materials marked with <!>.

Reagents and Solutions

Please see Appendix 1 for components of stock solutions, buffers, and reagents. Dilute stock solutions to the appropriate concentrations.

ATP (10 mM)
> Not required if 10x ligation buffer contains ATP

dNTP solution (containing all four dNTPs, each at a concentration of 1 mM) (pH 8.0, PCR grade)

Ethanol

Ethanol (70%)

Phenol:chloroform (1:1, v/v) <!>

Sodium acetate (3 M, pH 5.2)

Enzymes and Buffers

Bacteriophage T4 DNA ligase and 10x ligation buffer

Restriction enzymes *PstI* or *NsiI* and 10x buffers

Taq DNA polymerase and 10x amplification buffer
> 500 mM KCl
> 100 mM Tris (pH 8.3, room temperature)
> 15 mM $MgCl_2$

Nucleic Acids/Oligonucleotides

Oligonucleotide cassette (5.0 OD_{260}/ml [~8.5 µM]) in TE (pH 7.6)
> 5'CATGCTCGGTCGGGATAGGCACTGGTCTAGAGGGTTAGGTTCCTGCTACATCTCCAGC CTTGCA3'
>
> This 64-nucleotide single-stranded cassette is designed for ligation to target DNAs carrying termini generated by *PstI* or *NsiI*. The four 3'-terminal nucleotides (underlined) are complementary to the protruding termini of fragments of DNA generated by cleavage with these enzymes. If another restriction enzyme is used, the nucleotides at the 3' end of the linker must be changed so as to complement the protruding terminus generated by the enzyme.
>
> Before use, the oligonucleotide should be purified by C_{18} chromatography or electrophoresis through a 12% polyacrylamide gel. The 5' terminus of the oligonucleotide should not be phosphorylated.

Oligonucleotide (linker) primer (5.0 OD_{260}/ml [~17 µM]) in TE (pH 7.6)
> 5'CATGCTCGGTCGGGATAGGCACTGGTCTAGAG3'
>
> This oligonucleotide is identical in sequence to the 32 nucleotides at the 5' end of the oligonucleotide cassette and is used as an amplimer in the PCR.
>
> There is no need to purify or phosphorylate the deprotected oligonucleotide before use. Dissolve the oligonucleotide in TE (pH 7.6) at a concentration of 5.0 OD_{260}/ml solution (~17 µM).

Sequence-specific oligonucleotide (vector) primer (5.0 OD_{260}/ml [~17 µM]) in TE (pH 7.6)
> This primer is complementary to the vector when terminal sequences of a cloned segment of DNA are to be amplified or to cloned DNA sequences when a neighboring segment of genomic DNA is to be recovered. The primer should be 28–32 nucleotides in length and its

Adapted from Chapter 4, Protocol 14, p. 4.74 of MC3.

predicted melting temperature should be approximately equal to that of the 32-nucleotide oligonucleotide primer.

There is no need to purify or phosphorylate the deprotected oligonucleotide before use. Dissolve the oligonucleotide in TE (pH 7.6) at a concentration of 5.0 OD_{260}/ml solution (~17 µM).

Template DNAs

Template may be recombinant BAC, YAC, or cosmid DNA.

YACs can either be embedded in an agarose plug (2 µg in 100-µl plug) or in solution. Unless the yeast strain is carrying more than one YAC, there is no need to purify YAC DNA by PFGE before use in vectorette or single-site PCR.

DNAs should be purified by column chromatography using, e.g., Qiagen resin or GENECLEAN II (please see Protocol 1.9) and resuspended at a concentration of 1 µg/µl in TE (pH 7.6).

Gels/Loading Buffers

1.0% Agarose gels, cast and run in 0.5x TBE containing 0.5 mg/ml ethidium bromide <!>

Additional Items

Barrier tips for automatic pipettes
Heating-cooling block (15°C)
Microfuge tubes (0.5 ml, thin walled)
Thermal cycler, preferably with heated lid

Additional Information

Calculating melting temperatures of oligonucleotides	MC3, pp. 10.47–10.48
Enhancements of vectorette reactions when using genomic DNA templates	MC3, p. 4.81
Taq DNA polymerasae	MC3, pp. 8.108–8.109

METHOD

1. Digest approximately 5 µg of template DNA with *PstI* or *NsiI*. Check a small aliquot of the reaction by agarose gel electrophoresis to ensure that all of the DNA has been cleaved.

2. Extract the reaction mixture with phenol:chloroform and recover the DNA by standard precipitation with ethanol and subsequent washing in 70% ethanol. Store the open tube in an inverted position on a bed of paper towels to allow the last traces of ethanol to evaporate and then dissolve the damp pellet of DNA in 50 µl of H_2O.

3. In a sterile 0.5-ml microfuge tube, mix in the following order:

10x T4 DNA ligase buffer	10 µl
digested template DNA	20 µl
oligonucleotide cassette, 5.0 OD_{260}/ml	2 µl
T4 bacteriophage DNA ligase, 5 Weiss units/µl	2 µl
H_2O	to 100 µl

   Set up three control reactions as described above but without template DNA in one tube, without linker oligonucleotide in another, and without T4 DNA ligase in the third.

4. Incubate the test ligation reaction and controls for 12–16 hours at 15°C.

5. In a sterile 0.5-ml microfuge tube, mix in the following order:

10x amplification buffer	2 µl
1 mM solution of four dNTPs (pH 7.0)	2 µl
linker primer oligonucleotide, 5.0 OD_{260}/ml	1 µl
vector primer oligonucleotide, 5.0 OD_{260}/ml	1 µl
test DNA ligation reaction, from Step 4	1 µl
thermostable DNA polymerase, 5.0 units/µl	0.5 µl
H_2O	to 20 µl

Set up three control PCRs that contain 1 μl of the control ligation reactions instead of the test ligation reaction.

6. If the thermocycler is not fitted with a heated lid, overlay the reaction mixtures with 1 drop (~50 μl) of light mineral oil to prevent evaporation of the samples during repeated cycles of heating and cooling. Alternatively, place a bead of wax into the tube if using a hot start approach.

7. Place the PCR tubes in the thermocycler, programmed as follows, and start the amplification program.

Cycle Number	Denaturation	Annealing	Polymerization
1	2 min at 95°C		
2–35	30 sec at 94°C	30 sec at 60°C	3 min at 72°C
Last			5 min at 72°C

8. Analyze aliquots (25%) of each amplification reaction on an agarose gel.

A prominent DNA product visible by ethidium bromide staining should be present in the PCR containing the products of the test ligation. This DNA should be absent from the control reactions. The size of the product depends on the distance between the vector primer and the first cleavage site in the cloned insert. For *Pst*I and *Nsi*I, the amplified product is typically between 0.5 and 2 kb.

The amplified DNA can be sequenced directly, cloned, radiolabeled by random hexamer priming or PCR, and even used as a transcription template if a bacteriophage promoter is added to the linker or is present in the amplified segment of vector DNA.

CHAPTER 5

Gel Electrophoresis of DNA and Pulsed-field Agarose Gel Electrophoresis

BACKGROUND INFORMATION

Background information found in *Molecular Cloning: A Laboratory Manual,* 3rd edition (hereafter MC3) unless otherwise indicated.

Agarose gels	
Factors affecting rate of migration	MC3, pp. 5.4–5.9
Classes of agarose	MC3, pp. 5.6–5.7
Electrophoresis buffers	MC3, p. 5.8
Gel-loading buffers	MC3, p. 5.9
Alkaline agarose gels	MC3, p. 5.36
Autoradiography	MC3, pp. A9.9–A9.15
Polyacrylamide gels	
Polyacrylamide	MC3, pp. 5.40–5.42
Pulsed-field electrophoresis	MC3, pp. 5.55–5.60

Agarose Gel Electrophoresis

How to pour, load, and run an agarose gel.

MATERIALS

CAUTION: Please see Appendix 4 for appropriate handling of materials marked with <!>.

Reagents and Solutions

Please see Appendix 1 for components of stock solutions, buffers, and reagents. Dilute stock solutions to the appropriate concentrations.

Ethidium bromide (10 mg/ml) <!>
or
SYBR Gold solution (supplied commercially as a 10,000x concentrate in DMSO)

Nucleic Acids/Oligonucleotides

DNA samples
In TE or other buffer of low ionic strength and neutral pH
DNA size standards
Samples of DNAs of known size are typically generated by restriction enzyme digestion of a plasmid or bacteriophage DNA of known sequence. Alternatively, they may be produced by ligating a monomeric DNA of known size into a ladder of polymeric forms. Both types of size standards are available commercially.

Gels/Loading Buffers

Agarose solutions
Electrophoresis buffer
Almost always 0.5x TBE or 1x TAE
Gel-loading buffer (6x)
Almost always gel-loading buffer IV
0.25% bromophenol blue
40% (w/v) sucrose in H_2O

Additional Items

Equipment for agarose gel electrophoresis
Typically includes a horizontal electrophoresis tank with combs, glass plates, gel sealing tape (e.g., Time tape), and a power supply capable of delivering up to 500 V and 200 mA
Insulated gloves or tongs
Microwave oven or boiling-water bath
Transilluminator or other source of UV light
Water bath (50°C)

Additional Information

Agarose minigels	MC3, p. 5.13
SYBR Gold	MC3, pp. A9.7–A9.8, 5.15–5.16, and p. 191 of this volume

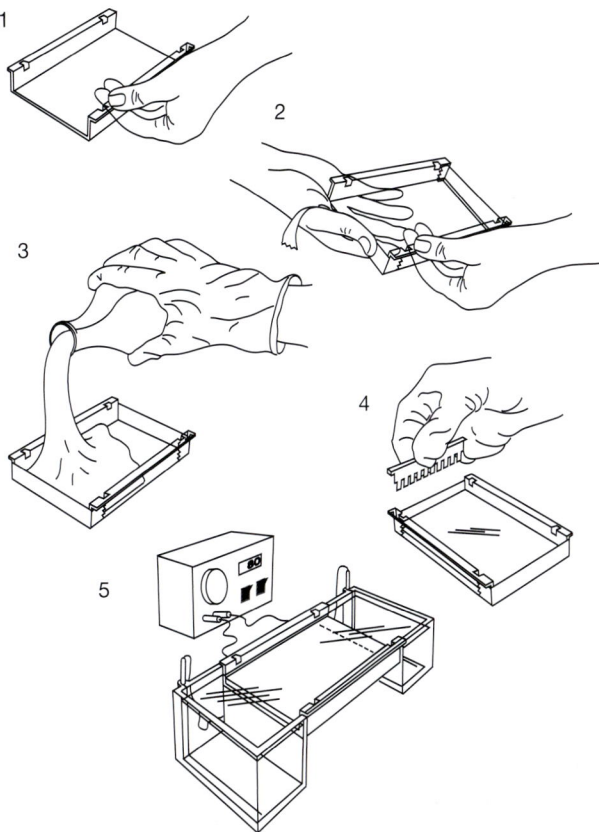

FIGURE 5-1 Pouring a Horizontal Agarose Gel

METHOD

1. Seal the edges of a clean, dry glass plate (or the open ends of the plastic tray supplied with the electrophoresis apparatus) with tape to form a mold (see Figure 5-1). Set the mold on a truly horizontal section of the bench.

2. Prepare sufficient electrophoresis buffer to fill the electrophoresis tank and to cast the gel.

   It is important to use the same batch of electrophoresis buffer in both compartments of the electrophoresis tank and the gel.

3. Prepare a solution of agarose in electrophoresis buffer at a concentration appropriate for separating the particular size fragments expected in the DNA sample(s): Add the correct amount of powdered agarose (please see Table 5-1 on the next page) to a measured quantity of electrophoresis buffer in an Erlenmeyer flask or a glass bottle.

   The buffer should occupy less than 50% of the volume of the flask or bottle.

4. Loosely plug the neck of the Erlenmeyer flask with Kimwipes. If using a glass bottle, make certain the cap is loose. Heat the slurry in a boiling-water bath or a microwave oven until the agarose dissolves.

   WARNING: The agarose solution can easily become superheated it if is heated too long or too quickly in the microwave oven. Heat the slurry for the minimum time required to allow all of the grains of agarose to dissolve.

5. Use insulated gloves or tongs to transfer the flask/bottle into a water bath at 50°C. When the molten gel has cooled, add ethidium bromide to a final concentration of 0.5 µg/ml. Mix the gel solution thoroughly by gentle swirling.

   IMPORTANT: SYBR Gold should not be added to the molten gel solution.

TABLE 5-1 Range of Separation in Gels Containing Different
Amounts of Standard Low-EEO Agarose

Agarose Concentration in Gel (% [w/v])	Range of Separation of Linear DNA Molecules (kb)
0.6	1–20
0.7	0.8–10
0.9	0.5–7
1.2	0.4–6
1.5	0.2–3
2.0	0.1–2

6. While the agarose solution is cooling, choose an appropriate comb for forming the sample slots in the gel. Position the comb 0.5–1.0 mm above the plate so that a complete well is formed when the agarose is added to the mold.

7. Pour the warm agarose solution into the mold.

> The gel should be between 3 and 5 mm thick. Check that no air bubbles are under or between the teeth of the comb. Air bubbles present in the molten gel can be removed easily by poking them with the corner of a Kimwipe.

8. Allow the gel to set completely (30–45 minutes at room temperature), pour a small amount of electrophoresis buffer on the top of the gel and carefully remove the comb. Pour off the electrophoresis buffer from the gel and carefully remove the tape. Mount the gel in the electrophoresis tank.

9. Add just enough electrophoresis buffer to cover the gel to a depth of approximately 1 mm.

10. Mix the samples of DNA with 0.20 volume of the desired 6x gel-loading buffer.

> The minimum amount of DNA that can be detected by photography of ethidium-bromide-stained gels is approximately 2 ng in a 0.5-cm-wide band (the usual width of a slot). More sensitive dyes such as SYBR Gold can detect as little as 20 pg of DNA in a band. Standard-size gel slots containing more than 1–2 µg of DNA may be overloaded, with consequent loss of resolution and blurry bands of DNA.

11. Slowly load the sample mixture into the slots of the submerged gel using a disposable micropipette, an automatic micropipettor, or a drawn-out pasteur pipette or glass capillary tube. Load size standards into slots on both the right and left sides of the gel.

> If possible, use a new pipette tip or a glass capillary tube for each DNA sample.

12. Close the lid of the gel tank and attach the electrical leads so that the DNA will migrate toward the positive anode (red lead). Apply a voltage of 1–5 V/cm (measured as the distance between the positive and negative electrodes). If the leads have been attached correctly, bubbles should be generated at the anode and cathode (due to electrolysis), and within a few minutes, the bromophenol blue should migrate from the wells into the body of the gel. Run the gel until the bromophenol blue and xylene cyanol FF have migrated an appropriate distance through the gel.

> The presence of ethidium bromide allows the gel to be examined by UV illumination at any stage during electrophoresis. The gel tray may be removed and placed directly on a transilluminator. Alternatively, the gel may be examined using a hand-held source of UV light. In either case, turn off the power supply before examining the gel!

13. When the DNA samples or dyes have migrated a sufficient distance through the gel, turn off the electric current and remove the leads and lid from the gel tank. If ethidium bromide is present in the gel and electrophoresis buffer, examine the gel by UV light and photograph the gel as described in Protocol 5.2. Otherwise, stain the gel by immersing it in electrophoresis buffer or H_2O containing ethidium bromide (0.5 µg/ml) for 30–45 minutes at room temperature or by soaking in a 1:10,000-fold dilution of SYBR Gold stock solution in electrophoresis buffer. Then photograph the gel as described in Protocol 5.2.

Nucleic acids that have been subjected to electrophoresis through agarose gels may be detected by staining and visualized by illumination with a 300-nm UV light. Methods for staining and visualization of DNA using either ethidium bromide or SYBR Gold are described here.

The most convenient and commonly used method to visualize DNA in agarose gels is staining with the fluorescent dye ethidium bromide. Ethidium bromide can be used to detect both single- and double-stranded nucleic acids (both DNA and RNA). However, the affinity of the dye for single-stranded nucleic acid is relatively low and the fluorescent yield is comparatively poor. In fact, most fluorescence associated with staining single-stranded DNA or RNA is attributable to binding of the dye to short intrastrand duplexes in the molecules.

Ethidium bromide is prepared as a stock solution of 10 mg/ml in H_2O, which is stored at room temperature in dark bottles or bottles wrapped in aluminum foil. The dye is usually incorporated into agarose gels and electrophoresis buffers at a concentration of 0.5 µg/ml. Note that polyacrylamide gels cannot be cast with ethidium bromide because the dye inhibits polymerization of the acrylamide. Acrylamide gels are therefore stained with the ethidium solution after the gel has been run (please see Protocol 5.10). Although the electrophoretic mobility of linear double-stranded DNA is reduced by approximately 15% in the presence of the dye, the ability to examine the agarose gels directly under UV illumination during or at the end of the run is a great advantage. However, sharper DNA bands are obtained when electrophoresis is performed in the absence of ethidium bromide. Staining is accomplished by immersing the gel in electrophoresis buffer or H_2O containing ethidium bromide (0.5 µg/ml) for 30–45 minutes at room temperature. Destaining is not usually required. However, detection of very small amounts (<10 ng) of DNA is made easier if the background fluorescence caused by unbound ethidium bromide is reduced by soaking the stained gel in H_2O or 1 mM $MgSO_4$ for 20 minutes at room temperature.

SYBR Gold is the trade name of an ultrasensitive dye with high affinity for DNA and a large fluorescence enhancement on binding to nucleic acid. The quantum yield of the SYBR Gold–DNA complex is greater than that of the equivalent ethidium bromide–DNA complex, and the fluorescence enhancement is >1000 times greater. As a result, <20 pg of double-stranded DNA can be detected in an agarose gel (up to 25 times less than the amount visible after ethidium bromide staining). In addition, staining of agarose or polyacrylamide gels with this dye can reveal as little as 100 pg of single-stranded DNA in a band or 300 pg of RNA. SYBR Gold shows maximum excitation at 495 nm and has a secondary excitation peak at 300 nm. Fluorescent emission occurs at 537 nm.

SYBR Gold is used to stain DNA by soaking the gel, after separation of the DNA fragments, in a 1:10,000-fold dilution of the stock dye solution. SYBR Gold should not be added to the molten agarose or to the gel before electrophoresis, because its presence in the hardened gel will cause severe distortions in the electrophoretic properties of the DNA and RNA. The greatest sensitivity is obtained when the gel is illuminated with UV light at 300 nm. Photography is carried out as described below with green or yellow filters. The dye is sensitive to fluorescent light, and working solutions containing SYBR Gold (1:10,000 dilution of the stock solution supplied by Molecular Probes) should be freshly made daily in electrophoresis buffer and stored at room temperature.

Photographs of ethidium-bromide-stained gels may be made using transmitted or incident UV light. Most commercially available devices (transilluminators) emit UV light at 302 nm (Figure 5-2). The fluorescent yield of ethidium bromide–DNA complexes is considerably greater at this wavelength than at 366 nm and slightly less than at short-wavelength (254 nm) light. However, the amount of nicking of the DNA is much less at 302 nm than at 254 nm. A further 10–20-fold

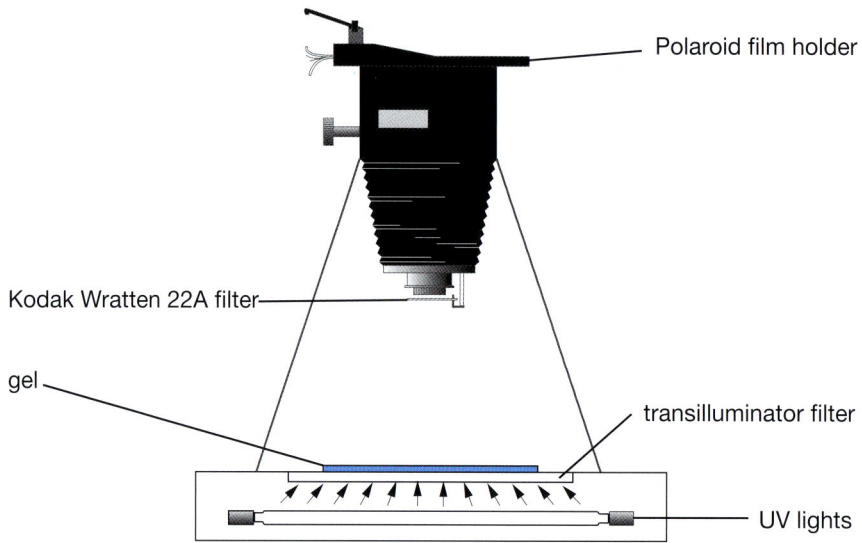

Polaroid film holder

Kodak Wratten 22A filter

gel

transilluminator filter

UV lights

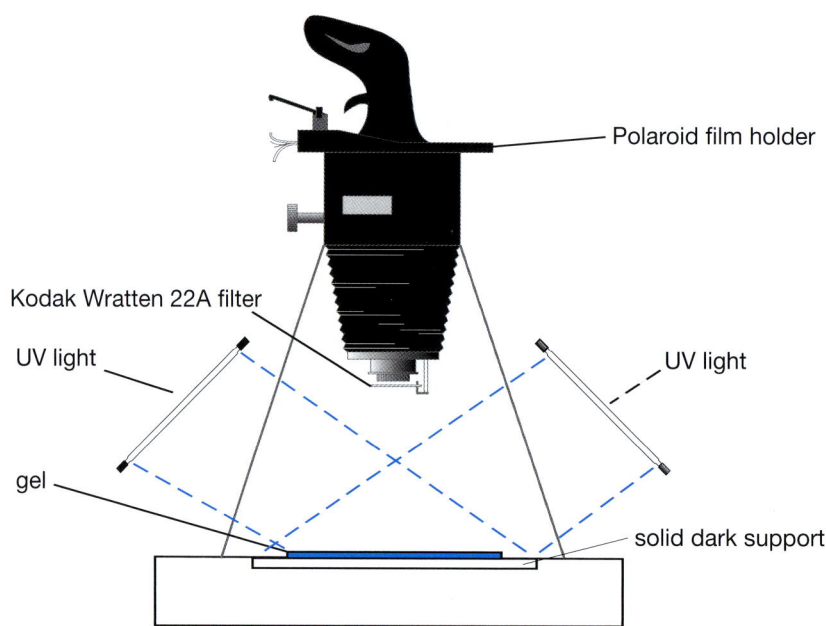

Polaroid film holder

Kodak Wratten 22A filter

UV light

UV light

gel

solid dark support

FIGURE 5-2 Photography of Gels by Ultraviolet Illumination

The top diagram shows the arrangement of the UV light source, the gel, and the camera that is used for photography by transmitted light. The bottom diagram shows the arrangement that is used for photography by incident light.

increase in the sensitivity of conventional photography can be obtained by staining DNA with SYBR Gold. Detection of DNAs stained with this dye requires the use of a yellow or green gelatin or cellophane filter (S-7569, available from Molecular Probes or Kodak) with the camera and illumination with 300-nm UV light.

Recovery of DNA from Agarose Gels: Electrophoresis onto DEAE-Cellulose Membranes

Recovery of bands of DNA from agarose gels by electrophoresis onto a sliver of DEAE-cellulose membrane can be performed simultaneously on many samples and reliably gives high yields of fragments between 500 bp and 5 kb in length.

MATERIALS

CAUTION: Please see Appendix 4 for appropriate handling of materials marked with <!>.

Reagents and Solutions

Please see Appendix 1 for components of stock solutions, buffers, and reagents. Dilute stock solutions to the appropriate concentrations.

Ammonium acetate (10 M)
DEAE high-salt elution buffer
    50 mM Tris-Cl (pH 8.0)
    1 M NaCl
    10 mM EDTA (pH 8.0)
DEAE low-salt wash buffer
    50 mM Tris-Cl (pH 8.0)
    0.15 M NaCl
    10 mM EDTA (pH 8.0)
EDTA (10 mM, pH 8.0)
Ethanol
Ethanol (70%)
NaOH (0.5 N) <!>
Phenol:chloroform (1:1, v/v) <!>
Sodium acetate (3 M, pH 5.2)
TE (pH 8.0)

Enzymes and Buffers

Restriction enzymes and 10x buffer(s)

Nucleic Acids/Oligonucleotides

DNA sample
DNA standards
Yeast tRNA (optional; may be used as carrier in Step 10)

Gels/Loading Buffers

Agarose gel, poured and run in 1x TAE containing ethidium bromide (0.5 µg/ml) <!>
    Recovery of DNA onto DEAE membranes is slightly more efficient when the gel is run in 1x TAE rather than 0.5x TBE.
Gel-loading buffer IV
1x TAE electrophoresis buffer

Additional Items

Blunt-ended forceps
DEAE-cellulose membranes
    May be obtained from Schleicher & Schuell or from Whatman
Handheld UV lamp, long wavelength (302 nm) <!>
Scalpel blade
Water bath or heating block (65°C)

Additional Information

Alternative protocol (Recovery of DNA from Agarose Gels Using Glass Beads)	MC3, p. 5.32
Other methods for DNA recovery (electroelution and anion-exchange chromatography)	Protocols 5.4 and 5.5, respectively

METHOD

1. Digest an amount of DNA that will yield at least 100 ng of the fragment(s) of interest. Separate the fragments by electrophoresis through an agarose gel of the appropriate concentration that contains 0.5 µg/ml ethidium bromide and locate the band of interest with a handheld, long-wavelength UV lamp.

2. Use a sharp scalpel to make an incision in the gel directly in front of the leading edge of the band of interest and approximately 2 mm wider than the band on each side.

   If DNA is to be eluted from an entire lane of an agarose gel (e.g., a restriction digest of mammalian genomic DNA), make the incision in the gel *parallel* to the lane of interest and place a single long piece of DEAE-cellulose membrane (prepared as in Step 3) into the incision. *Reorient* the gel so that the DNA can be transferred electrophoretically from the gel to the membrane. After electrophoresis, remove the membrane and cut it into segments. Elute DNA of the desired size from the appropriate segment(s) of the membrane as described in Steps 7–11.

3. Wearing gloves, cut a piece of DEAE-cellulose membrane that is the same width as the incision and slightly deeper (1 mm) than the gel. Soak the membrane in 10 mM EDTA (pH 8.0) for 5 minutes at room temperature. To activate the membrane, replace the EDTA with 0.5 N NaOH, and soak the membrane for an additional 5 minutes. Wash the membrane six times in sterile H_2O.

4. Use blunt-ended forceps or tweezers to hold apart the walls of the incision on the agarose gel and insert the membrane into the slit. Remove the forceps and close the incision, being careful not to trap air bubbles.

   Minimize the chance of contamination with unwanted species of DNA by either

   - cutting out a segment of gel containing the band of interest and transferring it to a hole of the appropriate size cut in another region of the gel far from any other species of DNA

   or

   - inserting a second piece of membrane above the band of interest to trap unwanted species of DNA.

5. Resume electrophoresis (5 V/cm) until the band of DNA has just migrated onto the membrane. Follow the progress of the electrophoresis with a handheld, long-wavelength (302 nm) UV lamp.

   Electrophoresis should be continued for the minimum time necessary to transfer the DNA from the gel to the membrane. Extended electrophoresis can result in cross-contamination with other DNA fragments (see above) or unnecessary accumulation of contaminants from the agarose.

6. When all of the DNA has left the gel and is trapped on the membrane, turn off the electric current. Use blunt-ended forceps to recover the membrane and rinse it in 5–10 ml of DEAE low-salt wash buffer at room temperature to remove any agarose pieces from the membrane.

    Do not allow the membrane to dry; otherwise, the DNA becomes irreversibly bound.

7. Transfer the membrane to a microfuge tube. Add enough DEAE high-salt elution buffer to cover the membrane completely. The membrane should be crushed or folded gently, but not tightly packed. Close the lid of the tube and incubate it for 30 minutes at 65°C.

8. While the DNA is eluting from the membrane, photograph the gel as described in Protocol 5.2 to establish a record of which bands were isolated.

9. Transfer the fluid from Step 7 to a fresh microfuge tube. Add a second aliquot of DEAE high-salt elution buffer to the membrane and incubate the tube for an additional 15 minutes at 65°C. Combine the two aliquots of DEAE high-salt elution buffer.

10. Extract the high-salt eluate once with phenol:chloroform. Transfer the aqueous phase to a fresh microfuge tube and add 0.2 volume of 10 M ammonium acetate and 2 volumes of ethanol at 4°C. Store the mixture for 10 minutes at room temperature and recover the DNA by centrifugation at maximum speed for 10 minutes at room temperature in a microfuge. Carefully rinse the pellet with 70% ethanol, store the open tube on the bench for a few minutes to allow the ethanol to evaporate, and then redissolve the DNA in 3–5 μl of TE (pH 8.0).

    The addition of 10 μg of carrier RNA before precipitation may improve the recovery of small amounts of DNA. However, before adding the RNA, make sure that the presence of RNA will not compromise any subsequent enzymatic reactions in which the DNA is used as a substrate or template.

11. If exceptionally pure DNA is required (e.g., for microinjection of fertilized mouse eggs or electroporation of cultured cells), reprecipitate the DNA with ethanol as follows.

    a. Suspend the DNA in 200 μl of TE (pH 8.0), add 25 μl of 3 M sodium acetate (pH 5.2), and precipitate the DNA once more with 2 volumes of ethanol at 4°C.

    b. Recover the DNA by centrifugation at maximum speed for 5–15 minutes at 4°C in a microfuge.

    c. Carefully rinse the pellet with 70% ethanol. Store the open tube on the bench for a few minutes to allow the ethanol to evaporate and then dissolve the DNA in 3–5 μl of TE (pH 8.0).

12. Check the amount and quality of the DNA by gel electrophoresis. Mix a small aliquot (~10–50 ng) of the final preparation of the fragment with 10 μl of TE (pH 8.0) and add 2 μl of the desired gel-loading buffer. Load and run an agarose gel of the appropriate concentration, using as markers restriction digests of known quantities of the original DNA and the appropriate DNA size standards. The isolated fragment should comigrate with the correct fragment in the restriction digest. Examine the gel carefully for the presence of faint fluorescent bands that signify the presence of contaminating species of DNA.

    It is often possible to estimate the amount of DNA in the final preparation from the relative intensities of fluorescence of the fragment and the markers.

Protocol 5.4 — Recovery of DNA from Agarose and Polyacrylamide Gels: Electroelution into Dialysis Bags

A messy but reliable technique that works well for DNAs ranging in size from 200 bp to >50 kb.

MATERIALS

CAUTION: Please see Appendix 4 for appropriate handling of materials marked with <!>.

Reagents and Solutions

Please see Appendix 1 for components of stock solutions, buffers, and reagents. Dilute stock solutions to the appropriate concentrations.

Ethanol
Ethanol (70%)
Ethidium bromide (10 mg/ml) <!>
or
SYBR Gold solution (supplied commercially as a 10,000x concentrate in DMSO)
Phenol:chloroform (1:1, v/v) <!>
Sodium acetate (3 M, pH 5.2)

Enzymes and Buffers

Restriction enzyme(s) and 10x buffer(s)

Nucleic Acids/Oligonucleotides

DNA samples
DNA size standards

Gels/Loading Buffers

Agarose gel or polyacrylamide gel suitable for separating the fragments of interest in the
    DNA sample (before preparing the gel, read the note to Step 1)
0.25x TBE electrophoresis buffer
0.25x TBE electrophoresis buffer containing 0.5 µg/ml ethidium bromide <!>

Additional Items

Blunt-ended forceps or thin spatula
Dialysis clips
Dialysis tubing
    Molecular biology grade is available from VWR.
Handheld UV lamp, long wavelength (302 nm) <!>
Scalpel blade
Step 3 of this protocol requires the reagents listed in Protocol 5.2.
Step 10 of this protocol requires the reagents listed in Protocol 5.5.

Additional Information

Alternative protocol (Recovery of DNA from Agarose Gels Using Glass Beads)	MC3, p. 5.32
Other methods for DNA recovery (electroelution and anion-exchange chromatography)	Protocols 5.4 and 5.5, respectively

   Adapted from Chapter 5, Protocol 4, p. 5.23 of MC3.

METHOD

1. Digest an amount of the sample DNA that will yield at least 100 ng of the fragment(s) of interest. Separate the fragments by electrophoresis through an agarose or polyacrylamide gel of the appropriate concentration, stain with 0.5 μg/ml ethidium bromide or SYBR Gold, and locate the band(s) of interest with a handheld, long-wavelength UV lamp.

   > Agarose gels may be cast with ethidium bromide or run and subsequently stained either with ethidium bromide or with SYBR Gold (please see Protocol 5.2). If the DNA is separated by electrophoresis through acrylamide, the gel is subsequently stained either with ethidium bromide or with SYBR Gold (please see Protocols 5.9 and 5.10). Excitation of the ethidium bromide–DNA complex may cause photobleaching of the dye and single-strand breaks. Using a source that emits at 302 nm instead of 254 nm will minimize both effects.

2. Use a sharp scalpel to cut out a slice of agarose or polyacrylamide containing the band of interest and place it on a square of Parafilm wetted with 0.25x TBE. Cut the smallest slice of gel possible to reduce the amount of contamination of DNA with inhibitors, to minimize the distance that the DNA must migrate to exit the gel, and to ensure an easy fit into the dialysis tubing on hand.

3. After excising the band, photograph the gel as described in Protocol 5.2 to establish a record of which band was removed.

4. Wearing gloves, seal one end of a piece of dialysis tubing with a secure knot. Fill the dialysis bag to overflowing with 0.25x TBE. Holding the neck of the bag and slightly squeezing the tubing to open it, use a thin spatula to transfer the gel slice into the buffer-filled bag.

5. Allow the gel slice to sink to the bottom of the bag. Squeeze out most of the buffer, leaving just enough to keep the gel slice in constant contact with the buffer (Figure 5-3). Place a dialysis clip just above the gel slice to seal the bag. Avoid trapping air bubbles and clipping the gel slice itself. Use a permanent felt-tipped marker to label the dialysis clip with the name of the DNA fragment.

6. Immerse the bag in a shallow layer of 0.25x TBE in a horizontal electrophoresis tank. Use a glass rod or pipette to prevent the dialysis bag from floating and to maintain the gel fragment in an orientation that is parallel to the electrodes. Pass an electric current through the bag (7.5 V/cm) for 45–60 minutes. Use a handheld, long-wavelength UV lamp to monitor the movement of the DNA fragment out of the gel slice.

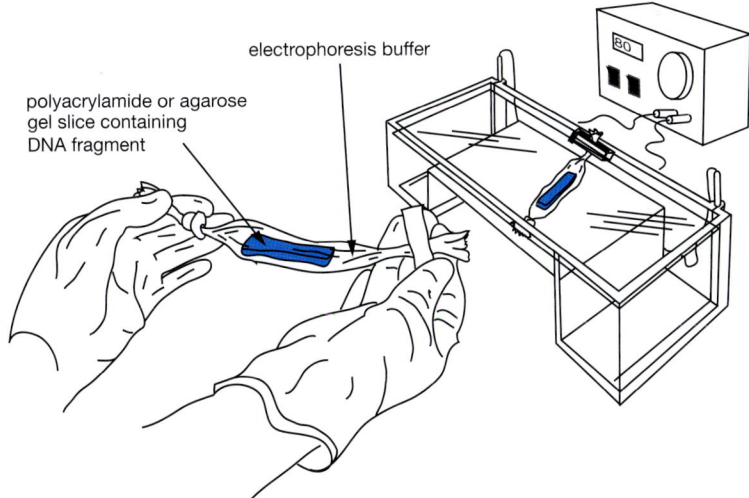

electrophoresis buffer

polyacrylamide or agarose
gel slice containing
DNA fragment

FIGURE 5-3 Electroelution of DNA from the Gel Slice

Typically, 45–60 minutes at 7.5 V/cm in 0.25x TBE is sufficient to electroelute approximately 85% of a DNA fragment of 0.1–2.0 kb from the gel slice. If electrophoresis is prolonged, the DNA may become attached to the wall of the dialysis bag. Other buffers, larger DNA fragments, and gels containing high concentrations of agarose or polyacrylamide require different electrophoresis times.

7. Reverse the polarity of the current for 20 seconds to release the DNA from the wall of the bag. Turn off the electric current and recover the bag from the electrophoresis chamber. Gently massage the bag to mix the eluted DNA into the buffer.

8. After the reverse electrophoresis, remove the dialysis clip and transfer the buffer surrounding the gel slice to a plastic tube. Remove the gel slice from the bag and stain it as described in Step 9. Use a pasteur pipette to wash out the empty bag with a small quantity of 0.25x TBE after the initial transfer and add the wash to the tube.

9. Stain the gel slice by immersing it in 0.25x TBE containing ethidium bromide (0.5 µg/ml) for 30–45 minutes at room temperature. Examine the stained slice by UV illumination to confirm that all of the DNA has eluted.

10. Purify the DNA either by passage through DEAE-Sephacel (please see Protocol 5.5), by chromatography on commercial resins, or by extraction with phenol:chloroform and standard ethanol precipitation.

Purification of DNA Recovered from Agarose and Polyacrylamide Gels by Anion-exchange Chromatography

Fragments of DNA recovered from agarose gels are sometimes poor templates or substrates in subsequent enzymatic reactions. This problem can be solved by binding the DNA to a positively charged matrix, such as DEAE-Sephadex or DEAE-Sephacel, in buffers of low ionic strength. After washing the matrix, the DNA is eluted by raising the strength of the buffer.

MATERIALS

CAUTION: Please see Appendix 4 for appropriate handling of materials marked with <!>.

Reagents and Solutions

Please see Appendix 1 for components of stock solutions, buffers, and reagents. Dilute stock solutions to the appropriate concentrations.

Ethanol
Ethanol (70%)
Isopropanol
Phenol:chloroform (1:1, v/v) <!>
Solutions of TE (pH 7.6) containing 0.1, 0.2, 0.3, or 0.6 M NaCl
    These buffers should be sterilized by autoclaving or filtration.
TE (pH 7.6 and pH 8.0)

Enzymes and Buffers

Restriction enzyme(s) and 10x buffer(s)

Nucleic Acids/Oligonucleotides

DNA samples (in TE, pH 7.6)

Additional Items

DEAE-Cellulose or DEAE-Sephadex (preswollen)
    Commercially available from, e.g., Sigma-Aldrich; usually supplied in buffers containing anti-bacterial agents that must be removed before the resin is used.
Column (e.g., Bio-Rad Dispo columns) or barrel of 2-cc syringe containing a small circle of filter paper to retain the DEAE resin
Step 10 of this protocol requires the reagents listed in Protocol 5.12.

Additional Information

Alternative protocol (Recovery of DNA from Agarose
    Gels Using Glass Beads   MC3, p. 5.32
Other methods of DNA recovery (electroelution
    onto DEAE membranes or into dialysis bags)   Protocols 5.2 and 5.4, respectively

METHOD

1. Suspend the DEAE resin in 20 volumes of TE (pH 7.6) containing 0.6 M NaCl. Allow the resin to settle and then remove the supernatant by aspiration. Add another 20 volumes of TE (pH 7.6) containing 0.6 M NaCl and gently resuspend the resin. Allow the resin to settle once more and then remove most of the supernatant by aspiration. Store the equilibrated resin at 4°C.

2. Pack 0.6 ml (sufficient to bind 20 μg of DNA) of the slurry of DEAE resin into a small column or into the barrel of a 2-cc syringe.

3. Wash the column as follows:

TE (pH 7.6) containing 0.6 M NaCl	3 ml
TE (pH 7.6)	3 ml
TE (pH 7.6) containing 0.1 M NaCl	3 ml

4. Mix the DNA (in TE at pH 7.6) with an equal volume of TE (pH 7.6) containing 0.2 M NaCl. Load the mixture directly onto the column. Collect the flow-through and reapply it to the column.

5. Wash the column twice with 1.5 ml of TE (pH 7.6) containing 0.3 M NaCl.

6. Elute the DNA with three 0.5-ml washes of TE (pH 7.6) containing 0.6 M NaCl.

7. Extract the eluate once with phenol:chloroform.

8. Divide the aqueous phase equally between two microfuge tubes and add an equal volume of isopropanol to each tube. Store the mixtures for 15 minutes at room temperature and then recover the DNA by centrifugation at maximum speed for 10 minutes at 4°C in a microfuge.

9. Wash the pellets carefully with 70% ethanol, store the open tube on the bench for a few minutes to allow the ethanol to evaporate, and then redissolve the DNA in a small volume (3–5 μl) of TE (pH 7.6).

10. Check the amount and quality of the fragment by polyacrylamide or high-resolution agarose gel electrophoresis.

   a. Mix a small aliquot (~20 ng) of the final preparation of the fragment with 10 μl of TE (pH 8.0) and add 2 μl of the desired gel-loading buffer.

   b. Load and run a polyacrylamide or high-resolution agarose gel of the appropriate concentration, using as markers restriction digests of known quantities of the original DNA. The isolated fragment should comigrate with the correct fragment in the restriction digest.

   c. Examine the gel carefully for the presence of faint fluorescent bands that signify the presence of contaminating species of DNA. It is often possible to estimate the amount of DNA in the final preparation from the relative intensities of fluorescence of the fragment and the markers.

Recovery of DNA from Low-melting-temperature Agarose Gels: Organic Extraction

DNA fragments separated by electrophoresis through gels cast with low-melting-temperature agarose are recovered by melting the agarose and extracting the resulting solution with phenol:chloroform. The protocol works best for DNA fragments ranging in size from 0.5 to 5 kb.

MATERIALS

CAUTION: Please see Appendix 4 for appropriate handling of materials marked with <!>.

Reagents and Solutions

Please see Appendix 1 for components of stock solutions, buffers, and reagents. Dilute stock solutions to the appropriate concentrations.

Ammonium acetate (10 M)
Chloroform <!>
Ethanol
Ethanol (70%)
Ethidium bromide (10 mg/ml) <!>
or
SYBR Gold solution (supplied commercially as a 10,000x concentrate in DMSO)
LMT elution buffer
    2 mM Tris-Cl (pH 8.0)
    1 mM EDTA (pH 8.0)
Phenol, H_2O saturated and equilibrated to pH 8.0
Phenol:chloroform (1:1, v/v) <!>
TE (pH 8.0)

Enzymes and Buffers

Restriction enzyme(s) and 10x buffer(s)

Nucleic Acids/Oligonucleotides

DNA samples

Gels/Loading Buffers

Gel cast with low-melting-temperature agarose, poured and run in 1x TAE
Gel-loading buffer IV

Additional Items

Scalpel blade
Steps 4 and 6 of this protocol require the reagents listed in Protocol 5.2.

Additional Information

Alternative protocol (Recovery of DNA from Agarose Gels Using Glass Beads)	MC3, p. 5.32
Low-melting-temperature agarose	MC3, pp. 5.6–5.7

METHOD

1. Prepare a gel containing the appropriate concentration of low-melting-temperature agarose in 1x TAE buffer.

2. Cool the gel to room temperature and then transfer it and its supporting glass plate to a horizontal surface in a gel box.

3. Mix the samples of DNA with gel-loading buffer IV, load them into the slots of the gel, and perform electrophoresis at 3–6 V/cm.

4. If needed, stain the agarose gel with ethidium bromide or with SYBR Gold as described in Protocol 5.2 and locate the DNA band of interest using a handheld, long-wavelength (302 nm) UV lamp.

5. Use a sharp scalpel to cut out a slice of agarose containing the band of interest and transfer it to a clean, disposable plastic tube.

6. After cutting out the band, photograph the gel as described in Protocol 5.2 to record which band of DNA was removed.

7. Add approximately 5 volumes of LMT elution buffer to the slice of agarose, close the top of the tube, and melt the gel by incubation for 5 minutes at 65°C.

8. Cool the solution to room temperature and then add an equal volume of equilibrated phenol. Vortex the mixture for 20 seconds and then recover the aqueous phase by centrifugation at 4000*g* (5800 rpm in a Sorvall SS-34 rotor) for 10 minutes at 20°C.

   The white substance at the interface is agarose.

9. Extract the aqueous phase once with phenol:chloroform and once with chloroform.

10. Transfer the aqueous phase to a fresh centrifuge tube. Add 0.2 volume of 10 M ammonium acetate and 2 volumes of absolute ethanol at 4°C. Store the mixture for 10 minutes at room temperature and then recover the DNA by centrifugation, for example, at 5000*g* (6500 rpm in a Sorvall SS-34 rotor) for 20 minutes at 4°C.

11. Wash the DNA pellet with 70% ethanol and dissolve in an appropriate volume of TE (pH 8.0).

    DNA fragments recovered from low-melting-temperature agarose gels can be purified by anion-exchange chromatography (Protocol 5.5).

Recovery of DNA from Low-melting-temperature Agarose Gels: Enzymatic Digestion with Agarase

A fragment of gel containing a band of DNA is excised and digested with agarase, which hydrolyzes the polymer to disaccharide subunits. The released DNA is then purified by phenol extraction and ethanol precipitation. The method works well for DNAs ranging in size from <5 to >20 kb.

MATERIALS

CAUTION: Please see Appendix 4 for appropriate handling of materials marked with <!>.

Reagents and Solutions

Please see Appendix 1 for components of stock solutions, buffers, and reagents. Dilute stock solutions to the appropriate concentrations.

Ethanol
Ethanol (70%)
Ethidium bromide (10 mg/ml) <!>
or
SYBR Gold solution (supplied commercially as a 10,000x concentrate in DMSO)
Gel equilibration buffer
10 mM Bis Tris-Cl (pH 6.5) (a zwitterionic buffer, commercially available from Sigma-Aldrich)
5 mM EDTA (pH 8.0)
0.1 M NaCl
NaCl (5 M)
Phenol, H_2O saturated and equilibrated to pH 8.0
TE (pH 8.0)

Enzymes and Buffers

Agarase (commercially available from New England BioLabs, Calbiochem, and others)

Nucleic Acids/Oligonucleotides

DNA samples

Centrifuges/Rotors/Tubes

Microfuge (4°C)

Additional Items

Dialysis clips
Dialysis tubing
    Molecular biology grade is available from VWR.
Handheld UV lamp, long wavelength (302 nm) <!>
Scalpel blade
Step 1 of this protocol requires the reagents listed in Protocol 5.6.
Step 3 of this protocol requires the reagents listed in Protocol 5.2.
Water baths (40°C and 76°C)

Additional Information

Alternative protocol (Recovery of DNA from Agarose Gels Using Glass Beads)	MC3, p. 5.32
Low-melting-temperature agarose	MC3, pp. 5.6–5.7
Minimizing damage to large DNA molecules	MC3, p. 2.110
Zwitterionic buffers	MC3, pp. A1.3–A1.4

METHOD

1. Follow Protocol 5.6, Steps 1–4 to prepare a gel cast with low-melting-temperature agarose, to load the DNA sample, and to perform electrophoresis.

2. Excise a segment of gel containing the DNA of interest and incubate the gel slice for 30 minutes at room temperature in 20 volumes of gel equilibration buffer.

3. After cutting out the band, photograph the gel as described in Protocol 5.2 to record which band of DNA was removed.

4. Transfer the segment of gel to a fresh tube containing a volume of gel equilibration buffer approximately equal to that of the gel slice.

5. Melt the gel slice by incubation for 10 minutes at 65°C. Cool the solution to 40°C and add DNase-free agarase, using 1–2 units of agarase per 200-µl gel slice. Incubate the sample for 1 hour at 40°C.

6. Purify and concentrate the DNA.

   To purify small DNA fragments (<20 kb)

   a. Extract the DNA solution twice with equilibrated phenol.

   b. After the second extraction, transfer the aqueous phase to a fresh tube and add 2 volumes of TE (pH 8.0).

   c. Add 0.05 volume of 5 M NaCl followed by 2 volumes of ethanol. (Here, 1 volume is equal to the volume of DNA at the end of Step b.) Incubate the tube for 15 minutes at 0°C and then collect the precipitate by centrifugation at maximum speed for 15 minutes at 4°C in a microfuge.

   d. Carefully remove the ethanol and add 0.5 ml of 70% ethanol at room temperature. Vortex the mixture and then centrifuge as described in Step c.

   e. Remove the supernatant and store the open tube on the bench for a few minutes at room temperature to allow the ethanol to evaporate. Dissolve the DNA in an appropriate volume of TE (pH 8.0).

   To purify large DNA fragments (>20 kb)

   a. Transfer the agarase-digested sample to a dialysis bag, seal, and place the bag in a beaker or flask containing 100 ml of TE (pH 8.0).

   b. Dialyze the sample for several hours at 4°C.

Alkaline Agarose Gel Electrophoresis

Alkaline agarose gels are run at a pH that is sufficiently high to denature double-stranded DNA. The denatured DNA is maintained in a single-stranded state and migrates through the alkaline gel as a function of its size. Alkaline agarose gels are used chiefly to measure the size of first and second strands of cDNA (Protocol 1.11) and to analyze the size of the DNA strand after digestion of DNA-RNA hybrids with nucleases such as S1.

MATERIALS

CAUTION: Please see Appendix 4 for appropriate handling of materials marked with <!>.

Reagents and Solutions

Please see Appendix 1 for components of stock solutions, buffers, and reagents. Dilute stock solutions to the appropriate concentrations.

Agarose
10x Alkaline agarose gel electrophoresis buffer
500 mM NaOH <!>
10 mM EDTA
Dilute the 10x alkaline agarose gel electrophoresis buffer with H_2O to generate a 1x working solution immediately before use in Step 3. Use the same stock of 10x alkaline agarose gel electrophoresis buffer to prepare the alkaline agarose gel and the 1x working solution.
Ethanol
Ethanol (70%)
Ethidium bromide (10 mg/ml) <!>
or
SYBR Gold solution (supplied commercially as a 10,000x concentrate in DMSO)
Neutralizing solution for alkaline agarose gels
1 M Tris-Cl (pH 7.6)
1.5 M NaCl
Sodium acetate (3 M, pH 5.2)
TAE electrophoresis buffer

Nucleic Acids/Oligonucleotides

DNA samples (usually radiolabeled) <!>

Gels/Loading Buffers

Alkaline agarose gel
6x Alkaline gel-loading buffer
300 mM NaOH <!>
6 mM EDTA
18% (w/v) Ficoll (Type 400, Pfizer)
0.15% (w/v) bromocresol green
0.25% (w/v) xylene cyanol

Additional Items

Boiling-water bath or microwave oven
Glass plate just large enough to cover the gel
Step 3 of this protocol requires the special equipment listed in Protocol 5.1.
Water bath (55°C)

> **Additional Information**
>
> Additional protocol (Autoradiography of Alkaline Agarose Gels)   MC3, p. 5.39
> Additional information on autoradiography and phosphorimaging MC3, pp. A9.11–A9.15

METHOD

1. Prepare the agarose solution by adding the appropriate amount of powdered agarose (please see Protocol 5.1) to a measured quantity of H_2O in an Erlenmeyer flask or a glass bottle.

2. Loosely plug the neck of the Erlenmeyer flask with Kimwipes. When using a glass bottle, make sure that the cap is loose. Heat the slurry in a boiling-water bath or a microwave oven until the agarose dissolves.

   Heat the slurry for the minimum time required to allow all of the grains of agarose to dissolve. Check that the volume of the solution has not been decreased by evaporation during boiling; replenish with H_2O if necessary.

3. Cool the clear solution to 55°C. Add 0.1 volume of 10x alkaline agarose gel electrophoresis buffer and immediately pour the gel as described in Protocol 5.1. After the gel is completely set, mount it in the electrophoresis tank and add freshly made 1x alkaline electrophoresis buffer until the gel is just covered.

   Do not add ethidium bromide because the dye will not bind to DNA at high pH.

   The addition of NaOH to a hot agarose solution causes hydrolysis of the polysaccharide. For this reason, the agarose is first melted in H_2O and then made alkaline by the addition of NaOH just before the gel is poured.

4. Collect the DNA samples by standard precipitation with ethanol. Dissolve the damp precipitates of DNA in 10–20 μl of 1x gel buffer. Add 0.2 volume of 6x alkaline gel-loading buffer.

   It is important to chelate all Mg^{++} with EDTA before adjusting the electrophoresis samples to alkaline conditions. In solutions of high pH, Mg^{++} forms insoluble $Mg(OH)_2$ precipitates that entrap DNA.

   It is not strictly necessary to denature the DNA with base before electrophoresis. The exposure of the samples to the alkaline conditions in the gel is usually enough to render the DNA single stranded.

5. Load the DNA samples dissolved in 6x alkaline gel-loading buffer into the wells of the gel as described in Protocol 5.1. Start the electrophoresis at <3.5 V/cm and, when the bromocresol green has migrated into the gel approximately 0.5–1 cm, turn off the power supply and place a glass plate on top of the gel. Continue electrophoresis until the bromocresol green has migrated approximately two-thirds the length of the gel.

   Alkaline gels draw more current than neutral gels at comparable voltages and heat up during the run. Alkaline agarose electrophoresis should therefore be carried out at <3.5 V/cm. A glass plate placed on top of the gel after the run is started slows the diffusion of the bromocresol green dye out of the gel and prevents the gel from detaching and floating in the buffer.

6. Process the gel according to one of the procedures described below, as appropriate for the goal of the experiment.

   ### Southern hybridization

   a. Soak the gel in neutralizing solution for 45 minutes at room temperature and transfer the DNA to an uncharged nitrocellulose or nylon membrane as described in Protocol 6.8.

      Alternatively, transfer the DNA directly (without soaking the gel) from the alkaline agarose gel to a charged nylon membrane (please see Protocol 6.8).

   b. Detect the target sequences in the immobilized DNA by hybridization to an appropriate labeled probe (please see Protocol 6.10).

Staining

a. Soak the gel in neutralizing solution for 45 minutes at room temperature.

b. Stain the neutralized gel with 0.5 μg/ml ethidium bromide in 1x TAE or with SYBR Gold.

> A band of interest can be sliced from the gel and subsequently eluted by one of the procedures described in Protocol 5.3 or 5.4.

Neutral Polyacrylamide Gel Electrophoresis

How to pour and run a neutral polyacrylamide gel.

MATERIALS

CAUTION: Please see Appendix 4 for appropriate handling of materials marked with <!>.

Reagents and Solutions

Please see Appendix 1 for components of stock solutions, buffers, and reagents. Dilute stock solutions to the appropriate concentrations.

Acrylamide:bisacrylamide (29:1) (% w/v) <!>
> WARNING: Wear gloves when working with acrylamide.

Ammonium persulfate (10% w/v), freshly prepared <!>

Ethanol

Ethanol (70%)

KOH/methanol solution <!>

Siliconizing fluid (e.g., Sigmacote or Acrylease)

TEMED <!>
> Available from Sigma-Aldrich and others

Nucleic Acids/Oligonucleotides

DNA samples

Gels/Loading Buffers

Gel-loading buffer IV

5x TBE electrophoresis buffer

Additional Items

Binder or "bulldog" clips (6–8 per gel), 5 cm wide

Electrophoresis apparatus including glass plates, comb, and spacers

Gel-sealing tape
> Time tape or equivalent

Micropipette with drawn-out plastic tip (e.g., Research Products International) *or* Hamilton syringe

Petroleum jelly

Scalpel blade

Side-arm flash (200 ml) (optional; see Step 6)

Syringe barrel (50 ml)

METHOD

1. If necessary, clean the glass plates and spacers with KOH/methanol.

2. Wash the glass plates and spacers in warm detergent solution and rinse them well, first in tap water and then in deionized H_2O. Hold the plates by the edges or wear gloves, so that oils

 Adapted from Chapter 5, Protocol 9, p. 5.40 of MC3.

from the hands do not become deposited on the working surfaces of the plates. Rinse the plates with ethanol and set them aside to dry.

> The glass plates must be free of grease spots to prevent air bubbles from forming in the gel.

3. (*Optional*) Treat one surface of one of the two plates with siliconizing fluid (e.g., Sigmacote or Acrylease): Place the glass on a pad of paper in a chemical fume hood and pour a small quantity of siliconizing fluid onto the surface. Wipe the fluid over the surface of the plate with a pad of Kimwipes and then rinse the plate in deionized H_2O. Dry the plate with paper towels.

   > This treatment prevents the gel from sticking tightly to one plate and reduces the possibility that the gel will tear when the mold is dismantled after electrophoresis.

4. Assemble the glass plates with spacers:

   a. Lay the larger (or unnotched) plate flat on the bench and arrange the spacers at each side parallel to the two edges.

   b. Apply minute dabs of petroleum jelly to keep the spacer bars in position during the next steps.

   c. Lay the inner (notched) plate in position, resting on the spacer bars.

   d. Clamp the plates together with binder or "bulldog" paper clips and bind the entire length of the two sides and the bottom of the plates with gel-sealing tape to make a watertight seal.

   > Take particular care with the bottom corners of the plates, as these are the places where leaks often occur. An extra band of tape around the bottom of the plates can help to prevent leaks.

   > There are many types of electrophoresis apparatuses available commercially, and the arrangement of the glass plates and spacers differs slightly by manufacturer. Whatever the design, the aim is to form a watertight seal between the plates and the spacers so that the unpolymerized gel solution does not leak out. Several manufacturers also sell precast polyacrylamide gels, which are foolproof but expensive and often can be used only in the manufacturer's gel apparatus.

5. Taking into account the size of the glass plates and the thickness of the spacers, calculate the volume of gel required. Prepare the gel solution with the desired polyacrylamide percentage according to Table 5-2, which gives the amount of each component required to make 100 ml.

TABLE 5-2 Volume of Reagents Used to Cast Polyacrylamide Gels of Indicated Concentrations in 1x TBE[a]

Polyacrylamide Gel (%)	29% Acrylamide plus 1% *N,N'*-Methylenebisacrylamide[b] (ml)	H_2O (ml)	5x TBE (ml)	10% Ammonium Persulfate (ml)
3.5	11.6	67.7	20.0	0.7
5.0	16.6	62.7	20.0	0.7
8.0	26.6	52.7	20.0	0.7
12.0	40.0	39.3	20.0	0.7
20.0	66.6	12.7	20.0	0.7

[a]Some investigators prefer to run acrylamide gels in 0.5x TBE. In this case, adjust the volumes of 5x TBE and H_2O accordingly.

[b]Stock solutions other than 29:1 (% w/v) acrylamide:bisacrylamide can be used to cast polyacrylamide gels. However, it is then necessary to recalculate the appropriate amount of stock solution to use. Gels can be cast with acrylamide solutions containing different acrylamide:bisacrylamide (cross-link) ratios, such as 19:1 and 37.5:1, in place of the 29:1 ratio recommended here. The mobility of DNA and dyes in such gels will be different from those given in this protocol.

6. (*Optional*) Place the required quantity of acrylamide:bis solution in a clean sidearm flask with a magnetic stir bar. De-aerate the solution by applying vacuum, gently at first. Swirl the flask during de-aeration until no more air bubbles are released.

De-aeration of the acrylamide solution is not essential, but it does reduce the chance that air bubbles will form when thick gels (>1 mm) are poured, as well as reduce the amount of time required for polymerization.

7. Perform the following manipulations over a tray so that any spilled acrylamide:bis solution will not spread over the bench. Wear gloves. Work quickly to complete the gel before the acrylamide polymerizes.

   a. Add 35 μl of TEMED for each 100 ml of acrylamide:bis solution and mix the solution by gentle swirling.

      Gels can be cast with as much as 1 μl of TEMED per milliliter of gel solution to increase the rate of polymerization.

   b. Draw the solution into the barrel of a 50-cc syringe. Invert the syringe and expel any air that has entered the barrel. Introduce the nozzle of the syringe into the space between the two glass plates. Expel the acrylamide gel solution from the syringe, filling the space almost to the top.

      Keep the remaining acrylamide solution at 4°C to reduce the rate of polymerization. If the plates have been well cleaned and well sealed, there should be no trapped air bubbles and no leaks. If air bubbles form, they can sometimes be coaxed to the top of the mold by gentle tapping or may be snagged with a bubble hook made of thin polypropylene tubing. If these methods fail, empty the gel mold, thoroughly clean the glass plates, and pour a new gel.

   c. Place the glass plates against a test tube rack at an angle of approximately 10° to the benchtop.

8. Immediately insert the appropriate comb into the gel, being careful not to allow air bubbles to become trapped under the teeth. The tops of the teeth should be slightly higher than the top of the glass. Clamp the comb in place with bulldog paper clips. If necessary, use the remaining acrylamide gel solution to fill the gel mold completely. Make sure that no acrylamide solution is leaking from the gel mold.

9. Allow the acrylamide to polymerize for 30–60 minutes at room temperature, adding more acrylamide:bis gel solution if the gel retracts significantly.

10. After polymerization is complete, surround the comb and the top of the gel with paper towels that have been soaked in 1x TBE. Then seal the entire gel in Saran Wrap and store it at 4°C until needed.

11. When ready to proceed with electrophoresis, squirt 1x TBE buffer around and on top of the comb and carefully pull the comb from the polymerized gel. Use a syringe to rinse out the wells with 1x TBE. Remove the gel-sealing tape from the bottom of the gel with a scalpel.

12. Attach the gel to the electrophoresis tank, using large bulldog clips on the sides or clamps built into the apparatus. The notched plate should face inward toward the buffer reservoir.

13. Fill the reservoirs of the electrophoresis tank with electrophoresis buffer prepared from the same batch of 5x TBE used to cast the gel. Use a bent pasteur pipette or syringe needle to remove any air bubbles trapped beneath the bottom of the gel.

    It is important to use the same batch of electrophoresis buffer in both of the reservoirs and in the gel. Small differences in ionic strength or pH produce buffer fronts that can greatly distort the migration of DNA.

14. Use a pasteur pipette or a syringe to flush out the wells once more with 1x TBE. Mix the DNA samples with the appropriate amount of 6x gel-loading buffer. Load the mixture into the wells using a Hamilton syringe or a micropipette equipped with a drawn-out plastic tip.

    Usually, approximately 20–100 μl of DNA sample is loaded per well depending on the size of the slot. Do not attempt to expel all of the sample from the loading device, as this almost always produces air bubbles that blow the sample out of the well. In many cases, the same device can be used to load many samples, provided it is thoroughly washed between each loading. However, it is important not to take too long to complete loading the gel; otherwise, the samples will diffuse from the wells.

15. Connect the electrodes to a power pack (positive electrode connected to the bottom reservoir), turn on the power, and begin the electrophoresis run.

   Nondenaturing polyacrylamide gels are usually run at voltages between 1 and 8 V/cm. If electrophoresis is performed at a higher voltage, differential heating in the center of the gel may cause bowing of the DNA bands or even melting of the strands of small DNA fragments. Therefore, with higher voltages, gel boxes that contain a metal plate or extended buffer chamber should be used to distribute the heat evenly. Many types of gel apparatuses are equipped with thermal sensors that monitor the temperature of the gel during the run. These are particularly useful when striving to minimize variation from one gel run to the next. Alternatively, use a gel-temperature-monitoring strip.

16. Run the gel until the marker dyes have migrated the desired distance. Turn off the electric power, disconnect the leads, and discard the electrophoresis buffer from the reservoirs.

17. Detach the glass plates, and use a scalpel to remove the gel-sealing tape. Lay the glass plates on the bench (siliconized plate uppermost). Use a spacer or plastic wedge to lift a corner of the upper glass plate. Check that the gel remains attached to the lower plate. Pull the upper plate smoothly away. Remove the spacers.

18. Use one of the methods described in Protocol 5.10 or 5.11 to detect the positions of bands of DNA in the polyacrylamide gel.

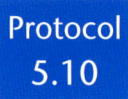

Detection of DNA in Polyacrylamide Gels by Staining

Staining bands of DNA in polyacrylamide gels with ethidium bromide, SYBR Gold, and methylene blue.

MATERIALS

CAUTION: Please see Appendix 4 for appropriate handling of materials marked with <!>.

Reagents and Solutions

Please see Appendix 1 for components of stock solutions, buffers, and reagents. Dilute stock solutions to the appropriate concentrations.
Ethidium bromide (0.5 µg/ml) <!>
Methylene blue (0.001–0.0025% w/v in 1x TAE electrophoresis buffer)

Gels/Loading Buffers

Polyacrylamide gel containing bands of DNA of interest

Additional Items

Glass tray large enough to hold gel and backing plate
UV transilluminator

Additional Information

Staining nucleic acids

MC3, pp. A9.3–A9.8

METHOD

1. Gently submerge the gel and its attached glass plate in the appropriate staining solution. Use just enough staining solution to cover the gel completely and stain the gel for 30–45 minutes at room temperature.

2. Remove the gel from the staining solution, using the glass plate as a support, rinse the gel with water, and carefully blot excess liquid from the surface of the gel with a pad of Kimwipes.

    IMPORTANT: Do not use absorbent paper; it will stick to the gel.

3. Cover the gel with a piece of Saran Wrap. Smooth out any air bubbles or folds in the Saran Wrap with the broad end of a slot comb or a crumpled Kimwipe.

4. Place a piece of Saran Wrap on the surface of a UV transilluminator. Invert the gel and place it on the transilluminator. Remove the glass plate, leaving the gel on the Saran Wrap.

5. Photograph the gel as described in Protocol 5.2.

Detection of DNA in Polyacrylamide Gels by Autoradiography

Detection of radioactive DNA in polyacrylamide gels by autoradiography.

MATERIALS

CAUTION: Please see Appendix 4 for appropriate handling of materials marked with <!>.

Buffers and Solutions

Please see Appendix 1 for components of stock solutions, buffers, and reagents. Dilute stock solutions to the appropriate concentrations.
  Acetic acid (7% v/v) <!>

Gels/Loading Buffers

  Polyacrylamide gel containing bands of DNA of interest

Additional Items

  Aluminum foil (heavy weight)
  Cellophane tape (Scotch tape)
  Gel dryer
  Glass backing plate to support gel
  Glass tray large enough to hold gel and backing plate
  Chemiluminescent markers of radioactive ink (see MC3, Appendix 8) <!>
  Whatman 3MM paper sheets
  X-ray film (Kodak XAR-5 or equivalent) with intensifying screen (optional; see Step 6)
  or
  Phosphorimager

Additional Information

  Autoradiography   MC3, pp. A9.9–A9.15

METHOD

1. Immerse the gel, together with its attached glass plate, in 7% acetic acid for 5 minutes. Remove the gel from the fixative by carefully lifting the glass plate from the fluid.

2. Rinse the gel briefly in deionized H$_2$O. Remove excess fluid from the surface of the gel with a pad of Kimwipes.

   IMPORTANT: Do not use absorbent paper; it will stick to the gel.

3. (*Optional*) Dry the gel onto a piece of Whatman 3MM paper using a commercial gel dryer.

   Drying the gel is generally necessary only when the gel contains DNA labeled with weak β-emitting isotopes such as ^{35}S or such small amounts of ^{32}P-labeled DNA that long exposures (longer than 24 hours) are necessary to obtain an adequate autoradiographic image.

4. Wrap the gel, together with its supporting glass plate, in Saran Wrap. Smooth out any air bubbles or folds in the Saran Wrap with the broad end of a slot comb or a crumpled Kimwipe.

If the DNA samples separated through the gel have been labeled with ^{35}S, it is better not to use Saran Wrap because the plastic film will block weak β particles. Make sure that the gel is very dry (in Step 3) and proceed to Step 5.

5. To align the gel and the film, attach adhesive dot labels marked with radioactive ink or with chemiluminescent markers to the surface of the Saran Wrap. Cover the radioactive ink labels with cellophane tape to prevent contamination of the film holder or intensifying screen.

6. Invert the gel and expose it to X-ray film (e.g., Kodak XAR-5 or its equivalent) as follows:

   a. In a darkroom, tape the sealed gel to a piece of X-ray film cut to the same size as the glass plate.

   > The plate serves as a weight to ensure good contact between the Saran Wrap and the X-ray film.

   b. Wrap the gel and film in light-tight aluminum foil.

   > Do not use a metal film cassette; it may break the glass plate and crush the gel. If the gel has been dried onto a piece of Whatman 3MM paper (Step 3), a metal film cassette may be used.

   c. Expose the film for the appropriate period of time at room temperature or at –70°C with or without an intensifying screen.

   d. Develop, fix, and dry the X-ray film as recommended by the manufacturer.

Isolation of DNA Fragments from Polyacrylamide Gels by the Crush and Soak Method

The "crush and soak" method, which works best for DNAs <1 kb in size, can be used to recover both single- and double-stranded DNAs from polyacrylamide gels. The yield of eluted DNA varies from <30% to >90% depending on the size of the DNA fragment.

MATERIALS

CAUTION: Please see Appendix 4 for appropriate handling of materials marked with <!>.

Reagents and Solutions

Please see Appendix 1 for components of stock solutions, buffers, and reagents. Dilute stock solutions to the appropriate concentrations.

Acrylamide gel elution buffer
  0.5 M ammonium acetate
  10 mM magnesium acetate tetrahydrate
  1 mM EDTA (pH 8.0)
  0.1% (w/v) SDS (required only when working with small amounts of DNA [<20 ng, >1 kb in size])
Chloroform <!>
Ethanol
Ethanol (70%)
Phenol:chloroform (1:1, v/v) <!> (optional; see Step 8)
Sodium acetate (3 M, pH 5.2)
TE (pH 8.0)

Gels/Loading Buffers

Polyacrylamide gel containing bands of DNA of interest
Gel-loading buffer IV

Additional Reagents

Column (disposable plastic), e.g., Quik-Sep column, Isolabs Inc. *or* a syringe barrel plugged with a Whatman GF/C filter or siliconized glass wool (optional; see Step 8)
Handheld UV lamp, long wavelength (302 nm)
Rotating wheel (37°C)
Scalpel blade
Steps 1 and 12 of this protocol require the reagents listed in Protocols 5.9 and 5.10 or Protocols 5.9 and 5.11.

METHOD

1. Perform polyacrylamide gel electrophoresis of the DNA sample and markers as described in Protocol 5.9. Locate the DNA of interest by autoradiography (Protocol 5.11) or by examination of ethidium bromide– or SYBR Gold–stained gels in long-wavelength (302 nm) UV light (Protocol 5.10).

2. Use a clean, sharp scalpel to cut out the segment of the gel containing the band of interest, keeping the size of the polyacrylamide slice as small as possible. This can be achieved by any of the following methods:

   - While the DNA is illuminated with UV light, cut through both the gel and the Saran Wrap and then peel the small piece of gel containing the DNA from the Saran Wrap.

- Use a permanent felt-tipped marker (e.g., Sharpie pen) to outline the DNA band on the back of the glass plate while the gel is illuminated from below with UV light. Invert the gel, remove the Saran Wrap, and cut out the band using the marker outline as a guide.

- In the case of a fragment of DNA identified by autoradiography, place the exposed autoradiographic film on the Saran Wrap and align it with the gel. Use a permanent marker to outline the position of the desired DNA fragment on the back of the glass plate. Remove the exposed film and Saran Wrap and cut out the band.

Photograph or autoradiograph the gel after the bands of DNA have been excised to produce a permanent record of the experiment.

3. Transfer the gel slice to a microfuge tube or a polypropylene tube. Use a disposable pipette tip or inoculating needle to crush the polyacrylamide gel against the wall of the tube.

4. Calculate the approximate volume of the slice and add 1–2 volumes of acrylamide gel elution buffer to the microfuge tube.

5. Close the tube and incubate it at 37°C on a rotating wheel or rotary platform.

    At this temperature, small fragments of DNA (<500 bp) are eluted in 3–4 hours; larger fragments take 12–16 hours.

6. Centrifuge the sample at maximum speed for 1 minute at 4°C in a microfuge. Transfer the supernatant to a fresh microfuge tube, being extremely careful to avoid transferring fragments of polyacrylamide (a drawn-out pasteur pipette works well).

7. Add an additional 0.5 volume of acrylamide gel elution buffer to the pellet of polyacrylamide, vortex briefly, and centrifuge again. Combine the supernatants.

8. (*Optional*) Remove any remaining fragments of polyacrylamide by passing the supernatant through a disposable plastic column.

    The eluted DNA can be extracted with phenol:chloroform and chloroform to remove SDS, which can inhibit subsequent enzymatic manipulation of the DNA. Precipitate the extracted DNA with ethanol as described in Step 9 and continue with the remainder of the protocol.

9. Add 2 volumes of ethanol at 4°C to the flow-through and store the solution on ice for 30 minutes. Recover the DNA by centrifugation at maximum speed for 10 minutes at 4°C in a microfuge.

    Even small quantities of DNA are efficiently precipitated by ethanol in this method. However, 10 μg of carrier RNA can be added before precipitation, which may improve even further the recovery of small amounts of DNA. Before adding the RNA, make sure that the presence of RNA will not compromise subsequent reactions with the DNA. (For preparation of carrier RNA, please see Protocol 5.3.)

10. Dissolve the DNA in 200 μl of TE (pH 8.0), add 25 μl of 3 M sodium acetate (pH 5.2), and again precipitate the DNA with 2 volumes of ethanol as described in Step 9.

11. Carefully rinse the pellet once with 70% ethanol and dissolve the DNA in TE (pH 8.0) to a final volume of 10 μl.

12. Check the amount and quality of the fragment by polyacrylamide or high-resolution agarose gel electrophoresis:

    a. Mix a small aliquot (~20 ng) of the final preparation of the fragment with 10 μl of TE (pH 8.0) and add 2 μl of gel-loading buffer.

    b. Load and run a polyacrylamide or high-resolution agarose gel of the appropriate concentration, using as markers restriction digests of known quantities of the original DNA. The isolated fragment should comigrate with the correct fragment in the restriction digest.

    c. Examine the gel carefully for the presence of faint fluorescent bands that signify the presence of contaminating species of DNA. It is often possible to estimate the amount of DNA in the final preparation from the relative intensities of fluorescence of the fragment and the markers.

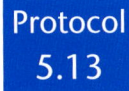
Preparation of DNA for Pulsed-field Gel Electrophoresis: Isolation of DNA from Mammalian Cells and Tissues

Genomic DNAs from mammalian cells are prepared for pulsed-field gel electrophoresis by lysing cells in situ in an agarose plug. Following digestion with an appropriate restriction enzyme, the plug is loaded directly into the well of a pulsed-field gel or it can be melted before loading.

MATERIALS

CAUTION: Please see Appendix 4 for appropriate handling of materials marked with <!>.

Reagents and Solutions

Please see Appendix 1 for components of stock solutions, buffers, and reagents. Dilute stock solutions to the appropriate concentrations.

L Buffer (4°C)
    0.1 M EDTA (pH 8.0)
    0.01 M Tris-Cl (pH 7.6)
    0.02 M NaCl

L Buffer (room temperature) containing
    Sarkosyl (1% w/v)
    Proteinase K (100 µg/ml), added just before use in Step 6

PBS (room temperature for white blood cells and 4°C for all other types of cells)

Red blood cell lysis buffer, used to isolate white blood cells
    155 mM NH_4Cl
    0.1 mM EDTA
    12 mM $NaHCO_3$

TE (pH 7.6)

TE (pH 7.6) containing 40 µg/ml PMSF (phenylmethylsulfonyl fluoride) <!>

Cells, Tissues/Culture Media

Cell or tissue samples (this protocol describes methods for dealing with cultured cell lines, fresh or frozen tissue samples, and white blood cells)

Enzymes and Buffers

Proteinase K

Gels/Loading Buffers

Low-melting-temperature agarose (1%), e.g., SeaKem GTG

Centrifuges/Rotors/Tubes

Sorvall SS-34 rotor (4°C), required only for processing white blood cells

Additional Items

Cheesecloth, required only for tissue samples
Glass homogenizer with tight-fitting pestle (0°C)
Hemocytometer, required only for fresh tissue samples
LeucoPREP cell separation tubes (Becton Dickinson), required only to isolate white blood cells

Mortar and pestle (–70°C), required only to process frozen tissue
Preformed Plexiglas molds (10–100 ml; Bio-Rad) *or* a length of Tygon
  tubing (1/8-inch or 3.2-mm internal diameter) *or* plastic syringe barrel (1 ml)
Scalpel blade
Water baths (42°C and 50°C)

Additional Information

Low-melting-temperature agarose	MC3, pp. 5.6–5.7
Phenylmethylsulfonyl fluoride (PMSF)	MC3, pp. A5.1 and 15.41

METHOD

1. Prepare cells or tissue samples.

   ### For cultured cells

   a. Wash cells that have been growing in culture three times in ice-cold PBS.

   b. Harvest the cells by scraping into a small volume of ice-cold PBS using a sterilized rubber policeman. Collect the cells by low-speed centrifugation.

   c. Resuspend the cells at a concentration of approximately 2×10^7 cells/ml in ice-cold L buffer.

   ### For fresh tissue samples

   a. In a petri dish, use a clean scalpel to mince freshly excised tissue into small cubes (1–2 mm^3) and then homogenize the cubes in ice-cold PBS in a chilled glass homogenizer with a tight-fitting pestle.

   b. Remove fragments of connective tissue by filtration through two layers of cheesecloth.

   c. Wash the suspended cells three times in ice-cold PBS and resuspend them at a concentration of 2×10^7 cells/ml in ice-cold L buffer.

      Use a hemocytometer to count the cells.

   ### For frozen tissue samples

   a. Grind frozen tissue to a fine powder using a mortar and pestle chilled to –70°C and suspend the powdered tissue in ice-cold PBS.

   b. Remove fragments of connective tissue by filtration through two layers of cheesecloth.

   c. Wash the suspended cells three times in ice-cold PBS and resuspend them at a concentration of 2×10^7 cells/ml in ice-cold L buffer.

   ### For white blood cells

   a. Fractionate 5–10 ml of starting blood by centrifugation in LeucoPREP cell separation tubes to grossly separate white and red blood cells.

   b. To the white blood cell layer (buffy coat), add 4 volumes of red blood cell lysis buffer and gently mix by two to three inversions of the tube.

   c. Incubate the cells in buffer for 5 minutes at room temperature and then centrifuge the tube at 3000*g* (5000 rpm in a Sorvall SS-34 rotor) for 5 minutes at room temperature.

   d. Resuspend the pellet in 1 ml of PBS at room temperature.

2. Prepare a volume of 1% low-melting-temperature agarose in L buffer that is equal to the volume of the cell preparation in Step 1. Cool the melted agarose to 42°C.

3. When the agarose has cooled to 42°C, warm the cell suspension (Step 1) to the same temperature. Mix the melted agarose with the suspended cells. Stir the mixture with a sealed pasteur pipette to ensure that the cells are evenly dispersed throughout the agarose.

4. Pipette or pour the molten mixture into preformed Plexiglas molds (50–100 μl), or draw the mixture into an appropriate length of Tygon tubing (1/8-inch or 3.2-mm internal diameter) or a 1-ml plastic syringe barrel. Store the molds for 15 minutes at room temperature and then transfer them to 4°C for 15–30 minutes.

5. When the agarose has set, gently collect the plugs from the Plexiglas molds or gently blow out the agarose from the Tygon tubing or syringe barrel into a petri dish. Cut the cylindrical plugs into 1-cm sections.

   Each 1-cm length of agarose (45 μl) should contain approximately 0.5 x 10^6 cells and yield approximately 2–5 μg of DNA. The migration of large DNA fragments is significantly slowed when pulsed-gel electrophoresis is performed at high DNA concentrations. This slowing will lead to an overestimate of the size of an individual DNA fragment.

6. Transfer the plugs to 3 volumes of L buffer containing 0.1 mg/ml proteinase K and 1% (w/v) Sarkosyl. Incubate the plugs for 3 hours at 50°C. Replace the original digestion mixture with two volumes of fresh digestion mixture and continue the incubation for 12–16 hours at 50°C.

   Be careful not to scar the agarose plugs when changing buffer solutions.

7. Incubate the plugs at room temperature in 50 volumes of TE (pH 7.6) with three to five changes of buffer over a period of 3 hours.

8. Remove the TE and replace with 2 volumes of TE (pH 7.6) containing 40 μg/ml PMSF. Incubate for 30 minutes at 50°C.

9. Incubate the plugs at room temperature in 50 volumes of TE (pH 7.6) with three to five changes of buffer over a period of 3 hours.

Yeast cells are first treated enzymatically to break down the cell walls and then resuspended in low-melting-temperature agarose plugs. The DNA is liberated by infusing the plugs with lysis buffer and proteases. This method is used to prepare both conventional and artificial yeast chromosomes.

MATERIALS

CAUTION: Please see Appendix 4 for appropriate handling of materials marked with <!>

Reagents and Solutions

Please see Appendix 1 for components of stock solutions, buffers, and reagents. Dilute stock solutions to the appropriate concentrations.

EDTA (0.05 M, pH 8.0)
L Buffer (4°C)
    0.1 M EDTA (pH 8.0)
    0.01 M Tris-Cl (pH 7.6)
    0.02 M NaCl
L Buffer (room temperature) containing
    Sarkosyl (1% w/v)
    Proteinase K (100 μg/ml), added just before use in Step 6
TE (pH 7.6) containing 40 μg/ml PMSF (phenylmethylsulfonyl fluoride) <!>
Yeast cell wash buffer (4°C)
    0.01 M Tris-Cl (pH 7.6)
    0.05 M EDTA (pH 8.0)
Yeast lysis buffer (4°C)
    0.01 M Tris-Cl (pH 7.6)
    0.5 M EDTA (pH 8.0)
    β-mercaptoethanol (1% v/v) <!>, added just before use

Cells, Tissues/Culture Media

Yeast suspension culture of a volume and cell density required to yield the required number of plugs of embedded DNA

Enzymes and Buffers

Zymolyase 5000 (Kirin Brewery), 2.0 mg/ml of 0.01 M sodium phosphate containing 50% glycerol (prepare just before use)
or
Lyticase (Sigma-Aldrich), 67 mg/ml (900 units/ml) in 0.01 M sodium phosphate containing 50% glycerol (prepared just before use)

Gels/Loading Buffers

Low-melting-temperature agarose (1%), e.g., SeaKem GTG

Centrifuges/Rotors/Tubes

Sorvall SLC-1500 rotor (4°C)

Adapted from Chapter 5, Protocol 14, p. 5.65 of MC3.

Additional Items

Preformed Plexiglas molds (10–100 ml; Bio-Rad) *or* a length of Tygon
  tubing (1/8-inch or 3.2-mm internal diameter) *or* plastic syringe barrel (1 ml)
Scalpel blade
Water baths (42°C and 50°C)

Additional Information

Chromosome size in yeasts and bacteria	MC3, p. 5.65
Low-melting-temperature agarose	MC3, pp. 5.6–5.7
Phenylmethylsulfonyl fluoride (PMSF)	MC3, pp. A5.1 and 15.41

METHOD

1. Collect yeast cells growing in suspension by centrifugation at 3000*g* (4300 rpm in a Sorvall SLC-1500 rotor) for 5 minutes at 4°C. Wash the cell pellet twice with yeast cell wash buffer.

2. Resuspend the cells at a concentration of 3×10^9 cells/ml in 0.05 M EDTA (pH 8.0) at 0°C.

3. Prepare an equal volume of 1% low-melting-temperature agarose in L buffer. Cool the melted agarose to 42°C.

4. Add 75 µl of zymolyase or lyticase solution to the cell suspension of Step 2. Mix well.

5. Warm the cell suspension to 42°C. Mix 5 ml of the melted agarose with 5 ml of the suspended cells. Stir the mixture with a sealed pasteur pipette to ensure that the cells are evenly dispersed throughout the agarose.

6. Pipette or pour the molten mixture into preformed Plexiglas molds (50–100 µl) or draw the mixture into an appropriate length of Tygon tubing (3/32-inch internal diameter) or a 1-ml plastic syringe barrel. Store the molds for 15 minutes at room temperature and then transfer them to 4°C for 15–30 minutes.

7. When the agarose has set, collect the plugs from the Plexiglas molds or blow out the agarose from the Tygon tubing or syringe barrel into a petri dish. Cut the cylindrical plugs into 1-cm blocks.

   Each block (50 µl) should contain approximately 5 µg of yeast chromosomal DNA.

8. Incubate the blocks in a chemical fume hood for 3 hours at 37°C in 3 volumes of yeast lysis buffer.

9. Add 3 volumes of L buffer containing 0.1 mg/ml proteinase K and 1% (w/v) Sarkosyl into a fresh petri dish. Transfer the blocks to this buffer and incubate them for 3 hours at 50°C. Replace the original digestion mixture with an equal volume of fresh digestion mixture and continue incubation for 12–16 hours at 50°C.

   Be careful not to scar the agarose plugs when changing buffer solutions.

10. Incubate the plugs at 50°C in 50 volumes of TE (pH 7.6) containing 40 µg/ml PMSF. After 1 hour, replace the original rinse buffer (TE containing PMSF) with an equal volume of fresh rinse buffer and continue incubation for another hour at 50°C.

11. Remove the rinse buffer (TE containing PMSF), replace with an equal volume of fresh TE (pH 7.6), and continue incubation for another hour at room temperature.

Restriction Endonuclease Digestion of DNA in Agarose Plugs

Genomic DNA isolated from mammalian, yeast, or bacterial cells can be digested with restriction endonucleases by incubating agarose plugs containing the DNA in the presence of the desired enzyme. After digestion, the DNA can be fractionated by pulsed-field gel electrophoresis and either isolated from the gel or analyzed by Southern hybridization.

MATERIALS

Reagents and Solutions

Please see Appendix 1 for components of stock solutions, buffers, and reagents. Dilute stock solutions to the appropriate concentrations.
   Spermidine (0.1 M), dissolved in H_2O just before use
   TE (pH 7.6)

Enzymes and Buffers

   Restriction enzymes and 10x buffers

Nucleic Acids/Oligonucleotides

   Genomic DNA embedded in plugs cast with low-melting-temperature agarose (see
      Protocols 5.13 and 5.14)

Gels/Loading Buffers

   Pulsed-field gel, cast with 1% agarose as described in Protocols 5.17 and 5.18

Additional Information

   Choosing restriction endonucleases for PFGE   MC3, pp. 5.68–5.69

METHOD

1. If plugs have not been stored in TE (e.g., plugs received through the mail or those that have been stored in 0.5 M EDTA [pH 8.0]), incubate them in 50 volumes of TE (pH 7.6) for 30 minutes at room temperature. Transfer the plugs to 50 volumes of fresh TE (pH 7.6) and continue incubation for an additional 30 minutes. Otherwise, proceed directly to Step 2.

2. Transfer the plugs to individual microfuge tubes and add 10 volumes of the appropriate 1x restriction enzyme buffer to each tube. Incubate the tubes for 30 minutes at room temperature.

   Individual buffers should be supplemented with spermidine to enhance the efficiency of restriction digestion. High-salt buffers (containing 100–150 mM salt) should be supplemented to a final concentration of 10 mM spermidine, medium-salt buffers (50–100 mM salt) to 5 mM spermidine, and low-salt buffers (<50 mM salt) to 3 mM spermidine. It is best to supplement the buffers from a 0.1 M spermidine stock (dissolved in H_2O) just before use.

3. Remove the buffer and replace it with 2–3 volumes of fresh 1x restriction enzyme buffer. Add 20–30 units of the appropriate restriction enzyme to each tube and incubate the tubes at the

optimal temperature for the restriction enzyme: 3 hours if YAC DNA is used or 5–6 hours if mammalian DNA is used.

4. If the DNA is to be digested with only one restriction enzyme, soak the plugs in 20 volumes of TE (pH 7.6) at 4°C. After 1 hour, proceed with Step 7. If the DNA is to be treated with more than one enzyme, skip the incubation in TE and proceed to Step 5.

5. If the DNA is to be digested with more than one restriction enzyme, reequilibrate the plug in buffer before adding the second enzyme. To reequilibrate, use automatic pipetting devices to remove as much as possible of the first restriction enzyme buffer from each tube and replace it with 1 ml of TE (pH 7.6). Remove the TE and replace it with a fresh 1 ml of TE. Store the plug for 30–60 minutes at room temperature. Gently remove the TE buffer.

6. Add 10 volumes of the appropriate second 1x restriction enzyme buffer to each tube. Incubate the tubes for 30 minutes at room temperature. Repeat the restriction enzyme digestion as described in Step 3. Finally, soak each plug in 20 volumes of TE (pH 7.6) for 1 hour at 4°C.

7. Use a disposable pipette tip to push the blocks directly into the slots of a pulsed-field gel and separate the fragments of DNA by electrophoresis (Protocols 5.17 and 5.18).

Protocol 5.16 Markers for Pulsed-field Gel Electrophoresis

Markers for pulsed-field gel electrophoresis can be generated by ligation of linear monomers of bacteriophage λ DNA (48.5 kb) into a nested series of concatemers. This procedure yields a series of concatemers that contain up to 20 tandemly arranged copies of bacteriophage DNA.

MATERIALS

CAUTION: Please see Appendix 4 for appropriate handling of materials marked with <!>.

Reagents and Solutions

Please see Appendix 1 for components of stock solutions, buffers, and reagents. Dilute stock solutions to the appropriate concentrations.
ATP (0.1 M)
Dithiothreitol (1.0 M)
EDTA (0.5 M, pH 8.0)
1x Ligation buffer with PEG
    50 mM Tris-Cl (pH 7.6)
    1 mM ATP
    10 mM dithiothreitol
    10 mM MgCl$_2$
    2% (w/v) PEG 8000 <!>
Low-melting-temperature (LMT) agarose buffer
    100 mM Tris-Cl (pH 7.6)
    20 mM MgCl$_2$
MgCl$_2$ (1.0 mM)
PEG 8000 (8% w/v) <!>
TE (pH 7.6)

Enzymes and Buffers

Bacteriophage T4 DNA ligase

Nucleic Acids/Oligonucleotides

Purified bacteriophage λ DNA (see Protocols 2.11 and 2.12)

Gels/Loading Buffers

Low-melting-temperature agarose, e.g., SeaKem GTG or equivalent

Additional Items

Boiling-water bath or microwave oven
Tygon tubing (3/32-inch internal diameter)
Water baths (37°C and 56°C)

Additional Information

Bacteriophage T4 ligase	MC3, p. A4.31
Use of condensing and crowding reagents in ligation reactions	MC3, p. 1.152

METHOD

1. Dissolve 0.1 g of low-melting-temperature agarose in 10 ml of LMT buffer by heating it in a boiling-water bath or by microwaving. Cool the solution to 37°C.

2. Dissolve 10 µg of purified bacteriophage λ DNA in 172.5 µl of TE (pH 7.6) and heat the solution to 56°C for 5 minutes.

3. Cool the solution to 37°C and rapidly add the following reagents in the following order:

8% PEG 8000	62.5 µl
20 mM MgCl$_2$	5.0 µl
0.1 M ATP	5.0 µl
1 M dithiothreitol	5.0 µl
bacteriophage T4 DNA ligase	0.5 Weiss unit
1% LMT agarose solution (Step 1)	250 µl

   Polyethylene glycol acts like a crowding agent and increases the efficiency of ligation.

4. Draw the mixture into a short length of Tygon tubing and chill the tubing on ice until the agarose has completely set.

5. Blow the agarose plug into a sterile, disposable plastic tube containing at least 3 volumes of 1x ligation buffer with PEG.

6. Incubate the plug in ligation buffer for 24 hours at room temperature and then transfer it to a tube containing 10 volumes of 20 mM EDTA (pH 8.0).

7. Incubate the plug in EDTA for 1 hour, transfer the plug to a tube containing 10 volumes of fresh 20 mM EDTA (pH 8.0), and store at 4°C until needed for electrophoresis in Protocols 5.17 and 5.18.

Pulsed-field Gel Electrophoresis via Transverse Alternating Field Electrophoresis (TAFE)

In this form of pulsed-field gel electrophoresis, electrodes are positioned on opposite sides of a vertically oriented gel. The DNA moves first toward one electrode and then toward the other, forming a zigzag pattern. The vector of this oscillation is a straight line from the loading well to the base of the gel. Variation in voltage and pulse time allows separation of DNAs ranging in size from 2 to >6000 kb. This protocol describes the resolution of genomic DNA by TAFE, followed by blotting and hybridization.

MATERIALS

CAUTION: Please see Appendix 4 for appropriate handling of materials marked with <!>.

Reagents and Solutions

Please see Appendix 1 for components of stock solutions, buffers, and reagents. Dilute stock solutions to the appropriate concentrations.
  Denaturation solution
      0.5 N NaOH <!>
      1.5 M NaCl
  Ethidium bromide (10 mg/ml) <!>
  or
  SYBR Gold solution (supplied commercially as a 10,000x concentrate in DMSO)
  TAFE gel electrophoresis buffer
      20 mM Tris-acetate (pH 8.2)
      0.5 M EDTA
          This buffer must be cooled to 14°C before filling the gel apparatus.
  TE (pH 8.0)

Nucleic Acids/Oligonucleotides

  DNA size standards
      See Protocols 5.14 and 5.16. Standards can also be purchased from New England BioLabs, Bio-Rad, and others.
  Genomic DNA of interest

Gels/Loading Buffers

  Agarose
      PFGE grade; e.g., SeaKem LE (BioWhittaker) and LMP (Invitrogen).

Additional Reagents

  Power supply capable of delivering electrical pulses of variable time and amperage. See Table 5-3.
  Step 2 of this protocol requires the reagents listed in Protocols 5.13 or 5.14, 5,15, and 5.16.
  Step 11 of this protocol requires the reagents listed in Protocol 6.10.
  Steps 9 and 10 of this protocol require the reagents listed in Protocol 6.8.
  TAFE gel apparatus attached to circulating water bath (14°C)

Additional Information

Alternative protocol to Southern hybridization and
  autoradiography (Silver Staining PFGE Gels)   MC3, p. 5.77
Staining agarose gels   MC3, pp. 5.14–5.15
Staining nucleic acids   MC3, pp. A9.3–A9.8
TAFE gels   MC3, pp. 5.55–5.59 and 5.74–5.75
Troubleshooting problems with TAFE gels   MC3, p. 5.78

METHOD

1. Cast a 1% agarose gel in 1x TAFE gel buffer without ethidium bromide and allow the gel to set. Pour the gel with the same buffer solution to be used to fill the gel apparatus.

2. Prepare agarose plugs containing the DNA of interest (please see Protocol 5.13 for preparation of mammalian DNA or Protocol 5.14 for preparation of yeast DNA embedded in plugs) and perform digestion with restriction enzymes as described in Protocol 5.15. Prepare and embed the appropriate DNA size standards as described in Protocols 5.14 and 5.16.

3. Rinse all of the plugs in 10 volumes of TE (pH 8.0) for 30 minutes with two changes of buffer.

4. Embed the digested and rinsed DNA plugs in individual wells of the gel. Seal the plugs in the wells with molten 1% agarose in 1x TAFE gel buffer.

5. Place the gel in the TAFE apparatus filled with 1x TAFE gel buffer previously cooled to 14°C.

6. Connect the gel apparatus to the appropriate power supply set to deliver 4-second pulses at a constant current of 170–180 mA for 30 minutes. This setting forces the DNA to enter the gel rapidly. After this time period, decrease the current input to 150 mA, change the pulse time to a setting optimum for the size range of DNAs to be resolved (please see Table 5-3), and continue electrophoresis for 12–18 hours.

   In a Tris-acetate/EDTA buffer, a pulse time of 15 seconds will separate DNA fragments in the 50–400-kb size range. This same range of fragments can be separated in Tris-borate/EDTA buffer (0.045 M Tris-borate [pH 8.2], 0.01 M EDTA) using a program of 8-second pulses at 350 mA for 12 hours followed by 15-second pulses at 350 mA for an additional 12 hours.

TABLE 5-3 TAFE Gel Conditions for Separating DNAs of Various Sizes

Size Range (kb)	Pulse Time (seconds)	Time (hours)
5–50	1	10
20–100	3	10
50–250	10	14
100–400	20	14
200–1000	45	18
200–1600	70	20
Larger DNAs	see text	see text

7. Disconnect the power supply, dismantle the gel apparatus, and stain the gel in 1x TAFE buffer containing 0.5 µg/ml ethidium bromide or an appropriate dilution of SYBR Gold. Photograph the gel under UV light.

   The standard technique used to detect DNAs separated by PFGE is staining with ethidium bromide or SYBR Gold (please see Protocol 5.2). To facilitate the detection of minor species of DNA stained with ethidium bromide, the gels may be destained in H_2O for up to 1 hour before photography. For details of methods that can be used to maximize the photographic detection of DNA stained with these dyes, please see Protocol 5.2.

8. Rinse the stained gel twice with H_2O. Pour off the second H_2O rinse and replace with denaturation solution. Incubate with gentle shaking for 30 minutes. Change the denaturation solution and incubate for an additional 30 minutes.

> In general, standard methods of Southern blotting as described in Protocols 6.8 and 6.9 can be used to detect hybridizing gene sequences in TAFE gels. Some investigators find that the very large DNA fragments typically analyzed by PFGE transfer to nylon membranes more efficiently after partial hydrolysis of the DNA by acid treatment. This treatment causes partial depurination and nicking of larger DNA fragments and in so doing enhances capillary transfer. To include an acid hydrolysis step, after the electrophoresis is complete, rinse the gel twice with H_2O in Step 8, pour off the second H_2O rinse, and replace with 25 mM HCl. Soak the gel with gentle agitation for 3–5 minutes. Rinse the gel with H_2O and continue with the denaturation protocol of Step 8 and onward.

> When acid treatments are used in the protocol, it is important to stain the agarose gel after transfer of the DNA to the nylon membrane. A slight residual smear of DNA should be visible. If no residual DNA is detected, the acid treatment may have been too harsh (i.e., too long or too strong). Too much depurination and nicking increases the transfer of the DNA but also tends to reduce the intensity of subsequent hybridization signals.

9. Transfer the DNA directly to a nylon membrane by capillary blotting in denaturation solution (for details, please see Protocol 6.8).

> Some investigators find that transfer of larger DNA fragments is enhanced when 6x SSC buffer is used in capillary blotting rather than the more standard 10x SSC solution.

10. After transfer, affix the DNA to the nylon membrane by baking for 2 hours at 80°C, by UV cross-linking, or by microwaving (see Table 6-2, p. 261).

11. Perform prehybridization and hybridization with labeled probes in a formamide-containing buffer (for details, please see Protocol 6.10).

> Typically, ^{32}P-radiolabeled single-stranded bacteriophage M13 probes are used at a concentration of 5×10^6 to 6×10^6 cpm/ml of hybridization buffer to detect a single-copy gene in a complex mammalian genome. A membrane hybridized for 12–16 hours in this fashion is washed in 500 ml of 2x SSC containing 1% (w/v) SDS for 15 minutes at room temperature, scrubbed gently with a sponge, washed in 1 liter of 0.5x SSC containing 1% (w/v) SDS for 2 hours at 68°C, and then subjected to autoradiography.

In CHEF gels, the electric field is generated from multiple electrodes, arranged in a square of hexagonal contour around the horizontal gel, and clamped to predetermined potentials. Using a combination of low field strengths, low concentrations of aragose, long switching intervals, and extended periods of electrophoresis, DNAs up to 5000 kb can be resolved. This protocol describes the resolution of genomic DNA by CHEF, followed by blotting and hybridization.

MATERIALS

CAUTION: Please see Appendix 4 for appropriate handling of materials marked with <!>.

Reagents and Solutions

Please see Appendix 1 for components of stock solutions, buffers, and reagents. Dilute stock solutions to the appropriate concentrations.

> Denaturation solution
>> 0.5 N NaOH <!>
>> 1.5 M NaCl
> 0.5x TBE gel electrophoresis buffer
> Ethidium bromide (10 mg/ml) <!>
> *or*
> SYBR Gold solution (supplied commercially as a 10,000x concentrate in DMSO)

Nucleic Acids/Oligonucleotides

> DNA size standards
>> See Protocols 5.14 and 5.16. Standards can also be purchased from New England BioLabs, Bio-Rad, and others.
> Genomic DNA of interest

Gels/Loading Buffers

> Agarose
>> PFGE grade, e.g., SeaKem GTG or Seakem Gold (BioWhittaker), and Agarose-LE (Beckman), among others

Additional Items

> Power supply capable of delivering electrical pulses of variable time and amperage (see Table 5-11)
> Step 3 of this protocol requires the reagents listed in Protocols 5.13 or 5.14, 5.15, and 5.16.
> Step 12 of this protocol requires the reagents listed in Protocol 6.10.
> Steps 10 and 11 of this protocol require the reagents listed in Protocol 6.8.
> TAFE gel apparatus attached to circulating water bath (14°C)

Additional Information

Alternative protocol to Southern hybridization and autoradiography (Silver Staining PFGE Gels)	MC3, p. 5.77
CHEF gels	MC3, pp. 5.55–5.59 and 5.79
Staining agarose gels	MC3, pp. 5.14–5.15
Staining nucleic acids	MC3, pp. A9.3–A9.8

METHOD

1. Cast an agarose gel of the appropriate concentration (usually 0.6%) in 0.5x TBE buffer as described in Protocol 5.1. Use a bubble level to ensure that the casting tray is completely flat on the laboratory bench. Allow the gel to harden for 1 hour at room temperature.

2. Place the agarose gel in the CHEF apparatus, add just enough 0.5x TBE to cover the gel, and cool the remaining buffer to 14°C.

3. Prepare agarose plugs containing the DNA of interest (please see Protocol 5.13 for preparation of mammalian DNA or Protocol 5.14 for preparation of yeast DNA embedded in plugs) and perform digestion with restriction enzymes as described in Protocol 5.15. Prepare and embed the appropriate DNA size standards as described in Protocol 5.14 or 5.16.

4. Gently place the plugs in individual microfuge tubes and add 200 μl of 0.5x TBE to each. Incubate the plugs for 15 minutes at room temperature.

5. Embed the digested and rinsed DNA plugs in individual wells of the gel. Seal the plugs in the wells with the same solution of molten agarose used to pour the gel.

6. Allow the sealed plugs to harden in the gel for approximately 5 minutes and then add additional 0.5x TBE buffer (previously cooled to 14°C) to the apparatus to cover the agarose gel completely.

7. Start the buffer circulating and begin the electrophoresis run using power supply settings as described in Table 5-4.

TABLE 5-4 **Conditions for Separating DNAs of Various Sizes in CHEF Gels**

% Agarose	Size Range (kb)	Switch Times	V/cm	Time (hours)
1% Fast Lane	6–500	ramped, 3–80 seconds	6	18
	10–800	ramped, 6–80 seconds	6	20
	100–1000	60 seconds,	6	15
		then 90 seconds	6	9
0.8% Fast Lane	150–2000	ramped, 30–180 seconds	5	24

All gels are run in 0.5x TBE. When very high resolution is required in the 800–2000-kb range, a lower voltage and a longer run time than those indicated above are used. If the gels are cast with SeaKem GTG agarose, add 10% to the electrophoresis time indicated.

8. After electrophoresis, stain the gel in a solution of ethidium bromide (1 μg/ml) or an appropriate dilution of SYBR Gold for 30 minutes at room temperature. Destain the gel in H_2O for 30 minutes and photograph the gel under UV light.

9. After photography, incubate the gel in 250 ml of denaturation solution with gentle shaking for 30 minutes. Change the denaturation solution and incubate for an additional 30 minutes.

10. Transfer the DNA directly to a nylon membrane by capillary blotting in denaturation solution (for details, please see Protocol 6.8).

   Enhanced transfer of larger DNA fragments has also been noted when 6x SSC buffer is used in capillary blotting rather than the more standard 10x SSC solution.

   In general, standard methods of Southern blotting as described in Protocols 6.8 and 6.9 can be used to detect hybridizing gene sequences in TAFE gels. Some investigators find that the very large DNA fragments typically analyzed by PFGE transfer to nylon membranes more efficiently after partial hydrolysis of the DNA by acid treatment. This treatment causes partial depurination and nicking of larger DNA fragments and in so doing enhances capillary transfer. To include an acid hydrolysis step, after the electrophoresis is complete, rinse the gel twice with H_2O in Step 8, pour off the second H_2O rinse, and replace with 25 mM HCl. Soak the gel with gentle agitation for 3–5 minutes. Rinse the gel with H_2O and continue with the denaturation protocol of Step 8 and onward.

When acid treatments are used in the protocol, it is important to stain the agarose gel after transfer of the DNA to the nylon membrane. A slight residual smear of DNA should be visible. If no residual DNA is detected, the acid treatment may have been too harsh (i.e., too long or too strong). Too much depurination and nicking increases the transfer of the DNA but also tends to reduce the intensity of subsequent hybridization signals.

11. After transfer, affix the DNA to the nylon membrane by baking for 2 hours at 80°C, by UV cross-linking, or by microwaving (see Table 6-2, p. 261).

12. Perform prehybridization and hybridization with labeled probes in a formamide-containing buffer (for details, please see Protocol 6.10).

Typically, ^{32}P-radiolabeled single-stranded bacteriophage M13 probes are used at a concentration of 5×10^6 to 6×10^6 cpm/ml of hybridization buffer to detect a single-copy gene in a complex mammalian genome. A membrane hybridized for 12–16 hours in this fashion is washed in 500 ml of 2x SSC containing 1% (w/v) SDS for 15 minutes at room temperature, scrubbed gently with a sponge, washed in 1 liter of 0.5x SSC containing 1% (w/v) SDS for 2 hours at 68°C, and then subjected to autoradiography.

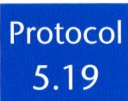

Direct Retrieval of DNA Fragments from Pulsed-field Gels

A gel slice containing a fragment of DNA resolved by pulsed-field gel electrophoresis is treated with agarose. The released DNA can be used as a substrate for ligation or restriction without additional purification.

MATERIALS

CAUTION: Please see Appendix 4 for appropriate handling of materials marked with <!>.

Reagents and Solutions

Please see Appendix 1 for components of stock solutions, buffers, and reagents. Dilute stock solutions to the appropriate concentrations.

Denaturation solution
0.5 N NaOH <!>
1.5 M NaCl
Ethidium bromide (10 mg/ml) <!>
or
SYBR Gold solution (supplied commercially as a 10,000x concentrate in DMSO)
Isopropanol (optional; see Step 9)
Phenol:chloroform (1:1, v/v) <!>
Sodium acetate (0.3 M, pH 5.2) (optional; see Step 9)

Enzymes and Buffers

Agarase, e.g., GELase (Epicentre Technologies) or β-agarase I (New England BioLabs)
Agarase buffer
Use the buffer supplied by the manufacturer, supplemented with 100 mM NaCl, 70 µM spermidine.

Nucleic Acids/Oligonucleotides

DNA size standards
See Protocols 5.14 and 5.16. Standards can also be purchased from New England BioLabs, Bio-Rad, and others.
Genomic DNA embedded in low-melting-temperature agarose plugs (Protocol 5.13 or 5.14)

Gels/Loading Buffers

Agarose, PFGE grade, e.g., SeaKem GTG or Seakem Gold (BioWhittaker), and Agarose-LE (Beckman), among others

Additional Reagents

Scalpel blades
Steps 1 and 2 of this protocol require the reagents listed in Protocol 5.17 or 5.18.
Water baths (65–68°C and the temperature recommended by the manufacturer for agarase digestion)

Additional Information

Staining nucleic acids   MC3, pp. A9.3–A9.8

METHOD

1. Prepare a preparative low-melting-point agarose PFGE (Protocols 5.17 or 5.18) that will provide optimum resolution of the DNA fragment or size fraction of interest.

2. Load the DNA size standards and single plugs of the digested genomic DNA in lanes located on both sides of the preparative slot. Load the preparative sample into the preparative slot. Perform electrophoresis as described in Protocol 5.17 (for TAFE gel) or Protocol 5.18 (for CHEF gel).

3. After electrophoresis, cut the lanes containing the size standards and single genomic DNA plugs from the gel and stain them with ethidium bromide or SYBR Gold for 30 minutes at room temperature. If necessary, destain the gel slices in H_2O for 30 minutes. Do not stain the preparative lane.

4. Examine the stained slices by UV illumination and make notches on the slices to mark the locations of markers flanking the position of DNA of interest.

5. Reassemble the gel with the lanes containing the stained size standards and single plugs and locate the approximate region of the unstained preparative lane containing the DNA of interest. Carefully excise this region with a scalpel and transfer the gel slice to a snap-cap polypropylene tube.

6. Cover the gel fragment with agarase buffer and incubate for 1 hour at room temperature with occasional agitation. Discard the buffer and repeat the procedure twice more.

7. After the buffer exchange is complete, melt the agarose slice containing the fractionated DNA at 65–68°C. During the melting step, swirl the tube to ensure complete melting.

8. Add an appropriate amount of agarase to the melted gel and digest the gel at the temperature recommended by the manufacturer.

   Digestions are usually carried out at temperatures between 37°C and 45°C and for times ranging from 1 hour to overnight.

9. After digestion, inactivate the agarase by heating (according to manufacturer instructions) or remove the enzyme by extraction with phenol:chloroform.

   Heat inactivation is preferred to avoid possible shearing of the DNA during extraction with organic solvents. Ligation or restriction enzyme digestion of the released DNA can be carried out in the presence of the agarose monomers produced by the agarase. Alternatively, the DNA can be precipitated in the presence of 0.3 M sodium acetate (pH 5.2) and 2 volumes of isopropanol.

Retrieval of DNA Fragments from Pulsed-field Gels following DNA Concentration

DNA contained in a slice of low-melting-temperature agarose is first concentrated by electrophoresis into a high-percentage agarose gel and then isolated by treatment with agarase. The resulting DNA preparation is purified by microdialysis.

MATERIALS

CAUTION: Please see Appendix 4 for appropriate handling of materials marked with <!>.

Reagents and Solutions

Please see Appendix 1 for components of stock solutions, buffers, and reagents. Dilute stock solutions to the appropriate concentrations.

Ethidium bromide (10 mg/ml) <!>

or

SYBR Gold solution (supplied commercially as a 10,000x concentrate in DMSO)

Equilibration buffer

1x TBE
100 mM NaCl
30 µM spermine
70 µM spermidine

Injection/transfection buffer

10 mM Tris-Cl (pH 7.5)
0.1 mM EDTA (pH 8.0)
100 mM NaCl
30 µM spermine
70 µM spermidine

Enzymes and Buffers

Agarase, e.g., GELase (Epicentre Technologies) or β-agarase I (New England BioLabs)

Agarase digestion buffer

Use the buffer supplied by the manufacturer, supplemented with 100 mM NaCl, 70 µM spermidine.

Nucleic Acids/Oligonucleotides

DNA size standards

See Protocols 5.14 and 5.16. Standards can also be purchased from New England BioLabs, Bio-Rad, and others.

Genomic DNA embedded in low-melting-temperature agarose plugs (Protocols 5.13 and 5.14)

Gels/Loading Buffers

Minigel, poured in a taped gel mold and cast without a comb using 5% NuSieve GTG agarose (Cambrex and other suppliers)

Additional Items

Membranes for drop dialysis (0.05-µm pore size), available from Millipore

Scalpel blades

Steps 1 and 2 of this protocol require the reagents listed in either Protocol 5.17 or 5.18.

Water baths (65–68°C and the temperature recommended by the manufacturer for agarase digestion)

Adapted from Chapter 5, Protocol 20, p. 5.86 of MC3.

METHOD

1. Prepare a preparative low-melting-point agarose PFGE that will provide optimum resolution of the DNA fragment or size fraction of interest.

2. Load the DNA size standards and single plugs of the digested genomic DNA in lanes located on both sides of the preparative slot. Load the preparative sample into the preparative slot. Perform electrophoresis as described in Protocol 5.17 (TAFE gel) or Protocol 5.18 (for CHEF gel).

3. After electrophoresis, cut the lanes containing the size standards and single genomic DNA plugs from the gel and stain them with ethidium bromide or SYBR Gold for 30 minutes at room temperature. If necessary, destain the gel slices in H$_2$O for 30 minutes. Do not stain the preparative lane.

4. Examine the stained slices by UV illumination and make notches on the slices to mark the locations of markers flanking the position of DNA of interest.

5. Under normal illumination, reassemble the gel with the lanes containing the stained size standards and single plugs, and locate the approximate region of the unstained preparative lane containing the DNA of interest. Carefully excise this region with a scalpel and transfer the gel slice to a snap-cap polypropylene tube.

6. Equilibrate the gel slice containing the size-fractionated DNA in 40 ml of PFGE equilibration buffer for 20–30 minutes at room temperature. Agitate the mixture gently throughout this period.

7. Pour off the buffer and melt the gel slice at 65–68°C, gently swirling the tube periodically to ensure complete melting. Record the volume of the melted gel slice.

8. While the 5% NuSieve GTG minigel is still in the taped casting tray, slice off enough of the top of the gel to accommodate the volume of the melted gel slice that contains DNA.

9. Pour the melted gel slice into the casting tray and allow it to harden. Concentrate the size-fractionated DNA in the 5% gel by applying 60 V for 12 minutes per millimeter of low-melting-temperature gel.

10. When electrophoresis is complete, slice a very thin section from the *center* of the gel and stain it with ethidium bromide. Determine how far into the gel the DNA has migrated (usually ~2 mm).

    It is important to stain a sliver from the center of the gel as some smiling occurs during electrophoresis.

11. Remove the low-melting portion of the gel and trim as small a portion as possible of the 5% gel containing the concentrated DNA.

12. Equilibrate the gel slice containing the DNA in 12 ml of 1x agarase digestion buffer containing 100 mM NaCl, 30 μM spermine, and 70 μM spermidine. Incubate the gel slice in this buffer for 20 minutes at room temperature with gentle agitation.

13. Drain off the buffer, transfer the gel slice to a microfuge tube, and melt the DNA slice at 65–68°C. Transfer the melted gel to a water bath set at a temperature optimal for the agarase preparation (recommended by the manufacturer).

14. Incubate the melted gel slice for 15 minutes and add an appropriate amount of agarase to digest the starting volume of 5% gel.

15. After digestion, centrifuge the tube at maximum speed for 5 minutes in a microfuge to pellet debris and transfer the supernatant to a fresh microfuge tube.

16. Set up a drop dialysis of the supernatant using membranes with a 0.05-μm pore size.

    a. Spot the supernatant onto the center of the membrane, floating shiny side up on 100 ml of injection/transfection buffer.

b. Dialyze for 1 hour at room temperature. Replace the original buffer with 100 ml of fresh injection/transfection buffer and dialyze for an additional hour.

c. Transfer the DNA to a clean microfuge tube.

> Drop dialysis removes agarase and the digested carbohydrates released from the agarose.
>
> After drop dialysis, the concentration of the DNA can be estimated by gel electrophoresis.
>
> The DNA can be injected directly or combined with a lipofection reagent for transfection into cultured cells.

CHAPTER 6

Preparation and Analysis of Eukaryotic Genomic DNA

BACKGROUND INFORMATION

Background information found in *Molecular Cloning: A Laboratory Manual,* 3rd edition (hereafter MC3) unless otherwise indicated.

Minimizing damage to large DNA molecules	MC3, p. 2.110
Phenol	MC3, pp. A1.23–A1.24
Southern blotting and hybridization	MC3, pp. 6.33–6.39
Spooling DNA	MC3, p. 6.61

Isolation of High-molecular-weight DNA from Mammalian Cells Using Proteinase K and Phenol

This is the method of choice when large amounts of mammalian DNA are required, for example, for Southern blotting (Protocols 6.8–6.10) or for construction of genomic libraries. Approximately 200 µg of mammalian DNA, 100–150 kb in length, is obtained from 5 x 10^7 cultured aneuploid mammalian cells (e.g., HeLa cells). The usual yield of DNA from 20 ml of normal blood is approximately 250 µg.

The initial stages of the procedure vary, depending on the type of samples used (cells growing in monolayer, cells growing in suspension, freshly drawn blood, or frozen blood). The protocol therefore contains four alternative versions of Step 1, each of which requiring different reagents and equipment.

MATERIALS

CAUTION: Please see Appendix 4 for appropriate handling of materials marked with <!>.

WARNING: Primate tissues and priming cultures of cells require special handling precautions.

Reagents and Solutions

Please see Appendix 1 for components of stock solutions, buffers, and reagents. Dilute stock solutions to the appropriate concentrations.

    Ammonium acetate (10 M) (optional; see Step 9)

    Dialysis buffer (optional; see Step 9)

        50 mM Tris-Cl (pH 8.0)

        10 mM EDTA (pH 8.0)

    Ethanol (optional; see Step 9)

    Phenol, equilibrated with 0.1 M Tris-Cl (pH 8.0) <!>

    TE (pH 8.0)

Cells, Tissues/Culture Media

Monolayers or suspensions of mammalian cells, fresh tissue, or blood samples

Enzymes and Buffers

Proteinase K (20 mg/ml)

Nucleic Acids/Oligonucleotides

Bacteriophage λ DNA

or

Commercially available DNA markers (50–200 kb in size)

Gel-loading buffer IV

Gels/Loading Buffers

Pulsed-field gel

or

Conventional horizontal 0.6% agarose gel

Centrifuges/Rotors/Tubes

Sorvall SS-34 and H-1000B rotors (room temperature)

 Adapted from Chapter 6, Protocol 1, p. 6.4 of MC3.

Additional Items

Dialysis tubing and closures (optional; see Step 9)
Dog nail clippers
Tube mixer or roller apparatus
Wide-bore pipettes (0.3-cm internal diameter orifice)

Additional Information

Additional protocol (Estimating the Concentration of DNA by Fluorometry)	MC3, p. 6.12
Proteinase K	MC3, p. A4.50
Quantitation of nucleic acids using fluorescent dyes	MC3, pp. A8.19–A8.23
Spectrophotometric methods of measuring the concentration of DNA	MC3, pp. A8.19–A8.21

METHOD

1A. Cells growing in monolayer cultures

Reagents and equipment

Bed of crushed ice
Erlenmayer flask (see Step f)
Lysis buffer
    10 mM Tris-Cl (pH 8.0)
    0.1 M EDTA (pH 8.0)
    0.5% (w/v) SDS
    20 mg/ml DNase-free RNase
Rubber policeman
Sorvall H-1000B rotor (4°C)
TE (pH 8.0)
Tris-buffered saline (TBS), ice cold

Work with batches of 10–12 culture dishes at a time. Store the remaining culture dishes in the incubator until they are required.

a. Take one batch of culture dishes (containing cells grown to confluency) from the incubator and immediately remove the medium by aspiration. Working quickly, wash the monolayers of cells twice with ice-cold TBS. This is most easily accomplished by gently pipetting approximately 10 ml of TBS onto the first monolayer. Swirl the dish gently for a few seconds and then tip the fluid into a 2-liter beaker. Add another 10 ml of ice-cold TBS and store the dish on a bed of ice. Repeat the procedure until the entire batch of monolayers has been processed.

b. Tip the fluid from the first monolayer into the 2-liter beaker. Remove the last traces of TBS from the culture dish by aspiration. Add 1 ml of fresh ice-cold TBS and store the dish on a bed of ice. Repeat the procedure until the entire batch of monolayers has been processed.

c. Use a rubber policeman to scrape the cells from the first culture dish into the 1 ml of TBS. Use a pasteur pipette to transfer the cell suspension to a centrifuge tube on ice. Immediately wash the culture dish with 0.5 ml of ice-cold TBS and combine the washings with the cell suspension in the centrifuge tube. Process the remaining monolayers in the same way.

d. Recover the cells by centrifugation at 1500*g* (2700 rpm in a Sorvall H-1000B rotor and swinging buckets) for 10 minutes at 4°C.

e. Resuspend the cells in 5–10 volumes of ice-cold TBS and repeat the centrifugation.

f. Resuspend the cells in TE (pH 8.0) at a concentration of 5×10^7 cells/ml. Transfer the solution to an Erlenmeyer flask.

> For 1 ml of cell suspension, use a 50-ml flask; for 2 ml, use a 100-ml flask, and so on.

g. Add 10 ml of lysis buffer for each milliliter of cell suspension. Incubate the suspension for 1 hour at 37°C and then proceed immediately to Step 2.

> Make sure that the cells are well dispersed over the inner surface of the Erlenmeyer flask when the lysis buffer is added.

1B. Cells growing in suspension cultures

Reagents and equipment

Erlenmayer flask (see Step c)
Lysis buffer
>   10 mM Tris-Cl (pH 8.0)
>   0.1 M EDTA (pH 8.0)
>   0.5% (w/v) SDS
>   20 mg/ml DNase-free RNase

Sorvall H-1000B rotor (4°C)
TE (pH 8.0)
Tris-buffered saline (TBS), ice cold

a. Transfer the cells to a centifuge tube or bottle and recover them by centrifugation at 1500g (2700 rpm in a Sorvall H-1000B rotor and swinging buckets) for 10 minutes at 4°C. Remove the supernatant medium by aspiration.

b. Wash the cells by resuspending them in a volume of ice-cold TBS equal to the volume of the original culture. Repeat the centrifugation. Remove the supernatant by aspiration and then gently resuspend the cells once more in ice-cold TBS. Recover the cells by centrifugation.

c. Remove the supernatant by aspiration and gently suspend the cells in TE (pH 8.0) at a concentration of 5×10^7 cells/ml. Transfer the suspension to an Erlenmeyer flask.

> For 1 ml of cell suspension, use a 50-ml flask; for 2 ml, use a 100-ml flask, and so on.

d. Add 10 ml of lysis buffer for each milliliter of cell suspension. Incubate the solution for 1 hour at 37°C and then proceed immediately to Step 2.

> Make sure that the cells are well dispersed over the inner surface of the Erlenmeyer flask when the lysis buffer is added.

1C. Tissue samples

Reagents and equipment

Liquid nitrogen
Lysis buffer
>   10 mM Tris-Cl (pH 8.0)
>   0.1 M EDTA (pH 8.0)
>   0.5% (w/v) SDS
>   20 mg/ml DNase-free RNase

Polypropylene tubes (50 ml)
Waring blender, equipped with a stainless-steel container
or
Mortar and pestle, prechilled in liquid nitrogen

a. Drop approximately 1 g of freshly excised tissue into liquid nitrogen in the stainless-steel container of a Waring blender. Blend at top speed until the tissue is ground to a powder.

Alternatively, smaller quantities of tissue can be snap-frozen in liquid nitrogen and then pulverized to a powder using a mortar and pestle precooled with liquid nitrogen.

b. Allow the liquid nitrogen to evaporate and add the powdered tissue little by little to approximately 10 volumes (w/v) of lysis buffer in a beaker. Allow the powder to spread over the surface of the lysis buffer and then shake the beaker to submerge the material.

c. When all of the material is in solution, transfer the suspension to a 50-ml centrifuge tube. Incubate the tube for 1 hour at 37°C and proceed to Step 2.

1D. Blood cells

Human blood must be collected by a trained phlebotomist under sterile conditions.

To collect cells from freshly drawn blood

Reagents and equipment

ACD Blood tubes
These tubes, which are widely used in phlebotomy, are commercially available and contain 3.5 ml of acid citrate dextrose solution as an anticoagulant. Tubes containing EDTA as a coagulant may also be used, although some loss of integrity of high-molecular-weight DNA can occur.
Lysis buffer
10 mM Tris-Cl (pH 8.0)
0.1 M EDTA (pH 8.0)
0.5% (w/v) SDS
20 mg/ml DNase-free RNase
Sorvall H-1000B rotor (4°C)

a. Collect approximately 20 ml of fresh blood in tubes containing 3.5 ml of either ACD (acid citrate dextrose solution B) or EDTA.
   The blood may be stored for several days at 0°C or indefinitely at −70°C before the DNA is prepared. Blood should not be collected into heparin, which is an inhibitor of the polymerase chain reaction.

b. Transfer the blood to a centrifuge tube and centrifuge at 1300g (2500 rpm in a Sorvall H-1000B rotor and 50-ml swinging buckets) for 15 minutes at 4°C.

c. Remove the supernatant fluid by aspiration. Use a pasteur pipette to transfer the buffy coat carefully to a fresh tube and repeat the centrifugation. Discard the pellet of red cells.
   The buffy coat is a broad band of white blood cells of heterogeneous density.

d. Remove residual supernatant from the buffy coat by aspiration. Resuspend the buffy coat in 15 ml of lysis buffer. Incubate the solution for 1 hour at 37°C and proceed to Step 2.

To collect cells from frozen blood samples

Reagents and equipment

ACD Blood tubes
These tubes, which are widely used in phlebotomy, are commercially available and contain 3.5 ml of acid citrate dextrose solution as an anticoagulant. Tubes containing EDTA as a coagulant may also be used, although some loss of integrity of high-molecular-weight DNA can occur.
Lysis buffer
10 mM Tris-Cl (pH 8.0)
0.1 M EDTA (pH 8.0)
0.5% (w/v) SDS
20 mg/ml DNase-free RNase
Sorvall SS-34 rotor (room temperature)
Water bath (37°C)

a. Collect approximately 20 ml of fresh blood in tubes containing 3.5 ml of either ACD or EDTA.

> The blood may be stored for several days at 0°C or indefinitely at –70°C before the DNA is prepared.

b. Thaw the blood in a water bath at room temperature and then transfer it to a centrifuge tube. Add an equal volume of phosphate-buffered saline at room temperature.

c. Centrifuge the blood at 3500*g* (5400 rpm in a Sorvall SS-34 rotor) for 15 minutes at room temperature.

d. Remove the supernatant, which contains lysed red cells, by aspiration. Resuspend the pellet in 15 ml of lysis buffer. Incubate the solution for 1 hour at 37°C and proceed to Step 2.

2. Transfer the lysate to one or more centrifuge tubes that fit into a Sorvall SS-34 rotor or its equivalent. The tubes should not be more than one-third full.

3. Add proteinase K (20 mg/ml) to a final concentration of 100 µg/ml. Use a glass rod to mix the enzyme solution gently into the viscous lysate of cells.

4. Incubate the lysate in a water bath for 3 hours at 50°C. Swirl the viscous solution from time to time.

5. Cool the solution to room temperature and add an equal volume of phenol equilibrated with 0.1 M Tris-Cl (pH 8.0). Gently mix the two phases by slowly turning the tube end-over-end for 10 minutes on a tube mixer or roller apparatus. If the two phases have not formed an emulsion at this stage, place the tube on a roller apparatus for 1 hour.

6. Separate the two phases by centrifugation at 5000*g* (6500 rpm in a Sorvall SS-34 rotor) for 15 minutes at room temperature.

7. Use a wide-bore pipette (0.3-cm diameter orifice) to transfer the viscous aqueous phase to a fresh centrifuge tube.

> When transferring the aqueous (upper) phase, it is essential to draw the DNA into the pipette very slowly to avoid disturbing the material at the interface and to minimize hydrodynamic shearing forces. If the DNA solution is so viscous that it cannot easily be drawn into a wide-bore pipette, use a long pipette attached to an aspirator to remove the organic (lower) phase.

8. Repeat the extraction with phenol twice more and pool the aqueous phases.

9. Isolate DNA by one of the following two methods.

To isolate DNA in the size range of 150–200 kb

a. Transfer the pooled aqueous phases to a dialysis bag. Close the top of the bag with a dialysis tubing clip, allowing room in the bag for the sample volume to increase 1.5- to 2-fold during dialysis.

b. Dialyze the solution at 4°C against 4 liters of dialysis buffer. Change the buffer three times at intervals of ≥6 hours.

> Because of the high viscosity of the DNA solution, dialysis generally takes ≥24 hours to complete.

To isolate DNA that has an average size of 100–150 kb

a. After the third extraction with phenol, transfer the pooled aqueous phases to a fresh centrifuge tube and add 0.2 volume of 10 M ammonium acetate. Add 2 volumes of ethanol at room temperature and swirl the tube until the solution is thoroughly mixed.

b. The DNA immediately forms a precipitate. Remove the precipitate in one piece from the ethanolic solution with a Shepherd's crook (a pasteur pipette whose end has been sealed and shaped into a U; please see Steps 5–7 of Protocol 6.3). Contaminating oligonucleotides remain in the ethanolic phase.

c. If the DNA precipitate becomes fragmented, abandon the Shepherd's crook and collect the precipitate by centrifugation at 5000*g* (6500 rpm in a Sorvall SS-34 rotor) for 5 minutes at room temperature.

d. Wash the DNA precipitate twice with 70% ethanol and collect the DNA by centrifugation as described in Step c.

e. Remove as much of the 70% ethanol as possible, using an aspirator. Store the pellet of DNA in an open tube at room temperature until the last visible traces of ethanol have evaporated.

> Do not allow the pellet of DNA to dry completely; desiccated DNA is very difficult to dissolve.

f. Add 1 ml of TE (pH 8.0) for each 0.1 ml of cells (Step 1). Place the tube on a rocking platform and gently rock the solution for 12–24 hours at 4°C until the DNA has completely dissolved. Store the DNA solution at 4°C.

10. Measure the concentration of the DNA by absorbance at 260 nm or by fluorometry.

> It is often difficult to measure the concentration of high-molecular-weight DNA by standard methods such as absorbance at 260 nm. This is because the DNA solution is so viscous that it is impossible to withdraw a representative sample. This problem can be minimized by withdrawing a large sample (10–20 μl) using an automatic pipettor equipped with a cutoff yellow tip. The sample is then diluted with 0.5 ml of TE (pH 8.0) and vortexed vigorously for 1–2 minutes. The absorbance of the diluted sample can then be read in the standard way.

11. Analyze the quality of the preparation of high-molecular-weight DNA by pulsed-field gel electrophoresis (Protocol 5.17 or 5.18) or by electrophoresis through a conventional 0.6% agarose gel (Protocol 5.1). Use unit-length and/or linear concatemers of λ DNA as markers. A method to generate linear concatemers of λ DNA is described in Protocol 5.16.

> Do not be concerned if some of the DNA remains in the well, since DNA molecules >250 kb have difficulty entering the gel. This problem can usually be solved by embedding the DNA in a small amount of melted agarose (at 55°C) and transferring the molten solution to the well of a preformed agarose gel.

Isolation of High-molecular-weight DNA from Mammalian Cells Using Formamide

This procedure, although lengthy, generates preparations of DNA that are large enough (200 kb) to be used for the construction of genomic libraries in high-capacity vectors and for analysis by pulsed-field electrophoresis. However, the concentration of DNA in the final preparation is low (~10 μg/ml), as is the yield (~1 mg of DNA/10^8 cultured aneuploid mammalian cells, such as HeLa cells).

MATERIALS

CAUTION: Please see Appendix 4 for appropriate handling of materials marked with <!>.

Reagents and Solutions

Please see Appendix 1 for components of stock solutions, buffers, and reagents. Dilute stock solutions to the appropriate concentrations.

Dialysis buffer 1 (4°C)
  20 mM Tris-Cl (pH 8.0)
  0.1 M NaCl
  10 mM EDTA (pH 8.0)
Dialysis buffer 2 (4°C)
  10 mM Tris-Cl (pH 8.0)
  10 mM NaCl
  0.5 mM EDTA (pH 8.0)
Formamide denaturation buffer
  20 mM Tris-Cl (pH 8.0)
  0.8 M NaCl
  80% (v/v) formamide <!>
TE (pH 8.0)

Cells, Tissues/Culture Media

Monolayers or suspensions of mammalian cells, fresh tissue, or blood samples

Nucleic Acids/Oligonucleotides

Bacteriophage λ DNA
or
Commercially available DNA markers (50–200 kb in size)

Gels/Loading Buffers

Pulsed-field gel (see Protocols 5.17 and 5.18)
or
Conventional horizontal 0.6% agarose gel (see Protocol 5.1)
Gel-loading buffer IV

Additional Items

Collodion bags (supplied by Sartorius in 20% ethanol)
Dialysis tubing and closures
Dog nail clippers to make cutoff yellow tips for automatic micropipette

Adapted from Chapter 6, Protocol 2, p. 6.13 of MC3.

> Glass rod
> Materials and equipment used in Protocol 6.1, Steps 1–4
> Tube mixer or roller apparatus
>
> **Additional Information**
>
> | | |
> |---|---|
> | Additional protocol (Estimating the Concentration of DNA by Fluorometry) | MC3, p. 6.12 |
> | Quantitation of nucleic acids using fluorescent dyes | MC3, pp. A8.19–A8.23 |
> | Spectrophotometric methods of measuring the concentration of DNA | MC3, pp. A8.19–A8.21 |

METHOD

1. Prepare lysates of cell suspensions (or frozen cell powders) as described in Protocol 6.1, Steps 1–4.

2. Cool the solution containing lysed cells and lysis buffer to 15°C. For every 1 ml of cell lysate, add 1.25 ml of formamide denaturation buffer and mix the solution gently using a glass rod. Store the solution for 12 hours at 15°C.

3. Rinse one or more collodion bags in dialysis buffer 2 and then immerse the bags in 100 ml of the buffer for 30 minutes. Pour the viscous DNA solution into one or more collodion bags. Secure the open end of the bag with a dialysis clip and dialyze the solution for 45 minutes at 4°C in 2 liters of dialysis buffer 1. Replace the buffer with fresh dialysis buffer 1 and continue the dialysis for at least 4 hours, followed by an additional 4 hours in a third 2-liter aliquot of dialysis buffer 1. Then dialyze the DNA against 2 liters of fresh dialysis buffer 2, three times, for at least 4 hours each.

   Dialysis intervals should be 45 minutes for the first buffer change and 4 hours for all subsequent changes.

   A total dialysis time of 24 hours is required to remove proteins from the DNA effectively.

4. Measure the concentration of the DNA by absorbance at 260 nm or by fluorometry.

   It is often difficult to measure the concentration of high-molecular-weight DNA by standard methods such as absorbance at 260 nm. This is because the DNA solution is so viscous that it is impossible to withdraw a representative sample. This problem can be minimized by withdrawing a large sample (10–20 μl) using an automatic pipettor equipped with a cutoff yellow tip. The sample is then diluted with 0.5 ml of TE (pH 8.0) and vortexed vigorously for 1–2 minutes. The absorbance of the diluted sample can then be read in the standard way.

5. Analyze the quality of the preparation of high-molecular-weight DNA by pulsed-field gel electrophoresis (Protocol 5.17 or 5.18) or by electrophoresis through a conventional 0.6% agarose gel (Protocol 5.1). Use unit-length and linear concatemers of λ DNA as markers (please see Protocol 5.16). The genomic DNA should be more than 200 kb in size.

   Do not be concerned if some of the DNA remains in the well, since DNA molecules >250 kb have difficulty entering the gel. This problem can usually be solved by embedding the DNA in a small amount of melted agarose (at 55°C) and transferring the molten solution to the well of a preformed agarose gel.

Isolation of DNA from Mammalian Cells by Spooling

This procedure is used to prepare DNA simultaneously from many different types of samples or tissues. Although the DNA is generally too small (~80 kb) for efficient construction of genomic DNA libraries, it gives excellent results in Southern hybridizations and PCRs. Cultured aneuploid mammalian cells (2×10^7, e.g., HeLa cells) yield 100 µg of DNA in a volume of 1 ml.

MATERIALS

CAUTION: Please see Appendix 4 for appropriate handling of materials marked with <!>.

Reagents and Solutions

Please see Appendix 1 for components of stock solutions, buffers, and reagents. Dilute stock solutions to the appropriate concentrations.
    Ethanol (room temperature)
    Mammalian cell lysis solution
        6 M guanidinium hydrochloride <!>
        0.1 M sodium acetate (pH 5.5)
    TE (pH 8.0)

Cells, Tissues/Culture Media

Monolayers or suspensions of mammalian cells, fresh tissue, or blood samples

Nucleic Acids/Oligonucleotides

Bacteriophage λ DNA
or
Commercially available DNA markers (30–100 kb in size)

Gels/Loading Buffers

Pulsed-field gel (see Protocols 5.17 and 5.18)
or
Conventional horizontal 0.6% agarose gel (see Protocol 5.1)
Gel-loading buffer IV

Additional Items

Materials and equipment used in Protocol 6.1, Step 1
Rocking platform (4°C)
Shepherd's crooks (sealed pasteur pipettes with bent ends)
Wide-bore pipettes (0.3-cm internal diameter orifice)

Additional Information

Additional protocol (Estimating the Concentration of DNA by Fluorometry)	MC3, p. 6.12
Quantitation of nucleic acids using fluorescent dyes	MC3, pp. A8.19–A8.23
Spectrophotometric methods of measuring the concentration of DNA	MC3, pp. A8.19–A8.21
Spooling DNA	MC3, p. 6.61

Adapted from Chapter 6, Protocol 3, p. 6.16 of MC3.

METHOD

1. Prepare cell suspensions (or frozen cell powders) as described in Protocol 6.1, Step 1.

2. Lyse the cells by one of the following two methods.

    ### For lysis of cells from suspensions

    a. Transfer the cell suspensions to disposable 50-ml polypropylene centrifuge tubes.

    b. Add 7.5 volumes of mammalian cell lysis solution.

    ### For lysis of cells from tissues

    a. Add the frozen cell powders little by little to approximately 7.5 volumes of cell lysis solution in beakers. Allow the powders to spread over the surface of the lysis solution and then shake the beakers to submerge the material.

    b. When all of the material is in solution, transfer the solutions to 50-ml disposable centrifuge tubes.

3. Close the tubes and incubate them for 1 hour at room temperature on a rocking platform.

4. Dispense 18 ml of ethanol at room temperature into each of a series of disposable 50-ml polypropylene centrifuge tubes. Use wide-bore pipettes to layer the cell suspensions carefully *under* the ethanol.

5. Recover the DNA from each tube by slowly stirring the interface between the cell lysate and the ethanol with a Shepherd's crook. The DNA will adhere to the crook, forming a gelatinous mass. Continue stirring until the ethanol and the aqueous phase are thoroughly mixed.

6. Transfer each Shepherd's crook, with its attached DNA, to a separate polypropylene tube containing 5 ml of ethanol at room temperature. Leave the DNA submerged in the ethanol until all of the samples have been processed.

7. Remove each crook, with its attached DNA, and allow as much ethanol as possible to drain away. By this stage, the DNA should have shrunk into a tightly packed, dehydrated mass; it is then possible to remove most of the free ethanol by capillary action by touching the U-shaped end of the crook to a stack of Kimwipes. Before all of the ethanol has evaporated from the DNA, transfer the crook into a fresh polypropylene tube containing 5 ml of ethanol at room temperature.

8. When all of the samples have been processed, again remove as much ethanol as possible (see Step 7).

    Do not allow the DNA pellets to dry completely or they will be very difficult to dissolve.

9. Transfer each pipette to a fresh polypropylene tube containing 1 ml of TE (pH 8.0). Allow the DNAs to rehydrate by storing the tubes overnight at 4°C.

10. During rehydration, the DNAs become highly gelatinous but remain attached to their pipettes. Use fresh Shepherd's crooks as scrapers to free the pellets of DNA gently from their pipettes. Discard the pipettes, leaving the DNA pellets floating in the TE. Close the tubes and incubate them at 4°C on a rocking platform until the pellets are completely dissolved (~24–48 hours).

    The level of contamination by RNA is kept within acceptable limits if 1.5 ml or more of lysis solution is used per 10^7 cells. However, the amount of RNA contaminating the DNA sample can be further reduced by adding DNase-free RNase (final concentration of 1 μg/ml) to the solution DNA.

11. Analyze an aliquot by pulsed-field gel electrophoresis (Protocol 5.17 or 5.18) or by electrophoresis through a 0.6% agarose gel (Protocol 5.1). Store the DNA at 4°C.

    The DNA should be approximately 80 kb in size and migrate more slowly than monomers of bacteriophage λ DNA.

    Because DNA made by this procedure is always contaminated with a small amount of RNA, it is necessary to estimate the concentration of DNA in the final preparation either by fluorometry or by gel electrophoresis and staining with ethidium bromide (Protocol 5.2).

Isolation of DNA from Mammalian Cells Grown in 96-well Microtiter Plates

Each well yields sufficient genomic DNA for several standard PCRs or for analysis in a single lane of a Southern hybridization.

MATERIALS

Reagents and Solutions

Please see Appendix 1 for components of stock solutions, buffers, and reagents. Dilute stock solutions to the appropriate concentrations.

Ethanol (room temperature)
Ethanol (70% at room temperature)
Microtiter cell lysis solution
    10 mM Tris-Cl (pH 7.5)
    10 mM NaCl
    10 mM EDTA (pH 8.0)
    0.5% (w/v) Sarkosyl (usually supplied as a 30% solution in H_2O)
NaCl/ethanol solution
    Add 150 µl of 5 M NaCl per 10 ml of absolute ethanol. Store the solution at –20°C.
Phosphate-buffered saline (PBS)
TE (pH 8.0)

Cells, Tissues/Culture Media

Mammalian cells growing in 96-well microtiter plates

Enzymes and Buffers

DNase-free pancreatic RNase (RNase A)
Restriction enzymes and buffers

Nucleic Acids/Oligonucleotides

Bacteriophage λ DNA
or
Commercially available DNA markers (30–100 kb in size)

Gels/Loading Buffers

Pulsed-field gel (see Protocols 5.17 and 5.18)
or
Conventional horizontal 0.6% agarose gel (see Protocol 5.1)
Gel-loading buffer IV

Additional Items

Materials and equipment used in Protocol 6.1, Step 1
Multichannel pipettor, 8 or 12 channels
Oven (60°C and 37°C)
Rocking platform
Tupperware containers, tough enough to withstand incubation at 60°C

> **Additional Information**
>
> Additional protocol (Optimizing Genomic DNA Isolation for PCR)   MC3, p. 6.22

METHOD

1. Remove the medium from confluent cultures of cells growing in individual wells of 96-well plates by aspiration through a blue pipette tip or a pasteur pipette.

2. Rinse the monolayers of cells in the individual wells twice with 100 µl of phosphate-buffered saline.

3. Use a multichannel pipettor to add 50 µl of microtiter cell lysis solution to each well of the microtiter plate. Place several wet paper towels in a polypropylene box (e.g., a Tupperware box) and then stack the microtiter plates containing the lysis buffer and cells on top of the towels. Seal the box tightly with the lid.

4. Incubate the sealed box for 12–16 hours in a 60°C oven.

5. Remove the box from the oven, place the plates on a flat benchtop, and allow them to cool for a few minutes before adding 100 µl of NaCl/ethanol solution per well. Store the plates for 30 minutes at room temperature without mixing. A stringy precipitate of nucleic acid should be visible at the end of the incubation.

6. Slowly invert each plate over a sink to decant the ethanolic solution. The precipitated nucleic acid should remain attached to the base of the wells. Place each plate in an upside-down position on a bed of dry paper towels and allow the remaining ethanol to drain from the plate.

7. Add 150 µl of 70% ethanol to each well, being careful not to dislodge the precipitate of nucleic acid. Discard the 70% ethanol by inverting the plate as in Step 6. Blot the excess liquid on a bed of paper towels. Rinse the precipitates of DNA twice more with 70% ethanol.

8. Allow the plates to dry at room temperature until the last traces of ethanol have evaporated. If the genomic DNA is to be analyzed by PCR, then proceed to Step 9. If the DNA is to be analyzed by Southern hybridization, proceed to Steps 10–12.

9. Add 30–50 µl of TE (pH 8.0) to each well and allow the DNA to dissolve during gentle rocking for 12–16 hours at room temperature.

   > Dissolution of the DNA can be accelerated by placing the microtiter dishes on the heating block of a thermal cycler that is programmed to cycle 10 times between 80°C and 50°C (1 minute at each temperature).

   > The DNA may now be used as template for standard PCR (please see Protocol 8.1).

10. If the DNA is to be analyzed by Southern blotting, make up the following restriction enzyme mixture; 40 µl of the mixture will be required for each well.

H_2O	0.8 volume
10x restriction enzyme buffer	0.1 volume
DNase-free RNase	10 µg/ml

    Just before use, add 10 units of restriction enzyme for each 40 µl of mixture.

11. Use a multichannel pipettor to add 40 µl of the restriction enzyme mixture to each well. Mix the contents of the wells by pipetting up and down several times, taking care to avoid air bubbles. Incubate the reactions at the appropriate digestion temperature for 12–16 hours in a humidified sealed Tupperware box as described in Step 3.

12. Stop the reactions by adding 5–10 µl of sucrose gel-loading buffer and analyze the digested DNA by Southern blotting and hybridization as described in Protocols 6.8–6.10.

Preparation of Genomic DNA from Mouse Tails and Other Small Samples

This simple protocol is used to extract DNA from small numbers of cultured cells and from fragments of soft or bony tissues. The method is used chiefly to genotype transgenic and knockout mice. Each 6–10-mm snippet of mouse tail yields 50–100 μg of DNA that can be used in dot or slot blotting to detect a transgene of interest, in Southern hybridization to detect DNA fragments that are <20 kb in size, and as a template in PCRs.

MATERIALS

CAUTION: Please see Appendix 4 for appropriate handling of materials marked with <!>.

Reagents and Solutions

Please see Appendix 1 for components of stock solutions, buffers, and reagents. Dilute stock solutions to the appropriate concentrations.

Ethanol (70%)

Isopropanol

Phenol:chloroform (1:1, v/v) <!>

Phosphate-buffered saline (PBS)

SNET

  20 mM Tris-Cl (pH 8.0)

  5 mM EDTA (pH 8.0)

  400 mM NaCl

  1% SDS

Cells, Tissues/Culture Media

Cultured mammalian cells

Monolayer cultures, grown to confluence or semiconfluence in 100-mm dishes, should be washed twice with ice-cold phosphate-buffered saline and then immediately lysed by addition of 1 ml of SNET containing 400 μg/ml proteinase K, as described in Step 1.

Cells growing in suspension should be recovered by centrifugation, washed twice in ice-cold phosphate-buffered saline, and then resuspended in TE (pH 8.0) at a concentration of 5 x 10^7/ml. Aliquots of the suspension (0.2 ml) are then transferred to a series of 17 x 100-mm Falcon polypropylene tubes and the cells are immediately lysed with SNET containing 400 μg/ml proteinase K, as described in Step 1.

Mouse tails

Samples of mouse tails are generally cut from 10-day old suckling animals or at the time of weaning (~3 weeks of age). In the former case, the distal third of the tail is removed and transferred into a microfuge. In the latter case, 6–10 mm of the tail is removed under anesthesia and transferred to a 17 x 100-mm Falcon polypropylene tube. Under rare circumstances, where obtaining a result rapidly is of paramount importance, the entire tail can be removed from newborn animals and transferred to a microfuge tube.

Mouse tissue

To isolate DNA from mouse tissue (other than tail snippets), transfer approximately 100 mg of the freshly dissected tissue to a 17 x 100-mm Falcon polypropylene tube.

All experiments carried out on laboratory mice, including removing sections of tail, require prior authorization from the appropriate institutional ethics committee.

Enzymes and Buffers

Proteinase K (20 mg/ml)

Centrifuges/Rotors/Tubes

Sorvall H-1000B and SH-3000 rotors (room temperature and 4°C)

Additional Items

Rocking platform (55°C, room temperature, and 4°C)

Additional Information

Alternative protocols (Isolation of DNA from Mouse Tails without Extraction by Organic Solvents, One-tube Isolation of DNA from Mouse Tails, and DNA Extraction from Paraffin Blocks)	MC3, pp. 6.26–6.27
Minimizing damage to large DNA molecules	MC3, p. 2.110
Proteinase K	MC3, p. A4.50

METHOD

1. Prepare the appropriate amount of lysis buffer (see Table 6-1) by adding proteinase K to a final concentration of 400 μg/ml in SNET. Add lysis buffer to the mouse tails or other tissues.

   This procedure also can be used to isolate DNA from monolayers of cultured mammalian cells. In this case, 1 ml of SNET containing 400 μg/ml proteinase K is added directly to 100-mm monolayers that have been rinsed twice in phosphate-buffered saline. The viscous cell slurry is scraped from the dish with a rubber policeman and transferred to a 17 x 100-mm Falcon polypropylene tube.

   Cells growing in suspension that have been washed twice in phosphate-buffered saline are resuspended in TE and lysed with SNET containing 400 μg of proteinase K (1 ml per 10^9 cells).

TABLE 6-1 SNET Lysis Buffer Volumes

Age of Mouse	Amount of Tissue	Type of Tube	Volume of SNET Lysis Buffer (ml)
Newborn	entire tail (1 cm)	microfuge	0.5
10 days old	distal one-third	microfuge	0.5
Weanling (3–4 weeks)	6–10 mm	17 x 100-mm polypropylene	4.0
Any age	100 mg of fresh tissue	17 x 100-mm polypropylene	4.0

2. Incubate the tube overnight at 55°C in a horizontal position on a rocking platform or with agitation in a shaking incubator.

   It is important that the sample be mixed adequately during digestion. After overnight incubation, the tissue/tails should no longer be visible and the buffer should be a milky-gray color.

3. Add an equal volume of phenol:chloroform:isoamyl alcohol, seal the top of the tube, and place it on a rocking platform for 30 minutes at room temperature.

4. Separate the organic and aqueous phases by centrifugation. Centrifuge the samples in 17 x 100-mm polypropylene tubes at 666*g* (1800 rpm in a Sorvall H-1000B rotor with swinging buckets or 1600 rpm in a Sorvall SH-3000 swinging-bucket rotor) for 5 minutes at room temperature. Alternatively, for smaller sample volumes, centrifuge the samples in microfuge tubes at maximum speed for 5 minutes at room temperature in a microfuge. Transfer the upper aqueous phase to a fresh Falcon or microfuge tube.

5. Precipitate the DNA by adding an equal volume of isopropanol. Collect the precipitated DNA by centrifugation at 13,250g (8000 rpm in a Sorvall SH-3000 swinging-bucket rotor or maximum speed in a microfuge) for 15 minutes at 4°C.

6. Carefully remove the isopropanol. Rinse the pellet of DNA with 1 ml of 70% ethanol. If the pellets are loose, centrifuge the samples again for 5 minutes. Remove the 70% ethanol and allow the pellets to air dry for 15–20 minutes at room temperature.

   Do not allow the DNA pellets to dry completely or they will be very difficult to dissolve.

7. Dissolve the nucleic acid pellet by rocking it gently overnight in 0.5 ml of TE (pH 8.0) at 4°C.

8. Transfer the solution to a microfuge tube and store it at room temperature.

   Between 100 and 250 µg of genomic DNA is typically isolated from 1 cm (~100 mg) of mouse tail.

   The addition of bovine serum albumin at a concentration of 100 µg/ml to restriction enzyme digests of genomic DNA prepared by this method will absorb residual SDS and reduce the possibility of incomplete digestions. If problems persist, reextract the samples once more with phenol:chloroform and precipitate the DNA with 2 volumes of ethanol.

Rapid Isolation of Mammalian DNA

Mammalian DNA prepared from blood or tissues as described in this protocol is 20–50 kb in size and suitable for use as a template in PCRs. The yields of DNA vary between 0.5 and 3.0 µg/mg of tissue or 5 and 15 µg per 300 µl of whole blood.

MATERIALS

CAUTION: Please see Appendix 4 for appropriate handling of materials marked with <!>.

Reagents and Solutions

Please see Appendix 1 for components of stock solutions, buffers, and reagents. Dilute stock solutions to the appropriate concentrations.

Ethanol (70%)

Isopropanol

Potassium acetate solution (5 M) (store at room temperature)

> 60 ml of 5 M potassium acetate solution
>
> 11.5 ml of glacial acetic acid <!>
>
> 28.5 ml of H_2O
>
> The resulting solution is 5 M with respect to potassium ions and 3 M with respect to acetate ions.

Red blood cell lysis buffer

> 20 mM Tris-Cl (pH 7.6)

Cells, Tissues/Culture Media

Mouse tissue (10–20 mg)

or

60 µl of fresh whole blood

Enzymes and Buffers

DNase-free pancreatic RNase (RNase A)

Proteinase K (20 mg/ml) (optional; see Step 3)

Centrifuges/Rotors/Tubes

Sorvall H-1000B and SH-3000 rotors (room temperature and 4°C)

Additional Items

Pestle to fit microfuge tube (Sigma-Aldrich)

Mortar and pestle (–70°C)

Water bath (55°C or 65°C) (optional; see note to Step 10)

Additional Information

Alternative protocols (Isolation of DNA from Mouse
 Tails without Extraction by Organic Solvents,
 One-tube Isolation of DNA from Mouse Tails,
 and DNA Extraction from Paraffin Blocks)   MC3, pp. 6.26–6.27

Minimizing damage to large DNA molecules   MC3, p. 2.110

Proteinase K   MC3, p. A4.50

METHOD

1. Prepare tissue or whole blood for genomic DNA isolation.

   For tissue

   a. Dissect 10–20 mg of tissue.

   b. Either mince the tissue finely with a razor blade/scalpel or freeze the tissue in liquid nitrogen and then grind it to a powder in a mortar prechilled with liquid nitrogen, as described in Protocol 6.1.

   For blood

   a. Transfer 300-μl aliquots of whole blood to each of two microfuge tubes. Add 900 μl of 20 mM Tris-Cl (pH 7.6) to each tube and invert the capped tubes to mix the contents. Incubate the solution at room temperature for 10 minutes, occasionally inverting the tubes.

   b. Centrifuge the tubes at maximum speed for 20 seconds at room temperature in a microfuge.

   c. Discard all but 20 μl of each supernatant.

   d. Resuspend the pellets of white cells in the small amount of supernatant left in each tube. Combine the resuspended cell pellets in a single tube.

2. Transfer the minced tissue or the resuspended white blood cell pellets to a microfuge tube containing 600 μl of ice-cold SNET buffer. Homogenize the suspension quickly with 30–50 strokes of a microfuge pestle.

   The SDS will precipitate from the ice-cold SNET buffer, producing a cloudy solution. This precipitation will not affect isolation of DNA.

3. (*Optional*) Add 3 μl of proteinase K solution to the lysate to increase the yield of genomic DNA. Incubate the digest for at least 3 hours but no more than 16 hours at 55°C.

4. Allow the digest to cool to room temperature and then add 3 μl of 4-mg/ml DNase-free RNase A. Incubate the digest for 15–60 minutes at 37°C.

5. Allow the sample to cool to room temperature. Add 200 μl of potassium acetate solution and mix the contents of the tube by vortexing vigorously for 20 seconds.

6. Pellet the precipitated protein/SDS complex by centrifugation at maximum speed for 3 minutes at 4°C in a microfuge.

   A pellet of protein should be visible at the bottom of the microfuge tube after centrifugation. If not, incubate the lysate for 5 minutes on ice and repeat the centrifugation step.

7. Transfer the supernatant to a fresh microfuge tube containing 600 μl of isopropanol. Mix the solution well and then recover the precipitate of DNA by centrifuging the tube at maximum speed for 1 minute at room temperature in a microfuge.

8. Remove the supernatant by aspiration and add 600 μl of 70% ethanol to the DNA pellet. Invert the tube several times and centrifuge the tube at maximum speed for 1 minute at room temperature in a microfuge.

9. Carefully remove the supernatant by aspiration and allow the DNA pellet to air dry for 15 minutes.

10. Redissolve the pellet of DNA in 100 μl of TE (pH 7.6).

    The solubilization of the genomic DNA pellet can be faciltated by incubation for 16 hours at room temperature or for 1 hour at 65°C.

Rapid Isolation of Yeast DNA

This method is used to isolate genomic yeast DNA or shuttle plasmids that replicate in both *S. cerevisiae* and *E. coli*. The DNA can be used as a template for PCR and for transformation.

MATERIALS

CAUTION: Please see Appendix 4 for appropriate handling of materials marked with <!>.

Reagents and Solutions

Please see Appendix 1 for components of stock solutions, buffers, and reagents. Dilute stock solutions to the appropriate concentrations.

> Ethanol
> Ethanol (70%)
> Phenol:chloroform (1:1, v/v) <!>
> Sodium acetate (3 M, pH 5.2)
> STES buffer
>> 200 mM Tris-Cl (pH 7.6)
>> 500 mM NaCl
>> 0.1% (w/v) SDS
>> 10 mM EDTA
> TE (pH 7.6)

Cells, Tissues/Culture Media

> Yeast cells, freshly grown, either as colonies on an agar plate or as an overnight suspension culture

Additional Items

> Acid-washed glass beads (0.4 mm)

METHOD

1. Prepare the yeast cells for lysis.

   For yeast colonies on plates

   Use a sterile inoculating loop to transfer one or more large, freshly grown colonies to a microfuge tube containing 50 µl of STES buffer.

   For yeast grown in liquid culture

   a. Transfer 1.5 ml from an overnight culture of yeast cells to a microfuge tube.

   b. Pellet the cells by centrifuging at maximum speed for 1 minute at room temperature in a microfuge.

   c. Remove the culture medium by aspiration and resuspend the pellet in 50 µl of STES buffer.

2. Add approximately 50 μl of acid-washed glass beads to each tube containing the resuspended yeast. Add 20 μl of TE (pH 7.6) to each tube.

3. Add 60 μl of phenol:chloroform, cap the tubes, and mix the organic and aqueous phases by vortexing for 1 minute.

4. Centrifuge the tubes at maximum speed for 5 minutes at room temperature in a microfuge.

5. Transfer the upper aqueous phase to a fresh microfuge tube. Collect the DNA by standard precipitation with ethanol for 15 minutes at 0°C.

6. Recover the precipitate of nucleic acids by centrifugation at maximum speed for 10 minutes at 4°C in a microfuge.

7. Remove the supernatant by aspiration and rinse the pellet with 100 μl of 70% ethanol in H_2O. Centrifuge the tubes at maximum speed for 1 minute at room temperature in a microfuge.

8. Remove the supernatant by aspiration and allow the pellet to air dry for 15 minutes. Redissolve the pellet in 40 μl of TE (pH 7.6).

> Use 1–10 μl of the solution of DNA as a template in PCRs. Shuttle plasmids can be recovered by transforming preparations of competent *E. coli* with 1 μl of the DNA.

Southern Blotting: Capillary Transfer of DNA to Membranes

This protocol describes the first stages of Southern blotting: digestion of genomic DNA with one or more restriction enzymes, separation of the resulting fragments by electrophoresis through an agarose gel, and transfer of the denatured fragments to a membrane by downward capillary transfer.

MATERIALS

CAUTION: Please see Appendix 4 for appropriate handling of materials marked with <!>.

Reagents and Solutions

Please see Appendix 1 for components of stock solutions, buffers, and reagents. Dilute stock solutions to the appropriate concentrations.

Ethidium bromide (10 mg/ml) <!>

or

SYBR Gold solution (supplied commercially as a 10,000x concentrate in DMSO)

Gel-loading buffer (6x)

    Almost always gel-loading buffer IV

        0.25% bromophenol blue

        40% (w/v) sucrose in H_2O

6x SSC

TE (pH 8.0)

For alkaline transfer of DNA to nylon membranes

    Alkaline transfer buffer

        0.4 N NaOH <!>

        1 M NaCl

    Neutralization buffer II

        0.5 M Tris-Cl (pH 7.2)

        1 M NaCl

HCl (0.2 N) (optional; used to depurinate DNA before transfer. See note to Step 6) <!>

For neutral transfer of DNA to nitrocellulose or nylon membranes

    Denaturation solution

        1.5 M NaCl

        0.5 M NaOH <!>

    Neutral transfer buffer

        10x SSC *or* 10x SSPE

    Neutralization buffer I

        1 M Tris (ph 7.4)

        1.5 M NaCl

Enzymes and Buffers

Appropriate restriction enzymes and 10x restriction buffers

Nucleic Acids/Oligonucleotides

DNA size markers, e.g., 1-kb ladder (New England BioLabs) *or* HindIII digest of bacteriophage λ DNA

Genomic DNA, for mammalian DNA (use ~10 μg/gel slot)

Adapted from Chapter 6, Protocol 8, p. 6.39 of MC3.

Gels/Loading Buffers

0.7% Agarose gel, cast and run in 0.5x TBE or 1x TAE in the absence of ethidium bromide
   or SYBR Gold (see MC3, pp. A9.3–A9.7)

Additional Items

Blunt-ended forceps
Cross-linking device for fixation of DNA to membrane (see Table 6-2)
Dog nail clippers (Scientific)
Glass baking dishes, large enough to hold gel
Glass rod
Neoprene stoppers
Nylon or nitrocellulose membranes (see MC3, pp. 6.37–6.38)
Plexiglas sheets or glass plates
Rotary platform shaker
Scalpel or paper cutter
Transparent ruler with fluorescent markings
Weight (~400 g)
Whatman 3MM paper sheets

Additional Information

Depurination of DNA before transfer	MC3, pp. 6.34–6.35
Options and choice of membranes for Southern blotting and hybridization	MC3, pp. 6.34–6.38
Setting up restriction digestions for Southern analysis	MC3, pp. 6.39–6.40

METHOD

1. Digest an appropriate amount of genomic DNA with one or more restriction enzymes.

   Use large-bore yellow pipette tips to transfer high-molecular-weight DNA.

2. If necessary, concentrate the DNA fragments at the end of the digestion by ethanol precipitation. Dissolve the DNAs in approximately 25 µl of TE (pH 8.0).

   Make sure that the ethanol is removed from the DNA solution before it is loaded on the gel. If significant quantities of ethanol remain, the DNA "crawls" out of the slot of the gel. Heating the solution of dissolved DNA to 70°C in an open tube for 10 minutes is usually sufficient to drive off most of the ethanol. This treatment also disrupts base pairing between cohesive termini of restriction fragments.

3. Measure the concentrations of the digested DNAs by fluorometry or by the ethidium bromide or SYBR Gold spot test (see MC3, p. A8.24). Transfer the appropriate amount of each digest to a fresh microfuge tube. Add 0.15 volume of 6x sucrose gel-loading buffer and separate the fragments of DNA by electrophoresis through an agarose gel. Maintain a low voltage through the gel (about <1 V/cm) to allow the DNA to migrate slowly.

   If the digested DNAs have been stored at 4°C, they should be heated to 56°C for 2–3 minutes before they are applied to the gel. This heating disrupts any base pairing that may have occurred between protruding cohesive termini.

4. After electrophoresis is complete, stain the gel with ethidium bromide or SYBR Gold and photograph the gel as described in Protocol 5.2. Place a transparent ruler alongside the gel so that the distance that any band of DNA has migrated can be read directly from the photographic image.

5. Denature the DNA and transfer it from the agarose gel to a nitrocellulose or a neutral or charged-nylon membrane using one of the methods described below.

Preparation of the Gel for Transfer

6. After fractionating the DNA by gel electrophoresis, transfer the gel to a glass baking dish. Use a razor blade to trim away unused areas of the gel, including the section of gel above the wells. Be sure to leave enough of the wells attached to the gel so that the positions of the lanes can be marked on the membrane after transfer of DNA. Cut a small triangular piece from the bottom left-hand corner of the gel to simplify orientation during the succeeding operations.

   Cut off the lanes containing the molecular-weight markers.

   If the fragments of interest are larger than approximately 15 kb, transfer may be improved by nicking the DNA by brief depurination before denaturation. After Step 6, soak the gel in several volumes of 0.2 N HCl until the bromophenol blue turns yellow and the xylene cyanol turns yellow/green. Immediately place the 0.2 N HCl in a hazardous-waste container and then rinse the gel several times with deionized H_2O.

   Depurination is best avoided when the size of the target fragments is <20 kb. However, for Southern analysis of higher-molecular-weight DNA separated by conventional or pulsed-field electrophoresis, depurination/nicking is advisable, if not essential.

7. Denature the DNA by soaking in a denaturing (alkaline) solution as follows.

   #### For transfer to uncharged membranes

   a. Soak the gel for 45 minutes at room temperature in 10 gel volumes of denaturation solution with constant *gentle* agitation (e.g., on a rotary platform) at room temperature.

   b. Rinse the gel briefly in deionized H_2O and then neutralize it by soaking for 30 minutes at room temperature in 10 gel volumes of neutralization buffer I with constant gentle agitation. Change the neutralization buffer and continue soaking the gel for an additional 15 minutes.

   #### For transfer to charged nylon membranes

   a. Soak the gel for 15 minutes at room temperature in several volumes of alkaline transfer buffer with constant *gentle* agitation (e.g., on a rotary platform) at room temperature.

   b. Change the solution and continue to soak the gel for an additional 20 minutes with gentle agitation.

      If the gel floats to the surface of the liquid, weigh it down with several pasteur pipettes.

8. Use a fresh scalpel or paper cutter to cut a piece of nylon or nitrocellulose membrane approximately 1 mm larger than the gel in each dimension. Also cut two sheets of thick blotting paper to the same size as the membrane.

   IMPORTANT: Use appropriate gloves and blunt-ended forceps (e.g., Millipore) to handle the membrane. A membrane that has been touched by oily hands will not wet!

9. Float the membrane on the surface of a dish of deionized H_2O until it wets completely (no spots or other blemishes) from beneath and then immerse the membrane in the appropriate transfer buffer for at least 5 minutes. Use a clean scalpel blade to cut a corner from the membrane to match the corner cut from the gel.

10. While the DNA is denaturing, place a piece of thick blotting paper on a sheet of Plexiglas or a glass plate to form a support that is longer and wider than the gel. The ends of the blotting paper should drape over the edges of the plate. Place the support inside a large baking dish. The support can be placed on top of four neoprene stoppers to elevate it from the bottom of the dish.

11. Fill the dish with the appropriate transfer buffer until the level of the liquid reaches almost to the top of the support. When the blotting paper on the top of the support is thoroughly wet, smooth out all air bubbles with a glass rod or pipette.

    Neutral transfer buffer (10x SSC *or* 10x SSPE) is used to transfer DNA to uncharged membranes. Alkaline transfer buffer (0.4 N NaOH with 1 M NaCl) is used to transfer DNA to charged nylon membranes.

12. Remove the gel from the solution in Step 7 and invert it so that its underside is now upper-most. Place the inverted gel on the support so that it is centered on the wet blotting paper.

    Make sure that there are no air bubbles between the blotting paper and the gel.

13. Surround, but do not cover, the gel with Saran Wrap or Parafilm.

14. Wet the top of the gel with the appropriate transfer buffer. Place the wet membrane on top of the gel so that the cut corners are aligned. To avoid bubbles, touch one corner of the membrane to the gel and gently lower the membrane onto the gel. One edge of the membrane should extend just over the edge of the line of slots at the top of the gel.

    IMPORTANT: Do not move the membrane once it has been applied to the surface of the gel. Make sure that there are no air bubbles between the membrane and the gel.

15. Wet the two pieces of thick blotting paper in the appropriate transfer buffer and place them on top of the wet membrane. Roll a pipette across the surface of the membrane to smooth away any air bubbles.

16. Cut or fold a stack of paper towels (5–8 cm high) just smaller than the blotting papers. Place the towels on the blotting papers. Put a glass plate on top of the stack and weigh it down with a 400-g weight.

17. Allow the transfer of DNA to proceed for 8–24 hours. Replace the paper towels as they become wet. Try to prevent the entire stack of towels from becoming wet with buffer.

18. Remove the paper towels and the blotting papers above the gel. Turn the gel and the attached membrane over and lay them, gel side up, on a dry sheet of blotting paper. Mark the positions of the gel slots on the membrane with a very soft-lead pencil or a ballpoint pen.

19. Peel the gel from the membrane and discard the gel.

    Before discarding the gel, it can be stained (45 minutes) in a 0.5-μg/ml solution of ethidium bromide in H_2O and visualized on a UV transilluminator to gauge the success of the DNA transfer.

20. Soak the membrane in *one* of the following solutions as appropriate (see Table 6-2).

    The sequence of steps from immobilization of DNA to the membrane to subsequent hybridization depends on the type of membrane, the method of transfer, and the method of fixation. Because alkaline transfer results in covalent attachment of DNA to positively charged nylon membranes, there is no need to fix the DNA to the membrane before hybridization. DNA transferred to uncharged nylon membranes in neutral transfer buffer should be fixed to the membrane by baking under vacuum, heating in a microwave oven, or cross-linking to the membrane by UV irradiation.

TABLE 6-2 Fixing DNA to the Membrane for Hybridization

Type of Membrane	Type of Transfer	Method of Fixation	Sequence of Steps
Positively charged nylon	alkaline	alkaline transfer	1. Soak membrane in neutralization buffer II. 2. Proceed to prehybridization.
Uncharged nylon or positively charged nylon	neutral	UV irradiation (please see Step 21 for details)	1. Soak membrane in 6x SSC. 2. Fix the DNA by UV irradiation. 3. Proceed to prehybridization.
Uncharged nylon or positively charged nylon	neutral	baking in vacuum oven or microwave oven (please see Step 21 for details)	1. Soak membrane in 6x SSC. 2. Bake the membrane. 3. Proceed to prehybridization.

For neutral transfer

6x SSC for 5 minutes at room temperature.

For alkaline transfer

Neutralization buffer II (0.5 M Tris-Cl [pH 7.2] with 1 M NaCl) for 15 minutes at room temperature.

> This rinse removes any pieces of agarose sticking to the membrane and, in the latter case, also neutralizes the membrane.

21. Immobilize the DNA that has been transferred to uncharged membranes.

> Because alkaline transfer results in covalent attachment of DNA to positively charged nylon membranes, there is no need for additional steps to fix the DNA to the membrane.

To fix by baking in a vacuum oven

a. Remove the membrane from the 6x SSC and allow excess fluid to drain away. Place the membrane flat on a paper towel to dry for at least 30 minutes at room temperature.

b. Sandwich the membrane between two sheets of dry blotting paper. Bake for 30 minutes to 2 hours at 80°C in a vacuum oven.

> Overbaking can cause nitrocellulose membranes to become brittle. If the gel was not completely neutralized before the DNA was transferred, nitrocellulose membranes will turn yellow or brown during baking and chip very easily. The background of nonspecific hybridization also increases dramatically.

To fix by baking in a microwave oven

a. Place the damp membrane on a dry piece of blotting paper.

b. Heat the membrane for 2–3 minutes at full power in a microwave oven (750–900 W).

> Proceed directly to hybridization (Protocol 6.10) or dry the membrane and store it between sheets of blotting paper until it is needed.

To cross-link by UV irradiation

a. Place the damp membrane on a dry piece of blotting paper.

b. Irradiate at 254 nm to cross-link the DNA to the membrane.

> Make sure that the side of the membrane carrying the DNA faces the UV light source. Most manufacturers advise that damp membranes be exposed to a total of 1.5 J/cm^2 and that dry membranes be exposed to 0.15 J/cm^2. However, we recommend performing a series of preliminary experiments to determine empirically the amount of irradiation required to produce the maximum hybridization signal.

22. Proceed directly to hybridization of immobilized DNA to a probe (Protocol 6.10).

> Any membranes not used immediately in hybridization reactions should be thoroughly dried, wrapped loosely in aluminum foil or blotting paper, and stored at room temperature, preferably under vacuum.

Southern Blotting: Simultaneous Transfer of DNA from a Single Agarose Gel to Two Membranes

DNA can be simultaneously transferred from opposite sides of a single agarose gel to two membranes. Bidirectional transfer occurs rapidly at first, but soon slows down as the gel becomes dehydrated. Because the efficiency of transfer is low, the method works best when the target sequences are present in high concentration, for example, when analyzing restriction digests of cloned DNAs or less complex genomes. This protocol describes bidirectional transfer of denatured DNA from an agarose gel to nitrocellulose or uncharged nylon membranes. It can be easily adapted for use with charged nylon membranes.

MATERIALS

CAUTION: Please see Appendix 4 for appropriate handling of materials marked with <!>.

Reagents and Solutions

Please see Appendix 1 for components of stock solutions, buffers, and reagents. Dilute stock solutions to the appropriate concentrations.

    Ethidium bromide (10 mg/ml) <!>

    or

    SYBR Gold solution (supplied commercially as a 10,000x concentrate in DMSO)

    Denaturation solution

        1.5 M NaCl

        0.5 M NaOH <!>

    Gel-loading buffer (6x)

        Almost always gel-loading buffer IV

            0.25% bromophenol blue

            40% (w/v) sucrose in H_2O

    HCl (0.2 N) (optional; used to depurinate DNA before transfer. See note to Step 6) <!>

    Neutral transfer buffer

        10x SSC *or* 10x SSPE

    Neutralization buffer I

        1 M Tris (pH 7.4)

        1.5 M NaCl

    6x SSC

    TE (pH 8.0)

Enzymes and Buffers

Appropriate restriction enzymes and 10x restriction buffers

Nucleic Acids/Oligonucleotides

DNA size markers, e.g., 1-kb ladder (New England BioLabs) *or* HindIII digest of bacteriophage λ DNA

Genomic DNA, for mammalian DNA (use ~10 µg/gel slot)

Gels/Loading Buffers

0.7% Agarose gel, cast and run in 0.5x TBE or 1x TAE in the absence of ethidium bromide *or* SYBR Gold (see MC3, pp. A9.3–A9.7)

Additional Items

Blunt-ended forceps

Cross-linking device for fixation of DNA to membrane (see Table 6-2, p. 261)

Dog nail clippers (Scientific)

Glass baking dishes, large enough to hold gel

Glass plates

Glass rod

Neoprene stoppers

Nylon or nitrocellulose membranes (see MC3, pp. 6.37–6.38)

Rotary platform shaker

Scalpel or paper cutter

Transparent ruler with fluorescent markings

Weight (~400 g)

Whatman 3MM paper sheets

Additional Information

Depurination of DNA before transfer	MC3, pp. 6.34–6.35
Options and choice of membranes for Southern blotting and hybridization	MC3, pp. 6.34–6.38
Setting up restriction digestions for Southern analysis	MC3, pp. 6.39–6.40

METHOD

1. Digest the DNA and fractionate it by gel electrophoresis according to Protocol 6.8, Steps 1–3.

2. After fractionating the DNA by gel electrophoresis, stain the gel with ethidium bromide or SYBR Gold and photograph as described in Protocol 5.2. Place a transparent ruler alongside the gel so that the distance that any band has migrated can be read directly from the photographic image. Prepare the gel for transfer under neutral conditions (Protocol 6.8, Steps 6 and 7).

3. Use a fresh scalpel or paper cutter to cut two pieces of nylon or nitrocellulose membrane approximately 1–2 mm larger than the gel in each dimension. Cut a corner from the membranes to match the corner cut from the gel. Also cut four sheets of thick blotting paper to the same size as the membranes.

   IMPORTANT: Use appropriate gloves and blunt-ended forceps (e.g., Millipore) to handle the membranes. A membrane that has been touched by oily hands will not wet!

   To retain small fragments of DNA (<300 nucleotides), use nitrocellulose membranes with a small pore size (0.2 μm) or nylon membranes.

4. Float the membranes on the surface of a dish of deionized H_2O until they wet completely from beneath and then immerse the membranes in 10x SSC for at least 5 minutes.

5. Roll a moistened pipette over each layer as it is assembled to ensure that no air bubbles are trapped, especially between the membranes and the gel sides. Place one of the membranes on two pieces of dampened blotting paper. Lay the gel on top of the membrane, aligning the cut corner of the gel with the cut corner of the membrane. Without delay, place the second membrane on the other side of the gel, followed by two sheets of dampened blotting paper.

6. Transfer the entire sandwich of blotting papers, membranes, and gel onto a 2–4-inch stack of paper towels. Cover the sandwich with a second stack of paper towels. Put a glass plate on top of the entire stack and weigh it down with a 400-g weight.

7. After 2–4 hours, remove the paper towels and blotting papers. Transfer the gel and membrane sandwich to a dry sheet of blotting paper and mark the approximate positions of the gel slots with a very soft lead pencil or a ballpoint pen.

8. Immobilize the DNA onto the membranes by completing Protocol 6.8, Steps 19–21.

9. Proceed directly to hybridization of immobilized DNA to a probe (Protocol 6.10).

Southern Hybridization of Radiolabeled Probes to Nucleic Acids Immobilized on Membranes

This protocol describes how to perform Southern hybridizations at high stringency in phosphate-SDS buffers. Although a wide variety of formats are available, most Southern hybridizations are performed in heat-sealable bags, roller bottles, or plastic boxes, as described here.

MATERIALS

CAUTION: Please see Appendix 4 for appropriate handling of materials marked with <!>.

Reagents and Solutions

Please see Appendix 1 for components of stock solutions, buffers, and reagents. Dilute stock solutions to the appropriate concentrations.

Phosphate-SDS washing solution 1
- 40 mM sodium phosphate buffer (pH 7.2)
- 1 mM EDTA (pH 8.0)
- 5% (w/v) SDS
- 0.5% (w/v) Fraction-V-grade bovine serum albumin

Phosphate-SDS washing solution 2
- 40 mM sodium phosphate buffer (pH 7.2)
- 1 mM EDTA (pH 8.0)
- 1% (w/v) SDS

Prehybridization/hybridization solutions

Solution for hybridization in aqueous buffer
- 6x SSC (or 6x SSPE)
- 5x Denhardt's reagent
- 0.5% (w/v) SDS
- 1 µg/ml poly(A)
- 100 µg/ml salmon sperm DNA

Solution for hybridization in formamide buffers
- 6x SSC (or 6x SSPE)
- 5x Denhardt's reagent
- 0.5% (w/v) SDS
- 1 µg/ml poly(A)
- 100 µg/ml salmon sperm DNA
- 50% (v/v) formamide <!>

After thorough mixing, filter the solution through a 0.45-µm disposable cellulose acetate membrane (Schleicher & Schuell Uniflo syringe membrane or its equivalent).

Solution for hybridization in phosphate-SDS buffer
- 0.5 M sodium phosphate (pH 7.2)
- 1 mM EDTA (pH 8.0)
- 7% (w/v) SDS
- 1% (w/v) bovine serum albumin

Use an electrophoresis grade of bovine serum albumin. No blocking agents or hybridization rate enhancers are required with this particular prehybridization/hybridization solution.

Sodium phosphate (1 M, pH 7.2)
0.1x SSC
0.1x SSC with 0.1% (w/v) SDS
2x SSC with 0.1% (w/v) SDS
2x SSC with 0.5% (w/v) SDS
6x SSC or 6x SSPE

Nucleic Acids/Oligonucleotides

DNA immobilized on membrane

Poly(A) RNA (10 mg/ml) in sterile H_2O (optional, for hybridization buffers)

> Prepare solution by dissolving poly(A) RNA in sterile H_2O and store in 100-μl aliquots.

Radiolabeled probe <!>

> For Southern analysis of mammalian genomic DNA, where each lane of the gel contains 10 μg of DNA, use 10–20 ng/ml radiolabeled probe DNA or RNA (sp. act. ≥10^9 cpm/μg). For Southern analysis of cloned DNA fragments, where each band of the restriction digest contains 10 ng of DNA or more, much less probe is required. When analyzing cloned DNA, hybridization is performed for 6–8 hours using 1–2 ng/ml radiolabeled probe (sp. act. 10^6 to 10^9 cpm/μg). Labeling should be performed according to the methods described in Chapter 9 or 10.

Salmon sperm DNA (~10 mg/ml)

> For advice on preparing a solution of salmon sperm DNA see MC3, p. 6.52.

Additional Items

Adhesive dots marked with radioactive ink <!> or commercially available phosphorescent adhesive labels (e.g., Glogos [Stratagene])

Boiling-water bath

Containers for hybridization, e.g., roller bottles, heat-sealable bags, etc.

Incubator or commercial hybridization device, set at the appropriate temperature for hybridization

Additional Information

Background and how to avoid it	MC3, p. 6.56
Hybridization at low stringency	MC3, p. 6.58
Formamide and its uses	MC3, pp. 6.59–6.60
Options and choice of membranes for Southern blotting and hybridization	MC3, pp. 6.34–6.38
Rapid hybridization buffers	MC3, pp. 6.61–6.62
Stripping probes from membranes	MC3, p. 6.57

METHOD

1. Float the membrane containing the target DNA on the surface of a tray of 6x SSC (or 6x SSPE) until the membrane becomes thoroughly wetted from beneath. Submerge the membrane for 2 minutes.

2. Prehybridize the membrane by one of the following methods.

For hybridization in a heat-sealable bag

a. Slip the wet membrane into a heat-sealable bag (e.g., Rival Seal-a-Meal or equivalent), and add 0.2 ml of prehybridization solution for each square centimeter of membrane. Squeeze as much air as possible from the bag.

b. Seal the open end of the bag with a heat sealer and then make a second seal. Test the strength and integrity of the seal by gently squeezing the bag. Incubate the bag for 1–2 hours submerged in a water bath set to the appropriate temperature (68°C for aqueous solvents; 42°C for solvents containing 50% formamide; 65°C for phosphate-SDS solvents).

For hybridization in a roller bottle

a. Gently roll the wetted membrane into the shape of a cylinder and place it inside a hybridization roller bottle together with the plastic mesh provided by the manufacturer.

Add 0.1 ml of prehybridization solution for each square centimeter of membrane. Close the bottle tightly.

b. Place the hybridization tube inside a prewarmed hybridization oven at the appropriate temperature (68°C for aqueous solvents; 42°C for solvents containing 50% formamide; 65°C for phosphate-SDS solvents).

For hybridization in a plastic container

a. Place the wet membrane in a plastic (e.g., Tupperware) container and add 0.2 ml of pre-hybridization solution for each square centimeter of membrane.

b. Seal the box with the lid and place the box on a rocking platform in an air incubator set at the appropriate temperature (68°C for aqueous solvents; 42°C for solvents containing 50% formamide; 65°C for phosphate-SDS solvents).

3. If the radiolabeled probe is double-stranded DNA, denature it by heating for 5 minutes at 100°C. Chill the probe rapidly in ice water.

> Alternatively, denature DNA probes by adding 0.1 volume of 3 N NaOH. After 5 minutes at room temperature, chill the probe to 0°C in an ice-water bath and add 0.05 volume of 1 M Tris-Cl (pH 7.2) and 0.1 volume of 3 N HCl. Store the probe in ice water until it is needed.
>
> Single-stranded DNA and RNA probes need not be denatured.

4. To hybridize the probe to a blot containing genomic DNA, perform one of the following methods.

For hybridization in a heat-sealable bag

a. Working quickly, remove the bag containing the membrane from the water bath. Open the bag by cutting off one corner with scissors and pour off the prehybridization solution.

b. Add the denatured probe to an appropriate amount of fresh prehybridization solution and deliver the solution into the bag. Squeeze as much air as possible from the bag.

c. Reseal the bag with the heat sealer; make sure that as few bubbles as possible are trapped in the bag. To avoid radioactive contamination of the water bath, seal the resealed bag inside a second, noncontaminated bag. Incubate the bag submerged in a water bath set at the appropriate temperature for the required period of hybridization.

For hybridization in a roller bottle

a. Pour off the prehybridization solution from the hybridization bottle and replace with fresh hybridization solution containing probe.

b. Seal bottle and replace in hybridization oven. Incubate for the required period of hybridization.

For hybridization in a plastic container

a. Transfer the membrane from the container to a sealable bag or a hybridization bottle.

b. Immediately treat as described above.

5. After hybridization, wash the membrane.

For hybridization in a heat-sealable bag

a. Wearing gloves, remove the bag from the water bath, remove the outer bag, and immediately cut off one corner of the inner bag. Pour out the hybridization solution into a container suitable for disposal of radioactivity and cut the bag along the length of three sides.

b. Remove the membrane and immediately submerge it in a tray containing several hundred milliliters of 2x SSC and 0.5% SDS (i.e., ~1 ml/cm² membrane) at room temperature. Agitate the tray gently on a slowly rotating platform.

For hybridization in a roller bottle

a. Remove the membrane from the hybridization bottle and briefly drain excess hybridization solution from the membrane by holding the corner of the membrane to the lip of the bottle or container.

b. Place the membrane in a tray containing several hundred milliliters of 2x SSC and 0.5% SDS (i.e., ~1 ml/cm² membrane) at room temperature. Agitate the tray gently on a slowly rotating platform.

> When hybridizing in phosphate-SDS solution, remove the membrane from the hybridization chamber as described in Step 5 and place it in several hundred milliliters (i.e., ~1 ml/cm² membrane) of phosphate-SDS washing solution 1 at 65°C. Agitate the tray. Repeat this rinse once.
>
> IMPORTANT: Do not allow the membrane to dry out at any stage during the washing procedure.

6. After 5 minutes, pour off the first rinse solution into a radioactivity disposal container and add several hundred milliliters of 2x SSC and 0.1% SDS to the tray. Incubate for 15 minutes at room temperature with occasional gentle agitation.

> If hybridization was performed in a phosphate-SDS buffer, rinse the membrane a total of eight times for 5 minutes each in several hundred milliliters of phosphate-SDS washing solution 2 at 65°C. Skip to Step 9 after the eighth rinse.

7. Replace the rinse solution with several hundred milliliters of fresh 0.1x SSC with 0.1% SDS. Incubate the membrane for 30 minutes to 4 hours at 65°C with gentle agitation.

> During the washing step, periodically monitor the amount of radioactivity on the membrane using a handheld minimonitor. The parts of the membrane that do not contain DNA should not emit a detectable signal. Do not expect to pick up a signal on the minimonitor from membranes containing mammalian DNA that has been hybridized to single-copy probes.

8. Briefly wash the membrane with 0.1x SSC at room temperature.

9. Remove most of the liquid from the membrane by placing it on a pad of paper towels. Place the damp membrane on a sheet of Saran Wrap. Apply adhesive dot labels marked with radioactive ink (or phosphorescent dots) to several asymmetric locations on the Saran Wrap. These markers serve to align the autoradiograph with the membrane. Cover the labels with Scotch tape. This prevents contamination of the film holder or intensifying screen with the radioactive ink.

> Alternatively, dry the membrane in the air and glue it to a piece of 3MM paper using a water-soluble glue.

10. Cover the membrane with a sheet of Saran Wrap and expose the membrane to X-ray film for 16–24 hours at –70°C with an intensifying screen to obtain an autoradiographic image.

> Alternatively, cover the hybridized and rinsed membrane with Sarap Wrap and expose it to a phosphorimager plate. An exposure time of 1–4 hours is usually long enough to detect single-copy gene sequences in a Southern blot of mammalian genomic DNA.

CHAPTER 7

Extraction, Purification, and Analysis of mRNA from Eukaryotic Cells

BACKGROUND INFORMATION

Background information found in *Molecular Cloning: A Laboratory Manual,* 3rd edition (hereafter MC3) unless otherwise indicated.

Bacterial artificial chromosomes (BACs)	MC3, pp. 4.48–4.51
Diethylpyrocarbonate	MC3, p. 7.84
How to win the battle with RNase	MC3, p. 7.82
Inhibitors of RNases	MC3, p. 7.83
Yeast artificial chromosomes (YACs)	MC3, pp. 4.58–4.65

Purification of RNA from Cells and Tissues by Acid Phenol–Guanidinium Thiocyanate–Chloroform Extraction

In this single-step technique, cells are homogenized in guanidinium thiocyanate, and the RNA is purified from the lysate by extraction with phenol:chloroform at reduced pH. Many samples can be processed simultaneously and speedily. The yield of total RNA depends on the tissue or cell source and is generally in the range of 4–7 µg/ml starting tissue or 5–10 µg/10^6 cells.

IMPORTANT: Prepare all reagents used in this protocol with DEPC-treated H_2O.

MATERIALS

CAUTION: Please see Appendix 4 for appropriate handling of materials marked with <!>.

Reagents and Solutions

Please see Appendix 1 for components of stock solutions, buffers, and reagents. Dilute stock solutions to the appropriate concentrations.

Chloroform:isoamyl alcohol (49:1, v/v) <!>

Ethanol (75%)

Formamide, stabilized <!> (optional; may be used for storage of RNA [see MC3, p. 7.8]).
> Stabilized formamide is available commercially from Fisher Scientific under the trade name Formazol.

Isopropanol

Liquid nitrogen <!> (required only for processing of tissues)

Phenol <!>

Phosphate-buffered saline (PBS; required only for cultured cells)

Sodium acetate (2 M, pH 4.0)

Solution D (denaturing solution)
> 4 M guanidinium thiocyanate <!>
> 25 mM sodium citrate · 2H_2O
> 0.5% (w/v) sodium lauryl sarcosinate
> 0.1 M β-mercaptoethanol <!>
> Dissolve 250 g of guanidinium thiocyanate in 293 ml of H_2O, 17.6 ml of 0.75 M sodium citrate (pH 7.0), and 26.4 ml of 10% (w/v) sodium lauryl sarcosinate. Add a magnetic bar and stir the solution on a combination heater-stirrer at 65°C until all ingredients are dissolved. Store solution D at room temperature and add 0.36 ml of 14.4 M stock β-mercaptoethanol per 50 ml of solution D just before use. Solution D may be stored for months at room temperature but is sensitive to light. Note that guanidinium will precipitate at low temperatures. Table 7-1 presents the amounts of solution D required to extract RNA from various sources.
> WARNING: Solution D is very caustic. Wear appropriate gloves, a laboratory coat, and eye protection when preparing, handling, or working with the solution.

Cells, Tissues/Culture Media

Cells or tissue samples for RNA isolation

Centrifuges/Rotors/Tubes

Sorvall SS-34 and H-1000B rotors (4°C)

Additional Items

Cuvettes, for measuring absorbance at 260 and 280 nm
Mortar and pestle, washed in DEPC-treated H_2O (see MC3, p. 7.84)

Spectrophotometer
Tissue homogenizer (e.g., Tissumizer [Tekmar-Dohrmann] *or* Polytron [Brinkman])
Water bath (65°C) (optional; see Step 10)

Additional Information

Chaotropic agents	MC3, p. 15.60
Guanidinium salts	MC3, p. 7.85
Quantitation of nucleic acids	MC3, pp. A8.19–A8.24
Storage of RNA	MC3, p. 7.8

METHOD

1. Prepare cells or tissue samples for isolation of RNA. Table 7-1 describes the amounts of solution D required for each type of sample.

   Table 7-1 Amounts of Solution D Required to Extract RNA from Cells and Tissues

Amount of Tissue or Cells	Amount of Solution D
100 mg of tissue	3 ml
T-75 flask of cells	3 ml
60-mm plate of cells	1 ml
90-mm plate of cells	2 ml

   For tissues

   a. Isolate the desired tissues by dissection and immediately place them in liquid nitrogen.

   b. Transfer approximately 100 mg of the frozen tissue to a mortar containing liquid nitrogen and pulverize the tissue using a pestle. The tissue can be kept frozen during pulverization by the addition of liquid nitrogen.

   c. Transfer the powdered tissue to a polypropylene snap-cap tube containing 3 ml of solution D.

   d. Homogenize the tissue for 15–30 seconds at room temperature with a Tissumizer or Polytron homogenizer.

   For mammalian cells grown in suspension

   a. Harvest the cells by centrifugation at 200–1900g (1000–3000 rpm in a Sorvall RT600 using the H-1000B rotor) for 5–10 minutes at room temperature in a benchtop centrifuge.

   b. Remove the medium by aspiration and resuspend the cell pellets in 1–2 ml of sterile ice-cold PBS.

   c. Harvest the cells by centrifugation, remove the PBS completely by aspiration, and add 2 ml of solution D per 10^6 cells.

   d. Homogenize the cells with a Tissumizer or Polytron homogenizer for 15–30 seconds at room temperature.

   For mammalian cells grown in monolayers

   a. Remove the medium and rinse the cells once with 5–10 ml of sterile ice-cold PBS.

   b. Remove PBS and lyse the cells in 2 ml of solution D per 90-mm culture dish (1 ml per 60-mm dish).

   c. Transfer the cell lysates to a polypropylene snap-cap tube.

   d. Homogenize the lysates with a Tissumizer or Polytron homogenizer for 15–30 seconds at room temperature.

2. Transfer the homogenate to a fresh polypropylene tube and sequentially add 0.1 ml of 2 M sodium acetate (pH 4.0), 1 ml of phenol, and 0.2 ml of chloroform:isoamyl alcohol per milliliter of solution D. After addition of each reagent, cap the tube and mix the contents thoroughly by inversion.

3. Vortex the homogenate vigorously for 10 seconds. Incubate the tube for 15 minutes on ice to permit complete dissociation of nucleoprotein complexes.

4. Centrifuge the tube at 10,000g (9000 rpm in a Sorvall SS-34 rotor) for 20 minutes at 4°C and then transfer the upper aqueous phase containing the extracted RNA to a fresh tube.

5. Add an equal volume of isopropanol to the extracted RNA. Mix the solution well and allow the RNA to precipitate for 1 hour or more at –20°C.

6. Collect the precipitated RNA by centrifugation at 10,000g (9000 rpm in a Sorvall SS-34 rotor) for 30 minutes at 4°C.

7. Carefully decant the isopropanol and dissolve the RNA pellet in 0.3 ml of solution D for every 1 ml of this solution used in Step 1.

   IMPORTANT: Pellets are easily lost. Decant the supernatant into a fresh tube. Do not discard it until the pellet has been checked.

8. Transfer the solution to a microfuge tube, vortex it well, and precipitate the RNA with 1 volume of isopropanol for 1 hour or more at –20°C.

9. Collect the precipitated RNA by centrifugation at maximum speed for 10 minutes at 4°C in a microfuge. Wash the pellet twice with 75% ethanol, centrifuge again, and remove any remaining ethanol with a disposable pipette tip. Store the open tube on the bench for a few minutes to allow the ethanol to evaporate. Do not allow the pellet to dry completely.

10. Add 50–100 µl of DEPC-treated H_2O. Store the RNA solution at –70°C.

   Addition of SDS to 0.5% followed by heating to 65°C may assist dissolution of the pellet.

11. Estimate the concentration of the RNA by measuring the absorbance at 260 nm of an aliquot of the final preparation.

A Single-step Method for the Simultaneous Preparation of DNA, RNA, and Protein from Cells and Tissues

This protocol, a variation of the method described in Protocol 7.1, involves lysis of cells in a monophasic solution of guanidine isothiocyanate and phenol. Addition of chloroform generates a second (organic) phase into which DNA and proteins are extracted, leaving RNA in the aqueous supernatant. The yield of total RNA depends on the tissue or cell source, but it is generally in the range of 4–7 μg/mg starting tissue or 5–10 μg/10^6 cells.

IMPORTANT: Prepare all reagents used in this protocol with DEPC-treated H_2O.

MATERIALS

CAUTION: Please see Appendix 4 for appropriate handling of materials marked with <!>.

Reagents and Solutions

Please see Appendix 1 for components of stock solutions, buffers, and reagents. Dilute stock solutions to the appropriate concentrations.

Chloroform <!>

Ethanol (75%)

Formamide, stabilized <!> (optional; may be used for storage of RNA [see MC3, p. 7.8]).

> Stabilized formamide is available commercially from Fisher Scientific under the trade name Formazol.

Isopropanol

Monophasic lysis reagents

> These reagents, which contain phenol, chaotropic agents such as guanidine salts or ammonium thiocyanate, and stabilizing agents, are available commercially.

TABLE 7-2 Monophasic Lysis Reagents

Reagent	Commercial Supplier
Trizol Reagent	Invitrogen www.invitrogen.com
TRI Reagent	Molecular Research Center www.mrcgene.com
Isogen	Nippon Gene, Toyama, Japan www.Nippongene.jp
RNA STAT-60	Tel-Test www.tel-test.com

> When using commercial reagents for the simultaneous isolation of RNA, DNA, and protein, we recommend following the manufacturer's instructions. In most cases, these differ little from the generic instructions given below.

Phosphate-buffered saline (PBS; required only for cultured cells)

RNA precipitation solution

> 1.2 M NaCl
> 0.8 M sodium citrate·$15H_2O$

Sodium acetate (2 M, pH 4.0; 3 M, pH 5.2)

Cells, Tissues/Culture Media

Cells or tissue samples for RNA isolation

Centrifuges/Rotors/Tubes

Sorvall SS-34 and H-1000B rotors (4°C)

Additional Items

Cuvettes, for measuring absorbance at 260 and 280 nm
Liquid nitrogen <!>, required only for processing tissues
Mortar and pestle, washed in DEPC-treated H_2O (see MC3, p. 7.84)
Tissue homogenizer (e.g., Polytron [Brinkman] *or* Tissumizer [Tekmar-Dohrmann])
Water bath (65°C) (optional; see Step 7)

Additional Information

Chaotropic agents	MC3, p. 15.60
Guanidinium salts	MC3, p. 7.85
Quantitation of nucleic acids	MC3, pp. A8.19–A8.24
Storage of RNA	MC3, p. 7.8

METHOD

1. Prepare cells or tissue samples for isolation of RNA.

   For tissues

   a. Isolate the desired tissues by dissection and immediately place them in liquid nitrogen.

   b. Transfer approximately 100 mg of the frozen tissue to a mortar containing liquid nitrogen and pulverize the tissue using a pestle. The tissue can be kept frozen during pulverization by the addition of liquid nitrogen.

   c. Transfer the powdered tissue to a polypropylene snap-cap tube containing 1 ml of ice-cold monophasic lysis reagent.

   d. Homogenize the tissue with a Tissumizer or Polytron homogenizer for 15–30 seconds at room temperature.

   For mammalian cells grown in suspension

   a. Harvest the cells by centrifugation at 200–1900g (1000–3000 rpm in a Sorvall H-1000B rotor) for 5–10 minutes at room temperature in a benchtop centrifuge.

   b. Remove the medium by aspiration and resuspend the cell pellets in 1–2 ml of sterile ice-cold PBS.

   c. Harvest the cells by centrifugation, remove the PBS completely by aspiration, and add 1 ml of monophasic lysis reagent per 10^6 cells.

   d. Homogenize the cells with a Tissumizer or Polytron homogenizer for 15–30 seconds at room temperature.

   For mammalian cells grown in monolayers

   a. Remove the medium and rinse the cells once with 5–10 ml of sterile ice-cold PBS.

   b. Remove PBS and lyse the cells in 1 ml of monophasic lysis reagent per 90-mm culture dish (0.7 ml per 60-mm dish).

   c. Transfer the cell lysates to a polypropylene snap-cap tube.

   d. Homogenize the lysates with a Tissumizer or Polytron homogenizer for 15–30 seconds at room temperature.

2. Incubate the homogenates for 5 minutes at room temperature to permit complete dissociation of nucleoprotein complexes.

3. Add 0.2 ml of chloroform per milliliter of monophasic lysis reagent. Mix the samples by vigorous shaking or vortexing.

   > The composition of the monophasic lysis reagent used for the simultaneous isolation of RNA, DNA, and proteins has not been published. However, a large number of commercial reagents, with a variety of names, are available (please see Table 7-2). These reagents are all monophasic solutions containing phenol, guanidine, or ammonium thiocyanate and solubilizing agents.

4. Separate the mixture into two phases by centrifuging at 10,000*g* (12,000 rpm in a Sorvall SS-34 rotor) for 15 minutes at 4°C. Transfer the upper aqueous phase to a fresh tube.

5. Precipitate the RNA from the aqueous phase: For each initial milliliter of monophasic lysis reagent, add 0.25 volume of isopropanol and 0.25 volume of RNA precipitation solution. After thorough mixing, store the final solution for 10 minutes at room temperature.

6. Collect the precipitated RNA by centrifugation at maximum speed for 10 minutes at 4°C in a microfuge. Wash the pellet twice with 75% ethanol and centrifuge again. Remove any remaining ethanol with a disposable pipette tip. Store the open tube on the bench for a few minutes to allow the ethanol to evaporate. Do not allow pellet to dry completely.

7. Add 50–100 μl of DEPC-treated H_2O. Store the RNA solution at –70°C.

   > Addition of SDS to 0.5% followed by heating to 65°C may assist dissolution of the pellet.

8. Estimate the concentration of the RNA by measuring the absorbance at 260 nm of an aliquot of the final preparation.

Selection of Poly(A)+ RNA by Oligo(dT)-Cellulose Chromatography

Chromatography on oligo(dT) columns is the preferred method for large-scale purification (>25 µg) of poly(A)+ RNA extracted from mammalian cells. Typically, between 1% and 10% of the RNA applied to the oligo(dT) column is recovered as poly(A)+ RNA. Because the method can be frustratingly slow, it is not recommended for purification of poly(A)+ RNA from multiple samples. For this purpose, batch elution (Protocol 7.4) is the better choice.

IMPORTANT: Prepare all reagents used in this protocol with DEPC-treated H_2O.

MATERIALS

CAUTION: Please see Appendix 4 for appropriate handling of materials marked with <!>.

Reagents and Solutions

Please see Appendix 1 for components of stock solutions, buffers, and reagents. Dilute stock solutions to the appropriate concentrations.

Ethanol

Ethanol (70%)

NaCl (5 M)

NaOH (10 N) <!>

Elution buffer

> 10 mM Tris-Cl (pH 7.6)
> 1 mM EDTA (pH 8.0)
> 0.05% SDS
>
> The stock solutions of Tris-Cl and EDTA used to make elution buffer should be freshly autoclaved (15 minutes at 15 psi [1.05 kg/cm²] on liquid cycle) and then diluted with the appropriate amount of sterile DEPC-treated H_2O. Then add the SDS from a concentrated stock solution (10% or 20%) made in sterile DEPC-treated H_2O.
>
> **IMPORTANT:** Do not attempt to sterilize elution buffer by autoclaving because it froths excessively.

2x Oligo(dT)-cellulose loading buffer

> 40 mM Tris-Cl (pH 7.6)
> 1 M NaCl
> 2 mM EDTA (pH 8.0)
> 0.2% (w/v) sodium lauryl sarcosinate
> Prepare as described below.
>
> Make up Tris-Cl (pH 7.6) from a fresh bottle in autoclaved, DEPC-treated H_2O. Prepare NaCl and EDTA in 0.1% DEPC <!> in H_2O. Store for at least 12 hours at 37°C and autoclave the mixture for 15 minutes at 15 psi (1.05 kg/cm²) on liquid cycle. To prepare sterile loading buffer, mix appropriate amounts of RNase-free stock solutions of Tris-Cl (pH 7.6), NaCl, and EDTA (pH 8.0) in an RNase-free container. Allow the solution to cool to approximately 65°C and then add sodium lauryl sarcosinate from a 10% stock solution that has been heated to 65°C for 30 minutes.
>
> Alternatively, substitute 0.05 M sodium citrate for Tris-Cl and treat the sodium citrate/NaCl/EDTA mixture and sodium lauryl sarcosinate with DEPC (please see Protocol 7.1). Store loading buffer at room temperature.

Sodium acetate (3 M, pH 5.2)

Sterile, DEPC-treated H_2O (see MC3, p. 7.84)

Nucleic Acids/Oligonucleotides

Total cellular RNA

> Prepare as described in either Protocol 7.1 or 7.2.

Adapted from Chapter 7, Protocol 3, p. 7.13 of MC3.

Centrifuges/Rotors/Tubes

Sorvall SS-34 rotor (4°C)

Additional Items

Cuvettes, for measuring absorbance at 260 nm

Dispo column (Bio-Rad) or pasteur pipette plugged with sterile glass wool

> Before use, treat Dispo column with DEPC (see MC3, p. 7.84). Sterilize the plugged pasteur pipette by baking for 4 hours at 300°C.

Oligo(dT)-cellulose, usually purchased commercially as a dried powder (e.g., Type VII oligo(dT)-cellulose [Pfizer])

> Columns of oligo(dT)-cellulose can be stored at 4°C and reused many times. Between uses, the column should be regenerated by sequential washing with NaOH, H_2O, and loading buffer as described in Steps 1–3 of this protocol. Spun columns of oligo(dT)-cellulose are available from several manufacturers.

pH paper (pH test strips [Sigma-Aldrich])

Water bath (65°C)

Additional Information

Additional methods to select poly(A)⁺ RNA	Protocol 7.4 and MC3, p. 7.20
Oligo(dT)-cellulose	MC3, p. 7.14
Quantitation of nucleic acids	MC3, pp. A8.19–A8.24
Storage of RNA	MC3, p. 7.8

METHOD

1. Suspend 0.5–1.0 g of oligo(dT)-cellulose in 0.1 N NaOH.

2. Pour a column of oligo(dT)-cellulose (0.5–1.0-ml packed volume) in a DEPC-treated Dispo column (or a pasteur pipette, plugged with sterile glass wool and sterilized by baking for 4 hours at 300°C). Wash the column with 3 column volumes of sterile DEPC-treated H_2O.

3. Wash the column with sterile 1x oligo(dT)-cellulose-loading buffer (dilute from 2x stock using sterile DEPC-treated H_2O) until the pH of the effluent is <8.0. Use pH paper for this measurement.

4. Dissolve the RNA in sterile H_2O and heat the solution to 65°C for 5 minutes. Cool the solution to room temperature quickly and add 1 volume of 2x oligo(dT)-cellulose-loading buffer.

5. Apply the solution of RNA to the column and, in a sterile tube, immediately begin to collect the material flowing through the column. When all of the RNA solution has entered the column, wash the column with 1 column volume of 1x oligo(dT)-cellulose-loading buffer while continuing to collect the flow-through.

6. When all of the liquid has emerged from the column, heat the collected flow-through to 65°C for 5 minutes and reapply it to the top of the column. Again collect the material flowing through the column.

7. Wash the column with 5–10 column volumes of 1x oligo(dT)-cellulose-loading buffer, collecting 1-ml fractions into sterile plastic tubes (e.g., microfuge tubes).

8. Use quartz or disposable methacrylate cuvettes to measure the absorbance at 260 nm of a 1:20 dilution of each fraction collected from the column using 1x oligo(dT)-cellulose loading buffer as a blank.

9. Precipitate the fractions containing a majority of the OD_{260} material by the addition of 2.5 volumes of ethanol at –20°C.

10. Elute the poly(A)$^+$ RNA from the oligo(dT)-cellulose with 2–3 column volumes of sterile, RNase-free elution buffer. Collect fractions equivalent in size to one third of the column volume.

11. Use quartz or disposable methacrylate cuvettes to measure the absorbance at 260 nm of each fraction collected from the column. Pool the fractions containing the eluted RNA.

12. To purify poly(A)$^+$ RNA further, heat the preparation of RNA to 65°C for 3 minutes and then cool it quickly to room temperature. Adjust the concentration of NaCl in the eluted RNA to 0.5 M using 5 M NaCl and carry out a second round of chromatography on the same column of oligo(dT)-cellulose (i.e., repeat Steps 3 and 5–11).

    The material obtained after a single round of chromatography on oligo(dT)-cellulose usually contains approximately equal amounts of polyadenylated and nonpolyadenylated species of RNA. Polyadenylated RNA may be further purified as described by a second round of chromatography on oligo(dT)-cellulose.

13. To the poly(A)$^+$ RNA eluted from the second round of oligo(dT)-cellulose chromatography, add 3 M sodium acetate (pH 5.2) to a final concentration of 0.3 M. Mix well. Add 2.5 volumes of ice-cold ethanol, mix, and store the solution for at least 30 minutes on ice.

14. Recover the poly(A)$^+$ RNA by centrifugation at 10,000g (9000 rpm in a Sorvall SS-34 rotor) for 15 minutes at 4°C. Carefully discard the supernatant, and wash the pellet (which is often invisible) with 70% ethanol. Recentrifuge briefly, remove the supernatant by aspiration, and store the open tube in an inverted position for a few minutes to allow most of the residual ethanol to evaporate. Do not allow the pellet to dry.

15. Redissolve the damp pellet of RNA in a small volume of sterile, DEPC-treated H$_2$O. Use quartz or disposable methacrylate cuvettes to measure the absorbance at 260 nm of each fraction collected from the column. Pool the fractions that contain RNA.

16. Store the preparation of poly(A)$^+$ RNA (see MC3, p. 7.8).

Protocol 7.4

Selection of Poly(A)⁺ RNA by Batch Chromatography

When many RNA samples are to be processed or when working with small amounts (<50 µg) of total mammalian RNA, the technique of choice is batch chromatography on oligo(dT)-cellulose. The method described in this protocol uses a combination of temperature and ionic strength to maximize binding and recovery of polyadenylated RNA.

IMPORTANT: Prepare all reagents used in this protocol with DEPC-treated H₂O.

MATERIALS

CAUTION: Please see Appendix 4 for appropriate handling of materials marked with <!>.

Reagents and Solutions

Please see Appendix 1 for components of stock solutions, buffers, and reagents. Dilute stock solutions to the appropriate concentrations.

Absorption/washing buffer
> TES, containing 0.5 M NaCl

Ammonium acetate (10 M) <!>

Ethanol

Ethanol (70%)

NaCl (5 M)

TES
> 10 mM Tris-Cl (pH 7.5)
> 1 mM EDTA (pH 8.0)
> 0.1% SDS

Sterile, DEPC-treated H₂O (see MC3, p. 7.84)

Nucleic Acids/Oligonucleotides

Total cellular RNA
> Prepare as described in either Protocol 7.1 or 7.2.

Additional Items

Cuvettes, for measuring absorbance at 260 nm

Microfuge fitted with variable speed control

Oligo(dT)-cellulose, purchased commercially as a dried powder (e.g., Type III oligo(dT)-cellulose [Pfizer] [binding capacity, 100 OD₂₆₀/g])
> Before use, equilibrate the oligo(dT) in absorption/washing buffer as in Protocol 7.3, Steps 1–3. Oligo(dT)-cellulose can be stored at 4°C and reused many times.

Rotating wheel

Water bath (65°C and 55°C)

Additional Information

Additional methods to select poly(A)⁺ RNA	Protocol 7.3 and MC3, p. 7.20
Oligo(dT)-cellulose	MC3, p. 7.14
Quantitation of nucleic acids	MC3, pp. A8.19–A8.24
Storage of RNA	MC3, p. 7.8

METHOD

1. In a series of sterile microfuge tubes, adjust the volume of each sample of total RNA (up to 1 mg) to 600 µl with TES. Heat the sealed tubes to 65°C for 10 minutes and then cool them quickly in ice to 0°C. Add 75 µl (0.1 volume) of 5 M NaCl to each sample.

2. Add 50 mg (500 µl) of equilibrated oligo(dT)-cellulose to each tube and incubate the closed tubes on a rotating wheel for 15 minutes at room temperature.

3. Centrifuge the tubes at 600–800*g* (~1500–2500 rpm) for 2 minutes at room temperature in a microfuge.

4. Transfer the supernatants to a series of fresh microfuge tubes. Store the tubes on ice.

5. To the pellets of oligo(dT) remaining in the first set of tubes, add 1 ml of ice-cold absorption/washing buffer. Disperse the pellets of oligo(dT) by gentle vortexing. Incubate the closed tubes on a rotating wheel for 2 minutes at room temperature.

6. Centrifuge the tubes at 600–800*g* (~1500–2500 rpm) for 2 minutes at room temperature in a microfuge. Discard the supernatants and then repeat Steps 5 and 6 twice more.

7. Resuspend the pellets of oligo(dT) in 0.4 ml of *ice-cold*, double-distilled, autoclaved H_2O by gentle vortexing. Immediately centrifuge the tubes for 2 minutes at 4°C in a microfuge.

8. Remove the supernatants by careful aspiration.

9. Recover the bound poly(A)[+] RNA by resuspending the pellets of oligo(dT)-cellulose in 400 µl of double-distilled, autoclaved H_2O. Incubate the suspensions for 5 minutes at 55°C and then centrifuge the tubes for 2 minutes at 4°C in a microfuge.

10. Transfer the supernatants to a series of fresh tubes and repeat Step 9 twice, pooling the recovered supernatants.

11. Add 0.2 volume of 10 M ammonium acetate and 2.5 volumes of ethanol to the supernatants. Store the tubes for 30 minutes at –20°C.

12. Recover the precipitated poly(A)[+] RNAs by centrifugation at maximum speed for 15 minutes at 4°C in a microfuge. Carefully discard the supernatants and wash the pellets (which are often invisible) with 70% ethanol. Centrifuge briefly, remove the supernatants by aspiration, and store the open tubes in an inverted position for a few minutes to allow most of the residual ethanol to evaporate.

13. Dissolve the RNA in a small volume of sterile H_2O.

14. Estimate the concentration of the RNA by absorbance at 260 nm.

15. Store the preparations (see MC3, p. 7.8).

Separation of RNA According to Size: Electrophoresis of Glyoxylated RNA through Agarose Gels

Separation of RNAs according to size is the first stage in northern blotting and hybridization. The method described in this protocol uses glyoxal to denature the RNA, ethidium bromide to stain it, and agarose gel electrophoresis to separate the resulting glyoxal-RNA-ethidium adducts.

IMPORTANT: Prepare all reagents used in this protocol with DEPC-treated H_2O.

MATERIALS

CAUTION: Please see Appendix 4 for appropriate handling of materials marked with <!>.

Reagents and Solutions

Please see Appendix 1 for components of stock solutions, buffers, and reagents. Dilute stock solutions to the appropriate concentrations.

Glyoxal reaction mixture
> 6 ml of DMSO <!>, HPLC grade or better
> 2 ml of Glyoxal, deionized
> 1.2 ml of 10x BPTE electrophoresis buffer
> 0.6 ml of 80% glycerol in H_2O
> 0.2 ml of ethidium bromide (10 mg/ml in H_2O) <!>
> Store in small aliquots at –70°C.

Nucleic Acids/Oligonucleotides

RNA samples (~10 µg of RNA in a volume of 1–2 µl of DEPC-treated H_2O)
RNA size markers
> RNA ladders are available commercially from New England BioLabs and others.

Gels/Loading Buffers

10x BPTE electrophoresis buffer (pH ~6.5)
> 100 mM PIPES
> 300 mM Bis-Tris
> 10 mM EDTA (pH 8.0)
> Add 3 g of PIPES (free acid), 6 g of Bis-Tris (free base), and 2 ml of 0.5 M EDTA to 90 ml of distilled H_2O. Treat the solution with 0.1% DEPC for 1 hour at 37°C before autoclaving.

Horizontal agarose gel, cast in 1x BPTE electrophoresis buffer
> A 1.5% gel is suitable for resolving RNAs in the 0.5–8.0-kb range. Separate large RNAs on 10% or 1.2% gels.

RNA gel-loading buffer
> 95% (w/v) deionized formamide <!>
> 0.025% (w/v) bromophenol blue
> 0.025% (w/v) xylene cyanol FF
> 5 mM EDTA (pH 8.0)
> 0.025% (w/v) SDS

Additional Items

Gel electrophoresis tank
> Reserve a particular electrophoresis apparatus for RNA analysis. Before each use, clean the tank and combs with detergent, rinse in H_2O, dry with ethanol, and treat with a solution of 3% H_2O_2 for 10 minutes. Finally, rinse the tank and combs in H_2O treated with 0.1% DEPC.

Ruler, transparent
UV transilluminator
Water bath (55°C)

Additional Information

Checking the quality of preparations of RNA	MC3, p. 7.30
Diethylpyrocarbonate (DEPC)	MC3, p. 7.84
Glyoxal	MC3, pp. 7.27–7.28
How to win the battle with RNase	MC3, p. 7.82
Inhibitors of RNases	MC3, p. 7.83
Markers used in gels to fractionate RNA	MC3, p. 7.23
Separation of RNA according to size	MC3, pp. 7.21–7.22

METHOD

1. Set up the glyoxal denaturation reaction. In sterile microfuge tubes, mix:

RNA (up to 10 μg)	1–2 μl
glyoxal reaction mixture	10 μl

   Up to 10 μg of RNA may be analyzed in each lane of the gel. Abundant mRNAs (0.1% or more of the mRNA population) can usually be detected by northern analysis of 10 μg of total cellular RNA. Detection of rare RNAs requires at least 1.0 μg of poly(A)$^+$ RNA. Samples containing RNA size markers should be prepared in the glyoxal reaction mixture in the same way as the RNA samples under test.

2. Close the tops of the microfuge tubes and incubate the RNA solutions for 60 minutes at 55°C. Chill the samples for 10 minutes in ice water and then centrifuge them for 5 seconds to deposit all of the fluid in the bottom of the microfuge tubes.

3. While the samples are incubating, install the agarose gel in a horizontal electrophoresis box. Add sufficient 1x BPTE electrophoresis buffer to cover the gel to a depth of approximately 1 mm.

4. Add 1–2 μl of RNA gel-loading buffer to the glyoxylated RNA samples and, without delay, load the glyoxylated RNA samples into the wells of the gel, leaving the two outermost lanes on each side of the gel empty. Load the RNA size markers in the outside lanes of the gel.

5. Perform electrophoresis at 5 V/cm until the bromophenol blue has migrated approximately 8 cm.

6. Visualize the RNAs by placing the gel on a piece of Saran Wrap on a UV transilluminator. Align a transparent ruler with the stained gel and photograph the gel under UV illumination.

7. Use the photograph to measure the distance from the loading well to each of the bands of RNA. Plot the $\log_{10}$ of the size of the fragments of RNA against the distance migrated. Use the resulting curve to calculate the sizes of the RNA species detected by blot hybridization.

8. Proceed with immobilization of RNA onto a solid support by upward or downward capillary transfer (please see Protocol 7.7).

Separation of RNA According to Size: Electrophoresis of RNA through Agarose Gels Containing Formaldehyde

Separation of RNAs according to size is the first stage in northern blotting and hybridization. The method described in this protocol uses formaldehyde to denature the RNA, ethidium bromide to stain it, and electrophoresis through agarose gels containing 2.2 M formamide to separate the resulting formaldehyde-RNA-ethidium adducts.

IMPORTANT: Prepare all reagents used in this protocol with DEPC-treated H_2O.

MATERIALS

CAUTION: Please see Appendix 4 for appropriate handling of materials marked with <!>.

Reagents and Solutions

Please see Appendix 1 for components of stock solutions, buffers, and reagents. Dilute stock solutions to the appropriate concentrations.

Ethidium bromide (200 µg/ml in DEPC-treated H_2O) <!>

Formaldehyde (37–40%, w/v; 12.3 M) <!>

If the pH of the formaldehyde solution is less than 4.0 or if the solution is yellow, deionize by treatment with a mixed bed resin such as Bio-Rad AG-501-X8 or Dowex HG8.

Formamide (deionized) <!>

Nucleic Acids/Oligonucleotides

RNA samples (~20 µg of RNA in a volume of 1–2 µl of DEPC-treated H_2O)

RNA size markers

RNA ladders are available commercially from New England BioLabs and others.

Gels/Loading Buffers

Agarose gel (1.5%) containing 2.2 M formaldehyde

A 1.5% gel is suitable for resolving RNAs in the 0.5–8.0-kb range. Separate larger RNAs on 10% or 1.2% gels. Dissolve 1.5 g of agarose in 72 ml of H_2O. Cools, the solution to 55°C and add 10 ml of 10x MOPS electrophoresis buffer and 18 ml of deionized formamide. Allow the gel to set for at least 1 hour at room temperature. Then cover the gel with Saran Wrap until the samples are ready to be loaded.

10x Formaldehyde gel-loading buffer

50% glycerol, diluted in DEPC-treated H_2O

10 mM EDTA (pH 8.0)

0.25% (w/v) bromophenol blue

0.25% (w/v) xylene cyanol FF

10x MOPS electrophoresis buffer <!>

0.2 M MOPS (pH 7.0) <!>

20 mM sodium acetate

10 mM EDTA (pH 8.0)

Dissolve 41.8 g of MOPS (3-[N-morpholino]propanesulfonic acid) in 700 ml of sterile DEPC-treated H_2O. Adjust the pH to 7.0 with 2 N NaOH. Add 20 ml of DEPC-treated 1 M sodium acetate and 20 ml of DEPC-treated 0.5 M EDTA. Adjust the volume to 1 liter with DEPC-treated H_2O. Sterilize the solution by filtration through a 0.45-µm filter and store it at room temperature, protected from light.

Additional Items

Gel electrophoresis tank

Reserve a particular electrophoresis apparatus for RNA analysis. Before each use, clean the tank and combs with detergent, rinse in H_2O, dry with ethanol, and treat with a solution of 3% H_2O_2 for 10 minutes. Finally, rinse the tank and combs in H_2O treated with 0.1% DEPC.

Ruler, transparent

UV transilluminator

Water bath (55°C or 85°C)

Additional Information

Checking the quality of preparations of RNA	MC3, p. 7.30
Diethylpyrocarbonate (DEPC)	MC3, p. 7.84
Formaldehyde	MC3, p. 7.31
How to win the battle with RNase	MC3, p. 7.82
Inhibitors of RNases	MC3, p. 7.83
Markers used in gels to fractionate RNA	MC3, p. 7.23
Separation of RNA according to size	MC3, pp. 7.21–7.22

METHOD

1. Set up the denaturation reaction. In sterile microfuge tubes, mix:

RNA (up to 20 μg)	2.0 μl
10x MOPS electrophoresis buffer	2.0 μl
formaldehyde	4.0 μl
formamide	10.0 μl
ethidium bromide (200 μg/ml)	1.0 μl

   As much as 20 μg of RNA may be analyzed in each lane of the gel. Abundant mRNAs (0.1% or more of the mRNA population) can usually be detected by northern analysis of 10 μg of total cellular RNA. For detection of rare RNAs, at least 1.0 μg of poly(A)⁺ RNA should be applied to each lane of the gel. Samples containing RNA size markers should be prepared in the same way as the RNA samples under test.

2. Close the tops of the microfuge tubes and incubate the RNA solutions for 60 minutes at 55°C. Chill the samples for 10 minutes in ice water and then centrifuge them for 5 seconds to deposit all of the fluid in the bottom of the microfuge tubes.

3. Add 2 μl of 10x formaldehyde gel-loading buffer to each sample and return the tubes to an ice bucket.

4. Install the agarose/formaldehyde gel in a horizontal electrophoresis box. Add sufficient 1x MOPS electrophoresis buffer to cover the gel to a depth of approximately 1 mm. Run the gel for 5 minutes at 5 V/cm and then load the RNA samples into the wells of the gel, leaving the two outermost lanes on each side of the gel empty. Load the RNA size standards in the outside lanes of the gel.

5. Run the gel submerged in 1x MOPS electrophoresis buffer at 4–5 V/cm until the bromophenol blue has migrated approximately 8 cm (4–5 hours).

6. Visualize the RNAs by placing the gel on a piece of Saran Wrap on a UV transilluminator. Align a transparent ruler with the stained gel and photograph under UV illumination.

7. Use the photograph to measure the distance from the loading well to each of the bands of RNA. Plot the $\log_{10}$ of the size of the fragments of RNA against the distance migrated. Use the resulting curve to calculate the sizes of the RNA species detected by blot hybridization.

8. Proceed with immobilization of RNA onto a solid support by upward or downward capillary transfer (please see Protocol 7.7).

Protocol 7.7

Transfer and Fixation of Denatured RNA to Membranes

This protocol describes the transfer of RNA from agarose gels to neutral or positively charged nylon membranes, using upward capillary flow of neutral or alkaline buffers. RNA becomes covalently fixed to positively charged nylon membranes during transfer in alkaline buffers. However, treatment by UV irradiation or heating is required to fix RNA to neutral membranes.

IMPORTANT: Prepare all reagents used in this protocol with DEPC-treated H_2O.

MATERIALS

CAUTION: Please see Appendix 4 for appropriate handling of materials marked with <!>.

Reagents and Solutions

Please see Appendix 1 for components of stock solutions, buffers, and reagents. Dilute stock solutions to the appropriate concentrations.

Ammonium acetate (0.1 M) with 0.5 µg/ml ethidium bromide <!> (optional; see Step 13)

Methylene blue (0.02% [w/v] in 0.3 M sodium acetate [pH 5.5]) (optional; see Step 15)

Soaking solution (for charged nylon membranes, use 0.01 N NaOH <!> combined with
  3 M NaCl; for uncharged membranes, use 0.05 N NaOH <!>)

20x SSC

0.2x SSC with 1% (w/v) SDS

Transfer buffer (for alkaline transfers to charged membranes, use 0.01 N NaOH <!> with
  3 M NaCl; for neutral transfers to uncharged membranes, use 20x SSC)

Nucleic Acids/Oligonucleotides

RNA, fractionated through an agarose gel (see Protocols 7.5 and 7.6)

Additional Items

Blotting paper (e.g., Sigma-Aldrich P9039)

Cross-linking device (e.g., Stratalinker [Stratagene]) *or* microwave oven *or* vacuum oven

Glass baking dish

Nylon membranes (either positively charged or uncharged)

Platform shaker

Plexiglas or glass plate

Scalpel blade

Thick blotting paper (Whatman 3MM)

Visible-spectrum light box

Weight (~400 g)

Yellow filter for photography

Additional Information

Alternative protocol (Capillary Transfer by Downward Flow)	MC3, p. 7.41
Background in Southern and northern hybridizations and how to avoid it	MC3, p. 7.45, Table 7-5
Checking the quality of preparations of RNA	MC3, p. 7.30
How to win the battle with RNase	MC3, p. 7.82
Inhibitors of RNases	MC3, p. 7.83
Markers used in gels to fractionate RNA	MC3, p. 7.23
Membranes used for northern hybridization	MC3, pp. 7.23–7.25

METHOD

1. (*Optional*) Partially hydrolyze the RNA sample, fractionated through agarose, by soaking the gel in the appropriate soaking solution as described below.

For transfer to uncharged nylon membranes

a. Rinse the gel with DEPC-treated H_2O.

b. Soak the gel for 20 minutes in 5 gel volumes of 0.05 N NaOH.

c. Transfer the gel into 10 gel volumes of 20x SSC for 40 minutes.

d. Without delay, proceed directly with Step 2 to transfer the partially hydrolyzed RNA to an uncharged nylon membrane by capillary action.

For transfer to charged nylon membranes

a. Rinse the gel with DEPC-treated H_2O.

b. Soak the gel for 20 minutes in 5 gel volumes of 0.01 N NaOH/3 M NaCl.

c. Without delay, proceed directly with Step 2 to transfer the partially hydrolyzed RNA to a positively charged nylon membrane by capillary action.

2. Move the gel containing fractionated RNA to a glass baking dish and use a sharp scalpel to trim away unused areas of the gel. Cut along the slot line to allow the top of the trimmed gel to be aligned with the top of the membrane during transfer. Cut off a small triangular piece from the bottom left-hand corner of the gel to simplify orientation during the succeeding operations.

3. Place a piece of thick blotting paper on a sheet of Plexiglas or a glass plate to form a support that is longer and wider than the trimmed gel. Make sure that the ends of the blotting paper drape over the edges of the plate. Place the support inside a large baking dish (please see Figure 7-1).

4. Fill the dish with the appropriate RNA transfer buffer (0.01 N NaOH/3 M NaCl for positively charged membranes and 20x SSC for uncharged membranes) until the level of the liquid reaches almost to the top of the support. When the blotting paper on the top of the support is thoroughly wet, smooth out all air bubbles with a glass rod or pipette.

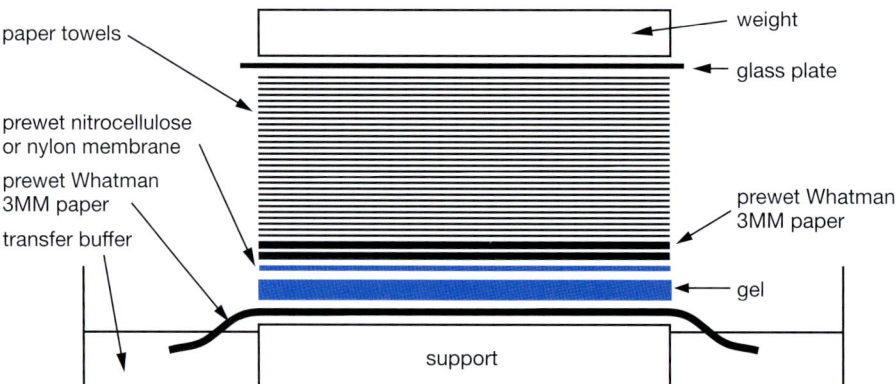

FIGURE 7-1 Upward Capillary Transfer

Capillary transfer of nucleic acids from an agarose gel to solid supports is achieved by drawing the transfer buffer from the reservoir upward though the gel into a stack of paper towels. The nucleic acid is eluted from the gel by a moving stream of buffer and is deposited onto a nitrocellulose filter or nylon membrane. A weight applied to the top of the paper towels helps to ensure a tight connection between the layers of material used in the transfer system.

Alkaline transfer buffer (0.01 N NaOH/3 M NaCl) is used to transfer RNA to positively charged nylon membranes. Neutral transfer buffer (20x SSC) is used to transfer RNA to uncharged nylon membranes.

5. Use a fresh scalpel or a paper cutter to cut a piece of the appropriate nylon membrane approximately 1 mm larger than the gel in both dimensions.

6. Float the nylon membrane on the surface of a dish of deionized H_2O until it saturates completely from beneath and then immerse the membrane in 10x SSC for at least 5 minutes. Use a clean scalpel blade to cut a corner from the membrane to match the corner cut from the gel.

7. Carefully place the gel on the support in an inverted position so that it is centered on the wet blotting paper.

8. Surround (but do not cover) the gel with Saran Wrap or Parafilm.

9. Wet the top of the gel with the appropriate transfer buffer (please see Step 4). Place the wet nylon membrane on top of the gel so that the cut corners are aligned. One edge of the membrane should extend just beyond the edge of the line of slots at the top of the gel.

    IMPORTANT: Do not move the membrane once it has been applied to the surface of the gel. Make sure that there are no air bubbles between the membrane and the gel.

10. Wet two pieces of thick blotting paper (cut to exactly the same size as the gel) in the appropriate transfer buffer and place them on top of the wet nylon membrane. Smooth out any air bubbles with a glass rod.

11. Cut or fold a stack of paper towels (5–8 cm high) just smaller than the blotting papers. Place the towels on the blotting papers. Put a glass plate on top of the stack and weigh it down with a 400-g weight.

12. Allow upward transfer of RNA to occur for no more than 4 hours in neutral transfer buffer and approximately 1 hour in alkaline transfer buffer.

13. Dismantle the capillary transfer system. Mark the positions of the slots on the membrane with a ballpoint pen through the gel. Transfer the membrane to a glass tray containing approximately 300 ml of 6x SSC at 23°C. Place the tray on a platform shaker and agitate the membrane very slowly for 5 minutes.

14. Remove the membrane from the 6x SSC and allow excess fluid to drain away. Lay the membrane, RNA side upward, on a dry sheet of blotting paper for a few minutes.

15. If the RNA is to be fixed by UV irradiation, proceed first to Step 16; otherwise, stain the membrane.

    a. Transfer the damp membrane to a glass tray containing methylene blue solution. Stain the membrane for just enough time to visualize the rRNAs (~3–5 minutes).

       The order of steps during staining and fixation depends on the type of transfer, the type of membrane, and the method of fixation. Because alkaline transfer results in covalent attachment of RNA to positively charged nylon membranes, there is no need to fix the RNA to the membrane before staining. RNA transferred to uncharged nylon membranes in neutral transfer buffer should be stained and then fixed to the membrane by baking under vacuum or heating in a microwave oven. If the RNA is to be cross-linked to the membrane by UV irradiation, the staining step should follow fixation (please see Table 7-3).

    b. Photograph the stained membrane under visible light with a yellow filter.

    c. After photography, destain the membrane by washing in 0.2x SSC and 1% (w/v) SDS for 15 minutes at room temperature.

       For RNA transferred to positively charged nylon using alkaline transfer, proceed directly to hybridization (Protocol 7.8). Any membranes not used immediately in hybridization reactions should be thoroughly dry, wrapped loosely in aluminum foil or blotting paper, and stored at room temperature, preferably under vacuum.

TABLE 7-3 Sequence of Staining RNA and Fixing to the Membrane

Type of Membrane	Method of Fixation	Order of Steps
Positively charged nylon	alkaline transfer	1. Stain with methylene blue. 2. Proceed to prehybridization.
Uncharged nylon or positively charged nylon (nonalkaline transfer)	UV irradiation (please see Step 16 for details)	1. Fix the RNA by UV irradiation. 2. Stain with methylene blue. 3. Proceed to prehybridization.
Uncharged nylon or positively charged nylon (nonalkaline transfer)	baking in vacuum oven or microwave oven (please see Step 16 for details)	1. Stain with methylene blue. 2. Bake the membrane. 3. Proceed to prehybridization.

16. Fix the RNA to the uncharged nylon membrane or to positively charged nylon (nonalkaline transfer).

To fix by baking

• Allow the membrane to air dry and then bake for 2 hours between two pieces of blotting paper under vacuum at 80°C in a vacuum oven.

  or

• Place the damp membrane on a dry piece of blotting paper and heat for 2–3 minutes at full power in a microwave oven (750–900 W).

   Proceed directly to hybridization (Protocol 7.8). Any membranes not used immediately in hybridization reactions should be thoroughly dry, wrapped loosely in aluminum foil or blotting paper, and stored at room temperature, preferably under vacuum.

To cross-link by UV irradiation

a. Place the damp, unstained membrane on a piece of dry blotting paper, transfer it to a cross-linking device, and irradiate at 254 nm for 1 minute 45 seconds at 1.5 J/cm².

b. After irradiation, stain the membrane with methylene blue as described in Step 15.

   Proceed directly to hybridization (Protocol 7.8). Any membranes not used immediately in hybridization reactions should be thoroughly dry, wrapped loosely in aluminum foil or blotting paper, and stored at room temperature, preferably under vacuum.

Northern Hybridization

This protocol describes how to perform northern hybridization at high stringency in phosphate-SDS buffers. Although a wide variety of formats are available, hybridization is usually performed in heat-sealable bags, roller bottles, or plastic boxes, as described here.

IMPORTANT: Prepare all reagents used in this protocol with DEPC-treated H_2O.

MATERIALS

CAUTION: Please see Appendix 4 for appropriate handling of materials marked with <!>.

Reagents and Solutions

Please see Appendix 1 for components of stock solutions, buffers, and reagents. Dilute stock solutions to the appropriate concentrations.

HCl (3 N) (optional; see Step 2)
NaOH (3 N) <!> (optional; see Step 2)
Prehybridization/hybridization solution
     0.5 M sodium phosphate (pH 7.2)
     7% (w/v) SDS
     1 mM EDTA (pH 8.0)
SSC (0.5x, 1x, and 2x)
SSC (0.1x) with 0.1% (w/v) SDS (optional; see Step 4)
Tris-Cl (1 M, pH 7.2) (optional; see Step 2)

Nucleic Acids/Oligonucleotides

Probe DNA or RNA, labeled in vitro with ^{32}P to high specific activity (>2 x 10^8 cpm/μg).
    The highest sensitivity is obtained from single-stranded probes (for preparation of probes, see protocols in Chapter 9).
RNA, immobilized on a nylon membrane (Protocol 7.7)

Radioactive Compounds

[α-^{32}P]dNTP or rNTP (highest specific activity available) <!>

Additional Items

Blotting paper (e.g., Whatman 3MM)
Boiling-water bath
Containers for prehybridization, hybridization, and washing (e.g., Seal-a-Meal bags, plastic boxes, rotating wheels)
Materials for autoradiography/phosphorimaging
Water bath (68°C)
Water bath (temperature of hybridization)

Additional Information

Alternative protocol (Capillary Transfer by Downward Flow)	MC3, p. 7.41
Autoradiography and phosphorimaging	MC3, pp. A9.11–A9.15
Background in Southern and northern hybridizations and how to avoid it	MC3, p. 7.45, Table 7-5

METHOD

1. Incubate the membrane for 2 hours at 68°C in 10–20 ml of prehybridization solution.

2. If using a double-stranded probe, denature the ^{32}P-labeled double-stranded DNA by heating for 5 minutes at 100°C. Chill the probe rapidly in ice water.

   Alternatively, denature the probe by adding 0.1 volume of 3 N NaOH. After 5 minutes at room temperature, transfer the probe to ice water and add 0.05 volume of 1 M Tris-Cl (pH 7.2) and 0.1 volume of 3 N HCl. Store the probe in ice water until it is needed.

   Single-stranded probes need not be denatured.

3. Add the denatured or single-stranded radiolabeled probe directly to the prehybridization solution. Continue incubation for 12–16 hours at the appropriate temperature.

   To detect low-abundance mRNAs, use at least 0.1 μg of probe whose specific activity exceeds 2×10^8 cpm/μg. Low-stringency hybridization, in which the probe is not homologous to the target gene, is best carried out at lower temperatures (37–42°C) in a hybridization buffer containing 50% deionized formamide, 0.25 M sodium phosphate (pH 7.2), 0.25 M NaCl, and 7% SDS.

4. After hybridization, remove the membrane from the container and transfer it as quickly as possible to a plastic box containing 100–200 ml of 1x SSC, 0.1% SDS at room temperature. Place the closed box on a platform shaker and agitate the fluid gently for 10 minutes.

   IMPORTANT: Do not allow the membrane to dry out at any stage during the washing procedure.

   Increase the concentration of SDS in the washing buffer to 1% if single-stranded probes are used. Following low-stringency hybridization in formamide-containing buffers, rinse the membrane in 2x SSC at 23°C and then successively wash in 2x SSC, 0.5x SSC with 0.1% SDS, and 0.1x SSC with 0.1% SDS for 15 minutes each at 23°C. Perform a final wash containing 0.1x SSC and 1% SDS at 50°C.

5. Transfer the membrane to another plastic box containing 100–200 ml of 0.5x SSC, 0.1% SDS, prewarmed to 68°C. Agitate the fluid gently for 10 minutes at 68°C.

6. Repeat the washing in Step 5 twice more for a total of three washes at 68°C.

7. Dry the membrane on blotting paper and establish an autoradiograph by exposing the membrane for 24–48 hours to X-ray film (Kodak XAR-5 or its equivalent) at –70°C with an intensifying screen.

   Tungstate-based intensifying screens are more effective than the older rare-earth screens. Alternatively, an image of the membrane can be obtained by scanning in a phosphorimager.

Dot and Slot Hybridization of Purified RNA

Dot blotting of RNA is best performed using purified preparations of RNA that are denatured with glyoxal or formaldehyde immediately before loading onto a nylon membrane through a vacuum manifold.

IMPORTANT: Prepare all reagents used in this protocol with DEPC-treated H_2O.

MATERIALS

CAUTION: Please see Appendix 4 for appropriate handling of materials marked with <!>.

Reagents and Solutions

Please see Appendix 1 for components of stock solutions, buffers, and reagents. Dilute stock solutions to the appropriate concentrations.

NaOH (10 N) <!>

Prehybridization/hybridization solution

    0.5 M sodium phosphate (pH 7.2)

    7% (w/v) SDS

    1 mM EDTA (pH 8.0)

RNA denaturation solution

    660 µl of formamide <!> (deionized; see Protocol 7.8)

    210 µl of formaldehyde (37% w/v) <!> (deionized; see Protocol 7.8 or p. 707)

    130 µl of 10x MOPS electrophoresis buffer (pH 7.0) <!> (see Protocol 7.8 or p. 706)

SSC (0.5x, 1x, 2x, and 20x)

SSC (0.1x) with 0.1% or 1.0% (w/v) SDS

Nucleic Acids/Oligonucleotides

Probe DNA or RNA, labeled in vitro with ^{32}P to high specific activity (>2 x 10^8 cpm/µg)

    The highest sensitivity is obtained from single-stranded probes, which are capable of detecting mRNAs of medium to high abundance when 5 µg of total cellular DNA is loaded into a slot. RNAs of the lowest abundance (1–5 copies/cell) are detected by loading at least 1 µg of purified poly(A)$^+$ RNA per slot and hybridizing with strand-specific probes of high specific activity (for preparation of probes, see Chapter 9).

RNA samples, standards, and controls

    Prepare samples by one of the methods described in Protocols 7.1–7.4. All slots of the dot-blot apparatus should contain equivalent amounts of RNA dissolved in 10 µl of DEPC-treated H_2O. Standards are generated by mixing varying quantities of unlabeled sense-strand RNA synthesized in vitro (Chapter 9) to aliquots of control DNA that lacks sequences complementary to the radiolabeled probe.

Radioactive Compounds

[α-^{32}P]dNTP or rNTP (highest specific activity available) <!>

Additional Items

Blotting manifold for dot or slot blotting

Blotting paper (e.g., Whatman 3MM)

Boiling-water bath

Containers for prehybridization, hybridization, and washing (e.g., Seal-a-Meal bags, plastic boxes, rotating wheels)

Cross-linking device (e.g., Stratalinker [Stratagene]) or vacuum oven or microwave oven

Glass dish
Materials for autoradiography/phosphorimaging
Nylon membrane (positively charged)
Plastic box (for posthybridization washing)
Water bath or hybridization chamber (65°C and 50°C)

Additional Information

Autoradiography and phosphorimaging	MC3, pp. A9.11–A9.15
Background in Southern and northern hybridizations and how to avoid it	MC3, p. 7.45, Table 7-5
Checking the quality of preparations of RNA	MC3, p. 7.30
Dot and slot hybridization of purified RNA	MC3, pp. 7.46–7.47
How to win the battle with RNase	MC3, p. 7.82
Inhibitors of RNases	MC3, p. 7.83
Membranes used for northern hybridization	MC3, pp. 7.23–7.25

METHOD

1. Cut a piece of positively charged nylon membrane to a suitable size. Mark the membrane with a soft pencil or ballpoint pen to indicate the orientation. Wet the membrane briefly in H_2O and soak it in 20x SSC for 1 hour at room temperature.

2. While the membrane is soaking, clean the blotting manifold carefully with 0.1 N NaOH and then rinse it well with sterile H_2O.

3. Wet two sheets of thick blotting paper with 20x SSC and place them on top of the vacuum unit of the apparatus.

4. Place the wet nylon membrane on the bottom of the sample wells cut into the upper section of the manifold. Roll a pipette across the surface of the membrane to smooth away any air bubbles trapped between the upper section of the manifold and the nylon membrane.

5. Clamp the two parts of the manifold together and connect the unit to a vacuum line.

6. Fill all of the slots/dots with 10x SSC and apply gentle suction until the fluid has passed through the nylon membrane. Turn off the vacuum and refill the slots with 10x SSC.

7. Mix each of the RNA samples (dissolved in 10 μl of H_2O) with 30 μl of RNA denaturation solution.

8. Incubate the mixture for 5 minutes at 65°C and then cool the samples on ice.

9. Add an equal volume of 20x SSC to each sample.

10. Apply gentle suction to the manifold until the 10x SSC in the slots has passed through the membrane. Turn off the vacuum.

11. Load all of the samples into the slots and then apply gentle suction. After all of the samples have passed through the membrane, rinse each of the slots twice with 1 ml of 10x SSC.

12. After the second rinse has passed through the nylon membrane, continue suction for 5 minutes to dry the membrane.

13. Remove the membrane from the manifold and fix the RNA to the membrane by either UV irradiation, baking, or microwaving, as described in Protocol 7.7, Step 16.

    Before setting up the prehybridization and hybridization reactions, please see Protocol 7.8.

14. Incubate the membrane for 2 hours at 68°C in 10–20 ml of prehybridization solution in a baking dish or hybridization chamber.

15. Add the denatured radiolabeled probe directly to the prehybridization solution. Continue the incubation for 12–16 hours at the appropriate temperature.

    To detect low-abundance mRNAs, use at least 0.1 μg of probe whose specific activity exceeds 5×10^8 cpm/μg. Low-stringency hybridization, in which the probe is not homologous to the target gene, is best performed at lower temperatures (37–42°C) in a hybridization buffer containing 50% deionized formamide, 0.25 M sodium phosphate (pH 7.2), 0.25 M NaCl, and 7% SDS.

16. After hybridization, remove the membrane from the plastic bag and transfer it as quickly as possible to a plastic box containing 100–200 ml of 1x SSC, 0.1% SDS at room temperature. Place the closed box on a platform shaker and agitate the fluid gently for 10 minutes.

    IMPORTANT: Do not allow the membrane to dry out at any stage during the washing procedure.

    Increase the concentration of SDS in the washing buffer to 1% if single-stranded probes are used. Following low-stringency hybridization in formamide-containing buffers, rinse the membrane in 2x SSC at 23°C and then successively wash in 2x SSC, 0.5x SSC with 0.1% SDS, and 0.1x SSC with 0.1% SDS for 15 minutes each at 23°C. Perform a final wash containing 0.1x SSC and 1% SDS at 50°C.

17. Transfer the membrane to another plastic box containing 100–200 ml of 0.5x SSC, 0.1% SDS, prewarmed to 68°C. Agitate the fluid gently for 10 minutes at 68°C.

18. Repeat the washing in Step 17 twice more for a total of three washes at 68°C.

19. Dry the membrane on filter paper and establish an autoradiograph by exposing the membrane for 24–48 hours to X-ray film (Kodak XAR-5 or its equivalent) at –70°C with an intensifying screen.

    Tungstate-based intensifying screens are more effective than the older rare-earth screens. Alternatively, an image of the membrane can be obtained by scanning in a phosphorimager.

Mapping RNA with Nuclease S1

Preparations of RNA containing an mRNA of interest are hybridized to a complementary single-stranded DNA probe. At the end of the reaction, nuclease S1 is used to degrade unhybridized regions of the probe, and the surviving DNA-RNA hybrids are then separated by gel electrophoresis and visualized by either autoradiography or Southern hybridization. The method can be used to quantitate RNAs, to map the positions of introns, and to identify the locations of 5′ and 3′ ends of mRNAs on cloned DNA templates.

IMPORTANT: Prepare all reagents used in this protocol with DEPC-treated H_2O.

MATERIALS

CAUTION: Please see Appendix 4 for appropriate handling of materials marked with <!>.

Reagents and Solutions

Please see Appendix 1 for components of stock solutions, buffers, and reagents. Dilute stock solutions to the appropriate concentrations.

Acrylamide solution (40% w/v) (acrylamide:bisacrylamide 19:1) <!>

Ammonium persulfate (10%) <!> (freshly made)

10x Annealing buffer

  100 mM Tris (pH 7.5)
  100 mM MgCl$_2$
  0.5 M NaCl
  100 mM dithiothreitol

dNTP solution, containing all four dNTPs at a concentration of 20 mM

Ethanol

Ethanol (70%)

Gel elution buffer

  0.5 M ammonium acetate <!>
  1 mM EDTA
  0.1% SDS

Hybridization buffer without formamide (for RNA)

  40 mM PIPES (pH 6.4)
  0.1 mM EDTA (pH 8.0)
  0.4 M NaCl

  Use the disodium salt of PIPES (piperazine-N,N′-bis[2-ethanesulfonic acid]) and adjust the pH to 6.4 with 1 N HCl.

Nuclease S1 stop mixture

  4 M ammonium acetate <!>
  50 mM EDTA
  50 µg/ml carrier RNA

Phenol:chloroform (1:1, v/v) <!>

Sodium acetate (3 M, pH 5.2)

TE (pH 7.6)

TEMED (N,N,N′,N′-tetramethylethylene diamine) <!>

Trichloroacetic acid (TCA) (1% and 10%) <!>

Urea

Enzymes and Buffers

Bacteriophage T4 polynucleotide kinase and 10x buffer

Klenow fragment of *E. coli* DNA polymerase and 10x buffer

Nuclease S1

Titrate nuclease S1 every time a new batch of enzyme, probe, or RNA preparation is used.

Nuclease S1 digestion buffer

0.28 M NaCl

0.05 sodium acetate (pH 4.5)

4.5 mM $ZnSO_4 \cdot 7H_2O$

Restriction enzymes and 10x buffers

Nucleic Acids/Oligonucleotides

Carrier RNA (yeast tRNA)

Dissolve commercially available yeast RNA at a concentration of 10 mg/ml in sterile 0.1 M NaCl. Extract the solution twice with phenol:chloroform and twice with chloroform alone. Recover the RNA by precipitation with 2.5 volumes of ethanol at room temperature and dissolve the RNA at a concentration of approximately 10 mg/ml in sterile TE (pH 7.6). Store aliquots of the preparation at –20°C.

Probe DNA, labeled in vitro with ^{32}P to high specific activity (>2 x 10^8 cpm/µg) and prepared as described in this protocol (Steps 1–15)

RNA, for use as a standard

Synthesize in vitro by transcription of the appropriate strand of a recombinant plasmid containing the DNA sequences of interest and a bacteriophage promoter (Protocol 9.6).

RNA samples, standards, and controls

Prepare samples by one of the methods described in Protocols 7.1–7.4. All slots of the dot-blot apparatus should contain equivalent amounts of RNA dissolved in 10 µl of DEPC-treated H_2O. Standards are generated by mixing varying quantities of unlabeled sense-strand RNA synthesized in vitro (Chapter 9) to aliquots of control DNA that lacks sequences complementary to the radiolabeled probe.

Synthetic oligonucleotide (10 pmoles/µl in H_2O)

Use to prime synthesis of a probe from a single-stranded DNA template. The oligonucleotide should be 20–25 nucleotides in length and complementary to the RNA strand to be analyzed. It should hybridize to the template DNA strand 250–500 nucleotides 3' of the position that will be cleaved by the chosen restriction enzyme.

Template DNA (1 µg/µl) (single stranded)

Prepare from a recombinant bacteriophage M13 carrying the target DNA strand in the same sense as the test RNA.

Test RNA

Poly(A)+ or total cellular RNA (Protocols 7.1–7.4)

Radioactive Compounds

$[\gamma\text{-}^{32}P]ATP$ (10 mCi/ml, 3000 Ci/mmole) <!>

Gels/Loading buffers

Denaturing polyacrylamide gel containing 8 M urea, cast in 1x TBE electrophoresis buffer

For most 5'- and 3'-end mapping experiments, a denaturing gel composed of 5% or 6% polyacrylamide and containing 8 M urea nicely resolves protected DNA fragments. A typical gel is 1.5 mm thick. However, "thin" or "sequencing" gels (0.4-mm thickness) can also be used (see Chapter 12). If thin gels are used to resolve protected DNA fragments, it is usually not necessary to fix the gel in trichloroacetic acid (TCA), as described in Steps 30–32, before drying the gel. Fixing sharpens and increases the resolution of thicker gels. In many cases, a miniprotein gel apparatus (e.g., Bio-Rad Mini-Protean) (13 cm x 13 cm x 1 mm) can be used both to prepare the radiolabeled single-stranded DNA or RNA probe and to analyze the products of nuclease S1 digestion. Table 7-4 shows the percentage of polyacrylamide used to purify DNA fragments of various sizes; Table 7-5 shows the expected mobilities of tracking dyes in these gels. The method used to prepare the polyacrylamide gel is described in Step 1 of this protocol; for additional details, please see Protocol 12.8.

TABLE 7-4 Percentage of Gel for Purifying Various DNA Fragments

% Polyacrylamide/Urea Gel	Size of Band (nt)
4	>250
6	60–250
8	40–120
10	20–60
12	10–50

Formamide gel-loading buffer
    80% (w/v) deionized formamide <!>
    10 mM EDTA (pH 8.0)
    0.25% (w/v) bromophenol blue
    0.25% (w/v) xylene cyanol FF
RNA gel-loading buffer
    95% (w/v) deionized formamide <!>
    5 mM EDTA (pH 8.0)
    0.25% (w/v) bromophenol blue
    0.25% (w/v) xylene cyanol FF

Additional Items

Conventional apparatus for polyacrylamide gel electrophoresis
Gel dryer
Materials for autoradiography/phosphorimaging
Polyacrylamide minigel apparatus (e.g., Bio-Rad Mini-Protean)
Water baths (37°C, 65°C, 85°C, 95°C, the appropriate digestion temperature [Step 22], and the desired hybridization temperature [Step 21])
Whatman 3MM paper

Additional Information

Autoradiography and phosphorimaging	MC3, pp. A9.11–A9.15
Markers used in gels to fractionate RNA	MC3, p. 7.23
Nuclease S1	MC3, pp. 7.51–7.55, 7.61, and 7.86

METHOD

1. Prepare a polyacrylamide minigel containing 8 M urea (13 cm x 15 cm x 1 mm) (e.g., Bio-Rad Mini-Protean).

    a. Mix the following reagents:

        7.2 g of urea
        1.5 ml of 10x TBE

    Add the appropriate amounts of 40% acrylamide (acrylamide:bisacrylamide 19:1; please see Table 7-5) to generate a gel containing the desired concentration of polyacrylamide.

    b. Add H_2O to a final volume of 15 ml.

TABLE 7-5 Volumes of Polyacrylamide Required to Cast Minigels of Various Percentages

% Gel	Volume of 40% Acrylamide (ml)
4	1.5
5	1.875
6	2.25
8	3.0

c. Stir the mixture at room temperature on a magnetic stirrer until the urea dissolves. Then add:

120 μl of 10% ammonium persulfate
16 μl of TEMED

Mix the solution quickly and then pour the gel into the mold of a minigel apparatus.

2. While the gel is polymerizing, mix the following reagents:

10 pmoles (1 μl) of unlabeled oligonucleotide
10 μl of $[\gamma\text{-}^{32}P]$ATP (3000 Ci/mmole, 10 mCi/ml)
2 μl of 10x polynucleotide kinase buffer
6 μl of H_2O
10 units (1 μl) of polynucleotide kinase

Incubate the reaction mixture for 45 minutes at 37°C and then for 3 minutes at 95°C to inactivate the polynucleotide kinase.

3. Add to the kinase reaction:

12 μl (2 μg) of single-stranded DNA template
4 μl of 10x annealing buffer
14 μl of H_2O

Incubate the reaction mixture for 10 minutes at 65°C and then allow it to cool to room temperature.

4. Add to the reaction mix from Step 3:

4 μl of dNTP mixture
1 μl (10 units) of the Klenow fragment of *E. coli* DNA polymerase I

Incubate the reaction mixture for 15 minutes at room temperature and then inactivate the DNA polymerase by incubation for 3 minutes at 65°C.

5. Adjust the ionic composition and pH of the reaction mixture to suit the restriction enzyme. Add 20 units of restriction enzyme and incubate the reaction mixture for 2 hours at the appropriate temperature.

6. Add to the restriction endonuclease digestion reaction:

2 μl of carrier RNA
5 μl of 3 M sodium acetate (pH 5.2)

Recover the DNA probe by standard precipitation with ethanol.

7. Dissolve the DNA in 20 μl of formamide gel-loading buffer. Heat the solution to 95°C for 5 minutes to denature the DNA and then cool the DNA quickly to 0°C.

8. While the DNA is incubating at 95°C, wash the loading slots of the gel to remove urea and then, without delay, load the probe into one of the slots of the gel.

9. Run the gel until the bromophenol blue reaches the bottom of the gel (200 mA for ~30 minutes).

10. Dismantle the gel apparatus, leaving the gel attached to the bottom glass plate. Wrap the gel and plate in a piece of plastic wrap (e.g., Saran Wrap). Make sure that no bubbles are between the gel and the plastic film.

11. Expose the gel to X-ray film. Mark the location of the corners and sides of the plate on the film with a permanent marker. Also mark the position of the bromophenol blue and xylene cyanol.

12. Realign the glass plate with the film and excise the radiolabeled band with a scalpel. Reexpose the mutilated gel to a fresh piece of film to ensure that the region of the gel containing the band of the correct molecular weight has been accurately excised.

13. Transfer the fragment of gel to a fresh sterile microfuge tube and add just enough gel elution buffer to cover the fragment (250–500 μl). Incubate the closed tube on a rotating wheel overnight at room temperature.

14. Centrifuge the tube at maximum speed for 5 minutes in a microfuge.

15. Taking care to avoid the pellet of polyacrylamide, use an automatic pipetting device to transfer the supernatant to a fresh microfuge tube. The labeled probe should emit approximately 1×10^4 cpm/μl as measured by liquid scintillation spectroscopy.

16. Store the probe at –70°C.

17. Transfer 0.5–150-μg aliquots of RNA (test and standard) into sterile microfuge tubes. Add an excess of uniformly labeled single-stranded DNA probe to each tube.

18. Precipitate the RNA and DNA by adding 0.1 volume of 3 M sodium acetate (pH 5.2) and 2.5 volumes of ice-cold ethanol. After storage for 30 minutes at 0°C, recover the nucleic acids by centrifugation at maximum speed for 15 minutes at 4°C in a microfuge. Discard the ethanolic supernatant, rinse the pellet with 70% ethanol, and centrifuge the sample. Carefully remove all of the ethanol, and store the pellet containing RNA and DNA at room temperature until the last visible traces of ethanol have evaporated.

19. Dissolve the nucleic acid pellet in 30 μl of hybridization buffer without formamide. Pipette the solution up and down many times to ensure that the pellet is completely dissolved.

20. Close the lid of the tube tightly and incubate the hybridization reaction in a water bath set at 85°C for 10 minutes to denature the nucleic acids.

21. Rapidly transfer the tube to a water bath set at the desired hybridization temperature (usually 65°C). Do not allow the tube to cool below the hybridization temperature during transfer. Hybridize the DNA and RNA for 12–16 hours at the chosen temperature.

22. Taking care to keep the body of the tube submerged, open the lid of the hybridization tube. Rapidly add 300 μl of ice-cold nuclease S1 digestion buffer and immediately remove the tube from the water bath. Quickly mix the contents of the tube by vortexing gently and then transfer the tube to a water bath set at the temperature appropriate for digestion with nuclease S1. Incubate for 1–2 hours, depending on the degree of digestion desired.

23. Chill the reaction to 0°C. Add 80 μl of nuclease S1 stop mixture and vortex the tube to mix the solution.

24. Extract the reaction once with phenol:chloroform. After centrifugation at maximum speed for 2 minutes at room temperature in a microfuge, transfer the aqueous supernatant to a fresh tube. Add 2 volumes of ethanol, mix, and store the tube for 1 hour at –20°C.

25. Recover the nucleic acids by centrifugation at maximum speed for 15 minutes at 4°C in a microfuge. Carefully remove all of the supernatant, and store the open tube at room temperature until the last visible traces of ethanol have evaporated.

26. Dissolve the pellet in 4 μl of TE (pH 7.6). Add 6 μl of formamide gel-loading buffer and mix well.

27. Heat the nucleic acids for 5 minutes at 95°C and then immediately transfer the tube to an ice bath. Centrifuge the tubes briefly in a microfuge to consolidate the samples at the bottoms of the tubes.

28. Analyze the radiolabeled DNA by electrophoresis through a polyacrylamide/8 M urea gel.

29. After the tracking dyes have migrated an appropriate distance through the gel (please see Table 7-6), turn off the power supply and disassemble the electrophoresis setup. Gently pry up one corner of the larger glass plate and slowly remove the plate from the gel. Cut off one corner of the gel for orientation purposes.

TABLE 7-6 Expected Mobilities of Tracking Dyes

% Polyacrylamide/Urea Gel	Xylene Cyanol (nt)	Bromophenol Blue (nt)
4	155	30
6	110	25
8	75	20
10	55	10

Tracking dyes can serve as useful size standards on denaturing polyacrylamide gels. The table indicates the approximate sizes of tracking dyes (in nucleotides) on gels of different polyacrylamide concentrations.

30. Transfer the glass plate containing the gel to a tray containing an excess of 10% TCA. Gently rock or rotate the tray for 10 minutes at room temperature.

31. Pour off the 10% TCA solution and replace with an excess of 1% TCA. Gently rock or rotate the tray for 5 minutes at room temperature.

32. Pour off the 1% TCA solution and briefly rinse the fixed gel with distilled deionized H_2O. Lift the glass plate together with the gel out of the tray and place them on a flat benchtop. Use paper towels or Kimwipes to remove excess H_2O.

33. Cut a piece of Whatman 3MM filter paper (or equivalent) that is 1 cm larger than the gel on all sides. Transfer the gel to the filter paper by laying the paper on top of the gel and inverting the glass plate.

34. Remove the plate and dry the gel on a gel dryer for 1.0–1.5 hours at 60°C.

35. Establish an autoradiographic image of the dried gel. Scan the image by densitometry or phosphorimaging, or excise the segments of the gel containing the fragments and count them by liquid scintillation spectroscopy.

Ribonuclease Protection: Mapping RNA with Ribonuclease and Radiolabeled RNA Probes

Preparations of RNA containing an mRNA of interest are hybridized to a radiolabeled single-stranded RNA probe. At the end of the reaction, a mixture of RNase A and RNase T1 is used to degrade unhybridized regions of the probe, and the surviving molecules are then separated by denaturing gel electrophoresis and visualized by autoradiography. The method can be used to quantitate RNAs, to map the positions of introns, and to identify the locations of 5′ and 3′ ends of mRNAs on cloned DNA templates.

IMPORTANT: Prepare all reagents used in this protocol with DEPC-treated H_2O.

MATERIALS

CAUTION: Please see Appendix 4 for appropriate handling of materials marked with <!>.

Reagents and Solutions

Please see Appendix 1 for components of stock solutions, buffers, and reagents. Dilute stock solutions to the appropriate concentrations.

Acrylamide solution (40% w/v) (acrylamide:bisacrylamide 19:1) <!>
Ammonium acetate (10 M) <!>
Ammonium persulfate (10%) <!> (freshly made)
Dithiothreitol (200 mM)
Ethanol
Ethanol (70%)
Ethanol (75%)
Hybridization buffer without formamide (for RNA)
    40 mM PIPES (pH 6.4)
    0.1 mM EDTA (pH 8.0)
    0.4 M NaCl
    Use the disodium salt of PIPES (piperazine-N,N′-bis[2-ethanesulfonic acid]) and adjust the pH of the buffer to 6.4 with 1 N HCl.
Phenol:chloroform (1:1, v/v) <!>
RNase digestion mixture
    300 mM NaCl
    10 mM Tris-Cl (pH 7.4)
    5 mM EDTA (pH 7.5)
    40 µg/ml RNase A
    2 µg/ml T1 RNase
    Prepare 10 mg/ml of ribonuclease A (bovine pancreatic RNase) with 10 mM Tris-Cl (pH 7.5), 15 mM NaCl. Prepare ribonuclease T1 separately in the same buffer. Add the RNases to the buffer just before digestion.
SDS (10%)
Sodium acetate (3 M, pH 5.2)
TE (pH 7.6)
TEMED (N,N,N′,N′-tetramethylethylene diamine) <!>
Trichloroacetic acid (TCA) (1% and 10%) <!>
Urea

Enzymes and Buffers

Bacteriophage-encoded DNA-dependent–RNA polymerase and 10x transcription buffer supplied by the manufacturer

The choice of polymerase depends on the plasmid vector and the strand of DNA to be transcribed.

DNase I (1 mg/ml), free of RNase (e.g., RQ1 [Promega])

Protein inhibitor of RNase (sold under various trade names: RNasin, Promega; Prime Inhibitor, 5 Prime→3 Prime, etc. See Inhibitors of RNases [MC3, p. 7.83].)

Proteinase K (10 mg/ml)

Restriction enzyme and 10x buffer

Nucleic Acids/Oligonucleotides

Carrier RNA (yeast tRNA)

> Dissolve commercially available yeast RNA at a concentration of 10 mg/ml in sterile 0.1 M NaCl. Extract the solution twice with phenol:chloroform and twice with chloroform alone. Recover the RNA by precipitation with 2.5 volumes of ethanol at room temperature and dissolve the RNA at a concentration of approximately 10 mg/ml in sterile TE (pH 7.6). Store aliquots of the preparation at –20°C.

Plasmid DNA or linearized target DNA for use as templates

RNA, for use as standards

> RNA standards are synthesized in vitro by transcription of the appropriate strand of a recombinant plasmid equipped with bacteriophage promoters and containing target DNA sequences (see Protocol 9.6).

rNTP solution, containing GTP, CTP, and ATP, each at a concentration of 5 mM

Test RNA

> Poly(A)$^+$ or total RNA (Protocols 7.1–7.4)

UTP (100 µM)

Radioactive Compounds

[α-^{32}P]UTP (10 mCi/ml, 800 Ci/mmole) <!>

Gels/Loading Buffers

1% Agarose minigel, cast and run in 1x TBE

Denaturing polyacrylamide gel containing 8 M urea, cast in 1x TBE electrophoresis buffer

> For most 5′- and 3′-end mapping experiments, a denaturing gel composed of 5% or 6% polyacrylamide and containing 8 M urea nicely resolves protected DNA fragments. A typical gel is 1.5 mm thick. However, "thin" or "sequencing" gels (0.4-mm thickness) can also be used (see Chapter 12). If thin gels are used to resolve protected DNA fragments, it is usually not necessary to fix the gel in trichloroacetic acid (TCA), as described in Steps 22–24, before drying the gel. Fixing sharpens and increases the resolution of thicker gels. In many cases, a miniprotein gel apparatus (e.g., Bio-Rad Mini-Protean) (13 cm x 13 cm x 1 mm) can be used both to prepare the radiolabeled single-stranded DNA or RNA probe and to analyze the products of nuclease S1 digestion. Table 7-4 shows the percentage of polyacrylamide used to purify DNA fragments of various sizes; Table 7-5 shows the expected mobilities of tracking dyes in these gels. The method used to prepare the polyacrylamide gel is described in Step 1 of this protocol; for additional details, please see Protocol 12.8.

RNA gel-loading buffer

> 95% (w/v) deionized formamide <!>
> 5 mM EDTA (pH 8.0)
> 0.025% (w/v) SDS
> 0.025% (w/v) bromophenol blue
> 0.025% (w/v) xylene cyanol FF

Additional Items

Conventional apparatus for polyacrylamide gel electrophoresis

Gel dryer

Materials for autoradiography/phosphorimaging

Water baths (30°C, 37°C, 85°C, and 95°C)

Whatman 3MM paper

Additional Information	
Autoradiography and phosphorimaging	MC3, pp. A9.11–A9.15
Bacteriophage RNA polymerases	MC3, pp. A4.28–A4.29
Markers used in gels to fractionate RNA	MC3, p. 7.23
Promoter sequences recognized by bacteriophage-encoded RNA polymerases	MC3, p. 7.87

METHOD

1. Prepare the linearized template DNA.

To prepare template from plasmid DNA

a. Linearize 5–20 μg of plasmid DNA by digestion with a fivefold excess of an appropriate restriction enzyme that cleaves either within the cloned DNA sequence or downstream from the DNA sequence. The distance from the promoter to the newly created terminus should be 200–400 bp. Make sure not to use an enzyme that separates the promoter from the sequence of interest. Because bacteriophage-encoded RNA polymerases may initiate transcription at 3′-protruding termini, choose a restriction enzyme that generates a blunt terminus or a 5′ extension.

b. At the end of the digestion, analyze an aliquot (~200 ng) of the reaction by agarose gel electrophoresis. No trace of circular plasmid DNA should be visible. If necessary, add more restriction enzyme and continue digestion until no more circular plasmid DNA can be detected.

c. Purify the linear DNA by extracting twice with phenol:chloroform and then recover the DNA by standard precipitation with ethanol. After washing the precipitate with 70% ethanol, dissolve the DNA in TE (pH 7.6) at a concentration of 1 μg/μl.

To prepare template by amplification of target DNA

a. Perform PCR to synthesize double-stranded DNA templates, 100–400 bp in length (please see Protocol 8.1 for details).

   The template should be either linearized plasmid DNA or a DNA fragment encoding the sequence of interest. Either one or both of the oligonucleotide primers are designed to contain the consensus sequence of a bacteriophage promoter at their 5′ termini. Amplification in PCR yields double-stranded DNA fragments carrying bacteriophage promoters at one or both ends. See Promoter Sequences Recognized by Bacteriophage-encoded RNA Polymerases in MC3, p. 7.87.

b. Analyze the products of the PCR by electrophoresis through an agarose or a polyacrylamide gel to ensure that a DNA fragment of the appropriate size has been amplified.

c. Purify the linear amplification product by extracting twice with phenol:chloroform and then recover the DNA by standard precipitation with ethanol. After washing with 70% ethanol, dissolve the DNA in TE (pH 7.6) at a concentration of 1 μg/μl.

2. Mix the following in order, prewarmed to room temperature except when noted otherwise:

   > 0.5 μg of linearized template DNA (from Step 1)
   > 1 μl of 0.2 M dithiothreitol
   > 2 μl of ribonucleotide solution
   > 1 μl of 100 μM UTP
   > 50–100 μCi [α-^{32}P]UTP (800 Ci/mmole, 10 mCi/ml)
   > H_2O to a volume of 16 μl
   > 2 μl of 10x transcription buffer
   > 24 units of protein inhibitor of RNase (on ice)

15–20 units of the appropriate bacteriophage-encoded RNA polymerase (on ice)

Adding the reagents in the order shown at room temperature prevents both precipitation of the DNA by spermidine and Mg++ in the transcription buffer and inactivation of the RNase inhibitor by high concentrations of dithiothreitol.

If, as is often the case, the bacteriophage RNA polymerase supplied by the manufacturer is highly concentrated, prepare an appropriate dilution of the enzyme in polymerase dilution buffer.

Incubate the reaction mixture for 60 minutes at 37°C.

The specific activity of the RNA synthesized in the reaction will be high (~10^9 cpm/µg) because 60–80% of the radiolabeled UTP will be incorporated. The total yield of RNA should be approximately 100 ng.

3. At the end of the incubation period, add 1 unit of RNase-free DNase equivalent to approximately 1 µg of the enzyme and continue incubation for an additional 10 minutes at 37°C.

4. Perform Steps 4 and 5 simultaneously. Dilute the reaction mixture to 100 µl with TE (pH 7.6) and measure the total radioactivity and the amount of TCA-precipitable radioactivity in 1-µl aliquots of the diluted mixture. From the fraction of radioactivity incorporated in TCA-precipitable material, calculate the weight and specific activity of the RNA probe synthesized in the reaction.

5. After removing 1-µl aliquots in Step 4, add 1 µl of 1-mg/ml carrier RNA to the remainder of the diluted reaction mixture. Extract the diluted reaction mixture once with phenol:chloroform. Transfer the aqueous phase to a fresh tube and precipitate the RNA by adding 10 µl of 10 M ammonium acetate and 300 µl of ethanol. Store the tube at –20°C until Step 4 has been completed.

A solution of carrier RNA (1 mg/ml) is prepared by diluting the stock solution 1:10 with DEPC-treated H_2O.

6. Recover the RNA by centrifugation at maximum speed for 10 minutes at 4°C in a microfuge. Wash the RNA pellet with 75% ethanol and centrifuge again. Remove the supernatant and allow the pellet of RNA to air dry until no visible trace of ethanol remains. Dissolve the RNA in 20 µl of RNA gel-loading buffer if the probe is to be purified by gel electrophoresis (Step 7) or in 20 µl of TE (pH 7.6) if the probe is to be used without further purification.

7. Following the instructions given in Protocol 7.10, Steps 8–16, purify the probe by electrophoresis using the previously prepared polyacrylamide/8 M urea gel.

8. Combine each of the test RNAs and RNA standards with the riboprobe (2×10^5 to 10×10^5 cpm, 0.1–0.5 ng). Add 0.1 volume of 3 M sodium acetate (pH 5.2) and 2.5 volumes of ice-cold ethanol. Store the mixtures for 10 minutes at –20°C and then recover the RNAs by centrifugation at maximum speed for 10 minutes at 4°C in a microfuge. Wash the pellet in 75% ethanol. Carefully remove all of the ethanol and store the pellet at room temperature until the last visible traces of ethanol have evaporated. See Setting Up Hybridizations for Ribonuclease Protection Assays in MC3, p. 7.72.

9. Dissolve the RNAs in 30 µl of hybridization buffer. Pipette the solution up and down numerous times to ensure that the pellet is completely dissolved.

10. Incubate the hybridization mixture for 10 minutes at 85°C to denature the RNAs. Quickly transfer the hybridization mixture to an incubator or water bath set at the annealing temperature. Incubate the mixture for 8–12 hours.

11. Cool the hybridization mixture to room temperature and add 300 µl of RNase digestion mixture. Digest the hybridization reaction for 60 minutes at 30°C.

12. Add 20 µl of 10% SDS and 10 µl of 10 mg/ml proteinase K to stop the reaction. Incubate the reaction mixture for 30 minutes at 37°C.

13. Add 400 µl of phenol:chloroform, vortex the mixture for 30 seconds, and separate the phas-

es by centrifugation at maximum speed for 5 minutes at room temperature in a microfuge.

14. Transfer the upper aqueous phase to a fresh tube, carefully avoiding the interface between the organic and aqueous phases.

15. Add 20 μg of carrier RNA and 750 μl of ice-cold ethanol. Mix the solution well by vortexing and then store the solution for 30 minutes at –20°C.

16. Recover the RNA by centrifugation at maximum speed for 15 minutes at 4°C in a microfuge. Carefully remove the ethanol and wash the pellet with 500 μl of 75% ethanol. Centrifuge as before.

17. Carefully remove all of the ethanol and store the open tube at room temperature until the last visible traces of ethanol have evaporated.

18. Resuspend the precipitate in 10 μl of gel-loading buffer.

19. Heat the nucleic acids for 5 minutes at 95°C and then immediately transfer the tube to an ice bath. Centrifuge the tubes briefly in a microfuge to consolidate the samples at the bottom of the tubes.

20. Analyze the radiolabeled RNA by electrophoresis through a "thin" polyacrylamide/8 M urea gel.

21. After the tracking dyes have migrated an appropriate distance through the gel, turn off the power supply and dismantle the electrophoresis setup. Gently pry up one corner of the larger glass plate and slowly remove the plate from the gel. Cut off one corner of the gel for orientation purposes.

22. Transfer the glass plate containing the gel to a tray containing an excess of 10% TCA. Gently rock or rotate the tray for 10 minutes at room temperature.

23. Pour off the 10% TCA solution and replace with an excess of 1% TCA. Gently rock or rotate the tray for 5 minutes at room temperature.

24. Pour off the 1% TCA solution and briefly rinse the fixed gel with distilled deionized H_2O. Lift the glass plate together with the gel out of the tray and place them on a flat benchtop. Apply paper towels or Kimwipes to the sides of the gel to remove excess H_2O.

25. Cut a piece of Whatman 3MM filter paper (or equivalent) that is 1 cm larger than the gel on all sides. Transfer the gel to the filter paper by laying the paper on top of the gel and inverting the glass plate.

26. Remove the plate and dry the gel for 1.0–1.5 hours at 60°C on a gel dryer.

27. Establish an autoradiographic image of the dried gel. Scan the image by densitometry or phosphorimaging, or excise the segments of the gel containing the fragments and count them by liquid scintillation spectroscopy.

Analysis of RNA by Primer Extension

Primer extension is used chiefly to map the 5′ termini of mRNAs. A preparation of polyadenylated mRNA is first hybridized with an excess of a single-stranded oligodeoxynucleotide primer, which is complementary to the target RNA and radiolabeled at its 5′ terminus. Reverse transcriptase is then used to extend the 3′ end of the primer. The size of the resulting cDNA, measured by denaturing polyacrylamide gel electrophoresis, is equal to the distance between the 5′ end of the priming oligonucleotide and the 5′ terminus of the target mRNA.

IMPORTANT: Prepare all reagents used in this protocol with DEPC-treated H_2O.

MATERIALS

CAUTION: Please see Appendix 4 for appropriate handling of materials marked with <!>.

Reagents and Solutions

Please see Appendix 1 for components of stock solutions, buffers, and reagents. Dilute stock solutions to the appropriate concentrations.

Acrylamide solution (40% w/v) (acrylamide:bisacrylamide 19:1) <!>
Ammonium acetate (10 M) <!>
Ammonium persulfate (10%) <!> (freshly made)
Chloroform <!>
Dithiothreitol (200 mM)
Ethanol
Ethanol (70%)
Ethanol (75%)
Hybridization buffer without formamide (for RNA)
    40 mM PIPES (pH 6.4)
    0.1 mM EDTA (pH 8.0)
    0.4 M NaCl
    Use the disodium salt of PIPES (piperazine-*N,N*′-bis[2-ethanesulfonic acid]) and adjust the pH of the buffer to 6.4 with 1 N HCl.
KCl (1.25 M)
Phenol:chloroform (1:1, v/v) <!>
Primer extension mixture
    20 mM Tris-Cl (pH 8.4, room temperature)
    10 mM $MgCl_2$
    dNTP solution containing all four dNTPs, each at a concentration of 1.6 mM
    50 μg/ml actinomycin D
    Dissolve actinomycin D in methanol at a concentration of 5 mg/ml. Store aliquots of the stock solution in light-proof containers at −20°C.
Sodium acetate (3 M, pH 5.2)
TE (pH 7.6)
TEMED (*N,N,N*′,*N*′-tetramethylethylene diamine) <!>
Trichloroacetic acid (TCA) (1% and 10%) <!>
Urea

Enzymes and Buffers

Polynucleotide kinase and 10x buffer
Protein inhibitor of RNase (sold under various trade names: RNasin, Promega; Prime Inhibitor, 5 Prime→3 Prime, etc. See Inhibitors of RNases [MC3, p. 7.83].)

Reverse transcriptase

> A cloned version of reverse transcriptase encoded by the Moloney murine leukemia virus (Mo-MLV) is the enzyme of choice, e.g., StrataScript (Stratagene).

Nucleic Acids/Oligonucleotides

Carrier RNA (yeast tRNA)

> Dissolve commercially available yeast RNA at a concentration of 10 mg/ml in sterile 0.1 M NaCl. Extract the solution twice with phenol:chloroform and twice with chloroform alone. Recover the RNA by precipitation with 2.5 volumes of ethanol at room temperature and dissolve the RNA at a concentration of approximately 10 mg/ml in sterile TE (pH 7.6). Store aliquots of the preparation at –20°C.

DNA markers for gel electrophoresis <!>

Oligonucleotide primer

> The primer should be 20–30 nucleotides in length and purified through Sep-Pak chromatography and gel electrophoresis (Protocol 10.1). Resuspend the purified oligonucleotide at a concentration of 60 ng/μl (5–7 μm/μl) in TE (pH 7.6).

Test RNA

> Preparations of poly(A)+ RNA give the best results.

Radioactive Compounds

[γ-^{32}P]ATP (10 mCi/ml, 7000 Ci/mmole) <!>

Gels/Loading Buffers

Denaturing polyacrylamide gel containing 8 M urea, cast in 1x TBE electrophoresis buffer

> In many cases, a miniprotein gel apparatus (e.g., Bio-Rad Mini-Protean) can be used to analyze radiolabeled primer extension products (see Tables 7-4 and 7-5 and the note to the entry for polyacrylamide gel electrophoresis in Protocol 7.10). The method used to prepare the minidenaturing polyacrylamide gel is described in Protocol 7.10, Step 1.

RNA gel-loading buffer

> 95% (w/v) deionized formamide <!>
> 5 mM EDTA (pH 8.0)
> 0.025% (w/v) SDS
> 0.025% (w/v) bromophenol blue
> 0.025% (w/v) xylene cyanol FF

Additional Items

Gel dryer
Materials for autoradiography/phosphorimaging
Minigel polyacrylamide gel apparatus
Water baths (37°C, 42°C, and 95°C)
Whatman 3MM paper

Additional Information

Actinomycin D	MC3, p. 7.88
Autoradiography and phosphorimaging	MC3, pp. A9.11–A9.15
Markers used in gels to fractionate RNA	MC3, p. 7.23

METHOD

1. Phosphorylate the oligonucleotide primer in a reaction containing:

oligonucleotide primer (5–7 pmoles or 60 ng)	1 µl
distilled deionized H_2O	6.5 µl
10x kinase buffer	1.5 µl
polynucleotide kinase (~10 units)	1 µl
[γ-^{32}P]ATP (7000 Ci/mmole)	2 µl

   Incubate the reaction for 60 minutes at 37°C.

   The final concentration of radiolabeled ATP in the reaction should be approximately 30 nM.

2. Stop the kinase reaction with the addition of 500 µl of TE (pH 7.6). Add 25 µg of carrier RNA.

3. Add 400 µl of equilibrated phenol (pH 8.0) and 400 µl of chloroform. Vortex vigorously for 20 seconds. Separate the aqueous and organic phases by centrifugation for 2 minutes in a microfuge.

4. Transfer the aqueous layer to a fresh sterile microfuge tube and extract with 800 µl of chloroform. Vortex vigorously for 20 seconds. Separate the aqueous and organic phases by centrifugation for 2 minutes in a microfuge. Again transfer the aqueous layer to a fresh sterile microfuge tube.

5. Repeat Step 4.

6. Add 55 µl of 3 M sodium acetate (pH 5.2) and 1 ml of ethanol to the aqueous layer from Step 5. Mix by vortexing and store the solution for at least 1 hour at –70°C.

7. Collect the precipitated oligonucleotide primer by centrifugation at maximum speed for 15 minutes at 4°C in a microfuge. Remove and discard the radioactive supernatant. Wash the pellet in 70% ethanol and centrifuge again. Discard the supernatant and dry the precipitate in the air. Dissolve the precipitate in 500 µl of TE (pH 7.6).

8. Count 2 µl of radiolabeled oligonucleotide primer in 10 ml of scintillation fluid in a liquid scintillation counter. Calculate the specific activity of the radiolabeled primer assuming 80% recovery. The specific activity should be approximately 2×10^6 cpm/pmole of primer.

9. Mix 10^4 to 10^5 cpm (20–40 fmoles) of the DNA primer with 0.5–150 µg of the RNA to be analyzed. Add 0.1 volume of 3 M sodium acetate (pH 5.2) and 2.5 volumes of ethanol. Store the solution for 60 minutes at –70°C and then recover the RNA by centrifugation at maximum speed for 10 minutes at 4°C in a microfuge. Wash the pellet with 70% ethanol and centrifuge again. Carefully remove all of the ethanol and store the pellet at room temperature until the last visible traces of ethanol have evaporated.

10. Resuspend the pellets in 8 µl of TE (pH 7.6) per tube. Pipette the samples up and down several times to dissolve pellets.

11. Add 2.2 µl of 1.25 M KCl. Vortex the samples gently and then deposit the fluid in the base of the tubes by centrifuging for 2 seconds in a microfuge.

12. Place the oligonucleotide/RNA mixtures in a water bath set at the appropriate annealing temperature. Incubate the samples for 15 minutes at the optimum temperature, as determined in preliminary experiments.

13. While the oligonucleotide and RNA are annealing, supplement an aliquot of primer extension mix with dithiothreitol and reverse transcriptase as follows: Thaw a 300-µl aliquot of primer extension mix on ice and then add 3 µl of 1 M dithiothreitol and reverse transcriptase to a concentration of 1–2 units/µl. Add 0.1 unit/µl of protein inhibitor of RNase, gently mix by inverting the tube several times, and store it on ice.

14. Remove the tubes containing the oligonucleotide primer and RNA from the water bath and deposit the fluid in the base of the tubes by centrifuging for 2 seconds in a microfuge.

15. Add 24 µl of the supplemented primer extension mix to each tube. Gently mix the solution in the tubes and again deposit the liquid at the tube bottoms by centrifugation.

16. Incubate the tubes for 1 hour at 42°C to allow the primer extension reaction to proceed.

17. Terminate the primer extension reactions by the addition of 200 µl of TE (pH 7.6), 200 µl of phenol:chloroform. Vortex for 20 seconds. Separate aqueous and organic phases by centrifugation for 4 minutes at room temperature in a microfuge.

18. Precipitate the nucleic acids by the addition of 50 µl of 10 M ammonium acetate and 700 µl of ethanol. Mix well by vortexing and incubate ethanol precipitations for at least 1 hour at –70°C.

19. Collect the precipitated nucleic acids by centrifugation for 10 minutes at 4°C in a microfuge. Carefully rinse the pellets with 400 µl of 70% ethanol. Centrifuge again for 5 minutes at 4°C and remove the 70% ethanol rinse with a pipette. Store the open tubes at room temperature until all visible traces of ethanol have evaporated.

20. Dissolve the nucleic acid precipitates in 10 µl of formamide loading buffer. Pipette the samples up and down to assist resuspension.

21. Heat the samples for 8 minutes at 95°C. Then plunge the tubes into an ice-water bath and immediately analyze the primer extension products by electrophoresis through a denaturing polyacrylamide gel.

22. After the tracking dyes have migrated an appropriate distance through the gel, turn off the power supply and dismantle the electrophoresis setup. Gently pry up one edge of the larger glass plate and slowly remove the plate from the gel. Cut off one corner of the gel for orientation purposes.

23. If a polyacrylamide gel 1.0 mm in thickness was used, fix the gel in TCA. Transfer the glass plate containing the gel to a tray containing an excess of 10% TCA. Gently rock or rotate the tray for 10 minutes at room temperature.

     This step is not necessary if a thin gel (0.4-mm thickness) was used. In this case, proceed to Step 26.

24. Pour off the 10% TCA solution and replace it with an excess of 1% TCA. Gently rock or rotate the tray for 5 minutes at room temperature.

25. Pour off the 1% TCA solution and briefly rinse the fixed gel with distilled deionized H_2O. Lift the glass plate together with the gel out of the tray and place them on a flat benchtop. Apply paper towels or Kimwipes to the sides of the gel to remove excess H_2O.

26. Cut a piece of Whatman 3MM filter paper (or equivalent) that is 1 cm larger than the gel on all sides. Transfer the gel to the filter paper by laying the paper on top of the gel and inverting the glass plate.

27. Remove the plate and dry the gel on a heat-assisted vacuum-driven gel dryer for 1.0–1.5 hours at 60°C.

28. Establish an image of the gel using autoradiography or phosphorimaging.

In Vitro Amplification of DNA by the Polymerase Chain Reaction

BACKGROUND INFORMATION

Background information found in *Molecular Cloning: A Laboratory Manual*, 3rd edition (hereafter MC3) unless otherwise indicated.

Oligonucleotide primers	MC3, pp. 8.13–8.16
Parameters that affect PCRs	MC3, pp. 8.4–8.6
PCR in theory	MC3, p. 8.12
Thermostable DNA polymerases	MC3, pp. 8.6–8.8 and 8.10–8.11

The Basic Polymerase Chain Reaction

This protocol describes how to amplify a segment of double-stranded DNA in a chain reaction catalyzed by a thermostable DNA polymerase. It is the foundation for all subsequent variations of the polymerase chain reaction.

MATERIALS

CAUTION: Please see Appendix 4 for appropriate handling of materials marked with <!>.

Reagents and Solutions

Please see Appendix 1 for components of stock solutions, buffers, and reagents. Dilute stock solutions to the appropriate concentrations.

dNTP solution (pH 8.0) containing all four deoxynucleotide triphosphates, each at a concentration of 20 mM

Enzymes and Buffers

Thermostable DNA polymerase and 10x amplificaton buffer as supplied by manufacturer or homemade

500 mM KCl
100 mM Tris-Cl (pH 8.3, room temperature)
15 mM $MgCl_2$

Nucleic Acids/Oligonucleotides

DNA markers
Forward primer (20 µM in H_2O)
Positive- and negative-control DNAs that, respectively, do or do not contain the target sequences
Reverse primer (20 µM in H_2O)

Primers should be designed according to standard rules, i.e., they should be complementary to sequences spaced 100–400 nucleotides apart on opposite strands of the target DNA. They should be 20–25 nucleotides in length and contain approximately equal numbers of the four bases, with a balanced distribution, G and C residues, and a low propensity to form secondary structures.

Template DNA

The amount of target sequence required for an amplification reaction to proceed efficiently is approximately 1 fmole. The mass of template DNA added to the PCR therefore varies according to its complexity. Preparations of mammalian genomic DNA used as templates in PCRs typically contain 100 µg DNA/ml; yeast genomic DNA, 1 µg/ml; bacterial genomic DNA, 0.1 µg/ml; and plasmid DNA, 1–5 ng/ml.

Gels/Loading Buffers

Polyacrylamide <!> or agarose gel, used to analyze products of PCR

Additional Items

Barrier tips with automatic pipettor
Light mineral oil or wax bead (optional; see Step 2)
Microtiter plates or microfuge tubes, 0.5 ml and thin walled
Positive displacement pipette

Step 4 of this protocol may require the reagents listed in Protocol 6.10 and/or 12.6.
Thermal cycler, programmed with desired amplification protocol

Additional Information

Equipment required for PCR	MC3, pp. 8.18–8.20
Minimizing contamination	MC3, pp. 8.16–8.17
Primer design	MC3, pp. 8.13–8.16
Troubleshooting	MC3, pp. 8.23–8.24

METHOD

1. In a sterile 0.5-ml microfuge tube, amplification tube, or the well of a sterile microtiter plate, mix in the following order:

10x amplification buffer	5 µl
20 mM solution of four dNTPs (pH 8.0)	1 µl
20 µM forward primer	2.5 µl
20 µM reverse primer	2.5 µl
1–5 units/µl thermostable DNA polymerase	1–2 units
H₂O	28–33 µl
template DNA	5–10 µl
Total volume	50 µl

   The table below provides standard reaction conditions for PCR.

Mg++	KCl	dNTPs	Primers	DNA polymerase	Template DNA
1.5 mM	50 mM	200 µM	1 µM	1–5 units	1 pg to 1 µg

   The amount of template DNA required varies according to the complexity of its sequence. In the case of mammalian DNA, up to 1.0 µg is used per reaction. Typical amounts of yeast, bacterial, and plasmid DNAs used per reaction are 10 ng, 1 ng, and 10 pg, respectively.

2. If the thermal cycler is not fitted with a heated lid, overlay the reaction mixtures with 1 drop (~50 µl) of light mineral oil. Alternatively, place a bead of wax into the tube if using a hot start protocol. Place the tubes or the microtiter plate in the thermal cycler.

3. Amplify the nucleic acids using the denaturation, annealing, and polymerization times and temperatures listed below.

Cycle Number	Denaturation	Annealing	Polymerization
30 cycles	30 sec at 94°C	30 sec at 55°C	1 min at 72°C
Last cycle	1 min at 94°C	30 sec at 55°C	1 min at 72°C

   Times and temperatures may need to be adapted to suit the particular reaction conditions. Polymerization should be carried out for 1 minute for every 1000 bp of length of the target DNA.

4. Withdraw a sample (5–10 µl) from the test reaction mixture and the four control reactions, analyze them by electrophoresis through an agarose or polyacrylamide gel, and stain the gel with ethidium bromide or SYBR Gold to visualize the DNA.

   A successful amplification reaction should yield a readily visible DNA fragment of the expected size. The identity of the band can be confirmed by DNA sequencing (please see Protocol 12.6), Southern hybridization (please see Protocol 6.10), and/or restriction mapping.

   If all has gone well, lanes of the gel containing samples of the two positive controls (Tubes 1 and 2) and the template DNA under test should contain a prominent band of DNA of the appropriate molecular weight. This band should be absent from the lanes containing samples of the negative controls (Tubes 3 and 4).

5. If mineral oil was used to overlay the reaction (Step 2), remove the oil from the sample by extraction with 150 μl of chloroform.

   The aqueous phase, which contains the amplified DNA, will form a micelle near the meniscus. The micelle can be transferred to a fresh tube with an automatic micropipette.

   IMPORTANT: Do not attempt chloroform extractions in microtiter plates. The plastic used in these plates is not resistant to organic solvents.

Purification of PCR Products in Preparation for Cloning

The residual enzymatic activity of thermostable DNA polymerases that survive the rigors of PCR can compromise subsequent enzymatic reactions. This protocol describes how to use proteinase K to destroy thermostable enzymes and to purify amplified DNA in preparation for cloning.

MATERIALS

CAUTION: Please see Appendix 4 for appropriate handling of materials marked with <!>.

Reagents and Solutions

Please see Appendix 1 for components of stock solutions, buffers, and reagents. Dilute stock solutions to the appropriate concentrations.

Ammonium acetate (10 M)
Chloroform <!>
Ethanol
Ethanol (70%)
Phenol:chloroform (1:1, v/v) <!>
TE (pH 8.0)

Enzymes and Buffers

Proteinase K (20 mg/ml)
Proteinase K buffer
    100 mM Tris-Cl (pH 8.0)
    50 mM EDTA (pH 8.0)
    500 mM NaCl

Nucleic Acids/Oligonucleotides

DNA markers
DNA(s), amplified by PCR

Gels/Loading Buffers

Polyacrylamide <!> or agarose gel

Additional Items

Step 1 may require the reagents listed in Prococol 5.6.
Step 6 may require the reagents listed in Protocol 1.9.
Water baths (37°C and 75°C)

Additional Information

Proteinase K

MC3, p. A4.50

METHOD

1. Pool up to eight PCRs (400 µl) containing 1 µg of the desired amplification products.

   If nonspecific amplification products are present at significant levels (e.g., detectable by gel electrophoresis), purify the desired product by electrophoresis through low-melting-temperature agarose before proceeding (please see Protocol 5.6).

   If mineral oil was used to prevent evaporation during PCR, centrifuge the pooled samples briefly and transfer the lower (aqueous) phase to a fresh microfuge tube.

2. Add 0.2 volume of 5x proteinase K buffer and proteinase K to a final concentration of 50 µg/ml. Incubate the mixture for 60 minutes at 37°C.

3. Inactivate the proteinase K by heating to 75°C for 20 minutes.

4. Extract the reaction mixture once with phenol:chloroform and once with chloroform.

5. Add 0.2 volume of 10 M ammonium acetate and 2.5 volumes of ethanol. Mix the solution well and store it for 30 minutes at 4°C.

6. Recover the DNA by centrifugation at maximum speed for 5 minutes at 4°C in a microfuge. Discard the supernatant and then wash the pellet with 70% ethanol. Centrifuge again, remove the supernatant, and allow the DNA to dry.

   The DNA may be further purified by chromatography or by gel electrophoresis. This step is recommended when primers have been used to add restriction sites to the ends of the amplified DNA. Unused primers and primer-dimers should be removed before digesting the DNA with the appropriate restriction enzymes (please see Protocol 8.3).

7. Dissolve the pellet in TE (pH 8.0). Assume that the recovery of amplified DNA is 50–80% and dissolve the DNA in TE (pH 8.0) at an estimated concentration of 25 µg/ml (25 ng/µl).

8. Analyze approximately 25 ng of the purified DNA by agarose or polyacrylamide gel electrophoresis, using markers of an appropriate size. Check that the amplified band fluoresces with the intensity expected of approximately 25 ng of DNA.

Removal of Oligonucleotides and Excess dNTPs from Amplified DNA by Ultrafiltration

In this protocol, ultrafiltration through Centricon or Microcon concentrators is used to remove unused primers, primer-dimers, and NTPs from preparations of amplified DNA.

MATERIALS

CAUTION: Please see Appendix 4 for appropriate handling of materials marked with <!>.

Reagents and Solutions

Please see Appendix 1 for components of stock solutions, buffers, and reagents. Dilute stock solutions to the appropriate concentrations.

Chloroform <!> (optional; see Step 1)
Ethanol
Ethanol (70%)
Sodium acetate (3 M, pH 5.2)
TE (pH 8.0)

Nucleic Acids/Oligonucleotides

DNA(s), amplified by PCR

Centrifuges/Rotors/Tubes

Preparative centrifuge with angle rotor, to hold Centricon or Microcon concentrators

Additional Items

Concentrators, Centricon-100 or Microcon-100 (Millipore)

Additional Information

Other methods to remove oligonucleotides and dNTPs   MC3, p. 8.27

METHOD

1. Place 2 ml of TE (pH 8.0) in the reservoir chamber of a Centricon-100 unit. Carefully separate the amplification reaction products from the upper mineral oil layer by pipetting or by extraction with chloroform. Transfer the amplification reaction products to the reservoir chamber of the Centricon-100 unit.

2. Place the entire unit into an appropriate rotor of a preparative centrifuge (e.g., a fixed-angle rotor). Insert the microconcentrator into the centrifuge with the filtrate cup (translucent portion) toward the bottom of the rotor.

   IMPORTANT: Do not touch the membrane with pipette or pipette tips when loading the microconcentrator.

3. Centrifuge the loaded concentrator at 1000g for 30 minutes at a temperature between 4°C and 25°C.

4. Remove the concentrator from the centrifuge and discard the filtrate cup. Invert the unit and replace it in the centrifuge (i.e., the retentate tube should now be placed toward the bottom of the rotor). Centrifuge at 300–1000g for 2 minutes.

5. Remove the concentrator from the centrifuge; remove the retentate cup and discard the rest of the device. Transfer the fluid in the retentate cup to a fresh microfuge tube.

6. If necessary, precipitate the sample by adding one-tenth volume of 3 M sodium acetate and 2–3 volumes of ethanol. The amplified sample is now ready for subsequent manipulation (DNA sequencing and ligation).

> A single purification step is usually sufficient for most subsequent manipulation steps. If necessary, trace oligonucleotide primers can be further removed by performing a second 30-minute centrifugation step. At Step 4 above, empty the translucent filtrate cup, reassemble the device, and add another 2-ml aliquot of TE (pH 8.0) to the reservoir chamber. Repeat Steps 2–4.

Blunt-end Cloning of PCR Products

Incubation of a blunt-end ligation reaction in the presence of an excess amount of an appropriate restriction enzyme can dramatically increase the yield of recombinant plasmids. The role of the restriction enzyme is to cleave circular and linear concatemers at restriction sites that are re-formed when linear, blunt-ended plasmid molecules ligate to themselves. In almost all cases, ligation of the PCR product to the plasmid destroys the restriction site. The constant reclamation of vector molecules drives the equilibrium of the ligation reaction strongly in favor of the recombinants between vector and blunt-ended PCR product.

MATERIALS

CAUTION: Please see Appendix 4 for appropriate handling of materials marked with <!>.

Reagents and Solutions

Please see Appendix 1 for components of stock solutions, buffers, and reagents. Dilute stock solutions to the appropriate concentrations.

ATP (10 mM)
> Required only if 10x ligation buffer does not contain ATP

dNTP solution, containing all four deoxynucleotide triphosphates, each at a concentration of 2 mM

Vectors and Hosts

Plasmid DNA (50 µg/ml), closed circular
Transformation-competent *E. coli*

Enzymes and Buffers

Bacteriophage T4 DNA ligase and 10x ligation buffer
Bacteriophage T4 DNA polymerase and 10x universal KGB buffer
> 1 M potassium acetate
> 250 mM Tris-acetate (pH 7.6)
> 100 mM magnesium acetate
> 5 mM β-mercaptoethanol <!>
> 100 µg/ml bovine serum albumin

Restriction enzyme for cloning
> The restriction enzyme should generate blunt ends and cleave the vector at a single site, but it should not cleave the amplified DNA.

Restriction enzymes for analysis

Nucleic Acids/Oligonucleotides

DNA markers
Target DNA (25 µg/ml), amplified by PCR
> The PCR-amplified DNA should be prepared for ligation by extraction with phenol:chloroform and ultrafiltration (Protocol 8.3). When the PCR product contains more than one or two bands of amplified DNA, purify the target fragment by gel electrophoresis (Protocol 5.6).

Gels/Loading Buffers

Agarose gel

Additional Items

Step 3 of this protocol requires the reagents listed in Protocol 1.25, 1.26, or 1.27.
Step 4 of this protocol may require the reagents listed in Protocol 8.12.
Step 6 of this protocol requires the reagents listed in either Protocol 6.10 or 12.6.
Water bath or heating/cooling block (22°C)

Additional Information

Cloning PCR products   MC3, pp. 8.30–8.31

METHOD

1. In a microfuge tube, mix the following in the order shown:

50 µg/ml closed circular plasmid vector	1 µl
25 µg/ml amplified target DNA	8 µl
10x universal KGB buffer	2 µl
H_2O (please see note below)	5 µl
10 mM ATP	1 µl
2 mM dNTPs	1 µl
restriction enzyme for cloning	2 units
T4 DNA polymerase	1 unit
T4 DNA ligase	3 units

   Adjust the amount of H_2O added so that the final reaction volume is 20 µl.

   Set up a control reaction that contains all of the reagents listed above except for the amplified target DNA.

2. Incubate the ligation mixture for 4 hours at 22°C.

3. Dilute 5 µl of each of the two ligation mixtures with 10 µl of H_2O and transform a suitable strain of competent *E. coli* to antibiotic resistance as described in Protocol 1.25 or 1.26. Plate the transformed cultures on media containing IPTG, X-gal (please see Protocol 1.27), and the appropriate antibiotic.

4. Calculate the number of colonies obtained from each of the ligation mixtures. Pick a number of colonies obtained by transformation with the ligation reaction containing the target DNA. Confirm the presence of the amplified fragment by (i) isolating the plasmid DNAs and digesting them with restriction enzymes whose sites flank the insert in the multiple cloning site or (ii) colony PCR (Protocol 8.12).

5. Fractionate the restricted DNA by electrophoresis through an agarose gel using appropriate DNA size markers. Measure the size of the cloned fragments.

6. Confirm the identity of the cloned fragments by DNA sequencing, restriction mapping, or Southern hybridization.

Cloning PCR Products into T Vectors

This method of direct cloning takes advantage of the unpaired adenosyl residue added to the 3′ terminus of amplified DNAs by *Taq* and other thermostable DNA polymerases.

MATERIALS

Reagents and Solutions

ATP (10 mM)
    Required only if 10x ligation buffer does not contain ATP

Vectors and Hosts

T vectors (see MC3, p. 8.35), commercially available as components of cloning kits (e.g., pGEM-T [Promega])
Transformation-competent *E. coli*

Enzymes and Buffers

Bacteriophage T4 DNA ligase and 10x ligation buffer

Nucleic Acids/Oligonucleotides

DNA markers
Target DNA (25 µg/ml), amplified by PCR
    The PCR-amplified DNA should be prepared for ligation by extraction with phenol:chloroform and ultrafiltration (Protocol 8.3). When the PCR product contains more than one or two bands of amplified DNA, purify the target fragment by gel electrophoresis (Protocol 5.6).

Gels/Loading Buffers

Agarose gel

Additional Items

Step 3 of this protocol requires the reagents listed in Protocol 1.25, 1.26, or 1.27.
Step 6 of this protocol requires the reagents listed in either Protocol 6.10 or 12.6.
Water bath or heating/cooling block (14°C)

Additional Information

Cloning PCR products	MC3, pp. 8.30–8.31
T vectors	MC3, p. 8.35

METHOD

1. In a microfuge tube, set up the following ligation mixture:

25 µg/ml amplified target DNA	1 µl
T-tailed plasmid	20 ng
10x ligation buffer	1 µl
bacteriophage T4 DNA ligase	3 units
H_2O	to 10 µl

   If necessary, add ATP to a final concentration of 1 mM. A 1:5 molar ratio of vector:amplified DNA fragment is recommended.

   Set up a control reaction that contains all of the reagents listed above except for the amplified target DNA.

2. Incubate the ligation mixture for 4 hours at 14°C.

3. Dilute 5 µl of each of the two ligation mixtures with 10 µl of H_2O and transform a suitable strain of competent *E. coli* to antibiotic resistance as described in Protocol 1.25 or 1.26. Plate the transformed cultures on media containing IPTG, X-gal (please see Protocol 1.27), and the appropriate antibiotic.

4. Calculate the number of colonies obtained from each of the ligation mixtures. Pick a number of white colonies obtained by transformation with the ligation reaction containing the target DNA. Confirm the presence of the amplified fragment by (i) isolating the plasmid DNAs and digesting them with restriction enzymes whose sites flank the insert in the multiple cloning site or (ii) colony PCR (Protocol 8.12).

   The ratio of blue:white colonies varies between 1:5 and 2:1.

5. Fractionate the restricted DNA by electrophoresis through an agarose gel using appropriate DNA size markers. Measure the size of the cloned fragments.

6. Confirm the identity of the cloned fragments by DNA sequencing, restriction mapping, or Southern hybridization.

Cloning PCR Products by Addition of Restriction Sites to the Termini of Amplified DNA

Pairs of oligonucleotide primers used in PCR are often designed with restriction sites in their 5' regions. In many cases, the sites are different in the two primers. In this case, amplification generates a target fragment whose termini now carry new restriction sites that can be used for directional cloning into plasmid vectors. The purified fragment and the vector are digested with the appropriate restriction enzymes, ligated together, and transformed into *E. coli*.

MATERIALS

CAUTION: Please see Appendix 4 for appropriate handling of materials marked with <!>.

Reagents and Solutions

Please see Appendix 1 for components of stock solutions, buffers, and reagents. Dilute stock solutions to the appropriate concentrations.

ATP (10 mM)
> Required only if 10x ligation buffer does not contain ATP

Chloroform <!>
EDTA (0.5 M, pH 7.5)
Ethanol
Ethanol (70%)
Phenol:chloroform (1:1, v/v) <!>
Sodium acetate (3 M, pH 5.2)
TE (pH 7.5)

Vectors and Hosts

Plasmid DNA, cleaved with the appropriate restriction enzyme(s)
> If the linearized plasmid vector carries compatible termini that can be ligated to one another, use alkaline phosphatase to remove the 5'-phosphate groups (see Protocol 1.20).

Transformation-competent *E. coli*

Enzymes and Buffers

Bacteriophage T4 DNA ligase and 10x ligation buffer
Restriction enzymes and 10x buffers

Nucleic Acids/Oligonucleotides

DNA markers
Forward and reverse primers, each 20 μM in H_2O
Target DNA (25 μg/ml), amplified by PCR
> When the PCR product contains more than one or two bands of amplified DNA, purify the target fragment by gel electrophoresis (Protocol 5.6). If not purified by gel electrophoresis, the PCR-amplified DNA should be prepared for ligation by extraction with phenol:chloroform and ultrafiltration (Protocol 8.3).

Gels/Loading Buffers

Agarose gel

Additional Items

Step 1 requires the reagents listed in Protocol 8.1.
Step 10 requires the reagents listed in Protocol 1.25, 1.26, or 1.27.
Step 11 may require the reagents listed in Protocol 8.12.
Step 13 requires the reagents listed in either Protocol 6.10 or 12.6.
Water bath or heating/cooling block (16°C)
Water bath, set to optimal temperature for restriction enzyme digestion

Additional Information

Cloning PCR products	MC3, pp. 8.30–8.31
Designing primers that contain restriction sites	MC3, pp. 8.37–8.38

METHOD

1. Design and synthesize the appropriate oligonucleotide primers. Use these forward and reverse primers to set up and carry out four identical amplification reactions (50-μl volume) to amplify the target fragment (please see Protocol 8.1). Combine the four PCRs, which, in aggregate, should contain 200–500 ng of the desired amplification product.

   Design forward and reverse primers carrying the appropriate restriction sites. The 3′ end of each primer should be an exact complement of approximately 15 consecutive bases at a selected site in the target DNA. The 5′ terminus of each primer serves as a clamp to hold together the termini of the amplified DNA and to provide a landing site for the restriction enzyme. The clamp should be 3–10 nucleotides in length. The midportion of the primer contains the recognition site for the restriction enzyme. Each primer should therefore be 24–31 nucleotides in length and contain approximately equal numbers of the four bases, with a balanced distribution of G and C residues and a low propensity to form stable secondary structures.

2. If the PCR mixture contains more than one or two bands of amplified DNA, purify the target fragment by electrophoresis through low-melting-/-gelling-temperature agarose (please see Protocol 5.6). If not purified by gel electrophoresis, prepare PCR-amplified DNA for ligation by extraction with phenol:chloroform and ultrafiltration through a Centricon-100 filter (please see Protocol 8.3). Dissolve in TE (pH 7.5) at a concentration of 25 μg/ml.

3. In a reaction volume of 20 μl, digest approximately 100 ng of purified PCR product with 1.0–2.0 units of the relevant restriction enzyme(s). Incubate the reactions for 1 hour at the optimum temperature for digestion.

4. At the end of the digestion, adjust the volume to 100 μl with H_2O and add 0.5 M EDTA to a final concentration of 5 mM. Extract once with phenol:chloroform and once with chloroform.

5. Transfer the aqueous phase to a fresh tube and add 3 M sodium acetate (pH 5.2) to achieve a final concentration of 0.3 M. Add 2 volumes of ethanol. Store the mixture for 30 minutes at 0°C.

6. Recover the precipitated DNA by centrifugation at maximum speed for 5 minutes at 4°C in a microfuge. Discard the supernatant and then wash the pellet with 70% ethanol. Centrifuge again, remove the supernatant, and allow the DNA to dry.

7. Dissolve the DNA in 10 μl of H_2O.

8. In a microfuge tube, set up the following ligation mixture:

25 μg/ml amplified target DNA	1.0 μl (25~s)
plasmid DNA	20 ng
10x ligation buffer	1.0 μl
T4 DNA ligase	1 unit
H_2O	to 10 μl

If necessary, add ATP to a final concentration of 1 mM.

When directional cloning is used, the ligation mixture should contain an approximately 1:1 molar ratio of purified target DNA to cleaved plasmid vector.

Set up a control reaction that contains all of the reagents listed above except for the amplified target DNA.

9. Incubate the ligation mixtures for 4 hours at 16°C.

10. Dilute 5 µl of each of the two ligation mixtures with 10 µl of H$_2$O and transform a suitable strain of competent *E. coli* to antibiotic resistance as described in Protocol 1.25 or 1.26. Plate the transformed cultures on media containing IPTG, X-gal (please see Protocol 1.27), and the appropriate antibiotic.

11. Calculate the number of colonies obtained from each of the ligation mixtures. Pick a number of white colonies obtained by transformation with the ligation reaction containing the target DNA. Confirm the presence of the amplified fragment by (i) isolating the plasmid DNAs and digesting them with restriction enzymes whose sites flank the insert in the multiple cloning site or (ii) colony PCR (Protocol 8.12).

In different experiments, the ratio of blue:white colonies can vary between 1:5 and 2:1.

12. Fractionate the restricted DNA by electrophoresis through an agarose gel using appropriate DNA size markers. Measure the size of the cloned fragments.

13. Confirm the identity of the cloned fragments by DNA sequencing, restriction mapping, or Southern hybridization.

Genetic Engineering with PCR

This method describes how to modify the termini of PCR products by introducing restriction sites and other features.

MATERIALS

CAUTION: Please see Appendix 4 for appropriate handling of materials marked with <!>.

Reagents and Solutions

Please see Appendix 1 for components of stock solutions, buffers, and reagents. Dilute stock solutions to the appropriate concentrations.

ATP (10 mM)

> Required only if 10x ligation buffer does not contain ATP

Chloroform <!> (optional; see Step 6)

dNTP solution (pH 8.0), containing all four deoxynucleotide triphosphates, each at a concentration of 20 mM

Vectors and Hosts

Plasmid DNA, cleaved with the appropriate restriction enzyme(s)

> If the linearized plasmid vector carries compatible termini that can be ligated to one another, use alkaline phosphatase to remove the 5'-phosphate groups (see Protocol 1.20).

Transformation-competent *E. coli*

Enzymes and Buffers

Bacteriophage T4 DNA ligase and 10x ligation buffer

Restriction enzymes

Thermostable DNA polymerase and 10x amplification buffer as supplied by manufacturer or homemade

> 500 mM KCl
> 100 mM Tris-Cl (pH 8.3, room temperature)
> 15 mM $MgCl_2$

Nucleic Acids/Oligonucleotides

DNA markers

Oligonucleotide primers 1 and 2, each 10 µM in H_2O

Positive-control DNA that contains the target sequences

Target DNA (25 µg/ml), amplified by PCR

> The PCR-amplified DNA should be prepared for ligation by extraction with phenol:chloroform and ultrafiltration (Protocol 8.3). When the PCR product contains more than one or two bands of amplified DNA, purify the target fragment by gel electrophoresis (Protocol 5.6).

Gels/Loading Buffers

Agarose or polyacrylamide gel <!>

Additional Items

Barrier tips for automatic pipettor

Light mineral oil or wax bead (optional; see Step 3)

Microtiter plates or microfuge tubes, 0.5 ml and thin walled

Adapted from Chapter 8, Protocol 7, p. 8.42 of MC3.

Positive displacement pipette
Step 5 requires the reagents listed in either Protocol 6.10 or 12.6.
Step 6 requires the reagents listed in Protocol 8.3.
Step 7 requires the reagents listed in Protocol 8.3.
Step 9 requires the reagents listed in Protocol 8.6.
Thermal cycler, programmed with desired amplification protocol
Water bath (optimal temperature for restriction enzyme digestion)

Additional Information

Cloning PCR products	MC3, pp. 8.30–8.31
Designing primers that contain restriction sites	MC3, pp. 8.37–8.38

METHOD

1. Design and synthesize the appropriate oligonucleotide primers for the end modifications desired.

   In this example, two primers derived from the 5′ sequence (5′ dATCATATGGCTCTGGATGAAC TGTGCCTGCTGGACATGCT 3′) and the 3′ sequence (5′ dATAAGCTTTTATTAAGACAGACTCAGCT CATGGGAGGCAA 3′) of the starting cDNA template are used to introduce an NdeI (CATATG) site at the 5′ end of the cDNA and to change several codons to those preferentially used in *E. coli*. The underlined nucleotides indicate differences between the oligonucleotide primers and the cDNA template. The number of perfectly matched nucleotides required at the 3′ end of the oligonucleotide primers for a successful amplification has not been rigorously determined; however, eight to ten generally work well.

2. In a sterile 0.5-ml microfuge tube, amplification tube, or the well of a sterile microtiter plate, mix in the following order:

100 ng template DNA	10 μl
10x amplification buffer	5 μl
20 mM solution of four dNTPs	5 μl
10 μM primer 1 (50 pmoles)	5 μl
10 μM primer 2 (50 pmoles)	5 μl
1–2 units of thermostable DNA polymerase	1 μl
H_2O	to 50 μl

   Set up two control reactions. In one reaction, include all of the above additions, except for the template DNA. In the other reaction, include a DNA template that has previously yielded a positive result in the PCR. Carry the controls through all subsequent steps of the protocol.

3. If the thermal cycler is not fitted with a heated lid, overlay the reaction mixtures with 1 drop (approximately 50 μl) of light mineral oil. This prevents evaporation of the samples during repeated cycles of heating and cooling. Alternatively, place a bead of wax into the tube if using hot start PCR. Place the tubes or the microtiter plate in the thermal cycler.

4. Amplify the nucleic acids using the denaturation, annealing, and polymerization times and temperatures listed in the table.

Cycle Number	Denaturation	Annealing/Polymerization
20 cycles	1 min at 94°C	3 min at 68°C
Last cycle	1 min at 94°C	15 min at 68°C

   Times and temperatures may need to be adapted to suit the particular reaction conditions.

5. Analyze 5–10% of the amplification on an agarose or polyacrylamide gel and estimate the concentration of the amplified target DNA. Include DNA markers of an appropriate size. Stain the gel with ethidium bromide or SYBR Gold to visualize the DNA.

A successful amplification reaction should yield a readily visible DNA fragment of the expected size. The identity of the band can be confirmed by DNA sequencing (please see Chapter 12), Southern hybridization (please see Chapter 6), and/or restriction mapping.

6. If mineral oil was used to overlay the reaction (Step 3), remove the oil from the sample by extraction with 150 µl of chloroform.

   The aqueous phase, which contains the amplified DNA, will form a micelle near the meniscus. The micelle can be transferred to a fresh tube with an automatic micropipette.

   IMPORTANT: Do not attempt chloroform extractions in microtiter plates. The plastic used in these plates is not resistant to organic solvents.

7. For subsequent cloning, cleave the DNA fragment at the restriction sites placed (or located) at the 5′ ends of the primers (with *Nde*I and *Hind*III in the above example). Purify the digested fragment using gel electrophoresis or ultrafiltration (please see Protocol 8.3).

8. Set up the appropriate ligation reaction with the desired vector DNA. Use a molar ratio of insert to a vector of 3:1 in the ligation reactions.

   Because of the error rate of thermostable DNA polymerases, it is very important to verify the sequence of the amplified DNA after cloning into the expression vector.

9. Use the products of the ligation reaction to transform *E. coli* as described in Protocol 8.6, Steps 10–13.

Amplification of cDNA Generated by Reverse Transcription of mRNA

In this method, an oligodeoxynucleotide primer hybridized to mRNA is extended by an RNA-dependent DNA polymerase to create a cDNA copy that can be amplified by PCR. Depending on the purpose of the experiment, the primer for first-strand cDNA synthesis can be specifically designed to hybridize to a particular target gene, or less frequently a general primer such as oligo(dT) that can be used to prime cDNA synthesis from essentially all mammalian mRNAs. Similarly, the reverse primer used in the subsequent amplification reaction can be gene specific or general (e.g., random hexamers).

MATERIALS

CAUTION: Please see Appendix 4 for appropriate handling of materials marked with <!>.

Reagents and Solutions

Please see Appendix 1 for components of stock solutions, buffers, and reagents. Dilute stock solutions to the appropriate concentrations.

Chloroform <!>

dNTP solution (pH 8.0), containing all four deoxynucleotide triphosphates, each at a concentration of 20 mM

Ethanol

Ethanol (70%)

$MgCl_2$ (1 M)

Phenol:chloroform (1:1, v/v) <!>

Placental RNase inhibitor (20 units/µl)

Sodium acetate (3 M, pH 5.2)

TE (pH 7.6)

Enzymes and Buffers

Reverse transcriptase (RNA-dependent DNA polymerase) and 10x reverse transcriptase buffer as supplied by the manufacturer or homemade

500 mM Tris-Cl (pH 8.3)
750 mM KCl
30 mM $MgCl_2$

Thermostable DNA polymerase and 10x amplification buffer as supplied by the manufacturer or homemade

500 mM KCl
100 mM Tris-Cl (pH 8.3, room temperature)
15 mM $MgCl_2$

Nucleic Acids/Oligonucleotides

DNA markers for gel electrophoresis

Gene-specific oligonucleotide primer (10 µM in H_2O), complementary to a known sequence in the target mRNA, used to prime synthesis of cDNA

Depending on the experiment, random hexanucleotides or oligo(dT)$_{12-18}$ can be used in Step 1 in place of a gene-specific primer (see MC3, pp. 8.46–8.49).

Sense and antisense primers for amplification of cDNAs by PCR (20 µM in H_2O)

The primer used to generate cDNA may also be used as the antisense primer in the amplification stage of standard RT-PCR. However, the specificity of amplification can be improved by using an antisense primer that binds to an upstream sequence in the target transcript. Both sense and anti-sense primers are gene-specific synthetic oligonucleotides that should be 20–30 nucleotides in length and contain approximately equal numbers of the four bases, with a balanced distribution of G and C residues and a low propensity to form stable secondary structures (see MC3, pp. 8.13–8.16).

Template mRNA

The template mRNA may be total cellular RNA (100 µg/ml in H_2O) or poly(A)$^+$ RNA (10 µg/ml in H_2O), which is preferred when the target mRNA is expressed at low abundance.

Gels/Loading Buffers

Agarose or polyacrylamide gel <!>

Additional Items

Barrier tips for automatic pipettor
Concentrators, Microcon- or Centricon-100 (Millipore) (optional; see Step 4)
Light mineral oil or wax bead (optional; see Step 6)
Microtiter plates or microfuge tubes, 0.5 ml and thin walled
Positive displacement pipette
Step 8 requires the reagents listed in Protocol 5.2, 6.10, or 12.6.
Step 10 requires the reagents listed in Protocols 8.3 and 8.4, 8.5, or 8.6.
Thermal cycler, programmed with desired amplification protocol
Water baths (75°C, 37°C, and 95°C)

Additional Information

Design of oligonucleotide primers	MC3, pp. 8.13–8.16
Inhibitors of RNases	MC3, p. 7.83
Reverse transcriptase	MC3, pp. A4.24–A4.26
Reverse transcriptases used in RT-PCR	MC3, p. 8.48
Terminal transferase	MC3, pp. A1.11, A4.27, and 8.111
Troubleshooting RT-PCR	MC3, p. 8.53

METHOD

1. Transfer 1 pg to 100 ng of poly(A)$^+$ mRNA or 10 pg to 1 µg of total RNA to a fresh microfuge tube. Adjust the volume to 10 µl with H_2O. Denature by heating for 5 minutes at 75°C, followed by chilling on ice.

2. To the denatured RNA, add:

10x amplification buffer	2 µl
20 mM solution of four dNTPs (pH 8.0)	1 µl
gene-specific oligonucleotide primers (10 mM)	1 µl
approximately 20 units/µl placental RNase inhibitor	1 µl
50 mM $MgCl_2$	1 µl
100–200 units/µl reverse transcriptase	1 µl
H_2O	to 20 µl

    Incubate the reaction for 60 minutes at 37°C.

    Depending on the experiment, oligo(dT)$_{12-18}$, random hexanucleotides, or gene-specific anti-sense oligonucleotides can be used as primers for synthesis of first-strand cDNA.

    The optimum ratio of primer to template should be ascertained empirically for each preparation of RNA. As a starting point for optimization, we recommend adding varying amounts of primers to 20-µl reactions:

synthetic oligonucleotide complementary to the target RNA	5–20 pmoles
oligo(dT)$_{12-18}$	0.1–0.5 µg
random hexanucleotides	1–5 µg

Set up three negative-control reactions. In one reaction, include all components of the first-strand reaction except for the RNA template. In another reaction, include all components except for the reverse transcriptase. Omit primers from the third reaction. Carry the controls through all subsequent steps of the protocol.

3. Inactivate the reverse transcriptase and denature the template-cDNA complexes by heating the reaction to 95°C for 5 minutes or by phenol extraction and ethanol precipitation.

4. Adjust the reaction mixture so that it contains 20 pmoles of the sense and antisense primers.

5. Add to the reaction mixture:

1x amplification buffer (or volume required to bring reaction mixture to 99 µl)	77 µl
1–2 units thermostable DNA polymerase	1 µl

6. If the thermal cycler is not fitted with a heated lid, overlay the reaction mixtures with 1 drop (~50 µl) of light mineral oil. Alternatively, place a bead of paraffin wax into the tube if using a hot start protocol. Place the tubes or the microtiter plate in the thermal cycler.

7. Amplify the nucleic acids using the denaturation, annealing, and polymerization times and temperatures listed in the table.

Cycle Number	Denaturation	Annealing	Polymerization
35 cycles	45 sec at 94°C	45 sec at 55°C	1 min 15 sec at 72°C
Last cycle	1 min at 94°C	45 sec at 55°C	1 min 15 sec at 72°C

Times and temperatures may need to be adapted to suit the particular reaction conditions.
Polymerization should be performed for 1 minute for every 1000 bp of length of the target DNA.

8. Withdraw a sample (5–10 µl) from the test reaction mixture and from the four control reactions and analyze them by electrophoresis through an agarose or polyacrylamide gel. Include DNA markers of an appropriate size. Stain the gel with ethidium bromide or SYBR Gold to visualize the DNA.

   A successful amplification reaction should yield a readily visible DNA fragment of the expected size. The identity of the band can be confirmed by DNA sequencing (please see Protocol 12.6), Southern hybridization (please see Protocol 6.10), and/or restriction mapping.

   If no product is visible after 30 cycles of amplification, add fresh thermostable DNA polymerase and continue the amplification reaction for an additional 15–20 cycles.

9. If mineral oil was used to overlay the reaction (in Step 6), remove the oil from the sample before cloning by extraction with 150 µl of chloroform.

   The aqueous phase, which contains the amplified DNA, will form a micelle near the meniscus. The micelle can then be transferred to a fresh tube with an automatic micropipette.

   IMPORTANT: Do not attempt chloroform extractions in microtiter plates. The plastic used in these plates is not resistant to organic solvents.

10. Clone the amplified products into an appropriately prepared vector by any of the methods described in Protocol 8.4, 8.5, or 8.6. Before cloning, separate the amplified DNA from the residual thermostable DNA polymerase and dNTPs (please see Protocol 8.3). The DNA can then be ligated to a blunt-ended or T vector, or it can be digested with restriction enzymes and ligated to a vector with compatible termini.

Amplification of 5′ cDNA (5′-RACE)

This method is used to extend partial cDNA clones by amplifying the 5′ sequences of the corresponding mRNAs. The technique requires knowledge of a small region of sequence within the partial cDNA clone. During PCR, the thermostable DNA polymerase is directed to the appropriate target RNA by a single primer derived from the region of known sequence; the second primer required for PCR is complementary to a general feature of the target—in the case of 5′-RACE, to a homopolymeric tail added (via terminal transferase) to the 3′ termini of cDNAs transcribed from a preparation of mRNA. This synthetic tail provides a primer-binding site upstream of the unknown 5′ sequence of the target mRNA. The products of the amplification reaction are cloned into a plasmid vector for sequencing and subsequent manipulation.

MATERIALS

CAUTION: Please see Appendix 4 for appropriate handling of materials marked with <!>.

Reagents and Solutions

Please see Appendix 1 for components of stock solutions, buffers, and reagents. Dilute stock solutions to the appropriate concentrations.

Chloroform <!>

dATP (1 mM) disodium salt

dNTP solution (pH 8.0), containing all four deoxynucleotide triphosphates, each at a concentration of 20 mM

Ethanol

Ethanol (70%)

Phenol:chloroform (1:1, v/v) <!>

Placental RNase inhibitor (20 units/µl)

Sodium acetate (3 M, pH 5.2)

TE (pH 7.6)

Trichloroacetic acid (10%) (optional; see Step 2)

Enzymes and Buffers

Reverse transcriptase (RNA-dependent DNA polymerase) and 10x reverse transcriptase buffer as supplied by the manufacturer or homemade

500 mM Tris-Cl (pH 8.3)
750 mM KCl
30 mM MgCl$_2$

Terminal transferase buffer and 5x terminal transferase buffer

500 mM potassium cacodylate
10 mM CoCl$_2$
1 mM dithiothreitol
For advice on making this buffer, see MC3, p. A1.11.

Thermostable DNA polymerase and 10x amplification buffer as supplied by the manufacturer or homemade

500 mM KCl
100 mM Tris-Cl (pH 8.3, room temperature)
15 mM MgCl$_2$

Nucleic Acids/Oligonucleotides

Adapter-primer (10 µM in H$_2$O) (5′ GACTCGAGTCGACATCG 3′)
Used in conjunction with a gene-specific sense primer to amplify a target cDNA

DNA markers for gel electrophoresis

(dT)$_{17}$ Adapter-primer (10 μM in H$_2$O) (5′ GACTCGAGTCGACATCGA(T)$_{17}$ 3′)

> The adapter-primer binds to the poly(A)$^+$ tract added to the 3′-terminal sequences of cDNAs by terminal transferase. In this protocol, 3′ termini are modified by addition of recognition sites for *Xho*I, *Sal*I, *Acc*I, *Hinc*II, and *Cla*I.

Gene-specific oligonucleotide primer (10 μM in H$_2$O)

> Complementary to a known sequence in the target mRNA; used to primer synthesis of cDNA. The gene-specific primer should contain approximately equal numbers of the four bases, with a balanced distribution of G and C residues and a low propensity to form stable secondary structures (see MC3, pp. 8.13–8.16)
>
> Random hexanucleotides (1 mg/ml) may be used in Step 1 in place of gene-specific primer (see MC3, p. 8.57).

Template RNA

> The template mRNA may be total cellular RNA (100 μg/ml in H$_2$O) or poly(A)$^+$ RNA (10 μg/ml in H$_2$O), which is preferred when the target mRNA is expressed at low abundance.

Radiolabeled Compounds

[^{32}P]dCTP (sp. act. 3000 Ci/mmole) <!> (optional; see Step 2)

Gels/Loading Buffers

Agarose or polyacrylamide gel <!>

Centrifuges/Rotors/Tubes

Preparative centrifuge with angle rotor, to hold Microcon or Centricon concentrators
Sorval SS-34 rotor

Additional Items

Barrier tips for automatic pipettor
Concentrators, Microcon- or Centricon-100 (Millipore)
Light mineral oil or wax bead (optional; see Step 6)
Microtiter plates or microfuge tubes, 0.5 ml and thin walled
Positive displacement pipette
Rotary evaporator (optional; see Step 3)
Step 9 requires the reagents listed in Protocol 5.2, 6.10, or 12.6.
Step 11 requires the reagents listed in Protocols 8.3 and 8.4, 8.5, or 8.6.
Thermal cycler, programmed with desired amplification protocol
Water baths (75°C, 37°C, and 80°C)

Additional Information

Design of oligonucleotide primers	MC3, pp. 8.13–8.16
Inhibitors of RNases	MC3, p. 7.83
Reverse transcriptase	MC3, pp. A4.24–A4.26
Reverse transcriptases used in RT-PCR	MC3, p. 8.48
Terminal transferase	MC3, pp. A1.11, A4.27, and 8.111
Troubleshooting 5′-RACE	MC3, p. 8.60

METHOD

1. Transfer 1 pg to 100 ng of poly(A)$^+$ mRNA or 10 pg to 1 μg of total RNA to a fresh microfuge tube. Adjust the volume to 9 μl with H$_2$O. Denature the RNA by heating for 5 minutes at 75°C, followed by rapid chilling on ice.

2. To the denatured RNA, add:

5x reverse transcriptase buffer	4 µl
20 mM solution of four dNTPs (pH 8.0)	1 µl
10 µM gene-specific antisense primer 1	4 µl
approximately 20 units/µl placental RNase inhibitor	1 µl
100–200 units/µl reverse transcriptase	1 µl
H_2O	to 20 µl

Incubate the reaction for 60 minutes at 37°C. Set up three negative-control reactions. In one reaction, include all of the components of the first-strand reaction, except for the RNA template. In another reaction, include all of the components, except for the reverse transcriptase. Omit primer from the third reaction. Carry the controls through all subsequent steps of the protocol.

> *Total* cDNA synthesis can be estimated from the proportion of trichloroacetic acid (TCA)-precipitable radioactivity incorporated in reverse transcription reactions supplemented with 10–20 µCi of [^{32}P]dCTP (sp. act. 3000 Ci/mmole). Be sure to set up a control reaction that contains no oligonucleotide primer.

> The success or failure of 5'-RACE is determined here. If the reverse transcription step works efficiently, the chance of isolating clones that contain the 5'-terminal sequences of the target mRNA is high. On the other hand, no amount of work on the later steps of the protocol can compensate for ineffective reverse transcription. It is therefore worthwhile to take the time to optimize the reverse transcriptase reaction by determining the optimum ratio of primer to template for each preparation of RNA and by varying the concentration of Mg^{++} in the reaction.

3. Remove excess primer by diluting the reverse transcriptase reaction to a final volume of 2 ml with H_2O and then applying the solution to a Centricon-100 microconcentrator (see Protocol 8.3). Centrifuge the solution at 500–1100g (2000–3000 rpm in a Sorvall SS-34 rotor) for 20 minutes at a temperature between 4°C and 25°C. Repeat the dilution step and centrifuge again. Transfer the retentate to a fresh 0.5-ml microfuge tube and reduce the volume to approximately 10 µl in a rotary vacuum evaporator.

> Alternatively, remove the dNTPs and unused primers by precipitating the cDNA twice in 2.5 M ammonium acetate and 3 volumes of ethanol.

4. To the cDNA in a volume of 10 µl, add:

5x terminal transferase buffer	4 µl
1 mM dATP	4 µl
terminal transferase	10–25 units

Incubate the reaction for 15 minutes at 37°C.

> The tailing reaction can be optimized by setting up mock reactions containing approximately 50 ng of a control DNA fragment, 100–200 nucleotides in length. After tailing, the size of the fragment should increase by 20–100 nucleotides as measured by electrophoresis through a 1% neutral agarose gel.

5. Inactivate the terminal transferase by heating the reaction for 3 minutes at 80°C. Dilute the dA-tailed cDNA to a final volume of 1 ml with TE (pH 7.6).

6. In a sterile 0.5-ml microfuge tube, amplification tube, or the well of a sterile microtiter plate, set up a series of PCRs containing the following:

diluted cDNA	0–20 µl
10x amplification buffer	5 µl
20 mM solution of four dNTPs	5 µl
10 µM (dT)$_{17}$ adapter-primer (16 pmoles)	1.6 µl
10 µM adapter-primer (32 pmoles)	3.2 µl
10 µM gene-specific primer 2 (32 pmoles)	3.2 µl
1–2 units of thermostable DNA polymerase	1 µl
H_2O	to 50 µl

It is essential to set up a series of amplification reactions to find the amount of tailed cDNA that generates the largest quantity of amplified 5′ termini. The control reaction containing no cDNA template serves as a control for contamination.

7. If the thermal cycler is not fitted with a heated lid, overlay the reaction mixtures with 1 drop (~50 μl) of light mineral oil. Alternatively, place a bead of paraffin wax into the tube if using a hot start protocol. Place the tubes or the microtiter plate in the thermal cycler.

8. Amplify the nucleic acids using the denaturation, annealing, and polymerization times and temperatures listed in the table.

Cycle Number	Denaturation	Annealing	Polymerization
First cycle	5 min at 94°C	5 min at 50–58°C	40 min at 72°C
Subsequent cycles (30)	40 sec at 94°C	1 min at 50–58°C	3 min at 72°C
Last cycle	40 sec at 94°C	1 min at 50–58°C	15 min at 72°C

Times and temperatures may need to be adapted to suit the particular reaction conditions.

9. Withdraw a sample (5–10 μl) from the test reaction mixture and from the four control reactions and analyze them by electrophoresis through an agarose or polyacrylamide gel. Include DNA markers of an appropriate size. Stain the gel with ethidium bromide or SYBR Gold to visualize the DNA.

A successful amplification reaction should yield a readily visible DNA fragment of the expected size. The identity of the band can be confirmed by DNA sequencing (please see Protocol 12.6), Southern hybridization (please see Protocol 6.10), and/or restriction mapping.

10. If mineral oil was used to overlay the reaction (in Step 7), remove the oil from the sample before cloning by extraction with 150 μl of chloroform.

The aqueous phase, which contains the amplified DNA, will form a micelle near the meniscus. The micelle can be transferred to a fresh tube with an automatic micropipette.

IMPORTANT: Do not attempt chloroform extractions in microtiter plates. The plastic used in these plates is not resistant to organic solvents.

11. Separate the amplified DNA from the residual thermostable DNA polymerase and dNTPs (please see Protocol 8.3).

The DNA can now be ligated to a blunt-ended or T vector, or it can be digested with restriction enzymes and ligated to a vector with compatible termini (please see Protocols 8.4–8.6).

Amplification of 3′ cDNA (3′-RACE)

3′-RACE reactions are used to isolate unknown 3′ sequences or to map the 3′ termini of mRNAs onto a gene sequence. 3′-RACE requires knowledge of a small region of sequence within either the target RNA or a partial clone of cDNA. A population of mRNAs is transcribed into cDNA with an adapter-primer consisting at its 3′ end of a poly(T) tract and at its 5′ end of an arbitrary sequence of 30–40 nucleotides. Reverse transcription is usually followed by two successive PCRs. The first is primed by a gene-specific sense oligonucleotide and an antisense primer complementary to the arbitrary sequence in the (dT) adapter-primer. If necessary, the products of the first PCR can be used as templates for a second nested PCR, which is primed by a gene-specific sense oligonucleotide internal to the first, and a second antisense oligonucleotide complementary to the central region of the (dT) adapter-primer. The products amplified in the second PCR are isolated from an agarose gel, cloned, and characterized.

MATERIALS

CAUTION: Please see Appendix 4 for appropriate handling of materials marked with <!>.

Reagents and Solutions

Please see Appendix 1 for components of stock solutions, buffers, and reagents. Dilute stock solutions to the appropriate concentrations.

Chloroform <!>

dNTP solution (pH 8.0), containing all four deoxynucleotide triphosphates, each at a concentration of 20 mM

Ethanol

Ethanol (70%)

Phenol:chloroform (1:1, v/v) <!>

Placental RNase inhibitor (20 units/µl)

Sodium acetate (3 M, pH 5.2)

TE (pH 7.6)

Trichloroacetic acid (10%) (optional; see Step 2)

Enzymes and Buffers

Reverse transcriptase (RNA-dependent DNA polymerase) and 10x reverse transcriptase buffer as supplied by the manufacturer or homemade
  500 mM Tris-Cl (pH 8.3)
  750 mM KCl
  30 mM MgCl$_2$
Thermostable DNA polymerase and 10x amplification buffer as supplied by the manufacturer or homemade
  500 mM KCl
  100 mM Tris-Cl (pH 8.3, room temperature)
  15 mM MgCl$_2$

Nucleic Acids/Oligonucleotides

Adapter-primer (10 µM in H$_2$O) (5′ GACTCGAGTCGACATCG 3′)
  Used in conjunction with a gene-specific sense primer to amplify a target cDNA
DNA markers for gel electrophoresis
(dT)$_{17}$ Adapter-primer (10 µM in H$_2$O) (5′ GACTCGAGTCGACATCGA(T)$_{17}$ 3′)

Adapted from Chapter 8, Protocol 10, p. 8.61 of MC3.

The oligo(dT) region of the adapter-primer binds to the poly(A)$^+$ tract added to the 3′-terminal sequences mRNA, leaving the adapter sequences unpaired. During amplification, the adapter sequences are converted to double-stranded cDNA that is equipped with recognition sites for *Xho*I, *Sal*I, *Acc*I, *Hinc*II, and *Cla*I.

Gene-specific oligonucleotide primer (10 μM in H$_2$O)

Used to prime synthesis of double-stranded cDNA. The gene-specific primer should contain approximately equal numbers of the four bases, with a balanced distribution of G and C residues and a low propensity to form stable secondary structures (see MC3, pp. 8.13–8.16). Random hexanucleotides (1 mg/ml) may be used in Step 1 in place of a gene-specific primer (see MC3, p. 8.57).

Template RNA

The template mRNA may be total cellular RNA (100 μg/ml in H$_2$O) or poly(A)$^+$ RNA (10 μg/ml in H$_2$O), which is preferred when the target mRNA is expressed at low abundance.

Radiolabeled Compounds

[^{32}P]dCTP (sp. act. 3000 Ci/mmole) <!> (optional; see Step 2)

Gels/Loading Buffers

Agarose or polyacrylamide gel <!>

Additional Items

Barrier tips for automatic pipettor
Light mineral oil or wax bead (optional; see Step 5)
Microtiter plates or microfuge tubes, 0.5 ml and thin walled
Positive displacement pipette
Step 6 requires the reagents listed in Protocol 5.2, 6.10, or 12.6.
Step 8 requires the reagents listed in Protocols 8.3 and 8.4, 8.5, or 8.6.
Thermal cycler, programmed with desired amplification protocol
Water baths (75°C, 37°C, and 80°C)

Additional Information

Theory of 3′-RACE   MC3, p. 8.61

METHOD

1. Transfer 1 pg to 100 ng of poly(A)$^+$ mRNA or 10 pg to 1 μg of total RNA to a fresh microfuge tube. Adjust the volume to 10 μl with H$_2$O. Denature the RNA by heating for 5 minutes at 75°C, followed by rapid chilling on ice.

2. To the denatured RNA, add:

5x reverse transcriptase buffer	10 μl
20 mM solution of four dNTPs	1.5 μl
10 μM (dT)$_{17}$ adapter-primer (80 pmoles)	8.0 μl
approximately 20 units/μl placental RNase inhibitor	1 μl
100–200 units/μl reverse transcriptase	1 μl
H$_2$O	to 50 μl

   Incubate the reaction for 60 minutes at 37°C. Set up three negative-control reactions. In one reaction, include all of the components of the first-strand reaction, except for the RNA template. In another reaction, include all of the components, except for the reverse transcriptase. Omit primer from the third reaction. Carry the controls through all subsequent steps of the protocol.

   Total cDNA synthesis can be estimated from the proportion of TCA-precipitable radioactivity incorporated in reverse transcriptase reactions supplemented with 10–20 μCi of [^{32}P]dCTP (sp. act. 3000 Ci/mmole). Be sure to set up a control reaction that contains no (dT)$_{17}$ adapter antisense primer.

The success or failure of 3′-RACE is determined here. If the reverse transcription step works efficiently, the chance of isolating clones that contain the 3′-terminal sequences of the target mRNA is high. On the other hand, no amount of work on the later steps of the protocol can compensate for ineffective reverse transcription. It is therefore worthwhile to take the time to optimize the reverse transcriptase reaction by determining the optimum ratio of primer to template for each preparation of RNA and by varying the concentration of Mg^{++} in the reaction.

3. Dilute the reverse transcriptase reaction (cDNA) to a final volume of 1 ml with TE (pH 7.6).

   It may be necessary to remove excess oligonucleotide and random hexamer primers from the cDNA preparation and then to optimize the concentrations of the sense and antisense primers in the amplification reaction (please see Protocol 8.9, Step 3).

4. In a sterile 0.5-ml microfuge tube, amplification tube, or the well of a sterile microtiter plate, set up a series of PCRs containing the following:

diluted cDNA	0–20 µl
10x amplification buffer	5 µl
20 mM solution of four dNTPs	5 µl
10 µM (dT)$_{17}$ adapter-primer (16 pmoles)	1.6 µl
10 µM adapter-primer (32 pmoles)	3.2 µl
10 µM gene-specific sense oligonucleotide primer (32 pmoles)	3.2 µl
1–2 units thermostable DNA polymerase	1 µl
H$_2$O	to 50 µl

   It is essential to set up a series of amplification reactions to find the amount of cDNA that generates the largest quantity of amplified 3′ termini. The control containing no cDNA template serves as a control for contamination.

5. If the thermal cycler is not fitted with a heated lid, overlay the reaction mixtures with 1 drop (~50 µl) of light mineral oil. Alternatively, place a bead of paraffin wax into the tube if using a hot start protocol. Place the tubes or the microtiter plate in the thermal cycler. Amplify the nucleic acids using the denaturation, annealing, and polymerization times and temperatures listed in the table.

Cycle Number	Denaturation	Annealing	Polymerization
First cycle	5 min at 94°C	5 min at 50–58°C	40 min at 72°C
Subsequent cycles (20)	40 sec at 94°C	1 min at 50–58°C	3 min at 72°C
Last cycle	40 sec at 94°C	1 min at 50–58°C	15 min at 72°C

   Times and temperatures may need to be adapted to suit the particular reaction conditions.

6. Withdraw a sample (5–10 µl) from the test reaction mixture and from the four control reactions and analyze them by electrophoresis through an agarose or polyacrylamide gel. Include DNA markers of an appropriate size. Stain the gel with ethidium bromide or SYBR Gold to visualize the DNA.

   A successful amplification reaction should yield a readily visible DNA fragment of the expected size. The identity of the band can be confirmed by DNA sequencing (please see Protocol 12.6), Southern hybridization (please see Protocol 6.10), and/or restriction mapping.

7. If mineral oil was used to overlay the sample, remove the oil from the sample before cloning by extraction with 150 µl of chloroform.

   The aqueous phase, which contains the amplified DNA, will form a micelle near the meniscus. The micelle can be transferred to a fresh tube with an automatic micropipette.

   IMPORTANT: Do not attempt chloroform extractions in microtiter plates. The plastic used in these plates is not resistant to organic solvents.

8. Separate the amplified DNA from the residual thermostable DNA polymerase and dNTPs (please see Protocol 8.3).

   The DNA can now be ligated to a blunt-ended or T vector, or it can digested with restriction enzymes and ligated to a vector with compatible termini (please see Protocols 8.4–8.6).

Mixed Oligonucleotide-primed Amplification of cDNA (MOPAC)

In MOPAC, the amino-terminal and carboxy-terminal sequences of a peptide are used to design two redundant families of oligonucleotides encoding the amino- and carboxy-terminal sequences of the peptide. The primers are used to either amplify a segment of cDNA prepared by RT-PCR from a tissue known to express the protein or amplify a segment of DNA from an established genomic or cDNA library. Because the length of the peptide is known, the size of the expected PCR product can be predicted exactly. After gel electrophoresis to resolve the amplification products, DNAs of the correct size are isolated, cloned, and sequenced. At least some of the clones should contain a DNA segment of the correct length that specifies the sequence of the starting peptide. Once identified, the entire cloned segment or the unique sequence lying between the two oligonucleotide primers is used as a probe to screen a cDNA library.

MATERIALS

CAUTION: Please see Appendix 4 for appropriate handling of materials marked with <!>.

Reagents and Solutions

Please see Appendix 1 for components of stock solutions, buffers, and reagents. Dilute stock solutions to the appropriate concentrations.

>   dNTP solution (pH 8.0), containing all four deoxynucleotide triphosphates, each at a
>   concentration of 20 mM
>
>   TE (pH 7.6)

Enzymes and Buffers

>   Thermostable DNA polymerase and 10x amplification buffer as supplied by the
>   manufacturer or homemade
>
>>   500 mM KCl
>>   100 mM Tris-Cl (pH 8.3, room temperature)
>>   15 mM $MgCl_2$

Nucleic Acids/Oligonucleotides

>   DNA markers for gel electrophoresis
>   Families of degenerate sense and antisense oligonucleotide primers (10 µM in TE [pH 8.0];
>   see MC3, pp. 8.66–8.69)
>   Template DNA (100 µg/ml) in TE (pH 8.0)
>>   The template mRNA may be uncloned cDNA, genomic DNA, or a cDNA library that contains
>>   the target sequence. Before performing MOPAC, it is essential to remove any primers (such as
>>   oligo[dT]) that were used to prime synthesis of uncloned cDNA templates (see Protocol 8.3).

Gels/Loading Buffers

>   Polyacrylamide gel <!>

Additional Items

>   Barrier tips for automatic pipettor
>   Light mineral oil or wax bead (optional; see Step 5)
>   Microtiter plates or microfuge tubes, 0.5 ml and thin walled
>   Positive displacement pipette
>   Step 4 requires the reagents listed in Protocol 5.2, 6.10, or 12.6.
>   Thermal cycler, programmed with desired amplification protocol

Additional Information

Band-stab PCR	MC3, p. 8.71
Design of degenerate oligonucleotide primers	MC3, pp. 8.13–8.16
Theory of MOPAC reactions	MC3, pp. 8.66–8.68

METHOD

1. In a sterile 0.5-ml microfuge tube, amplification tube, or the well of a sterile microtiter plate, mix in the following order:

0.5 µg of template DNA	5 µl
10x amplification buffer	5 µl
20 mM solution of four dNTPs	5 µl
10 µM sense primer family (70 pmoles)	7 µl
10 µM antisense primer family (70 pmoles)	7 µl
1–2 units thermostable DNA polymerase	1 µl
H₂O	to 50 µl

   Set up several control reactions. In one reaction, include all of the above components, except for the template DNA. In two other reactions, include all the components minus one or the other of the oligonucleotide primers.

2. If the thermal cycler is not fitted with a heated lid, overlay the reaction mixtures with 1 drop (~50 µl) of light mineral oil. Alternatively, place a bead of paraffin wax into the tube if using a hot start protocol. Place the tubes or the microtiter plate in the thermal cycler.

3. Amplify the nucleic acids using the denaturation, annealing, and polymerization times and temperatures listed in the table.

Cycle Number	Denaturation	Annealing	Polymerization
First cycle	5 min at 94°C	2.5 min at 40°C	5 min at 50°C
Subsequent cycles (35)	1.5 min at 94°C	2.5 min at 40°C	5 min at 50°C

   Times and temperatures may need to be adapted to suit the particular reaction conditions.

   The amplification conditions described above are based on 32-member degenerate families of oligonucleotide primers that are 17–20 bases in length. Higher annealing and polymerization temperatures can be used with longer and/or less complex mixtures of oligonucleotides.

   The specificity of MOPAC can be increased by using (i) a thermal cycler program that includes a temperature ramping protocol or (ii) touchdown PCR (see MC3, p. 8.112).

4. Withdraw a sample (5–10 µl) from the test reaction mixture and from the control reactions and analyze them by electrophoresis through a neutral polyacrylamide gel. Include DNA markers of an appropriate size. Stain the gel with ethidium bromide or SYBR Gold to visualize the DNA (see Band-stab PCR, MC3, p. 8.71).

   If a single contiguous amino acid sequence was used to design both sense and antisense primers for the MOPAC reaction, the exact size of the desired product will be known. In most cases, thin polyacrylamide gels (≤6%) should allow size fractionation at nucleotide resolution, and recovery and unambiguous identification of the desired product. Once identified, the product can be end-labeled (Protocol 9.14) and subjected to chemical sequencing (Protocol 12.7) to determine a unique sequence linking the two families of oligonucleotide probes. This unique sequence can in turn be synthesized and used as a probe to isolate longer cDNAs. Alternatively, the MOPAC product itself can be radiolabeled during a second round of PCR and used as a probe.

5. If mineral oil was used to overlay the reaction (Step 2), remove the oil from the sample by extraction with 150 µl of chloroform.

> The aqueous phase, which contains the amplified DNA, will form a micelle near the meniscus. The micelle can be transferred to a fresh tube with an automatic micropipette.

> IMPORTANT: Do not attempt chloroform extractions in microtiter plates. The plastic used in these plates is not resistant to organic solvents.

Protocol 8.12   Rapid Characterization of DNAs Cloned in Prokaryotic Vectors

In this method, sequences cloned in standard bacteriophage or plasmid vectors are amplified in PCRs containing primers targeted to flanking vector sequences. The amplified fragments can be analyzed by gel electrophoresis, DNA sequencing, and/or restriction mapping. Many colonies or plaques can be assayed simultaneously.

MATERIALS

Reagents and Solutions

Please see Appendix 1 for components of stock solutions, buffers, and reagents. Dilute stock solutions to the appropriate concentrations.

dNTP solution (pH 8.0), containing all four deoxynucleotide triphosphates, each at a concentration of 20 mM

Tris-Cl (10 mM, pH 7.6)

Vectors and Hosts

Bacterial colonies or bacteriophage λ plaques (see Template DNAs below)

Enzymes and Buffers

Thermostable DNA polymerase and 10x amplification buffer as supplied by the manufacturer or homemade

500 mM KCl

100 mM Tris-Cl (pH 8.3, room temperature)

15 mM MgCl₂

Nucleic Acids/Oligonucleotides

DNA markers for gel electrophoresis

Forward and reverse primers, each at a concentration of 20 μM in H₂O

Use commerically available oligonucleotide primers for sequencing the 5′ and 3′ regions of DNA segments cloned into the most popular vectors. These oligonucleotides can also be used as primers in PCR reactions. For details, see MC3, pp. 8.113–8.114.

Template DNAs

This method uses unpurified templates obtained by cracking bacteriophage λ particles, transformed bacterial cells, or yeast.

Gels/Loading Buffers

Agarose gel

Additional Items

Barrier tips for automatic pipettor

Boiling-water bath

Light mineral oil or wax bead (optional; see Step 8)

Microtiter plates or microfuge tubes, 0.5 ml and thin walled

Positive displacement pipette

Step 9 requires the reagents listed in Protocols 5.2 and 6.10.

Thermal cycler, programmed with desired amplification protocol

340   Adapted from Chapter 8, Protocol 12, p. 8.72 of MC3.

Additional Information

Screening bacterial libraries by PCR	MC3, p. 8.76
Screening yeast colonies by PCR	MC3, p. 8.75
Troubleshooting	MC3, p. 8.75

METHOD

1. Calculate the number of bacterial colonies or bacteriophage λ plaques that are to be screened. Prepare the appropriate amount of master mix; analysis of each colony or plaque requires 25 μl of master mix. *One ml of master mix contains:*

10x amplification buffer	100 μl
20 mM solution of four dNTPs	50 μl
forward primer	1 nmole
reverse primer	1 nmole
H_2O	to 1 ml

2. Dispense 25-μl aliquots of the master mix into the appropriate number of amplification tubes.

3. Use a sterile 200-μl pipette tip (NOT a toothpick) to touch each bacterial colony or bacteriophage λ plaque. Working quickly, wash the pipette tip in 25 μl of master mix.

4. Close the caps of the tubes. Incubate the closed tubes in a boiling water bath for 10 minutes (bacterial colonies) or 2 minutes (bacteriophage λ plaques).

5. Dilute the required amount of thermostable DNA polymerase to a concentration of 1 unit/μl in 10 mM Tris (pH 7.6). Store the diluted enzyme on ice.

6. Allow the samples (from Step 4) to cool to room temperature. Centrifuge the tubes briefly and then add 1 μl of the diluted thermostable DNA polymerase to each tube.

7. Set up two control reactions. In one reaction, include all of the components, except for the template DNA. In the other reaction, include a recombinant bacteriophage λ plaque or transformed bacterial lysate that has previously produced a positive result in this assay.

8. If the thermal cycler does not have a heated lid, overlay the reaction mixtures with 1 drop (~50 μl) of light mineral oil. Place the tubes in the thermal cycler. Amplify the nucleic acids using the denaturation, annealing, and polymerization times and temperatures listed in the table.

Cycle Number	Denaturation	Annealing	Polymerization
30 cycles	1 min at 94°C	2 min at 50°C	2 min at 72°C

   Times and temperatures may need to be adjusted to achieve maximum amplification.

9. Withdraw a sample (5–10 μl) from the test reaction mixture and from the control reactions and analyze them by electrophoresis through an agarose gel. Include DNA markers of an appropriate size. Stain the gel with ethidium bromide or SYBR Gold to visualize the DNA.

   A successful amplification reaction should yield a readily visible DNA fragment. Nonrecombinant PCR products will be equal to the length of DNA between the locations of the 5′ termini of the two primers in the cloning vector. Recombinant PCR products will be the sum of (i) the length of the insert and (ii) the distance between the 5′ termini of two primers in the vector. If necessary, the identity of the band can be confirmed by restriction mapping and Southern hybridization.

Long PCR

Whereas standard PCRs yield amplified DNAs 1–2 kb in length, the following protocol uses a combination of thermostable DNA polymerases to amplify DNA up to 25 kb in length.

MATERIALS

CAUTION: Please see Appendix 4 for appropriate handling of materials marked with <!>.

Reagents and Solutions

Please see Appendix 1 for components of stock solutions, buffers, and reagents. Dilute stock solutions to the appropriate concentrations.

Chloroform <!>

dNTP solution (pH 8.0), containing all four deoxynucleotide triphosphates, each at a concentration of 20 mM

Enzymes and Buffers

10x Long PCR buffer, as supplied by the manufacturer or homemade
  500 mM Tris-Cl (pH 9.0, room temperature)
  160 mM ammonium sulfate
  25 mM $MgCl_2$
  1.5 mg/ml bovine serum albumin

Mixture of thermostable DNA polymerase mix
  This protocol uses a homemade mixture of KlenTaq1 (Ab Peptides), an efficient but error-prone DNA polymerase, and *Pfu* (Stratagene), which provides a thermostable 3′-5′ exonuclease that resects mismatched termini of duplex DNA. A typical mixture contains 0.187 unit of *Pfu* and 33.7 units of KlenTaq1 in a total volume of 1.2 μl. Other cocktails of enzymes are available commercially, e.g., TaqPlus Long PCR System (Stratagene).

Nucleic Acids/Oligonucleotides

DNA markers for gel electrophoresis

Forward and reverse primers, each at a concentration of 20 μM in H_2O
  Oligonucleotide primers used for long PCR are generally slightly longer (25–30 nucleotides) than those used for standard PCR. When designing the oligonucleotides, make sure that the calculated melting temperatures of the forward and reverse primers do not differ by more than 1°C.

Template DNAs
  Long PCR works well on a variety of templates including recombinant PACs, BACs, cosmids, and bacteriophage λ clones, as well as high-molecular-weight genomic DNAs. For success, the quality of the DNA is paramount and the median length of the template should be at least 3x the length of the desired PCR product. Use between 100 pg and 2 μg of template DNA per rection, depending on the complexity of the target sequences.

Gels/Loading Buffers

Agarose gel

Additional Items

Barrier tips for automatic pipettor
Light mineral oil or wax bead (optional; see Step 2)
Microtiter plates or microfuge tubes, 0.5 ml and thin walled

Positive displacement pipette
Step 4 requires the reagents listed in Protocols 5.2 and 6.10.
Thermal cycler, programmed with desired amplification protocol

Additional Information

Modified thermostable DNA polymerase	MC3, pp. 12.46–12.47
Why long PCR works	MC3, p. 8.77

METHOD

1. In a thin-walled amplification tube, add and mix in the following order:

10x long PCR buffer	5 μl
20 mM solution of four dNTPs	5 μl
20 mM forward primer	1 μl
20 mM reverse primer	1 μl
thermostable DNA polymerase mix	0.2 μl
template DNA	100 pg to 2 μg
H_2O	to 50 μl

   Templates purified from individual recombinant clones constructed in bacteriophage λ, cosmid, bacteriophage P1, PAC, and BAC vectors should be used in amounts ranging from 100 pg to 300 ng. Larger amounts of total genomic DNAs are required, usually between 100 ng and 1 μg per reaction. The optimum amount of template and the optimum ratio of primers:template should be ascertained empirically for each new preparation of DNA.

2. If the thermal cycler is not fitted with a heated lid, overlay the reaction mixtures with 1 drop (~50 μl) of light mineral oil. Alternatively, place a bead of paraffin wax into the tube if using a hot start PCR. Place the tubes or the microtiter plate in the thermal cycler. Amplify the nucleic acids using the denaturation, annealing, and polymerization times and temperatures listed in the table.

Cycle Number	Denaturation	Annealing	Polymerization
24 cycles	1 min at 94°C	1 min at 60–67°C	5–20 min at 68°C

   Times and temperatures may need to be adapted to suit the particular reaction conditions.

   The temperature used for the annealing step depends on the melting temperature of the oligonucleotide primers. Because the primers used in long PCR are generally 27–30 nucleotides in length, the annealing temperatures used in long PCR can be considerably higher than those used in standard PCR.

3. If mineral oil was used to overlay the reaction (Step 2), remove the oil from the sample by extraction with 150 μl of chloroform.

   The aqueous phase, which contains the amplified DNA, will form a micelle near the meniscus. The micelle can be transferred to a fresh tube with an automatic micropipette.

   IMPORTANT: Do not attempt chloroform extractions in microtiter plates. The plastic used in these plates is not resistant to organic solvents.

4. Analyze an aliquot of the aqueous phase by electrophoresis through an agarose gel using markers of an appropriate size. In many cases, the amount of amplified product may be too small to be detected by conventional staining with ethidium bromide. In this case, stain the DNA in the gel with SYBR Gold or transfer to a nylon or nitrocellulose filter and probe by Southern hybridization (please see Protocol 6.10).

Inverse PCR

Inverse PCR is used to amplify and clone unknown DNA that flanks one end of a known DNA sequence and for which no primers are available. The technique involves digestion by a restriction enzyme of a preparation of DNA containing the known sequence and its flanking region. The individual restriction fragments (many thousands in the case of total mammalian genomic DNA) are converted into circles by intramolecular ligation and the circularized DNA is then used as a template in the PCR. The unknown sequence is amplified by two primers that bind specifically to the known sequence and point in opposite directions. The product of the amplification reaction is a linear DNA fragment containing a single site for the restriction enzyme originally used to digest the DNA. Inverse PCR is used chiefly to generate end-specific probes for chomosome walking.

MATERIALS

CAUTION: Please see Appendix 4 for appropriate handling of materials marked with <!>.

Reagents and Solutions

Please see Appendix 1 for components of stock solutions, buffers, and reagents. Dilute stock solutions to the appropriate concentrations.

ATP (10 mM)
> Required only if 10x ligation buffer does not contain ATP

Chloroform <!>

dNTP solution (pH 8.0), containing all four deoxynucleotide triphosphates, each at a
> concentration of 20 mM

Ethanol

Ethanol (70%)

Phenol:chloroform (1:1, v/v) <!>

Sodium acetate (3 M, pH 5.2)

TE (pH 8.0)

Tris-Cl (10 mM, pH 7.6)

Enzymes and Buffers

Bacteriophage T4 DNA ligase and 10x buffer

Restriction enzymes and 10x buffers

Thermostable DNA polymerase and 10x amplification buffer as supplied by the
manufacturer or homemade
> 500 mM KCl
> 100 mM Tris-Cl (pH 8.3, room temperature)
> 15 mM MgCl$_2$

Nucleic Acids/Oligonucleotides

DNA markers for gel electrophoresis

Oligonucleotide primers 1 and 2 (each 20 μM in H$_2$O)
> There is nothing unusual about the primers used in inverse PCR, except for their placement on the circular template. The standard rules for primer design apply (see MC3, pp. 8.13–8.16).

Template DNA in 10 mM Tris-Cl (pH 7.6)
> Inverse PCR requires a circular DNA as template. Steps 1–4 of this protocol describe how such templates can be generated from conventional preparations of linear DNAs, which can range in complexity from a small fragment of cloned DNA to a preparation of total mammalian

Adapted from Chapter 8, Protocol 14, p. 8.81 of MC3.

genomic DNA. The amount of DNA required for inverse PCR depends on its complexity. A reasonable amount for an inverse PCR experiment, when using total mammalian DNA as template, would be 2–5 µg in a volume of approximately 100 µl.

Gels/Loading Buffers

Polyacrylamide <!> or agarose gel

Additional Items

Barrier tips for automatic pipettor
Light mineral oil or wax bead (optional; see Step 6)
Microtiter plates or microfuge tubes, 0.5 ml and thin walled
Positive displacement pipette
Step 7 requires the reagents listed in Protocol 5.2, and in Protocol 6.10 or 12.6.
Thermal cycler, programmed with desired amplification protocol
Water block or heat/cooling block (16°C)

Additional Information

Design of oligonucleotide primers for basic PCR	MC3, pp. 8.13–8.16
Inverse PCR, including a schematic representation of the technique	MC3, pp. 8.81–8.82

METHOD

1. Design and synthesize oligonucleotide primers 1 and 2 based on a known sequence of DNA.

2. Digest 2–5 µg of DNA template (sequence complexity <10^9 bp) with an appropriate restriction enzyme that generates complementary cohesive termini. Extract the digested DNA with phenol:chloroform and then with chloroform alone. Precipitate the DNA with 0.1 volume of 3 M sodium acetate (pH 5.2) and 2.5 volumes of ethanol. Recover the precipitated DNA by centrifugation and dissolve it in TE (pH 8.0) at a concentration of 100 µg/ml.

   Alternatively, heat the digested DNA to 65°C for 15–20 minutes to inactivate the restriction enzyme.

3. In sterile 0.5-ml microfuge tubes, amplification tubes, or the wells of a sterile microtiter plate, set up a series of ligation reactions containing cleaved template DNA at concentrations ranging from 0.1 to 1 µg/ml.

template DNA	10–100 ng
10x ligation buffer	10 µl
bacteriophage T4 DNA ligase	4 units
10 mM ATP	10 µl
H_2O	to 100 µl

   Incubate the reactions for 12–16 hours at 16°C.

4. Extract the ligated DNA with phenol:chloroform, and then with chloroform alone. Precipitate the DNA with 0.1 volume of 3 M sodium acetate (pH 5.2) and 2.5 volumes of ethanol. Recover the precipitated DNA by centrifugation and dissolve it in 10 mM Tris (pH 7.6) or H_2O at a concentration of 100 µg/ml.

5. In a sterile 0.5-ml thin-walled amplification tube, add and mix in the following order:

10x amplification buffer	5 µl
20 mM solution of four dNTPs (pH 8.0)	1 µl
20 µM oligonucleotide primer 1	2.5 µl
20 µM oligonucleotide primer 2	2.5 µl

1–5 units/μl thermostable DNA polymerase	1.0 μl
H$_2$O	28–33 μl
ligated template DNA	5–10 μl
Total volume	50 μl

Set up two control reactions. In one reaction, include all of the above reagents, except for the template DNA. In the other reaction, replace the template with a plasmid of known size, containing the DNA insert from which the oligonucleotide primers were derived. Carry each control reaction through all subsequent steps of the protocol.

6. If the thermal cycler is not fitted with a heated lid, overlay the reaction mixtures with 1 drop (~50 μl) of light mineral oil. Alternatively, place a bead of wax into the tube if using hot start PCR. Place the tubes or the microtiter plate in the thermal cycler. Amplify the nucleic acids using the denaturation, annealing, and polymerization times and temperatures listed in the table.

Cycle Number	Denaturation	Annealing	Polymerization
30 cycles	30 sec at 94°C	30 sec at 60°C	2.5 min at 72°C
Last cycle	30 sec at 94°C	30 sec at 60°C	10 min at 72°C

Times and temperatures may need to be adapted to suit the particular reaction conditions.

The exact annealing temperature should be established empirically for the primer pairs used in a given amplification reaction. An extended polymerization time (up to 10 minutes per cycle) should be tried if the target DNA is long (>4 kb). Alternatively, the use of mutant thermostable DNA polymerases that lack and/or contain a low level of 3′-exonuclease activity may produce longer templates (see Protocol 8.13).

7. Withdraw a sample (5–10 μl) from the test and control reactions and analyze them by electrophoresis through an agarose or polyacrylamide gel. Include DNA markers of an appropriate size. Stain the gel with ethidium bromide or SYBR Gold.

A successful amplification reaction should yield a readily visible DNA. The identity of the band can be confirmed by DNA sequencing (please see Protocol 12.6), restriction mapping, and/or Southern hybridization using probes homologous to the known DNA sequence (please see Protocol 6.10). If low yields are a problem, or if the amplification reactions yield multiple fragments, nested PCR should be used in a second round of PCR.

Quantitative PCR

Quantitative PCR involves co-amplification of two templates: a constant amount of a preparation containing the desired target sequence and varying amounts of a reference template. After amplification, the concentration of the target sequence in the preparation of nucleic acid under test is established by interpolation into a standard curve. Quantitation of nucleic acids by PCR is best performed by real-time PCR (see MC3, pp. 8.94–8.95). However, the following robust protocol, which uses radioactivity to quantify PCR products, remains useful when a real-time instrument is unavailable. The method can be easily adapted to other methods of quantification such as fluorometry.

MATERIALS

CAUTION: Please see Appendix 4 for appropriate handling of materials marked with <!>.

Reagents and Solutions

Please see Appendix 1 for components of stock solutions, buffers, and reagents. Dilute stock solutions to the appropriate concentrations.

Chloroform <!>

dNTP solution (pH 8.0), containing all four deoxynucleotide triphosphates, each at a concentration of 20 mM

$MgCl_2$ (1 M)

Placental RNase inhibitor (20 units/µl)

Enzymes and Buffers

Appropriate restriction enzymes and 10x buffers

Bacteriophage T4 DNA ligase and 10x buffer

Reverse transcriptase, required only if RNA is used as a template (see MC3, p. 8.48)

Thermostable DNA polymerase and 10x amplification buffer as supplied by the manufacturer or homemade

    500 mM KCl

    100 mM Tris-Cl (pH 8.3, room temperature)

    15 mM $MgCl_2$

Nucleic Acids/Oligonucleotides

DNA markers for gel electrophoresis

Externally added reference (either DNA or RNA) of known concentration

Use a DNA reference to measure the concentration of DNA sequences and, if possible, an RNA reference for RNA targets. A method to construct reference RNA is described in Protocol 15.2.

Sense and antisense primers, each 20 µM in H_2O

There is nothing unusual about the primers used in quantitative PCR. The standard rules for primer design apply (see MC3, pp. 8.13–8.16).

Target nucleic acid

The target can be a preparation of DNA or RNA, either total or poly(A)+. Dissolve preparations of total RNA in H_2O at a concentration of 0.5–1.0 mg/ml and preparations of poly(A)+ RNA at 10–100 µg/ml. Dissolve DNA targets in 10 mM Tris-Cl (pH 7.6) at the following concentrations: mammalian genomic DNA, 100 µg/ml; yeast genomic DNA, 1 µg/ml; bacterial genomic DNA, 0.1 µg/ml; and plasmid DNA, 1–5 ng/ml.

Radiolabeled Compounds

[α-^{32}]dCTP (sp. act. 3000 Ci/mmole at 10 mCi/ml) <!>

Gels/Loading Buffers

Polyacrylamide <!> or agarose gel

Additional Items

Barrier tips for automatic pipettor
Fluorometer (optional; see Step 1 and MC3, pp. A8.22–A8.24)
Light mineral oil or wax bead (optional; see Step 5)
Materials for autoradiography or phosphorimaging
Microtiter plates or microfuge tubes, 0.5 ml and thin walled
Positive displacement pipette
Thermal cycler, programmed with desired amplification protocol
Water baths (94°C and, for RNA templates only, 75°C)

Additional Information

Inhibitors of RNases	MC3, p. 7.83
PCR in theory	MC3, p. 8.12
Real-time PCR	MC3, pp. 8.94–8.95
Reverse transcriptases used in RT-PCR	MC3, p. 8.48
Theory of quantitative PCR	MC3, pp. 8.86–8.89

METHOD

1. Design and prepare a reference template suitable for the task at hand. Measure the concentration of the reference template as carefully as possible, preferably by fluorometry. Alternatively, estimate the amount of reference template after gel electrophoresis and ethidium bromide staining.

2. Make a series of tenfold dilutions (in H_2O) containing concentrations of the reference template ranging from 10^{-6} to 10^{-12} M. After using the dilutions (Step 3), they should be stored at –70°C for later use in Step 8.

3. If starting from RNA, denature the target RNA by incubating aliquots for 5 minutes at 75°C, followed by rapid chilling in ice water. Then, without delay, set up a series of reverse transcription reactions containing increasing amounts of reference template in sterile 0.5-ml microfuge tubes. For each reaction in the series, prepare the following:

10x amplification buffer	2 µl
20 mM solution of four dNTPs (pH 8.0)	1 µl
20 µM antisense primer	2.5 µl
approximately 20 units/µl placental RNase inhibitor	1 µl
50 mM $MgCl_2$	1 µl
denatured target RNA	10 pg to 1.0 µg
100–200 units/µl reverse transcriptase	1 µl
tenfold dilution of reference template	1 µl
H_2O	to 20 µl

Incubate the reaction for 60 minutes at 37°C and then denature the reverse transcriptase by heating to 95°C for 20 minutes.

4. In sterile 0.5-ml microfuge tubes, amplification tubes, or the wells of a sterile microtiter plate, set up amplification reactions with each reaction in the series from Step 3:

reverse transcriptase reaction (Step 3) or target DNA	5 µl

20 μM sense primer	1.5 μl
20 μM antisense primer	1.25 μl
10x amplification buffer	5 μl
[α-^{32}P]dCTP (3000 Ci/mmole)	10 μCi
20 mM solution of four dNTPs	1 μl
thermostable DNA polymerase	2 units
H$_2$O	to 50 μl

IMPORTANT: Do not reduce the concentration of unlabeled dCTP in the reaction mixture to increase the specific activity of the precursor pool. There is a danger that the amount of the nucleotide could become limiting at late stages in the amplification reaction.

5. If the thermal cycler is not fitted with a heated lid, overlay the reaction mixtures with 1 drop (~50 μl) of light mineral oil. Alternatively, place a bead of wax into the tube if using hot start PCR. Place the tubes or the microtiter plate in the thermal cycler.

6. Amplify the nucleic acids using the denaturation, annealing, and polymerization times and temperatures listed in the table.

Cycle Number	Denaturation	Annealing	Polymerization
30 cycles	30 sec at 95°C	30 sec at 55°C	1 min at 72°C
Last cycle	1 min at 94°C	30 sec at 55°C	1 min at 72°C

Times and temperatures may need to be adapted to suit the particular reaction conditions.

7. Analyze and quantitate the amplified products.

When using a reference template that differs from the target sequence in size

a. Analyze the sizes of the amplified products in a 20-μl aliquot of each of the reactions by gel electrophoresis and autoradiography.

b. Excise the amplified bands of the control template and target sequences from the gel and measure the amount of radioactivity in each band in a liquid scintillation counter. Alternatively, scan the gel with the appropriate detector (e.g., GE Healthcare scanner or phosphorimager).

c. Calculate the relative amounts of the two radiolabeled DNAs in each of the PCRs.
   Correct the amount of radioactivity to allow for differences in the molecular weights of the two radiolabeled DNAs.

When using a reference template that contains a novel restriction site or lacks a naturally occurring site

a. Heat the samples to 94°C for 5 minutes following the final round of amplification.

b. Allow the samples to cool gradually to room temperature and then digest a 20-μl aliquot of each of the reactions with the appropriate restriction enzyme.

c. Analyze the sizes of the amplified DNA fragments by gel electrophoresis and autoradiography or phosphorimaging.

d. Excise the amplified bands of the control template and target sequences from the gel and measure the amount of radioactivity in each band in a liquid scintillation counter. Alternatively, scan the gel with the appropriate detector (e.g., GE Healthcare scanner or phosphorimager).

e. Calculate the relative amounts of the two radiolabeled DNAs in each of the PCRs.
   Correct the amount of radioactivity to allow for differences in the molecular weights of the two radiolabeled DNAs.

8. Examine the results to determine the concentration of reference template that yields approximately the same amount of amplified product as the target sequence. Set up a second series of amplification reactions (please see Step 4) containing a narrower range of concentrations of reference template.

   It is best to generate this series of dilutions from the appropriate tenfold dilution of the reference template (Step 2).

9. Repeat Steps 5–7. For each amplification reaction, measure the ratio of the yield of amplified reference template to the yield of amplified target sequence. Plot this ratio against the amount of reference template added to each amplification reaction. From the resulting straight line, determine the equivalence point (i.e., the amount of reference template that gives exactly the same quantity of amplified product as the target sequence in the reaction). Calculate the concentration of the target sequence in the original sample.

This method is included in this book for completeness rather than for practical value. Differential display PCR (DD-PCR) had its heyday in the mid-1990s and was therefore included in the third edition of *Molecular Cloning*, published in 2001. These days, anyone wishing to catalog the mRNAs expressed in cells or tissues would be better off using DNA microarrays rather than DD-PCR, which was always problematic in principle and difficult in performance.

As its name implies, DD-PCR is used to amplify and display many cDNAs derived from the mRNAs of a given cell or tissue type. The method relies on two different types of synthetic oligonucleotides: anchored antisense primers and arbitrary sense primers. A typical anchored primer is complementary to approximately 13 nucleotides of the poly(A) tail of mRNA and the adjacent two nucleotides of the transcribed sequence. Anchored primers therefore anneal to the junction between the poly(A) tail and the 3'-untranslated region of mRNA templates, from where they can prime synthesis of first-strand cDNA. A second primer, an arbitrary sequence of approximately 10 nucleotides, is then added to the reaction mixture, and double-stranded cDNAs are produced by conventional PCR, carried out at low stringency. The products of the amplification reaction are separated by electrophoresis through a denaturing polyacrylamide gel and visualized by autoradiography. By comparing the banding patterns of cDNA products derived from two different cell types, or from the same cell type grown under different conditions, it is sometimes possible to identify the products of differentially expressed genes. Bands of interest can then be recovered from the gel, amplified further, and cloned and/or used as probes to screen northern blots, cDNA libraries, etc.

MATERIALS

CAUTION: Please see Appendix 4 for appropriate handling of materials marked with <!>.

Reagents and Solutions

Please see Appendix 1 for components of stock solutions, buffers, and reagents. Dilute stock solutions to the appropriate concentrations.

Chloroform <!> (optional; see Step 21)

Dithiothreitol (100 mM)

dNTP solution (pH 8.0), containing all four deoxynucleotide triphosphates, each at a concentration of 20 mM

Placental RNase inhibitor (20 units/µl)

Enzymes and Buffers

RNase-free DNase I (optional; see Step 3)

Reverse transcriptase deficient in RNase H (e.g., SuperScript [Stratagene]) and 5x DD-PCR reverse transcriptase buffer

250 mM Tris-Cl (pH 8.3)

375 mM KCl

15 mM $MgCl_2$

Thermostable DNA polymerase and 10x amplification buffer as supplied by the manufacturer or homemade

500 mM KCl

100 mM Tris-Cl (pH 8.3, room temperature)

15 mM $MgCl_2$

Nucleic Acids/Oligonucleotides

Anchoring 3'-oligonucleotide primers (300 μg/ml) in 10 mM Tris-Cl (pH 7.6), 0.1 mM EDTA
> The anchoring 3' oligonucleotides are a family of 12 primers with the general structure 5'-d(T)$_{12}$VN-3', where V is C, A, or G and N is C, T, A, or G.

Arbitrary 5'-oligonucleotide primers (50 μg/ml) in 10 mM Tris-Cl (pH 7.6), 0.1 mM EDTA
> Sixteen arbitrary 5'-oligonucleotide primers are required, each 10 nucleotides in length. The sequence of each primer is chosen at random, with a balanced distribution of G and C residues and a low propensity to form secondary structures.

Total RNA (100 μg/ml), extracted from cells with chaotropic agents (Protocol 7.1)

Radioactive Compounds

[α-^{33}P or α-^{35}S]ATP (sp. act. 3000 Ci/mmole at 10 mCi/ml) <!>

Gels/Loading Buffers

1% Agarose gel
Electrolyte gradient DNA-sequencing gel and 5x formamide loading buffer (see Protocol 12.10)

Additional Items

Barrier tips for automatic pipettor
Light mineral oil or wax bead (optional; see Step 7)
Materials for autoradiography or phosphorimaging
Microtiter plates or microfuge tubes, 0.5 ml and thin walled
Positive displacement pipette
Razor blades
Small-gauge hypodermic needle
Step 22 requires the reagents listed in Protocol 8.5.
Step 23 requires the reagents listed in Protocol 8.12.
Step 24 requires the reagents listed in Protocol 7.8 or 7.11 and in Protocol 8.15.
Steps 10 and 11 require the reagents listed in Protocols 12.10 and 12.11.
Thermal cycler, programmed with desired amplification protocol
Water baths (94°C, 65°C, and 37°C)
Whatman 3MM paper

Additional Information

Differential display	MC3, pp. 8.96–8.100
Reverse transcriptases used in RT-PCR	MC3, p. 8.48

METHOD

1. In sterile 0.5-ml microfuge tubes, set up a series of trial reactions to establish the optimum concentrations of "control" and "test" RNAs required to produce a pattern of 100–300 amplified cDNA bands after gel electrophoresis and autoradiography. Make fivefold serial dilutions in H$_2$O of the RNA preparations to produce concentrations of between 1 and 100 μg/ml.

2. Choose one or more primers from the collection of anchored 3' oligonucleotides and set up a series of annealing reactions that contain different amounts of diluted RNA templates:

template RNA	8.0 μl
anchored 3'-oligonucleotide primer	2.0 μl

   Incubate the reactions for 10 minutes at 65°C and then place them in a 37°C water bath.
   > The total amount of RNA in the annealing reactions should vary between 8 and 800 ng.

3. Add the following to the annealing reactions:

5x DD-PCR reverse transcriptase buffer	4 μl
100 mM dithiothreitol	2 μl
200 μM solution of four dNTPs	2 μl
approximately 25 units/μl placental RNase inhibitor	0.25 μl
200 units/μl reverse transcriptase	0.25 μl
H$_2$O	to 20 μl

   Incubate the tubes for 1 hour at 37°C.

   > To test for contaminating genomic DNA, set up one or more control reactions that contain no reverse transcriptase and carry these through Step 10 of the protocol. If necessary, the RNA preparation can be treated with RNase-free DNase I either as a separate step during purification or in the same reaction tube that will later be used to synthesize cDNA.

4. Inactivate the reverse transcriptase by incubating the reaction mixtures for 10 minutes at 94°C.

   > Steps 3 and 4 can be performed in a thermal cycler programmed with a single cycle of 37°C for 1 hour/94°C for 10 minutes, followed by a 4°C hold.

5. Set up two series of eight 0.5-ml amplification tubes. Each tube should contain:

10x amplification buffer	2 μl
anchored 3'-oligonucleotide primer	2 μl
20 mM solution of four dNTPs (pH 8.0)	1 μl
[α-^{33}P]dATP or [α-^{35}S]dATP (3000 Ci/mmole)	1 μl
H$_2$O	9 μl
5 units/μl thermostable DNA polymerase	1 unit

   To each tube, add 2 μl of a different arbitrary 5' primer. Mix the contents by tapping the sides of the tubes.

   > Wherever possible, use the 10x amplification buffer supplied by the manufacturer of the DNA polymerase.

6. Into one series of eight tubes, dispense approximately 3-μl aliquots of the reverse transcriptase reaction containing the test RNA. Into the other series of eight tubes dispense approximately 3-μl aliquots of the reverse transcriptase reaction containing the preparation of control RNA. Close the tubes and mix the contents gently.

7. If the thermal cycler is not fitted with a heated lid, overlay the reaction mixtures with 1 drop (~50 μl) of light mineral oil. Place the tubes in the thermal cycler.

8. Amplify the nucleic acids using the denaturation, annealing, and polymerization times and temperatures listed in the table.

Cycle Number	Denaturation	Annealing	Polymerization
30 cycles	15 sec at 94°C	30 sec at 42°C	15 sec at 72°C
Last cycle	15 sec at 94°C	30 sec at 42°C	2 min at 72°C

   > Times and temperatures may need to be adapted to suit the particular reaction conditions.

9. At the end of the program, remove the tubes from the thermal cycler and add 5 μl of 5x formamide loading buffer to each.

10. Separate the radiolabeled products of the reactions by electrophoresis through an electrolyte gradient polyacrylamide gel of the type used for DNA sequencing. Electrophoresis is carried at constant electrical power until the xylene cyanol tracking dye has migrated about two-thirds the length of the gel (please see Protocols 12.10 and 12.11). Dry the gel and expose it to autoradiographic film or phosphorimaging.

11. Examine the pattern of DNA bands arising from reactions containing different concentrations of control and test RNAs. A good differential display contains between 100 and 250 well-

resolved bands. Select the concentration of test and control RNAs that work well with the largest number of primer pairs.

12. Repeat the annealing, reverse transcriptase, and amplification reactions using all combinations of primer pairs and the optimum amount of RNA templates. Set up the reactions in 96-well microtiter plates designed for use in a thermal cycler.

13. Separate the products of the amplification reactions by electrophoresis through polyacrylamide sequencing gels, as in Steps 9 and 10.

    Load the reactions generated with each primer pair in adjacent lanes on the gel, i.e., load the reaction obtained with primer pair A + B from one RNA preparation next to the reaction obtained with primer pair A + B from the other RNA preparation.

14. Compare the patterns of bands obtained with each primer pair from the different RNA populations.

    When a differentially expressed band is identified, it is advisable to repeat the experiment to make sure that the initial finding is reproducible. Ideally, different batches of the two RNAs should be used, although this precaution may not always be practicable.

15. Recover target bands from the dried polyacrylamide gel. Lay the autoradiogram on top of the gel and use a soft pencil to lightly mark the position of the desired band on the autoradiogram. Cutting through the autoradiogram with a clean razor blade, excise each target band and the attached Whatman 3MM paper. Soak each sliver of dried gel/paper overnight at room temperature in a separate 0.5-ml microfuge tube containing 50 μl of sterile H_2O.

16. Puncture the bottom of each 0.5-ml tube with a small-gauge needle. Place each punctured tube inside a 1.5-ml microfuge tube. Centrifuge for 20 seconds to transfer the eluate to the larger tube. Discard the amplification tube containing the residue of the Whatman 3MM paper and polyacrylamide.

17. Amplify the eluted fragment in a reaction containing the following:

10x amplification buffer	2 μl
DNA eluted from polyacrylamide gel	3 μl
arbitrary 5'-oligonucleotide primer	2 μl
anchoring 3'-oligonucleotide primer	2 μl
20 mM solution of four dNTPs (pH 8.0)	1 μl
H_2O	9.5 μl
5 units/μl thermostable DNA polymerase	2 units

    Wherever possible, use the 10x amplification buffer supplied by the manufacturer of the DNA polymerase.

18. If the thermal cycler is not fitted with a heated lid, overlay the reaction mixtures with 1 drop (~50 μl) of light mineral oil. Place the tubes in a thermal cycler.

19. Amplify the nucleic acids using the denaturation, annealing, and polymerization times and temperatures listed in the table.

Cycle Number	Denaturation	Annealing	Polymerization
30 cycles	15 sec at 94°C	30 sec at 42°C	15 sec at 72°C
Last cycle	15 sec at 94°C	30 sec at 42°C	2 min at 72°C

    Polymerization should be performed for 1 minute for every 1000 bp of length of the target DNA.

20. Estimate the concentration of the reamplified DNA fragment by electrophoresis of 5–10% of the reaction through a 1% (w/v) agarose gel.

21. If mineral oil was used to overlay the reaction (Step 18), remove the oil from the sample by extraction with 150 μl of chloroform.

The aqueous phase, which contains the amplified DNA, will form a micelle near the meniscus. The micelle can be transferred to a fresh tube with an automatic micropipette.

IMPORTANT: Do not attempt chloroform extractions in microtiter plates. The plastic used in these plates is not resistant to organic solvents.

22. Ligate the DNA into a vector that has been tailed with dT (e.g., pGEM T vector from Promega) (please see Protocol 8.5) and transform *E. coli* with aliquots of the ligation reaction.

23. Isolate plasmid DNA from six or more recombinants and compare the sizes of the inserts released by restriction enzyme digestion.

    The sequence of the insert DNA can be established by using universal primers that bind to the flanking regions of the vector. These oligonucleotides can also be used as primers to check the size of the inserts by PCR (please see Protocol 8.12).

    It is important to isolate and sequence more than one plasmid recombinant from the ligation reaction. Compare the cDNA sizes and sequences to one another and to those in the various databases.

24. Confirm the differential expression of a candidate cDNA/mRNA in as many ways as possible, including northern hybridization (Protocol 7.8), RNase protection (Protocol 7.11), or quantitative PCR (Protocol 8.15). In situ mRNA hybridization can be used to localize the transcript to a diseased or developing tissue.

CHAPTER 9

Preparation of Radiolabeled DNA and RNA Probes

BACKGROUND INFORMATION

Background information found in *Molecular Cloning: A Laboratory Manual*, 3rd edition (hereafter MC3) unless otherwise indicated.

Comparison of methods used to radiolabel nucleic acids in vitro   MC3, p. 9.3

Random Priming: Radiolabeling of Purified DNA Fragments by Extension of Random Oligonucleotides

Short oligonucleotides of random sequence can serve as primers for the initiation of DNA synthesis at multiple sites on single-stranded DNA templates. The newly synthesized DNA is labeled by using one [α-^{32}P]dNTP and three unlabeled dNTPs as precursors, generating probes with specific activities of 5×10^8 to 5×10^9 dpm/μg.

MATERIALS

CAUTION: Please see Appendix 4 for appropriate handling of materials marked with <!>.

Reagents and Solutions

Please see Appendix 1 for components of stock solutions, buffers, and reagents. Dilute stock solutions to the appropriate concentrations.

Ammonium acetate (10 M) (optional; see Step 5)

Dithiothreitol (1 M), freshly made

dNTP solution (pH 8.0), containing three unlabeled deoxynucleoside triphosphates, each at a concentration of 5 mM

> The composition of this solution depends on the [α-^{32}P]dNTP to be used for radiolabeling. For example, if radiolabeled dATP is used, the mix should contain dCTP, dTTP, and dGTP.

Ethanol (optional; see Step 5)

NA stop/storage buffer

50 mM Tris-Cl (pH 7.5)

50 mM NaCl

5 mM EDTA (pH 8.0)

0.5% (w/v) SDS

Trichloroacetic acid (10%) (optional; see Step 4)

Enzymes and Buffers

Klenow fragment of *E. coli* DNA polymerase I and 5x random priming buffer

250 mM Tris-Cl (pH 8.0)

25 mM MgCl$_2$

100 mM NaCl

10 mM dithiothreitol (from freshly made 1 M stock)

1 M HEPES (adjusted to pH 6.6 with 4 M NaOH)

Nucleic Acids/Oligonucleotides

Random oligonucleotide primers, six or seven bases in length (e.g., Roche Applied Science)

Template DNA (5–25 ng/ml in TE [pH 7.6])

> Purified by one of the methods described in Chapter 5. Protocol 9.1 works optimally when 25 ng of template DNA is used per reaction. Larger amounts of DNA yield probes of lower specific activity; smaller amounts require longer reaction times.

Radioactive Compounds

[α-^{32}P]dNTP (sp. act. 3000 Ci/mmole at 10 mCi/ml) <!>

Gels/Loading Buffers

Alkaline agarose or denaturing polyacrylamide gel <!>

     Adapted from Chapter 9, Protocol 1, p. 9.4 of MC3.

Additional Items

Spun column of Sephadex G-50, equilibrated in TE (pH 7.6)
Water bath or heating block (100°C)

Additional Information

Spun-column chromatography   MC3, pp. A8.30–A8.31

METHOD

1. In a 0.5-ml microfuge tube, combine template DNA (25 ng) in 30 µl of H_2O with 1 µl of random deoxynucleotide primers (~125 ng). Close the top of the tube tightly and place the tube in a 95°C to 100°C bath for 2 minutes.

2. Remove the tube and place it on ice for 1 minute. Centrifuge the tube for 10 seconds at 4°C in a microfuge. Return the tube to the ice bath.

3. To the mixture of primer and template, add:

5 mM dNTP solution	1 µl
5x random primer buffer	10 µl
10 mCi/ml [α-^{32}P]dNTP (sp. act. 3000 Ci/mmole)	5 µl
H_2O	to 50 µl

4. Add 5 units (~1 µl) of the Klenow fragment. Mix the components gently. Centrifuge the tube at maximum speed for 1–2 seconds in a microfuge. Incubate the reaction mixture for 60 minutes at room temperature.

   To label larger amounts of DNA, assemble reaction mixtures as described in Steps 3 and 4 and then incubate the reaction for 60 minutes. To label smaller amounts of DNA, incubate the reactions for times that are inversely proportional to the amount of template added.

   To monitor the course of the reaction, measure the proportion of radiolabeled dNTPs that is either incorporated into material precipitated by trichloroacetic acid (TCA) or that adheres to a DE81 filter (see Appendix 3, p. 745).

   Under these reaction conditions, the length of the radiolabeled product is approximately 400–600 nucleotides, as determined by electrophoresis through an alkaline agarose gel (Protocol 5.8) or a denaturing polyacrylamide gel (Protocol 12.8).

5. Add 10 µl of NA stop/storage buffer to the reaction and proceed with one of the following options as appropriate.

   • Store the radiolabeled probe at –20°C until it is needed for hybridization

   or

   • Separate the radiolabeled probe from unincorporated dNTPs by either spun-column chromatography or selective precipitation of the radiolabeled DNA with ammonium acetate and ethanol (please see MC3, Appendix 8). This step is generally not required if >50% of the radiolabeled dNTP has been incorporated during the reaction.

Random Priming: Radiolabeling of Purified DNA Fragments by Extension of Random Oligonucleotides in the Presence of Melted Agarose

A variation of the technique described in Protocol 9.1 can be used to radiolabel DNA in slices cut from gels cast with low-melting-temperature agarose.

MATERIALS

CAUTION: Please see Appendix 4 for appropriate handling of materials marked with <!>.

Reagents and Solutions

Please see Appendix 1 for components of stock solutions, buffers, and reagents. Dilute stock solutions to the appropriate concentrations.

Ammonium acetate (10 M) (optional; see Step 5)

Bovine serum albumin (10 mg/ml)

Dithiothreitol (1 M), freshly made

dNTP solution (pH 8.0), containing three unlabeled deoxynucleoside triphosphates, each at a concentration of 20 mM

> The composition of this solution depends on the [α-^{32}P]dNTP to be used for radiolabeling. For example, if radiolabeled dATP is used, the mix should contain dCTP, dTTP, and dGTP.

Ethanol (optional; see Step 5)

Ethidium bromide (10 mg/ml) <!>

NA stop/storage buffer

> 50 mM Tris-Cl (pH 7.5)
> 50 mM NaCl
> 5 mM EDTA (pH 8.0)
> 0.5% (w/v) SDS

Trichloroacetic acid (10%) (optional; see Step 4)

Enzymes and Buffers

Klenow fragment of *E. coli* DNA polymerase I and 5x oligonucleotide labeling buffer

> 250 mM Tris-Cl (pH 8.0)
> 25 mM MgCl$_2$
> 2 mM of each of the three unlabeled dNTPs
> 20 mM dithiothreitol (from freshly made 1 M stock)
> 1 M HEPES (adjusted to pH 6.6 with 4 M NaOH)

Nucleic Acids/Oligonucleotides

Random oligonucleotide primers, six or seven bases in length (e.g., Roche Applied Science)

Template DNA (5–25 ng/µl in TE [pH 7.6])

> The DNA to be labeled is purified by electrophoresis through a low-melting-temperature agarose gel, cast and run in 1x TAE electrophoresis buffer. Protocol 9.2 works optimally when 25 ng of template DNA is used per reaction. Larger amounts of DNA yield probes of lower specific activity; smaller amounts require longer reaction times.

Radioactive Compounds

[α-^{32}P]dNTP (sp. act. 3000 Ci/mmole at 10 mCi/ml) <!>

Gels/Loading Buffers

Akaline agarose or denaturing polyacrylamide gel <!>

Low-melting-temperature agarose gel

Additional Items

Spun column of Sephadex G-50, equilibrated in TE (pH 7.6)
Water bath or heating block (100°C and 37°C)

Additional Information

Spun-column chromatography   MC3, pp. A8.30–A8.31

METHOD

1. After electrophoresis, stain the gel with ethidium bromide (final concentration 0.5 µg/ml) or SYBR Gold and excise the desired band, eliminating as much extraneous agarose as possible.

2. Place the band in a preweighed microfuge tube and measure its weight. Add 3 ml of H_2O for every gram of agarose gel.

3. Place the microfuge tube in a boiling-water bath for 7 minutes to melt the gel and denature the DNA.

   If radiolabeling is to be performed immediately, store the tube at 37°C until the template is required. Otherwise, store the tube at –20°C. After each removal from storage, reheat the DNA/gel slurry to 100°C for 3–5 minutes and then store at 37°C until the radiolabeling reaction is initiated.

4. To a fresh microfuge tube in a 37°C water bath or heating block, add in the following order:

5x oligonucleotide labeling buffer	10 µl
10 mg/ml bovine serum albumin solution	2 µl
DNA in a volume no greater than 32 µl	20–50 ng
10 mCi/ml [α-^{32}P]dNTP	
(sp. act. 3000 Ci/mmole)	5 µl
Klenow fragment (5 units)	1 µl
H_2O	to 50 µl

   Mix the components completely with a micropipettor. Incubate the reaction for 2–3 hours at room temperature or for 60 minutes at 37°C.

   To label larger amounts of DNA, adjust the volume of the reaction mixture proportionately and incubate the reaction for 60 minutes. To label smaller amounts of DNA, incubate the reactions for times that are inversely proportional to the amount of template added.

   To monitor the course of the reaction, measure the proportion of radiolabeled dNTPs that is either incorporated into material precipitated by TCA or that adheres to a DE81 filter (see Appendix 3, p. 745).

   Under these reaction conditions, the length of the radiolabeled product is approximately 400–600 nucleotides, as determined by electrophoresis through an alkaline agarose gel (Protocol 5.8) or a denaturing polyacrylamide gel (Protocol 5.9).

5. Add 50 µl of NA stop/storage buffer to the reaction, and proceed with one of the following options as appropriate:

   • Store the radiolabeled probe at –20°C until it is needed for hybridization

   or

   • Separate the radiolabeled probe from unincorporated dNTPs by either spun-column chromatography or selective precipitation of the radiolabeled DNA with ammonium acetate and ethanol (please see MC3, Appendix 8). This step is not required if >50% of the radiolabeled dNTP has been incorporated during the reaction.

Radiolabeling of DNA Probes by the Polymerase Chain Reaction

In this protocol, double-stranded DNA probes, labeled in each strand, are produced in conventional PCRs containing three unlabeled dNTPs at concentrations exceeding the K_m, and one radiolabeled dNTP at a concentration at or slightly above the K_m (2–3 μM) of thermostable DNA polymerase.

MATERIALS

CAUTION: Please see Appendix 4 for appropriate handling of materials marked with <!>.

Reagents and Solutions

Please see Appendix 1 for components of stock solutions, buffers, and reagents. Dilute stock solutions to the appropriate concentrations.

Ammonium acetate (10 M)

Carrier (glycogen [50 mg/ml in H_2O] or yeast tRNA [10 mg/ml in H_2O])

> Used to improve the efficiency of precipitation of DNA with ethanol (see MC3, p. A8.13)

Chloroform <!>

dCTP solution (0.1 mM)

> Dilute 1 volume of a 10-mM stock solution of dCTP with 99 volumes of 10 mM Tris-Cl (pH 8.0).

dNTP solution containing three unlabeled deoxynucleoside triphosphates (dATP, dGTP, and dTTP), each at a concentration of 10 mM

Ethanol

TE (pH 7.6)

Enzymes and Buffers

Thermostable DNA polymerase and 10x amplification buffer as supplied by manufacturer or homemade

> 500 mM KCl
>
> 100 mM Tris-Cl (pH 8.3, room temperature)
>
> 15 mM $MgCl_2$

Nucleic Acids/Oligonucleotides

Forward and reverse primers, each at a concentration of 20 μM in H_2O

> Primers should be designed according to standard rules, i.e., they should be complementary to sequences spaced 100–400 nucleotides apart on opposite strands of the target DNA. They should be 20–25 nucleotides in length and contain approximately equal numbers of the four bases, with a balanced distribution of G and C residues and a low propensity to form secondary structures.

Template DNA

> A variety of templates can be used in this protocol, including crude preparations of plasmid or bacteriophage λ DNA, purified DNA fragments recovered from gels, and genomic DNAs. The minimum amount of target sequence required for the reaction to proceed efficiently is approximately 1 fmole. The amount of template DNA added to the PCR therefore varies according to its complexity. Preparations of mammalian genomic DNA used as templates in PCRs typically contain 100 μg DNA/ml; yeast genomic DNA, 1 μg/ml; bacterial genomic DNA, 0.1 μg/ml; and plasmid DNA, 1–5 ng/ml.

Radioactive Compounds

[α-^{32}P]dCTP (sp. act. 3000 Ci/mmole at 10 mCi/ml) <!>

Additional Items

Barrier tips for automatic pipettor
Light mineral oil or wax bead (optional; see Step 2)
Microtiter plates or microfuge tubes, 0.5 ml and thin walled
Positive displacement pipette
Spun column of Sephadex G-75, equilibrated in TE (pH 7.6)
Thermal cycler, programmed with desired amplification

Additional Information

Generating asymmetric probes	MC3, p. 9.18
Spun-column chromatography	MC3, pp. A8.30–A8.31

METHOD

1. In a 0.5-ml thin-walled microfuge tube, set up an amplification/radiolabeling reaction containing:

10x amplification buffer	5.0 μl
10 mM dNTP solution (dATP, dGTP, and dTTP)	1.0 μl
0.1 mM dCTP	1.0 μl
20 μM forward oligonucleotide primer	2.5 μl
20 μM reverse oligonucleotide primer	2.5 μl
template DNA (2–10 ng or ~1 fmole)	5–10 μl
10 mCi/ml [α-^{32}P]dCTP	
(sp. act. 3000 Ci/mmole)	5.0 μl
H$_2$O	to 48 μl

   Add 2.5 units of thermostable DNA polymerase to the reaction mixture. Gently tap the side of the tube to mix the ingredients.

   > If more than one DNA fragment is to be radiolabeled using a single pair of primers, make up and dispense a master mix consisting of all of the reaction components except the DNA templates to the PCR tubes. Individual DNA templates can then be added to each tube just before addition of enzyme and initiation of the reaction.

2. If the thermal cycler is not fitted with a heated lid, overlay the reaction mixture with 1 drop (50 μl) of light mineral oil or a bead of paraffin wax. Place the tubes in a thermal cycler.

3. Amplify the samples using the denaturation, annealing, and polymerization times listed in the following table.

Cycle Number	Denaturation	Annealing	Polymerization
30 cycles	30–45 sec at 94°C	30–45 sec at 55–60°C	1–2 min at 72°C
Last cycle	1 min at 94°C	30 sec at 55°C	1 min at 72°C

   > Times and temperatures may need to be adapted to suit the particular reaction conditions.
   > Polymerization should be performed for 1 minute for every 1000 bp of length of the target DNA.

4. Remove the tubes from the thermal cycler. Use a micropipettor to remove as much mineral oil from the top of the reaction mixture as possible. Extract the reaction mix with 50 μl of chloroform. Separate the aqueous and organic layers by centrifugation for 1 minute at room temperature in a microfuge.

5. Remove the upper, aqueous layer to a fresh microfuge tube, add carrier tRNA (10–100 µg) or glycogen (5 µg), and precipitate the DNA with an equal volume of 4 M ammonium acetate and 2.5 volumes of ethanol. Store the tube for 1–2 hours at –20°C or for 10–20 minutes at –70°C. Collect the precipitated DNA by centrifugation at maximum speed for 5–10 minutes at 4°C.

6. Dissolve the DNA in 20 µl of TE (pH 7.6) and remove remaining unincorporated dNTPs and the oligonucleotide primers by spun-column chromatography through Sephadex G-75.

7. Use a liquid scintillation counter to measure the amount of radioactivity in 1.0 µl of the void volume of the spun column. Store the remainder of the radiolabeled DNA at –20°C until required.

Synthesis of Single-stranded DNA Probes of Defined Length from Bacteriophage M13 Templates

A synthetic oligonucleotide annealed to single-stranded DNA derived from a recombinant bacteriophage M13 or phagemid template is used to prime the synthesis of complementary radiolabeled DNA. Synthesis is catalyzed by the Klenow fragment of *E. coli* DNA polymerase I, which extends the annealed primer for various distances along the single-stranded template DNA. The products of the reaction, which are heterogeneous both in length and in the amount of incorporated radiolabeled dNTPs, are digested with a restriction enzyme to create double-stranded DNA fragments of uniform length that are subsequently purified by agarose gel electrophoresis.

The oligonucleotide primer is usually complementary to a region of the *lac* gene immediately 3′ to the polycloning site in the mp series of bacteriophage M13 vectors. This "universal" primer, which is sold by several companies, has the advantage that it can be used to prepare probes complementary to any segment of DNA that has been cloned into any restriction site in the polylinker of the vector. However, custom-made oligonucleotides complementary to specific sequences within the cloned DNA can also be used to prepare probes that represent only a portion of the cloned sequence.

MATERIALS

CAUTION: Please see Appendix 4 for appropriate handling of materials marked with <!>.

Reagents and Solutions

Please see Appendix 1 for components of stock solutions, buffers, and reagents. Dilute stock solutions to the appropriate concentrations.

Ammonium acetate (10 M) (optional; see Step 12)

dATP (40 μM and 20 mM)

Dithiothreitol (1 M), freshly made

dNTP solution (pH 8.0), containing three unlabeled deoxynucleoside triphosphates, each at a concentration of 20 mM

> The composition of this solution depends on the $[\alpha\text{-}^{32}P]dNTP$ to be used for radiolabeling. If, as in this case, radiolabeled dATP is used, the mix should contain dCTP, dTTP, and dGTP.

EDTA (0.5 M, pH 8.0)

Ethanol

NaCl (5 M)

Phenol:chloroform (1:1, v/v) <!>

5x TBE (optional; see Step 16)

Trichloroacetic acid or DE81 filters

Tris-Cl (1 M, pH 7.6) (optional; see Step 16)

Enzymes and Buffers

Klenow fragment of *E. coli* DNA polymerase I and 10x Klenow basic buffer

500 mM Tris-Cl (pH 8.0)

100 mM MgSO₄

Restriction enzyme(s)

> The restriction enzyme used to digest the products of the labeling reaction should cleave between 200 and 1 kb downstream from the primer-binding site.

Nucleic Acids/Oligonucleotides

Oligonucleotide primer
> Use a generic primer such as a commercial bacteriophage M13 universal primer (see MC3, p. 8.113) *or* a custom-synthesized primer, specific for the target sequences of interest.

Template DNA (~100 μg/ml in TE [pH 7.6])
> Purified single-stranded bacteriophage M13 or phagemid DNA (Protocol 3.4 or 3.8)

Radioactive Compounds

[α-^{32}P]dATP (sp. act. 3000 Ci/mmole at 10 mCi/ml) <!>

Gels/Loading Buffers

Alkaline agarose or denaturing polyacrylamide gel <!> (optional; see Step 13)

Additional Items

Materials for autoradiography or phosphorimaging
Spun column of Sephadex G-50, equilibrated in TE (pH 7.6)
Step 17 of this protocol requires the reagents listed in Protocol 5.4 or 5.12.
Thermal cycler
Water baths or heating blocks (68°C and an appropriate temperature for digestion with a restriction enzyme)

Additional Information

E. coli DNA polymerase I and the Klenow fragment	MC3, pp. 9.82–9.86
Spun-column chromatography	MC3, pp. A8.30–A8.31
Troubleshooting	MC3, p. 9.24
Universal primers	MC3, pp. 8.113–8.117

METHOD

1. In a 0.5-ml microfuge tube, mix:

single-stranded template (bacteriophage M13 or phagemid DNA) (~0.5 pmole)	1 μg
oligonucleotide primer	5 pmoles
10x Klenow basic buffer	3 μl
H$_2$O	to 20 μl

2. Heat the mixture to 85°C for 5 minutes in a heating block and then let it cool slowly to 37°C.

   Slow cooling using a thermal cycler programmed to heat at 85°C for 5 minutes before ramping down to 37°C over a 30-minute period can be achieved by floating the microfuge tube in a piece of Styrofoam in a 250-ml beaker filled with water equilibrated to 85°C. Place the beaker containing the primer-template mix at room temperature until the temperature of the water falls to 37°C (~30 minutes).

3. To the tube of annealed primer and template, add:

0.1 M dithiothreitol	2 μl
10 mCi/ml [α-^{32}P]dATP (sp. act. 3000 Ci/mmole)	5 μl
40 μM dATP	1 μl
20 mM solution of dTTP, dCTP, and dGTP	1 μl

Mix the reagents by gently tapping the side of the tube. Centrifuge the tube at maximum speed for 1–2 seconds in a microfuge to transfer all of the liquid to the bottom.

4. Transfer 0.5 µl of the mixture to a microfuge tube containing 15 µl of 20 mM EDTA (pH 8.0). Store the tube on ice.

5. Add 1 µl (5 units) of the Klenow fragment to the remainder of the mixture. Mix the components of the reaction by gently tapping the side of the tube. Incubate the reaction for 30 minutes at room temperature.

   Approximately 5 units of the Klenow fragment are required for each labeling reaction.

6. Transfer 0.5 µl of the reaction to a fresh microfuge tube containing 20 µl of 0.5 M EDTA (pH 8.0). Store the tube on ice.

7. Add 1 µl of 20 mM unlabeled dATP to the remainder of the reaction. Mix by gently tapping the side of the tube. Centrifuge the tube at maximum speed for 1–2 seconds in a microfuge to transfer all of the liquid to the bottom. Incubate the reaction mixture for an additional 20 minutes at room temperature.

8. During the 20-minute incubation (Step 7), measure the fraction of radioactivity in the samples stored in Steps 4 and 6 that has either become insoluble in 10% TCA or that adheres to a DE81 filter (see MC3, pp. A8.25–A8.26).

9. Heat the reaction to 68°C for 10 minutes to inactivate the Klenow fragment.

10. Adjust the concentration of NaCl in the reaction to achieve optimal conditions for cleavage of the product by the selected restriction enzyme.

    The concentration of NaCl in the primer-extension reaction is 50 mM.

    If the restriction enzyme works best in the absence of NaCl, or if the enzyme requires unusual conditions (e.g., high concentrations of Tris or the presence of detergent), transfer the radiolabeled DNA to the appropriate buffer.

11. Add 20 units of the desired restriction enzyme and incubate the reaction for 1 hour at the appropriate temperature.

12. Purify the DNA by standard extraction with phenol:chloroform and remove unincorporated dNTPs by spun-column chromatography or differential precipitation with 2.5 M ammonium acetate and ethanol. Add 0.5 M EDTA (pH 8.0) to a final concentration of 10 mM.

    IMPORTANT: It is essential that free magnesium ions not be present when the DNA is exposed to alkali in the next step. Otherwise, the DNA will form an insoluble complex with $Mg(OH)_2$.

13. Isolate the radiolabeled DNA by electrophoresis through a denaturing polyacrylamide gel (please see Protocol 12.8) or an alkaline agarose gel (please see Protocol 5.8), depending on the size of the fragment.

14. After electrophoresis, prepare the gel for phosphorimaging or autoradiography. Expose the gel to X-ray film for 5–10 minutes.

15. Align the developed film with the gel, using the images from the phosphorescent dots (Glogos [Stratagene]) or radioactive ink spots. Tape the film to the gel and mark the position of the primer-extended radioactive DNA fragment on the back of the gel plate. Remove the film and cut out the segment of the gel containing the desired fragment of DNA.

16. If a polyacrylamide gel was used to separate the DNA fragment from the template, proceed to Step 17. If an alkaline agarose gel was used, neutralize the gel by gentle shaking for 45 minutes in 0.5 M Tris-Cl (pH 7.6), followed by shaking for an additional 45 minutes in 0.5x TBE before the radiolabeled DNA is eluted.

17. Extract the DNA from the gel by electroelution or by crushing and soaking the polyacrylamide gel slice in an appropriate buffer (please see Protocols 5.4 and 5.12, respectively).

    Check the efficiency of the elution process with a handheld minimonitor.

    The eluted DNA is ready for use as a probe and does not need to be denatured.

Synthesis of Single-stranded DNA Probes of Heterogeneous Length from Bacteriophage M13 Templates

The following technique yields a heterogeneous population of short radiolabeled molecules 200–300 nucleotides in length. These probes are synthesized, as in Protocol 9.4, by extension of an oligonucleotide primer on a single-stranded DNA template. The radiolabeled products of the reaction are then separated from the template by electrophoresis through a denaturing gel from which they are eluted directly into hybridization buffer. The method is useful for synthesizing single-stranded DNA probes of very high specific activity for Southern analysis of single-copy genes in complex genomes or for northern analysis of rare mRNAs.

MATERIALS

CAUTION: Please see Appendix 4 for appropriate handling of materials marked with <!>.

Reagents and Solutions

Please see Appendix 1 for components of stock solutions, buffers, and reagents. Dilute stock solutions to the appropriate concentrations.

EDTA (0.5 M, pH 8.0)

Elution buffer (1x SSPE containing 0.5% SDS)

$MgCl_2$ (1 M)

NaOH (10 N) <!>

3x Probe synthesis buffer

    62 µM each dATP, dGTP, and dTTP

    0.5 mg/ml bovine serum albumin (Fraction V [Sigma-Aldrich])

    625 µM dithiothreitol

    32 mM Tris-Cl (pH 7.5)

    6.5 mM $MgSO_4$

Enzymes and Buffers

Klenow fragment of *E. coli* DNA polymerase I

Nucleic Acids/Oligonucleotides

Oligonucleotide primer, 5 pmole/µl

    Use a generic primer such as a commercial bacteriophage M13 universal primer (see MC3, p. 8.113) *or* a custom-synthesized primer, specific for the target sequences of interest.

Template DNA (~0.1 µg/µl)

    Purified single-stranded bacteriophage M13 or phagemid DNA (Protocols 3.4 and 3.8)

Radioactive Compounds

[α-^{32}P]dCTP (sp. act. 800–3000 Ci/mmole at 10 mCi/ml) <!>

Gels/Loading Buffers

Denaturing polyacrylamide gel <!> (7% polyacrylamide containing 7 M urea in 1x TBE; see Protocol 12.8)

Formamide loading buffer

    80% deionized formamide

    1 mg/ml bromophenol blue

    1 mg/ml xylene cyanol

    5 mM EDTA

    0.025% SDS

Centrifuges/Rotors/Tubes

Desktop clinical centrifuge
Plastic snap-top tubes (12 x 17 mm)

Additional Items

Air incubator or shaking water bath (50°C)
Boiling-water bath
Fritted column (e.g., Quik-sep column [Isolab])
Heating blocks (57°C and 37°C)
Materials for autoradiography or phosphorimaging
Razor blades

Additional Information

E. coli DNA polymerase I and the Klenow fragment	MC3, pp. 9.82–9.86
Troubleshooting	MC3, p. 9.24
Universal primers	MC3, pp. 8.113–8.117

METHOD

1. In a 0.5-ml microfuge tube, mix:

single-stranded template (bacteriophage M13 or phagemid DNA) (~0.15 pmole, ~0.1 µg/µl)	0.3 µg
oligonucleotide primer (5 pmoles/µl)	1 µl
25 mM MgCl$_2$	1 µl

   If possible, include as a positive control in each experiment a reaction tube containing a control DNA template and an oligonucleotide that has worked well previously.

2. Cap the tube and incubate the reaction mixture for 5–10 minutes at 57°C in a heating block.

3. Transfer the tube from the heating block to a bucket of ice and add to the reaction mixture:

3x probe synthesis buffer	4 µl
10 mCi/ml [α-^{32}P]dCTP (sp. act. 800–3000 Ci/mmole)	3 µl
Klenow fragment (~2.5 units)	0.5 µl

   Mix the reagents by gently tapping the side of the tube. Centrifuge the tube at maximum speed for 1–2 seconds in a microfuge. Incubate the reaction at 37°C for 40 minutes.

4. While the primer reaction is incubating, pour a 5% polyacrylamide gel containing 7 M urea in 1x TBE buffer (please see Protocol 12.8). The gel should be 1.5-mm thick and at least 15 cm long, with 2.5-cm wide slots. Prerun the gel for 15 minutes at 20–25 V/cm of gel length to remove ammonium persulfate from the wells.

5. Stop the primer-extension reaction by adding 25 µl of formamide loading buffer and heating the reaction mixture in a boiling-water bath for 3–5 minutes. Transfer the tube to an ice bucket and add 1 µl of 1 N NaOH.

6. Wash urea and loose fragments of polyacrylamide from the wells of the gel using a syringe loaded with 1x TBE and immediately load the DNA samples. Separate the radiolabeled probe from the template DNA by electrophoresis for 30 minutes at 20–25 V/cm of gel length.

   It is unnecessary to include DNA size standards on the polyacrylamide gel because the xylene cyanol dye migrates to a position comparable to that of a DNA fragment of approximately 125 bases.

7. After 30 minutes, separate the gel plates, cover the gel with Saran Wrap, and locate the radiolabeled DNA by autoradiography. Expose the gel to X-ray film for approximately 1 minute.

The majority of the radiolabeled DNA should have migrated slightly slower than the xylene cyanol (green) tracking dye. Although a continuous smear of radiolabeled material is the expected result, in practice, between three and ten bands of discrete length are usually detected that may be spaced over a distance of 2–3 cm of the gel. The cause of the discrete bands is presumed to be secondary structures in the particular single-stranded DNA template used in the reaction.

8. Cut the radioactive bands out of the gel with a clean razor blade. Place them in the bottom of a 12 × 17-mm plastic snap-cap tube and crush with a disposable inoculating stick. Add 1–2 ml of elution buffer and shake the fragments of gel for >3 hours at 50°C.

    Elution of the radioactive probe is more efficient if individual bands are cut from the gel, rather than excising a wide swath of polyacrylamide that spans all the bands.

9. Place the eluate in a fritted column, place the column in a 12 × 17-mm plastic snap-cap tube, and centrifuge in a desktop clinical centrifuge for 1–2 minutes. Measure the amount of radioactivity in 1 µl of the radioactive probe by liquid scintillation spectroscopy.

Protocol 9.6

Synthesis of Single-stranded RNA Probes by In Vitro Transcription

This protocol describes a procedure for synthesizing RNA probes of high specific activity (1 x 10^9 dpm/mg) from double-stranded linear DNAs containing promoters for bacteriophage-encoded RNA polymerases.

MATERIALS

CAUTION: Please see Appendix 4 for appropriate handling of materials marked with <!>.

Reagents and Solutions

Please see Appendix 1 for components of stock solutions, buffers, and reagents. Dilute stock solutions to the appropriate concentrations.

Ammonium acetate (10 M) (optional; see Step 8)
Bovine serum albumin (2 mg/ml; Fraction V [Sigma-Aldrich])
Dithiothreitol (1 M), freshly made
Ethanol
Phenol:chloroform (1:1, v/v) <!>
Placental RNase inhibitor (20 units/µl), e.g., RNasin (Promega)
rGTP (0.5 mM) (optional; see Step 5)
RNase-free H$_2$O (see MC3, p. 7.84)
rNTP solution, containing rATP, rCTP, and rUTP, each at a concentration of 5 mM
Sodium acetate (3 M, pH 5.2)

Vectors and Hosts

Commercially available plasmid containing bacterial promoters on both sides of the polycloning sequence (e.g., pGEM [Promega] or pBlueScript [Stratagene])
    Purify the superhelical plasmid by one of the methods described in Chapter 1.

Enzymes and Buffers

Appropriate restriction enzymes, used to linearize plasmid DNA in preparation for in vitro transcription
Bacteriophage T4 DNA polymerase (optional; see Step 2)
DNA-dependent RNA polymerase of bacteriophage T3, T7, or SP6 (commercially available)
    Usually supplied at concentrations of 10–20 units/µl together with 10x transcription buffer
Pancreatic DNase I, free of RNase (e.g., RQ1 [Promega])

Nucleic Acids/Oligonucleotides

Template DNA
    The DNA fragment to be used as a template for transcription should be cloned into one of the commercially available plasmids containing bacterial promoters on both sides of the poly-cloning sequence (e.g., pGEM [Promega] or pBlueScript [Stratagene]).

Radioactive Compounds

[α-^{32}P]rGTP (sp. act. 800–3000 Ci/mmole at 10 mCi/ml) <!>

Gels/Loading Buffers

0.8% Agarose gel, cast and run in 1x TBE
6% Neutral polyacrylamide gel <!> (optional; see Step 8)
    See Protocol 5.9 for directions on acrylamide gel electrophoresis.

Additional Items

Spun column of Sephadex G-50 in 10 mM Tris-Cl (pH 7.5)
    The column should be autoclaved before use in Step 8.
Water baths (37°C or 40°C)

Additional Information

Additional protocol (Using PCR to Add Promoters for Bacteriophage-encoded RNA Polymerases to Fragments of DNA)	MC3, p. 9.36
Bacteriophage RNA polymerases	MC3, pp. A4.28–A4.29
How to win the battle with RNase	MC3, p. 7.82
In vitro transcription systems	MC3, pp. 9.87–9.88
Plasmid vectors used for in vitro transcription	MC3, p. 9.29
Promoter sequences recognized by bacteriophage-encoded RNA polymerases	MC3, p. 9.87
Spun-column chromatography	MC3, pp. 8.30–8.31
Troubleshooting	MC3, p. 9.36

METHOD

1. Prepare 5 pmoles of linear template DNA by complete digestion of superhelical plasmid DNA with a suitable restriction enzyme. Analyze an aliquot (100 ng) of the digested DNA by agarose gel electrophoresis.

   It is essential that plasmid molecules be cleaved to completion and that no trace of circular DNA is visible.

2. If restriction enzymes, such as *Pst*I or *Sst*I, that generate protruding 3' termini must be used, treat the digested DNA with bacteriophage T4 DNA polymerase in the presence of all four dNTPs to remove the 3' protrusion (please see Protocol 9.11).

3. Purify the template DNA by extraction with phenol:chloroform and standard precipitation with ethanol. Dissolve the DNA in H_2O at a concentration of 100 nM (i.e., 200 μg/ml for a 3-kb plasmid).

4. Warm the first six components listed below to room temperature and in a sterile 0.5-ml microfuge tube, mix *in the following order at room temperature:*

template DNA	0.2 pmole (400 ng for a 3-kb plasmid)
RNase-free H_2O	to 6 μl
5 mM rNTP solution	2 μl
100 mM dithiothreitol	2 μl
10x transcription buffer (commercial)	2 μl
2 mg/ml bovine serum albumin	1 μl
10 mCi/ml [α-^{32}P]rGTP (sp. act. 800–3000 Ci/mmole)	5 μl

   Mix the components of the mixture by gently tapping the outside of the tube. Then add:

placental RNase inhibitor (10 units)	1 μl
bacteriophage DNA-dependent RNA polymerase (~10 units)	1 μl

Mix the reagents by gently tapping the outside of the tube. Centrifuge the tube for 1–2 seconds to transfer all of the liquid to the bottom. Incubate the reaction for 1–2 hours at 37°C (bacteriophages T3 and T7 DNA-dependent RNA polymerases) or 40°C (bacteriophage SP6 DNA-dependent RNA polymerase).

> The reaction may be scaled from 20 to 50 µl to accommodate more dilute reagents.

> When the reaction is carried out as described above, 80–90% of the radiolabel will be incorporated into RNA. The yield of RNA will be approximately 20 ng (sp. act. 4.7×10^9 dpm/µg) when the specific activity of the $[\alpha\text{-}^{32}\text{P}]$GTP is 3000 Ci/mmole and approximately 75 ng (sp. act. ~1.0 $\times 10^8$ dpm/µg) when the specific activity of the precursor is 800 Ci/mmole.

5. (*Optional*) If full-length transcripts are desired, add 2 µl of 0.5 mM rGTP and incubate the reaction mixture for an additional 60 minutes at the temperature appropriate for the polymerase.

6. Terminate the in vitro transcription reaction by adding 1 µl of 1 mg/ml RNase-free pancreatic DNase I to the reaction tube. Mix the reagents by gently tapping the outside of the tube. Incubate the reaction mixture for 15 minutes at 37°C.

7. Add 100 µl of RNase-free H_2O and purify the RNA by extraction with phenol:chloroform.

> If the probe will be used in experiments where length is important (e.g., RNase protection), purify the radiolabeled RNA by polyacrylamide gel electrophoresis (please see Protocol 5.9).

8. Transfer the aqueous phase to a fresh microfuge tube and separate the radiolabeled RNA from undesired small RNAs and rNTPs by one of three methods:

To purify RNA by ethanol precipitation

a. Add 30 µl of 10 M ammonium acetate to the aqueous phase. Mix and then add 250 µl of ice-cold ethanol to the tube. After storage for 30 minutes on ice, collect the RNA by centrifugation at maximum speed for 10 minutes at 4°C in a microfuge.

b. Remove as much of the ethanol as possible by gentle aspiration and leave the open tube on the bench for a few minutes to allow the last visible traces of ethanol to evaporate. Dissolve the RNA in 100 µl of RNase-free H_2O.

c. Add 2 volumes of ice-cold ethanol to the tube and store the RNA at –70°C until needed.

> To recover the RNA, transfer an aliquot of the ethanolic solution to a fresh microfuge tube. Add 0.25 volume of 10 M ammonium acetate, mix, and then store the tube for at least 15 minutes at –20°C. Centrifuge the solution at maximum speed for 10 minutes at 4°C in a microfuge. Remove the ethanol by aspiration and dissolve the RNA in the desired volume of the appropriate RNase-free buffer.

To purify RNA by spun-column chromatography

a. Prepare a Sephadex G-50 spun column that has been autoclaved in 10 mM Tris-Cl (pH 7.5).

b. Purify the RNA by spun-column chromatography.

c. Store the eluate in a microfuge tube at –70°C until the RNA is needed.

To purify RNA by gel electrophoresis

a. Prepare a neutral polyacrylamide gel according to Protocol 5.9.

b. Add the appropriate gel-loading buffer to the aqueous phase and purify the RNA by gel electrophoresis.

c. Locate the RNA by autoradiography according to Protocol 5.11.

d. Purify the RNA from the gel slice using the crush and soak method according to Protocol 5.12.

e. Store the RNA at –70°C until needed.

Synthesis of cDNA Probes from mRNA Using Random Oligonucleotide Primers

This protocol describes the generation of cDNA probes from poly(A)$^+$ mRNA using random oligonucleotide primers. Probes of this type are used for differential screening of cDNA libraries.

MATERIALS

CAUTION: Please see Appendix 4 for appropriate handling of materials marked with <!>.

Reagents and Solutions

Please see Appendix 1 for components of stock solutions, buffers, and reagents. Dilute stock solutions to the appropriate concentrations.

Ammonium acetate (10 M) (optional; see Step 7)
Dithiothreitol (1 M), freshly made
dCTP (125 μM)
dNTP solution, containing dATP, dGTP, and dTTP, each at a concentration of 20 mM
EDTA (0.5 M, pH 8.0)
Ethanol
HCl (2.5 N) <!>
NaOH (3 N) <!>
Phenol:chloroform (1:1, v/v) <!>
Placental RNase inhibitor (20 units/μl), e.g., RNasin (Promega)
RNase-free H$_2$O (see MC3, p. 7.84)
SDS (10% w/v)
Trichloroacetic acid (10%) or DE81 filters (see MC3, pp. A8.25–A8.26)
Tris-Cl (1 M, pH 7.4)

Enzymes and Buffers

Reverse transcriptase (RNA-dependent DNA polymerase) and 10x reverse transcriptase
buffer as supplied by the manufacturer or homemade
500 mM Tris-Cl (pH 8.3)
750 mM KCl
30 mM MgCl$_2$
A cloned version of reverse transcriptase encoded by Mo-MLV is the enzyme of choice, e.g.,
StrataScript (Stratagene) or SuperScript II (Invitrogen); see MC3, p. 8.48.

Nucleic Acids/Oligonucleotides

Random oligonucleotide primers, six or seven bases in length (e.g., Roche Applied
Science)
Dissolve random primers at a concentration of 0.125 μg/μl in H$_2$O.
Template poly(A)$^+$ mRNA
Prepare as described in Protocol 7.3 or 7.4 and dissolve in RNase-free H$_2$O at a concentration
of approximately 250 μg/ml.

Radioactive Compounds

[α-^{32}P]dCTP (sp. act. 3000 Ci/mmole at 10 mCi/ml) <!>

Gels/Loading Buffers

0.8% Agarose gel, cast and run in 1x TBE

6% Neutral polyacrylamide gel <!>
    See Protocol 5.9 for directions on acrylamide gel electrophoresis.

Additional Items

pH paper
Spun column of Sephadex G-50 in 10 mM Tris-Cl (pH 7.5) (optional; see Step 7)
    The column should be autoclaved before use in Step 7.
Water baths or heating blocks (45°C, 68°C, and 70°C)

Additional Information

How to win the battle with RNase	MC3, p. 7.82
Spun-column chromatography	MC3, pp. 8.30–8.31

METHOD

1. Transfer 1 μg of poly(A)$^+$ RNA to a sterile microfuge tube. Adjust the volume of the solution to 4 μl with RNase-free H$_2$O. Heat the closed tube for 5 minutes at 70°C and then quickly transfer the tube to an ice-water bath.

2. To the chilled solution in the microfuge tube, add:

10 mM dithiothreitol	2.5 μl
placental RNase inhibitor	20 units
random deoxyoligonucleotide primers	5 μl
10x reverse transcriptase buffer	2.5 μl
20 mM solution of dATP, dGTP, and dTTP	1 μl
125 μM solution of dCTP	1 μl
10 mCi/ml [α-^{32}P]dCTP (sp. act. 3000 Ci/mmole)	10 μl
RNase-free H$_2$O	to 24 μl
reverse transcriptase (200 units)	1 μl

   IMPORTANT: Add the reverse transcriptase last.

   Reverse transcriptase supplied by different manufacturers varies in its activity per unit. When using a new batch of enzyme, set up a series of extension reactions containing equal amounts of poly(A)$^+$ RNA and oligonucleotide primer and different amounts of enzyme. If possible, the primer should be specific for an mRNA present at moderate abundance in the preparation of poly(A)$^+$ RNA. Assay the products of each reaction by gel electrophoresis as described in this protocol. Use the minimal amount of enzyme required to produce the maximum yield of extension product. The units used in this protocol work well with most batches of StrataScript and SuperScript II.

   Mix the components by gently tapping the side of the tube. Remove bubbles by brief centrifugation in a microfuge. Incubate the reaction mixture for 1 hour at 45°C.

   As an alternative, [α-^{32}P]dCTP of specific activity 800 Ci/mmole can be substituted in this reaction. If this substitution is made, omit the 125 μM dCTP from the reaction mixture.

3. Stop the reaction by adding:

0.5 M EDTA (pH 8.0)	1 μl
10% (w/v) SDS	1 μl

   Mix the reagents in the tube completely.

4. Add 3 μl of 3 N NaOH to the reaction tube. Incubate the mixture for 30 minutes at 68°C to hydrolyze the RNA.

5. Cool the reaction mixture to room temperature. Neutralize the solution by adding 10 μl of 1 M Tris-Cl (pH 7.4), mixing well, and then adding 3 μl of 2.5 N HCl. Check the pH of the solution by spotting a very small amount on pH paper.

6. Purify the cDNA by extraction with phenol:chloroform.

7. Separate the radiolabeled probe from the unincorporated dNTPs by either spun-column chromatography or selective precipitation by ethanol in the presence of 2.5 M ammonium acetate.

8. Measure the proportion of radiolabeled dNTPs that is either incorporated into material precipitated by TCA or that adheres to a DE81 filter.

> If a larger amount of radiolabeled cDNA is required, scale up the reaction by increasing the volumes of all components proportionally. It is important to maintain a ratio of 200 units of reverse transcriptase/μg of input mRNA to ensure maximum yield.
>
> The purified radiolabeled cDNA can be used for hybridization without denaturation. Use 5×10^7 dpm of radiolabeled cDNA for each 150-mm filter and 5×10^6 to 1×10^7 dpm for each 90-mm filter.

Synthesis of Radiolabeled, Subtracted cDNA Probes Using Oligo(dT) as Primer

This protocol describes the preparation of subtracted cDNA probes by hybridization to an mRNA driver, followed by purification of the single-stranded radiolabeled cDNA by hydroxyapatite chromatography. For additional information, please see MC3, p. 9.89. Before preparing the probe, it is a good idea to have filters (which contain the cDNA library to be screened) ready to hybridize.

MATERIALS

CAUTION: Please see Appendix 4 for appropriate handling of materials marked with <!>.

Reagents and Solutions

Please see Appendix 1 for components of stock solutions, buffers, and reagents. Dilute stock solutions to the appropriate concentrations.

Ammonium acetate (10 M)

Dithiothreitol (1 M), freshly made

dCTP (125 µM)

dNTP solution, containing dATP, dGTP, and dTTP, each at a concentration of 20 mM

EDTA (0.5 M, pH 8.0)

Ethanol

HCl (2.5 N) <!>

Isobutanol <!>

NaOH (3 N) <!>

Phenol:chloroform (1:1, v/v) <!>

Placental RNase inhibitor (20 units/µl), e.g., RNasin (Promega)

RNase-free H_2O (see MC3, p. 7.84)

SDS (10% w/v)

SDS/EDTA solution
    30 mM EDTA (pH 8.0)
    1.2% (w/v) SDS

Sodium phosphate buffer (2 M, pH 6.8)

SPS buffer
    0.12 M sodium phosphate buffer (pH 6.8)
    0.1% (w/v) SDS

Trichloroacetic acid (10%) or DE81 filters (see MC3, pp. A8.25–A8.26)

Tris-Cl (1 M, pH 7.4)

Enzymes and Buffers

Reverse transcriptase (RNA-dependent DNA polymerase) and 10x reverse transcriptase
    buffer as supplied by the manufacturer or homemade
        500 mM Tris-Cl (pH 8.3)
        750 mM KCl
        30 mM $MgCl_2$
        A cloned version of reverse transcriptase encoded by Mo-MLV is the enzyme of choice, e.g.,
        StrataScript (Stratagene) or SuperScript II (Invitrogen); see MC3, p. 8.48.

Nucleic Acids/Oligonucleotides

Driver mRNA

Oligo(dT)$_{12-18}$ (1 mg/ml in TE [pH 7.6])

Random oligonucleotide primers can be substituted in the cDNA synthesis reaction (see Protocol 9.7). The choice of primer depends on the method used to construct the cDNA library that is to be screened. Use oligo(dT) in this protocol if the library was constructed from cDNA whose synthesis had been primed by oligo(dT). Use random primers if the library was constructed in any other way.

Template poly(A)+ mRNA

Prepare as described in Protocol 7.3 or 7.4 and dissolve in RNase-free H$_2$O at a concentration of approximately 250 µg/ml.

Radioactive Compounds

[α-^{32}P]dCTP (sp. act. 3000 Ci/mmole at 10 mCi/ml) <!>

Additional Items

Apparatus for separation of single- and double-stranded nucleic acids by hydroxyapatite chromatography at 60°C (see MC3, pp. A8.32–A8.35)
Boiling-water bath
Glass capillary tubes (20 µl), siliconized
Ice-water bath
Microfuge tubes (1.5 ml), siliconized
pH paper
Sephadex G-50 column (5 ml), equilibrated in TE (pH 8.0)
Spun column of Sephadex G-50 in 10 mM Tris-Cl (pH 7.5) containing 0.1% (w/v) SDS
Water baths or heating blocks (45°C, 68°C, and 70°C)

Additional Information

How to win the battle with RNase	MC3, p. 7.82
Hydroxyapatite chromatography	MC3, pp. A8.32–A8.34
Sargent T.D. 1987. Isolation of differentially expressed genes. *Meth. Enzymol.* **152**: 423–432.	
Spun-column chromatography	MC3, pp. A8.30–A8.31
Trichloroacetic acid and DE81 filters	MC3, pp. A8.25–A.8.26

METHOD

1. Transfer 5–10 µg of poly(A)+ RNA to a sterile microfuge tube. Adjust the volume of the solution to 40 µl with RNase-free H$_2$O. Heat the closed tube to 70°C for 5 minutes and then quickly transfer the tube to an ice-water bath.

2. To the chilled microfuge tube, add:

0.1 M dithiothreitol	2.5 µl
placental RNase inhibitor	200 units
oligo(dT)$_{12-18}$	10 µl
10x reverse transcriptase buffer	25 µl
20 mM solution of dATP, dGTP, and dTTP	10 µl
125 µM dCTP	10 µl
10 mCi/ml [α-^{32}P]dCTP (sp. act. 3000 Ci/mmole)	100 µl
RNase-free H$_2$O	to 240 µl
reverse transcriptase (2000 units)	10 µl

IMPORTANT: Add the reverse transcriptase last.

Mix the components by gently tapping the side of the tube. Collect the reaction mixture in the bottom of the tube by brief centrifugation in a microfuge. Incubate the reaction for 1 hour at 45°C.

3. Stop the reaction by adding:

0.5 M EDTA (pH 8.0)	10 μl
10% (w/v) SDS	10 μl

   Mix the reagents in the tube well.

4. Add 30 μl of 3 N NaOH to the reaction tube. Incubate the mixture for 30 minutes at 68°C to hydrolyze the RNA.

5. Cool the mixture to room temperature. Neutralize the solution by adding 100 μl of 1 M Tris-Cl (pH 7.4), mixing well, and then adding 30 μl of 2.5 N HCl. Check the pH of the solution by spotting <1 μl on pH paper.

6. Purify the cDNA by extraction with phenol:chloroform.

7. Measure the proportion of radiolabeled dNTPs that is either incorporated into material precipitated by TCA or that adheres to a DE81 filter (see MC3, pp. A8.25–A8.26). Calculate the yield of cDNA as follows:

   In a reaction containing 1.5 nmoles of the limiting dNTP:

   $$\frac{\text{cpm incorporated}}{\text{total cpm}} \times 1.5 \text{ nmoles dCTP} \times 330 \text{ ng/nmole} \times 160 = \text{ng of cDNA synthesized}$$

8. Separate the radiolabeled probe from the unincorporated dNTPs by chromatography through a 5-ml column of Sephadex G-50.

   IMPORTANT: Perform this step and all subsequent steps with siliconized tubes.

9. To the radiolabeled cDNA, add tenfold excess by weight of the driver RNA that will be used to subtract the cDNA probe, 0.2 volume of 10 M ammonium acetate, and 2.5 volumes of ice-cold ethanol. Incubate the mixture for 10–15 minutes at 0°C and then recover the nucleic acids by centrifugation at maximum speed for 5 minutes at 4°C in a microfuge.

10. Remove all of the ethanol by aspiration and store the open tube on the bench for a few minutes. Dissolve the nucleic acids in 6 μl of RNase-free H_2O.

11. To the dissolved nucleic acids, add:

2 M sodium phosphate (pH 6.8)	2 μl
SDS/EDTA solution	2 μl

12. Draw the mixture into a siliconized, disposable 20-μl glass capillary tube and seal the ends of the tube in the flame of a Bunsen burner.

13. Place the microfuge tube or sealed capillary tube in a boiling-water bath for 5 minutes. Transfer to a water bath set at 68°C and allow the nucleic acids to hybridize to $C_{r_0}t = 1000$ moles seconds/liter.

    To calculate the time required to reach $C_{r_0}t$, solve the following equation for *t*:

    $$D/D_o = e^{-kC_{r_0}t}$$

    where *D* is the remaining single-stranded cDNA at time *t*, D_o is the total amount of input cDNA, *e* is the natural logarithm, *k* is a rate constant for the formation of RNA-DNA hybrids that is dependent on the complexity of the mRNA population and may be assumed to be approximately 6.7×10^{-3} liters/mol-sec, and C_{r_0} is the initial concentration of the RNA driver (which does not change appreciably during the hybridization reaction).

14. Remove the microfuge or capillary tube from the water bath. Use a drawn-out pipette tip attached to a micropipettor to remove the hybridization solution from the microfuge tube, or open the ends of the capillary tube with a file or diamond pen. Transfer the hybridization mixture into a tube containing 1 ml of SPS buffer.

15. Separate the single- and double-stranded nucleic acids by chromatography on hydroxyapatite at 60°C.

Measure the amount of radioactivity in each fraction by liquid scintillation counting. At least 90% of the input [^{32}P]cDNA should have hybridized to the mRNA and be present in the >0.36 M sodium phosphate wash.

16. Pool the fractions containing the single-stranded cDNA and concentrate them by *repeated* extractions with isobutanol as follows: Add an equal volume of isobutanol. Mix the two phases by vortexing and centrifuge the mixture at maximum speed for 2 minutes at room temperature in a microfuge. Discard the upper (organic) phase. Repeat the extraction with isobutanol until the volume of the aqueous phase is <100 µl.

17. Remove salts from the cDNA by spun-column chromatography through Sephadex G-50 equilibrated in TE (pH 8.0) containing 0.1% SDS.

    IMPORTANT: Do not use ethanol precipitation to concentrate the cDNA because the presence of phosphate ions interferes with precipitation. Do not use dialysis to remove phosphate ions because the cDNA will stick to the dialysis bag.

18. Measure the amount of radioactivity in the sample and calculate the weight of DNA in the subtracted probe.

19. Repeat Steps 9–18.

    Between 10% and 30% of the cDNA will form hybrids with the driver RNA during the second round of hybridization.

    It is not necessary to concentrate or remove salts from the final preparation of cDNA if it is to be used to probe a cDNA library. The radiolabeled cDNA can be used for hybridization without denaturation. The subtractive hybridizations should be carried out as rapidly as practicable, and the probe should be used without delay. Use 5×10^7 dpm of radiolabeled cDNA for each 150-mm filter and 5×10^6 to 1×10^7 dpm for each 90-mm filter.

    If a genomic DNA library is screened with the radiolabeled subtracted probe, oligo(dA) can be added to the prehybridization and hybridization reactions at 1 µg/ml to prevent nonspecific hybridization between the oligo(dT) tails of the cDNA and oligo(dA) tracts in the genomic DNA.

In this procedure, synthesis of cDNA is performed in the presence of saturating concentrations of all four dNTPs and trace amounts of a single radiolabeled dNTP. After subtraction hybridization, the enriched single-stranded cDNA is radiolabeled to high specific activity in a second synthetic reaction by extension of random oligonucleotide primers using the Klenow fragment of *E. coli* DNA polymerase I. Because the concentrations of dNTP in the first reaction are nonlimiting, both the amounts and size of cDNA generated are greater than those achieved in standard labeling protocols.

MATERIALS

CAUTION: Please see Appendix 4 for appropriate handling of materials marked with <!>.

Reagents and Solutions

Please see Appendix 1 for components of stock solutions, buffers, and reagents. Dilute stock solutions to the appropriate concentrations.

Dithiothreitol (1 M), freshly made

dNTP solution, containing all four dNTPs, each at a concentration of 5 mM

dNTP solution, containing three dNTPs (dCTP, dGTP, and dTTP), each at a concentration of 5 M

EDTA (0.5 M, pH 8.0)

Ethanol

HCl (2.5 N) <!>

Isobutanol

Phenol:chloroform (1:1, v/v) <!>

Placental RNase inhibitor (20 units/μl), e.g., RNasin (Promega)

RNase-free H_2O (see MC3, p. 7.84)

SDS (20% w/v)

Sodium acetate (3 M, pH 5.2)

Trichloroacetic acid (10%) or DE81 filters (see MC3, pp. A8.25–A8.26)

Tris-Cl (1 M, pH 7.4)

Enzymes and Buffers

Klenow fragment of *E. coli* DNA polymerase I and 5x random priming buffer

250 mM Tris-Cl (pH 8.0)

25 mM $MgCl_2$

100 mM NaCl

10 mM dithiothreitol

1 M HEPES (adjusted to pH 6.6 with 4 M NaOH)

Reverse transcriptase (RNA-dependent DNA polymerase) and 10x reverse transcriptase buffer as supplied by the manufacturer or homemade

500 mM Tris-Cl (pH 8.3)

750 mM KCl

30 mM $MgCl_2$

A cloned version of reverse transcriptase encoded by Mo-MLV is the enzyme of choice, e.g., StrataScript (Stratagene) or SuperScript II (Invitrogen); see MC3, p. 8.48.

Nucleic Acids/Oligonucleotides

Driver mRNA

Oligo(dT)$_{12-18}$ (1 mg/ml in TE [pH 7.6])

Random oligonucleotide primers, 6–7 nucleotides in length (0.125 µg/µl in H$_2$O)

Template poly(A)$^+$ mRNAs

Prepare two sets of template RNAs:

(1) poly(A)$^+$ RNA, prepared from cells or tissues that express the genes of interest

(2) poly(A)$^+$ RNA, prepared from cells or tissues that do not express the genes of interest

Both preparations should be enriched for poly(A)$^+$ species by two passes through an oligo(dT) column (see Protocol 7.3).

Radioactive Compounds

[α-^{32}P]dATP (10 mCi/ml, sp. act. 3000 Ci/mmole) <!>

[α-^{32}P]dCTP (10 mCi/ml, sp. act. 800–3000 Ci/mmole) <!>

Additional Items

Apparatus for separation of single- and double-stranded nucleic acids by hydroxyapatite chromatography (60°C) (see MC3, pp. A8.32–A8.35)

Boiling-water bath

Glass capillary tubes (20 µl), silconized

Ice-water bath

Microfuge tubes (1.5 ml), siliconized

pH paper

Sephadex G-50 column (5 ml), equilibrated in TE (pH 8.0)

Spun column of Sephadex G-50 in 10 mM Tris-Cl (pH 7.5), containing 0.1% (w/v) SDS

Water baths or heating blocks (45°C, 60°C, and 68°C)

Steps 8 and 9 of this protocol require the reagents listed in Protocol 9.8

Additional Information

How to win the battle with RNase	MC3, p. 7.82
Spun-column chromatography	MC3, pp. A8.30–A8.31
Trichloroacetic acid and DE81 filters	MC3, pp. A8.25–A.8.26

METHOD

1. To synthesize first-strand cDNA, mix the following ingredients at 4°C in a sterile microfuge tube:

template RNA (1 mg/ml)	10 µl
oligo(dT)$_{12-18}$ (1 mg/ml)	10 µl
5 mM dNTP solution (complete)	10 µl
50 mM dithiothreitol	1 µl
10x reverse transcriptase buffer	5 µl
10 mCi/ml [α-^{32}P]dCTP (sp. act. 800–3000 Ci/mmole)	5 µl
placental RNase inhibitor	25 units
RNase-free H$_2$O	to 46 µl
reverse transcriptase (~800 units)	4 µl

IMPORTANT: Add the reverse transcriptase last.

Mix the components by gently tapping the side of the tube. Collect the reaction mixture in the bottom of the tube by brief centrifugation in a microfuge. Incubate the reaction for 1 hour at 45°C.

[α-^{32}P]dCTP is used as a tracer to measure the synthesis of the first strand of cDNA.

2. Measure the proportion of radiolabeled dNTPs that is either incorporated into material precipitated by TCA or that adheres to a DE81 filter. Calculate the yield of cDNA using the equation below.

   In a reaction containing 50 nmoles of each dNTP:

   $$\frac{\text{cpm incorporated}}{\text{total cpm}} \times 200 \text{ nmoles dNTP} \times 330 \text{ ng/nmole} = \text{ng of cDNA synthesized}$$

3. Stop the reaction by adding:

0.5 M EDTA (pH 8.0)	2 µl
20% (w/v) SDS	2 µl

   Mix the reagents in the tube completely.

   > The single-stranded, radiolabeled cDNA is quite sticky and adheres nonspecifically to glass, filters, and some plastics. For this reason, it is important to maintain a minimum of 0.05% (w/v) SDS in Step 3 and 0.1–1.0% SDS in hybridization buffers.

4. Add 5 µl of 3 N NaOH to the reaction tube. Incubate the mixture for 30 minutes at 68°C to hydrolyze the RNA.

5. Cool the mixture to room temperature. Neutralize the solution by adding 10 µl of 1 M Tris-Cl (pH 7.4), mixing well, and then adding 5 µl of 2.5 N HCl. Check the pH of the solution by spotting <1 µl on pH paper.

6. Purify the cDNA by extraction with phenol:chloroform.

7. Separate the radiolabeled probe from the unincorporated dNTPs by chromatography through a spun column of Sephadex G-50.

   IMPORTANT: Perform this step and all subsequent steps with siliconized tubes.

8. Perform two rounds of subtractive hybridization as described in Protocol 9.8, Steps 9–19.

9. Concentrate the final preparation of cDNA by sequential extractions with isobutanol and remove salts by chromatography on Sephadex G-50 as described in Protocol 9.8, Steps 16 and 17.

10. Recover the cDNA by standard precipitation with ethanol. Dissolve the cDNA in H_2O at a concentration of 15 ng/µl.

    IMPORTANT: Do not attempt to precipitate the cDNA with ethanol before removing the phosphate ions by spun-column chromatography.

    > The subtracted cDNA prepared through Step 10 of this protocol can be converted into double-stranded DNA (please see Protocol 11.1 Stage 2) and cloned into a bacteriophage or plasmid vector to produce a subtracted cDNA library.

11. To radiolabel the subtracted cDNA to high specific activity, mix the following in a 0.5-ml microfuge tube:

subtracted cDNA	5 µl
random deoxynucleotide primers (125 µg/ml)	5 µl

12. Heat the mixture to 60°C for 5 minutes and then cool to 4°C.

13. To the primer:cDNA template mixture, add:

5x random primer buffer	10 µl
5 mM dNTP solution of dCTP, dGTP, and dTTP	5 µl
10 mCi/ml [α-^{32}P]dATP (sp. act. 3000 Ci/mmole)	25 µl
Klenow fragment (12.5 units)	2.5 µl
H_2O	to 50 µl

    Incubate the reaction for 4–6 hours at room temperature.

    > 10–15 units of the Klenow fragment are required in each random priming reaction.

14. Stop the reaction by adding:

0.5 M EDTA (pH 8.0)	1 μl
20% (w/v) SDS	2.5 μl

15. Separate the radiolabeled cDNA from the unincorporated dNTPs by spun-column chromatography through Sephadex G-50.

    The radiolabeled cDNA should be denatured by heating to 100°C for 5 minutes before it is used for hybridization. Use 5×10^7 dpm of radiolabeled cDNA for each 138-mm filter and 5×10^6 to 1×10^7 dpm for each 82-mm filter. Once radiolabeled, use the probe immediately to avoid damage by radiochemical decay.

Labeling 3′ Termini of Double-stranded DNA Using the Klenow Fragment of *E. coli* DNA Polymerase I

Templates for end-labeling are produced by digesting DNA with a restriction enzyme that creates recessed 3′ termini. The Klenow fragment of *E. coli* DNA polymerase I is then used to catalyze the incorporation of one or more $[\alpha\text{-}^{32}P]$dNTPs into the recessed 3′ termini in an end-filling reaction. Because the Klenow enzyme works well under a wide variety of conditions, there is usually no need to change buffers or to purify the DNA fragment at the end of restriction reaction.

The choice of which $[\alpha\text{-}^{32}P]$dNTP(s) to use as substrates for the reaction depends on the sequence of the protruding 5′ terminus. One or both ends of a linear double-stranded DNA can be labeled depending on the nature of the termini and on the radiolabeled nucleotide chosen as substrate. In most circumstances, it is best to include unlabeled dNTPs in the reaction. This allows the recessed 3′ termini to be filled completely and to be labeled at any position. In addition, unlabeled dNTPs incorporated downstream from the labeled dNTP shield the labeled nucleotide from the indolent 3′→5′ exonuclease of the Klenow fragment. For additional information on end-filling, see MC3, p. 9.52.

MATERIALS

CAUTION: Please see Appendix 4 for appropriate handling of materials marked with <!>.

Reagents and Solutions

Please see Appendix 1 for components of stock solutions, buffers, and reagents. Dilute stock solutions to the appropriate concentrations.
Ammonium acetate (10 M) (optional; see Step 4)
dNTP solution, containing the appropriate dNTPs, each at a concentration of 1 mM
Ethanol (optional; see Step 4)

Enzymes and Buffers

Klenow fragment of *E. coli* DNA polymerase I

Nucleic Acids/Oligonucleotides

Template DNA (0.1–5 µg), carrying at least one recessed 3′ terminus

Radioactive Compounds

$[\alpha\text{-}^{32}P]$dNTP (10 mCi/ml, sp. act. 800–3000 Ci/mmole) <!>

Additional Items

Sephadex G-50 spun column, equilibrated in TE (pH 7.6)
Water bath (75°C)

Additional Information

E. coli DNA polymerase I and the Klenow fragment	MC3, pp. 9.82–9.86
Methods of labeling the ends of DNA	MC3, pp. 9.55–9.56
Spun-column chromatography	MC3, pp. A8.30–A8.31

METHOD

1. Digest 0.1–5 µg of template DNA with the desired restriction enzyme in 25–50 µl of the appropriate restriction enzyme buffer.

   > The labeling reaction may be performed immediately after digesting the DNA with a restriction enzyme. It is not usually necessary to change buffers.

2. To the completed restriction digest, add:

10 mCi/ml [α-^{32}P]dNTP (sp. act. 800–3000 Ci/mmole)	2–50 µCi
unlabeled dNTPs	to a final concentration of 100 µM
Klenow fragment	1–5 units

   Incubate the reaction for 15 minutes at room temperature.

   > Approximately 0.5 unit of the Klenow enzyme is required for each µg of template DNA (1 µg of a 1000-bp fragment is equivalent to ~3.1 pmoles of termini of double-stranded DNA).

   > Reverse transcriptase (1–2 units) can be used in place of the Klenow enzyme in this protocol. However, reverse transcriptase is not as forgiving of buffer conditions as the Klenow enzyme and is therefore used chiefly to label purified DNA fragments in reactions containing conventional reverse transcriptase buffer.

   > If the labeled DNA is to be used for sequencing by the Maxam-Gilbert technique (please see Protocol 12.7) or for mapping mRNA by the nuclease S1 method (please see Protocol 7.10), the concentration of labeled dNTP in the reaction should be increased to the greatest level that is practicable. After the reaction has been allowed to proceed for 15 minutes at room temperature, add all four unlabeled dNTPs to a final concentration of 0.2 mM for each dNTP, and continue the incubation for an additional 5 minutes at room temperature.

3. Stop the reaction by heating it for 10 minutes at 75°C.

4. Separate the radiolabeled DNA from unincorporated dNTP by spun-column chromatography through Sephadex G-50 or by two rounds of precipitation with ethanol in the presence of 2.5 M ammonium acetate.

Labeling 3′ Termini of Double-stranded DNA with Bacteriophage T4 DNA Polymerase

Bacteriophage T4 DNA polymerase, unlike the Klenow enzyme, rapidly digests 3′-protruding termini and then continues at a slower pace to remove 3′ nucleotides from the double-stranded portion of the DNA substrate. Consequently, in the absence of dNTPs, the enzyme will degrade double-stranded molecules to about half-length single strands. However, in the presence of high concentrations of dNTPs, recessed 3′-hydroxyl termini generated by exonucleolytic activity act as primers for template-directed addition of mononucleotides by the 5′→3′ polymerase. Because the synthetic capacity of bacteriophage T4 DNA polymerase exceeds its exonucleolytic abilities, protruding 3′ termini are converted to termini with flush ends. The reaction therefore consists of cycles of removal and replacement of the 3′-terminal nucleotides from recessed or blunt-ended DNA. If one of the four dNTPs is radiolabeled, the resulting blunt-ended double-stranded molecules will be labeled at or near their 3′ termini.

MATERIALS

CAUTION: Please see Appendix 4 for appropriate handling of materials marked with <!>.

Reagents and Solutions

Please see Appendix 1 for components of stock solutions, buffers, and reagents. Dilute stock solutions to the appropriate concentrations.

Ammonium acetate (10 M) (optional; see Step 6)

dNTP solution, containing one unlabeled dNTP at a concentration of 2 mM
> The dNTP used in this solution should contain the same base as the [α-^{32}P]dNTP.

dNTP solution, containing the three unlabeled dNTPs, each at a concentration of 2 mM
> The three dNTPs to be included depend on the sequence of the restriction site and on the [α-^{32}P]dNTP used in the labeling reaction.

Ethanol (optional; see Steps 1 and 6)

Phenol:chloroform (1:1, v/v) <!> (optional; see Step 1)

Enzymes and Buffers

Appropriate restriction enzyme(s)

Bacteriophage T4 DNA polymerase and 10x buffer
> 300 mM Tris acetate (pH 8.0)
> 600 mM potassium acetate
> 100 mM magnesium acetate
> 5 mM dithiothreitol
> 1 mg/ml bovine serum albumin (Fraction V [Sigma-Aldrich])

Nucleic Acids/Oligonucleotides

Template DNA (0.1–5 µg), carrying the appropriate blunt or protruding 3′ terminus

Radioactive Compounds

[α-^{32}P]dNTP (10 mCi/ml, sp. act. 800–3000 Ci/mmole) <!>

Additional Items

Sephadex G-50 spun column, equilibrated in TE (pH 7.6) (optional; see Step 6)

Water baths or heating blocks (37°C and 70°C)

Additional Information

Bacteriophage T4 DNA polymerase	MC3, pp. A4.18–A4.21
Methods of labeling the ends of DNA	MC3, pp. 9.55–9.56
Spun-column chromatography	MC3, pp. A8.30–A8.31

METHOD

1. Digest 0.1–5.0 µg of the template DNA with a restriction enzyme(s) that generates a blunt or 3′-protruding end.

    In many cases, digestion can be performed in T4 DNA polymerase buffer, allowing the digestion and labeling reactions to be carried out sequentially without an intermediate extraction with phenol:chloroform and precipitation with ethanol.

2. If necessary, purify the DNA by extraction with phenol:chloroform and precipitation with ethanol in the presence of 2.5 M ammonium acetate.

3. Dissolve the DNA pellet in:

10x bacteriophage T4 DNA polymerase buffer	2 µl
2 mM solution of three unlabeled dNTPs	1 µl
10 mCi/ml [α-^{32}P]dNTP (800–3000 Ci/mmole)	1 µl
bacteriophage T4 DNA polymerase (2.5 units/µl)	1 µl
H$_2$O	to 20 µl

    Incubate the reaction for 5 minutes at 37°C.

4. Add 1 µl of a 2 mM solution of the unlabeled fourth dNTP. Continue the incubation for an additional 10 minutes.

5. Stop the reaction by heating it to 70°C for 5 minutes.

6. Separate the labeled DNA from unincorporated dNTPs by spun-column chromatography through Sephadex G-50 or by two rounds of precipitation with ethanol in the presence of 2.5 M ammonium acetate.

End-labeling Protruding 3′ Termini of Double-stranded DNA with [α-³²P]Cordycepin 5′-Triphosphate or [α-³²P]DideoxyATP

The 3′-protruding termini of DNA, generated by cleavage with restriction enzymes such as *Pst*I or *Sac*I, can be labeled using calf thymus terminal transferase to catalyze the transfer of [α-³²P]dideoxyATP or [α-³²P]cordycepin triphosphate. Because neither of these nucleotide analogs carries a 3′-hydroxyl group, no additional nucleotides can be added to the modified protruding 3′ terminus.

MATERIALS

CAUTION: Please see Appendix 4 for appropriate handling of materials marked with <!>.

Reagents and Solutions

Please see Appendix 1 for components of stock solutions, buffers, and reagents. Dilute stock solutions to the appropriate concentrations.

Ammonium acetate (10 M) (optional; see Step 6)
Ethanol (optional; see Step 6)
Phenol:chloroform (1:1, v/v) <!>

Enzymes and Buffers

Appropriate restriction enzyme(s)
Calf thymus terminal transferase and 5x buffer
500 mM potassium cacodylate
10 mM CoCl₂ · 6H₂O
1 mM dithiothreitol
For advice on assembling this buffer, see MC3, p. A1.11.

Nucleic Acids/Oligonucleotides

Template DNA (0.1–5 μg), carrying the appropriate protruding 3′ terminus

Radioactive Compounds

[α-³²P]Cordycepin 5′-triphosphate (10 mCi/ml, sp. act. 5000 Ci/mmole) <!>
or
[α-³²P]DideoxyATP dNTP (10 mCi/ml, sp. act. 3000 Ci/mmole) <!>

Additional Items

Sephadex G-50 spun column, equilibrated in TE (pH 7.6) (optional; see Step 6)
Water bath or heating block (37°C)

Additional Information

Calf thymus deoxynucleotidyl terminal transferase	MC3, pp. A1.11 and A4.27
Cordycepin (3′ adenosine), an antibiotic, is isolated from the culture fluids of *Cordyceps militaris* (a club-shaped mushroom)	
Methods of labeling the ends of DNA	MC3, pp. 9.55–9.56
Spun-column chromatography	MC3, pp. A8.30–A8.31

METHOD

1. Digest 0.1–5 µg of template DNA with the appropriate restriction enzyme.

2. Purify the DNA by extraction with phenol:chloroform and standard precipitation with ethanol.

3. Dissolve the digested DNA in 10 µl of 5x terminal transferase buffer and 34 µl of H_2O.

4. Add 5 µl of 10 mCi/ml [α-^{32}P]cordycepin 5'-triphosphate (5000 Ci/mmole) or [α-^{32}P]dideoxy-ATP (3000 Ci/mmole) and 1 µl of calf thymus terminal transferase (~20 units).

   20 units of calf thymus terminal transferase are required to catalyze the radiolabeling of approximately 10 pmoles of protruding 3′ termini.

5. Incubate the reaction for 1 hour at 37°C.

6. Separate the labeled DNA from unincorporated [α-^{32}P]cordycepin 5'-triphosphate (or [α-^{32}P]dideoxyATP) by spun-column chromatography through Sephadex G-50 or by two rounds of precipitation with ethanol in the presence of 2.5 M ammonium acetate.

Dephosphorylation of DNA Fragments with Alkaline Phosphatase

Essentially any alkaline phosphatase (e.g., bacterial alkaline phosphatase [BAP], calf intestinal phosphatase [CIP], placental alkaline phosphatase, and shrimp alkaline phosphatase [SAP]) will catalyze the removal of 5′-phosphate residues from nucleic acid templates. Both CIP and SAP can easily be inactivated and they are the most widely used phosphatases in molecular cloning. Although CIP is cheaper per unit of activity, SAP has the advantage of being readily inactivated in the absence of chelators.

MATERIALS

CAUTION: Please see Appendix 4 for appropriate handling of materials marked with <!>.

Reagents and Solutions

Please see Appendix 1 for components of stock solutions, buffers, and reagents. Dilute stock solutions to the appropriate concentrations.

Chloroform <!>
Carrier (glycogen [50 mg/ml in H_2O] or linear polyacrylamide [1 mg/ml in H_2O])
  (optional; see Step 2e)
EDTA (0.5 M, pH 8.0)
EGTA (0.5 M, pH 8.0) (optional; see Step 2d)
Ethanol
Phenol:chloroform (1:1, v/v) <!>
SDS (10% w/v) (if using CIP)
Sodium acetate (3 M, pH 7.0 [if using CIP] and pH 5.2 [if using SAP])
TE (pH 7.6)
Tris-Cl (1 M, pH 8.5)

Enzymes and Buffers

Alkaline phosphatase (CIP or SAP) and 10x buffers
  10x Dephosphorylation buffer (CIP)
    100 mM Tris-Cl (pH 8.3)
    10 mM $MgCl_2$
    10 mM $ZnCl_2$
  10x Dephosphorylation buffer (SAP)
    100 mM Tris-Cl (pH 8.8)
    100 mM $MgCl_2$
    10 mM $ZnCl_2$
Proteinase K
Restriction enzyme(s) and 10x buffer(s)

Nucleic Acids/Oligonucleotides

Substrate DNA (1.0–10 μg [10–100 pmoles])

Additional Items

Water baths or heating blocks (37°C, 56°C, 65°C, 75°C, or 70°C)

Additional Information

Alkaline phosphatase	MC3, pp. 9.92–9.93
Calculating the amount of 5′ ends in a DNA sample	MC3, p. 9.63
Carriers used during precipitation of DNA	MC3, p. A8.13
Proteinase K	MC3, p. A4.50

METHOD

1. Use the restriction enzyme of choice to digest to completion 1–10 μg (10–100 pmoles) of the DNA to be dephosphorylated.

   CIP and SAP will dephosphorylate DNA at a slightly reduced efficiency in restriction buffers that have been adjusted to pH 8.5 with 10x CIP or 10x SAP buffer. Alternatively, the restricted DNA may be purified by extraction with phenol:chloroform and standard precipitation with ethanol and then dissolved in a minimal volume of 10 mM Tris-Cl (pH 8.5).

2. Dephosphorylate the 5′ ends of the restricted DNA with either CIP or SAP.

To dephosphorylate DNA using CIP

a. Add to the DNA:

10x CIP dephosphorylation buffer	5 μl
H₂O	to 48 μl

b. Add the appropriate amount of CIP.

   One unit of CIP will dephosphorylate approximately 1 pmole of 5′-phosphorylated termini (5′-recessed or blunt-ended DNA) or approximately 50 pmoles of 5′-protruding termini. These amounts may vary slightly from one manufacturer to the next.

c. Incubate the reaction for 30 minutes at 37°C, add a second aliquot of CIP, and continue incubation for an additional 30 minutes.

d. To inactivate CIP at the end of the incubation period, add SDS and EDTA (pH 8.0) to final concentrations of 0.5% and 5 mM, respectively. Mix the reagents well and add proteinase K to a final concentration of 100 μg/ml. Incubate for 30 minutes at 56°C.

   Alternatively, CIP can be inactivated by heating to 65°C for 30 minutes (or 75°C for 10 minutes) in the presence of 10 mM EGTA (pH 8.0).
   IMPORTANT: Use EGTA, not EDTA.

e. Cool the reaction to room temperature and purify the DNA by extracting it twice with phenol:chloroform and once with chloroform alone.

   Proteinase K and SDS used to inactivate and digest CIP must be completely removed by extraction with phenol:chloroform before subsequent enzymatic treatments (phosphorylation by polynucleotide kinase, ligation, etc.).

   A carrier (glycogen or linear polyacrylamide) can be added before phenol:chloroform extraction if small amounts of DNA (<100 ng) were used in the reaction. Do not add carrier nucleic acid (tRNA, salmon sperm DNA, etc.) because it will compete with the dephosphorylated DNA for the radiolabeled ATP during any subsequent kinasing reactions.

To dephosphorylate DNA using SAP

a. Add to the DNA:

10x SAP dephosphorylation buffer	5 μl
H₂O	to 48 μl

b. Add the appropriate amount of SAP.

> One unit of SAP will dephosphorylate approximately 1 pmole of 5'-phosphorylated termini (3'-recessed or 5'-recessed DNA) or approximately 0.2 pmole of blunt-ended DNA. These amounts may vary slightly from one enzyme manufacturer to the next.

c. Incubate the reaction for 1 hour at 37°C.

d. To inactivate SAP, transfer the reaction to 70°C, incubate for 20 minutes, and cool to room temperature.

3. Transfer the aqueous phase to a clean microfuge tube and recover the DNA by standard ethanol precipitation in the presence of 0.1 volume of 3 M sodium acetate (pH 5.2) if SAP was used or 0.1 volume of 3 M sodium acetate (pH 7.0) if CIP was used.

4. Allow the precipitate to dry at room temperature before dissolving it in TE (pH 7.6) at a DNA concentration of >2 nmoles/ml.

Phosphorylation of DNA Molecules with Protruding 5′-Hydroxyl Termini

Bacteriophage T4 polynucleotide kinase catalyzes the transfer of [γ-^{32}P]ATP to the terminal 5′-hydroxyl groups of single- or double-stranded nucleic acids. When [γ-^{32}P]ATP of high specific activity (3000–7000 Ci/mmole) is used as a substrate, approximately 40–50% of the protruding 5′ termini in the reaction becomes radiolabeled.

MATERIALS

CAUTION: Please see Appendix 4 for appropriate handling of materials marked with <!>.

Reagents and Solutions

Please see Appendix 1 for components of stock solutions, buffers, and reagents. Dilute stock solutions to the appropriate concentrations.

Ammonium acetate (10 M) (optional; see Step 3)

EDTA (0.5 M, pH 8.0)

Ethanol (optional; see Step 3)

Enzymes and Buffers

Bacteriophage T4 polynucleotide kinase lacking 3′-phosphatase activity (e.g., Roche Applied Science)

10x Polynucleotide kinase buffer

700 mM Tris-Cl (pH 7.4)

100 mM $MgCl_2$

50 mM dithiothreitol

Nucleic Acids/Oligonucleotides

Substrate DNA (10–50 pmoles), dephosphorylated

Radioactive Compounds

[γ-^{32}P]ATP (10 mCi/ml, sp. act. 3000–7000 Ci/mmole) <!>

Additional Items

Liquid scintillation spectrometer, capable of quantifying ^{32}P by Cerenkov radiation

Sephadex G-50 spun column, equilibrated in TE (pH 7.6) (optional; see Step 3)

or

Sephadex G-50 column (1 ml), equilibrated in TE (pH 7.6) (optional; see Step 3)

Water bath or heating block (37°C)

Additional Information

Bacteriophage T4 polynucleotide kinase	MC3, pp. A4.30 and A4.35–A4.36
Calculating the amount of 5′ ends in a DNA sample	MC3, p. 9.63
Labeling the 5′ termini of DNA with bacteriophage T4 polynucleotide kinase	MC3, p. 9.67
Methods of labeling the ends of DNA	MC3, pp. 9.55–9.56
Spun-column chromatography	MC3, pp. A8.30–A8.31

METHOD

1. In a microfuge tube, mix the following reagents:

dephosphorylated DNA	10–50 pmoles
10x bacteriophage T4 polynucleotide kinase buffer (10–20 units)	5 μl
10 mCi/ml [γ-^{32}P]ATP (sp. act. 3000–7000 Ci/mmole)	50 pmoles
bacteriophage T4 polynucleotide kinase	10 units
H$_2$O	to 50 μl

   Incubate the reaction for 1 hour at 37C.

   Ideally, ATP should be in a fivefold molar excess over DNA 5' ends, and the concentration of DNA termini should be ≥0.4 μM. The concentration of ATP in the reaction should therefore be >2 μM, but this is rarely achievable in practice. To increase the specific activity of the radiolabeled DNA product, increase the amount of [γ-^{32}P]ATP used in the phosphorylation reaction. Decrease the volume of H$_2$O to maintain a reaction volume of 50 μl.

2. Terminate the reaction by adding 2 μl of 0.5 M EDTA (pH 8.0). Measure the total radioactivity in the reaction mixture by Cerenkov counting in a liquid scintillation counter.

3. Separate the radiolabeled probe from unincorporated dNTPs by

   - spun-column chromatography through Sephadex G-50

   or

   - conventional size-exclusion chromatography through 1-ml columns of Sephadex G-50 (equilibrated in TE)

   or

   - two rounds of selective precipitation of the radiolabeled DNA with ammonium acetate and ethanol

4. Measure the amount of radioactivity in the probe preparation by Cerenkov counting. Calculate the efficiency of transfer of the radiolabel to the 5' termini by dividing the amount of radioactivity in the probe by the total amount in the reaction mixture (Step 2).

Phosphorylation of DNA Molecules with Dephosphorylated Blunt Ends or Recessed 5′ Termini

Blunt ends and recessed 5′ termini of DNA are labeled less efficiently in reactions catalyzed by T4 DNA kinase than are protruding 5′ termini. However, the efficiency of phosphorylation of blunt-ended DNAs greater than 300 bp in length can be increased by including a condensing reagent such as polyethylene glycol in the reaction.

MATERIALS

CAUTION: Please see Appendix 4 for appropriate handling of materials marked with <!>.

Reagents and Solutions

Please see Appendix 1 for components of stock solutions, buffers, and reagents. Dilute stock solutions to the appropriate concentrations.
Ammonium acetate (10 M) (optional; see Step 5)
EDTA (0.5 M, pH 8.0)
Ethanol (optional; see Step 3)

Enzymes and Buffers

Bacteriophage T4 polynucleotide kinase lacking 3′-phosphatase activity (e.g., Roche Applied Science)
10x Imidazole buffer
    500 mM imidazole·Cl (pH 6.4)
    180 mM $MgCl_2$
    50 mM dithiothreitol
    1 mM spermidine HCl
    1 mM EDTA
Polyethylene glycol (24% [w/v] PEG 8000) in H_2O

Nucleic Acids/Oligonucleotides

Substrate DNA (10–50 pmoles), dephosphorylated as described in Protocol 9.13

Radioactive Compounds

[γ-^{32}P]ATP (10 mCi/ml, sp. act. 3000 Ci/mmole) <!>

Additional Items

Liquid scintillation spectrometer, capable of quantifying ^{32}P by Cerenkov radiation
Sephadex G-50 spun column, equilibrated in TE (pH 7.6) (optional; see Step 3)
or
Sephadex G-50 column (1 ml), equilibrated in TE (pH 7.6) (optional; see Step 3)
Water bath or heating block (37°C)

Additional Information

Bacteriophage T4 polynucleotide kinase	MC3, pp. A4.30 and A4.35–A4.36
Calculating the amount of 5′ ends in a DNA sample	MC3, p. 9.63
Condensing and crowding agents	MC3, p. 1.152

Labeling the 5′ termini of DNA with bacteriophage	
T4 polynucleotide kinase	MC3, p. 9.67
Methods of labeling the ends of DNA	MC3, pp. 9.55–9.56
Polyethylene glycol	MC3, p. A1.28
Spun-column chromatography	MC3, pp. A8.30–A8.31

METHOD

1. In a microfuge tube, mix *in the following order:*

dephosphorylated DNA	10–50 pmoles
10x imidazole buffer	4 µl
H_2O	to 15 µl
24% (w/v) PEG	10 µl

2. Add 40 pmoles of $[\gamma\text{-}^{32}P]ATP$ (10 mCi/ml, sp. act. 3000 Ci/mmole) to the tube and bring the final volume of the reaction to 40 µl with H_2O.

   Ideally, ATP should be in a fivefold molar excess over DNA 5′ ends, and the concentration of DNA termini should be ≥0.4 µM. The concentration of ATP in the reaction should therefore be >2 µM, but this is rarely achievable in practice. To increase the specific activity of the radiolabeled DNA product, increase the amount of $[\gamma\text{-}^{32}P]ATP$ used in the phosphorylation reaction. Decrease the volume of H_2O to maintain a reaction volume of 40 µl.

3. Add 40 units of bacteriophage T4 polynucleotide kinase to the reaction. Mix the reagents gently by tapping the side of tube and incubate the reaction for 30 minutes at 37°C.

4. Terminate the reaction by adding 2 µl of 0.5 M EDTA (pH 8.0). Measure the total radioactivity in the reaction mixture by Cerenkov counting in a liquid scintillation counter.

5. Separate the radiolabeled probe from unincorporated dNTPs by

   • spun-column chromatography through Sephadex G-50

   or

   • conventional size-exclusion chromatography through 1-ml columns of Sephadex G-50 (equilibrated in TE)

   or

   • two rounds of selective precipitation of the radiolabeled DNA with ammonium acetate and ethanol

6. Measure the amount of radioactivity in the probe preparation by Cerenkov counting. Calculate the efficiency of transfer of the radiolabel to the 5′ termini by dividing the amount of radioactivity in the probe by the total amount in the reaction mixture (Step 4).

Phosphorylation of DNA Molecules with Protruding 5′ Termini by the Exchange Reaction

In the exchange reaction, an excess of ADP causes bacteriophage T4 polynucleotide kinase to transfer the terminal 5′ phosphate from phosphorylated DNA to ADP; the DNA is then rephosphorylated by transfer of a radiolabeled γ phosphate from [γ-^{32}P]ATP. Although the exchange reaction catalyzed by bacteriophage T4 polynucleotide kinase does not require that the 5′ termini of DNA substrates be dephosphorylated, the efficiency of the exchange reaction is poor unless crowding reagents such as polyethylene glycol are included in the reaction mixture.

MATERIALS

CAUTION: Please see Appendix 4 for appropriate handling of materials marked with <!>.

Reagents and Solutions

Please see Appendix 1 for components of stock solutions, buffers, and reagents. Dilute stock solutions to the appropriate concentrations.
ADP (1 mM in 25 mM Tris-Cl [pH 8.0])
Ammonium acetate (10 M) (optional; see Step 3)
ATP (10 mM in 25 mM Tris-Cl [pH 8.0])
EDTA (0.5 M, pH 8.0)
Ethanol (optional; see Step 3)

Enzymes and Buffers

Bacteriophage T4 polynucleotide kinase lacking 3′-phosphatase activity (e.g., Roche Applied Science)
10x Imidazole buffer
500 mM imidazole·Cl (pH 6.4)
180 mM MgCl$_2$
50 mM dithiothreitol
1 mM spermidine HCl
1 mM EDTA
Polyethylene glycol (24% [w/v] PEG 8000) in H$_2$O

Nucleic Acids/Oligonucleotides

Substrate DNA (10–50 pmoles), dephosphorylated in a volume of <11 μl

Radioactive Compounds

[γ-^{32}P]ATP (10 mCi/ml, sp. act. 3000–7000 Ci/mmole) <!>

Additional Items

Liquid scintillation spectrometer, capable of quantifying ^{32}P by Cerenkov radiation
Sephadex G-50 spun column, equilibrated in TE (pH 7.6) (optional; see Step 3)
or
Sephadex G-50 column (1 ml), equilibrated in TE (pH 7.6) (optional; see Step 3)
Water bath or heating block (37°C)

Additional Information

Bacteriophage T4 polynucleotide kinase   MC3, pp. A4.30 and A4.35–A4.36

Adapted from Chapter 9, Protocol 16, p. 9.73 of MC3.

METHOD

1. Mix the following reagents in a microfuge tube *in the following order:*

DNA with 5′ terminal phosphates	10–50 pmoles
10x imidazole buffer	5 μl
1 mM ADP	5 μl
50 nM ATP	1 μl
10 mCi/ml [γ-^{32}P]ATP (sp. act. 3000–7000 Ci/mmole)	20–100 pmoles
H$_2$O	to 40 μl
24% (w/v) PEG	10 μl
bacteriophage T4 polynucleotide kinase (20 units)	1 μl

   Mix reagents gently by tapping the side of tube and incubate the reaction for 30 minutes at 37°C.

2. Terminate the reaction by adding 2 μl of 0.5 M EDTA (pH 8.0). Measure the total radioactivity in the reaction mixture by Cerenkov counting in a liquid scintillation counter.

3. Separate the radiolabeled probe from unincorporated dNTPs by

   - spun-column chromatography through Sephadex G-50

   or

   - conventional size-exclusion chromatography through 1-ml columns of Sephadex G-50 (equilibrated in TE)

   or

   - two rounds of selective precipitation of the radiolabeled DNA with ammonium acetate and ethanol

4. Measure the amount of radioactivity in the probe preparation by Cerenkov counting. Calculate the efficiency of transfer of the radiolabel to the 5′ termini by dividing the amount of radioactivity in the probe by the total amount in the reaction mixture (Step 2).

CHAPTER 10

Working with Synthetic Oligonucleotide Probes

BACKGROUND INFORMATION

Background information found in *Molecular Cloning: A Laboratory Manual*, 3rd edition (hereafter MC3) unless otherwise indicated.

Methods used to purify synthetic oligonucleotides	MC3, pp. 10.48–10.49
Oligonucleotide synthesis	MC3, pp. 10.42–10.46
Working with synthetic oligonucleotide probes	MC3, pp. 10.1–10.10

As a rule of thumb, oligonucleotides >25 nucleotides in length should be purified by polyacrylamide gel electrophoresis. The oligonucleotide can then be eluted from the gel and concentrated by reversed-phase chromatography on Sep-Pak C_{18} columns.

MATERIALS

CAUTION: Please see Appendix 4 for appropriate handling of materials marked with <!>.

Reagents and Solutions

Please see Appendix 1 for components of stock solutions, buffers, and reagents. Dilute stock solutions to the appropriate concentrations.

Acetonitrile (HPLC grade) <!>

Ammonium acetate (10 M)

n-Butanol <!>

Methanol:H_2O solution <!>

    6 ml methanol

    4 ml filter-sterilized deionized H_2O

Oligonucleotide elution buffer

    0.5 M ammonium acetate

    10 mM magnesium acetate

    Some investigators include 0.1% SDS in their oligonucleotide elution buffers. This is not advisable if the eluted oligonucleotide is to be purified by Sep-Pak C_{18} chromatography.

TE (pH 8.0)

Nucleic Acids/Oligonucleotides

Crude preparation of synthetic oligonucleotide, usually supplied as a lyophilized powder after removal of protecting groups

    When opening a tube of oligonucleotide for the first time, vent the tube by opening it slowly to allow ammonia gas to escape (into a chemical hood). Otherwise much of the oligonucleotide may spray around the room.

    If the oligonucleotide is supplied in NH_4OH, transfer 0.5–1.0-ml aliquots to microfuge tubes and evaporate the NH_4OH to dryness on a centrifugal evaporator.

Gels/Loading Buffers

Denaturing polyacrylamide gel <!>, cast and run in 1x TBE

    The percentage of acrylamide in the solution used to form the gel and the conditions under which electrophoresis is performed vary according to the size of the oligonucleotides. Table 10-1 on p. 404 provides useful guidelines.

Formamide loading buffer without tracking dyes <!>

    Deionized formamide <!>

    10 mM EDTA

    Dyes (bromophenol blue and xylene cyanol FF) are omitted from the loading buffer because they may migrate through the gel at the same rate as the oligonucleotide and interfere with its detection by absorption of UV light. If desired, 0.2% orange G can be included in the loading buffer. This dye migrates with the buffer front and does not interfere with detection of the oligonucleotide.

Formamide-tracking dye mixture

    5 ml deionized formamide <!>

    2.5 ml 0.05% xylene cyanol FF

    2.5 ml 0.05% bromophenol blue

The dyes in the mixture are used as a size standards and are run in wells of the gel adjacent to those containing the oligonucleotide.

Additional Items

Centrifugal evaporator (Savant or equivalent)

Millex HV filter (Millipore, 0.45-μm pore size)

Parafilm

or

Fluorescent thin-layer chromatographic plate (F_{254}, 20 x 20 cm) (Brinkmann or Merck)

Razor blade

Saran Wrap and plastic-backed bench paper

Sep-Pak C_{18} classic columns, short body (Waters Division of Millipore)

> These columns contain 360 mg of a hydrophobic reversed-phase chromatography resin. The oligonucleotide binds to the column when the polarity of the solvent is high and elutes when the polarity is reduced. One column is required per 10 OD_{260} units of oligonucleotide loaded onto the polyacrylamide gel.

Shaker incubator (37°C)

Steps 7–9 require the reagents listed in Protocols 12.8 and 12.11.

Syringes (5 and 10 cc), polypropylene

Ultraviolet lamp (260 nm, handheld) <!>

Water bath or heating block (55°C)

Additional Information

Methods used to purify synthetic oligonucleotides	MC3, pp. 10.48–10.49
Oligonucleotide synthesis	MC3, pp. 10.42–10.46
Visualizing oligonucleotides in polyacrylamide gels	MC3, p. 10.16

METHOD

1. In a sterile microfuge tube, prepare a 10-μM solution of the crude oligonucleotide in sterile, filtered H_2O (Milli-Q or its equivalent). Vortex the solution thoroughly.

   > The solution is often slightly cloudy because of the presence of insoluble benzamides generated during the synthesis of the oligonucleotide.

2. Centrifuge the tube at maximum speed for 5 minutes at room temperature in a microfuge. Transfer the supernatant to a fresh, sterile microfuge tube.

3. Extract the solution three times in succession with 400 μl of *n*-butanol. Discard the upper (organic) phase after each extraction.

4. Evaporate the solution to dryness in a centrifugal evaporator (Savant SpeedVac or its equivalent). The tube should contain a yellowish pellet and a creamy-white powder.

5. Dissolve the pellet and powder in 200 μl of sterile filtered H_2O (Milli-Q or its equivalent).

6. Estimate the amount of oligonucleotide in the preparation as follows: Add 1 μl of the solution to 1 ml of H_2O. Mix the solution well and read the OD_{260}. Calculate the oligonucleotide concentration as follows.

   > Calculate the millimolar extinction coefficient of the oligonucleotide (ε) from the following equation:
   >
   > $$\varepsilon = A(15.2) + G(12.01) + C(7.05) + T(8.4)$$
   >
   > where *A*, *G*, *C*, and *T* are the number of times each nucleotide is represented in the sequence of the oligonucleotide. The numbers in parentheses are the molar extinction coefficients for each deoxynucleotide at pH 8.0.
   >
   > For example, a 19-mer containing five dA residues, four dG residues, four dC residues, and six dT residues would have a millimolar extinction coefficient of

$$(5 \times 15.2) + (4 \times 12.01) + (4 \times 7.05) + (6 \times 8.4) = 202.64 \text{ mM}^{-1} \text{ cm}^{-1}$$

Calculate the concentration (c) of the undiluted solution of oligonucleotide from the following equation:

$$c = (OD_{260})(1000)/\varepsilon$$

7. Pour a denaturing polyacrylamide gel (as described in Protocol 12.8) of the appropriate concentration (see Table 10-1). The loading slots in the gel should be approximately 1 cm in length.

TABLE 10-1 Range of Resolution of Gels Containing Different Concentrations of Acrylamide

Acrylamide (%)	Size of Oligonucleotides (in Bases)
20–30	2–8
15–20	8–25
13–15	15–35
10–13	35–45
8–10	45–70
6–8	70–300

8. Run the gel at constant wattage (50–70 W) for approximately 45 minutes or until the temperature of the gel reaches 45–50°C. Turn off the power supply and disconnect the electrodes.

   Prerunning the gel in this way causes ammonium persulfate to migrate from the wells and, more importantly, warms the gel to a temperature optimal for electrophoresis of DNA.

9. Without delay, load approximately 2 OD$_{260}$ units of oligonucleotide (in a volume of 10 μl or less for maximum resolution) onto one or more slots of the gel as follows:

   a. Add an equal volume of formamide loading buffer lacking dyes to the oligonucleotide solution. Mix the reagents well by vortexing and then heat the mixture to 55°C for 5 minutes to disrupt secondary structure.

   b. Flush out the urea from the wells with 1× TBE.

   c. Load the heated oligomer into the slots. Load 5 μl of formamide-tracking dye mixture into an unused slot.

   For further details on loading polyacrylamide gels, please see Protocol 12.11.

10. Run the gel at 1500 V until the oligonucleotide has migrated approximately two thirds of the length of the gel.

    The position of the oligonucleotide may be estimated from the positions of the tracking dyes as detailed in Table 10-2. Note that a synthetic oligonucleotide carrying a hydroxyl residue at its 5′ terminus migrates more slowly through a denaturing polyacrylamide gel than does a phosphorylated oligonucleotide of equivalent length. Furthermore, the electrophoretic mobility of an oligonucleotide is dependent on its base composition and sequence. Thus, there may not be an exact correspondence between the predicted and observed positions of the oligonucleotide in the polyacrylamide gel.

TABLE 10-2 Approximate Lengths of Oligonucleotides Comigrating with Tracking Dyes

Polyacrylamide (%)	Xylene Cyanol FF	Bromophenol Blue
20	22	6
15	30	9–10
12	40	~15

11. Lay the gel mold flat on plastic-backed protective bench paper with the smaller (notched) plate uppermost. Allow the gel to cool to <37°C before proceeding.

12. Remove any remaining pieces of electrical tape from the gel plates. Use a spacer or a plate-separating tool to slowly and gently pry apart the plates of the mold. The gel should remain attached to the longer (nonsiliconized) glass plate.

If the gel adheres to both plates, replace the partially dislodged, smaller, or notched plate back on the gel, invert the plates, and try again.

13. Place a piece of Saran Wrap on the gel, turn the glass plate over, and transfer the gel to the Saran Wrap. Place a piece of Parafilm or a fluorescent thin-layer chromatographic plate under the gel where the oligonucleotide is predicted to be.

14. Use a handheld UV lamp to examine the gel by illumination from above at 260 nm.

    The DNA in the gel absorbs the UV radiation and appears as dark blue bands against a uniform fluorescent background contributed by the Parafilm or chromatographic plate. If the DNA is difficult to visualize, take the gel into a darkened room and illuminate it with the handheld UV lamp.

15. Recover the desired oligonucleotide, which should be the slowest-migrating band (i.e., closest to the top of the gel), by excising each DNA band with a sharp, clean scalpel or razor blade. Avoid taking UV-absorbing material smaller in length than the desired oligonucleotide.

16. Transfer the gel slices to three or four microfuge tubes. Add 1 ml of oligonucleotide elution buffer to each tube. Crush the slices with a disposable pipette tip, using a circular motion and pressing the fragments of gel against the sides of the tubes. Seal the tubes well. Incubate the tubes for 12 hours at 37°C in a shaker incubator.

17. Centrifuge the tubes at maximum speed for 5 minutes at room temperature in a microfuge. Pool the supernatants, transfer them to a 5-cc disposable syringe, and pass them through a Millex-HV filter (Millipore). Collect the effluent in a 15-ml polypropylene tube.

18. Prepare a Sep-Pak C_{18} reversed-phase column as follows:

    a. Attach the barrel of a disposable 10-cc polypropylene syringe to the longer end of a Sep-Pak C_{18} classic column.

    b. Add 10 ml of acetonitrile to the barrel and slowly push it through the column with the plunger of the syringe.

    c. Remove the syringe from the Sep-Pak column and then take the plunger out of the barrel. This prevents air from being pulled back into the column. Reattach the barrel to the column.

    d. Add 10 ml of sterile filtered H_2O (Milli-Q or its equivalent) to the barrel and slowly push it through the column with the plunger. Repeat Step c.

    e. Add 2 ml of 10 mM ammonium acetate to the barrel and push it slowly through the column. Again remove the syringe, remove the barrel, and reattach the barrel to the column. The column is now ready for use.

19. Add the solution containing the gel-purified oligonucleotide (from Step 17) to the barrel and slowly push it through the column with the plunger. Collect the effluent in a sterile 50-ml polypropylene tube. Repeat Step 18c.

20. Add 10 ml of H_2O to the barrel and push it slowly through the column with the plunger. Repeat this wash step twice more.

21. Elute the bound oligonucleotide from the Sep-Pak column with three aliquots of 1 ml of methanol:H_2O solution. Repeat Step 18c after each elution. Collect each effluent in a separate microfuge tube. Read the OD_{260} of the solution in each of the three microfuge tubes, using the methanol:H_2O solution as a blank. More than 90% of the oligonucleotide applied to the column should elute in the first fraction.

22. Evaporate the solution containing the oligonucleotide to dryness in a centrifugal evaporator.

23. Dissolve the oligonucleotide in a total volume of 200 µl of H_2O or TE (pH 8.0).

24. Transfer 5 µl of the solution to a cuvette containing 995 µl of H_2O. Mix the contents of the cuvette and read the OD_{260} of the diluted sample. Calculate the amount of oligonucleotide present in the total solution (Step 23) as described in Step 6 of this protocol.

Synthetic oligonucleotides lacking phosphate groups at their 5′ termini are easily radiolabeled by transfer of the γ-^{32}P from [γ-^{32}P]ATP in a reaction catalyzed by bacteriophage T4 polynucleotide kinase. The reaction described below is designed to label 10 pmoles of an oligonucleotide to high specific activity. Labeling of different amounts of oligonucleotide can easily be achieved by increasing or decreasing the size of the reaction while keeping the concentrations of all components constant. When the reaction is performed efficiently, >50% of the oligonucleotide molecules in the reaction become radiolabeled. Similar reaction conditions can be used when adding a nonradiolabeled phosphate to the 5′ end of a synthetic oligonucleotide before its use in site-directed mutagenesis.

MATERIALS

CAUTION: Please see Appendix 4 for appropriate handling of materials marked with <!>.

Reagents and Solutions

Please see Appendix 1 for components of stock solutions, buffers, and reagents. Dilute stock solutions to the appropriate concentrations.

Tris-Cl (1 M, pH 8.0)

Enzymes and Buffers

Bacteriophage T4 polynucleotide kinase lacking 3′-phosphatase activity (e.g., Roche Applied Science)

10x Polynucleotide kinase buffer
700 mM Tris-Cl (pH 7.4)
100 mM MgCl$_2$
50 mM dithiothreitol

Nucleic Acids/Oligonucleotides

Synthetic oligonucleotide, preferably purified as described in Protocol 10.1
Crude preparations of oligonucleotides are labeled with lower efficiency. If using unpurified preparations, make sure that the last cycle of synthesis was programmed to "trityl-off." Otherwise the resident dimethyltrityl group will efficiently protect the 5′-hydroxyl group from phosphorylation by polynucleotide kinase.

Radioactive Compounds

[γ-^{32}P]ATP (10 mCi/ml, sp. act. 3000–7000 Ci/mmole) <!>
10 pmoles of [γ-^{32}P]ATP is required to label the 5′-hydroxyl termini of 10 pmoles of oligonucleotide.

Additional Items

DE81 filters
or
Sephadex G-15 or Biogel P-60 columns
Step 4 requires the reagents listed in Protocol 10.5 or 13.7.
Water bath or heating block (37°C and 68°C)

Additional Information

Bacteriophage T4 polynucleotide kinase	MC3, pp. A4.30, A4.35, and A4.36
Calculating the amount of 5′ ends in a DNA sample	MC3, p. 9.63
Labeling the 5′ termini of DNA with bacteriophage T4 polynucleotide kinase	MC3, p. 9.67
Methods of labeling the ends of DNA	MC3, pp. 9.55–9.56

METHOD

1. Set up a reaction mixture in a 0.5-ml microfuge tube containing:

synthetic oligonucleotide (10 pmoles/μl)	1 μl
10x bacteriophage T4 polynucleotide kinase buffer	2 μl
[γ-^{32}P]ATP (10 pmoles, sp. act. 3000–7000 Ci/mmole)	5 μl
H$_2$O	11.4 μl

   Mix the reagents well by gentle but persistent tapping on the outside of the tube. Place 0.5 μl of the reaction mixture in a tube containing 10 μl of 10 mM Tris-Cl (pH 8.0). Set aside the tube for use in Step 4.

   To label an oligonucleotide to the highest specific activity,

   - Increase the concentration of [γ-^{32}P]ATP in the reaction by a factor of 3 (i.e., use 15 μl of radio-label and decrease the volume of H$_2$O to 1.4 μl).

   - Decrease the amount of oligonucleotide to 3 pmoles.

   Under these circumstances, only about 10% of the radiolabel is transferred, but a high proportion of the oligonucleotide becomes radiolabeled.

2. Add 10 units (~1 μl) of bacteriophage T4 polynucleotide kinase to the remaining reaction mixture. Mix the reagents well and incubate the reaction mixture for 1 hour at 37°C.

3. At the end of the incubation period, place 0.5 μl of the reaction in a second tube containing 10 μl of 10 mM Tris-Cl (pH 8.0). Heat the remainder of the reaction for 10 minutes at 68°C to inactivate the polynucleotide kinase. Store the tube containing the heated reaction mixture on ice.

4. Before proceeding, determine whether the labeling reaction has worked well by measuring the fraction of the radiolabel that has been transferred to the oligonucleotide substrate in a small sample of the reaction mixture. Transfer a sample of the reaction (exactly 0.5 μl) to a fresh tube containing 10 μl of 10 mM Tris-Cl (pH 8.0). Use this sample (along with the two aliquots set aside in Steps 1 and 3) to measure the efficiency of transfer of the γ-^{32}P from ATP by one of the following methods:

   - Measure the proportion of the radiolabel that binds to DE81 filters. Oligonucleotides bind tightly to the positively charged filters, whereas [γ-^{32}P]ATP does not. For details of this method, please see Protocol 13.7.

     or

   - Measure the efficiency of the labeling reaction by estimating the fraction of label that migrates with the oligonucleotide during size-exclusion chromatography through Sephadex G-15 or Biogel P-60 columns. For details of this method, please see Protocol 10.5. In some ways, this is the easier of the two methods because the relative amounts of incorporated and unincorporated radioactivity can be estimated during chromatography on a handheld minimonitor.

5. If the specific activity of the oligonucleotide is acceptable, purify the radiolabeled oligonucleotide as described in Protocol 10.3, 10.4, 10.5, or 10.6.

   If the specific activity is too low, add an additional 8 units of polynucleotide kinase, continue incubation for an additional 30 minutes at 37°C (i.e., a total of 90 minutes), heat the reaction for 10 minutes at 68°C to inactivate the enzyme, and analyze the products of the reaction again, as described in Step 4.

Purification of Radiolabeled Oligonucleotides by Precipitation with Ethanol

If radiolabeled oligonucleotides are to be used only as probes in hybridization experiments, complete removal of unincorporated radiolabel is generally unnecessary. However, to reduce background, it is always advisable to separate the bulk of the unincorporated radioactivity from the oligonucleotide. Most of the residual radioactive precursors can be removed from the preparation by differential precipitation with ethanol if the oligonucleotide is more than 18 nucleotides in length (this protocol) or with cetylpyridinium bromide, regardless of the length of the oligonucleotide (Protocol 10.4). If complete removal of the unincorporated radiolabel is required (e.g., when the radiolabeled oligonucleotide will be used in primer extension reactions), chromatographic methods (Protocols 10.5 and 10.6) or gel electrophoresis (essentially as described in Protocol 10.1) should be used.

MATERIALS

CAUTION: Please see Appendix 4 for appropriate handling of materials marked with <!>.

Reagents and Solutions

Please see Appendix 1 for components of stock solutions, buffers, and reagents. Dilute stock solutions to the appropriate concentrations.

Ammonium acetate (10 M)
Ethanol (100% and 80% v/v)
TE (pH 7.6)

Nucleic Acids/Oligonucleotides

Radiolabeled oligonucleotide <!>

The starting material for this protocol is the reaction mixture (~20 µl) generated in Protocol 10.2 (Step 2 or 5), after it has been heated to inactivate bacteriophage T4 polynucleotide kinase.

METHOD

1. Add 40 µl of H_2O to the tube containing the radiolabeled oligonucleotide (~20 µl). After mixing, add 240 µl of a 5 M solution of ammonium acetate. Mix the reagents again and then add 750 µl of ice-cold ethanol. Mix the reagents once more and store the ethanolic solution for 30 minutes at 0°C.

2. Recover the radiolabeled oligonucleotide by centrifugation at maximum speed for 20 minutes at 4°C in a microfuge.

3. Use a micropipettor equipped with a disposable tip to remove all of the supernatant carefully from the tube.

4. Add 500 µl of 80% ethanol to the tube, tap the side of the tube to rinse the nucleic acid pellet, and centrifuge the tube again at maximum speed for 5 minutes at 4°C in a microfuge.

5. Use a micropipettor equipped with a disposable tip to remove the supernatant (which will contain appreciable amounts of radioactivity) carefully from the tube. Stand the open tube behind a Plexiglas screen until the residual ethanol has evaporated.

6. Dissolve the radiolabeled oligonucleotide in 100 μl of TE (pH 7.6).

> The radiolabeled oligonucleotide may be stored for a few days at –20°C. However, during prolonged storage, decay of ^{32}P causes radiochemical damage that can impair the ability of the oligonucleotide to hybridize to its target sequence.

Purification of Radiolabeled Oligonucleotides by Precipitation with Cetylpyridinium Bromide

Radiolabeled nucleic acids, including oligonucleotides, can be separated from unincorporated radiolabel by quantitative, differential precipitation with the cationic detergent cetylpyridinium bromide (CPB). The nucleic acids are first precipitated from aqueous solution with CPB. The detergent is then removed from the precipitate with ethanol (in which the nucleic acids are insoluble), and the nucleic acids are finally dissolved in the buffer of choice.

MATERIALS

CAUTION: Please see Appendix 4 for appropriate handling of materials marked with <!>.

Reagents and Solutions

Please see Appendix 1 for components of stock solutions, buffers, and reagents. Dilute stock solutions to the appropriate concentrations.

Cetylpyridinium bromide (CPB) (1%, w/v) <!>

> Dissolve 1 g of CPB (Sigma-Aldrich) in 100 ml of H_2O. To speed dissolution, heat the solution on a hot plate on a hot magnetic stirring plate. After all of the detergent has dissolved, cool the solution to room temperature and filter it through a nitrocellulose filter (0.45 μm).

EDTA-Tris (0.5 M, pH 6.0)

> Add 0.1 mole of EDTA to 60 ml of H_2O. While stirring the solution, slowly add Tris base (powder) until the pH of the solution reaches 6.0. By this stage, the concentration of Tris will be approximately 1.2 M. Adjust the volume of the solution to 200 ml with H_2O.

EDTA-Tris-DNA solution

> Dissolve carrier DNA (e.g., salmon sperm DNA) at a concentration of 50 μg/ml in 50 mM EDTA-Tris (pH 6.0) (i.e., a 1:10 dilution of the 0.5 M EDTA-Tris solution).

Ethanol–sodium acetate solution

> 80% (v/v) ethanol
> 20% (v/v) sodium acetate (0.1 M, pH 5.2)

TE (pH 7.6) or an appropriate prehybridization solution

Nucleic Acids/Oligonucleotides

Radiolabeled oligonucleotide <!>

> The starting material for this protocol is the reaction mixture (~20 μl) generated in Protocol 10.2 (Step 2 or 5), after it has been heated to inactivate bacteriophage T4 polynucleotide kinase.

Additional Items

Dry-ice/ethanol bath <!>

METHOD

1. Add 5–10 volumes of the EDTA-Tris-DNA solution to a microfuge tube containing the solution of radiolabeled oligonucleotide. Mix the reagents well.

2. Add sufficient 1% CPB to the tube to bring the concentration of the detergent in the mixture to 0.1%. Mix well.

3. Place the tube in a dry-ice/ethanol bath until the mixture is frozen. Remove the tube from the bath and allow the mixture to thaw at room temperature.

4. Centrifuge the solution at maximum speed for 5 minutes at 4°C in a microfuge. Use a pasteur pipette or a micropipettor equipped with a blue disposable tip to remove all of the supernatant carefully from the tube.

5. Add 500 μl of distilled H_2O to the tube, vortex the mixture for 20 seconds, and centrifuge the solution as in Step 4.

6. Remove the supernatant from the tube and add 500 μl of the ethanol–sodium acetate solution to the pellet. Vortex the mixture for 15 seconds and then centrifuge the solution at maximum speed for 2 minutes at room temperature in a microfuge.

7. Repeat Step 6.

8. Carefully remove the supernatant and stand the open tube on the bench behind a Plexiglas screen until the last visible traces of ethanol have evaporated. Dissolve the precipitated oligonucleotide in 20–50 μl of TE (pH 7.6) or in a small volume of prehybridization solution if the radiolabeled oligonucleotide is to be used as a probe.

Purification of Radiolabeled Oligonucleotides by Size-exclusion Chromatography

When radiolabeled oligonucleotides are to be used in enzymatic reactions such as primer extension, virtually all of the unincorporated radiolabel must be removed from the oligonucleotide. Chromato-graphic methods (Protocols 10.5 and 10.6) or gel electrophoresis (essentially as described in Protocol 10.1) are superior in this respect to differential precipitation of the oligonucleotide with ethanol or CPB.

MATERIALS

CAUTION: Please see Appendix 4 for appropriate handling of materials marked with <!>.

Reagents and Solutions

Please see Appendix 1 for components of stock solutions, buffers, and reagents. Dilute stock solutions to the appropriate concentrations.

    Chloroform <!> (optional; see Step 7)
    EDTA (0.5 M, pH 8.0)
    Ethanol (100% and 80%)
    Phenol:chloroform <!> (optional; see Step 7)
    Sodium acetate (3 M, pH 5.2) (optional; see Step 7)
    TE (pH 7.6)
    Tris-Cl (1 M, pH 8.0) (optional; see Step 7)
    Tris-SDS chromatography buffer
        10 mM Tris-Cl (pH 8.0)
        0.1% SDS

Nucleic Acids/Oligonucleotides

Radiolabeled oligonucleotide <!>
    The starting material for this protocol is the reaction mixture (~20 µl) generated in Protocol 10.2 (Step 2 or 5), after it has been heated to inactivate bacteriophage T4 polynucleotide kinase.

Additional Items

    Centrifugal evaporator (e.g., Savant SpeedVac) (optional; see note to Step 2a)
    DE81 filters (optional; see Step 6)
    Gel filtration resin (Biogel P-60 fine grade, available from Bio-Rad as a preswollen gel,
      or Sephadex G-15, which needs to be swollen and equilibrated before use)
    Glass wool, sterilized by autoclaving
    Liquid scintillation spectroscope, with the capacity to quantify ^{32}P by Cerenkov counting
    Handheld minimonitor

METHOD

1. Add 30 µl of 20 mM EDTA (pH 8.0) to the tube containing the radiolabeled oligonucleotide. Store the solution at 0°C while preparing a column of size-exclusion chromatography resin.

   For convenience, Biogel P-60 is used throughout this protocol as an example of a suitable resin; however, the method works equally well with Sephadex G-15.

2. Prepare a Biogel P-60 column in a sterile pasteur pipette.

   a. Equilibrate the slurry of Biogel P-60 supplied by the manufacturer in 10 volumes of Tris-SDS chromatography buffer.

   If a centrifugal evaporator (Savant SpeedVac or its equivalent) is available, the Biogel P-60 column may be poured and run in a solution of 0.1% ammonium bicarbonate. The pooled fractions containing the radiolabeled oligonucleotide (please see Step 6) can then be evaporated to dryness in a centrifugal evaporator, thereby eliminating the need for extraction of the oligonucleotide preparation with organic solvents and precipitation with ethanol.

   b. Tamp a sterile glass wool plug into the bottom of a sterile pasteur pipette.

   A glass capillary tube works well as a tamping device.

   c. With the plug in place, pour a small amount of Tris-SDS chromatography buffer into the column and check that the buffer flows at a reasonable rate (one drop every few seconds).

   d. Fill the pipette with the Biogel P-60 slurry. The column forms rapidly as the gel matrix settles under gravity and the buffer drips from the pipette. Add additional slurry until the packed column fills the pipette from the plug of glass wool to the constriction near the top of the pipette.

   e. Wash the column with 3 ml of Tris-SDS chromatography buffer.

   IMPORTANT: Do not allow the column to run dry. If necessary, seal the column by wrapping a piece of Parafilm around the bottom of the pipette.

3. Use a pipette to remove excess buffer from the top of the column and then rapidly load the radiolabeled oligonucleotide (in a volume of 100 µl or less) onto the column.

4. Immediately after the sample has entered the column, add 100 µl of Tris-SDS chromatography buffer to the top of the column. As soon as the buffer has entered the column, fill the pipette with buffer. Replenish the buffer as necessary so that it continuously drips from the column. Do not allow the column to run dry.

5. Use a handheld minimonitor to follow the progress of the radiolabeled oligonucleotide. When the radioactivity first starts to elute from the column, begin collecting two-drop fractions into microfuge tubes.

6. When nearly all of the radioactivity has eluted from the column, use a liquid scintillation counter to measure the radioactivity in each fraction by Cerenkov counting. If there is a clean separation of the faster-migrating peak (the radiolabeled oligonucleotide) from the slower peak of unincorporated [γ-^{32}P]ATP, pool the samples containing the radiolabeled oligonucleotide. If the peaks are not well separated, analyze approximately 0.5 µl of every other fraction either by adsorption to DE81 filters (please see Protocol 13.7, Step 3) or by thin-layer chromatography. Pool those fractions containing radiolabeled oligonucleotide that do not contain appreciable amounts of unincorporated [γ-^{32}P]ATP.

7. If the radiolabeled oligonucleotide is to be used in enzymatic reactions, proceed as follows. Otherwise, proceed to Step 8.

   a. Extract the pooled fractions with an equal volume of phenol:chloroform.

    b. Back-extract the organic phase with 50 µl of 10 mM Tris-Cl (pH 8.0) and combine the two aqueous phases.

    c. Extract the combined aqueous phases with an equal volume of chloroform.

    d. Add 0.1 volume of 3 M sodium acetate (pH 5.2), mix well, and add 3 volumes of ethanol. Incubate the sample for 30 minutes at 0°C and then centrifuge it at maximum speed for 20 minutes at 4°C in a microfuge. Use a micropipettor equipped with a disposable tip to remove the ethanol (which should contain very little radioactivity) from the tube.

8. Add 500 µl of 80% ethanol to the tube, vortex briefly, and centrifuge the tube again at maximum speed for 5 minutes in a microfuge.

9. Use a micropipettor equipped with a disposable tip to remove the ethanol from the tube. Stand the open tube behind a Plexiglas screen until the residual ethanol has evaporated.

10. Dissolve the precipitated oligonucleotide in 20 µl of TE (pH 7.6) and store it at –20°C.

Purification of Radiolabeled Oligonucleotides by Chromatography on a Sep-Pak C$_{18}$ Column

Radiolabeled oligonucleotides can be separated from unincorporated radiolabel by chromatography on silica gel resins. The protocol is suitable only for purifying oligonucleotides carrying 5′-phosphate groups. The method described in Protocol 10.1 should be used to purify oligonucleotides with free 5′-hydroxyl groups.

MATERIALS

CAUTION: Please see Appendix 4 for appropriate handling of materials marked with <!>.

Reagents and Solutions

Please see Appendix 1 for components of stock solutions, buffers, and reagents. Dilute stock solutions to the appropriate concentrations.

Acetonitrile HPLC grade (5%, 30%, and 100% in H$_2$O, freshly prepared) <!>
Ammonium bicarbonate (25 mM, pH 8.0)
Ammonium bicarbonate (25 mM, pH 8.0), containing 5% (v/v) acetonitrile
TE (pH 7.6)

Nucleic Acids/Oligonucleotides

Radiolabeled oligonucleotide <!>
The starting material for this protocol is the reaction mixture (~20 µl) generated in Protocol 10.2 (Step 2 or 5), after it has been heated to inactivate bacteriophage T4 polynucleotide kinase.

Additional Items

Centrifugal evaporator (e.g., Savant SpeedVac)
Sep-Pak C$_{18}$ classic columns, short body (Waters Division of Millipore)
These columns contain 360 mg of a hydrophobic reversed-phase chromatography resin. The oligonucleotide binds to the column when the polarity of the solvent is high and elutes when the polarity is reduced. A separate column is required for each phosphorylation reaction.
Syringe (10 cc), polypropylene

METHOD

1. Prepare a Sep-Pak C$_{18}$ reversed-phase column as follows:

   a. Attach a polypropylene syringe containing 10 ml of acetonitrile (100%) to a Sep-Pak C$_{18}$ column.

   b. Slowly push the acetonitrile through the Sep-Pak column.

   c. Remove the syringe from the Sep-Pak column and then take the plunger out of the barrel. This prevents air from being pulled back into the column. Reattach the barrel to the column.

   d. Flush out the organic solvent with two 10-ml aliquots of sterile H$_2$O. Repeat Step c after each wash.

2. Dilute the radiolabeled oligonucleotide preparation to 1.5 ml with sterile H_2O and apply the entire sample to the column through the syringe.

3. Wash the Sep-Pak column with the following four solutions. Repeat Step 1c after each wash.

> 10 ml of 25 mM ammonium bicarbonate (pH 8.0)
> 10 ml of 25 mM ammonium bicarbonate/5% acetonitrile
> 10 ml of 5% acetonitrile
> 10 ml of 5% acetonitrile

4. Elute the radiolabeled oligonucleotide with three 1-ml aliquots of 30% acetonitrile. Collect each fraction in a separate 1.5-ml microfuge tube. Repeat Step 1c after each elution.

5. Recover the oligonucleotide by evaporating the eluate to dryness in a centrifugal evaporator (Savant SpeedVac or its equivalent).

6. Dissolve the radiolabeled oligonucleotide in a small volume (10 μl) of TE (pH 7.6).

Labeling of Synthetic Oligonucleotides Using the Klenow Fragment of *E. coli* DNA Polymerase I

Probes of high specific activities can be obtained using the Klenow fragment of *E. coli* DNA polymerase I to catalyze synthesis of a strand of DNA complementary to a synthetic oligonucleotide. A short primer is hybridized to an oligonucleotide template whose sequence is the complement of the desired radiolabeled probe. The primer is then extended using the Klenow fragment of *E. coli* DNA polymerase I to incorporate $[\alpha\text{-}^{32}P]dNTPs$ in a template-directed manner. To keep the substrate concentration high, the reaction should be set up in as small a volume as possible. It is best to use radiolabeled dNTPs supplied in ethanol/H$_2$O rather than buffered aqueous solvents. Appropriate volumes of ethanolic radiolabeled dNTPs can then be mixed and evaporated to dryness in the microfuge tube that will be used to perform the reaction. After the reaction, the template and product are separated by denaturation, followed by electrophoresis through a polyacrylamide gel under denaturing conditions. With this method, it is possible to generate oligonucleotide probes that contain several radioactive atoms per molecule of oligonucleotide and to achieve specific activities as high as 2×10^{10} cpm/μg of probe.

MATERIALS

CAUTION: Please see Appendix 4 for appropriate handling of materials marked with <!>.

Reagents and Solutions

Please see Appendix 1 for components of stock solutions, buffers, and reagents. Dilute stock solutions to the appropriate concentrations.
   10% (w/v) Trichloroacetic acid (optional; see Step 4)

Enzymes and Buffers

Klenow fragment of *E. coli* DNA polymerase I and 10x Klenow basic buffer
   500 mM Tris-Cl (pH 8.0)
   100 mM MgSO$_4$

Nucleic Acids/Oligonucleotides

Oligonucleotide primer, purified as described in Protocol 10.1
   To ensure efficient labeling, the primer should be in three- to tenfold molar excess over the template oligonucleotide in the labeling reaction.
Template oligonucleotide, purified as described in Protocol 10.1

Radioactive Compounds

$[\alpha\text{-}^{32}P]dNTPs$ (sp. act. 3000 Ci/mmole at 10 mCi/ml) <!>

Gels/Loading Buffers

Denaturing polyacrylamide gel <!>, cast and run in 1x TBE
   The percentage of acrylamide in the solution used to form the gel and the conditions under which electrophoresis is performed vary according to the size of the oligonucleotides. Table 10-3 on the following page provides useful guidelines.
Formamide loading buffer
   80% (w/v) deionized formamide
   10 mM EDTA (pH 8.0)
   1 mg/ml xylene cyanol FF
   1 mg/ml bromophenol blue

Additional Items

Heating/cooling block (14°C)
Materials for autoradiography or phosphorimaging
Scotch tape
Step 9 of this protocol requires the reagents listed in Protocol 5.12.
Water bath or heating block (80°C)

Additional Information

E. coli DNA polymerase I and the Klenow fragment   MC3, pp. 9.82–9.86

METHOD

1. Transfer to a microfuge tube the calculated amounts of [α-³²P]dNTPs necessary to achieve the desired specific activity and sufficient to allow complete synthesis of all template strands.

   The concentration of dNTPs should not drop below 1 μM at any stage during the reaction. To keep the substrate concentration high, the extension reaction should be performed in as small a volume as possible.

2. Add to the tube the appropriate amounts of oligonucleotide primer and template oligonucleotide.

   To ensure efficient radiolabeling, the primer should be in three- to tenfold molar excess over the template DNA in the reaction mixture.

3. Add 0.1 volume of 10x Klenow basic buffer to the tube. Mix the reagents well.

4. Add 2–4 units of the Klenow fragment per 5 μl of reaction volume. Mix well. Incubate the reaction for 2–3 hours at 14°C.

   If desired, the progress of the reaction may be monitored by removing small (0.1 μl) aliquots and measuring the proportion of radioactivity that has become precipitable with 10% trichloroacetic acid (TCA).

5. Prepare a denaturing polyacrylamide gel (as described in Protocol 12.8) of the appropriate concentration (see Table 10-3). Dilute the reaction mixture with an equal volume of formamide-loading buffer, heat the mixture to 80°C for 3 minutes, and load the entire sample on the gel.

   TABLE 10-3 Percent Polyacrylamide Required to Resolve Oligonucleotides

Length of Oligonucleotide (nucleotides)	Polyacrylamide (%)
12–15	20
25–35	15
35–45	12
45–70	10

6. Following electrophoresis, disassemble the electrophoresis apparatus, leaving the polyacrylamide gel attached to one of the glass plates (for details, please see Protocol 12.11).

7. Wrap the gel and its backing plate in Saran Wrap. Note the position of the tracking dyes and use a handheld minimonitor to check the amount of radioactivity in the region of the gel that should contain the oligonucleotide. Attach a set of adhesive dot labels, marked with either very hot radioactive ink or phosphorescent spots, around the edge of the sample on the Saran Wrap. Cover the radioactive dots with Scotch tape to prevent contaminating the film holder or intensifying screen with the radioactive ink.

8. Expose the gel to autoradiographic film.

   Usually, the amount of radioactivity incorporated into the probe is so great that the time needed to obtain an image on film is no more than a few seconds.

9. After developing the film, align the images of the radioactive ink with the radioactive marks on the labels and locate the position of the probe in the gel. Excise the band and recover the radioactive oligonucleotide as described in Protocol 5.12.

Hybridization of Oligonucleotide Probes in Aqueous Solutions: Washing in Buffers Containing Quaternary Alkylammonium Salts

Finding conditions in which synthetic oligonucleotides hybridize with high specificity to their target sequences can be tricky. The variables that affect duplex formation and stability include the length of the oligonucleotide, its base composition, its base sequence, and the solvents and temperatures in which the hybridization reaction and posthybridization washings are performed. When a single oligonucleotide is used as a probe, acceptable conditions for hybridization can usually be established empirically, guided by the calculated melting temperature of perfect hybrids formed between the probe and its target (see MC3, pp. 10.2–10.4). Problems arise when degenerate pools of oligonucleotides that vary in base composition and content are used as probes. In this case, hybridization may be performed in the presence of quaternary alkylammonium salts, which greatly ameliorate the effects of G + C content on the melting temperature of oligonucleotides. In the following protocol, hybridization is performed in conventional aqueous solvents at a temperature well below the predicted melting temperature. Nonspecific hybrids are then removed by washing at high stringency in buffers containing quaternary salts. Tetramethylammonium chloride (TMACl) is used with probes that are 14–50 nucleotides in length, whereas tetraethylammonium chloride (TEACl) is used with longer oligonucleotides.

The graph in Figure 10-1 should be used to estimate a washing temperature in TMACl buffers for hybrids involving oligonucleotides of different lengths. When using TEACl buffers, subtract 33°C from the value obtained from Figure 10-1.

MATERIALS

CAUTION: Please see Appendix 4 for appropriate handling of materials marked with <!>.

Reagents and Solutions

Please see Appendix 1 for components of stock solutions, buffers, and reagents. Dilute stock solutions to the appropriate concentrations.

Oligonucleotide hybridization solution

> 6x SSC (*or* SSPE)
> 0.05 M sodium phosphate (pH 6.8)
> 1 mM EDTA (pH 8.0)
> 5x Denhardt's solution
> 100 µg/ml denatured, fragmented salmon sperm DNA (Pfizer)
> 100 mg/ml dextran sulphate

Oligonucleotide prehybridization solution

> As above, omitting dextran sulphate

Quaternary alkylammonium salts (3 M) (Sigma-Aldrich)

> Prepare a 5 M solution of TMACl or a 3 M solution of TEACl in H_2O. TMACl is used in post-hybridization washes when oligonucleotides 14–50 nucleotides in length are used as probes. TEACl is used in posthybridization washes for probes 50–200 nucleotides in length. To the solution of TMACl or TEACl, add activated charcoal to a final concentration of approximately 10%. Stir the solution for 20–30 minutes. Allow the charcoal to settle and then filter the solution through a Whatman No. 1 filter. Sterilize the solution by filtration through a nitrocellulose filter (0.45 µm). Measure the refractive index of the solution and calculate the precise concentration of the quaternary ammonium salts from the equation $C = (n - 1.331/0.018)$, where C is the molar concentration of the quaternary alkylammonium salts and n is the refractive index.

TEACl wash solution

> 2.4 M TEACl
> 50 mM Tris-Cl (pH 8.0)

0.2 mM EDTA (pH 8.0)

1 mg/ml SDS

Use this posthybridization wash solution for probes 50–200 nucleotides in length. It is essential to warm an aliquot of the solution to the appropriate temperature before use.

TMACl wash solution

3.0 M TMACl

50 mM Tris-Cl (pH 8.0)

0.2 mM EDTA (pH 8.0)

1 mg/ml SDS

Use this posthybridization wash solution for probes 14–50 nucleotides in length. It is essential to warm an aliquot of the solution to the appropriate temperature before use.

Nucleic Acids/Oligonucleotides

Nitrocellulose or nylon membranes containing the immobilized target nucleic acid(s) of interest

Radioactive Compounds

Radiolabeled oligonucleotide probes <!>

Prepare as described in Protocol 10.2 or 10.7 and purify by precipitation with cetylpyridinium bromide (Protocol 10.4).

Additional Items

Materials for autoradiography or phosphorimaging

Seal-a-Meal bags (Rival)

or

Plastic boxes with tight-fitting lids

Additional Information

Denhardt's solution	MC3, pp. A1.14–A1.15
Dextran sulphate as a crowding reagent	MC3, p. 6.58
Hybridization temperatures	MC3, pp. 10.2–10.4
Melting temperatures	MC3, pp. 10.47–10.48
Preparing phosphate buffers	MC3, p. A1.5
Quaternary alkylammonium salts	MC3, p. 10.6

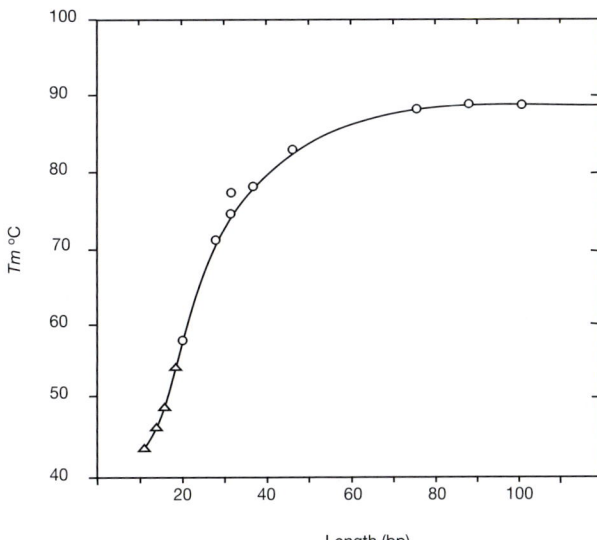

FIGURE 10-1 Estimating T_m in Buffers Containing 3.0 M TMACl

The graph depicts the increase in T_m as a function of probe length. (Adapted, with permission, from Wood et al. 1985 [*Proc. Natl. Acad. Sci.* **82**: 1585–1588].)

METHOD

1. Prehybridize the filters or membranes for 4–16 hours in oligonucleotide prehybridization solution at 37°C.

   > Prehybridization, hybridization, and washing of circular filters are best carried out in Rival Seal-a-Meal bags or plastic boxes with tight-fitting lids. For Southern and northern blots, a hybridization device equipped with sealable glass tubes may be used.

2. Discard the prehybridization solution and replace it with oligonucleotide hybridization solution containing a radiolabeled oligonucleotide probe at a concentration of 180 pM.

   > When hybridizing with several oligonucleotides simultaneously, each probe should be present at a concentration of 180 pM and the specific activity of the radiolabeled probe should be 5×10^5 to 1.5×10^6 cpm/pmole.

3. Incubate the filters for 12–16 hours at 37°C.

4. Discard the radiolabeled hybridization solution into an appropriate disposable container. Rinse the filters three times at 4°C with ice-cold 6x SSC or 6x SSPE to remove most of the dextran sulfate.

5. Wash the filters twice for 30 minutes at 4°C in ice-cold 6x SSC or 6x SSPE.

6. Rinse the filters at 37°C in two changes of the TMACl or TEACl wash solution.

   > The aim of this step is to replace the SSPE and SSC with the solution of quaternary alkylammonium salts. Unless this step is performed diligently, the full benefits of using TEACl or TMACl will not be realized.

7. Wash the filters twice for 20 minutes each in TMACl or TEACl wash solution at a temperature that is 2–4°C below the T_m indicated in Figure 10-1.

   > Note that the T_m of a hybrid is 33°C lower in a buffer containing TEACl than in a buffer containing TMACl. Make sure that the buffers are prewarmed to the desired temperature and that fluctuations in temperature are less then ±1°C.

8. Remove the filters from the washing solution. Blot them dry at room temperature and inspect them by autoradiography or phosphorimaging.

Empirical Measurement of Melting Temperature

The melting temperature (T_m) of an oligonucleotide may be determined empirically by measuring the temperature (T_i) at which dissociation of the double-stranded DNA becomes irreversible. The procedure requires a cloned target sequence that is complementary (perfectly or imperfectly, depending on the experiment) to the oligonucleotide probe. If a target sequence is not available from "natural" sources, it can be synthesized chemically or by PCR. The experiment is performed under nonequilibrium conditions that do not favor rehybridization of the released probe to the target.

MATERIALS

CAUTION: Please see Appendix 4 for appropriate handling of materials marked with <!>.

Reagents and Solutions

Please see Appendix 1 for components of stock solutions, buffers, and reagents. Dilute stock solutions to the appropriate concentrations.

Denaturation solution (required only for double-stranded DNA targets)
    1.5 M NaCl
    0.5 N NaOH <!>
Ethanol
Neutralizing solution (required only for double-stranded DNA targets)
    0.5 M Tris-Cl (pH 7.4)
    1.5 M NaCl
Oligonucleotide hybridization solution
    6x SSC (or SSPE)
    0.05 M sodium phosphate (pH 6.8)
    1 mM EDTA (pH 8.0)
    5x Denhardt's solution
    100 µg/ml denatured, fragmented salmon sperm DNA (Pfizer)
    0.5% (w/v) SDS
Phenol:chloroform (1:1, v/v) <!> (required only for double-stranded DNA targets)
Sodium acetate (3 M, pH 5.2) (required only for double-stranded DNA targets)
2x SSC

Enzymes and Buffers

Restriction enzyme to cleave target DNA (optional; see Step 3)

Nucleic Acids/Oligonucleotides

Control DNA
    Single- or double-stranded DNA whose sequence is unrelated to that of the target DNA
Target DNA
    Ideally, this sequence should be cloned into a bacteriophage M13 vector and isolated as single-stranded DNA. However, double-stranded DNA plasmids or PCR products can also be used after denaturation.

Radioactive Compounds

Radiolabeled oligonucleotide probes <!>, prepared as described in Protocol 10.2 or 10.7 and purified by precipitation with cetylpyridinium bromide (CPB) (Protocol 10.4)

> **Additional Items**
>
> Blunt-ended forceps
> Boiling-water bath
> Cross-linking device (e.g., Stratalinker [Stratagene] *or* microwave oven *or* vacuum oven)
> Handheld minimonitor
> Materials for liquid scintillation spectroscopy
> Nitrocellulose or nylon membrane
> Paper-hole punch
> Step 1 of this protocol requires the reagents listed in Protocols 10.2 and 10.4.
> Thick blotting paper (e.g., Whatman 3MM)
> Water bath, circulating, with precise temperature control
>
> **Additional Information**
>
> | | |
> |---|---|
> | Hybridization temperatures | MC3, pp. 10.2–10.4 |
> | Melting temperatures | MC3, pp. 10.47–10.48 |

METHOD

1. Radiolabel 1–10 pmoles of the oligonucleotide to be used as a probe by phosphorylation (Protocol 10.2) and remove excess unincorporated [γ-^{32}P]ATP by precipitation with CPB (Protocol 10.4).

2. Use a paper-hole punch to cut four small circles (diameter 3–4 mm) out of a nitrocellulose or nylon membrane for hybridization. Arrange the circles on a piece of Parafilm. Mark two of the membranes with a soft-lead pencil.

3. Apply target and control DNAs to the membrane circles as follows.

Single-stranded target DNA

a. Apply approximately 100 ng of target DNA in a volume of 1–3 µl of 2x SSC to each of the marked membranes.

b. Apply an equal amount of vector DNA (e.g., single-stranded bacteriophage M13 DNA without insert) to the unmarked membranes.

c. After the fluid has dried, use blunt-ended forceps (e.g., Millipore) to remove the two sets of membranes from the Parafilm and place them between sheets of thick blotting paper.

d. Fix the DNAs to the membranes by baking for 1–2 hours at 80°C in a vacuum oven.

   Alternatively, place the membranes on a sheet of blotting paper and fix the DNA by cross-linking using UV light.

Double-stranded target DNA

a. If the target DNA has been cloned into a plasmid, linearize both the recombinant plasmid and the vector by digestion with a restriction enzyme that does not cleave within the target sequence.

b. Purify the resulting double-stranded DNA or the PCR product by extraction with phenol:chloroform and standard precipitation with ethanol. Dissolve the DNA in 2x SSC at a concentration of 50–100 ng/µl.

c. Apply the solution of target DNA, as well as a control DNA, to the membranes prepared as described above. Use blunt-ended forceps to transfer the membranes to a sheet of thick blotting paper saturated with denaturation solution. Incubate the membranes for 5–10 minutes at room temperature.

d. Transfer the membranes to a fresh sheet of thick blotting paper saturated with neutralizing solution. Incubate the membranes for 10 minutes at room temperature.

e. Transfer the membranes to a dry sheet of thick blotting paper and leave them at room temperature until all of the fluid has evaporated. Immobilize the DNA to the membranes either by baking for 1–2 hours at 80°C in a vacuum oven or by cross-linking using UV light.

4. Use blunt-ended forceps to transfer all of the membranes to a polyethylene tube containing 2 ml of oligonucleotide prehybridization solution. Seal the tube and incubate, with occasional shaking, at a temperature estimated to be 25°C below the T_m for the solvent being used.

> Although the above protocol uses sodium salts in the hybridization solution, other solutes such as TMACl or TEACl can be substituted if desired to determine the T_i in these solvents.

5. After 2 hours, add the radiolabeled oligonucleotide to the prehybridization solution. The final concentration of oligonucleotide should be approximately 1 pmole/ml. Continue incubating at 25°C below the T_m for an additional 2–4 hours, with occasional shaking.

6. Remove the membranes from the hybridization solution and immediately immerse them in 2x SSC at room temperature. Agitate the fluid continuously. Replace the fluid every 5 minutes until the amount of radioactivity on the membranes remains constant (as measured with a handheld minimonitor).

7. Adjust the temperature of a circulating water bath to 25°C below the T_m. Dispense 5 ml of 2x SSC into each of 20 glass test tubes (17 x 100 mm). Monitor the temperature of the fluid in one of the tubes with a thermometer. Incubate the tubes in the water bath until the temperature of the 2x SSC is 25°C below the T_m.

8. Transfer the membranes individually to four empty glass tubes and add 1 ml of 2x SSC (from one of the tubes prepared in Step 7 and prewarmed to 25°C below the T_m) to each membrane. Place the tubes in the water bath for 5 minutes.

9. Remove the tubes containing membranes from the bath, transfer the liquid to scintillation vials, and wash the tubes and membranes with 1 ml of 2x SSC at room temperature. Add the wash solutions to the appropriate scintillation vials.

10. Increase the temperature of the water bath by 3°C and wait for the temperature of the 2x SSC in the tubes prepared in Step 7 to equilibrate.

11. Add 1 ml of 2x SSC at the higher temperature to each of the four tubes containing the membranes. Place the tubes in the water bath for 5 minutes.

12. Repeat Steps 9–11 at successively higher temperatures until a temperature of 30°C above the T_m is achieved.

13. Place the membranes in separate glass tubes (17 x 100 mm) containing 1 ml of 2x SSC and heat them to boiling for 5 minutes to remove any remaining radioactivity. Cool the solutions in ice and transfer them to scintillation vials. Wash the membranes and tubes used for boiling with 1 ml of 2x SSC and add the washing solutions to the appropriate scintillation vials.

14. Use a liquid scintillation counter to measure the radioactivity (by Cerenkov counting) in all of the vials. Calculate the proportion of the total radioactivity that has eluted at each temperature.

> For details on calculations, please see MC3, p. 10.41, Step 14.

> If the experiment has worked well, very little radioactivity should be associated with the membranes containing vector DNA alone. Furthermore, this radioactivity should be completely released from the membranes at temperatures much lower than the estimated T_m. On the other hand, considerable radioactivity should be associated with the membranes containing the target DNA; the elution of this radioactivity should show a sharp temperature dependence. Very little radioactivity should be released from the membranes until a critical temperature is reached, and then approximately 90% of the radioactivity should be released during the succeeding 6–9°C rise in temperature.

CHAPTER 11

Preparation of cDNA Libraries and Gene Identification

BACKGROUND INFORMATION

Background information found in *Molecular Cloning: A Laboratory Manual,* 3rd edition (hereafter MC3) unless otherwise indicated.

Overview of strategies used in cDNA cloning   MC3, pp. 11.2–11.35

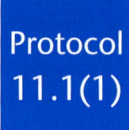

Construction of cDNA Libraries
Stage 1: Synthesis of First-strand cDNA Catalyzed by Reverse Transcriptase

This traditional protocol is based on a method described by Gubler-Hoffman (1983. *Gene* **25:** 263–269) and is divided into six stages:

- Stage 1: Synthesis of First-strand cDNA Catalyzed by Reverse Transcriptase
- Stage 2: Second-strand cDNA Synthesis
- Stage 3: Methylation of cDNA
- Stage 4: Attachment of Linkers or Adapters
- Stage 5: Fractionation of cDNA by Gel Filtration through Sepharose CL-4B
- Stage 6: Ligation of cDNA to Bacteriophage λ Arms

Protocol 11.1 Stage 1 describes the conversion of poly(A)$^+$ mRNA into first-strand cDNA in a reaction catalyzed by a murine RNase H$^-$ reverse transcriptase and primed by oligo(dT), random hexamers, or primer-adapters. The reaction conditions for first-strand synthesis using other enzymes are given in MC3, p. 11.38, Table 11-3.

MATERIALS

CAUTION: Please see Appendix 4 for appropriate handling of materials marked with <!>.

Reagents and Solutions

Please see Appendix 1 for components of stock solutions, buffers, and reagents. Dilute stock solutions to the appropriate concentrations.

Actinomycin D (5 mg/ml in H$_2$O), required only if using a wild-type version of Mo-MLV reverse transcriptase (i.e., an enzyme possessing RNase H activity). Shield the stock solution from light.

Dithiothreitol (1 M)

dNTP solution, containing all four dNTPs, each at a concentration of 5 mM

EDTA (0.5 M, pH 8.0)

KCl (1 M)

MgCl$_2$ (1 M)

Trichloroacetic acid (10%)

Tris-Cl (1 M, pH 8.3 at room temperature)

Enzymes and Buffers

Reverse transcriptase (Mo-MLV H$^-$ RT), a variant lacking RNase H activity, e.g., SuperScript (Invitrogen) or StrataScript (Stratagene)

RNase inhibitor (e.g., RNasin [Promega])

Nucleic Acids/Oligonucleotides

DNA markers for alkaline gel (optional; see Step 5)

Oligonucleotide primers for cDNA synthesis (1 mg/ml in H$_2$O)

Synthesis of first-strand cDNA is generally primed by random hexamers, oligo(dT)$_{12-18}$, by a mixture of the two, or by primer-adapters of the general structure $5'p(dX)$-(dR)-$(dT)_x$-$OH3'$, where X is a clamp composed of four nucleotides (usually GAGA), R is a recognition site for a restriction enzyme, and (dT)$_x$ is a homopolymeric run of approximately 15 (dT) residues. For additional information, see MC3, p. 11.39.

Poly(A)$^+$ RNA (100 µg/ml)

Approximately 5–10 µg of poly(A)$^+$ is required to synthesize enough cDNA to construct a large cDNA library.

 Adapted from Chapter 11, Protocol 1 Stage 1, p. 11.38 of MC3.

Radioactive Compounds

[α-^{32}P]dCTP (10 mCi/ml, 400 Ci/mmole) <!>
  A cDNA synthesis kit sold by Stratagene recommends the use of any label other than dCTP. This kit is unusual in that it includes 5-methyl dCTP as a substrate, which protects the cDNA against internal cleavage by *Xho*I, the enzyme used to create cohesive termini for ligation of cDNA into a vector. Unless using this kit, dCTP is the label of choice.

Gels/Loading Buffers

Alkaline agarose gel (optional; see Step 5)

Additional Items

Step 5 (optional) requires the reagents listed in Protocol 5.8.
Water baths (37°C and 70°C)

Additional Information

Actinomycin D	MC3, p. 7.88
Commercial kits for cDNA synthesis and library construction	MC3, pp. 11.107–11.108
Inhibitors of RNases	MC3, p. 7.83
Methods used to synthesize first-strand cDNA	MC3, pp. 11.11–11.15
Mo-MLV reverse transcriptase	MC3, pp. 11.109–11.110
Reverse transcriptases	MC3, p. 8.48
Troubleshooting	MC3, p. 11.42

METHOD

1. To synthesize first-strand cDNA, mix the following in a sterile microfuge tube on ice:

1 μg/μl poly(A)$^+$ RNA	10 μl
1 μg/μl oligonucleotide primer(s)	1 μl
1 M Tris-Cl (pH 8.0 at 37°C)	2.5 μl
1 M KCl	3.5 μl
250 mM MgCl$_2$	2 μl
solution of all four dNTPs, each at 5 mM	10 μl
0.1 M dithiothreitol	2 μl
RNase inhibitor (optional)	25 units
H$_2$O	to 48 μl

   Add the manufacturer's recommended amount of Mo-MLV H$^-$ RT to the reaction. Mix the reagents well by gentle vortexing.

   Mo-MLV RT is temperature sensitive and should be stored at –20°C until needed.

   IMPORTANT: If a type of RT is used other than Mo-MLV H$^-$, please see MC3, p. 11.38, Table 11-3 for reaction conditions.

2. After all of the components of the reaction have been mixed at 0°C, transfer 2.5 μl of the reaction to a fresh 0.5-ml microfuge tube. Add 0.1 μl of [α-^{32}P]dCTP (400 Ci/mmole, 10 mCi/ml) to the small-scale reaction.

3. Incubate the large- and small-scale reactions for 1 hour at 37°C.

   Higher temperatures (up to 55°C) can be used with some mutant forms of Mo-MLV RT.

4. At the end of the incubation period, add 1 μl of 0.25 M EDTA to the small-scale reaction containing the radioisotope. Transfer the small-scale reaction to ice. Heat the large-scale reaction to 70°C for 10 minutes and then transfer it to ice.

5. Measure both the total amount of radioactivity and the amount of trichloroacetic acid (TCA)-precipitable radioactivity in 0.5 μl of the small-scale reaction, as described in MC3, Appendix 8. In addition, it is worthwhile analyzing the products of the small-scale reaction through an alkaline agarose gel using appropriate DNA markers (please see Protocol 5.8).

    The remainder of the small-scale reaction can be stored at –20°C.

6. Calculate the amount of first-strand cDNA synthesized as follows:

    a. Since 10 μl of a solution containing all four dNTPs at a concentration of 5 mM each was used (i.e., 10 μl of 20 mmoles/liter of total dNTP), the large-scale reaction must contain

    20 nmoles/μl dNTP x 10 μl = 200 nmoles of dNTP

    b. Because the molecular weight of each dNMP incorporated into DNA is approximately 330 g/mole, the reaction is capable of generating a total of

    200 nmoles x 330 ng/nmole = 66 μg of DNA

    c. Therefore, from the results of the small-scale reaction,

    $$\frac{\text{cpm incorporated}}{\text{total cpm}} \times 66 \text{ μg} = \text{μg of first strand of cDNA synthesized}$$

7. Proceed as soon as it is feasible to the next stage in the synthesis of the cDNA.

Construction of cDNA Libraries
Stage 2: Second-strand cDNA Synthesis

Here, the DNA-RNA hybrids synthesized in Stage 1 are converted into full-length double-stranded cDNAs. The primers for synthesis of second-strand cDNA are created by RNase H, which introduces nicks into the RNA moiety of the cDNA-mRNA hybrids. *E. coli* DNA polymerase I extends the newly created 3'-hydroxyl termini, using the first-strand cDNA as a template and replacing segments of mRNA in the cDNA-mRNA hybrid with the newly synthesized second strand of DNA. Residual nicks are then repaired by *E. coli* DNA ligase, and the frayed termini of the double-stranded cDNA are polished by a DNA polymerase such as bacteriophage T4 DNA polymerase or *Pfu*. Finally, bacteriophage T4 polynucleotide kinase is used to catalyze the phosphorylation of 5'-hydroxyl groups on the ends of the cDNAs in preparation for ligation of linkers or adapters (Protocol 11.1 Stage 4).

MATERIALS

CAUTION: Please see Appendix 4 for appropriate handling of materials marked with <!>.

Reagents and Solutions

Please see Appendix 1 for components of stock solutions, buffers, and reagents. Dilute stock solutions to the appropriate concentrations.

Chloroform <!>
dNTP solution, containing all four dNTPs, each at a concentration of 10 mM
EDTA (0.5 M, pH 8.0)
Ethanol
Ethanol (70%)
$MgCl_2$ (1 M)
$(NH_4)_2SO_4$ (1 M)
β-Nicotinamide adenine dinucleotide (β-NAD) (50 mM)
Phenol:chloroform (1:1, v/v) <!>
Sodium acetate (3 M, pH 5.2)
TE (pH 7.6)
Trichloroacetic acid (10%)
Tris-Cl (2 M, pH 7.4)

Enzymes and Buffers

Bacteriophage T4 DNA polymerase (2.5 units/µl)
Bacteriophage T4 polynucleotide kinase (30 units/µl) and 10x buffer
  700 mM Tris-Cl (pH 7.4)
  100 mM $MgCl_2$
  50 mM dithiothreitol
  E. coli DNA ligase
E. coli DNA polymerase I (10,000 units/ml)
RNase H (~1,000 units/ml)

Nucleic Acids/Oligonucleotides

DNA markers for alkaline gel, e.g., end-labeled *Hin*dIII fragments of bacteriophage λ DNA
First-strand cDNA (Protocol 11.1 Stage 1)

Radioactive Compounds

[α-³²P]dCTP (10 mCi/ml, 400 Ci/mmole) <!>

Gels/Loading Buffers

Alkaline agarose gel (optional; see Step 5)

Additional Items

Step 5 (optional) requires the reagents listed in Protocol 5.8.
Water bath or heating/cooling block (16°C)
Whatman GF/C filters (25 mm)

Additional Information

Bacteriophage T4 DNA polymerase	MC3, pp. A4.18–A4.21
Bacteriophage T4 polynucleotide kinase	MC3, pp. A4.35–A4.36
E. coli DNA ligase	MC3, p. A4.33
E. coli DNA polymerase I	MC3, pp. A4.12–A4.14
Overview of synthesis of second-strand cDNA	MC3, pp. 11.14–11.20
RNase H	MC3, p. A4.38
Troubleshooting	MC3, p. 11.47

METHOD

1. Add the following reagents directly to the large-scale first-strand reaction mixture (Protocol 11.1 Stage 1, Step 4):

10 mM MgCl$_2$	70 µl
2 M Tris-Cl (pH 7.4)	5 µl
10 mCi/ml [α-³²P]dCTP (400 Ci/mmole)	10 µl
1 M (NH$_4$)$_2$SO$_4$	1.5 µl
RNase H (1000 units/ml)	1 µl
E. coli DNA polymerase I (10,000 units/ml)	4.5 µl

   Mix the reagents by gentle vortexing and centrifuge the reaction mixture briefly in a microfuge to eliminate any bubbles. Incubate the reaction for 2–4 hours at 16°C.

   Second-strand synthesis catalyzed by RNase H and DNA polymerase I results in loss of sequences (~20 nucleotides) from the extreme 5′ end of the mRNA template. The alternative to Step 1 (see the following paragraph) reduces the risk of losing 5′-terminal regions of cDNA clones. Unfortunately, the alternative method results in a reduced yield of double-stranded cDNA, and most investigators will therefore opt for the standard Step 1 method.

   Incubate the second-strand synthetic reaction (above) for 2 hours at 16°C *in the absence of RNase H*. Purify the resulting double-stranded cDNA by extraction with phenol:chloroform and precipitation with ethanol in the presence of 0.3 M sodium acetate (pH 5.2). Dissolve the precipitated DNA in 20 µl of 20 mM Tris-Cl (pH 7.6), 20 mM KCl, 0.1 mM EDTA (pH 8.0), 0.1 mM dithiothreitol. Digest the DNA for 20 minutes at 37°C with RNase H (0.5 unit). Proceed to Step 2.

2. At the end of the incubation, add the following reagents to the reaction mixture:

β-NAD (50 mM)	1 µl
E. coli DNA ligase (1000–4000 units/ml)	1 µl

   Incubate the reaction for 15 minutes at room temperature.

3. At the end of the incubation, add 1 µl of a mixture containing all four dNTPs each at a concentration of 10 mM and 2 µl (5 units) of bacteriophage T4 DNA polymerase. Incubate the reaction mixture for 15 minutes at room temperature.

4. Remove a small aliquot (3 µl) of the reaction. Measure the mass of second-strand DNA in the aliquot as described in Steps 7 and 8.

5. To the remainder of the reaction, add 5 µl of 0.5 M EDTA (pH 8.0). Extract the mixture once with phenol:chloroform and once with chloroform. Recover the DNA by precipitation with ethanol in the presence of 0.3 M sodium acetate (pH 5.2). Dissolve the DNA in 90 µl of TE (pH 7.6).

6. To the DNA, add:

10x T4 polynucleotide kinase buffer	10 µl
T4 polynucleotide kinase (3000 units/ml)	1 µl

   Incubate the reaction at room temperature for 15 minutes.

7. Use the small aliquot from Step 4 to determine the total amount of radioactivity and the TCA-precipitable counts in 1 µl of the second-strand synthesis reaction as described in MC3, Appendix 8.

8. Use the following equation to calculate the weight of the cDNA synthesized in the second-strand reaction, taking into account the amount of dNTPs already incorporated into the first strand of cDNA:

$$\frac{\text{cpm incorporated in the second-strand reaction}}{\text{total cpm}} \times (66\ \mu g - x\ \mu g)$$

$$= \mu g \text{ of second-strand cDNA synthesized}$$

   where x is the weight of the first strand of cDNA. The amount of second-strand cDNA synthesized is usually 70–80% of the weight of the first strand.

9. Extract the reaction containing phosphorylated cDNA (from Step 6) with an equal volume of phenol:chloroform.

10. Separate the unincorporated dNTPs from the cDNA by spun-column chromatography through Sephadex G-50 equilibrated in TE (pH 7.6) containing 10 mM NaCl.

11. Precipitate the eluted cDNA by adding 0.1 volume of 3 M sodium acetate (pH 5.2) and 2 volumes of ethanol. Store the sample on ice for at least 15 minutes. Recover the precipitated DNA by centrifugation at maximum speed for 15 minutes at 4°C in a microfuge. Use a hand-held minimonitor to check that all of the radioactivity has been precipitated.

12. Wash the pellet with 70% ethanol and centrifuge again.

13. Gently aspirate all of the fluid (check to see that none of the radioactivity is in the aspirated fluid) and allow the pellet to air dry.

14. Dissolve the cDNA in 80 µl of TE (pH 7.6) if it is to be methylated by *Eco*RI methylase (please see Protocol 11.1 Stage 3). Alternatively, if the cDNA is to be ligated directly to *Not*I or *Sal*I linkers, or an adapter oligonucleotide (please see Protocol 11.1 Stage 4), resuspend the cDNA in 29 µl of TE (pH 7.6).

Construction of cDNA Libraries
Stage 3: Methylation of cDNA

The goal of this stage is to introduce methyl groups that will modify and protect naturally occurring *Eco*RI sites in the double-stranded cDNA. Note that Stage 3 is obligatory if synthetic linkers are used to attach the cDNA to a vector. However, the entire stage should be omitted if adapters are used.

MATERIALS

CAUTION: Please see Appendix 4 for appropriate handling of materials marked with <!>.

Reagents and Solutions

Please see Appendix 1 for components of stock solutions, buffers, and reagents. Dilute stock solutions to the appropriate concentrations.

Chloroform <!>
EDTA (pH 8.0)
Ethanol
Ethanol (70%)
$MgCl_2$ (1 M)
NaCl (5 M)
Phenol:chloroform (1:1, v/v) <!>
S-Adenosylmethionine (20 mM in 5 mM H_2SO_4/10% ethanol)
Sodium acetate (3 M, pH 5.2)
TE (pH 8.0)
Tris-Cl (2 M, pH 8.0)

Enzymes and Buffers

*Eco*RI and 10x *Eco*RI buffer, supplied by enzyme manufacturer
*Eco*RI methylase

Nucleic Acids/Oligonucleotides

Double-stranded cDNA from Protocol 11.1 Stage 2
Linearized plasmid or bacteriophage λ DNA, used to check the efficiency of methylation of *Eco*RI sites
  The DNA should contain at least one *Eco*RI site located some distance from the 3′ and 5′ termini.

Gels/Loading Buffers

1% Agarose gel, cast and run in 0.5x TBE
10x Loading buffer

Additional Items

Handheld minimonitor
Water bath or heating/cooling block (37°C and 68°C)

Additional Information

*Eco*RI methylase   MC3, p. 11.48

METHOD

1. To the cDNA (from Protocol 11.1 Stage 2, Step 14), add:

2 M Tris-Cl (pH 8.0)	5 μl
5 M NaCl	2 μl
0.5 M EDTA (pH 8.0)	2 μl
20 mM S-adenosylmethionine	1 μl
H$_2$O	to 96 μl

2. Remove two 2-μl aliquots and place each in a separate 0.5-ml microfuge tube. Number the tubes 1 and 2, and store the numbered tubes on ice.

3. Add 2 μl of *Eco*RI methylase (80,000 units/ml) to the remainder of the reaction mixture and then store the reaction mixture at 0°C until Step 4 is completed.

4. Remove two additional aliquots (2 μl each) from the large-scale reaction and place each in a separate 0.5-ml microfuge tube. Number these tubes 3 and 4.

5. To each of the four small aliquots (Steps 2 and 4), add test DNA (100 ng of linearized plasmid DNA *or* 500 ng of bacteriophage λ DNA). These unmethylated DNAs are used as substrates in pilot reactions to assay the efficiency of methylation.

   IMPORTANT: Do not add any DNA to the large-scale reaction!

6. Incubate all four pilot reactions and the large-scale reaction for 1 hour at 37°C.

7. Heat the five reactions to 68°C for 15 minutes. Extract the large-scale reaction once with phenol:chloroform and once with chloroform.

8. To the large-scale reaction, add 0.1 volume of 3 M sodium acetate (pH 5.2) and 2 volumes of ethanol. Mix the reagents well and store the ethanolic solution at –20°C until the results of the pilot reactions are available.

9. Analyze the four pilot reactions as follows:

   a. To each control reaction, add:

0.1 M MgCl$_2$	2 μl
10x *Eco*RI buffer	2 μl
H$_2$O	to 20 μl

   b. Add 20 units of *Eco*RI to reactions 2 and 4.

   c. Incubate all four samples for 1 hour at 37°C and analyze them by electrophoresis through a 1% agarose gel.

10. Recover the precipitated cDNA (Step 8) by centrifugation at maximum speed for 15 minutes at 4°C in a microfuge. Remove the supernatant, add 200 μl of 70% ethanol to the pellet, and centrifuge again.

11. Use a handheld minimonitor to check that all of the radioactivity is recovered in the pellet. Remove the ethanol by gentle aspiration, dry the pellet in the air, and then dissolve the DNA in 29 μl of TE (pH 8.0).

12. Proceed as soon as is feasible to the next stage in the synthesis of the cDNA.

Construction of cDNA Libraries
Stage 4: Attachment of Linkers or Adapters

This stage achieves four goals: polishing the ends of double-stranded DNA, ligation of synthetic linkers or adapters, digestion of the attached linkers to create cohesive termini, and preparing the cDNA for cloning.

MATERIALS

CAUTION: Please see Appendix 4 for appropriate handling of materials marked with <!>.

Reagents and Solutions

Please see Appendix 1 for components of stock solutions, buffers, and reagents. Dilute stock solutions to the appropriate concentrations.

ATP (10 mM)
Bromophenol blue (0.25% w/v) in 50% glycerol
EDTA (0.5 M, pH 8.0)
Ethanol
Ethanol (70%)
Phenol:chloroform (1:1, v/v) <!>
Sodium acetate (3 M, pH 5.2)
TE (pH 8.0)
Tris-Cl (1 M, pH 8.0)

Enzymes and Buffers

Bacteriophage T4 DNA ligase
Bacteriophage T4 DNA polymerase and 10x repair buffer
    90 mM $(NH_4)_2SO_4$
    330 mM Tris-Cl (pH 8.3)
    50 mM β-mercaptoethanol <!>
Restriction enzyme(s)
    Depending on the type of linkers chosen, EcoRI, NotI, SalI, or other restriction enzymes may be required (see also MC3, p. 11.20).

Nucleic Acids/Oligonucleotides

Control DNA, a linear DNA (e.g., a linearized plasmid) containing the relevant restriction site is used to check the efficiency of restriction after addition of linkers
Double-stranded cDNA, either methylated or unmethylated (Protocol 11.1 Stage 2, Step 14 or Stage 3, Step 11)
Marker DNAs
Synthetic linkers or adapters, phosphorylated
    Choose linkers or adapters containing restriction sites that occur rarely in mammalian DNA; see MC3, p. A6.4, Table A6-3. Many of these adapters and linkers are available from commercial suppliers.

Gels/Loading Buffers

1% Agarose gel, cast and run in 0.5x TBE

Additional Items

Handheld minimonitor
Sephadex G-50 spun column
Water bath or heating/cooling block (68°C, 37°C, and 16°C)

Additional Information

Adapters	MC3, p. 1.160
Spun-column chromatography	MC3, pp. A8.30–A8.31
Use of linkers and adapters in cDNA cloning	MC3, pp. 11.20–11.21 and 11.51–11.52

METHOD

1. Heat the cDNA (Protocol 11.1 Stage 3) to 68°C for 5 minutes.

2. Cool the cDNA to 37°C and add the following to the tube:

5x bacteriophage T4 DNA polymerase repair buffer	10 μl
5 mM dNTP solution	5 μl
H_2O	to 50 μl

3. Add 1–2 units of bacteriophage T4 DNA polymerase (500 units/ml) and incubate the reaction for 15 minutes at 37°C.

4. Stop the reaction by adding 1 μl of 0.5 M EDTA (pH 8.0).

5. Extract the sample with phenol:chloroform and remove the unincorporated dNTPs by spun-column chromatography through Sephadex G-50 (please see MC3, Appendix 8).

6. Add 0.1 volume of 3 M sodium acetate (pH 5.2) and 2 volumes of ethanol to the column flow-through. Store the sample for at least 15 minutes at 4°C.

7. Recover the precipitated cDNA by centrifugation at maximum speed for 15 minutes at 4°C in a microfuge. Dry the pellet in the air and then dissolve it in 13 μl of 10 mM Tris-Cl (pH 7.6).

8. Add the following to the repaired DNA:

10x T4 DNA ligase buffer	2 μl
800–1000 ng of phosphorylated linkers or adapters	2 μl
10^5 Weiss units/ml bacteriophage T4 DNA ligase	1 μl
10 mM ATP	2 μl

   Mix and incubate for 8–12 hours at 16°C.

   It is essential to use a vast molar excess of linkers (>100-fold) to ensure that the ends of the cDNA become ligated to a linker and not to each other.

9. Withdraw 0.5 μl from the reaction and store the aliquot at 4°C. Inactivate the ligase in the remainder of the reaction by heating for 15 minutes at 68°C.

   If endonuclease digestion is necessary to permit ligation of the cDNA to the vector, proceed with Steps 10–12. However, when using certain adapter molecules, endonuclease digestion is not required. In these cases, skip to Step 13.

10. To the heated ligase reaction, add:

10x restriction enzyme buffer	20 μl
H_2O	150 μl
restriction enzyme	200 units

Mix the reagents at 0°C.

As a control, transfer 2 µl of the reaction to a 0.5-ml microfuge tube. To the 2-µl aliquot, add 100 ng (in a volume of no more than 0.5 µl) of control DNA, i.e., either a linearized plasmid or a preparation of cleaved bacteriophage λ DNA that contains an internal site for the particular restriction enzyme used.

11. Incubate the large-scale reaction and the control reaction for 2 hours at 37°C.

12. Analyze the DNA in the control sample and the 0.5-µl aliquot withdrawn at Step 9 by electrophoresis through a 1% agarose gel. Load one or two lanes of the gel with plasmid or bacteriophage λ marker DNA.

13. Purify the cDNA by extraction with phenol:chloroform and standard precipitation with ethanol. Dissolve the cDNA in 20 µl of TE (pH 8.0).

14. Add 2.5 µl of a solution of 0.25% bromophenol blue in 50% glycerol. This addition simplifies the next stage in the cDNA library—size-fractionation of the cDNA by chromatography through Sepharose CL-4B (Protocol 11.1 Stage 5).

Construction of cDNA Libraries
Stage 5: Fractionation of cDNA by Gel Filtration through Sepharose CL-4B

Unused linkers, adapter-linkers, and low-molecular-weight products created by digestion of linkers in Step 4 are removed by size-exclusion chromatography.

MATERIALS

Reagents and Solutions

Please see Appendix 1 for components of stock solutions, buffers, and reagents. Dilute stock solutions to the appropriate concentrations.

Ethanol
Sodium acetate (3 M, pH 5.2)
TE (pH 7.6)
TE (pH 7.6), containing 0.1 M NaCl
Tris-Cl (1 M, pH 7.6)

Nucleic Acids/Oligonucleotides

cDNA
  Prepare as in Protocol 11.1 Stage 4, Steps 13 and 14.
Marker DNAs
  Marker DNAs should be ^{32}P end-labeled fragments of DNA, between 200 bp and 5 kb in size.

Gels/Loading Buffers

1% Agarose gel, cast and run in 0.5x TBE

Additional Items

Disposable plastic pipette (1 ml)
Gel dryer
Handheld minimonitor
Hemostat or equivalent hose clamp
Hypodermic needle, with bent tip
Liquid scintillation spectroscope, capable of quantifying ^{32}P by Cerenkov counting
Materials for autoradiography or phosphorimaging
Polyvinyl chloride tubing, of the type normally used in polystaltic pumps
Sepharose CL-4B
  Wash the paste supplied by the manufacturer several times in sterile TE (pH 7.6) containing 0.1 M NaCl and proceed as described in Steps 1–3. Alternatively, use prepacked columns of Sepharose CL-4B (SizeSep 400 Spun Columns) from GE Healthcare.
Vinyl bubble tubing
Whatman 3MM paper

METHOD

1. Use a hypodermic needle with a bent end to pull the cotton wool pledget halfway out of the end of a sterile, disposable 1-ml pipette. Cut the pledget in half with sterile scissors. Discard

the loose piece of cotton wool. Use filtered compressed air to blow the remainder of the pledget to the narrow end of the pipette.

2. Attach a piece of sterile polyvinyl chloride tubing to the narrow end of the pipette. Dip the wide end of the pipette into a solution of TE (pH 7.6) containing 0.1 M NaCl. Attach the tubing to an Erlenmeyer flask connected to a vacuum line. Apply gentle suction until the pipette is filled with buffer. Close the tubing with a hemostat.

3. Attach a piece of vinyl bubble tubing to the wide end of the pipette. Fill the bubble tubing with Sepharose CL-4B equilibrated with TE (pH 7.6) containing 0.1 M NaCl. Allow the slurry to settle for a few minutes and then release the hemostat. The column will form as the buffer drips from the pipette. If necessary, add more Sepharose CL-4B until the packed matrix almost fills the pipette.

    The dimension of the packed column should be about 27 x 0.3 cm.

4. Wash the column with several column volumes of TE (pH 7.6) containing 0.1 M NaCl. After washing is completed, use a hemostat to close the tubing at the bottom of the column.

5. Use a pasteur pipette to remove the fluid above the Sepharose CL-4B. Apply the cDNA (in a volume of 50 µl or less) to the column. Release the hemostat and allow the cDNA to enter the gel matrix. Wash the microfuge tube used to store the cDNA with 50 µl of TE (pH 7.6) and apply this to the column. Fill the bubble tubing with TE (pH 7.6) containing 0.1 M NaCl.

    IMPORTANT: Do not allow the column to run dry at any stage!

6. Monitor the progress of the cDNA through the column using a handheld minimonitor. Begin collecting 2-drop fractions (~60 µl) in microfuge tubes when the radioactive cDNA has traveled two-thirds the length of the column. Continue collecting fractions until all of the radioactivity has eluted from the column.

7. Measure the radioactivity in each fraction by Cerenkov counting.

8. Analyze small aliquots of each fraction (~5 µl) by electrophoresis through a 1% agarose gel, using as markers end-labeled fragments of DNA of known size (200 bp to 5 kb). Store the remainder of the fractions at –20°C until the autoradiograph of the agarose gel is available.

9. After electrophoresis, transfer the gel to a piece of Whatman 3MM paper. Cover the gel with Saran Wrap and dry it on a commercial gel dryer. Heat the gel to 50°C for the first 20–30 minutes of drying and then turn off the heat. Continue drying the gel under vacuum for an additional 1–2 hours.

10. Expose the dried gel to X-ray film at –70°C with an intensifying screen.

11. Add 0.1 volume of 3 M sodium acetate (pH 5.2) and 2 volumes of ethanol to fractions containing cDNA molecules that are ≥500 bp in length. Allow the cDNA to precipitate for at least 15 minutes at 4°C. Recover the DNA by centrifugation at maximum speed for 15 minutes at 4°C in a microfuge.

12. Dissolve the DNA in a total volume of 20 µl of 10 mM Tris-Cl (pH 7.6).

13. Determine the amount of radioactivity present in a small aliquot and calculate the total amount of cpm available in the selected fractions. Calculate the total quantity of cDNA available for ligation to bacteriophage λ arms (for calculation of second-strand cDNA data, please see MC3, Protocol 11.1 Stage 2, Step 8):

$$\frac{\text{cpm available}}{\text{cpm incorporated into second strand}} \times 2x \text{ µg second-strand cDNA synthesized}$$

$$= \text{ µg of cDNA available for ligation}$$

If everything has gone well, 10 µg of poly(A)⁺ RNA should yield at least 250–400 ng, and possibly as much as 3 µg, of cDNA whose size is larger than 500 bp in length.

Construction of cDNA Libraries
Stage 6: Ligation of cDNA to Bacteriophage λ Arms

The final stage of cDNA library construction involves optimizing ligation of the size-fractionated cDNA to bacteriophage λ arms, followed by packaging and plating of the recombinant bacteriophages.

MATERIALS

Reagents and Solutions

Please see Appendix 1 for components of stock solutions, buffers, and reagents. Dilute stock solutions to the appropriate concentrations.

ATP (10 mM)
  Not required if 10x ligation buffer contains ATP
SM

Vectors and Hosts

Bacteriophage λ arms, made as described in Protocol 2.16 or purchased from commercial suppliers
E. coli, fresh overnight culture of an appropriate strain (see MC3, p. 11.66, Table 11-4)
Packaging extracts, available commercially

Enzymes and Buffers

Bacteriophage T4 DNA ligase (100 Weiss units/λ) and 10x buffer
Restriction enzyme

Nucleic Acids/Oligonucleotides

Bacteriophage λ DNA, used to determine the efficiency of the packaging extract
  This DNA is usually supplied with commercial packaging kits.
cDNA
  Prepare as described in Protocol 11.1 Stage 5, Step 12.
Marker DNAs
  Marker DNAs should be fragments of DNA, 200 bp to 5 kb in size.

Gels/Loading Buffers

1% Agarose gel, cast and run in 0.5x TBE

Additional Items

Step 7 of this protocol requires the reagents listed in Protocol 2.23.
Water bath or heating/cooling block (16°C)

Additional Information

Additional protocol (Amplification of cDNA Libraries)	MC3, p. 11.64
Alternative protocol (Ligation of cDNA into a Plasmid Vector)	MC3, p. 11.63
E. coli strains used to amplify cDNA libraries in bacteriophage λ vectors	MC3, p. 11.66
In vitro packaging	MC3, pp. 11.113–11.114
Ligation of cDNA to bacteriophage λ arms	MC3, pp. 11.59–11.60
Plating bacteriophage λ on E. coli strains	MC3, p. 11.62
Troubleshooting	MC3, p. 11.64

METHOD

1. Set up four test ligation/packaging reactions as follows:

Ligation	A	B	C	D
λ vector DNA (0.5 μg/μl)	1.0 μl	1.0 μl	1.0 μl	1.0 μl
10x T4 DNA ligase buffer	1.0 μl	1.0 μl	1.0 μl	1.0 μl
cDNA	0 ng	5 ng	10 ng	50 ng
Bacteriophage T4 DNA ligase (100 Weiss units/μl)	0.1 μl	0.1 μl	0.1 μl	0.1 μl
10 mM ATP (if required)	1.0 μl	1.0 μl	1.0 μl	1.0 μl
H₂O to a final volume of	10 μl	10 μl	10 μl	10 μl

   Incubate the ligation mixtures for 4–16 hours at 16°C. Store the unused portion of the cDNA at –20°C.

2. Package 5 μl of each ligation into bacteriophage λ particles following the directions provided by the manufacturer of the packaging extract.

3. After the packaging reaction is complete, add 0.5 ml of SM to each mixture.

4. Use the fresh overnight cultures of the appropriate strain(s) of *E. coli* to plate 10 μl and 100 μl of a 10^{-2} dilution of each packaging mixture on each strain (please see Protocol 2.1). Incubate the plates for 8–12 hours at the temperature required for the vector/host combination.

5. Count the number of recombinant and nonrecombinant plaques. Ligation A should yield no recombinant plaques, whereas ligations B, C, and D should yield increasing numbers of recombinant plaques.

6. From the number of recombinant plaques, calculate the efficiency of cloning of cDNA (pfu/ng cDNA). If all has gone well, the efficiency should be at least 2×10^4 pfu/ng cDNA. The total yield of recombinants from 5 μg of poly(A)$^+$ RNA should be in excess of 5×10^6.

7. Pick 12 recombinant bacteriophage λ plaques, grow small-scale lysates, and prepare DNAs for digestion with the appropriate restriction enzyme.

8. Analyze the size of the cDNA inserts by electrophoresis through a 1% agarose gel, using as markers fragments of DNA 500 bp to 5 kb in length.

Construction and Screening of Eukaryotic Expression Libraries
Stage 1: Construction and Screening of cDNA Libraries in Eukaryotic Expression Vectors

This protocol summarizes how to construct a cDNA library in a mammalian vector, using commercial kits.

MATERIALS

Reagents and Solutions

Please see Appendix 1 for components of stock solutions, buffers, and reagents. Dilute stock solutions to the appropriate concentrations.

cDNA synthesis kit (see Commercial Kits for cDNA Synthesis and Library Construction, MC3, pp. 11.107–11.108 and 11.71)

Vectors and Hosts

Bacteriophage λ arms and packaging extract, from commercial suppliers

Electrocompetent *E. coli*, for plasmid vectors

or

E. coli, fresh overnight culture of an appropriate strain for propagation of bacteriophage λ (see MC3, p. 11.66, Table 11-4)

Enzymes and Buffers

Restriction enzyme and 10x universal KGB buffer

Nucleic Acids/Oligonucleotides

Expression vector

The choice between a plasmid vector and a bacteriophage λ vector is dictated by the host system used for expression. If *Xenopus* oocytes are used, the expression vector should be a bacteriophage λ vector or a plasmid carrying promoters for RNA polymerase-encoded bacteriophages T3, T7, or SP6. If cultured cells are used, a plasmid-based expression vector should be used that contains a powerful pan-specific promoter and a strong transcriptional terminator. Examples include plasmids of the pCMV series (Stratagene) and pcDNA4 (Invitrogen).

Packaging extracts for bacteriophage λ vectors, available commercially

Poly(A)+ RNA, purified by two rounds of chromatography on oligo(dT)-cellulose (Protocol 7.3)

Additional Items

Step 2 of this protocol requires the reagents listed in Protocol 11.1 Stage 5.

Step 8 of this protocol may require the reagents listed in Protocol 1.26.

Steps 4 and 5 of this protocol require the reagents listed in Protocol 11.1 Stage 6.

Additional Information

METHOD

1. Use a commercial kit to synthesize blunt-ended, double-stranded cDNA and equip the termini with the appropriate linkers or adapters.

2. Fractionate the double-stranded cDNA according to size, using gel-filtration chromatography. For details, please see Protocol 11.1 Stage 5.

   If the size of the target mRNA is known, pool the column fractions that contain double-stranded cDNAs ranging in size from 1 kb smaller than the target mRNA to 1 kb larger than the target mRNA. If the size of the target mRNA is unknown, fractionate the preparation of double-stranded cDNA into three pools containing molecules of different sizes: 500–1500, 1500–3000, and >3000 bp.

3. Digest 10–25 μg of the plasmid or bacteriophage λ expression vector with two restriction enzymes whose recognition sequences occur in the linker-adapters placed at the 5′ and 3′ ends of the cDNA.

   IMPORTANT: Take care to ensure that both restriction enzymes digest the vector DNA to completion. Perform the digestions in sequence rather than simultaneously; purify the DNA by extraction with phenol:chloroform and precipitation with ethanol between digests, and where possible, use gel electrophoresis to check that both of the digests have gone to completion.

   To ensure complete cleavage of the expression vector before the cloning of cDNA, some investigators insert a short "stuffer" fragment of 200–300 bp between the *Sal*I and *Not*I (or other enzyme combination) sites of the polylinker.

4. Set up trial ligations using different ratios of cDNA to bacteriophage λ arms or plasmid DNA.

5. Package an aliquot of the products of each of the ligation reactions into bacteriophage λ particles. Determine the titer of infectious particles generated in each packaging reaction. Alternatively, use electroporation to transform *E. coli* with aliquots of each ligation reaction (please see Protocol 1.26).

6. Test six bacteriophage λ or plasmid recombinants for the presence of cDNA inserts of the appropriate size and determine the ratio of cDNA to vector DNA that generates the largest number of recombinant clones. Calculate the size of the library that can be generated from the ligation reactions containing the optimum ratio of cDNA to vector.

7. Using the optimum ratio of cDNA insert to vector, ligate as much of the cDNA as possible to the bacteriophage λ or plasmid DNA.

   It is usually better to set up many small ligation reactions, rather than one large reaction.

8. Prepare and analyze the recombinants using one of the methods below:

   If a bacteriophage λ vector is used

   Package the ligated cDNA into bacteriophage λ particles following the directions provided by the manufacturer of the packaging extract, measure the titer of the virus stock, and store the stock at 4°C.

   If a plasmid vector is used

   Measure the number of potential recombinants in the ligation reaction by electroporating small aliquots of the ligation mixture into *E. coli* cells (please see Protocol 1.26).

9. Proceed as soon as is feasible to the screening of the eukaryotic expression library.

This method requires a sensitive and specific biological assay for the protein encoded by the gene of interest (see MC3, p. 11.74). To minimize the amount of work involved in screening expression libraries, the cDNA library is divided into a series of pools, each containing a defined number of cDNA clones. Detection of the biological activity encoding a single target cDNA in a screening pool is a linear function of the number of clones in the pool and the sensitivity of the assay. Before screening an entire library, it is essential to set up a reconstruction experiment, in which a cloned cDNA encoding a biological activity is assayed in the presence of increasing numbers of irrelevant cDNA clones. The results of the reconstruction can be used to inform decisions about optimum size of the cDNA pools.

MATERIALS

Reagents and Solutions

Please see Appendix 1 for components of stock solutions, buffers, and reagents. Dilute stock solutions to the appropriate concentrations.

SM

Media and Antibiotics

TB agar plates containing the appropriate antibiotic, for plasmid vectors

Terrific Broth (or other rich medium), containing the appropriate antibiotic for plasmid vectors

Terrific Broth (or other rich medium), for propagation of bacteriophage λ

Enzymes and Buffers

Restriction enzymes and 10x buffers

Nucleic Acids/Oligonucleotides

cDNA library
Prepare as described in Protocol 11.2 Stage 2.

Transfection/injection controls
Whenever possible, choose a positive-control cDNA that encodes a biological activity similar to that of the desired target cDNA. The control cDNA is used (1) to establish the optimum pool size and (2) as an internal positive control to check that a given experiment has worked with the expected efficiency. The positive-control cDNA should be transferred into the appropriate strain of E. coli (if screening a plasmid-based cDNA library) or packaged in bacteriophage λ particles.

Two types of negative controls should be included in screening experiments using Xenopus oocytes: the plasmid vector alone and the in vitro transcription reaction mixture without added cDNA. In addition, Step 1 requires either an empty bacteriophage λ expression vector packaged into bacteriophage λ particles or E. coli transformed with the empty plasmid expression vector.

Additional Items

Step 1 of this protocol may require the reagents listed in Protocols 2.1 or 2.5 and in Protocols 2.23 or 2.24.

Step 1 of this protocol may require the reagents necessary to assay for the biological activity encoded by the positive-control cDNA.

Step 2 of this protocol may require the reagents listed in Protocol 1.26 and in Protocols 2.1 and 2.23 or 2.24.

Step 4 of this protocol requires the reagents necessary to assay the biological activity encoded by the target cDNA.

Steps 1 and 2 of this protocol may require a commercial kit for plasmid purification.

Steps 1 and 3 of this protocol may require the reagents from one of the transfection protocols in Chapter 16, the reagents listed in Protocol 9.6, and the reagents necessary to inject molecules into *Xenopus* oocytes.

METHOD

1. Set up a series of trial experiments to optimize the transfection and expression systems used to screen the library for cDNA clones of interest. This is best done by using a previously cloned cDNA encoding a biological activity for which reliable assays are available.

   If a eukaryotic plasmid expression vector and cultured cell host are used

   a. Inoculate a single colony of *E. coli* harboring a plasmid carrying the cDNA used as a positive control into 10 ml of rich medium (e.g., Terrific Broth containing a selective antibiotic) together with 10, 100, 1000, 10,000, or 100,000 colonies derived from electroporation of *E. coli* with an empty vector. Grow the cells to saturation in an overnight culture incubated with agitation at 37°C.

   b. Use a commercial kit to prepare plasmid DNA of a purity sufficient for efficient transfection of cultured mammalian cells (please see Protocol 1.9).

   c. Use one or more of the methods described in Chapter 16 to transfect the various plasmid preparations into cultures of eukaryotic cells and assay for the biological activity encoded by the positive-control cDNA.

   > When screening the cDNA library, use the transfection method that generates the maximum signal and an acceptably low level of background.

   If a bacteriophage RNA polymerase (e.g., T3, T7, or SP6) is used to transcribe cDNAs from a plasmid expression vector followed by *Xenopus* oocyte injection

   a. Follow Steps 1a and 1b above.

   b. Linearize the pooled, purified plasmid DNAs at the rare restriction site placed at the 3′ end of the cDNAs during library construction and transcribe the templates in vitro into mRNA (please see Protocol 9.6).

   c. Inject the mRNA prepared from the pooled cDNAs into *Xenopus* oocytes and assay the appropriate biological activity or transfect the mRNA into the appropriate cell line.

   If a bacteriophage λ vector containing bacteriophage-encoded RNA polymerase promoters was chosen as an expression vector

   a. Generate a set of bacteriophage suspensions containing different ratios (10:1, 100:1, 1000:1, etc.) of an empty bacteriophage λ vector to a recombinant bacteriophage λ harboring the control cDNA. Infect an appropriate strain of *E. coli* with a multiplicity of bacteriophage particles that yields near-confluent lysis of bacterial lawns or complete lysis of infected cells grown in liquid culture (please see Protocol 2.1 or 2.5).

   b. Prepare bacteriophage λ DNA from the plates (please see Protocol 2.23) or from the liquid cultures (please see Protocol 2.24).

c. Linearize the bacteriophage λ DNA at the rare restriction site placed at the 3′ end of the cDNAs during library construction and transcribe the cDNAs into mRNA as described in Protocol 9.6.

d. Inject the prepared mRNA into *Xenopus* oocytes and assay the mRNAs for their ability to encode the biological activity of the cDNA used as a positive control.

2. Using the results obtained in Step 1 as a rough guide, divide the cDNA library into pools of a suitable size for screening and transform or transfect *E. coli* with the expression library pools. Prepare plasmid or bacteriophage λ DNA that will be used to screen the pools for the presence of the target cDNA.

If the expression library is constructed in a plasmid vector and pools are grown on solid medium

a. Withdraw a sufficient number of aliquots (e.g., 50–100 aliquots, each capable of generating ~1000 recombinants) from the ligation mixture (Protocol 11.2 Stage 1, Step 8) and introduce each aliquot individually into *E. coli* by electroporation. Plate the entire contents of each electroporation cuvette onto a single dish of Terrific agar medium containing the appropriate antibiotic. Incubate the plates overnight at 37°C.

b. Estimate the number of colonies on each plate. Scrape the colonies from each plate into 5 ml of rich medium (e.g., Terrific broth containing a selectable antibiotic). Grow the bacterial suspensions to saturation and then prepare purified plasmid DNA for screening.

If the expression library is constructed in a plasmid vector and pools are grown in liquid medium

The advantages of this procedure are that the plating step is omitted, thus saving time and effort, and there is less chance that any individual cDNA will be lost because of poor growth on agar or because of mechanical damage.

a. Withdraw a sufficient number of aliquots (e.g., 50–100 aliquots, each capable of generating ~1000 recombinants) from the ligation mixture (Protocol 11.1 Stage 1, Step 8) and introduce them individually into separate cultures of *E. coli* by electroporation. The number of transformants should equal the desired pool size in each culture.

b. Add 1 ml of rich medium (e.g., Terrific broth) without antibiotics to each bacterial culture immediately after electroporation. Incubate the culture for 1 hour at 37°C with very gentle agitation.

c. Use an aliquot of each bacterial culture to inoculate separate 5-ml cultures. After growing the bacterial cultures to saturation, prepare purified plasmid DNA for screening.

If the expression library is constructed in a bacteriophage λ vector and pools are to contain 10,000 to 100,000 cDNAs

a. Plate an appropriate volume of packaging mixture to produce a semiconfluent lawn of recombinant bacteriophage plaques. Alternatively, use a sufficient amount of packaging mixture to obtain complete lysis of *E. coli* cells grown in liquid culture.

b. Isolate bacteriophage λ DNA by a plate lysis procedure or a liquid culture procedure (please see Protocol 2.23 or 2.24).

If the expression library is constructed in a bacteriophage λ vector and pools are to contain 10,000 or fewer cDNAs

a. Plate recombinant bacteriophages, equal in number to the pool size, at a density of 1000 pfu per 100-mm dish.

b. When plaque formation is complete, overlay each dish with 2–3 ml of SM to produce a low-titer plate lysate. Calculate the titer of the lysate.

c. From the low-titer plate lysate, produce a high-titer stock in liquid culture.

d. Purify bacteriophage λ DNA by a liquid culture procedure (please see Protocol 2.24).

3. Transfer the purified clone into the appropriate background to analyze for expression:

If the clone is purified plasmid DNA

Either transfect into cultured mammalian cells or use as a template for in vitro synthesis of RNA by a bacteriophage-encoded RNA polymerase, followed by injection into *Xenopus* oocytes.

If the clone is purified bacteriophage λ DNA

Linearize the DNA and use as a template for in vitro synthesis of RNA, followed by injection into *Xenopus* oocytes.

4. Assay for the biological activity encoded by the target cDNA.

5. After a positive pool(s) of cDNAs has been identified, subdivide and rescreen it in an iterative fashion until a single cDNA encoding the target activity has been identified. Carry out this goal, which is most easily accomplished using the preparation of purified plasmid or bacteriophage λ DNA corresponding to the positive pool(s) of recombinants, as follows:

a. Divide the positive (primary) pool of plasmid or bacteriophage λ DNA into aliquots that generate approximately one tenth of the original number of transformed colonies or plaques.

b. Repeat Steps 3 and 4 above. Include the original positive pool as a positive control in the rescreening assay.

c. Repeat the process of subdivision until a single recombinant is obtained that encodes the desired activity. Make sure that only single well-isolated plaques or bacterial colonies are used to prepare bacteriophage or plasmid DNA for the ultimate screen.

6. Characterize the single isolate that encodes the target activity by DNA sequencing (Chapter 12), expression (Chapter 15 or 16), and northern blotting (Chapter 7).

Exon Trapping and Amplification
Stage 1: Construction of the Library

This protocol describes how to construct and amplify a genomic DNA library in a plasmid vector equipped to express cloned exons in transfected mammalian cells. The plasmid vector used as an example is pSPL3, which is available from Invitrogen.

MATERIALS

CAUTION: Please see Appendix 4 for appropriate handling of materials marked with <!>.

Reagents and Solutions

Please see Appendix 1 for components of stock solutions, buffers, and reagents. Dilute stock solutions to the appropriate concentrations.

ATP (10 mM)
> Not required if 10x ligase buffer contains ATP

Ethanol
Phenol:chloroform (1:1, v/v) <!>
Sodium acetate (3 M, pH 5.2)
TE (pH 8.0)

Vectors and Hosts

E. coli strain HB101, competent for transformation (see Protocols 2.23–2.25)
pSPL3 (Invitrogen) (see Figure 11-1).

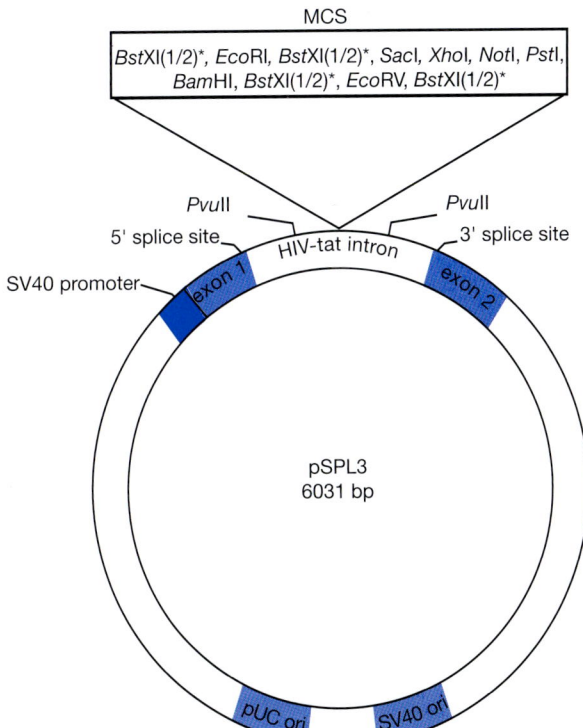

FIGURE 11-1 pSPL3

The SV40 promoter drives the transcription of a heterologous minigene. Two exons are interrupted by an intron derived from the HIV *tat* gene. A multiple cloning site (MCS) has been inserted into the intron. The vector carries the SV40 promoter and the origins of replication from pUC (pUC ori) and from SV40 (SV40 ori). (Adapted, with permission, from D.M. Church.)

Adapted from Chapter 11, Protocol 3 Stage 1, p. 11.81 of MC3.

449

Media and Antibiotics

LB agar plates containing 50 μg/ml ampicillin
LB broth containing 50 μg/ml ampicillin
SOC medium

Enzymes and Buffers

*Pvu*II and 10x buffer
Restriction enzyme(s) and appropriate 10x buffers

Nucleic Acids/Oligonucleotides

Cosmid, BAC, or PAC recombinant clone(s), encompassing the target region of genomic
   DNA, purified as described in Chapter 4
or
Genomic DNA, isolated as described in Chapter 6

Gels/Loading Buffers

0.9% Agarose gel, cast and run in 0.5x TBE

Additional Items

Step 1 of this protocol requires the reagents listed in Protocol 1.20.
Step 6 of this protocol requires the reagents listed in Protocols 1.23–1.25.
Steps 10 and 12 of this protocol require the reagents listed in Protocol 1.1 or 1.9.
Water bath or heating/cooling block (65°C, 15°C, and 37°C).

Additional Information

Overview of exon trapping   MC3, p. 11.79

METHOD

1. Digest pSPL3 with a restriction enzyme(s) that will permit insertion of the genomic DNA. Dephosphorylate the ends of the linearized vector and purify the vector by gel electrophoresis.

2. Digest 1–2 μg of genomic DNA or recombinant vector carrying genomic target DNA with a restriction enzyme(s) that is compatible with the enzyme used to prepare pSPL3 in Step 1.

3. Analyze 10% of the digested genomic DNA on a 0.9% agarose gel cast and run in TAE.

4. Inactivate the restriction enzyme from the digested genomic DNA by heating the reaction mixture to 65°C for 15 minutes. Extract the reaction mixture with phenol:chloroform and recover the DNA by standard ethanol precipitation. Resuspend the DNA at a concentration of 100 μg/ml in TE (pH 8.0).

5. Ligate the genomic DNA to the vector by combining the following:

digested genomic DNA	150 ng
digested, dephosphorylated pSPL3 vector	50 ng
10x ligation buffer	1 μl
bacteriophage T4 DNA ligase	10 Weiss units
H_2O	to 10 μl

Incubate the reaction mixture for 2–3 hours at room temperature or overnight at 15°C.

Be sure to include a control containing only the vector DNA. This control is important for assessing the quality of the library.

6. Transform 40 µl of competent HB101 cells with the ligation reaction.

   Be sure to include positive (vector alone DNA) and negative (no DNA) transformation controls to assess efficiency.

   IMPORTANT: It is imperative that HB101 be used to propagate the library and the vector, because pSPL3 is unstable in other bacterial strains.

7. After transformation, add 800 µl of SOC medium to the cells and incubate the culture for 30–45 minutes at 37°C to allow expression of the antibiotic resistance marker encoded on the plasmid.

8. Amplify the library by growing the bacteria overnight at 37°C in the presence of ampicillin.

If using a single cosmid

Plate 100 µl of the transformation mixture onto an LB agar plate containing 50 µg/ml ampicillin. Grow the remainder of the transformation in a 2-ml liquid culture of LB broth containing 50 µg/ml ampicillin.

If using a BAC, PAC, or pools of cosmids

Plate the entire contents of the transformation onto separate 150-mm LB agar plates containing 50 µg/ml ampicillin.

9. Estimate the efficiency of ligation by comparing the number of recombinant and nonrecombinant clones (i.e., colonies arising from the transformation of HB101 with the ligation mixture containing only pSPL3).

10. If the library was amplified on a 150-mm LB agar plate, proceed to Step 11. If the genomic library was amplified in liquid culture, purify plasmid DNA from the bacteria using a standard alkaline lysis procedure (please see Protocol 1.1) or use a commercial plasmid purification kit. Proceed to Step 13.

11. If the entire library was plated onto a large LB agar plate, flood the plate with 10 ml of LB broth and gently scrape the colonies from the surface of the agar. Minimize the amount of agar that is scraped into the LB broth.

12. Purify the plasmid DNA using a modification of the standard alkaline lysis minipreparation (Protocol 1.1): Triple the amounts of alkaline lysis solutions I (300 µl), II (600 µl), and III (450 µl) added to the bacterial pellet. To simplify the procedure, after addition of alkaline lysis solution I, transfer the bacterial slush to a 2-ml microfuge tube. The rest of the procedure can be performed in 2-ml tubes. After adding alkaline lysis solution III and centrifuging, divide the recovered, crude nucleic acid supernatant into two 1.5-ml tubes. Extract the resulting supernatant with phenol:chloroform and chloroform. Precipitate the supernatant with isopropanol and combine the two dissolved DNA pellets. Alternatively, use a commercial plasmid purification kit (Protocol 1.9).

13. Digest an aliquot of the purified DNA with *Pvu*II to determine the quality of the library.

Exon Trapping and Amplification
Stage 2: Electroporation of the Library into COS-7 Cells

In this stage, COS-7 cells are transfected with a library of plasmids containing the genomic DNA of interest.

MATERIALS

Cells, Tissues/Culture Media

COS-7 cells
COS-7 cells are grown to 75–85% confluency in Dulbecco's Modified Eagle's Medium (DMEM, containing 10% heat-inactivated calf serum) + 1% calf serum.
Phosphate-buffered saline (PBS) without Ca^{++} and Mg^{++}
Trypsin-EDTA

Nucleic Acids/Oligonucleotides

Library of plasmid DNAs for electroporation (~20 μg/ml in PBS lacking cations)
Prepare as described in Protocol 11.3 Stage 1.
Control plasmid

Centrifuges/Rotors/Tubes

Sorvall H-1000B rotor (4°C)

Additional Items

Electroporation device and cuvettes fitted with electrodes 0.4 cm apart
Tissue-culture paraphernalia

Additional Information

Electroporation	MC3, pp. 16.33–16.34 and 16.54–16.57
Overview of exon trapping	MC3, p. 11.79

METHOD

1. Wash the semiconfluent monolayer of COS-7 cells with PBS lacking divalent cations. Remove the cells from the surface of the dishes by treatment with trypsin-EDTA. Collect the cells by centrifugation at 250*g* (1100 rpm in a Sorvall H-1000B rotor) for 10 minutes at 4°C.
   IMPORTANT: Keep the COS-7 cells cold throughout the procedure.

2. Adjust the volume of the DNA sample to be used for transfection to a total volume of 100 μl in PBS without divalent cations.

3. Resuspend 4 x 10⁶ COS-7 cells in a volume of 700 μl of PBS without divalent cations. Mix the cells with the DNA in a cooled electroporation cuvette (0.4-cm chamber). Incubate the mixture for 10 minutes on ice.

4. Gently resuspend the cells and introduce the DNA into the cells by electroporation at 1.2 kV (3 kV/cm) and 25 μF. Immediately return the cuvettes to ice and store them for 10 minutes.

 Adapted from Chapter 11, Protocol 3 Stage 2, p. 11.85 of MC3.

5. Use a wide-bore pipette to transfer 1×10^6 of the electroporated cells into a series of 100-mm tissue culture dishes that contain 10 ml of DMEM containing 10% heat-inactivated fetal calf serum that has been warmed to 37°C. Incubate the cultures for 48–72 hours at 37°C in a humidified incubator with an atmosphere of 5–7% CO_2.

Standard techniques are used here to isolate cytoplasmic RNA from COS cells transfected with a library of recombinant plasmids containing the genomic DNA of interest.

MATERIALS

CAUTION: Please see Appendix 4 for appropriate handling of materials marked with <!>.

Reagents and Solutions

Please see Appendix 1 for components of stock solutions, buffers, and reagents. Dilute stock solutions to the appropriate concentrations.

DEPC-treated H_2O <!>
Ethanol
Ethanol (70%)
NaCl (5 M)
Phenol, equilibrated and H_2O saturated <!>
Phenol:chloroform (1:1, v/v) <!>
SDS (5% w/v)
TKM buffer
     10 mM Tris-Cl (pH 7.5)
     10 mM KCl
     1 mM $MgCl_2$
Triton X-100 (10% v/v) or Nonidet P-40 (4% v/v)

Cells, Tissues/Culture Media

COS-7 cells transfected with the recombinant and nonrecombinant forms of pSPL3 (see Protocol 11.3 Stage 2)
Phosphate-buffered saline (PBS) without Ca^{++} and Mg^{++}

Centrifuges/Rotors/Tubes

Sorvall H-1000B rotor (4°C)

Additional Items

Tissue-culture paraphernalia

Additional Information

How to win the battle with RNase	MC3, p. 7.82
Inhibitors of RNases	MC3, p. 7.83

METHOD

1. Rinse the tissue-culture plates of transfected COS-7 cells three times with ice-cold PBS lacking calcium and magnesium ions. Keep plates on a bed of ice between rinses.

2. Add 10 ml of ice-cold PBS to each plate. Place the plate on the bed of ice and gently scrape the cells off the plate.

Adapted from Chapter 11, Protocol 3 Stage 3, p. 11.87 of MC3.

3. Transfer the cell suspensions to chilled 15-ml polystyrene tubes.

4. Recover the cells by centrifugation at 300*g* (1200 rpm in a Sorvall H-1000B rotor) for 8 minutes at 4°C.

5. Decant as much of the supernatant as possible. Remove the residual supernatant with a pipette.

6. Resuspend the cell pellet from 1×10^6 to 2×10^6 COS cells in 300 µl of TKM buffer and store the suspension on ice for 5 minutes.

7. Add 15 µl of 10% Triton X-100 or 4% Nonidet P-40 and store the cell suspension on ice for an additional 5 minutes.

8. Recover the nuclei by centrifugation at 500*g* (1500 rpm in a Sorvall H-1000B rotor) for 5 minutes at 4°C.

9. Transfer the supernatant to a chilled 1.5-ml microfuge tube.

   IMPORTANT: Use extreme care when removing the supernatant. Do not touch the nuclear pellet. If the nuclear membranes burst, the sample will become too viscous to pipette because of the contamination with genomic DNA.

10. Add 20 µl of 5% SDS and 300 µl of phenol. Vortex the mixture vigorously and separate the organic and aqueous phases by centrifugation at maximum speed for 5 minutes at 4°C in a microfuge.

11. Transfer the aqueous layer to a 1.5-ml microfuge tube containing 300 µl of phenol:chloroform. Vortex the suspension vigorously and then separate the organic and aqueous phases by centrifugation at maximum speed for 3 minutes at room temperature in a microfuge.

12. Transfer the aqueous (upper) layer to a chilled 1.5-ml microfuge tube containing 12 µl of 5 M NaCl and 750 µl of ethanol. Allow the RNA to precipitate for 2–3 hours at –20°C or for 30 minutes at –80°C.

13. Recover the RNA by centrifugation at maximum speed for 10 minutes at 4°C in a microfuge.

14. Discard the supernatant and wash the pellet with 70% ethanol. Dry the pellet in air and redissolve it in 20 µl of DEPC-treated H_2O.

Exon Trapping and Amplification
Stage 4: Reverse Transcriptase–PCR

In this stage of the protocol, cDNA molecules generated from the RNA isolated in Protocol 11.3 Stage 3 are amplified by PCR. Amplified cDNAs containing functional 3′ and 5′ splice sites are selected by digestion with BstXI and cloned into a Bluescript vector.

MATERIALS

CAUTION: Please see Appendix 4 for appropriate handling of materials marked with <!>.

Reagents and Solutions

Please see Appendix 1 for components of stock solutions, buffers, and reagents. Dilute stock solutions to the appropriate concentrations.

Chloroform <!>
DEPC-treated H_2O <!>
Dithiothreitol (1 M)
dNTP solution, containing all four dNTPs, each at a concentration of 1.25 mM
Phenol, equilibrated and H_2O saturated <!>
RNasin or other placental RNase inhibitor

Vectors and Bacterial Strains

E. coli strain DH5α
pBlueScript II (pBSII) (KS or SK) (Stratagene)

Media and Antibiotics

LB agar plates, containing 50 µg/ml ampicillin

Enzymes and Buffers

Mo-MLV reverse transcriptase
Restriction enzymes BstXI and EcoRV, and 10x buffers
Thermostable DNA polymerase and 10x amplification buffer
500 mM KCl
100 mM Tris-Cl (pH 8.3, room temperature)
15 mM $MgCl_2$
Uracil DNA glycosylase (1 unit/µl)

Nucleic Acids/Oligonucleotides

Oligonucleotides primers (20 mM in TE [pH 8.0])
SA2   5′ ATC TCA GTG GTA TTT GTG AGC 3′
SD6   5′ TCT GAG TCA CCT GGA CCA CC 3′
SDDU   5′ AUA AGC UUG AUC UCA CAA GCT GCA CGC TCT AG 3′
SADU   5′ UUC GAG UAG UAC UTT CTA TTC CTT CGG GCC TGT 3′
BSD-U   5′ GAU CAA GCU UAU CGA TAC CGT CGA CCT 3′
BSA-U   5′ AGU ACU ACU CGA AUT CCT GCA GCC 3′
Please note that the 5′ regions of oligonucleotides SDDU, SADU, BSD-U, and BSA-U contain uracil residues to facilitate ligation-independent cloning (see MC3, pp. 11.121–11.124). Oligonucleotides SADU and SDDU are nested primers.
RNA isolated from COS-7 cells transfected with control and recombinant vectors (Protocol 11.3 Stage 3)

Gels/Loading Buffers

1.5% Agarose gel, cast and run in 0.5x TBE

Additional Items

Barrier tips for automatic pipettor
Light mineral oil or wax bead (optional; see Steps 3 and 10)
Microtiter plates or microfuge tubes, 0.5 ml and thin walled
Positive displacement pipette
Step 14 of this protocol requires the reagents listed in Protocol 1.25.
Thermal cycler, programmed with desired amplification protocol
Water baths (42°C, 55°C, and 65°C)

Additional Information

How to win the battle with RNase	MC3, p. 7.82
Inhibitors of RNases	MC3, p. 7.83

METHOD

1. Generate the first-strand cDNA from the cytoplasmic RNAs isolated in the previous stage (Stage 3), using the SA2 oligonucleotide as a primer. In an RNase-free 0.5-ml tube combine:

RNA	3 µl
10x amplification buffer	2.5 µl
dNTP solution at 1.25 mM	4 µl
0.1 M dithiothreitol	1 µl
3′ oligo (SA2, 20 µM)	1.25 µl
DEPC-treated H$_2$O	11.25 µl

   Heat the reaction mixture to 65°C for 5 minutes.

   IMPORTANT: To reduce formation of secondary structures in the RNAs, do not place the reaction mixture on ice!

   Then add:

RNase inhibitor	1 µl
Mo-MLV reverse transcriptase (200 units)	1 µl

   Incubate the reactions for 90 minutes at 42°C.

2. Use both the SA2 oligonucleotide primer and the SD6 oligonucleotide primer to perform a limited second-strand synthesis. Combine the following in a sterile thin-walled amplification tube:

reverse transcriptase reaction (Step 1)	12.5 µl
10x amplification buffer	4 µl
dNTP solution at 1.25 mM	6 µl
SA2 oligonucleotide primer (20 µM)	2 µl
SD6 oligonucleotide primer (20 µM)	2.5 µl
Taq DNA polymerase	1–2 units
DEPC-treated H$_2$O	to 40 µl

   Process the vector-only control in the same way. This will serve as an important control during the PCR analysis.

3. If the thermal cycler is not fitted with a heated lid, overlay the reaction mixtures with 1 drop (~50 µl) of light mineral oil. Alternatively, place a bead of paraffin wax into the tube if using a hot start PCR. Place the tubes in the thermal cycler.

4. Amplify the nucleic acids using the denaturation, annealing, and polymerization times and temperatures listed in the table.

Cycle Number	Denaturation	Annealing	Polymerization
1	5 min at 94°C		
5–6	30 sec at 94°C	30 sec at 62°C	3 min at 72°C
Last cycle			5 min at 72°C

Times and temperatures may need to be adapted to suit the particular reaction conditions.

The purpose of this PCR is to generate double-stranded material for *Bst*XI digestion.

Three-minute extensions are used in this initial PCR to increase the efficiency of capture of longer cDNA products. Limited cycling time is used to minimize the possibility of generating PCR artifacts.

Continue to process the material amplified from COS cells transfected with vector DNA alone (Protocol 11.3 Stage 2).

5. Add 30 units of *Bst*XI to the PCRs. Overlay the reaction mixtures with light mineral oil (if not already on the reaction) and incubate the mixtures overnight at 55°C.

6. Add an additional 20 units of *Bst*XI to the PCR and incubate the reaction for an additional 2–3 hours at 55°C.

7. Meanwhile, digest 1 μg of pBSII (KS or SK) to completion with *Eco*RV. Purify the DNA by extraction with phenol and precipitation with ethanol. Dissolve the digested DNA in H_2O at a concentration of 2 ng/μl.

8. In a 0.5-ml amplification tube, mix the following:

*Bst*XI-digested RT-PCR (Step 6)	5–10 μl
10x amplification buffer	4.5 μl
dNTP solution at 1.25 mM	7.5 μl
SADU oligonucleotide primer (20 μM)	2.5 μl
SDDU oligonucleotide primer (20 μM)	2.5 μl
Taq DNA polymerase	10–20 units
H_2O	to 45 μl

9. In another 0.5-ml amplification tube, mix the following:

pBSII (KS or SK) digested with *Eco*RV (2 ng/μl)	10 μl
10x amplification buffer	10 μl
dNTP solution at 1.25 mM	16 μl
BSD-U oligonucleotide primer (20 μM)	5 μl
BSA-U oligonucleotide primer (20 μM)	5 μl
Taq DNA polymerase	2–4 units
H_2O	to 100 μl

10. If the thermal cycler is not fitted with a heated lid, overlay the reaction mixtures (Steps 8 and 9) with 1 drop (~50 μl) of light mineral oil. Alternatively, place a bead of paraffin wax into the tubes if using a hot start PCR. Place the tubes in the thermal cycler.

11. Amplify the nucleic acids using the denaturation, annealing, and polymerization times and temperatures listed in the table.

Cycle Number	Denaturation	Annealing	Polymerization
1	5 min at 94°C		
25–30	30 sec at 94°C	30 sec at 62°C	3 min at 72°C
Last cycle			5 min at 72°C

Times and temperatures may need to be adapted to suit the particular reaction conditions.

The 5′ ends of the primers used in the second round of PCR amplification contain dUTP rather than dTTP. This facilitates ligation-independent cloning. The 5′ ends of BSD-U and BSA-U are complementary to the 5′ ends of SADU and SDDU.

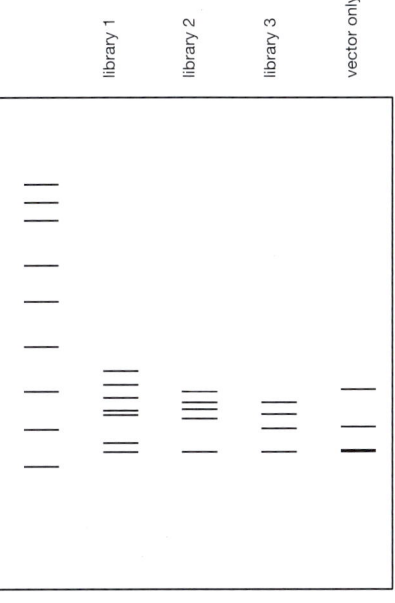

FIGURE 11-2 Representation of a Gel Demonstrating Possible Products Seen after Amplification of the Exon-trapped Products with SADU and SDDU, before Cloning

Amplification of nonrecombinant pSPL3 is shown in the right-hand lane (vector only). The lowest band represents the vector-only splice product, whereas the other two bands are derived from additional cryptic splice sites in the vector. The other three lanes (library 1, 2, 3) show representations of three typical amplification reactions. The vector-only band is present in all other lanes, but it is reduced in intensity. (Adapted, with permission, from D.M. Church and A.J. Butler, *Methods Enzymol.* **303:** 83–99.)

It is best to perform the uracil DNA glycosylase reaction (Step 13) the same day as the amplification. If this is not possible, purify the amplification reactions by extraction with phenol:chloroform, followed by ethanol precipitation, and store them at –20°C until ready for use. The dU-amplified vector (Step 9) can be stored for several weeks at –20°C.

12. Run an aliquot of the PCR on a 1.5% agarose gel in TBE to assess the quality of the reaction. There should be a smear representing several amplification products (see Figure 11.2).

13. In a 0.5-ml microfuge tube, mix the following:

*Bst*XI-digested cDNA (Step 8) amplified in Step 11	3 µl
*Eco*RV-digested pBSII DNA (Step 9) amplified in Step 11	1 µl
10x amplification buffer	1 µl
uracil DNA glycosylase (UDG; 1 unit/µl)	1 µl
H_2O	to 10 µl

Incubate the reaction mixture for at least 30 minutes at 37°C.

IMPORTANT: During this reaction, DNA containing dU residues is digested by uracil DNA glycosylase. In addition, the complementary termini of the plasmid and the amplifed cDNA anneal to form recombinant molecules. To prevent nonspecific annealing of the vector and PCR product, do not place the reaction on ice after the 37°C incubation. If Steps 14 and 15 cannot be performed immediately, leave the reaction at 37°C or store it at –20°C.

14. Use the entire UDG reaction mixture to transform 30–50 µl of DH5α *E. coli* cells using a $CaCl_2$ transformation procedure (please see Protocol 1.25).

15. Plate 100–500 µl of the transformation reaction onto LB agar plates containing 50 µg/ml ampicillin and incubate them overnight at 37°C.

Exon Trapping and Amplification
Stage 5: Analysis of Clones

The final stage of this protocol involves amplification and sequencing of the exon-trapped products.

MATERIALS

Reagents and Solutions

Please see Appendix 1 for components of stock solutions, buffers, and reagents. Dilute stock solutions to the appropriate concentrations.

10x Amplification buffer

dNTP solution, containing all four dNTPs, each at a concentration of 1.25 mM

Media and Antibiotics

LB broth containing 50 µg/ml ampicillin

LB broth containing 30% (v/v) glycerol

Enzymes and Buffers

Thermostable DNA polymerase and 10x amplification buffer

500 mM KCl

100 mM Tris-Cl (pH 8.3, room temperature)

15 mM MgCl$_2$

Nucleic Acids/Oligonucleotides

Exon-trapped products cloned into a *Bst*II vector (see Protocol 11.3 Stage 4)

Oligonucleotide primers (20 mM in TE [pH 8.0])

T3	5' AAT TAA CCC TCA CTA AAG GG 3'	
T7	5' GTA ATA CGA CTC ACT ATA GGG C 3'	
–20	5' GTA AAA CGA CGG CCA GT 3'	
REV	5' AGC GGA TAA CAA TTT CAC ACA GG 3'	

Gels/Loading Buffers

1.5% Agarose gel, cast and run in 0.5x TBE

Additional Items

Barrier tips for automatic pipettor

Light mineral oil or wax bead (optional; see Step 4)

Microtiter plates (96 well)

Multichannel pipette

Positive displacement pipette

Replicating device (96 prong)

Step 8 of this protocol requires the reagents listed in Protocols 12.3–12.5.

Thermal cycler, programmed with desired amplification protocol

Additional Information

Storage of bacterial cultures

MC3, p. A8.5

Adapted from Chapter 11, Protocol 3 Stage 5, p. 11.95 of MC3.

METHOD

1. Dispense 100 μl of LB broth containing 50 μg/ml of ampicillin into each well of a 96-well microtiter plate. Transfer one transformed bacterial colony (from Protocol 11.3 Stage 4, Step 15) at a time into individual wells. Cover the plate with Parafilm and grow the colonies for 3–4 hours at 37°C. It is not necessary to agitate the plate.

2. Prepare a master mix (sufficient for 100 wells) of PCRs by combining the following components together:

10x amplification buffer	250 μl
dNTP solution at 1.25 mM	400 μl
–20 oligonucleotide primer (20 μM)	125 μl
REV oligonucleotide primer(20 μM)	125 μl
H$_2$O	1475 μl
Taq DNA polymerase	75 units

   Aliquot 24 μl of the master mix into each well of fresh 96-well PCR plates.

3. Use a 96-prong replicating device to transfer bacterial cultures from the plate in Step 1 to the plate containing the PCR master mix.

4. If the thermal cycler is not fitted with a heated lid, overlay the reaction mixtures with 1 drop (~50 μl) of light mineral oil. Alternatively, place a bead of paraffin wax into the wells if using a hot start PCR. Place the 96-well plates in the thermal cycler.

5. Amplify the nucleic acids using the denaturation, annealing, and polymerization times and temperatures listed in the table.

Cycle Number	Denaturation	Annealing	Polymerization
1	5 min at 94°C		
25–30	30 sec at 94°C	30 sec at 62°C	30 sec at 72°C
Last cycle	30 sec at 94°C	30 sec at 62°C	5 min at 72°C

   Times and temperatures may need to be adapted to suit the particular reaction conditions.

6. Use a multichannel pipette to make a replica of the 96-well plate (from Step 1) containing the bacterial colonies and allow them to grow overnight at 37°C. The next day, add to each well 100 μl of LB broth containing 30% glycerol. Seal the plate with Parafilm and store it at –80°C.

7. Analyze the amplification products from Step 5 by gel electrophoresis on a 1.5% agarose gel.

8. Determine the sequence of each of the exon-trapped products (please see Protocols 12.3–12.5).

   If pBSIISK⁻ was used for cloning, sequencing with T7 primer will produce sequence from the 5′ end of the exon-trapped product. The exon-trapped product will be flanked by the following pSPL3-derived sequence:

   5′ GTCGACCCAGCA exon-trapped product sequence ACCTGGAGATCC 3′

Direct Selection of cDNAs with Large Genomic Clones

The goal of this method is to identify transcriptionally active genes in cloned segments of genomic DNA. The protocol uses hybridization and affinity purification to recover biotin-labeled cDNAs that bind to a 500-kb segment of human DNA cloned in a BAC vector. However, the method can be easily adapted to other clones of genomic DNAs cloned in high-capacity vectors.

MATERIALS

CAUTION: Please see Appendix 4 for appropriate handling of materials marked with <!>.

Reagents and Solutions

Please see Appendix 1 for components of stock solutions, buffers, and reagents. Dilute stock solutions to the appropriate concentrations.

ATP (10 mM)
> Not required if 10x ligase buffer contains ATP

Biotin-16-dUTP (0.4 mmole)

Chloroform <!>

dNTP solution, containing all four dNTPs, each at a concentration of 0.4 mM

dNTP solution, containing all four dNTPs, each at a concentration of 2.5 mM

Ethanol

2x Hybridization solution
> 1.5 M NaCl
> 40 mM Na phosphate buffer (pH 7.2)
> 10 mM EDTA (pH 8.0)
> 10x Denhardt's solution
> 0.2% SDS

NaOH (0.1 M) <!>

Phenol, equilibrated (pH 8.0) and H_2O saturated

Phenol:chloroform (1:1, v/v) <!>

SDS (10% w/v)

Sodium acetate (3 M, pH 5.2)

20x SSC

Streptavidin bead-binding buffer
> 10 mM Tris-Cl (pH 7.5)
> 1 mM EDTA (pH 8.0)
> 1 M NaCl

Tris-Cl (1 M, pH 7.5)

Enzymes and Buffers

Bacteriophage T4 DNA ligase and 10x ligation buffer, as supplied by manufacturer

10x Nick-translation buffer
> 500 mM Tris-Cl (pH 7.5)
> 100 mM $MgCl_2$
> 50 mM dithiothreitol

Nick-translation mixture
> *E. coli* DNA polymerase I (0.5 units/μl) + pancreatic DNase I (0.007 units/μl) (from BioNick [Invitrogen] or homemade)

Restriction enzymes and 10x buffers

Adapted from Chapter 11, Protocol 4, p. 11.98 of MC3.

Thermostable DNA polymerase and 10x amplification buffer + gelatin
    500 mM KCl
    100 mM Tris-Cl (pH 8.3, room temperature)
    15 mM MgCl$_2$
    0.01% (w/v) gelatin

Nucleic Acids/Oligonucleotides

BAC DNA encompassing the target region, purified as described in Protocol 4.8 or 4.9

Blunt-ended, random-primed cDNAs (see Protocol 11.1 Stage 1)

C$_0$t1 genomic DNA, available from Invitrogen

Cytoplasmic poly(A)$^+$ RNA, purified from tissue or cell line of interest

DNA markers for gel electrophoresis

Oligonucleotide primers (10 mM in TE [pH 8.0])
    Oligo3   5' CTC GAG AAT TCT GGA TCC TC 3'
    Oligo4   5' GAG GAT CCA GAA TTC TCG AGT T 3'

Positive-control DNA
    The positive-control DNA should be a part of a gene or expressed sequence tag (EST) known to be present in the starting segment of DNA. If no genes are known, a single-copy control DNA can be seeded into the genomic target before labeling (at a 1:1 molar ratio). Label the DNA sample using one of the protocols in Chapter 9. The choice of labeling protocol is determined by the experimental goal.

Radioactive Compounds

[α-^{32}P]dCTP (3000 Ci/mmole) <!>

Gels/Loading Buffers

1.0% Agarose gel, cast and run in 0.5x TBE

Additional Items

Barrier tips for automatic pipettor

Handheld minimonitor

Hybridization oven with rotating spindle *or* shaking water bath

Light mineral oil or wax bead (optional; see Step 16)

Magnetic separator for removing streptavidin-coated beads (Dynal Biotech)

Microfuge tubes, thin walled for amplification

Positive displacement pipette

Sephadex G-50 spun column, equilibrated in TE (pH 7.6)

Step 1 of this protocol requires the reagents listed in Protocol 10.2 and the reagents for labeling listed in Protocols 9.3, 9.5, and 9.6.

Step 2 of this protocol requires the reagents for the synthesis of (1) random-primed, blunt-ended cDNAs (Protocol 11.1 Stages 1 and 2) or (2) blunt-ended PCR-amplified inserts from a cDNA library.

Steps 19 and 20 require the reagents listed in Protocols 6.8 and 6.10.

Streptavidin-coated magnetic beads

Thermal cycler, programmed with desired amplification protocol

Water baths or heating/cooling blocks (100°C, 14°C, 65°C, and 4°C)

Additional Information

Magnetic beads	MC3, pp. 11.118–11.120
Nick translation	MC3, pp. 9.12–9.13
Overview of direct selection of cDNAs	MC3, pp. 11.98–11.100
Phosphate buffers	MC3, p. A1.5
Troubleshooting	MC3, p. 11.106

METHOD

1. Label oligonucleotides 3 and 4 separately by phosphorylation at their 5′ termini using polynucleotide kinase, as described in Protocol 10.2. Mix the labeled complementary oligonucleotides in equal molar ratios (~2 μg of each), denature them for 10 minutes at 100°C, and slowly cool them to room temperature. During this process, the oligonucleotides anneal to form an adapter. Adjust the concentration of the adapter to 1 μg/ml.

2. Prepare at least 2 μg of double-stranded cDNA from cytoplasmic polyadenylated RNA by random priming (please see Protocol 11.1).

3. Prepare the following ligation reaction in a sterile 0.5-ml microfuge tube:

double-stranded blunt-ended cDNA (Step 2)	2 μg
oligonucleotide amplification cassette mixture at 1 μg/μl, from Step 1	3 μl
10x T4 DNA ligase buffer	3 μl
10 mM ATP (if required)	3 μl
T4 DNA ligase at 1 unit/μl	3 μl
H$_2$O	to 30 μl

   Incubate the ligation for 16 hours at 14°C. Inactivate the T4 DNA ligase by a 10-minute incubation at 65°C.

4. Purify the products of the ligation reaction by phenol:chloroform extraction, spun-column chromatography through Sephadex G-50, and standard ethanol precipitation. Dry the pellet and resuspend it in 10 μl TE (pH 7.6).

5. Incorporate biotinylated residues into the BAC genomic DNA clone using nick translation. Prepare the following nick-translation reaction in a sterile 0.5-ml microfuge tube (label each BAC DNA separately):

purified BAC DNA (0.1 mg/ml)	1 μl
biotin-16-dUTP (0.04 mM)	1 μl
10x nick-translation buffer	2 μl
dNTP mix for nick translation (0.4 mM)	1 μl
[α-^{32}P]dCTP (3000 Ci/mmole)	1 μl
DNA polymerase/DNase I (5 units/μl)	1 μl
H$_2$O	to 20 μl

   Incubate the reaction for 2 hours at 4°C.

6. Purify the radiolabeled and biotinylated products of the nick-translation reaction by spun-column chromatography through Sephadex G-50 and standard ethanol precipitation. Resuspend the pellet in 10 μl of TE and store it at –20°C.

7. In a 1.5-ml sterile microfuge tube, wash 3 mg (300 μl) of beads with 500 μl of streptavidin bead-binding buffer three times. Following each wash, remove the beads from the binding buffer using a magnetic separator. Resuspend the beads at a concentration of 10 mg/ml in streptavidin bead-binding buffer.

8. Test an aliquot of each labeling reaction (from Step 6) for the ability to bind to streptavidin beads. Prepare the following binding reaction in a sterile 0.5-ml microfuge tube:

washed streptavidin-coated beads (Step 7)	20 μl
labeled genomic DNA (Step 6)	1 μl
streptavidin bead-binding buffer	29 μl

   Incubate the reaction for 15 minutes at room temperature with occasional gentle mixing. Remove the beads from the binding buffer using a magnetic separator and transfer the supernatant to a fresh, sterile 0.5-ml microfuge tube. Use a minimonitor to measure the radioactivity present on the beads and in the supernatant. If the ratio of bound-to-free cpm is >8:1, proceed with the selection.

> If the ratio of bound-to-free cpm is <8:1, it is likely that the DNA was resistant to proper labeling in Step 5. Before labeling, try purifying the BAC DNA further by several rounds of extraction with phenol:chloroform and passing it through a Sephadex G-50 spun column.

9. Block or "repeat suppress" repetitive sequences in the pool of cDNA (from Step 4) using C_0t1 DNA as follows:

   a. Prepare the following annealing reaction in a sterile 0.5-ml microfuge tube:

cDNA carrying linkers (Step 4)	5 μl (1 μg)
human genomic C_0t1 DNA	5 μl (1 μg)

   b. Overlay the reaction mixture with light mineral oil (~50 μl) to prevent evaporation and denature the DNA by heating for 10 minutes at 100°C. Cool the reaction mixture to 65°C and deliver 10 μl of 2x hybridization solution under the oil. Mix the components gently. Incubate the reaction mixture for 4 hours at 65°C.

10. After hybridization of the cDNA pools to the C_0t1 DNA is complete, set up the primary direct selection. Deliver 5 μl (50 ng) of biotinylated BAC DNAs from Step 6 into a fresh microfuge tube and overlay the solution with light mineral oil (~50 μl) to prevent evaporation. Denature the BAC DNA by heating to 100°C for 10 minutes. Cool the reaction to 65°C.

11. Prepare the following annealing reaction in a sterile 0.5-ml microfuge tube:

biotinylated BAC DNA from Step 10	5 μl (50 ng)
blocked cDNA from Step 9 (1 μg cDNA plus	20 μl
1 μg C_0t1 DNA)	
2x hybridization solution	5 μl

    Mix the reagents gently and incubate the reaction for >54 hours at 65°C in a rotating hybridization oven or shaking water bath.

12. To capture and wash the genomic DNA and hybridized cDNAs, prepare the following in a sterile 1.5-ml microfuge tube:

washed streptavidin-coated beads	100 μl
annealing reaction from Step 11	30 μl
streptavidin bead-binding buffer	100 μl

    Incubate the mixture for 15 minutes at room temperature with occasional gentle mixing. Remove the beads from the binding buffer using a magnetic separator and then remove and discard the supernatant. Wash the beads twice, for 15 minutes each time, in 1 ml of 1x SSC/0.1% SDS at room temperature followed by three washes, 15 minutes each, in 1 ml of 0.1x SSC/0.1% SDS at 65°C. After the final wash, transfer the beads to a fresh microfuge tube.

13. To elute the hybridizing cDNAs from the beads in Step 12:

    a. Add 100 μl of 0.1 M NaOH and incubate the reaction mixture for 10 minutes at room temperature.

    b. Add 100 μl of 1 M Tris-Cl (pH 7.5).

    c. Desalt the mixture by spun-column chromatography through Sephadex G-50.

14. Transfer three aliquots (1, 5, and 10 μl, respectively) from the 200 μl of eluted cDNA (Step 13) to 0.5-ml sterile amplification tubes.

15. To each tube, add the following:

primer oligo3 (10 mM)	5 µl
10x amplification buffer	2.5 µl
dNTP mixture for PCR (2.5 mM)	2.5 µl
thermostable DNA polymerase (5 units/µl)	0.2 µl
H_2O	to 25 µl

Set up two control reactions at the same time as the above test reactions. For the negative control, include all of the components listed above, but omit the template cDNA. For the second control, include 10 ng of the starting cDNA (Step 4).

16. If the thermal cycler is not fitted with a heated lid, overlay the reaction mixtures with 1 drop (~50 µl) of light mineral oil to prevent evaporation. Alternatively, place a bead of wax into the tube if using a hot start PCR. Place the tubes in the thermocycler.

17. Amplify the nucleic acids using the denaturation, annealing, and polymerization times and temperatures listed below.

Cycle Number	Denaturation	Annealing	Polymerization
30 cycles	30 sec at 94°C	30 sec at 55°C	1 min at 72°C

Times and temperatures may need to be adapted to suit the particular reaction conditions.

18. Analyze 10% of each amplification reaction on a 1% agarose gel cast and run in 0.5x TBE, including DNA markers of an appropriate size. Stain the gel with ethidium bromide or SYBR Gold to visualize the DNAs, estimate the concentration of the amplified cDNA, and determine which input cDNA concentration produces the highest yield of cDNA.

19. Onto a second 1% agarose gel, load the same amounts (~0.5 µg per lane) of the amplified products of each PCR, as well as the appropriate size markers. Also, load 0.5 µg of randomly primed cDNA from Step 4.

20. Transfer the separated DNA species to a membrane by Southern blotting (Protocol 6.8) and hybridize with the radiolabeled, positive-control cDNA.

21. Once the positive-control enrichment is confirmed, scale up the optimal amplification reaction to yield at least 1.5 µg of selected cDNAs. Extract the pooled reactions with phenol:chloroform and recover the DNA by standard ethanol precipitation. Resuspend the dried cDNA pellet in 7.5 µl of TE (200 µg/ml).

Proceed to perform the secondary selection under the same conditions as the primary selection, using 1 µg of the primary selected cDNA and 50 ng of the target DNA (total length 500 kb in this example).

22. Block repetitive sequences in the primary selected cDNA as described in Step 9 (with 1 µg of cDNA being used and 0.5 µg of cDNA being held in reserve for later analysis).

23. Set up the secondary selection as described in Steps 9–13.

24. Analyze the products of the second amplification by electrophoresis through a 1% agarose gel as in Step 18, including a lane with 0.5 µg of primary cDNA (held in reserve) and a lane with 0.5 µg of the starting cDNA.

25. Analyze the gel by Southern blotting and hybridization with the radiolabeled reporter probe as described in Step 20.

26. Once enrichment of the control is confirmed, clone the selected cDNAs into the appropriate vector(s). The restriction enzyme sites in the amplification cassette facilitate the cloning of the secondary selected cDNAs into bacteriophage or plasmid vectors.

DNA Sequencing

BACKGROUND INFORMATION

Background information found in *Molecular Cloning: A Laboratory Manual,* 3rd edition (hereafter MC3) unless otherwise indicated.

Strategies used in DNA sequencing   MC3, pp. 12.3–12.9

Generation of a Library of Randomly Overlapping DNA Inserts

Shotgun sequencing of a large segment of DNA involves random fragmentation of the target region into smaller segments that are subsequently cloned into a bacteriophage M13 vector. The goal is to create a library of overlapping clones that provide at least fivefold coverage over the entire length of the target fragment.

In this protocol, the target segment is sheared by sonication or nebulization. The termini of the resulting DNA fragments are repaired, phosphorylated, and fractionated according to size by gel electrophoresis. Molecules 0.8–1.5 kb in size are then recovered and cloned into the linearized, dephosphorylated DNA of a bacteriophage M13 vector.

MATERIALS

CAUTION: Please see Appendix 4 for appropriate handling of materials marked with <!>.

Reagents and Solutions

Please see Appendix 1 for components of stock solutions, buffers, and reagents. Dilute stock solutions to the appropriate concentrations.

Ammonium acetate (10 M)

ATP (20 mM)

dNTP solution containing all four dNTPs, each at a concentration of 2 mM (see MC3, p. 12.107)

Ethanol

Ethanol (70%)

Glycerol (sterile)

$MgCl_2$ (1 M)

$MgSO_4$ (1 M)

NaCl (5 M)

Phenol, equilibrated to pH 7.6 and saturated with H_2O <!>

Phenol:chloroform (1:1, v/v) <!>

Polyethylene glycol (30% [w/v] PEG 8000) <!> (optional; see Step 1)

Polyethylene glycol (20% [w/v] PEG 8000) <!> in 2.5 M NaCl

TE (pH 7.6)

10x TM buffer

    0.5 M Tris-Cl (pH 8.0)

    150 mM $MgCl_2$

TTE buffer

    0.5% (v/v) Triton X-100 in TE (pH 8.0)

Vectors and Hosts

Bacteriophage M13 vector DNA, equipped for blue-white screening, linearized, blunt-ended, and dephosphorylated

    Vector DNAs of this type are available from several commercial vendors or can be prepared as described in MC3, p. 12.14.

E. coli, male, competent for transformation or electroporation (e.g., XLF'-Blue or DH5α-F')

    Available commercially or may be prepared as described in Protocols 1.23–1.26

Media and Antibiotics

LB or 2YT medium containing 5 mM $MgCl_2$

Top agarose containing IPTG and X-gal

YT agar plates

Enzymes and Buffers

Bacteriophage T4 DNA ligase and 10x buffer
Bacteriophage T4 DNA polymerase and 10x buffer
Bacteriophage T4 polynucleotide kinase and 10x buffer
Klenow fragment of *E. coli* DNA polymerase

Nucleic Acids/Oligonucleotides

Bacteriophage λ or φX174 DNA, cleaved to completion with *Alu*I (50 µg/ml in TE, pH 8.0)
DNA size markers, 123-bp ladder (Invitrogen)
Target DNAi, usually prepared by releasing a cloned DNA from its vector by digestion with a
   restriction enzyme that does not cleave within the target sequences
      If possible, use a restriction enzyme that generates cohesive termini. This simplifies ligation of the
      target DNA into concatemers. After release from the vector, purify the target DNA by gel elec-
      trophoresis and dissolve the purified DNA in TE (pH 7.6) at a concentration of 1 mg/ml.

Gels/Loading Buffers

Standard 0.7% agarose gel, cast and run in 0.5x TBE
Standard 1% agarose gel, cast and run in 0.5x TBE
Low-melting-temperature 0.8% agarose gel, cast and run in 0.5x TBE
or
Neutral polyacrylamide gel (5%)

Centrifuges/Rotors/Tubes

Centrifuge capable of handling deep-well microplates and standard microtiter plates,
   e.g., Beckman GPR centrifuge with a Microplus carrier

Additional Items

Cup-horn sonicator (Misonix XL2000 Microson) with micro cup-horn
or
Nebulizer (IPI Medical Products, model 4207)
Deep-well microplates, lids, and caps, approximately 1.2-ml capacity, e.g., Beckman
   96-tube box, fitted with one-piece lid and cap
      Two lids and one cap are required to prepare 96 M13 DNA templates.
Microtiter plates (96 well, fitted with lids)
Multichannel pipettor (8 or 12 channel)
Multitube vortex mixer
Silver foil tape (3M), used for sealing 96-tube boxes
Toothpicks
Water baths (16°C, 37°C, 68°C, and 80°C)

Additional Information

Additional protocol (Preparation of Dephosphorylated Blunt-ended Bacteriophage M13 Vector DNA for Shotgun Cloning)	MC3, p. 12.24
Klenow fragment of *E. coli* DNA polymerase	MC3, pp. 12.101–12.102
Methods used to fragment DNA for shotgun sequencing	MC3, pp. 12.10–12.12
Polyethylene glycol as a crowding agent	MC3, p. 1.152
Preparation of stock solutions of dNTPs and ddNTPs for DNA sequencing	MC3, p. 12.107
Use of microtiter plates in DNA sequencing	MC3, p. 12.100

METHOD

Self-ligation is required to ensure that sequences at the ends of the target DNA are adequately represented in the population of fragments used to construct the library.

1. Transfer 5–10 μg of the purified target DNA (see Materials) to a fresh microfuge tube and add:

10x bacteriophage T4 DNA ligase buffer	2.5 μl
5 mM ATP	2.5 μl
30% (w/v) PEG 8000 (*optional*)	5.0 μl
bacteriophage T4 DNA ligase	0.5–2.0 Weiss units
H₂O	to 25 μl

   Incubate the mixture for 4 hours at 16°C and inactivate the ligase by heating the mixture for 15 minutes at 68°C.

   > If the target DNA is blunt-ended, include PEG in the reaction to increase the efficiency of ligation.
   >
   > Use 0.5 Weiss unit for cohesive termini and 2.0 Weiss units for blunt termini.
   >
   > ATP is not required if it is included in the 10x ligase buffer.

2. Add 175 μl of TE (pH 7.6) and purify the ligated DNA by extraction with phenol:chloroform. Recover the ligated DNA by precipitation with 3 volumes of ethanol in the presence of 2.0 M ammonium acetate. After centrifugation at maximum speed for 5 minutes in a microfuge, wash the pellet with 0.5 ml 70% ethanol at room temperature and centrifuge again.

3. Remove as much of the supernatant as possible and allow the last traces of ethanol to evaporate at room temperature. Dissolve the DNA in 25 μl of TE (pH 7.6) in a microfuge tube.

4. Fragment the target DNA into segments 0.8–1.5 kb in length by sonication or nebulization.

To fragment the DNA by sonication

a. Place ice water in the cup horn of the sonicator. Set the sonicator power switch to on, the timer to hold, and the power setting to 10. Apply two 40-second pulses and allow the sonicator to warm up.

b. Place the tube containing the DNA in the ice water such that the bottom of the tube is 1–2 mm above the hole in the center of the cup-horn probe and sonicate the DNA.

c. Centrifuge the tube briefly to collect the sonicated DNA sample at the bottom of the tube and place it on ice.

d. Analyze 1 μl of the sonicated DNA sample with the appropriate molecular-weight markers by electrophoresis through a 0.7% agarose gel. Keep the remainder of the DNA on ice while the gel is running.

To fragment the DNA by nebulization

a. Prepare the following DNA solution and place it in the nebulizer cup.

DNA sample (from Step 3)	5–10 μg
10x TM buffer (pH 8.0)	200 μl
sterile 100% glycerol	1 ml
sterile H₂O	to a final volume of 2 ml

b. Place the DNA sample in an ice-water bath and nebulize using the empirically determined optimal conditions.

c. Place the nebulizer in a suitable centrifuge rotor and cushion it with pieces of styrofoam. Centrifuge at 2000*g* (1000 rpm in a centrifuge equipped with a microplate adapter) briefly at 4°C to collect the DNA sample at the bottom of the nebulizer cup.

d. Divide the DNA sample into four equal aliquots in 1.5-ml microfuge tubes, perform standard ethanol precipitation of the DNA, and dry the DNA pellets under vacuum.

e. Dissolve each DNA pellet in 35 μl of TE (pH 7.6) and analyze 1 μl of the sheared DNA with appropriate molecular-weight markers by electrophoresis through a 0.7% agarose gel. Keep the remainder of the DNA at 4°C while the gel is running.

5. To the fragmented DNA (~25 μl), add:

10x bacteriophage T4 DNA polymerase buffer	4.0 μl
2.0 mM solution of four dNTPs	4.0 μl
bacteriophage T4 DNA polymerase	5 units
H₂O	to 40 μl

Incubate the reaction for 15 minutes at room temperature and then add approximately 5 units of the Klenow fragment. Continue incubation for an additional 15 minutes at room temperature. The combination of DNA polymerases results in efficient polishing of the termini of the fragmented DNA.

> The termini produced by nebulization and sonication are highly heterogeneous, consisting of blunt-ended and frayed ends, with and without phosphate residues. Because only a fraction of these molecules can be repaired by DNA polymerases, the efficiency with which hydrostatically sheared DNA can be cloned in bacteriophage M13 vectors is generally low. However, 5–10 μg of sonicated, repaired, and size-selected target DNA generally yields several thousand recombinant clones.

6. Purify the DNA by extraction with phenol:chloroform. Transfer the aqueous (upper) phase to a fresh tube, and adjust the solution to 0.1 M NaCl. Recover the DNA by precipitation with 2 volumes of ethanol. Wash the pellet with 70% ethanol.

7. Redissolve the precipitated DNA in 25 μl of TE (pH 7.6).

8. Combine the following in a microfuge tube:

fragmented DNA	23 μl
10x polynucleotide kinase buffer	3 μl
20 mM ATP	3 μl
bacteriophage T4 polynucleotide kinase	1 unit

> Bacteriophage T4 polynucleotide kinase catalyzes the phosphorylation of the 5′ termini of the blunt-ended DNA fragments. This step is not mandatory, but, in most cases, leads to more efficient ligation of the fragments to the vector.

9. Incubate the reaction for 30 minutes at 37°C.

10. Purify the fragments of DNA of the desired size (0.8–1.5 kb) by electrophoresis through a low-melting-temperature agarose gel (0.8%) or 5% neutral polyacrylamide gel (see Chapter 5).

11. Recover the target DNA from the gel by one of the methods described in Chapter 5. Dissolve the purified DNA in 25 μl of TE (pH 7.6).

12. Check the integrity and recovery of the purified DNA by analyzing an aliquot (1.0 μl) by electrophoresis through a 1% agarose gel.

13. Set up a series of test ligations containing 50 ng (~0.01 pmole) of linearized and dephosphorylated vector DNA and increasing concentrations of fragmented, blunt-ended, phosphorylated target DNA (please see Table 12-1).

14. Introduce aliquots of the test ligations into competent *E. coli* of the appropriate strain by electroporation or transformation (please see Protocols 1.23–1.26). Plate the bacteria on media containing IPTG and X-gal. Incubate the plates overnight at 37°C.

15. The following day, count the number of blue and colorless plaques.

16. Set up a large-scale ligation reaction using the minimum amount of fragmented, blunt-ended target that will yield sufficient recombinant clones to complete the sequencing project and transform *E. coli* with the ligated DNA. Incubate the plates overnight at 37°C.

17. The following day, collect the plates and store the resulting transformants under the appro-

TABLE 12-1 Test Ligation Reactions of Dephosphorylated Vector DNA

Tube	Type of DNA[a] A	B	C	H₂O (µl)	10x Ligation Buffer (µl)	ATP (5 mM) (µl)[b]	T4 DNA Ligase (Weiss units)	30% PEG (optional)[c] (µl)
1	+			5.0	1.0	1.0	2.0	1.6
2	+			6.0	1.0	1.0	—	1.6
3	+	+		3.0	1.0	1.0	2.0	1.6
4	+		+	3.0	1.0	1.0	2.0	1.6
5	+		+	3.0	1.0	1.0	2.0	1.6
6	+		+	3.0	1.0	1.0	2.0	1.6
7	+		+	3.0	1.0	1.0	2.0	1.6

[a]A = 50 ng of linearized, dephosphorylated vector (in a volume of 2 µl); B = 20 ng of bacteriophage λ DNA or φX174 RF DNA cleaved to completion with *AluI* (in a volume of 2 µl); C = fragmented, blunt-ended, phosphorylated target DNA. Tubes 4, 5, 6, and 7 should contain 10, 20, 50, and 100 ng of size-selected, fragmented target DNA, respectively (100 ng of size-selected fragments = ~ 0.1 pmole).
[b]Some commercial ligase buffers contain ATP. When using such buffers, omit the ATP.
[c]The efficiency of ligation can be increased by adding PEG 8000 (30% w/v) to a concentration of 5% in the final ligation mixture (see Protocol 1.19). It is important to warm the stock solution of PEG 8000 to room temperature before adding it as the last component of the ligation reactions. DNA can precipitate at cold temperatures from solutions containing PEG 8000.

priate conditions until required. Prepare template DNAs from a series of individual colorless plaques as described in Protocol 3.4.

18. For each set of 96 clones, inoculate 100 ml of LB medium or 2x YT medium in a 500-ml flask with a single colony of a suitable F' strain of *E. coli* (e.g., XL1-Blue, XL1-Blue MRF', or DH5αF'). Incubate the culture for 6–8 hours at 37°C with agitation at 300 rpm or until the culture enters the early log phase of growth.

19. Add MgSO₄ to the culture to a final concentration of 5 mM.

20. Use a multichannel pipettor to transfer 0.8-ml aliquots of the cells to individual tubes in a 96-tube box.

21. Wearing gloves, use sterile toothpicks to transfer individual well-separated, colorless bacteriophage M13 plaques into each tube of the 96-tube arrays. Stab the toothpick into the middle of a plaque and drop the toothpick into the culture tube.

22. To avoid confusion, leave the toothpicks in the tubes until all 96 tubes in the box have been inoculated.

23. When the last plaque has been picked, remove the toothpicks and seal the box. Label the box and place it in an orbital shaker set at 250–300 rpm and 37°C. Repeat Steps 20 and 21 as needed. Incubate the infected cultures for 8–12 hours.

24. Remove the boxes from the incubator and pellet the bacterial cells by centrifugation at 2400*g* (3000–3250 rpm in a centrifuge equipped with a microplate adapter) for 20 minutes.

25. Use a multichannel pipettor to transfer 120 µl of 20% PEG 8000 in 2.5 M NaCl to individual tubes in a fresh 96-tube box.

26. Carefully remove the tubes from the centrifuge; use the multichannel pipettor to transfer 0.6 ml of each supernatant to a tube containing the PEG/NaCl solution.

     IMPORTANT: Do not disturb the pellet of bacterial cells during this step. Inclusion of bacteria will drastically reduce the quality of DNA sequence obtained.

27. Place a 96-tube cap over the tubes containing the bacteriophage suspensions and PEG/NaCl solution. Make sure that a liquid-tight seal has formed and then mix the solutions by inverting the box several times. Incubate the box for 30 minutes at room temperature, followed by a 30–60-minute incubation on ice.

28. Collect the precipitated bacteriophage by centrifugation of the boxes at 2400*g* (3000–3250 rpm in a centrifuge equipped with a microplate adapter) for 30 minutes. Remove a row of tubes and drain the supernatant by inversion over a sink. A small white pellet should be visible on the bottom of each tube. Return the row of tubes to the box.

29. When all of the tubes have been emptied, invert the boxes on a paper towel and allow the last traces of supernatants to drain for a few minutes. Keeping the box in an upside-down position, replace the wet paper towel with a fresh dry towel. Transfer the inverted box-towel combinations to the centrifuge. Remove the last traces of PEG/NaCl solution from the bacteriophage M13 pellets by centrifugation at 300 rpm for 3–5 minutes.

30. Remove the boxes from the centrifuge, check that the pellets have remained in place, and add 20 µl of TTE to each tube.

31. Seal the tubes with 3M silver foil tape and shake the boxes vigorously on a multitube vortexer for 15–30 minutes.

32. Centrifuge the boxes briefly to bring the solution to the bottom of the tubes. Pry the base from each of the 96-tube boxes and place the tubes in an 80°C water bath for 10 minutes.

33. Remove the tubes from the water bath and allow them to cool to room temperature. Replace the bottoms of the boxes and briefly centrifuge each unit to bring the solutions to the bottom of the tubes.

34. Use a multichannel pipettor to transfer 70 µl of sterile H_2O to the individual wells of 96-well microtiter plates. Transfer the bacteriophage lysates from the tubes in Step 33 to the microtiter plate wells. Mix the two solutions by pipetting up and down. Seal the plates with a strip of 3M silver foil tape or with the plate lid if sequencing is to be performed within the next 24–48 hours. Label each plate and store it at –20°C.

    The yield of single-stranded M13 DNA should be 2.5–5 µg per starting culture.

35. Examine aliquots (5 µl) of DNA selected at random from a few wells by electrophoresis through a 1% agarose gel.

    Use between 2 and 7.5 µl of each DNA preparation in cycle sequencing reactions (see Protocol 12.6).

Preparing Denatured Double-stranded DNA Templates for Sequencing by Dideoxy-mediated Chain Termination

In this protocol, double-stranded plasmid DNA is denatured by alkali and annealed to an appropriate oligonucleotide primer, in preparation for dideoxy-sequencing reactions catalyzed by Sequenase (described in Protocol 12.3).

MATERIALS

CAUTION: Please see Appendix 4 for appropriate handling of materials marked with <!>.

Reagents and Solutions

Please see Appendix 1 for components of stock solutions, buffers, and reagents. Dilute stock solutions to the appropriate concentrations.

Ammonium acetate (5 M)

Chloroform:isoamyl alcohol (24:1, v/v) <!>

Ethanol

Ethanol (70%)

NaOH (2 N)/EDTA (2 mM) <!>
> Prepare just before use by diluting 10 N stock solution of NaOH.

Phenol, equilibrated to pH 7.6 and saturated with H_2O <!>

5x Sequenase reaction buffer
> 200 mM Tris-Cl (pH 7.5)
> 100 mM $MgCl_2$
> 125 mM NaCl

Nucleic Acids/Oligonucleotides

Oligonucleotide primers
> Oligonucleotide primers used to sequence denatured double-stranded DNA templates are frequently longer (20–30 nucleotides) than primers normally used to sequence single-stranded DNA templates. Longer primers give rise to fewer artifactual bands when used with denatured double-stranded DNA templates.
>
> For a list of "universal" primers that bind to vector sequences upstream of the target region, see MC3, pp. 8.113–8.117. For information on preparation of oligonucleotide primers for DNA sequencing, see MC3, p. 12.103.

Plasmid DNA, closed circular, double stranded
> Prepare plasmid DNA as described in Chapter 1 or in the additional protocol (Purification of Plasmid DNA from Small-scale Cultures by Precipitation with PEG) in MC3, p. 12.31. Five micrograms of DNA is required for each set of four sequencing reactions (ddA, ddG, ddC, and ddT) (see Protocol 12.3).

Additional Items

Dry-ice/ethanol bath

Water bath (65°C)

Additional Information

Additional protocol (Purification of Plasmid DNA from Small-scale Cultures by Precipitation with PEG)	MC3, p. 12.31
Additional protocol (Rapid Denaturation of Double-stranded DNA)	MC3, p. 12.30
Bacteriophage λ templates for DNA sequencing	MC3, p. 12.29
Troubleshooting	MC3, p. 12.29

METHOD

1. Transfer approximately 5 µg of purified plasmid DNA to a 1.5-ml microfuge tube. Adjust the volume to 50 µl with H_2O. Add 10 µl of freshly prepared 2 N NaOH/2 mM EDTA solution. Incubate the mixture for 5 minutes at room temperature.

2. Add 30 µl of 5 M ammonium acetate and mix the contents of the tube by vortexing.

3. Add 45 µl of equilibrated phenol and mix the contents of the tube by vortexing.

4. Add 135 µl of chloroform:isoamyl alcohol, mix the contents of the tube by vortexing, and separate the phases by centrifugation at maximum speed for 15 minutes at room temperature in a microfuge.

5. Carefully remove the upper aqueous phase to a fresh microfuge tube and precipitate the DNA by the addition of 330 µl of ice-cold ethanol. Mix the contents of the tube and place it in a dry ice–ethanol bath for 15 minutes.

6. Recover the denatured DNA by centrifugation at maximum speed for 15 minutes at 4°C in a microfuge.

7. Very carefully decant the supernatant without disturbing the DNA pellet that may or may not be visible on the side of the tube.

8. Recentrifuge the tube for 2 seconds and remove the last traces of ethanol with a drawn-out pipette tip without disturbing the DNA precipitate.

9. Allow the precipitate to dry at room temperature and resuspend the DNA in H_2O at a concentration of 1 µg/µl.

10. Anneal the primer DNA to the denatured template, using 5 µg (5 µl) of alkali-denatured plasmid DNA for each set of four DNA-sequencing reactions (i.e., 1.25 µg of denatured DNA for the reaction containing the ddA reaction, 1.25 µg for the reaction containing ddG, etc.). Set up annealing reactions as follows:

alkali-denatured DNA	5.0 µg (in 5 µl H_2O)
DMSO (*optional*)	2.0 µl
Sequenase reaction buffer	2.0 µl
sequencing primer	1.0 µl (10 ng)
H_2O	to 11.0 µl

    Heat the mixture to 65°C for 2 minutes and then allow it to cool slowly to room temperature. Store the annealed samples on ice until the elongation/chain termination reactions are set up.

11. Proceed to Step 3 in Protocol 12.3.

Dideoxy-mediated Sequencing Reactions Using Bacteriophage T7 DNA Polymerase (Sequenase)

This protocol describes DNA sequencing of single-stranded DNA templates in reactions catalyzed by Sequenase (version 2.0). All of the materials required for sequencing with Sequenase version 2.0 are contained in kits, sold by USB, and they include a detailed protocol describing the reagents and steps used in the sequencing reaction. This kit is invaluable for first-time or occasional sequencers, but it is expensive. After gaining experience, it may turn out to be more economical to purchase Sequenase version 2.0 in bulk and assemble the remaining reagents locally.

MATERIALS

CAUTION: Please see Appendix 4 for appropriate handling of materials marked with <!>.

Reagents and Solutions

Please see Appendix 1 for components of stock solutions, buffers, and reagents. Dilute stock solutions to the appropriate concentrations.
Dithiothreitol (100 mM)
dNTPs and ddNTPs, stock solutions (0.5 mM) of each of the four dNTPs and ddNTPs
TE (pH 7.6)

Enzymes and Buffers

5x Labeling mixture and ddNTP extension/termination mixtures, assembled by mixing the volumes of reagents shown in Table 12-2
5x Sequenase dilution buffer
10 mM Tris-Cl (pH 7.5)
5 mM dithiothreitol
0.5 mg/ml bovine serum albumin

TABLE 12-2 5x Labeling Mixture and ddNTP Extension/Termination Mixtures for Sequenase Reactions

Reaction Mixture	Stock Solutions of dNTPs and ddNTPs (all volumes in µl)								Other Reagents	
	dCTP 0.5 mM	dTTP 0.5 mM	dATP 0.5 mM	dGTP 0.5 mM	ddCTP 0.5 mM	ddTTP 0.5 mM	ddATP 0.5 mM	ddGTP 0.5 mM	5 M NaCl	H₂O
Labeling	15	15	—	15	—	—	—	—	—	955
ddCTP	160	160	160	160	16	—	—	—	10	334
ddTTP	160	160	160	160	—	16	—	—	10	334
ddATP	160	160	160	160	—	—	16	—	10	334
ddGTP	160	160	160	160	—	—	—	16	10	334

The 5x labeling mixture contains unlabeled dGTP, dCTP, and dTTP, each at a concentration of 7.5 mM. Radiolabeled dATP is added at Step 7 to a final concentration of approximately 30 µM. Each of the ddNTP extension/termination mixtures contains all four dNTPs at a concentration of 160 µM and a single ddNTP at a concentration of 0.16 µM. The 5x labeling mixture and extension/termination mixtures should be dispensed in 50-µl aliquots and stored frozen at –20°C.

When sequencing templates that are rich in GC structure, use base analogs such as 7-deaza-2'-dGTP or dITP in place of dGTP. If using 7-deaza-dGTP, simply replace the GTP in all of the ddNTP extension/termination mixtures with an equimolar amount of the base analog.

If using dITP, substitute 30 µl of a 0.5 mM stock solution of dITP for dGTP in the 5x labeling mixture and decrease the amount of H₂O to 940 µl. In the ddNTP extension/termination mixtures containing ddCTP, ddTTP, and ddATP, substitute 160 µl of a 0.5 mM stock dITP solution for dGTP. In the extension/termination mixture containing ddGTP, substitute 320 µl of 0.5 mM stock solution of dITP for dGTP, decrease the amount of ddGTP to 3.2 µl, and decrease the volume of H₂O to 187 µl.

Adapted from Chapter 12, Protocol 3, p. 12.32 of MC3.

5x Sequenase reaction buffer with $MgCl_2$
    200 mM Tris-Cl (pH 7.5)
    100 mM $MgCl_2$
    125 mM NaCl

5x Sequenase reaction buffer with $MnCl_2$
    200 mM Tris-Cl (pH 7.5)
    25 mM $MnCl_2$
    125 mM NaCl

    The presence of Mn^{++} increases the efficiency with which Sequenase utilizes ddNTPs, with marked improvement in the strength of the bands that run close to the oligonucleotide primer. To use Mn^{++}, add 1 μl of 5x Sequenase reaction buffer with $MnCl_2$ to the dideoxy-dNTP extension/termination reactions before adding Sequenase.

Sequenase version 2.0 (USB) (see MC3, pp. 12.104–12.105)

Yeast inorganic pyrophosphatase (GE Healthcare) (optional; see Step 6)

    The addition of yeast pyrophosphatase increases read-length by preventing buildup of pyrophosphate during extension/termination reactions. Pyrophosphatase can be mixed with Sequenase at a ratio of 1 unit of pyrophosphatase to 3 units of Sequenase. This is best done by mixing equal volumes of appropriate dilutions of the two enzymes.

Nucleic Acids/Oligonucleotides

Oligonucleotide primers (1 μg/ml in H_2O)

    Oligonucleotide primers used to sequence denatured double-stranded DNA templates are frequently longer (20–30 nucleotides) than primers normally used to sequence single-stranded DNA templates. Longer primers give rise to fewer artifactual bands when used with denatured double-stranded DNA templates.

    For a list of "universal" primers that bind to vector sequences upstream of the target region, please see MC3, pp 8.113–8.117. For information on preparation of oligonucleotide primers for DNA sequencing, see MC3, p. 12.103.

Template DNA (1 mg/ml in TE [pH 7.6])

    Use single-stranded DNA or double-stranded denatured plasmid DNA.

Radioactive Compounds

[α-^{35}S]dATP (1000 Ci/mmole, 10 mCi/ml) <!>

or

[α-^{33}P]dATP (1000–3000 Ci/mmole, 20 mCi/ml) <!>

or

[α-^{32}P]dATP (1000 Ci/mmole, 10 mCi/ml) <!>

    Instead of internal labeling with [^{32}P]dATP, use a 5′ ^{32}P-labeled oligonucleotide as a primer. In this case, 2 ml (~5 x 10^5 cpm) of the radiolabeled primer and 2 ml of H_2O are added in place of the unlabeled primer and [^{32}P]dATP. The 5′ termini of oligonucleotides can be labeled as described in Protocol 10.2.

Gels/Loading Buffers

Formamide gel-loading buffer <!>
    80% (w/v) deionized formamide
    10 mM EDTA (pH 8.0)
    0.25% bromophenol blue
    0.25% xylene cyanol FF

Centrifuges/Rotors/Tubes

Swing-out rotor and holder for microtiter plates

Additional Items

Dry-ice <!> /ethanol bath
Heating block or water bath (37°C)
Microtiter plates with 96 U-shaped wells
Water bath (65°C)

Additional Information

Additional protocol (Purification of Plasmid DNA from Small-scale Cultures by Precipitation with PEG)	MC3, p. 12.31
Additional protocol (Rapid Denaturation of Double-stranded DNA)	MC3, p. 12.30
Bacteriophage λ templates for DNA sequencing	MC3, p. 12.29
7-Deaza-dGTP	MC3, p. 12.111
Description of sequenase-mediated sequencing	MC3, p. 12.32
Inosine used to relieve compression in sequencing gels	MC3, p. 12.109
Preparation of stock solutions of dNTPs and ddNTPs for DNA sequencing	MC3, p. 12.107
Sequenase	MC3, pp. 12.104–12.105
Troubleshooting	MC3, pp. 12.29 and 12.38–12.39

METHOD

1. To a 0.5-ml microfuge tube or well of a microtiter plate, add:

single-stranded template DNA (~1 μg/μl)	1 μl
oligonucleotide primer (~1 ng/μl)	3 μl
5x Sequenase reaction buffer containing MgCl$_2$ or MnCl$_2$	2 μl
H$_2$O	to 10 μl

   IMPORTANT: When sequencing double-stranded plasmid DNAs that have already been denatured and annealed to a sequencing primer (Protocol 12.2), ignore the first two steps of this protocol and begin at Step 3.

2. Incubate the tightly closed tube for 2 minutes at 65°C. Remove the tube from the water bath and allow it to cool to room temperature over the course of 3–5 minutes.

3. While the primer and template are cooling, thaw the 5x labeling and ddNTP extension/termination mixtures and radiolabeled dATP. Store the thawed solutions on ice.

4. Transfer 2.5 μl of each ddNTP extension/termination mixture into separate 0.5-ml microfuge tubes or into individual wells of a microtiter plate color-coded or labeled C, T, A, and G.

5. Make a fivefold dilution of the 5x labeling mixture in ice-cold H$_2$O. A volume of 2 μl of the diluted labeling mixture is required for each template sequenced.

6. Dilute Sequenase in ice-cold Sequenase dilution buffer, with or without addition of yeast pyrophosphatase, as described in Materials.

   A volume of 2.0 μl, containing approximately 3.0 units of Sequenase enzyme, is required for each template sequenced. Store the diluted enzyme on ice at all times.

7. To perform the labeling reaction, add the following to the 10-μl annealing reaction of Step 2:

diluted labeling mixture (from Step 5)	2 μl
0.1 M dithiothreitol	1 μl
[α-^{33}P]dATP, [α-^{32}P]dATP, or [α-^{35}S]dATP	0.5 μl
diluted Sequenase (~1.6 units/μl)	2.0 μl

   Mix the components of the reaction by gently tapping the sides of the tube or microtiter plate and then incubate the reaction for 2–5 minutes at 20°C.

Store the diluted Sequenase in ice; do not allow it to warm to ambient room temperature. Store the concentrated stock of enzyme supplied by the manufacturer at –20°C. It may lose activity if stored in an ice bucket for hours.

8. Toward the end of the labeling reaction, prepare to set up the termination reactions by pre-warming the labeled microfuge tubes or microtiter plates to 37°C. *This step is important!* Then transfer the 3.5 μl of the labeling reaction to the walls of each of the prewarmed labeled microfuge tubes or to the sides of prewarmed microtiter wells containing the appropriate dideoxy terminator mixtures (Step 4 above).

9. Place the microfuge tubes in a microfuge *at room temperature* (use appropriate rotor or adapters for 0.5-ml tubes or place inside decapitated 1.5-ml tubes) or place the microtiter plates in a centrifuge *at room temperature* equipped with an appropriate adapter. Centrifuge the C, T, A, and G tubes or plates for a few seconds at 2000 rpm to mix the components of the reactions. Immediately transfer the reactions to a heating block or a water bath for 3–5 minutes at 37°C.

10. Stop the reactions by adding 4 μl of formamide loading buffer.

11. The reactions may be stored for up to 5 days at –20°C or analyzed directly by denaturing gel electrophoresis (Protocol 12.8 or 12.9). After heat denaturation (2 minutes at 100°C) and quick cooling on ice, load 3 μl of each of the C, T, A, and G reactions into individual wells of a sequencing gel.

Dideoxy-mediated Sequencing Reactions Using the Klenow Fragment of *E. coli* DNA Polymerase I and Single-stranded DNA Templates

The following protocol has its roots in methods devised in the Sanger laboratory in the 1970s, in the early days of dideoxy-DNA sequencing.

MATERIALS

CAUTION: Please see Appendix 4 for appropriate handling of materials marked with <!>.

Reagents and Solutions

Please see Appendix 1 for components of stock solutions, buffers, and reagents. Dilute stock solutions to the appropriate concentrations.

dATP (0.1 mM) (optional; see Step 4)

ddNTPs, four solutions (5 mM), each containing one of the four ddNTPs

dNTP solutions, four solutions (1 mM), each containing one of the four dNTPs

EDTA (10 mM, pH 8.0)

TM buffer
> 100 mM Tris-Cl (pH 8.5)
> 50 mM MgCl$_2$

Tris-Cl (1 M, pH 8.0)

Enzymes and Buffers

Klenow fragment of *E. coli* DNA polymerase I (~5 units/μl)
> Approximately 2.5 units of enzyme are required for each set of four sequencing reactions.

Extension/termination mixtures and chase mixture, assembled by mixing the volumes
> (in μl) of the solutions of dNTPs and ddNTPs shown in Table 12-3

Nucleic Acids/Oligonucleotides

Oligonucleotide primers, approximately 0.5 pmole/μl, in H$_2$O, equivalent to about 1 mg/ml
> For a list of "universal" primers that bind to vector sequences upstream of the target region, see MC3, pp. 8.113–8.117. For information on preparation of oligonucleotide primers for DNA sequencing, see MC3, p. 12.103.

TABLE 12-3 **ddNTP Extension/Termination Mixtures and Chase Mixtures for Klenow Sequencing**

Reaction Mixture	Stock Solutions of dNTPs (1 mM) (all volumes in μl)				Stock Solutions of ddNTPs (5 mM) (all volumes in μl)				Other Reagents		
	dCTP	dTTP	dATP	dGTP	ddCTP	ddTTP	ddATP	ddGTP	Tris-Cl (1 M, pH 8.0)	EDTA (10 mM, pH 8.0)	H$_2$O
ddCTP	5	100	—	100	30	—	—	—	10	10	745
ddTTP	100	5	—	100	—	25	—	—	10	10	675
ddATP	100	100	—	100	—	—	5	—	10	10	580
ddGTP	100	100	—	5	—	—	—	25	10	10	750
Chase	245	245	245	245	—	—	—	—	10	10	—
dATP	—	—	100	—	—	—	—	—	10	10	880

The ratios of ddNTP:dNTP in the extension/termination mixtures are 30:1, 25:1, and 25:1 for dCTP, dTTP, and dGTP, respectively. In the assembled sequencing reaction, the ratio of ddATP:dATP is approximately 45:1.

Dispense mixtures as 50-μl aliquots and store frozen at –20°C.

Template DNA, single stranded (~0.05 pmole/μl), equivalent to approximately 0.15 μg/μl of bacteriophage M13 single-stranded DNA

Radioactive Compounds

[α-^{32}P]dATP (1000 Ci/mmole, 10 mCi/ml) <!>

or

[α-^{33}P]dATP (1000–3000 Ci/mmole, 20 mCi/ml) <!>

or

[α-^{35}S]dATP (1000 Ci/mmole, 10 mCi/ml) <!>

Instead of internal labeling with [^{32}P]dATP, use a 5' ^{32}P-labeled oligonucleotide as a primer. In this case, 2 ml (~5 x 10^5 cpm) of the radiolabeled primer and 2 ml of H$_2$O are added in place of the unlabeled primer and [^{32}P]dATP. The 5' termini of oligonucleotides can be labeled as described in Protocol 10.2.

Gels/Loading Buffers

Formamide gel-loading buffer <!>
80% (w/v) deionized formamide
10 mM EDTA (pH 8.0)
0.25% bromophenol blue
0.25% xylene cyanol FF

Centrifuges/Rotors/Tubes

Swing-out rotor and holder for microtiter plates

Additional Items

Microtiter plates with 96 U-shaped wells
Water bath or heating block (55°C and 37°C)

Additional Information

Description of Klenow-mediated sequencing	MC3, p. 12.40
Klenow fragment of *E. coli* DNA polymerase I	MC3, pp. 12.101–12.102
Preparation of stock solutions of dNTPs and ddNTPs for DNA sequencing	MC3, p. 12.107
Troubleshooting	MC3, p. 12.44

METHOD

1. Transfer the following to a 0.5-ml microfuge tube or well of a microtiter plate:

single-stranded template DNA (0.1 μg/μl)	5 μl
oligonucleotide primer (1 μg/ml, ~160 pmoles/ml)	4 μl
TM buffer	1 μl

2. Close the top of the tube or seal the microtiter plate and anneal the oligonucleotide primer to the template DNA by incubating the reaction mixture for 5–10 minutes at 55°C.

   If necessary, the annealed template and primer may be stored for several months at –20°C.

3. Meanwhile, label four microfuge tubes or four contiguous wells of a 96-well disposable microtiter plate with the letters C, T, A, and G; then add 4 μl of the appropriate ddNTP extension/termination mixture to each tube (i.e., 4 μl of ddCTP mixture to tube/well labeled C, 4 μl of ddTTP mixture to tube/well labeled T, etc.).

4. Place the annealed primer-template solution from Step 2 on ice and add:

[α-^{32}P]dATP *or* [α-^{33}P]dATP *or* [α-^{35}S]dATP	1 µl
Klenow enzyme (~2.5 units)	1 µl
0.1 mM dATP (if [α-^{32}P] or [α-^{33}P] is used)	1 µl
or	
H$_2$O (if [α-^{35}S]dATP is used)	1 µl

Store the stock of Klenow enzyme at –20°C; do not allow it to warm to ambient room temperature. The enzyme may lose activity if stored in an ice bucket for hours.

5. Transfer 3 µl of the mixture from Step 4 to the walls of each of the C, T, A, and G tubes or to the sides of microtiter wells. Do not allow the radiolabeled mixture to come into contact with the ddNTP extension/termination mixture.

6. Place the microfuge tubes in a microfuge (use appropriate rotor or adapters for 0.5-ml tubes or place inside decapitated 1.5-ml tubes) or place the microtiter plates in a centrifuge equipped with an appropriate adapter. Centrifuge the C, T, A, and G tubes or plates for a few seconds at 2000 rpm to mix the components and initiate the extension/termination reactions. Incubate the reactions for 10–12 minutes at 37°C.

Extension/termination and chase reactions catalyzed by the Klenow enzyme work equally well at temperatures ranging from room temperature to 37°C.

7. After 9–11 minutes of incubation, transfer 1 µl of the chase mixture to the side of each C, T, A, and G tube or well. Do not allow the solution to slide into the polymerization reactions. When a total of 10–12 minutes has elapsed, centrifuge the tubes/plates for 2 seconds to introduce the chase mix into the extension/termination reactions. Incubate the reactions for an additional 10–12 minutes at room temperature.

8. After 9–11 minutes of the second incubation, transfer 6 µl of formamide loading buffer to the sides of each C, T, A, and G tube or well. Again, take care that the solution does not slide down into the reaction. At the end of the incubation period, centrifuge the tubes or plates to stop the sequencing reactions.

9. The DNA-sequencing reactions may be stored for up to 5 days at –20°C or analyzed directly by denaturing gel electrophoresis (Protocol 12.8 or 12.9). After heat denaturation (2 minutes at 100°C) and quick cooling on ice, load 3 µl of each of the C, T, A, and G reactions into the individual wells of a sequencing gel.

Protocol 12.5

Dideoxy-mediated Sequencing Reactions Using *Taq* DNA Polymerase

Because sequencing reactions catalyzed by thermostable DNA polymerases such as *Taq* are performed at elevated temperatures, problems caused by mismatched annealing of primers or templates rich in secondary structure are greatly alleviated.

MATERIALS

CAUTION: Please see Appendix 4 for appropriate handling of materials marked with <!>.

Reagents and Solutions

Please see Appendix 1 for components of stock solutions, buffers, and reagents. Dilute stock solutions to the appropriate concentrations.

dATP (0.1 mM) (optional; see Step 4)

ddNTPs, four solutions (5 mM), each containing one of the four ddNTPs

dNTP solutions four solutions (1 mM), each containing one of the four dNTPs

Enzymes and Buffers

AmpliTaq CS (5 units/µl) (Perkin-Elmer) or other exonuclease-deficient version of *Taq* DNA polymerase

> Thermostable enzymes purified from organisms other than *T. aquaticus* may require slightly different reaction conditions and dilution buffers. Please see the specifications provided by the manufacturer.

Taq Dilution buffer

> 25 mM Tris-Cl (pH 8.8)
> 0.01 mM EDTA (pH 8.0)
> 0.15% Tween-20
> 0.15% Nonidet P-40

10x Labeling mixture and extension/termination mixtures, assembled by mixing the volumes (in µl) of the solutions of dNTPs and ddNTPs shown in Table 12-4

> When sequencing templates that are rich in secondary structure, replace dCTP in the 10x labeling mixture and extension/termination mixtures with an equimolar amount of 7-deaza-2′-dGTP (see MC3, p. 12.111).

5x Reaction buffer

> 200 mM Tris-Cl (pH 8.8, room temperature)
> 25 mM $MgCl_2$

Nucleic Acids/Oligonucleotides

Oligonucleotide primers, approximately 1.0 pmole/µl, in H_2O, equivalent to about 6.5 µg/ml

> For a list of "universal" primers that bind to vector sequences upstream of the target region, please see MC3, pp. 8.113–8.117. For information on preparation of oligonucleotide primers for DNA sequencing, see MC3, p. 12.103.

Template DNA, single stranded (100 µg/ml in TE) (pH 7.6)

> Single-stranded DNA (500 ng) or 1.0 µg of denatured double-stranded plasmid DNA is required for each set of four sequencing reactions. Denature and anneal the linear double-stranded DNA to the sequencing primer by mixing the native DNA with a 20-fold molar excess of primer, heating the mixture in a boiling-water bath for approximately 2 minutes and then plunging the tube into an ice-water bath. Use the mixture immediately.

TABLE 12-4 10x Labeling Mixture and ddNTP Chain Extension/Termination Mixtures for Use in Conventional DNA-sequencing Reactions Catalyzed by *Taq* Polymerase

Reaction Mixture	Stock Solution of dNTP and ddNTPs (all volumes in μl)								Other Reagents		
	dCTP 1 mM	dTTP 1 mM	dATP 1 mM	dGTP 1 mM	ddCTP 5 mM	ddTTP 5 mM	ddATP 5 mM	ddGTP 5 mM	Tris-Cl (1 M, pH 8.0)	EDTA (10 mM, pH 8.0)	H₂O
Labeling	1.5	1.5	—	1.5	—	—	—	—	10	10	975
ddCTP	15	15	15	15	90	—	—	—	10	10	830
ddTTP	15	15	15	15	—	240	—	—	10	10	680
ddATP	15	15	7.5	15	—	—	120	—	10	10	800
ddGTP	15	15	15	15	—	—	—	9	10	10	911

The 10x labeling mixture contains dGTP, dCTP, and dTTP, each at a concentration of 1.5 μM.
The ratios of ddNTP:dNTP in the four ddNTP extension/termination mixtures are 30:1, 80:1, 80:1, and 3:1 for C, T, A, and G, respectively.
The 10x labeling mixture and ddNTP extension/termination mixtures should be dispensed in 50-μl aliquots and stored frozen at –20°C.

Radioactive Compounds

[α-^{32}P]dATP (1000 Ci/mmole, 10 mCi/ml) <!>
or
[α-^{33}P]dATP (1000–3000 Ci/mmole, 20 mCi/ml) <!>
or
[α-^{35}S]dATP (1000 Ci/mmole, 10 mCi/ml) <!>
Instead of internal labeling with [^{32}P]dATP, use a 5' ^{32}P-labeled oligonucleotide as a primer. In this case, 2 ml (~5 x 10^5 cpm) of the radiolabeled primer and 2 ml of H₂O are added in place of the unlabeled primer and [^{32}P]dATP. The 5' termini of oligonucleotides can be labeled as described in Protocol 10.2.

Gels/Loading Buffers

Formamide gel-loading buffer <!>
80% (w/v) deionized formamide
10 mM EDTA (pH 8.0)
0.25% bromophenol blue
0.25% xylene cyanol FF

Centrifuges/Rotors/Tubes

Swing-out rotor and holder for microtiter plates

Additional Items

Microtiter plates with 96 U-shaped wells
Water bath or heating block (65°C, 45°C, and 72°C)

Additional Information

7-Deaza-dGTP	MC3, p. 12.111
Preparation of stock solutions of dNTPs and ddNTPs for DNA sequencing	MC3, p. 12.107
Thermostable DNA polymerases used in DNA sequencing	MC3, pp. 12.45–12.47
Troubleshooting	MC3, pp. 12.56–12.59

METHOD

1. To a 0.5-ml microfuge tube or well of a microtiter plate, add:

single-stranded template DNA (250 fmoles) (100 ng/µl)	5.0 µl
oligonucleotide primer (0.5 pmole) (~3.3 ng/µl)	3.0 µl
5x dideoxy reaction buffer	2.0 µl

2. Incubate the tightly closed tube for 2 minutes at 65°C. Remove the tube from the water bath and allow it to cool to room temperature over the course of 3–5 minutes.

3. While the primer and template are cooling, thaw the 10x labeling and ddNTP extension/termination mixtures and radiolabeled dATP. Store the thawed solutions on ice.

4. For each annealing reaction, transfer 4 µl of each ddNTP termination/extension mixture into color-coded 0.5-ml microfuge tubes or into individual wells of a microtiter plate prelabeled C, T, A, and G (i.e., 4 µl of ddCTP mixture to tube/well labeled C, 4 µl of ddTTP mixture into tube/well labeled T, etc.). Store the tubes/microtiter plates on ice.

5. Dilute enough thermostable DNA polymerase 1:8 for all templates to be sequenced, for example,

Taq DNA polymerase (5–10 units/µl)	1 µl
Taq dilution buffer	7 µl

   2 µl (2 units) of diluted enzyme is needed for each set of four sequencing reactions. The final concentration of the enzyme should be approximately 1 unit/µl. Store the diluted enzyme on ice at all times.

6. Add the following to each annealing reaction (Step 2 above):

10x labeling mixture	2 µl
radiolabeled dATP	0.5 µl
diluted thermostable DNA polymerase (~1 unit/µl)	8 µl

   Mix the contents of the tube by vortexing and then incubate the reactions for 5 minutes at 45°C.

7. Transfer 4 µl of the labeling reaction to the sides of each of the C, T, A, and G tubes or to the sides of microtiter wells containing the appropriate dideoxy terminator mixtures (Step 4 above).

8. Place the microfuge tubes in a microfuge (use appropriate rotor or adapters for 0.5-ml tubes or place them inside decapitated 1.5-ml tubes) or place the microtiter plates in a centrifuge equipped with an appropriate adapter. Centrifuge the C, T, A, and G tubes or plates for a few seconds at 2000 rpm to mix the components of the reactions. Incubate the reactions for 5 minutes at 72°C.

   Incubate the tubes/plates in an efficient heating block or place them in contact with H_2O, oil, or other efficient conductor of heat. Tubes/plates placed in an air incubator set at 72°C will not reach optimum temperature for *Taq* DNA polymerase during the course of the brief incubation.

9. Stop the reactions by adding 4 µl of formamide loading buffer.

10. The reactions may be stored at –20°C for up to 5 days or analyzed directly by denaturing gel electrophoresis (Protocol 12.8 or 12.9). After heat denaturation (2 minutes at 100°C) and quick cooling on ice, load 3 µl of each of the C, T, A, and G reactions into individual wells of a sequencing gel.

Cycle Sequencing: Dideoxy-mediated Sequencing Reactions Using PCR and End-labeled Primers

In this protocol, asymmetric PCR is used to generate single-stranded DNA templates for dideoxy sequencing.

MATERIALS

CAUTION: Please see Appendix 4 for appropriate handling of materials marked with <!>.

Reagents and Solutions

Please see Appendix 1 for components of stock solutions, buffers, and reagents. Dilute stock solutions to the appropriate concentrations.

ddNTPs, four solutions (5 mM), each containing one of the four ddNTPs
dNTP solutions, four solutions (1 mM), each containing one of the four dNTPs
EDTA (0.05 M, pH 8.0)

Enzymes and Buffers

AmpliTaq CS, 5 units/μl (Perkin-Elmer), or other exonuclease-deficient version of *Taq* DNA polymerase

> Thermostable enzymes purified from organisms other than *T. aquaticus* may require slightly different reaction conditions and dilution buffers. Please see the specifications provided by the manufacturer. For advice on which enzyme to use, see MC3, pp. 12.45–12.47.

5x Cycle-sequencing buffer
> 200 mM Tris-Cl (pH 8.8, room temperature)
> 25 mM MgCl$_2$

ddNTP extension/termination mixtures, assembled by mixing the volumes (in μl) of the solutions of dNTPs and ddNTPs shown in Table 12-5

> When sequencing templates that are rich in secondary structure, replace dCTP in the 10x labeling mixture and extension/termination mixtures with an equimolar amount of 7-deaza-2'-dGTP (see MC3, p. 12.111).

Nucleic Acids/Oligonucleotides

Oligonucleotide primers, radiolabeled at the 5' terminus with ^{32}P or ^{33}P (Protocol 10.2) <!>
Template DNA
> Plasmids, cosmids, bacteriophage λ, and bacteriophage M13 DNAs, purified by any of the methods described in Chapters 1–4, can be used as templates.
> Table 12-6 shows the amounts of each type of template required.

TABLE 12-5 ddNTP Extension/Termination Mixtures for Use in Cycle-sequencing Reactions

ddNTP Reaction Mixture	Stock Solution of dNTP (1 mM) and ddNTPs (5 mM) (all volumes in μl)								Other Reagents		
	dCTP	dTTP	dATP	dGTP	ddCTP	ddTTP	ddATP	ddGTP	Tris-Cl (1 M, pH 8.0)	EDTA (10 mM, pH 8.0)	H$_2$O
ddCTP	20	20	20	20	80	—	—	—	10	10	820
ddTTP	20	20	20	20	—	160	—	—	10	10	740
ddATP	20	20	20	20	—	—	120	—	10	10	780
ddGTP	20	20	20	20	—	—	—	40	10	10	860

Dispense as 50-μl aliquots; store frozen at –20°C.

The ratios of ddNTP:dNTP in the four extension/temination mixtures are 20:1, 40:1, 30:1, and 10:1 for C, T, A, and G, respectively. These ratios were optimized for reactions catalyzed by AmpliTaq CS DNA polymerase. When using other thermostable DNA polymerases, the ratios may need to be reoptimized.

TABLE 12-6 Amounts of Various Types of Templates Required in Cycle-sequencing Reactions

Type of Template	Amount of Purified DNA Required for Each Set of Four Dideoxy-sequencing Reactions (fmoles)
Double-stranded plasmid DNA	20–200
Single-stranded bacteriophage M13 DNA	1–50
Double-stranded bacteriophage I DNA	20–200
Double-stranded cosmid DNA	50–200

Gels/Loading Buffers

Formamide gel-loading buffer <!>
   80% (w/v) deionized formamide
   10 mM EDTA (pH 8.0)
   0.25% bromophenol blue
   0.25% xylene cyanol FF

Centrifuges/Rotors/Tubes

Centrifuge with microplate adapters

Additional Items

Light mineral oil or wax bead (optional; see Step 4)
Microtiter plates with 96 U-shaped wells
Thermal cycler, programmed with desired amplification program
Water bath or heating block (65°C, 45°C, and 72°C)

Additional Information

Additional protocol (Cycle Sequencing Reactions Using PCR and Internal Labeling with [α-^{32}P]dNTPs)	MC3, p. 12.60
Cycle sequencing	MC3, pp. 12.51–12.52
7-Deaza-dGTP	MC3, p. 12.111
Preparation of stock solutions of dNTPs and ddNTPs for DNA sequencing	MC3, p. 12.107
Thermostable DNA polymerases used in DNA sequencing	MC3, pp. 12.45–12.47
Troubleshooting	MC3, pp. 12.55 and 12.56–12.59

METHOD

1. Transfer 4 μl of appropriate ddNTP extension/termination mixture (please see Table 12-5) into 0.5-ml color-coded microfuge tubes or into individual wells of a heat-stable microtiter plate prelabeled C, T, A, and G (i.e., 4 μl of ddCTP mixture into tube/well labeled C, 4 μl of ddTTP mixture into tube/well labeled T, etc.). Store the tubes/microtiter plates on ice.

2. To the side of each tube or well, add:

double-stranded template DNA	10–100 fmoles
1.5 pmoles of 5' ^{32}P end-labeled primer*	1.0 μl
5x cycle sequencing buffer	2.0 μl
H$_2$O	to 5.0 μl

   *If using a ^{33}P end-labeled primer: 1.5 μl of a standard preparation of ^{33}P-labeled oligonucleotide contains approximately 5.0 pmoles of primer. To maintain the stoichiometry of the components of the sequencing reaction, increase the amount of template DNA approximately threefold (30–300 fmoles) *without increasing the total volume of the reaction.*

3. Dilute an aliquot of the preparation of AmpliTaq CS DNA polymerase with 1x cycle sequencing buffer to a final concentration of 0.5–1.0 unit/µl. Add 1 µl of the diluted enzyme to the side of each tube or well.

4. Mix the reagents by flicking the tubes with a finger or shaking the microtiter plate. If necessary, overlay each reaction with a drop of light mineral oil, cap the tubes, and centrifuge them at 2000 rpm for 2 seconds in a microfuge or briefly in a centrifuge with microtiter plate adapters.

5. Load the tubes or plate into a thermocycler, preheated to 95°C. Begin thermal cycling according to the program outlined below.

    IMPORTANT: Do not delay in starting the program. Make sure that the samples are not exposed for >3 minutes to 95°C during the loading/preheating step. Otherwise, the DNA polymerase may be inactivated.

Cycle Number	Denaturation	Annealing	Polymerization
Preheating	60 sec at 95°C		
20–25 cycles	30 sec at 95°C	30 sec at 55°C	60 sec at 72°C
10 cycles	30 sec at 95°C	60 sec at 72°C	60 sec at 72°C

    Times and temperatures may need to be adapted to suit the particular reaction conditions.

6. Remove the tubes or plates from the thermocycler and add 5 µl of formamide loading buffer to each cycle sequencing reaction.

7. The reactions may be stored for up to 5 days at –20°C or analyzed directly by denaturing gel electrophoresis (Protocol 12.8 or 12.9). After heat denaturation (2 minutes at 100°C) and quick cooling on ice, load 3 µl of each of the C, T, A, and G reactions into individual wells of a sequencing gel.

This protocol is based on the classic Maxam-Gilbert base-specific cleavage method of DNA sequencing.

MATERIALS

CAUTION: Please see Appendix 4 for appropriate handling of materials marked with <!>.

Reagents and Solutions

Please see Appendix 1 for components of stock solutions, buffers, and reagents. Dilute stock solutions to the appropriate concentrations.

Acetic acid (1 M), freshly diluted from glacial acetic acid (17.4 M) <!>

Dimethylsulfate (DMS), gold label, 99% <!> (Sigma-Aldrich)

DMS (10% v/v) in ethanol <!>

DMS buffer
    50 mM sodium cacodylate (pH 7.0) <!>
    1 mM EDTA (pH 8.0)

DMS stop solution
    1.5 M sodium acetate (pH 7.0)
    1 M β-mercaptoethanol <!>
    250 µg/ml yeast tRNA

EDTA (0.5 M, pH 8.0)

Ethanol, chilled to –20°C

Formamide <!>

Hydrazine (95% [Eastman Kodak]) <!>
    Store in small aliquots in tightly closed tubes at –20°C.

Hydrazine stop solution
    0.3 M sodium acetate (pH 7.0)
    0.1 mM EDTA (pH 8.0)
    100 µg/ml yeast tRNA

NaCl (5 M)

NaOH (1.2 N), containing 1 mM EDTA <!>

Piperidine (1 M in H_2O) <!>, freshly made by mixing 1 volume of piperidine (10 M [Fisher Scientific]) with 9 volumes of H_2O

Piperidine formate (1 M), prepared by adjusting a 4% (v/v) solution of formic acid in H_2O to pH 2.0 with 10 M piperidine

Sodium acetate (3 M, pH 5.2)

Nucleic Acids/Oligonucleotides

Salmon sperm DNA, sheared (Sigma-Aldrich) (1 mg/ml in H_2O)

Target DNA, radiolabeled <!>
    Prepare at least 5×10^5 cpm of DNA, asymmetrically end-labeled with ^{32}P, and dissolve at a concentration of approximately 5000 cpm/µl in H_2O (see MC3, p. 12.73). The DNA *must* be free of salt.

Yeast tRNA (1 mg/ml in H_2O)

Gels/Loading Buffers

Formamide gel-loading buffer <!>
    80% (w/v) deionized formamide
    10 mM EDTA (pH 8.0)

0.25% bromophenol blue
0.25% xylene cyanol FF
Polyacrylamide DNA-sequencing gel <!> (Protocol 12.8)

Additional Items

Dry-ice/ethanol bath
Handheld minimonitor or liquid scintillation spectrometer for Cerenkov counting
Microfuge tubes of five different colors (one color for each base-specific cleavage reaction)
Rotary vacuum evaporator (Savant SpeedVac)
Round bath rack with screw-down pressure plate (Research Products International)
    (optional; see Step 4)
Water baths or heating blocks (37°C and 90°C)

Additional Information

Additional protocol (Preparation of End-labeled DNA for Chemical Sequencing)	MC3, p. 12.73
Alternative protocol (Rapid Maxam-Gilbert Sequencing)	MC3, p. 12.70
Chemical sequencing	MC3, pp. 12.61–12.63
Troubleshooting	MC3, pp. 12.67–12.69

METHOD

1. Subject the radiolabeled DNA to the base-modification procedures outlined in the flowchart in Table 12-7.

2. Resuspend each of the four or five lyophilized DNA samples containing the base-modified DNA by vortexing with 100 µl of 1 M piperidine.

3. Close the tops of the tubes securely. Mix the contents of the tubes by vortexing. If necessary, centrifuge the tubes briefly (2000 rpm) to deposit all of the fluid at the bottom.

4. Incubate the tubes for 30 minutes at 90°C. To prevent the tops of the tubes from popping open during heating, either place a heavy weight on the tubes or seal the tops with plastic tape. Alternatively, use a round bath rack with a screw-down pressure plate (e.g., Research Products International).

5. Allow the tubes to cool to room temperature. Open the lids of the tubes and seal the open tubes with Parafilm. Pierce several holes in the Parafilm with a 21-gauge needle and evaporate the contents of the tubes to dryness in a rotary vacuum evaporator (e.g., Savant SpeedVac). This step takes 1–4 hours, depending on the efficiency of the rotary evaporator.

6. Remove the tubes from the evaporator. Discard the Parafilm and add 20 µl of H_2O to each tube. Close the caps of the tubes and vortex the tubes for 30 seconds to dissolve the DNA. Centrifuge the tubes briefly to deposit all of the fluid at the bottom. Use a handheld minimonitor to check that all of the radiolabeled DNA has been washed from the walls of the tubes by the H_2O and is dissolved in the fluid.

7. Once again, evaporate all of the samples to dryness in a rotary vacuum evaporator (see Step 5 above). This step usually takes 15–30 minutes, depending on the efficiency of the evaporator.

8. Repeat Steps 6 and 7.

9. Estimate the amount of radioactivity remaining in each of the tubes by Cerenkov counting and dissolve the individual modification and cleavage reactions in sequencing gel-loading

TABLE 12-7 Maxam-Gilbert Sequencing

G	A+G	C+T	C	A>C
Mix 190 μl DMS buffer, 4 μl sonicated DNA, 5 μl ^{32}P-labeled DNA	Mix 10 μl H$_2$O, 4 μl sonicated DNA, 10 μl ^{32}P-labeled DNA	Mix 10 μl H$_2$O, 4 μl sonicated DNA, 10 μl ^{32}P-labeled DNA	Mix 15 μl 5 M NaCl, 4 μl sonicated DNA, 5 μl ^{32}P-labeled DNA	Mix 100 μl 1.2 N NaOH, 1 mM EDTA, 4 μl sonicated DNA, 5 μl ^{32}P-labeled DNA
Chill to 0°C.	Chill to 0°C.	Chill to 0°C.	Chill to 0°C.	Close top of tube tightly and secure with plastic tape.
Add 5 μl 10% DMS and mix.	Add 4 μl piperidine formate (pH 2.0) and mix.	Add 30 μl hydrazine and mix gently.	Add 30 μl hydrazine and mix gently.	Incubate 7 minutes at 90°C. Chill to 0°C.
Incubate 5 minutes at 20°C.	Incubate 15 minutes at 37°C.	Incubate 7 minutes at 20°C.	Incubate 5 minutes at 20°C.	Add 150 μl 1 M acetic acid, 5 μl tRNA (1 mg/ml) and mix.
Add 50 μl DMS stop solution (at 0°C) and mix.	Add 240 μl hydrazine stop solution (at 0°C) and mix.	Add 200 μl hydrazine stop solution (at 0°C) and mix.	Add 200 μl hydrazine stop solution (at 0°C) and mix.	Add 750 μl of ethanol (–20°C) and mix.
Add 750 μl of ethanol (–20°C) and mix.	Add 750 μl of ethanol (–20°C) and mix.	Add 750 μl of ethanol (–20°C) and mix.	Add 750 μl of ethanol (–20°C) and mix.	Store for 5 minutes at –70°C.
Store for 5 minutes at –70°C.	Store for 5 minutes at –70°C.	Store for 5 minutes at –70°C.	Store for 5 minutes at –70°C.	Centrifuge 5 minutes at 12,000g at 4°C.
Centrifuge 5 minutes at 12,000g at 4°C.	Centrifuge 5 minutes at 12,000g at 4°C.	Centrifuge 5 minutes at 12,000g at 4°C.	Centrifuge 5 minutes at 12,000g at 4°C.	Remove the supernatant.
Remove the supernatant carefully and dispose of it in the appropriate waste bottle.	Remove the supernatant.	Remove the supernatant carefully and dispose of it in the appropriate waste bottle.	Remove the supernatant carefully and dispose of it in the appropriate waste bottle.	

Add 300 μl 0.3 M sodium acetate (pH 5.2) at 0°C and mix by vortexing.
Add 900 μl ethanol (–20°C) and mix.
Store 5 minutes at –70°C.
Centrifuge 5 minutes at 12,000g at 4°C.
Remove supernatant carefully.
Add 1 ml ethanol (–20°C) and mix.
Centrifuge 2 minutes at 12,000g at 4°C.
Remove supernatant carefully.
Lyophilize pellet of DNA.
Add 100 μl 1 M piperidine. Secure the tops of the tubes with plastic tape and mix the contents of the tubes.
Incubate 30 minutes at 90°C.
Lyophilize.
Add 20 μl H$_2$O. Vortex for 30 seconds.
Centrifuge 10 seconds at room temperature.
Lyophilize.
Add 20 μl H$_2$O. Vortex for 30 seconds.
Centrifuge 10 seconds at room temperature.
Lyophilize.
Add 10 μl sequencing gel-loading buffer. Vortex for 20 seconds.
Centrifuge 10 seconds at room temperature.
Close the tops of the tubes tightly. Heat to 90°C for 1 minute.
Load 3 μl of each sample on a lane of a sequencing gel.

buffer. An overnight exposure on Kodak XAR-5 film requires approximately 25,000 cpm of reactions that cleave the DNA after only one base (i.e., the C and G reactions) and approximately 50,000 cpm of reactions that cleave after two bases (C+T, A+G, and A>C). Therefore, the modified and cleaved DNAs should be dissolved in sequencing gel-loading buffer so that the C and G reactions contain approximately 25,000 cpm/3 μl and the C+T, A+G, and A>C reactions contain approximately 50,000 cpm/3 μl. Vortex the tubes to dissolve the DNA fully. Centrifuge the tubes briefly to deposit all of the fluid at the bottom. If necessary, the samples may be stored at –20°C for a few hours while the sequencing gel is prepared.

10. Heat the tubes for 1 minute at 90°C to denature the DNA before quick cooling on ice. Analyze the reactions by electrophoresis through denaturing polyacrylamide gels as described in Protocol 12.8 or 12.9.

Preparation of Denaturing Polyacrylamide Gels

In 1978, Fred Sanger and Alan Coulson devised a method to pour and run thin polyacrylamide gels, which are still used to resolve the products of DNA-sequencing reactions.

MATERIALS

CAUTION: Please see Appendix 4 for appropriate handling of materials marked with <!>.

Reagents and Solutions

Please see Appendix 1 for components of stock solutions, buffers, and reagents. Dilute stock solutions to the appropriate concentrations.

Acrylamide solution (45% w/v) <!>

acrylamide (sequencing grade)	434 g
N,N'-methylenebisacrylamide	16 g
H_2O	to 600 ml

Heat the solution to 37°C to dissolve the chemicals. Adjust the volume to 1 liter with H_2O. Filter the solution through a 0.45-μm nitrocellulose filter and store the filtered solution in light-proof bottles at room temperature. During storage, acrylamide and methylenebisacrylamide are slowly deaminated. Before use, check that the pH of the solution is ≤ 7.0.

Ammonium persulfate (1.6% w/v in H_2O) <!>

Dishwashing detergent

Ethanol

KOH/methanol solution <!>

Dissolve 5 g of KOH pellets in 100 ml of methanol. Store the solution in a tightly capped glass bottle.

Silanizing fluid <!>, e.g., Gel Slick (FMC Bioproducts), Acrylease (Stratagene), and RainX (Unelko)

TEMED (N,N,N',N'-tetramethylethylene diamine) <!>, electrophoresis grade (Sigma-Aldrich or Bio-Rad)

An adjunct catalyst to ammonium persulfate for the polymerization of acrylamide

Urea (solid)

Gels/Loading Buffers

10x TBE electrophoresis buffer

TBE is used at a working strength of 1x (89 mM Tris-borate, 2 mM EDTA) for polyacrylamide gel electrophoresis. This is twice the strength usually used for agarose gel electrophoresis.

Additional Items

Bulldog binder clips (5-cm length, five to seven per gel)

Gel-drying racks

Gel-sealing tape, e.g., 3M Scotch stretchable tape, 3M Scotch yellow electrical tape #56, or 3M Scotch polytetrafluorethylene extruded film tape

Glass sequencing gel plates (matched pairs), commercially available

Gloves, talc-free

Petroleum jelly (optional; see Step 4)

Shark's tooth comb (0.4 mm thick, with 32, 64, or 96 teeth, depending on the capacity of the gel apparatus)

Side-arm flask (250 ml)

Spacers, two per gel, either constant thickness or wedge shaped, made of Teflon

Syringe (60 cc)
Water bath (55°C)

Additional Information

Acrylamide	MC3, p. A8.41
Ammonium persulfate	MC3, p. 12.75
Compressions in DNA-sequencing gels	MC3, pp. 12.82 and 12.109–12.110
Troubleshooting (air bubbles)	MC3, p. 12.79
Troubleshooting (leaking gels)	MC3, p. 12.80

METHOD

1. If necessary, remove old silanizing reagent from plates by swabbing them with KOH/ methanol solution.

   IMPORTANT: To prevent contamination of the glass surfaces by skin oils, wear talc-free gloves at all times and handle the plates by their edges.

2. Wash the plates and spacers in a warm, dilute solution of dishwashing liquid and then rinse them thoroughly in tap water, followed by deionized H_2O. Rinse the plates with absolute ethanol to prevent water spots and allow them ample time to dry.

3. Treat the inner surface of the smaller or notched plate with silanizing solution. Lay the plate, inner surface uppermost, on a pad of paper towels in a chemical fume hood and pour or spray a small quantity of silanizing fluid onto the surface of the plate. Wipe the fluid over the entire surface with a pad of Kimwipes and allow the fluid to air dry (1–2 minutes). Rinse the plate first with deionized H_2O and then with ethanol, and allow the plate ample time to dry.

4. Lay the larger (or unnotched) glass plate (clean side up) on an empty test-tube rack on the bench and arrange the spacers in place along each side of the glass plate so that they are flush with the bottom of the plate (please see Figure 12-1).

   Small dabs of petroleum jelly between the spacers and the larger (unnotched) plate will help keep the spacers in position during the next steps.

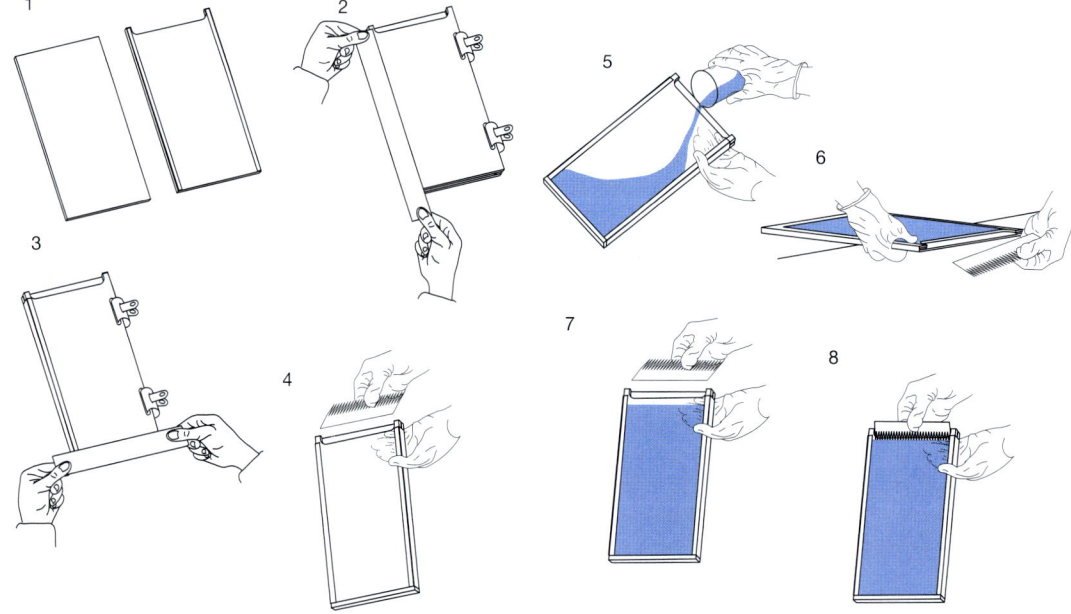

FIGURE 12-1 Preparation of a Sequencing Gel

TABLE 12-8 Acrylamide Solutions for Denaturing Gels

	Gel			
	4%	6%	8%	10%
Acrylamide:bis solution (45%)	8.9 ml	13.3 ml	17.8 ml	22.2 ml
10x TBE buffer	10 ml	10 ml	10 ml	10 ml
H₂O	45.8 ml	41.4 ml	36.9 ml	32.5 ml
Urea	42 g	42 g	42 g	42 g

5. Center the shorter (notched) plate, siliconized side down, on top of a larger (unnotched) plate. Make sure that the spacers remain in position at the very edges of the two plates.

6. Clamp the plates together on one side with two or three large (5-cm length) bulldog binder clips. Bind the entire length of the other side and the bottom of the plates with gel-sealing tape to make a watertight seal.

7. Remove the bulldog clips and seal this side of the gel plates with gel-sealing tape.

8. Place the flat side of a shark's tooth comb into the open end of the gel mold so that it fits snugly. Remove the comb and lay the empty gel mold on the test-tube rack.

9. Cover the working area of the bench with plastic-backed protective paper.

10. In a 250-ml side-arm flask, prepare a sequencing gel solution containing the desired concentration of acrylamide as specified in Table 12-8. The volumes given in the table are sufficient for a single 40 x 40-cm sequencing gel and can be proportionally adjusted to accommodate smaller or larger gels.

    IMPORTANT: The preparation of the gel must be completed without interruption from this point onward.

11. Combine all of the reagents and heat the solution in a 55°C water bath for 3 minutes to help dissolution of the urea.

12. Remove the solution from the water bath and allow it to cool at room temperature for 15 minutes. Swirl the mixture from time to time.

13. Attach the side-arm flask to a vacuum line and de-gas the solution.

14. Transfer the solution to a 250-ml glass beaker. Add 3.3 ml of freshly prepared 1.6% ammonium persulfate and swirl the gel solution gently to mix the reagents.

15. Add 50 μl of TEMED to the gel solution and swirl the solution gently to mix the reagents. Pour the gel solution into the mold directly from the beaker in which it has been prepared. Alternatively, draw approximately 40 ml of the solution into a 60-cc hypodermic syringe. Do not suck air bubbles into the syringe!

    Compared with polyacrylamide gels used to resolve proteins, a massive amount of TEMED is used to cast sequencing gels. The large amount of TEMED ensures that polymerization will occur rapidly and uniformly throughout the large surface area of the gel. Because the rate of polymerization is temperature dependent, cooling the gel solution allows more time for casting the gel. Experienced gel pourers can often cast two or more 40 x 40-cm gels from a single gel solution by judicious precooling.

    Work as quickly as possible from here onward because the gel solution will polymerize rapidly.

16. Allow a thin stream of gel solution to flow from the beaker or syringe into the top corner of the gel mold while holding the mold at an angle approximately 45° to the horizontal.

17. Lay the mold down on the test-tube rack (please see Figure 12-1).

18. Immediately insert the *flat side* of a shark's tooth comb approximately 0.5 cm into the gel solution. Insert both ends of the comb into the fluid to an equal depth so that the flat surface is level when the gel is standing in a vertical position.

19. Clamp the comb into position using bulldog binder clips. Use the remaining acrylamide/urea solution in the hypodermic syringe/pipette to form a bead of acrylamide across the top of the gel. Allow the gel to polymerize for at least 15 minutes at room temperature.

20. Wash out the 60-cc syringe so that it does not become clogged with polymerized acrylamide.

21. After 15 minutes of polymerization, examine the gel for the presence of a Schlieren line just underneath the flat surface of the comb. This is a sign that polymerization is occurring satisfactorily. When polymerization is complete (~1 hour after the gel was poured), remove the bulldog clips.

22. The gel can be used immediately (please see Protocol 12.11) or stored for up to 24 hours at room temperature or 48 hours at 4°C. To prevent dehydration during storage, leave the tape in place and surround the top of the gel with paper towels dampened with 1x TBE. Cover the paper towels with Saran Wrap. Do not remove the comb at this stage.

Preparation of Denaturing Polyacrylamide Gels Containing Formamide

The inclusion of formamide in sequencing gels eliminates secondary structure in the DNA during electrophoresis. Formamide gels are particularly useful and almost a necessity when sequencing DNA templates with a G/C content >55%.

MATERIALS

CAUTION: Please see Appendix 4 for appropriate handling of materials marked with <!>.

Reagents and Solutions

This protocol requires all the reagents listed in Protocol 12.8.
Formamide (100%) <!>

Additional Information

Acrylamide	MC3, p. A8.41
Compressions in DNA-sequencing gels	MC3, pp. 12.82 and 12.109–12.110
Troubleshooting (air bubbles)	MC3, p. 12.79
Troubleshooting (leaking gels)	MC3, p. 12.80

METHOD

1. Clean and assemble glass plates to form a gel mold as described in Protocol 12.8, Steps 1–8.

2. Cover the working area of the bench with plastic-backed protective paper.

3. In a 250-ml side-arm flask, prepare a sequencing gel solution containing the desired concentration of acrylamide as specified in Table 12-9. The volumes given in this table are sufficient for a single 40 x 40-cm sequencing gel and can be proportionally adjusted to accommodate smaller or larger gels.

   IMPORTANT: The preparation of the gel must be completed without interruption from this point onward.

4. Combine all of the reagents and then heat the solution in a 55°C water bath for 3 minutes to help dissolution of the urea.

5. Remove the solution from the water bath and allow it to cool for 15 minutes at room temperature. Swirl the mixture from time to time. Add H_2O to the solution bringing the final volume to 100 ml.

TABLE 12-9 Acrylamide Solutions for Denaturing Polyacrylamide Gels Containing Formamide

	Gel			
	4%	**6%**	**8%**	**10%**
Acrylamide:bis solution (45%)	8.9 ml	13.3 ml	17.8 ml	22.2 ml
10x TBE	10 ml	10 ml	10 ml	10 ml
H_2O	20.8 ml	16.4 ml	11.9 ml	7.5 ml
Formamide	25 ml	25 ml	25 ml	25 ml
Urea	42 g	42 g	42 g	42 g

6. Attach the side-arm flask to a vacuum line and de-gas the solution.

7. Transfer the solution to a 250-ml glass beaker. Add 3.3 ml of freshly prepared 1.6% ammonium persulfate and swirl the gel solution gently to mix the reagents.

8. Add 50 μl of TEMED to the gel solution and swirl the solution gently to mix the reagents. Draw approximately 40 ml of the solution into a 60-cc hypodermic syringe.

9. Pour the gel solution into the mold, as described in Protocol 12.8, Steps 16–22.

   IMPORTANT: Because formamide slows the polymerization reaction substantially, allow the gel to polymerize for 2–3 hours before clamping it in an electrophoresis apparatus.

Preparation of Electrolyte Gradient Gels

Electrolyte gradients are formed when buffers of different concentrations are used in the upper (low electrolyte concentration) and lower (high electrolyte concentration) chambers of the electrophoresis device. Fragments of DNA migrate more slowly as they travel anodically into regions of progressively higher ionic strength. The spacing between bands of DNA is therefore reduced at the bottom of the gel and increased at the top. In consequence, the number of bases that can be read is increased by approximately 30%.

MATERIALS

CAUTION: Please see Appendix 4 for appropriate handling of materials marked with <!>.

Reagents and Solutions

This protocol requires all the reagents listed in Protocol 12.8.
Formamide <!> (optional)
Electrolyte gradient gels are poured in 1x TBE buffer and can be run in the presence or absence of formamide to resolve compressions.
Sodium acetate (3 M, pH 7.0)

Additional Information

Acrylamide	MC3, p. A8.41
Compressions in DNA-sequencing gels	MC3, pp. 12.82 and 12.109–12.110
Troubleshooting (air bubbles)	MC3, p. 12.79
Troubleshooting (leaking gels)	MC3, p. 12.80

METHOD

1. Clean and assemble a set of glass plates to form a gel mold as described in Protocol 12.8, Steps 1–8.

2. Cover the working area of the bench with plastic-backed protective paper.

3. Prepare a denaturing polyacrylamide gel as described in Protocol 12.8 or 12.9.

4. Clamp the gel into the electrophoresis device. Fill the upper chamber with 0.5x TBE buffer and the lower chamber with a buffer composed of 2 parts 1x TBE and 1 part 3 M sodium acetate. Load the sequencing reactions and perform electrophoresis as described in Protocol 12.11.

Loading and Running DNA-sequencing Gels

The sets of nested DNA fragments generated by DNA-sequencing methods are resolved by electrophoresis through thin denaturing polyacrylamide gels. **WARNING:** Large voltages are passed through DNA-sequencing gels at substantial amperages. More than enough current is used in these gels to cause severe burns, ventricular fibrillation, central respiratory arrest, and asphyxia due to paralysis of the respiratory muscles. Make sure that the gel boxes used for electrophoresis are well insulated, all buffer chambers are covered, and the box is used on a stable bench top that is dry. Always turn off the power to the gel box before loading samples or dismantling the gel.

MATERIALS

CAUTION: Please see Appendix 4 for appropriate handling of materials marked with <!>.

Nucleic Acids/Oligonucleotides

DNA-sequencing reactions <!> (Protocols 12.3–12.7)

Gels/Loading Buffers

Denaturing polyacrylamide gel (Protocol 12.8 or 12.9)
Formamide gel-loading buffer <!>
    80% (w/v) deionized formamide
    10 mM EDTA (pH 8.0)
    0.25% bromophenol blue
    0.25% xylene cyanol FF
0.5x and/or 1.0x TBE

Additional Items

Automatic pipettor (20 ml), equipped with flat capillary tips (e.g., Multiflex tip [Research Products International]) or other loading device (see MC3, p. 12.88)
Bulldog binder clips (5-cm length, five to seven per gel)
Gel-temperature-monitoring strips, thermochromic liquid crystal indicators that change color as the temperature rises (available from BioWhittaker and other suppliers)
Plastic-covered metal clamps (optional; see Step 2)
Power pack, capable of delivering >75 W at constant power at up to 7 kV
Scalpel with disposable blades
Sequencing gel tanks, preferably fitted with heat-dispersing metal plates
Shark's tooth comb (0.4 mm thick, with 32, 64, or 96 teeth, depending on the capacity of the gel apparatus)
    Use the same comb that was used to pour the gel.
Syringe (10 cc) and 22-gauge hypodermic needles
Water bath (85°C) or heating block (100°C)

Additional Information

Loading-sequencing gels   MC3, p. 12.88

METHOD

1. Use damp paper towels or a wet sponge to wipe away any dried polyacrylamide/urea from the outside of the gel mold. Pipette several ml of 1x TBE buffer along the top of the smaller or notched glass plate and slowly remove the comb from the gel. Cut the electrical tape with a scalpel and strip it from the bottom of the gel mold.

   IMPORTANT: Do not remove the tape from the sides of the gel!

2. Attach the gel mold to the electrophoresis apparatus with bulldog binder clips, plastic-coated laboratory clamps, or, in the case of electrophoretic devices, with built-in screw clamps, according to manufacturer instructions. The smaller or notched plate should be in direct contact with the electrophoresis device. The larger, unnotched plate should face the investigator.

3. Fill the upper and lower reservoirs of the apparatus with the appropriate buffer.

   ### For standard or formamide-containing gels

   a. Fill the upper and lower buffer reservoirs with 1x TBE. Make sure that the level of the buffer in the lower chamber is well above the bottom of the plates. The level of the buffer in the upper chamber should be well above the level of the upper edge of the shorter or notched plate and in direct contact with the gel.

   b. Use a 10-ml syringe filled with 1x TBE to rinse the top of the gel. Make sure that excess polyacrylamide and urea are removed from the gel. If necessary, use a syringe needle to scrape off any polyacrylamide sticking to the glass plates. Remove air bubbles under the bottom of the glass plates in the lower reservoir using a pasteur pipette.

   c. Attach the electrodes to the electrophoresis apparatus and the power supply. The cathode (black lead) should be attached to the upper reservoir and anode (red lead) to the bottom. Attach the built-in thermal sensor (if available) or temperature-monitoring strip. Run the gel at constant wattage (50–70 W) for approximately 45 minutes or until the temperature of the gel reaches 45–50°C. Turn off the power supply and disconnect the electrodes.

   ### For electrolyte gradient gels

   a. Fill the upper reservoir with 0.5x TBE and the lower reservoir with a solution consisting of 2 volumes of 1x TBE plus 1 volume of 3 M sodium acetate (please see Protocol 12.10).

   b. Wash the well with 0.5x TBE and remove urea/polyacrylamide as described above. Do not pre-run electrolyte gradient gels.

4. If necessary, add 5 µl of formamide gel-loading buffer to each sample. Then, incubate the microfuge tubes containing the sequencing reactions for 2 minutes in a heating apparatus set at 100°C. If the reactions have been carried out in a microtiter plate, remove the cover from the plate and float the open plate in a water bath for 5 minutes at 85°C.

5. While the tubes or plates are incubating, fill a 10-cc syringe fitted with a 22-gauge hypodermic needle with 0.5x or 1x TBE, as appropriate. Squirt the TBE forcibly across the submerged loading surface of the gel to remove any remaining urea and fragments of polyacrylamide from the loading area. Continue squirting until no urea can be seen in the loading area.

6. Gently insert the shark's tooth comb (teeth downward) into the loading slot. Push the comb down until the points of the teeth just penetrate the surface of the gel.

7. Transfer the microfuge tubes or microtiter plate from the water bath or heating block to ice. Keep the samples at 0°C until they are loaded onto the gel.

8. Load 1–5 µl (please see Protocol 12.3 for recommended volumes) of each sequencing reaction into adjacent slots of the gel.

9. When all of the samples have been loaded, connect the electrodes to the power pack and the electrophoresis apparatus: cathode (black) to the upper reservoir and anode (red) to the bottom reservoir. Run the gel at sufficient constant power to maintain a temperature of 45–50°C.

Gel Size (cm)	Power (W)	Voltage (V)
20 x 40	35–40	~1700
40 x 40	55–75	~2–3000
40 x 40 (electrolyte gradient gel)	75	~2100

The time required to achieve optimal resolution of the sequence of interest must be determined empirically. Monitor the progress of the electrophoretic run by following the migration of the marker dyes in the formamide gel-loading buffer (please see Table 12-10).

TABLE 12-10 Migration Rates of Marker Dyes through Denaturing Polyacrylamide Gels

Polyacrylamide Gel	Bromophenol Blue[a]	Xylene Cyanol FF[a]
5%	35	130
6%	29	106
8%	26	76
10%	12	55
20%	8	28

In electrolyte gradient gels, bromophenol blue migrates progressively more slowly as it travels anodically through the gel. Migration essentially ceases when the dye nears the bottom of the gel. The xylene cyanol tracking dye behaves in a similar fashion. Typically, electrophoresis is continued until the xylene cyanol dye is within 5–10 cm of the bottom of the gel and the bromophenol blue dye is at the bottom.

[a]The numbers are the approximate sizes of DNA (in nucleotides) with which the indicated marker dye will comigrate in a standard DNA-sequencing gel.

10. Depending on the distance between the sequence of interest and the oligonucleotide primer, apply a second loading of the sequencing samples to a standard or formamide-containing denaturing polyacrylamide gel approximately 15 minutes after the bromophenol blue in the first set of samples has migrated to the bottom of the gel (1.5–2.0 hours). The sequence obtained from the first loading will be more distal to the primer, whereas that obtained from the second loading will be more proximal. The length of readable sequence can be extended by approximately 35% by reloading the samples into a fresh set of lanes 2 hours after the samples were first loaded.

   a. Turn off the power supply and disconnect the sequencing apparatus.

   b. Replace the buffer in the top and bottom reservoirs.

   c. Denature the samples by heating as described in Step 4 above.

   d. Load the samples.

   e. Reconnect the sequencing apparatus to the power supply.

   f. Run the gel, as before, at sufficient constant power to maintain a temperature of 45–50°C.

11. At the end of the run, follow the procedures described in Protocol 12.12 to dismantle the gel and perform autoradiography.

Autoradiography and Reading of Sequencing Gels

MATERIALS

CAUTION: Please see Appendix 4 for appropriate handling of materials marked with <!>.

Reagents and Solutions

Please see Appendix 1 for components of stock solutions, buffers, and reagents. Dilute stock solutions to the appropriate concentrations.

Gel-fixing solution (mix in the order shown)
300 ml methanol <!>
2.4 liters H_2O
300 ml glacial acetic acid (17.4 M) <!>

Gels/Loading Buffers

DNA-sequencing gels, prepared and run as described in Protocols 12.8–12.11

Additional Items

Autoradiography cassettes, metal, spring-loaded (35.6 x 43.2 cm)
Chemiluminescent markers (e.g., Gloglos [Stratagene]) or radioactive ink <!>
    For instructions on making radioactive ink, see MC3, p. A1.21.
Intensifying screens, e.g., Lightening Plus (Dupont) (optional; see Step 15)
Gel dryer (80°C)
Gloves, talc-free
Metal spatula
Mylar sheet (~5 cm larger than the gel in length and breadth)
Tray for fixing gel
Whatman 3MM CHR paper
Whatman 3MM paper
X-ray film, blue sensitive (35 x 43 cm), e.g., X-Omat AR (Eastman Kodak)
X-ray film processor

Additional Information

Autoradiography and phosphorimaging	MC3, pp. A9.11–A9.15
Intensifying screens	MC3, p. A9.11
Reading an autoradiograph	MC3, p. 12.113

METHODS

1. At the end of the electrophoretic run (Protocol 12.11), turn off the power and disconnect the sequencing apparatus from the power pack. Dispose of the electrophoresis buffer and remove the gel mold from the apparatus.

2. Lay the gel mold flat on plastic-backed protective bench paper with the smaller (notched) plate uppermost. Allow the gel to cool to <37°C before proceeding.

3. Remove any remaining pieces of gel-sealing tape. Use the end of a metal spatula to pry apart the plates of the mold slowly and gently. The gel should remain attached to the longer (non-siliconized) glass plate.

4. When the glass plates have been separated, cut off a bottom or top corner on the side of the gel that was loaded first. This landmark serves to orient the gel during subsequent manipulations.

5. Fix the gel in sequencing gel-fixing solution (methanol/acetic acid).

6. Transfer the gel (together with its supporting glass plate) to a shallow tray containing methanol: acetic acid fixing solution. Fix the gel for 30 minutes at room temperature. Do not agitate the fluid while the gel is being fixed.

7. After 30 minutes, lift the glass plate very, very slowly from the fixation solution. Try to keep the plate horizontal until most of the fixation solution has drained away. Lay the plate, gel side uppermost, on a stack of paper towels. Blot excess fixation solution from the glass plate with Kimwipes. Try not to touch the surface of the gel with the Kimwipes. Remove wrinkles and blemishes from the gel by gently caressing its surface with gloved fingers.

8. Prepare a piece of Whatman 3MM CHR paper (or its equivalent) that is slightly larger (2–3 cm) than the gel in both length and width. Hold the paper in a bow shape and touch the center of the bow to the center of the gel. Let the paper fall gently onto the surface of the gel and then apply gentle pressure so that the gel becomes firmly attached to the rough surface of the paper.

9. Hold the paper in place with one hand and pick up the supporting glass plate with the other hand. Quickly flip the sandwich over and lay it down on the bench top with the glass plate facing upward. Gently separate the 3MM CHR paper from the glass plate by lifting the plate upward. The gel will stick to the 3MM CHR paper as the glass plate is pulled away.

10. Lay the 3MM CHR paper (gel uppermost) on two pieces of Whatman 3MM paper of the same size. Cut a piece of Saran Wrap slightly longer and wider than the gel and lay it on top of the gel. Try to avoid creases and bubbles. This step is more easily accomplished with the help of another person. Hold the corners of the Saran Wrap and pull outward so that it is tightly stretched. Lower the stretched Saran Wrap onto the surface of the gel. Once the Saran Wrap has touched the gel, do not attempt to remove it, since this can cause the gel to tear. The flat end of an agarose gel comb, a Kimwipe, or a plastic card can be used to remove any bubbles of air trapped between the gel and wrap.

11. Use a paper cutter or sharp pair of scissors to trim all three pieces of Whatman paper and the Saran Wrap to approximately the same size as the gel.

12. Place the sandwich of paper, gel, and Saran Wrap on the gel dryer, with the plastic wrap uppermost. Use a sheet of Mylar to keep the sandwich flat during drying.

13. Following the instructions of the manufacturer of the gel dryer, dry the gel for 30–60 minutes under vacuum at 80°C.

14. Remove the gel from the dryer and peel off the Saran Wrap. The dried gel should feel smooth to the touch and not sticky. A quick remedy for stickiness is to turn a powdered latex glove inside out and use the talcum powder on the inside of the glove to dust the gel. To orient the gel, attach a small adhesive label marked with chemiluminescent or radioactive ink to the 3MM CHR paper in the space created by cutting the bottom corner of the gel (Step 4).

15. In a darkroom, place the dried gel (gel side up) on top of an intensifying screen (if desired) in a spring-loaded metal cassette. Cover the gel with a sheet of unexposed X-ray film. Close the cassette. Establish an autoradiograph by exposing the gel to the film for 16–24 hours at room temperature or –80°C.

16. Develop the autoradiograph according to the recommendations of the manufacturer of the film and read the sequence of the DNA.

Mutagenesis

BACKGROUND INFORMATION

Background information found in *Molecular Cloning: A Laboratory Manual*, 3rd edition (hereafter MC3) unless otherwise indicated.

Preparation of Uracil-containing Single-stranded Bacteriophage M13 DNA

This method of mutagenesis uses a genetic strategy to increase the proportion of mutant clones recovered after oligonucleotide-mediated site-directed mutagenesis (Kunkel T.A. 1985. Rapid and efficient site-specific mutagenesis without phenotypic selection. *Proc. Natl. Acad. Sci.* **52:** 488–492). Single-stranded templates of bacteriophage M13, generated in a strain of *E. coli* that lacks uracil-DNA glycosylase (*ung⁻ dut⁻*), carry 20–30 uracil residues in place of thymines. When this DNA is used in a classical in vitro mutagenesis procedure, the product is a heteroduplex with uracil in the wild-type strand and thymine in the strand synthesized in vitro. Transfection of this DNA into a wild-type (*ung⁺*) strain of *E. coli* results in removal of the uracil residues, destruction of the template strand, and suppression of the production of wild-type bacteriophages. A large proportion of the progeny bacteriophages are therefore derived from replication of the transfected uracil-free strand. Because the synthesis of this strand was primed in vitro by a mutagenic oligonucleotide, a high proportion of the progeny bacteriophages carry the desired mutation. This protocol describes how to generate uracil-containing, single-stranded bacteriophage M13 DNA for subsequent use in oligonucleotide-directed mutagenesis (Protocol 13.2).

MATERIALS

CAUTION: Please see Appendix 4 for appropriate handling of materials marked with <!>.

Reagents and Solutions

Please see Appendix 1 for components of stock solutions, buffers, and reagents. Dilute stock solutions to the appropriate concentrations.

Ethanol
Ethanol (70%)
NaCl (2.5 M), containing 15% (w/v) polyethylene glycol (PEG 8000) <!>
Phenol:chloroform (1:1, v/v) <!>
Phenol, equilibrated to pH 8.0 and saturated with H_2O <!>
Sodium acetate (3 M, pH 5.2)
TE (pH 7.6)

Vectors and Bacterial Strains

Bacteriophage M13 vector, e.g., M13mp18 or mp19 (see MC3, Figure 3-4 and Appendix 3)
Overnight culture of *E. coli* strain CJ236 (*dut⁻ ung⁻* F′)
Overnight culture of *E. coli* strains TG1, JM109, or their equivalents

Media and Antibiotics

2x YT medium
2x YT medium containing 0.25 µg/ml uridine

Enzymes and Buffers

0.8% and 0.9% Agarose gels (w/v) cast and run in 0.5x TBE
Marker DNA, single-stranded bacteriophage M13 DNA

Nucleic Acids/Oligonucleotides

DNA fragment (~500 bp), carrying the target sequence for mutagenesis

Centrifuges/Rotors/Tubes

Centrifuge bottles (250 ml)
Corex glass tubes (15 and 30 ml)
Sorvall SLC-1500 and SS-34 rotors (4°C)

Additional Items

Rotary shaker/incubator (37°C)
Spun column of Sephacryl S-400 (or equivalent resin)
Step 1 of this protocol requires the reagents listed in Protocols 3.3, 3.4, and 3.6.
Step 6 of this protocol requires the reagents listed in Protocol 3.1.
Water bath (60°C)

Additional Information

Commercial kits for site-directed mutagenesis	MC3, p. 13.89
dut and *ung*	MC3, p. 13.85
Selecting against wild-type DNA in site-directed mutagenesis	MC3, pp. 13.84–13.87
Spun-column chromatography	MC3, pp. A8.30–A8.31

METHOD

1. Prepare for mutagenesis.

   a. Clone a small fragment of DNA (<500 bp) carrying the target sequence into an appropriate bacteriophage M13 vector such as M13mp18 or mp19 (see Protocol 3.6).

   b. Isolate single-stranded template DNA and double-stranded replicative form DNA from a freshly grown plaque generated by the recombinant bacteriophage.

   c. Check the fidelity of the recombinant clone by restriction mapping of the replicative form DNA and by DNA sequencing of the single-stranded DNA.

2. Use a sterile pasteur pipette to transfer a single plaque produced by the bacteriophage M13 recombinant to a microfuge tube containing 1 ml of 2x YT medium.

3. Incubate the tube for 5 minutes at 60°C to kill bacterial cells. Vortex the tube vigorously for 30 seconds to release the bacteriophages trapped in the top agar. Remove dead bacterial cells and fragments of agar by centrifuging the tube at maximum speed for 2 minutes at 4°C in a microfuge.

4. Transfer 50 μl of the supernatant to a 500-ml flask containing 50 ml of 2x YT medium supplemented with 0.25 μg/ml uridine. Add 5 ml of a mid-log-phase culture of *E. coli* strain CJ236 (*dut⁻ ung⁻* F′). Incubate the culture with vigorous shaking (300 cycles/minute on a rotary shaker) for 6 hours at 37°C.

5. Pellet the cells by centrifugation at 5000*g* (6470 rpm in a Sorvall SS-34 rotor) for 30 minutes at 4°C. Transfer the supernatant to a fresh 250-ml centrifuge bottle that will fit into a Sorvall SLC-1500 rotor or its equivalent.

6. Determine the relative titer of the bacteriophage suspension on *E. coli* strain CJ236 (*dut⁻ ung⁻* F′) and a strain such as JM109 or TG1 (please see Protocol 3.1). The titer on strain CJ236 should be four to five orders of magnitude greater than that on the *dut⁺ ung⁺* strain of *E. coli*.

7. Measure the volume of the bacteriophage suspension and then add 0.25 volume of 2.5 M of NaCl containing 15% (w/v) PEG 8000. Mix the contents of the centrifuge bottle by swirling and store the bottle on ice for 1 hour.

8. Recover the precipitated bacteriophage particles by centrifugation at 5000*g* (5500 rpm in a Sorvall SLC-1500 rotor) for 20 minutes at 4°C. Remove the supernatant by aspiration and then invert the bottle to allow the last traces of supernatant to drain away. Use a pipette attached to a vacuum line to remove any drops of solution adhering to the walls of the bottle.

9. Resuspend the bacteriophage pellet in 4 ml of TE (pH 7.6). Transfer the suspension to a 15-ml Corex centrifuge tube and wash the walls of the centrifuge bottle with another 2 ml of TE (pH 7.6). Transfer the washing to the Corex tube. Vortex the suspension vigorously for 30 seconds and store the tube on ice for 1 hour.

10. Vortex the suspension vigorously for 30 seconds and pellet the bacterial debris by centrifugation at 5000*g* (6470 rpm in a Sorvall SS-34 rotor) for 20 minutes at 4°C.

12. Taking care not to disturb the pellet of bacterial debris, transfer the supernatant to a 15-ml polypropylene tube. Extract the suspension twice with phenol (pH 8.0) and once with phenol:chloroform. Separate the phases by centrifugation at 4000*g* (5800 rpm in a Sorvall SS-34 rotor) for 5 minutes at room temperature. Avoid transferring material from the interface.

13. Transfer the aqueous phase from the final extraction to a glass centrifuge tube (e.g., a 30-ml Corex tube). Measure the volume of the solution and add 0.1 volume of 3 M of sodium acetate (pH 5.2), followed by 2 volumes of ethanol at 0°C. Mix the contents of the tube thoroughly and then store the tube on ice for 30 minutes.

14. Recover the DNA by centrifugation at 5000*g* (6470 rpm in a Sorvall SS-34 rotor) for 20 minutes at 4°C. Carefully remove the supernatant. Add 10 ml of 70% ethanol at room temperature, vortex the solution briefly, and recentrifuge.

15. Carefully remove the supernatant by aspiration and store the tube in an inverted position at room temperature until the last traces of ethanol have evaporated. Dissolve the DNA in 200 μl of TE (pH 7.6).

16. Purify the resuspended single-stranded uracil-containing bacteriophage M13 DNA by spun-column chromatography using resins that exclude large DNAs (>100 nucleotides).

17. Measure the DNA spectrophotometrically at 260 nm (1 OD_{260} = 40 μg/ml). Analyze the size of an aliquot of the DNA (0.5 μg) by gel electrophoresis, using single-stranded DNA of the original bacteriophage M13 recombinant (Step 1) as a size marker.

18. Perform oligonucleotide-directed mutagenesis as described in Protocol 13.2.

Oligonucleotide-directed Mutagenesis of Single-stranded DNA

Uracil-containing bacteriophage M13 DNA, generated in Protocol 13.1, is used in the classic double-primer method of oligonucleotide-mediated mutagenesis to generate a heteroduplex molecule with uracil in the template strand and thymine in the strand synthesized in an in vitro reaction. Transformation of this DNA into an *ung+* strain of *E. coli* results in destruction of the template strand, with consequent suppression of production of wild-type bacteriophages. Up to 80% of the plaques are derived by replication of the uracil-free strand. Because synthesis of this strand has been primed by a mutagenic oligonucleotide, a large proportion of progeny bacteriophages carry the desired mutation.

MATERIALS

Reagents and Solutions

Please see Appendix 1 for components of stock solutions, buffers, and reagents. Dilute stock solutions to the appropriate concentrations.

> ATP (10 mM)
>
> dNTP solution, containing all four dNTPs, each at a concentration of 2 mM
>
> 10x PE1 buffer
>> 200 mM Tris-Cl (pH 7.5)
>> 100 mM MgCl$_2$
>> 500 mM NaCl
>> 10 mM dithiothreitol
>
> 10x PE2 buffer
>> 200 mM Tris-Cl (pH 7.5)
>> 100 mM MgCl$_2$
>> 100 mM dithiothreitol

Vectors and Hosts

> Overnight culture of *E. coli* strain, male, suitable for transfection (e.g., TG1; see MC3, Appendix 3)
>
> TG1 or other suitable strain, competent for transformation or electroporation (Protocol 1.25 or 1.26)

Media and Antibiotics

> 2x YT agar plates and top agar

Enzymes and Buffers

> Bacteriophage T4 DNA ligase
>
> Bacteriophage T4 polynucleotide kinase and 10x buffer
>> 700 mM Tris-Cl (pH 7.6)
>> 100 mM MgCl$_2$
>> 50 mM dithiothreitol
>
> T4 DNA polymerase
>> This is one of several DNA polymerases (e.g., Sequenase or the Klenow fragment of *E. coli* DNA polymerase) that can be used to catalyze the in vitro reaction primed by the mutagenic oligonucleotide.
>>
>> Sequenase and T4 DNA polymerase are preferred because they have the distinct advantage of being unable to displace the mutagenic oligonucleotide from its template.

Nucleic Acids/Oligonucleotides

Bacteriophage M13 universal sequencing primer (see MC3, p. 8.115)
Bacteriophage M13 recombinant DNA, containing target sequence for mutagenesis (1 µg of DNA in a volume of 2–3 µl of H_2O)
Mutagenic oligonucleotide primer, designed as described in MC3, pp. 13.82–13.83 and purified by Sep-Pak$_{18}$ column chromatography (Protocol 10.6)

Additional Items

Step 1 of this protocol requires the reagents listed in Protocol 13.1.
Step 7 of this protocol requires the reagents listed in Protocol 12.3 or 12.4 and 13.7.
Water bath/heating/cooling blocks (16°C, 42°C, 47°C, and 68°C)

Additional Information

Bacteriophage T4 DNA polymerase	MC3, pp. A4.18–A4.21
Calculating melting temperatures	MC3, pp. 10.47–10.48
Commercial kits for site-directed mutagenesis	MC3, p. 13.89
dut and *ung*	MC3, p. 13.85
Klenow fragment of *E. coli* DNA polymerase I	MC3, pp. A4.15–A4.17
Selecting against wild-type DNA in site-directed mutagenesis	MC3, pp. 13.84–13.87
Sequenase	MC3, p. 12.104
Strategies used in site-directed mutagenesis	MC3, pp. 13.2–13.10
Troubleshooting	MC3, p. 13.18

METHOD

1. Prepare the single-stranded bacteriophage M13 template as described in Protocol 13.1.

2. Phosphorylate the mutagenic oligonucleotide and the universal sequencing primer with bacteriophage T4 polynucleotide kinase. In separate microfuge tubes, mix:

synthetic oligonucleotide	100–200 pmoles
10x bacteriophage T4 polynucleotide kinase buffer	2 µl
10 mM ATP	1 µl
bacteriophage T4 polynucleotide kinase	4 units
H_2O	to 20 µl

   Incubate the reactions for 1 hour at 37°C and then heat them for 10 minutes at 68°C to inactivate the polynucleotide kinase.

3. Anneal the phosphorylated mutagenic oligonucleotide and universal sequencing primer to the single-stranded bacteriophage M13 DNA containing the target sequence. Mix:

single-stranded template DNA (~1 µg)	0.5 pmole
phosphorylated mutagenic oligonucleotide	10 pmoles
phosphorylated universal primer	10 pmoles
10x PE1 buffer	1 µl
H_2O	to 10 µl

   Heat the mixture for 5 minutes to 20°C above the theoretical T_m of a perfect hybrid formed by the mutagenic oligonucleotide, calculated from the formula $T_m = 4(G+C) + 2(A+T)$, where $(G+C)$ = the sum of G and C residues in the oligonucleotide and where $(A+T)$ = the sum of the A and T residues in the oligonucleotide. Transfer the tube containing the reaction mixture to a beaker containing H_2O at 20°C above the T_m. Stand the beaker on the bench and allow the reaction to cool slowly to room temperature (~20 minutes). Centrifuge the tube briefly

(5 seconds) in a microfuge to collect any fluid that has condensed on the walls of the tube.

4. While the annealing reaction cools to room temperature, mix the following reagents in a fresh 0.5-ml microfuge tube:

10x PE2 buffer	1.0 μl
2 mM dNTP solution	1.0 μl
10 mM ATP	1.0 μl
bacteriophage T4 DNA ligase	5 Weiss units
Klenow fragment or other DNA polymerase	2.5 units
H_2O	to 10 μl

Store the mixture on ice until needed.

5. Add 10 μl of the ice-cold reaction mixture from Step 4 to the reaction mixture containing single-stranded DNA and annealed oligonucleotides (Step 3). Incubate the final reaction mixture for 6–15 hours at 16°C.

> When using bacteriophage T4 DNA polymerase or Sequenase, increase the concentration of each of the four dNTPs to 500 μM and incubate the polymerization/extension reaction for 5 minutes at 0°C, 5 minutes at room temperature, and then 2 hours at 37°C. Incubation at low temperature optimizes initiation of DNA synthesis from the 3′ terminus of the mutagenic oligonucleotide, and the subsequent incubation at 37°C at high dNTP concentrations improves the efficiency of the extension reaction and suppresses the strong 3′-exonuclease activity of bacteriophage T4 DNA polymerase.

6. Transfect competent *E. coli* of an appropriate host strain (e.g., TG1) as follows:

   a. Prepare a series of dilutions of the reaction mixture (1:10, 1:100, and 1:500) in 10 mM Tris-Cl (pH 7.6).

   b. To a series of chilled (0°C) Falcon 2059 tubes, transfer 1 μl and 5 μl of (i) the original reaction mixture and (ii) each dilution. Add 200 μl of a preparation of competent TG1 cells to each tube.

   c. Store the mixtures on ice for 30 minutes and then transfer them for exactly 2 minutes to a water bath equilibrated at 42°C.

   d. Remove the transfected cultures from the water bath and add 100 μl of a standard overnight culture of TG1 cells. The addition of cells makes it easier to see bacteriophage M13 plaques in the lawn of bacterial cells.

   e. Add 2.5 ml of 2x YT top agar (melted and cooled to 47°C) to each tube and plate the resulting mixtures on separate YT agar plates. Incubate the plates for 12–16 hours at 37°C to allow plaques to form.

   > If electroporation is used, the polymerization/extension reaction should be diluted 1:100, 1:500, 1:1000, and 1:10,000 in H_2O. Between 1 and 10 ml of each dilution should be used for electroporation.

7. Screen plaques by sequencing preparations of single-stranded bacteriophage DNA (please see Protocol 12.3 or 12.4). If necessary, the plaques can be screened by hybridization with a radiolabeled oligonucleotide probe to detect mutants that arise at a low frequency (Protocol 13.7).

In Vitro Mutagenesis Using Double-stranded DNA Templates: Selection of Mutants with *Dpn*I

In this type of mutagenesis, sometimes called circular mutagenesis, two oligonucleotides are used to prime DNA synthesis catalyzed by a high-fidelity thermostable polymerase on a denatured plasmid template. The two oligonucleotides both contain the desired mutation and occupy the same starting and ending positions on opposite strands of the plasmid DNA. During several rounds of thermal cycling, both strands of the plasmid DNA are amplified in a linear fashion, generating a mutated plasmid containing staggered nicks on opposite strands.

The products of the linear amplification reaction are treated with the restriction enzyme *Dpn*I, which specifically cleaves fully methylated G^{Me6}ATC sequences. *Dpn*I will therefore digest the bacterially generated DNA used as a template for amplification but will not digest DNA synthesized during the course of the reaction in vitro. *Dpn*I-resistant molecules, which are rich in the desired mutants, are recovered by transforming *E. coli* to antibiotic resistance.

MATERIALS

CAUTION: Please see Appendix 4 for appropriate handling of materials marked with <!>.

Reagents and Solutions

Please see Appendix 1 for components of stock solutions, buffers, and reagents. Dilute stock solutions to the appropriate concentrations.

ATP (10 mM)
dNTP solution, containing all four dNTPs, each at a concentration of 5 mM
Ethanol
Ethanol (70%)
Phenol:chloroform (1:1, v/v) <!>
NaOH (1 M), containing EDTA (1 mM) <!>
Sodium acetate (3 M, pH 4.8)
TE (pH 8.0)

Vectors and Hosts

Transformation-competent preparation of an *E. coli* strain with an *hsdR17* genotype (e.g., XL1-Blue, XL1-Blue MRF', XL2-Blue MRF', or DH5α)

Media and Antibiotics

LB agar plates containing the appropriate antibiotic

Enzymes and Buffers

Bacteriophage T4 DNA ligase and 10x buffer
Bacteriophage T4 polynucleotide kinase and 10x buffer
*Dpn*I restriction endonuclease and 10x buffer
10x Long PCR buffer, used in reactions catalyzed by mixtures of thermostable DNA polymerases
    500 mM Tris-Cl (pH 9.0, room temperature)
    160 mM ammonium sulfate
    25 mM MgCl$_2$
    1.5 mg/ml bovine serum albumin

10x Mutagenesis buffer, when using DNA polymerases such as *Pfu*
  100 mM KCl
  100 mM ammonium sulfate
  200 mM Tris (pH 8.9, room temperature)
  20 mM MgSO₄
  1% Triton X-100
  1 mg/ml nuclease-free bovine serum albumin

Thermostable DNA polymerase (e.g., *Pfu* DNA polymerase)
  The conditions described in this protocol are optimized for *Pfu*Turbo DNA polymerase (Stratagene). However, they are easily adapted for use with other thermostable polymerases or mixtures of polymerases. It is important to choose an enzyme or mixture of enzymes that has an efficient proofreading activity, a low rate of incorporation of mismatched bases, and no untemplated terminal transferase activity. *Taq* does not fulfill these criteria, but *Pfu* and a number of other DNA polymerases do (see MC3, p. 13.20)

Nucleic Acids/Oligonucleotides

Closed circular plasmid (100–200 µg/ml), containing a cloned copy of the gene or sequence of interest
  In general, the smaller the plasmid, the better.

Mutagenic oligonucleotide primers, designed as described in MC3, pp. 13.19–13.20 and 13.82–13.83
  In brief, the mutagenic oligonucleotides must anneal to the complementary strands of the target sequence, be of equal length (32 bp ± 8); melt at the same temperature (>75°C), and terminate in a C or G residue. The oligonucleotides need not be phosphorylated but should be purified by Sep-Pak C₁₈ column chromatography or gel electrophoresis.

Gels/Loading Buffers

Agarose gel (1%), cast and run in 0.5x TBE containing ethidium bromide (0.5 µg/ml) <!>
DNA markers (1-kb ladders)

Additional Items

Barrier tips for automatic pipettor
Light mineral oil or wax bead (optional; see Step 5)
Microfuge tubes for PCR, 0.5 ml and thin walled
Step 14 of this protocol requires the reagents listed in Protocol 1.23.
Step 15 of this protocol requires the reagents listed in Protocol 1.1.
Step 16 of this protocol requires the reagents listed in Protocol 12.3, 12.4, or 12.5.
Thermal cycler, programmed with the desired amplification protocol
Water bath/heating/cooling blocks (16°C, 42°C, 47°C, and 68°C)

Additional Information

Commercially available kits for circular mutagenesis include QuikChange (Stratagene).

*Dpn*I	MC3, pp. 13.84 and A4.5–A4.6
hsdR17	MC3, pp. 3.10–3.13
Selecting against wild-type DNA in site-directed mutagenesis	MC3, pp. 13.84–13.87
Strategies used in site-directed mutagenesis	MC3, pp. 13.2–13.10
Troubleshooting	MC3, p. 13.25

METHOD

1. Denature the plasmid DNA template in a reaction containing 1–10 µg of plasmid DNA dissolved in 40 µl of H_2O plus 10 µl of 1 M NaOH/1 mM EDTA. Incubate the DNA in the denaturing solution for 15 minutes at 37°C.

2. Add 5 µl of 3 M sodium acetate (pH 4.8) to neutralize the solution. Precipitate the DNA with 150 µl of ice-cold ethanol.

3. Collect the denatured plasmid DNA by centrifugation for 10 minutes at 4°C in a microfuge. Carefully decant the ethanolic supernatant and rinse the pellet with 150 µl of 70% ethanol. Recentrifuge for 2 minutes, decant the supernatant, and allow the last traces of ethanol to evaporate at room temperature. Resuspend the DNA in 20 µl of H_2O.

4. In sterile 0.5-ml microfuge tubes, set up a series of reaction mixtures containing different amounts (e.g., 5, 10, 25, and 50 ng) of plasmid DNA and a constant amount of each of the two oligonucleotide primers.

10x mutagenesis buffer	5 µl
template plasmid DNA	5–50 ng
oligonucleotide primer 1 (20 mM)	2.5 µl
oligonucleotide primer 2 (20 mM)	2.5 µl
dNTP mix	2.5 µl
H_2O	to 50 µl

   It is important to add the reagents in the order shown.

   Add 2.5 units of *Pfu*Turbo DNA polymerase.

5. If the thermal cycler is not fitted with a heated lid, overlay the reaction mixtures with 1 drop (~50 µl) of light mineral oil or a bead of paraffin wax to prevent evaporation of the samples during repeated cycles of heating and cooling. Place the tubes in the thermal cycler.

6. Amplify the nucleic acids using the denaturation, annealing, and polymerization times and temperatures listed below.

Cycle Number	Denaturation	Annealing	Polymerization
1 cycle	1 min at 95°C		
2–18 cycles[a]	30 sec at 95°C	1 min at 55°C	2 min/kb of plasmid DNA at 68°C
Last cycle	1 min at 94°C	1 min at 55°C	10 min at 72°C

   Times and temperatures may need to be adapted to suit the particular reaction conditions.

   [a]For single-base substitutions, use 12 cycles of linear amplification; for substitution of one amino acid for another (usually two or three contiguous base substitutions), use 16 cycles; for insertions and deletions of any size, use 18 cycles.

   The rate of DNA synthesis is 1.5–2.0 times slower in amplification reactions catalyzed by *Pfu* than in reactions catalyzed by *Taq*.

7. After amplification of the DNA, place the reactions on ice.

8. Verify that the target DNA was amplified by analyzing 10 µl of each reaction by electrophoresis through a 1% agarose gel containing 0.5 µg/ml ethidium bromide. As standards, load 50 ng of unamplified linearized plasmid DNA and a 1-kb DNA ladder into the outer lanes of the gel.

9. Extract the amplified DNAs twice with phenol:chloroform and precipitate with ethanol.

   Steps 9–12 are generally used only when the efficiency of mutagenesis is expected to be low (e.g., when constructing insertions and deletions).

10. Resuspend the DNA pellets in the following:

10x bacteriophage T4 polynucleotide kinase buffer	5 µl
10 mM ATP	5 µl
bacteriophage T4 polynucleotide kinase	5 units
H$_2$O	to 50 µl

Incubate the reactions for 1 hour at 37°C. Inactivate the kinase enzyme by heating at 68°C for 10 minutes. Extract the phosphorylated DNAs twice with phenol:chloroform and collect the DNAs by ethanol precipitation.

11. Resuspend the pellets of phosphorylated DNA (~0.9 µg each) in 90 µl of TE. Set up a series of ligation reactions containing the phosphorylated DNAs at concentrations ranging from 0.1 to 1 µg/ml.

phosphorylated DNA	(10 to 100 ng)
10x bacteriophage T4 DNA ligase buffer	10 µl
10 mM ATP	10 µl
bacteriophage T4 DNA ligase	4 units
H$_2$O	to 100 µl

Incubate the reactions for 12–16 hours at 16°C.

12. Extract the ligated DNAs twice with phenol:chloroform and collect the DNA by ethanol precipitation. Resuspend each pellet in 45 µl of H$_2$O. Add 5 µl of 10x *Dpn*I buffer to each tube.

13. Digest the amplified DNAs by adding 10 units of *Dpn*I directly to the remainder of the amplification reactions (Step 7) or to the phosphorylated and ligated DNAs (Step 12). Mix the reagents by pipetting the solution up and down several times, centrifuge the tubes for 5 seconds in a microfuge, and then incubate them for 1 hour at 37°C.

14. Transform competent *E. coli* with 1, 2, and 5 µl of digested DNA according to the procedure described in Protocol 1.23.

15. Prepare plasmid DNA from at least 12 independent transformants. Screen the DNA preparations for mutations by DNA sequencing, oligonucleotide hybridization (see Protocol 13.7), or restriction digestion of small preparations of plasmid DNA if a site was created or destroyed by the introduced mutation, or if an insertion or deletion was introduced into the template.

16. Sequence the entire segment of target DNA to verify that the desired mutation has been generated and that no spurious mutations occurred during amplification (please see Protocol 12.3, 12.4, or 12.5).

Oligonucleotide-directed Mutagenesis by Elimination of a Unique Restriction Site (USE Mutagenesis)

Two oligonucleotide primers are hybridized to the same strand of a denatured double-stranded recombinant plasmid. One primer (the mutagenic primer) introduces the desired mutation into the target sequences, and the second primer carries a mutation that destroys a unique restriction site in the plasmid. Both primers are elongated in a reaction catalyzed by bacteriophage T4 or T7 DNA polymerase. Nicks in the strand of newly synthesized DNA are sealed with bacteriophage T4 DNA ligase. The product of the first part of the method is a heteroduplex plasmid consisting of a wild-type parental strand and a new full-length strand that carries the desired mutation but no longer contains the unique restriction site.

After residual wild-type plasmid molecules have been linearized by digestion with the appropriate restriction enzyme, the mixture of circular heteroduplex DNA and linear wild-type DNA is used to transform a strain of *E. coli* that is deficient in repair of mismatched bases. Because linear DNA transforms 10- to 1000-fold less efficiently than circular DNA, many of the wild-type molecules are unable to reestablish themselves in *E. coli*. The circular heteroduplex molecules, however, begin to replicate. Because the mismatched bases are not repaired, the first round of replication generates a wild-type plasmid that carries the original restriction site and a mutated plasmid that does not. DNA from the first set of transformants is recovered, digested once more with the same restriction enzyme to linearize the wild-type molecules, and then used to transform a standard laboratory strain of *E. coli*. This biochemical selection can be sufficiently powerful to ensure that a high proportion of the resulting transformants carry the desired mutation.

MATERIALS

CAUTION: Please see Appendix 4 for appropriate handling of materials marked with <!>.

Reagents and Solutions

Please see Appendix 1 for components of stock solutions, buffers, and reagents. Dilute stock solutions to the appropriate concentrations.
  10x Annealing buffer
      200 mM Tris-Cl (pH 7.5)
      100 mM MgCl$_2$
      500 mM NaCl
  10x Synthesis buffer
      100 mM Tris-Cl (pH 7.5)
      dTTP, dATP, dCTP, and dGTP, each at a concentration of 4 mM
      10 mM ATP
      20 mM dithiothreitol

Vectors and Hosts

  Transformation-competent preparation of an *E. coli* strain with a *mut*⁺ genotype
  Transformation-competent preparation of an *E. coli* strain with an *mutS* genotype (e.g., BMH 71-18)

Media and Antibiotics

  LB agar plates containing the appropriate antibiotic
  LB medium containing the appropriate antibiotic

Enzymes and Buffers

  Bacteriophage T4 DNA ligase and 10x buffer

Bacteriophage T4 DNA polymerase or Sequenase
Restriction enzyme that cleaves a unique site within the plasmid

Nucleic Acids/Oligonucleotides

Closed circular plasmid (100–200 µg/ml), containing a cloned copy of the gene or
sequence of interest and purified as described in Protocol 1.2 or 1.9
> In general, the smaller the plasmid, the better.

Mutagenic primer and selection primer, dissolved in H_2O at a concentration of
10 pmoles/µl and designed as described in MC3, pp. 13.82–13.83
> In brief, the primers must carry the engineered mutation in the center and be flanked on each
> side by 10–15 bases that pair perfectly with the template DNA. In addition, the selection
> primer and the mutagenic primer must be complementary to the same strand of the the plas-
> mid DNA. Both oligonucleotides should be phosphorylated and purified by FPLC or polyacryl-
> amide gel electrophoresis (Protocol 10.1 or 10.5).
>
> Several companies (e.g., BD Biosciences/CLONTECH) sell selection primers and also market
> pairs of "toggle" primers that are used for forward and reverse conversion of restriction sites
> and allow sequential rounds of mutagenesis without subcloning of the template.

Gels/Loading Buffers

1% Agarose gel, cast and run in 0.5x TBE containing ethidium bromide (0.5 µg/ml) <!>

Additional Items

Shaking incubator (37°C)
Step 17 of this protocol requires the reagents listed in Protocol 12.3, 12.4, or 12.5.
Steps 7 and 14 of this protocol require the reagents listed in Protocol 1.23, 1.24,
or 1.26 for the transformation of *E. coli*.
Steps 10 and 16 of this protocol require the reagents listed in Protocol 1.1 for
the minipreparation of plasmid DNA.
Water bath/heating blocks (70°C and at the optimal temperature for digestion with the
restriction enzyme)

Additional Information

Commercial kits for USE mutagenesis are available from several manufacturers, including
BD Biosciences/CLONTECH and Stratagene.

Selecting against wild-type DNA in site-directed mutagenesis	MC3, pp. 13.84–13.87
Troubleshooting	MC3, p. 13.30
USE mutagenesis	MC3, p. 13.85

METHOD

1. Mix the following components in a microfuge tube:

10x annealing buffer	2 µl
plasmid DNA	0.025 to 0.25 pmole
selection primer	25 pmoles
mutagenic primer	25 pmoles
H_2O	to 20 µl

   Incubate the tube in a boiling-water bath for 5 minutes.

2. Immediately chill the tube in ice for 5 minutes. Centrifuge the tube for 5 seconds in a
microfuge to deposit the fluid at the base.

3. To the tube of annealed primers and plasmid, add:

10x synthesis buffer	3 µl
bacteriophage T4 DNA polymerase (2–4 units/µl)	1 µl
bacteriophage T4 DNA ligase (4–6 units/µl)	1 µl
H₂O	5 µl

   Mix the reagents well by gentle up and down pipetting. Centrifuge the tube for 5 seconds in a microfuge to deposit the fluid at the base. Incubate the reaction for 1–2 hours at 37°C.

4. Stop the reaction by heating the tube for at least 5 minutes at 70°C to inactivate the enzymes. Store the tube on the bench to allow it to cool to room temperature.

5. Adjust the NaCl concentration of the reaction to a level that is optimal for the selected unique site restriction endonuclease. Use the 10x annealing buffer, a stock of NaCl, or the 10x buffer supplied with the restriction enzyme.

6. Add 20 units of the selective restriction endonuclease to the reaction mixture. Incubate the reaction for at least 1 hour at the appropriate digestion temperature.

   IMPORTANT: The volume of enzymes added to the reactions (including polymerase and ligase) should not exceed 10% of the total reaction volume. Adjust the reaction volume accordingly.

7. Transform a *mutS E. coli* strain such as BMH 71-18 with the plasmid DNAs contained in the digestion mixture, using one of the transformation procedures described in Protocols 1.23–1.26.

8. Spread 10, 50, and 250 µl of the transformation mixture onto LB agar plates containing the appropriate antibiotic. Incubate the plates overnight at 37°C. Perform Step 9 while the plates are incubating.

9. Amplify the plasmids by adding the remaining transformation mixture to 3 ml of LB medium containing the appropriate antibiotic. Incubate the culture overnight at 37°C with shaking.

10. The next day, prepare plasmid DNA from approximately 2.5 ml of the overnight culture (please see Protocol 1.1).

11. Digest the plasmid DNA prepared in Step 10 with the selective restriction enzyme:

plasmid DNA	~500 ng
10x restriction enzyme buffer	2 µl
unique site restriction endonuclease	20 units
H₂O	to 20 µl

   Incubate the reaction for 2 hours at the appropriate temperature.

12. Add an additional 10 units of the restriction enzyme and incubate for at least 1 additional hour.

13. Assess the extent of digestion by running 5–10 µl of the plasmid DNA on a 1% agarose gel containing 0.5 µg/ml ethidium bromide.

14. Transform competent *mutS⁺ E. coli* cells with either 2–4 µl of the digested plasmid DNA (~50–100 ng) for transformation of chemically treated competent cells or 1 µl of plasmid DNA diluted fivefold with sterile H₂O (~5 ng) for transformation by electroporation (please see Protocols 1.23–1.26).

15. Spread 10, 50, and 250 µl of the transformation mixture onto LB agar plates containing the appropriate antibiotic. Incubate the plates overnight at 37°C.

16. The next day, prepare minipreparations of plasmid DNA from at least 12 independent transformants. Screen the preparations by restriction endonuclease digestion and agarose gel electrophoresis to identify plasmids that are resistant to cleavage by the selective restriction enzyme.

17. Use DNA sequencing to confirm that the plasmids contain the desired mutation (please see Protocol 12.3, 12.4, or 12.5).

MATERIALS

CAUTION: Please see Appendix 4 for appropriate handling of materials marked with <!>.

Reagents and Solutions

Please see Appendix 1 for components of stock solutions, buffers, and reagents. Dilute stock solutions to the appropriate concentrations.

dNTP solution, containing all four dNTPs, each at a concentration of 20 mM

Enzymes and Buffers

10x Amplification buffer
500 mM KCl
100 mM Tris-Cl (pH 8.3, room temperature)
15 mM MgCl$_2$

Thermostable DNA polymerase
To avoid introducing erroneous bases, this protocol requires a highly processive thermostable DNA polymerase with 3′–5′ exonuclease proofreading ability. In addition, the enzyme, unlike *Taq*, must not catalyze the nontemplate addition of adenine residues. Thermostable polymerases with the appropriate properties include *Pwo* DNA polymerase (Roche Applied Science), *Pfu* DNA polymerase (Stratagene), *VentR* DNA polymerase (New England BioLabs), and *rTth* DNA polymerase XL (Perkin-Elmer).

Nucleic Acids/Oligonucleotides

Closed circular plasmid (10 µg/ml), purified as described in Protocol 1.1 and containing a cloned copy of the gene or sequence of interest

Oligonucleotide primers
Each primer should be 16–25 nucleotides in length and contain approximately equal numbers of the four bases, with a balanced distribution of G and C residues and a low propensity to form secondary structures. The sequences of the mutagenic oligonucleotide primers FM and RM (Figure 13-2) should share at least a 15-bp overlap, and the mismatched base pairs within these primers should be located at the center of the oligonucleotide. The 5′ sequences of primers R2 and F2 can incorporate unique restriction sites to aid in subsequent cloning.

Gels/Loading Buffers

1% Agarose gel, cast and run in 0.5x TBE containing ethidium bromide (0.5 µg/ml) <!>

Additional Items

Barrier tips for automatic pipettor
Light mineral oil or wax bead (optional; see Steps 4 and 9)
Microfuge tubes for PCR, 0.5 ml or thin walled
Step 11 of this protocol requires the reagents listed in Protocol 1.17 or 1.19.
Step 12 of this protocol requires the reagents listed in Protocol 12.3, 12.4, or 12.5.
Thermal cycler, programmed with the desired amplification protocol

Additional Information

Design of oligonucleotide primers	MC3, pp. 8.13–8.16
Mutagenic oligonucleotides	MC3, pp. 13.82–13.83

METHOD

1. Design and synthesize oligonucleotide primers FM, RM, R2, and F2 based on the known sequence of the DNA.

2. In a sterile 0.5-ml microfuge tube or amplification tube, set up PCR 1 by mixing the following reagents:

template DNA	~100 ng
10x amplification buffer	10 µl
20 mM mixture of four dNTPs	1.0 µl
5 µM primer FM (30 pmoles)	6.0 µl
5 µM primer R2 (30 pmoles)	6.0 µl
thermostable DNA polymerase	1–2 units
H_2O	to 100 µl

3. In a second sterile 0.5-ml microfuge tube or amplification tube, set up PCR 2 by mixing the following reagents:

template DNA	~100 ng
10x amplification buffer	10 µl
20 mM mixture of four dNTPs	1.0 µl
5 µM primer RM (30 pmoles)	6.0 µl
5 µM primer F2 (30 pmoles)	6.0 µl
thermostable DNA polymerase	1–2 units
H_2O	to 100 µl

4. If the thermocycler does not have a heated lid, overlay the PCRs with 1 drop (~50 µl) of light mineral oil. Place the tubes in a thermocycler.

5. Amplify the nucleic acids using the denaturation, annealing, and polymerization times and temperatures listed below.

Cycle Number	Denaturation	Annealing	Polymerization
20 cycles	1 min at 94°C	1 min at 50°C	1–3 min at 72°C
Last cycle	1 min at 94°C	10 min at 72°C	

Times and temperatures may need to be adapted to suit the particular reaction conditions.

The length of the polymerization step should be calculated from the polymerization rate of the thermostable DNA polymerase employed and the length of the DNA template to be amplified.

The temperature of the annealing reaction may have to be adjusted depending on the sequence of the mutagenic primers.

6. Analyze 5% of each of the two PCRs on an agarose or polyacrylamide gel and estimate the concentration of amplified target DNAs.

7. (*Optional*) Purify the two PCR products using one of the protocols described in Chapter 5. Including this purification step often increases the yield of the desired amplification product in Step 8 of the protocol and reduces the background of spurious amplification products.

8. In a sterile 0.5-ml microfuge tube or amplification tube, mix the following reagents in an amplification reaction to join the 5´ and 3´ ends of the target gene:

amplification product PCR 1 (Step 2)	~50 ng
amplification product PCR 2 (Step 3)	~50 ng
10x amplification buffer	10 µl
5 µM primer F2 (30 pmoles)	6.0 µl
5 µM primer R2 (30 pmoles)	6.0 µl
thermostable DNA polymerase	1–2 units
H_2O	to 100 µl

9. If the thermocycler does not have a heated lid, overlay the PCRs with 1 drop (~50 µl) of light mineral oil. Place the tubes in a thermocycler.

10. Amplify the nucleic acids using the denaturation, annealing, and polymerization times and temperatures listed above in Step 5.

11. Analyze 5% of the PCR on an agarose or polyacrylamide acrylamide gel and estimate the concentration of amplified target DNA.

12. Verify the complete sequence of the amplified DNA fragment after cloning to ensure that no mutations other than those in primers FM and RM were introduced during these manipulations.

Screening Recombinant Clones for Site-directed Mutants by Hybridization to Radiolabeled Oligonucleotides

Plaques formed by M13 bacteriophages or bacterial colonies transformed by plasmids carrying specific mutations can be detected by hybridization, using a radiolabeled oligonucleotide that forms a perfect duplex with the mutant sequence. Hybridization is carried out under conditions of low stringency that allow the radiolabeled oligonucleotide to anneal to both mutant and wild-type DNAs. A hybrid between the radiolabeled oligonucleotide and the wild-type sequence will contain one or more mismatched base pairs, whereas the hybrid formed between the newly created mutant and the oligonucleotide probe will be perfectly matched. These two types of hybrids usually differ in their thermal stabilities, with mismatched hybrids dissociating at a lower temperature than the corresponding perfect hybrid. Plaques or colonies that continue to hybridize to the probe after several cycles of washing at progressively higher temperatures are likely to contain recombinants carrying the desired mutation. The following protocol describes the screening of bacteriophage M13 recombinants. However, the technique can easily be adapted to screen bacterial colonies (MC3, p. 13.47).

MATERIALS

CAUTION: Please see Appendix 4 for appropriate handling of materials marked with <!>.

Reagents and Solutions

Please see Appendix 1 for components of stock solutions, buffers, and reagents. Dilute stock solutions to the appropriate concentrations.

Ammonium formate (0.2 M) <!>
Oligonucleotide hybridization solution
    6x SSC
    5x Denhardt's reagent
    10^6–10^7 dpm/ml radiolabeled oligonucleotide <!> (see Protocol 10.2)
Oligonucleotide prehybridization solution
    6x SSC
    5x Denhardt's reagent
    0.1% (w/v) SDS
6x SSC
TE (pH 7.6)

Vectors and Hosts

Bacterial lawns of E. coli TG-1 (or equivalent) containing plaques formed by (1) bacteriophage M13 recombinants that putatively carry the mutation of interest and (2) wild-type bacteriophage M13, for use as a negative control
Overnight culture of E. coli strain TG-1 (or equivalent)

Media and Antibiotics

YT agar plates
2x YT top agar

Enzymes and Buffers

Bacteriophage T4 polynucleotide kinase and 10x polynucleotide kinase buffer
   700 mM Tris-Cl (pH 7.6)
   100 mM MgCl$_2$
   50 mM dithiothreitol
Restriction enzymes and 10x buffers

Nucleic Acids/Oligonucleotides

Oligonucleotide (10 pmoles/μl in H$_2$O)
   The oligonucleotide used as a probe is generally 20–24 nucleotides in length and exactly complementary to the sequences in which the desired site-specific mutation is embedded. In most cases, the oligonucleotide probe is generated from the mutagenic oligonucleotide originally used to introduce the mutation into the target sequence.

Radioactive Compounds

[γ-^{32}P]ATP (>5000 Ci/mmole, 10 mCi/ml) <!>

Gels/Loading Buffers

1% Agarose gels, cast and run in 0.5x TBE containing ethidium bromide (0.5 μg/ml) <!>
   If the target sequence cloned into the bacteriophage M13 vector is small (<300 bp), it may be better to use a polyacrylamide gel rather than an agarose gel to recover the mutated fragment from bacteriophage M13.

Additional Items

Blunt-ended forceps (e.g., Millipore)
Hypodermic needle (18 gauge) and India ink
or
Chemiluminescent stickers, for marking and aligning autoradiographs, e.g., Glogos (Stratagene)
Incubator or hybridization chamber, set to appropriate temperature for hybridization
Material for autoradiography or phosphorimaging (MC3, pp. A9.11–A9.14)
Microwave oven, vacuum baking oven (preset to 80°C), *or* UV cross-linking device
Nitrocellulose (e.g., Millipore HATF) or nylon filters (82 mm)
Plastic box, for washing filters after hybridization
Rotating shaker
Step 4 of this protocol requires the reagents listed in Protocol 10.4.
Step 14 of this protocol requires the reagents listed in Protocol 3.4.
Step 17 of this protocol requires the reagents listed in Protocol 3.3.
Step 20 of this protocol requires the reagents listed in Protocols 6.8 and 6.10.
Steps 15 and 16 of this protocol require the reagents listed in Protocol 12.3, 12.4, or 12.5.
Water baths (65°C and 10°C below the calculated T_m of the oligonucleotide:target hybrid)
Whatman DEAE paper (DE81)

Additional Information

Alternative protocol (Detection of Defined Mutants by PCR)   MC3, p. 13.48
Denhardt's reagent   MC3, p. A1.15
Design of oligonucleotide primers for basic PCR   MC3, pp. 8.13–8.16
Mutagenic oligonucleotides   MC3, p. 13.82–13.83

METHOD

1. In a sterile microfuge tube, mix:

oligonucleotide (10 pmoles/µl) to be used as a probe	1 µl
10x bacteriophage T4 polynucleotide kinase buffer	1 µl
10 mCi/ml [γ-^{32}P]ATP (10–50 pmoles)	1 µl
H$_2$O	6 µl
5–10 units/µl bacteriophage T4 polynucleotide kinase	1 µl

   Incubate the reaction mixture for 30 minutes at 37°C.

2. Dilute the reaction mixture to 100 µl by the addition of 90 µl of TE (pH 7.6) and inactivate the polynucleotide kinase by heating for 10 minutes at 68°C.

3. Measure the efficiency of transfer of ^{32}P to the oligonucleotide and estimate its specific activity by chromatography on DE81 paper as follows:

   a. Cut a strip of Whatman DE81 paper approximately 1 cm wide and 7–10 cm long. With a soft-lead pencil, draw a fine line across the strip approximately 1.5 cm from one end. This line marks the origin of the chromatogram.

   b. Spot 1.0 µl of the diluted phosphorylation reaction at the origin. Fill a 250-ml beaker to a depth of approximately 0.5 cm with approximately 25–50 ml of 0.2 M ammonium formate. Place the DE81 strip vertically in a beaker so that the radioactive sample(s) at the origin is just above the buffer solution. Cover the beaker with a glass plate or aluminum foil and allow the chromatogram to develop until the solvent front has migrated almost to the top of the beaker.

   c. Wrap the strip of DE81 paper in Saran Wrap and subject it to a very brief period of autoradiography. Use the developed X-ray film as a guide to cut out the radioactive region at the solvent front, the origin, and any other region that contains radioactivity. Measure the amount of radioactivity in each section in a scintillation counter.

4. (*Optional*) Remove unincorporated radiolabel from the oligonucleotide by precipitation with cetylpyridinium bromide as described in Protocol 10.4.

   This step is necessary only when background hybridization is a persistent problem. Under normal circumstances, the unfractionated reaction mixture may be used as a probe.

5. Prepare replicas of the bacteriophage M13 plaques that are to be screened with the radiolabeled oligonucleotide as follows:

   a. Transfer plates containing 100–500 plaques to 4°C for at least 30 minutes.

      Include at least one plate that contains plaques of the original wild-type recombinant bacteriophage M13. This plate serves as a negative control in the hybridization and washing steps.

   b. When the plates are thoroughly chilled, remove them from the cold room and immediately lay a numbered, dry nitrocellulose or nylon filter on the agar surface of each plate. Use an 18-gauge hypodermic needle to make a series of holes in each filter and the underlying agar. These holes will later serve to key the filters to the plates.

   c. After 30 seconds to 4 minutes, use blunt-ended forceps to peel each filter carefully from its plate. Spread out all of the filters (plaque side up) on a pad of paper towels. Wrap the plates in Saran Wrap and store them at 4°C until they are needed.

   d. When the filters have dried (~30 minutes at room temperature), bake them for 1 hour at 80°C in a vacuum oven.

6. Transfer all of the filters to a heat-sealable plastic bag (e.g., Seal-a-Meal), an evaporating dish of the appropriate diameter, or a hybridization roller bottle. Add oligonucleotide prehybridization solution (~10 ml/82-mm filter in bags or dishes; 5 ml/82-mm filter in bottles). Seal

the bag, cover the evaporating dish with Saran Wrap, or cap the roller bottle and incubate the filters for 1–2 hours at 65°C.

7. Discard the prehybridization solution and replace it with oligonucleotide hybridization solution (~5 ml/82-mm filter). Reseal the bag, recover the evaporating dish or recap the hybridization roller bottle, and incubate the hybridization reaction for 4–6 hours at the appropriate temperature.

    Perform hybridization with the radiolabeled oligonucleotide at a temperature 5–10°C below the T_m estimated from the following formula:

    $$T_m = 4(G+C) + 2(A+T)$$

    where $(G+C)$ is the sum of G and C residues in the oligonucleotide and $(A+T)$ is the sum of the A and T residues in the oligonucleotide.

8. At the end of the hybridization period, quickly transfer the filters to a tray containing 200–300 ml of 6x SSC at room temperature. Cover the tray with Saran Wrap and place it on a rotating shaker for 15 minutes. Replace the washing fluid every 5 minutes. Meanwhile, transfer the remainder of the radioactive hybridization solution from the bag, dish, or roller bottle to a disposable plastic tube. Close the tube tightly and store the radioactive solution at –20°C until it is needed for rescreening positive plaques (Step 13).

9. At the end of the washing period, quickly transfer the filters to a piece of Saran Wrap stretched on the bench. Cover the filters with another piece of Saran Wrap. Fold the edges of the two pieces of Saran Wrap together to form a tight seal. Apply adhesive dot labels marked with radioactive or chemiluminescent ink to the outside of the package and generate an autoradiograph by exposing the package of filters to X-ray film for 1–2 hours at –70°C, using an intensifying screen.

    IMPORTANT: Do not allow the filters to dry on the Saran Wrap.

10. Compare the pattern of hybridization with the distribution of plaques. At this stage, it is normal to find that virtually every plaque hybridizes to the probe. Typically, however, some plaques hybridize more strongly than others, and these often turn out to be the plaques carrying the desired mutation.

11. Transfer the filters to a plastic box containing 100–200 ml of 6x SSC that has been warmed to 10°C below the calculated T_m of the nucleotide. Agitate the filters in the solution for 2 minutes and transfer them to a piece of Saran Wrap as described in Step 9. Establish another autoradiograph. At this stage, it is often possible to identify two types of plaques: those whose radioactive signal has decreased in intensity and those that show no change in intensity.

12. Repeat the cycles of washing and autoradiography, increasing the temperature of the washing solution by 2–10°C in each cycle. The aim is to find a temperature that does not markedly affect perfect hybrids but causes dissociation of mismatched hybrids (such as those formed between the mutagenic oligonucleotide and the original wild-type sequence).

13. Positively hybridizing plaques usually contain a mixture of both mutant and wild-type sequences. It is therefore *essential* to plaque-purify the bacteriophages from positively hybridizing plaques as follows:

    a. Touch the blunt end of a sterile, disposable wooden toothpick to the surface of a positively hybridizing plaque.

    b. Drop the toothpick into a sterile tube containing 1 ml of sterile TE (pH 7.6). Store the tube for 10–15 minutes at room temperature, shaking it from time to time to dislodge bacteriophage particles.

    c. Make a series of tenfold dilutions of the bacteriophage suspension with TE (pH 7.6). Mix 10 µl of the 10^{-3}, 10^{-4}, and 10^{-5} dilutions with 100-µl aliquots of an overnight culture of an appropriate strain of *E. coli* (e.g., TG1).

    d. Add 2.5 ml of 2x YT top agar (melted and cooled to 45°C) to each culture and plate the entire mixture on a single YT agar plate. Incubate the plate for 16 hours at 37°C to allow plaques to form.

    e. Rescreen the plaques with the radiolabeled oligonucleotide as described in Steps 5–12. In this second round of screening, there is no need to increase the temperature of the washing solution in a stepwise fashion. Instead, the filters can be transferred directly from the washing solution at room temperature (Step 8) to 6x SSC previously warmed to the discriminatory temperature found empirically in Step 12.

14. Pick two plaques from each of three independent putative mutants. Prepare single-stranded DNA from small-scale cultures infected with bacteriophages derived from each of these plaques as described in Protocol 3.4.

15. Carry out DNA sequencing by the dideoxy-mediated chain-termination method (please see Protocol 12.3 or 12.4) through the region containing the target sequence. Use either a universal sequencing primer or a custom-synthesized primer that binds 50–100 nucleotides upstream of the mutation site.

16. When the presence of the mutation has been confirmed, verify the sequence of the entire region of the target DNA cloned in the bacteriophage M13 vector to ensure that no adventitious mutations have been generated during propagation of the recombinant in bacteriophage M13. Often, this requires the synthesis of custom-designed sequencing primers that are complementary to segments of the target DNA spaced approximately 200–400 bp apart.

17. Isolate bacteriophage M13 replicative form DNA from a culture infected with plaque-purified recombinant bacteriophages (Step 14) that carry the desired mutation and show no other changes in sequence in the target region. For methods to isolate and purify bacteriophage M13 replicative form DNA, please see Protocol 3.3.

18. Recover the mutated target sequence by digestion of bacteriophage M13 replicative form DNA with the appropriate restriction enzyme(s) and preparative gel electrophoresis. Clone the target DNA into the desired vector.

19. Use several different restriction enzymes to digest aliquots of a recombinant that carries either the original (nonmutagenized) target sequence or the mutagenized target sequence.

20. Separate the resulting fragments by gel electrophoresis and transfer them to a solid support (e.g., nitrocellulose or nylon membrane) as described in Protocol 6.8. Perform Southern hybridization at 10°C below the T_m, using the ^{32}P-labeled mutagenic oligonucleotide as a probe. Wash the filter under the discriminatory conditions and establish an autoradiograph.

    The final autoradiograph should show hybridization only to the relevant fragments of the mutagenized target DNA.

Protocol 13.8

Detection of Mutations by Single-stranded Conformational Polymorphism (SSCP) Analysis

SSCP involves the following three steps:

- amplification by PCR of the target region
- denaturation of the PCR product
- electrophoresis of the single-stranded DNA through a gel at neutral pH

Single-stranded DNA molecules fold into complex three-dimensional structures as a result of intrastrand base pairing. Single strands of equal length can therefore vary considerably in electrophoretic mobility as a result of the looping and compaction caused by intrastrand pairing. Alteration of the sequence by as little as a single base can reshape the secondary structure, with consequent changes in electrophoretic mobility at neutral pH.

Because the mobility of single-stranded conformers varies considerably with temperature and ionic strength, it is usually necessary to analyze samples under two or more sets of conditions. Close to 100% of mutations in a target DNA can be detected if SSCP is performed in a range of buffers and temperatures (e.g., 0.5x TBE at 4°C and 15°C, or 0.5x TBE with and without compounds such as glycerol that are rich in hydroxyl groups). Accurate control of temperature during electrophoresis is essential for reproducibility of results, as is consistency in the structure of the gel matrix.

MATERIALS

CAUTION: Please see Appendix 4 for appropriate handling of materials marked with <!>.

Reagents and Solutions

Please see Appendix 1 for components of stock solutions, buffers, and reagents. Dilute stock solutions to the appropriate concentrations.

Ammonium persulfate (10% w/v), freshly made

10x Amplification buffer

dNTP solution, containing all four dNTPs, each at a concentration of 1 mM

Glycerol (e.g., UltraPure [Invitrogen])

TEMED (*N,N,N',N'*-tetramethylethylene diamine), electrophoresis grade (Sigma-Aldrich or Bio-Rad) <!>

Enzymes and Buffers

Restriction enzymes and 10x buffers (optional; see Step 8)

Thermostable DNA polymerase

Use an enzyme such as *Pfu* DNA polymerase (Stratagene) that, unlike *Taq*, does not catalyze the nontemplate addition of adenine residues. Use the 10x amplification buffer supplied by the manufacturer.

Nucleic Acids/Oligonucleotides

Control DNAs

Oligonucleotide primers

Forward and reverse primers should be designed according to the usual rules (see MC3, pp. 8.13–8.16) and should be spaced 150–200 bp apart on opposite strands of the target DNA.

Target DNA to be screened for mutation(s)

The mass of DNA required varies according to its complexity. This protocol describes the use of SSCP to detect a specific mutation in total human genomic DNA. Lesser amounts of DNA are required for less complex targets.

Radioactive Compounds

[α-^{32}P]dCTP (3000 Ci/mmole, 10 mCi/ml) <!>

Gels/Loading Buffers

Acrylamide:bisacrylamide (29:1, w/w) <!>
> Many investigators use premade commercial solutions of acrylamide and bisacrylamide (e.g., Acrylogel, BDH). Others prefer to use the Hydrolink series of gel matrices, which are sold under the trade name of MDE (Mutation Detection Enhancement) by FMC BioProducts.

Formamide loading buffer <!>
> 80% (w/v) deionized formamide
> 10 mM EDTA (pH 8.0)
> 0.25% bromophenol blue
> 0.25% xylene cyanol FF

Sucrose gel-loading buffer (Type 1)
> 0.25% bromophenol blue
> 0.25% xylene cyanol FF
> 40% (w/v) sucrose in H_2O

10x TBE electrophoresis buffer
> Use the same stock of 10x TBE to prepare both the gel and the running buffer

Additional Items

Barrier tips for automatic pipettor
Gel-drying apparatus
Hamilton syringe
Light mineral oil or wax bead (optional; see Step 2)
Materials required for autoradiography or phosphorimaging (MC3, pp. A9.9–A9.15)
Materials required to prepare and load polyacrylamide gels (glass plates, electrical tape, gel-loading tapes, etc.) (see Protocol 12.8)
Microfuge tubes for PCR, 0.5 ml and thin walled
Thermal cycler, programmed with the desired amplification protocol
Water bath/heating block (100°C and at temperature required for digestion of DNA with restriction enzymes)
Whatman 3MM paper

Additional Information

Autoradiography	MC3, pp. A9.9–A9.15
Background information of mutation detection	MC3, pp. 13.91–13.96
Background information on SSCP	MC3, pp. 13.49–13.52
Glycerol	MC3, p. 13.90
Troubleshooting	MC3, pp. 13.55 and 13.56

METHOD

1. In a sterile 0.5-ml microfuge tube, mix in the following order:

1 mM dNTP solution	1 μl
10x amplification buffer	2 μl
35 μM 5′-oligonucleotide solution	1 μl (35 pmoles)
35 μM 3′-oligonucleotide solution	1 μl
10 μCi/μl [α-^{32}P]dCTP	1 μl
human genomic DNA	10 μl (100 ng)
thermostable DNA polymerase	1–3 units
H_2O	to 20 μl

[α-^{32}P]dCTP is incorporated in the PCRs to label the amplified DNA uniformly. However, end-labeled DNA generated with ^{32}P-labeled primers works equally well in SSCP.

If possible, set up control reactions using two DNA samples known to contain alleles that differ in sequence by one or more base pairs and that are known to resolve on SSCP gels. In addition, set up a contamination control in which no template DNA is added to the reaction.

2. If the thermal cycler is not fitted with a heated lid, overlay the reaction mixtures with 1 drop (~50 µl) of light mineral oil to prevent evaporation of the samples during repeated cycles of heating and cooling. Alternatively, place a bead of paraffin wax into the tube if using a hot start protocol. Place the tubes in the thermal cycler.

3. Amplify the nucleic acids using the denaturation, annealing, and polymerization times and temperatures listed below. For advice on thermal cycler programs, please see Protocol 8.1.

Cycle Number	Denaturation	Annealing/Polymerization
30 cycles	5–30 sec at 94°C	0.5–1 min at 68°C
Last cycle	1 min at 94°C	5–10 min at 68°C

Times and temperatures may need to be adapted to suit the particular reaction conditions.

4. While the thermal cycler program is running, prepare a 5.5% polyacrylamide gel containing 10% (v/v) glycerol in 1x TBE gel buffer.

10x TBE gel buffer	10 ml
29:1% acrylamide:bisacrylamide solution	18 ml
10% ammonium persulfate	0.5 ml
glycerol	10 ml
H$_2$O	61.5 ml

Mix the reagents by gentle swirling or stirring.

This volume of gel solution is sufficient for one polyacrylamide gel of standard size (40 x 40-cm plates with 0.4-mm spacers). The volume of the gel solution can be increased or decreased as needed for other gel sizes.

Use the same stock of 10x TBE gel buffer to prepare enough 1x TBE gel buffer to fill the tanks of the electrophoresis apparatus.

5. Assemble and tape together two 40 x 40-cm glass electrophoresis plates with 0.4-mm spacers.

To obtain maximum resolution of single-stranded DNA conformers, it is important to use "thin-gel" spacers that are ≤ 0.4 mm in thickness.

6. Add 100 µl of TEMED to the gel solution. Mix the solution by gently swirling the flask and pour the gel.

Work quickly because the acrylamide solution will polymerize rapidly. For instructions on pouring thin gels, please see Protocol 12.8.

7. Assemble the polymerized gel into an electrophoresis apparatus at room temperature. Fill the buffer tanks with 1x TBE gel buffer made from the same stock as the gel solution.

8. (*Optional*) If the amplified DNA fragment is to be digested with a restriction enzyme, remove the PCR tubes from the thermal cycler at the end of the run and place them on ice. Set up the following restriction enzyme digestion:

PCR solution	5 µl
10x restriction enzyme buffer	4 µl
restriction enzyme (2–50 units)	2 µl
H$_2$O	29 µl

Incubate for 1–2 hours at the temperature appropriate for the restriction enzyme.

9. Dilute either 1.5 μl of the original PCR (from Step 3) or 5 μl of the restriction-enzyme-digested PCR (from Step 8) into 20 μl of sucrose gel-loading buffer. Dilute similar aliquots into 20 μl of formamide dye mix.

10. Heat the formamide-containing samples to 100°C for 6 minutes and then plunge the tubes directly into ice.

11. Use a pasteur pipette or a Hamilton syringe to wash out the wells of the polyacrylamide gel with 1x TBE gel buffer. With a drawn-out glass capillary tube or a micropipettor equipped with a gel-loading tip, load 2 μl of each sample on the polyacrylamide gel.

12. Apply 6–7 V/cm (~250 V [and 15 mA] for a 40 x 40-cm gel) to the gel for approximately 14 hours.

13. At the completion of electrophoresis, separate the glass plates and transfer the gel to a sheet of Whatman 3MM filter paper. Dry the gel on a vacuum dryer for 30–60 minutes.

14. Subject the dried gel to autoradiography for 4–16 hours at room temperature without an intensifying screen. The nondenatured PCR samples (i.e., those diluted into the sucrose gel-loading buffer) will migrate through the gel as double-stranded molecules. In contrast, the denatured samples will usually migrate as a mixture of double-stranded (faster-migrating) and single-stranded (slower-migrating) molecules. Only one single-stranded species will be detected if the two complementary strands of DNA fold into conformers that cannot be resolved under the conditions of electrophoresis used. Two single-stranded conformers will be detected if the complementary strands fold into resolvable conformers. The PCR product of a heterozygous allele from a diploid organism should generate at least four bands: two bands whose mobility is identical to those of the wild-type bands and two that are characteristic of the mutation. Often, however, there will be more than two bands, either as a consequence of polymorphisms or because the complementary DNA strands fold into more than one conformer. These ambiguities can be resolved by DNA sequencing.

Generation of Sets of Nested Deletion Mutants with Exonuclease III

Nested deletions lacking progressively more nucleotides from one end or the other of a target DNA are used to define boundaries of functional *cis*-acting control elements. The methods rely on nucleases such as exonuclease III and BAL 31 that digest DNA in a predictable fashion. In this protocol, the double-stranded DNA of recombinant plasmid, phagemid, or bacteriophage M13 replicative form DNA is digested with two restriction enzymes whose sites of cleavage both lie between one end of the target DNA and the binding site for universal primer. The enzyme that cleaves nearer the target sequence must generate either a blunt end or a recessed 3′ terminus; the other enzyme must generate a four-nucleotide protruding 3′ terminus. Because only the blunt or recessed 3′ terminus of the resulting linear DNA is susceptible to exonuclease III, digestion proceeds unidirectionally away from the site of cleavage and into the target DNA. The exposed single strands are then removed by digestion with nuclease S1 or mung bean nuclease, and the DNA is then recircularized. If desired, a synthetic linker can be inserted at the site of recircularization. Several commercial kits are available to construct deletions by this method.

MATERIALS

CAUTION: Please see Appendix 4 for appropriate handling of materials marked with <!>.

Reagents and Solutions

Please see Appendix 1 for components of stock solutions, buffers, and reagents. Dilute stock solutions to the appropriate concentrations.

dNTP solution, containing all four dNTPs, each at a concentration of 0.5 mM
Ethanol
Ethanol (70%)
Phenol:chloroform (1:1, v/v) <!>
Sodium acetate (3 M, pH 5.2)

Enzymes and Buffers

Exonuclease III and 10x exonuclease III buffer
    660 mM Tris-Cl (pH 8.0)
    66 mM MgCl$_2$
    100 mM β-mercaptoethanol <!>

Klenow mixture, sufficient for 30 samples (prepare just before Step 11)

H$_2$O	20 µl
1 M MgCl$_2$	6 µl
0.1 M Tris-Cl (pH 7.6)	3 µl
Klenow fragment	3 units

Ligase mixture, sufficient for 24 samples (prepare just before Step 13)

H$_2$O	20 µl
10x ligase buffer	100 µl
5 mM ATP	100 µl
PEG 8000 (30% w/v) <!>	250 µl
Bacteriophage T4 DNA ligase	5 Weiss units

Nuclease S1 and 10x S1 buffer
    0.28 M NaCl
    0.05 M sodium acetate (pH 5.6)
    4.5 mM ZnSO$_4$

Nuclease S1 reaction mixture (prepare just before Step 7)

H_2O	172 μl
10x S1 buffer	27 μl
nuclease S1	60 units

Nuclease S1 stop buffer
0.3 M Tris base
50 mM EDTA (pH 8.0)
Restriction enzymes (two)

Nucleic Acids/Oligonucleotides

Target DNA

Analyze the sequence of the target DNA for the presence of suitable restriction sites. Clone the segment of DNA to be digested by exonuclease III into a plasmid or bacteriophage vector and purify the recombinant DNA. (Note: the preparation of plasmid DNA used in this protocol must consist of >90% superhelical molecules.)

Additional Items

Microfuge tubes (0.5 ml and thin walled) or microtiter plates with U-shaped wells
Step 4 of this protocol, which is optional, uses the reagents listed in Protocol 10.4.
Step 14 of this protocol requires the reagents listed in Protocol 1.24, 1.25, or 1.26 (for transformation).
Step 16 of this protocol the requires reagents listed in Protocol 12.2, 12.3, or 12.5.
Water baths/heating blocks (30°C, 37°C, and 70°C)

Additional Information

Background information on digestion with exonuclease III	MC3, pp. 13.57–13.58 and 13.72–13.75
Nuclease S1	MC3, pp. 7.86 and A4.46

METHOD

1. Digest 10 μg of target DNA (recombinant bacteriophage M13 replicative form DNA, phagemid DNA, or plasmid DNA) with two restriction enzymes that cleave the polycloning site between the primer-binding site of the vector and the target DNA.

2. Purify the DNA by standard extraction with phenol:chloroform and precipitation with ethanol. Carefully remove the supernatant and add 0.5 ml of 70% ethanol to the pellet.

    Rinsing the pellet with ethanol is important because sodium ions inhibit exonuclease III.

3. Recover the washed pellet of DNA by centrifuging at maximum speed for 2 minutes at 4°C in a microfuge and then carefully remove the supernatant. Incubate the open tube on the bench to allow the last traces of ethanol to evaporate and dissolve the DNA in 60 μl of 1x exonuclease III buffer. Store the dissolved DNA on ice.

4. Place 7.5 μl of nuclease S1 reaction mixture in each of 25 0.5-ml microfuge tubes or in 25 wells of a 96-well microtiter plate with U-shaped wells. Store the microtiter plate or microfuge tubes on a bed of ice.

5. Incubate the DNA solution prepared in Step 3 for 5 minutes at 37°C. Transfer 2.5 μl of the solution to the first microfuge tube or well of the microtiter plate containing the nuclease S1 reaction mixture.

6. To the remainder of the DNA solution, add 150 units of exonuclease III per pmole of recessed 3′ termini (1 unit of exonuclease III will generate 1 nmole of acid-soluble total nucleotide in

30 minutes at 37°C). Tap the tube to mix the contents and immediately return the tube to the 37°C water bath.

7. At 30-second intervals, remove 2.5-μl samples of the DNA solution and place them in successive microfuge tubes or wells of the microtiter plate containing the nuclease S1 reaction mixture.

8. When all of the samples have been harvested, incubate the microfuge tubes or microtiter plate containing the nuclease S1 and digested plasmid DNA for 30 minutes at 30°C.

9. Add 1 μl of nuclease S1 stop mixture to each of the microfuge tubes or wells and incubate the reaction mixtures for 10 minutes at 70°C.

10. Transfer the microfuge tubes or microtiter plate to a bed of ice and analyze aliquots of each of the samples by agarose gel electrophoresis.

11. Pool the samples containing DNA fragments of the desired size. Add 1 μl of Klenow mixture for each 10 μl of pooled sample and incubate the reaction mixture for 5 minutes at 37°C.

12. For each 10 μl of pooled sample, add 1 μl of 0.5 mM dNTPs. Continue incubation for 15 minutes at room temperature.

13. Add 40 μl of T4 bacteriophage ligase mixture for each 10 μl of pooled sample. Mix and continue incubation for 2 hours at room temperature.

14. Transform the appropriate *E. coli* host with aliquots of the ligated DNA.

15. Prepare minipreparations of bacteriophage M13 replicative form DNA, plasmid, or phagemid DNA from at least 24 randomly selected plaques or colonies.

16. Linearize the DNAs by digestion with an appropriate restriction enzyme and analyze their sizes by electrophoresis through a 1% agarose gel. Include the original plasmid or bacteriophage DNA that has been linearized by restriction enzyme digestion as a marker. Choose clones of an appropriate size for sequencing (Chapter 12) or additional restriction enzyme mapping.

Generation of Sets of Deletion Mutants with BAL 31 Nuclease

In this method, the nuclease BAL 31 is used to make uni- or bidirectional deletions in a segment of cloned DNA. BAL 31 is a complex enzyme and tends to digest a population of double-stranded DNA targets in an asynchronous fashion. Deletions created by BAL 31 are therefore far more heterogeneous in size than those created by processive enzymes such as exonuclease III.

MATERIALS

CAUTION: Please see Appendix 4 for appropriate handling of materials marked with <!>.

Reagents and Solutions

Please see Appendix 1 for components of stock solutions, buffers, and reagents. Dilute stock solutions to the appropriate concentrations.

dNTP solution, containing all four dNTPs, each at a concentration of 0.5 mM
EGTA (0.5 M, pH 8.0)
Ethanol
Ethanol (70%)
Phenol:chloroform (1:1, v/v) <!>
Sodium acetate (3 M, pH 5.2)
TE

Enzymes and Buffers

Bacteriophage T4 DNA polymerase and 10x buffer
BAL 31 nuclease and 5x BAL 31 buffer
    2.5 M NaCl
    62.5 mM CaCl$_2$
    62.5 mM MgCl$_2$
    100 mM Tris-Cl (pH 8.0)
Klenow fragment of *E. coli* DNA polymerase I and 10x buffer
Restriction enzymes and 10x buffers

Nucleic Acids/Oligonucleotides

Target DNA
Marker DNAs for gel electrophoresis

Gels/Loading Buffers

Agarose gels
0.8% Agarose gels, cast and run in 0.5x TBE containing 0.5 µg/ml ethidium bromide <!>
Preparative agarose gel

Additional Items

Step 1 of this protocol requires the reagents listed in Protocols 1.17, 1.18, and 1.19, or Protocol 3.6.
Step 2 of this protocol requires the reagents listed in Protocol 1.9.
Step 27 of this protocol requires the reagents listed in Protocol 5.4, 5.5, 5.6, or 5.7.
Step 30 of this protocol requires the reagents listed in Protocol 1.24, 1.25, or 1.26 or Protocol 3.6 or 3.8.

Adapted from Chapter 13, Protocol 10, p. 13.62 of MC3.

Step 31 of this protocol requires the reagents listed in Protocol 3.3.
Step 34 of this protocol requires the reagents listed in Protocol 12.3, 12.4, or 12.5.
Stopwatch
Water baths/heating blocks (30°C, 65°C, and to the appropriate temperatures for digestion with restriction enzymes)

Additional Information

BAL 31   MC3, pp. 13.68–13.72

METHOD

1. Clone the target fragment into an appropriate plasmid or bacteriophage M13 vector.

   If deletion mutants are to be constructed from both termini of the target DNA, it will be necessary to clone the parental target DNA in both orientations with respect to the polycloning site in an appropriate vector.

2. Purify the closed circular recombinant DNA(s) by column chromatography on Qiagen columns (or their equivalent) and precipitation with ethanol. Redissolve the DNA in the smallest practical volume of Tris/EDTA.

   It is essential to use highly purified closed circular DNA (i) to minimize the contribution of contaminating RNA and small fragments of *E. coli* chromosomal DNA to the total concentration of termini in the reaction and (ii) to eliminate nicked circular molecules, which are degraded by BAL 31 from the site of the nick.

3. Digest 30 μg of the closed circular DNA to completion with a restriction endonuclease that cleaves at one end of the target DNA. This site defines the common point from which the nested deletions will begin. Use agarose gel electrophoresis to verify that digestion with the restriction endonuclease is complete.

4. Purify the DNA by extraction with an equal volume of phenol:chloroform. Separate the aqueous and organic phases by centrifugation at maximum speed for 3 minutes at 0°C in a microfuge and then transfer the aqueous phase to a fresh microfuge tube.

5. Add 0.1 volume of 3 M sodium acetate (pH 5.2) and 2 volumes of ice-cold ethanol. Store the tube for 10 minutes at 0°C and then recover the DNA by centrifugation at maximum speed for 10 minutes at 4°C in a microfuge.

6. Remove the supernatant and wash the pellet of DNA carefully with 70% ethanol at room temperature. Dry the pellet at room temperature and dissolve it in TE (pH 7.6) at a concentration of 1 μg/μl. Store the DNA at –20°C.

7. In a microfuge tube, mix:

linearized DNA (1 μg/μl)	4 μl
H₂O	48 μl
5x BAL 31 buffer	13 μl

   Dispense 9 μl of this mixture into each of seven separate microfuge tubes.

   Because the ratio of fast and slow forms of BAL 31 varies from preparation to preparation, it is essential to assay the activity of the particular batch of enzyme that will be used to generate deletions.

8. Make a series of seven twofold dilutions of BAL 31 (see MC3, p. 13.68) in 1x BAL 31 buffer. Enzyme dilution is best performed by placing seven aliquots (2 μl) of 1x BAL 31 buffer on the surface of a piece of Parafilm lying on a bed of ice or on a cold block. Use a disposable micropipette tip to mix 2 μl of the BAL 31 preparation under test with the first drop. Use a fresh tip to transfer 2 μl of the mixture to the next drop and again mix. Continue in this fash-

ion until the enzyme has been added to all of the drops. Working quickly, add 1 µl of each of the last six dilutions to six of the microfuge tubes containing the linear DNA being tested. Do not add enzyme to the seventh tube.

9. Incubate all of the microfuge tubes (including the tube that received no enzyme) for 30 minutes at 30°C.

10. Add 1 µl of 200 mM EGTA (pH 8.0) to each tube and then heat the tubes for 5 minutes at 65°C.

11. Mix each of the samples with 3 µl of agarose gel-loading buffer and analyze the size of the DNAs by electrophoresis through a 0.8% agarose gel containing 0.5 µg/ml ethidium bromide.

12. Examine the gel by UV illumination and determine the dilution of enzyme just sufficient to digest the DNA to the point where only a smear of small (200-bp) fragments is detectable. This dilution of BAL 31 will be used in the large-scale digestion (Step 15).

13. Mix:

linearized DNA (1 µg/µl)	20 µl
H$_2$O	240 µl
5x BAL 31 buffer	65 µl

Incubate the mixture in a water bath at 30°C.

14. While the mixture is warming to 30°C, prepare a set of eight microfuge tubes, each containing 5 µl of 200 mM EGTA (pH 8.0). Label the tubes 1.5 minutes, 3.0 minutes, 4.5 minutes, etc.

15. Add 36 µl of the appropriate dilution of BAL 31 (please see Step 12) to the reaction mixture prepared in Step 13. Quickly mix the enzyme by tapping the side of the tube, return the tube to the water bath set at 30°C, and start a stopwatch.

16. At 1.5-minute intervals, transfer 45 µl of the reaction mixture to the appropriately labeled microfuge tube. Store the tubes on ice until all of the samples have been collected.

17. Heat the tubes for 5 minutes at 65°C to inactivate the BAL 31.

18. Add 5 µl of 3 M sodium acetate (pH 5.2) to each tube, followed by 100 µl of ice-cold ethanol. Mix the solution by vortexing and store the tubes on ice for 20–30 minutes.

19. Recover the DNAs by centrifugation at maximum speed for 10 minutes at 4°C. Remove the supernatants and wash the pellets with 200 µl of ice-cold 70% ethanol. Centrifuge for an additional 2 minutes.

20. Carefully remove the supernatants and stand the open tubes at room temperature until all of the ethanol has evaporated. Dissolve each of the pellets in 23 µl of TE (pH 7.6).

21. Add to each of the DNA preparations:

0.5 mM dNTP solution	3 µl
10x polymerase buffer	3 µl
bacteriophage T4 DNA polymerase (~5 units)	1 µl

Incubate the reactions for 15 minutes at room temperature and then add approximately 1 µl (~5 units) of the Klenow fragment. Continue the incubation for an additional 15 minutes at room temperature.

22. Purify the DNAs by extraction with phenol:chloroform and then precipitate the DNAs with ethanol as described in Steps 18–20. Dissolve each of the DNAs in 16 µl of TE (pH 7.6).

23. To each DNA, add 2 µl of the appropriate 10x restriction enzyme buffer and 8 units of a restriction enzyme that will separate the target DNA from the vector. Incubate the reactions for 1 hour at the appropriate temperature.

24. At the end of the incubation, transfer an aliquot (3 μl) from each digest to a fresh microfuge tube. Store the remainder of the digests on ice until needed in Step 27.

25. Add 1 μl of sucrose gel-loading buffer to each 3-μl aliquot and load the contents of each tube into the wells of an agarose gel cast in 0.5x TBE and containing 0.5 μg/ml ethidium bromide. The wells at the sides of the gel should contain markers of the appropriate size.

26. Separate the target fragments from the vector DNA by electrophoresis. Examine the gel by UV illumination and determine which of the samples has been digested to an appropriate size by BAL 31.

27. Pool the samples (from Step 24) containing target DNA of the appropriate size and isolate the target fragments by preparative gel electrophoresis. Recover the target DNA fragments from the gel using one of the methods described in Protocols 5.4, 5.5, and 5.7.

28. Estimate the amount of purified target DNA from the intensity of ethidium-bromide-mediated fluorescence.

29. Ligate the deleted target fragments with a plasmid, phagemid, or bacteriophage M13 vector (please see Chapter 1 or 3) that carries one blunt end and one terminus that is compatible with the restriction enzyme used in Step 23.

30. Transform (plasmids or phagemids) or transfect (bacteriophage M13 replicative form DNA) competent *E. coli* of an appropriate strain with small aliquots or dilutions of the ligation mixture. The next day, grow small-scale cultures of 12 transformed colonies or bacteriophage M13 plaques, chosen at random.

31. Purify plasmid, phagemid, or bacteriophage M13 replicative form DNA from each of the 12 cultures by using one of the methods described in Chapter 1 or 3. Digest the DNAs with a restriction enzyme(s) that will liberate the target fragment from the vector.

32. Analyze the size of the target fragment liberated from each of the DNAs by agarose gel electrophoresis, using size markers of an appropriate size.

33. If the results are satisfactory (i.e., if the target fragments fall within the desired size range), pick a large number of individual transformed colonies or plaques and determine the size of the inserts as described above. Preserve those cultures that carry recombinants of the desired size.

34. Determine the exact end points of the deletion in each mutant by DNA sequencing (please see Protocol 12.3, 12.4, or 12.5).

Screening Expression Libraries

BACKGROUND INFORMATION

Background information found in *Molecular Cloning: A Laboratory Manual*, 3rd edition (hereafter MC3) unless otherwise indicated.

Classes of probes used for screening expression libraries	MC3, pp. 14.1–14.3
Plasmid and bacteriophage λ expression vectors	MC3, pp. 14.47–14.49
Using antibodies in immunological sceening	MC3, pp. 14.50–14.51

Screening Expression Libraries Constructed in Bacteriophage λ Vectors

In expression libraries constructed in bacteriophage λ vectors, such as λZAP, λZAPII, λZiplox, or their derivatives, cDNAs are embedded in the region of the *lacZ* gene coding for the α-complementation fragment of β-galactosidase. If the cDNA sequences are inserted in the correct orientation and reading frame, an inducible fusion protein is produced that consists of β-galactosidase sequences linked to polypeptide sequences encoded by the cDNA. Recombinants expressing target polypeptides that display antigenic epitopes can be detected by screening the library with specific antibody probes.

Bacteriophage λ vectors used for the construction of expression libraries usually carry a mutant copy of the λ repressor gene (*c*Its857) that encodes a temperature-sensitive repressor protein; many of the vectors have been engineered to allow recovery of cDNA inserts as plasmids. The properties of the most popular bacteriophage λ expression vectors are detailed in MC3, pp. 11.22–11.26 and 14.47. Information about their preferred *E. coli* hosts is available in MC3, p. 11.66.

In the screening experiment described in this protocol, a library constructed in a bacteriophage λ vector is plated on an appropriate *E. coli* host in the absence of isopropyl-β-D-thiogalactoside (IPTG). After 2–4 hours, the plates are moved from 42°C to 37°C to stabilize any fusion proteins that are temperature sensitive. Filters impregnated with IPTG are then laid on top of the developing plaques to collect and imprint the induced fusion proteins. After further incubation, the filters are removed and probed with an antibody specific to the target protein. Three alternative methods to detect bound antibodies are described: radiochemical screening, chromogenic screening, and chemiluminescent screening. The advantages and disadvantages of these methods are discussed in MC3, pp. 14.1–14.3.

MATERIALS

CAUTION: Please see Appendix 4 for appropriate handling of materials marked with <!>.

Reagents and Solutions

Please see Appendix 1 for components of stock solutions, buffers, and reagents. Dilute stock solutions to the appropriate concentrations.

Blocking buffer
- 10 mM Tris-Cl (pH 8.0)
- 150 mM NaCl
- 0.05% (v/v) Tween-20
- 20% (v/v) fetal bovine serum

Alternative blocking buffers that can be used if the backgrounds are unacceptably high are 5% (w/v) nonfat dried milk or TNT buffer containing 1% (w/v) gelatin and 3% (w/v) bovine serum albumin. Blocking buffers can be stored at 4°C and reused several times. In this case, sodium azide <!> should be added to a final concentration of 0.05% (w/v) to inhibit the growth of microorganisms.

Chloroform <!>

IPTG (10 mM) (prepare just before use in Step 4)

Reagents for detection of antigen-antibody complexes
1. Chromogenic screening with alkaline phosphatase–conjugated antibodies
   Alkaline phosphatase–conjugated anti-IgG antibodies
   AP buffer
   - 100 mM Tris-Cl (pH 9.5)
   - 100 mM NaCl
   - 5 mM MgCl$_2$

5-Bromo-4-chloro-3-indolyl phosphate (BCIP) (50 mg/ml in 100% dimethylformamide) <!>
> Store in the dark.

Nitro blue tetrazolium (50 mg/ml in 70% dimethylformamide) <!>
> Store in the dark.

2. Chromogenic screening with horseradish peroxidase–conjugated antibodies

4-Chloro-1-naphthol (60 mg dissolved in 20 ml of ice-cold methanol)

H_2O_2 (30%) <!>

HRP-conjugated anti-IgG antibodies

NaCl (5 M)

Tris-Cl (1 M, pH 7.5)

Tris-Cl (10 mM, pH 7.5) containing 150 mM NaCl

3. Materials for chemiluminescent screening

AP- or HRP-conjugated anti-IgG antibodies

Chemiluminescent substrates available from, e.g., Tropix

4. ^{125}I-labeled protein A or ^{125}I-labeled anti-IgG antibodies (available commercially)
> Secondary antibody can be labeled as described in MC3, p. 14.5. These reagents are required only if radiochemical screening is used.

SM

TNT buffer
> 10 mM Tris-Cl (pH 8.0)
> 150 mM NaCl
> 0.05% (v/v) Tween-20
> Approximately 1 liter of TNT is required to process ten filters.

Washing buffers
> TNT buffer containing 0.1% (w/v) bovine serum albumin or TNT buffer containing 0.1% (w/v) bovine serum albumin and 0.1% (v/v) Nonidet P-40.

Vectors and Hosts

Bacteriophage λ cDNA expression library
> These libraries are available commercially or can be constructed as described in Chapter 11.

E. coli (for information on appropriate strains, see MC3, p. 11.66)

Media and Antibiotics

LB agar plates
> Do not use fresh plates because the top agar will peel off with the nitrocellulose filters. Two-day-old plates work well. If using as hosts *E. coli* strains B44 or XL1-Blue, which carry a *tet*ᵣ marker on an F′ episome, include tetracycline (12.5 μg/ml) in the media (see MC3, p. 11.66).

LB top agarose

Radioactive Compounds

^{125}I-labeled protein A or ^{125}I-labeled anti-IgG antibodies <!>

Radioiodinated secondary antibody <!>
> These reagents are required only if radiochemical screening is used. ^{125}I-labeled protein A and ^{125}I-labeled anti-IgG are available commercially. Secondary antibody can be radioiodinated as described in MC3, p. 14.5.

Additional Items

Air incubators (42°C and 37°C)

Blunt-ended forceps (e.g., Millipore), for handling filters

Large glass petri dishes, crystallization dishes, or Pyrex baking dishes (see note to Step 10)

Materials for autoradiography or phosphorimaging (see MC3, Appendix A9)

Needle (18 gauge) and syringe loaded with India ink

Nitrocellulose filters (82 or 138 mm), free of Triton X-100, e.g., Millipore HATF

Primary antibody, directed against the target protein
    For advice on purifying primary and secondary antibodies, please see MC3, pp. 14.50–14.51.
Saran Wrap or Mylar sheets
Step 1 of this protocol requires the reagents listed in Protocol 2.1.
Step 17 of this protocol requires the reagents listed in Protocol 2.2.

Additional Information

Antibodies	MC3, pp. A9.25–A9.34
Autoradiography/phosphorimaging	MC3, pp. A9.9–A9.15
BCIP	MC3, pp. A9.41–A9.42
Chemiluminescence	MC3, pp. A9.16–A9.20
Enzyme-conjugated antibodies	MC3, p. A9.34
Horseradish peroxidase	MC3, p. A9.35
Methods of detecting labeled antibodies	MC3, pp. A8.54–A8.55
Protein A	MC3, p. A9.46
Radiolabeling of antibodies	MC3, p. A9.30
Troubleshooting	MC3, p. 14.13
Validation of clones isolated by immunological screening	MC3, p. 14.12

METHOD

1. Using a single colony of the appropriate strain of *E. coli* as inoculum, prepare a plating culture as described in Protocol 2.1.

2. Calculate the number of plates that will be required to screen the library, assuming 0.5×10^4 to 2×10^4 plaques per 90-mm plate or 0.5×10^4 to 5×10^4 plaques per 150-mm plate. Arrange a set of sterile tubes (13 × 100 mm) in a rack; use a fresh tube for each plate. In each tube, mix 0.1 ml of the plating bacteria with 0.1 ml of SM containing the desired number of plaque-forming units of the bacteriophage λ expression library. Incubate the infected bacteria for 20 minutes at 37°C.

3. Add to each tube 2.5 ml (90-mm plate) or 7.5 ml (150-mm plate) of molten top agarose and immediately pour the mixture onto an LB agar plate. Incubate the infected plates for 3.5 hours at 42°C.

4. Number the nitrocellulose filters with a soft-lead pencil or a ballpoint pen. Use gloves to handle the filters because skin oils will prevent the transfer of proteins. Soak the filters in 10 mM IPTG for a few minutes. Use blunt-ended forceps to remove the filters from the solution and allow them to dry at room temperature on a pad of Kimwipes.

5. Remove one plate at a time from the incubator and quickly overlay it with an IPTG-impregnated nitrocellulose filter. Do not move the filter once contact is made with the plate. Put the plate in the 37°C incubator and repeat the above procedure until all plates contain a nitrocellulose filter.

    It is important to place filters on the plates one at a time so that the temperature of the plates does not drop below 37°C. The growth of bacteriophages is severely retarded at temperatures <30°C.

6. Incubate the plates for at least 4 hours at 37°C.

7. Remove the lids from the plates and continue the incubation for an additional 20 minutes at 37°C. This step strengthens the bond between the soft agarose and the agar plate.

8. Move the plates in small batches to room temperature. Use an 18-gauge needle attached to a syringe containing waterproof black ink to mark each filter in at least three asymmetric locations by stabbing through it and into the agar underneath.

9. Use blunt-ended forceps to peel the filters from the top agarose and immediately immerse them in a large volume of TNT buffer. Rinse away any small remnants of agarose by gently agitating the filters in the buffer on a shaking platform. The speed of the shaking platform should be high enough to prevent the filters from sticking to one another.

   > If large areas of the top agarose stick to the nitrocellulose filters, chill the plates for 30 minutes at 4°C or 5 minutes at –20°C before peeling the filters off the surface of the agarose.

10. After all of the filters have been transferred to the TNT buffer, wrap the plates in plastic film and store them at 4°C until the results of the immunological screening are available.

    > IMPORTANT: Do not allow the filters to dry out during any of the subsequent steps. Antibodies bound nonspecifically and reversibly to wet filters become permanently attached if the filters dry out. It is also essential that the filters do not stick to one another when they are immersed in the various buffers and antibody solutions. This problem can be minimized by dividing the filters into small batches (e.g., five filters per batch) and using a separate large petri dish or crystallizing dish for each batch. The petri dishes can be stacked on top of one another on a slowly rotating platform shaker.

11. When all of the filters have been rinsed, transfer them one at a time to a fresh batch of TNT buffer. After transfer, agitate the buffer and filters gently for an additional 30 minutes at room temperature.

12. Use blunt-ended forceps to transfer the filters individually to glass dishes containing blocking buffer (7.5 ml for each 82-mm filter; 15 ml for each 138-mm filter). When all of the filters have been submerged, agitate the buffer and filters slowly on a rotary platform for 30 minutes at room temperature.

13. Use blunt-ended forceps to transfer the filters to fresh glass dishes containing the primary antibody diluted in blocking buffer (7.5 ml for each 82-mm filter; 15 ml for each 138-mm filter). Use the highest dilution of antibody that gives acceptable background yet still allows detection of 50–100 pg of denatured antigen (see MC3, p. 14.50). When all of the filters have been submerged, agitate the solution gently on a rotary platform for 2–4 hours at room temperature.

14. Wash the filters for 10 minutes in each of the buffers below, in the order given. Transfer the filters individually from one buffer to the next. Use 7.5 ml of each buffer for each 82-mm filter and 15 ml for each 138-mm filter.

    TNT buffer containing 0.1% bovine serum albumin
    TNT buffer containing 0.1% bovine serum albumin and 0.1% Nonidet P-40
    TNT buffer containing 0.1% bovine serum albumin

15. Detect the antigen-antibody complexes with the chosen radiochemical, chromogenic, or chemiluminescent reagent. For radiochemical screening, use approximately 1 μCi of ^{125}I-labeled protein A or ^{125}I-labeled immunoglobulin per filter. Radiolabeled protein A is available from commercial sources (sp. act. 2–100 μCi/μg). Radioiodinated secondary antibody is available commercially or can be prepared (MC3, p. A9.30). For chromogenic screening, antibodies coupled to horseradish peroxidase (HRP) or alkaline phosphatase (AP) that react with species-specific determinants on primary antibodies are available from commercial sources and should be used at the recommended dilution in accordance with manufacturer instructions. Typically, 5 μl of conjugated antiserum is used for each 82-mm filter in 7.5 ml of blocking buffer (without sodium azide). Chemiluminescence is the most sensitive method for detecting immunopositive plaques. Chemiluminescent detection is quick, produces a permanent record of the screened filter (an X-ray film or phosphorimage), and is sensitive (1–10 pg of antigen in a plaque can be detected).

Radiochemical screening

a. Dilute radiolabeled ligands in blocking buffer (7.5 ml for each 82-mm filter; 15 ml for each 138-mm filter).

b. Incubate the filters for 1 hour at room temperature and then wash them several times in TNT buffer before establishing autoradiographs.

Continue at Step 16.

Chromogenic screening with AP-conjugated antibodies

a. Gently agitate the filters for 1.5–2 hours at room temperature in the solution of AP-conjugated antibody.

b. Wash the filters as described in Step 14.

c. Prepare stock solutions of BCIP (50 mg/ml in 100% dimethylformamide) and NBT (50 mg/ml in 70% dimethylformamide). These solutions are stable when stored in the dark.

d. Prepare the BCIP/NBT developing solution just before use as follows:

   i. Add 33 µl of the NBT solution to 5 ml of AP buffer and mix well.

   ii. Add 16.5 µl of the stock solution of BCIP. Mix again. Protect the solution from strong light and use it within 1 hour.

e. Blot the washed filters on paper towels.

f. Incubate the filters (Step d) in the BCIP/NBT developing solution (7.5 ml for each 82-mm filter; 15 ml for each 138-mm filter) for several hours at room temperature.

g. Rinse the filters briefly in two changes of H_2O. An intense purple color will develop at the site of antigen-antibody complexes.

Continue at Step 16.

Chromogenic screening with HRP-conjugated antibodies

a. Gently agitate the filters in the HRP-conjugated antibody solution for 1.5–2 hours at room temperature.

b. Wash the filters as described in Step 14.

c. To prepare developing solution, dissolve 60 mg of 4-chloro-1-naphthol in 20 ml of ice-cold methanol. Just before use, mix the solution with 100 ml of 10 mM Tris-Cl (pH 7.5), 150 mM NaCl containing 60 µl of 30% H_2O_2.

d. Blot the washed filters on paper towels and wash them briefly in 10 mM Tris-Cl (pH 7.5) containing 150 mM NaCl.

e. Incubate the filters for 15–20 minutes at room temperature in the 4-chloro-1-naphthol developing solution (10 ml for each 82-mm filter; 25 ml for each 138-mm filter).

f. Wash the filters in two changes of H_2O. An intense purple color will develop at the site of antigen-antibody complexes.

Continue at Step 16.

Chemiluminescent screening

a. Gently agitate the filters for 1.5–2 hours at room temperature in the AP- or HRP-conjugated antibody solution.

b. Wash the filters as described in Step 14.

c. Prepare the chemiluminescent substrates (MC3, pp. A9.16–A9.20) according to manufacturer instructions.

d. Incubate the washed filters in the chemiluminescent substrates for 1–5 minutes (again, consult manufacturer recommendations for optimal exposure times).

e. Drain the excess solution from the filters and immediately wrap the filters in Saran Wrap. Do not allow the filters to dry out.

f. Establish an autoradiogram. Typically, the initial exposure is 1 minute. This interval provides enough information to establish the proper exposure time.

Continue at Step 16.

16. Identify the locations of positive plaques or, if made, compare the duplicate filters, searching for coincident signals. For screens involving radiolabeled or chemiluminescent probes, compare the resulting autoradiograms with the agar plates on a light box. For screens involving chromogenic reagents that leave a visible positive residue on the filter, perform the following steps:

    a. Lay a sheet of Saran Wrap or Mylar film over the filters.

    b. On the surface of the Saran Wrap, mark the locations of the holes in the filters and the locations of antigen-positive clones with different colored waterproof markers. Label the Saran Wrap to identify the plates from which the filters were derived.

    c. Place the sheet of Saran Wrap on a light box and align the plates containing the original bacteriophage λ plaques on top of it.

    d. Identify the area containing the positive plaque and remove a plug of agar from this area using the large end of a pasteur pipette. Transfer the plug to 1 ml of SM containing 2 drops of chloroform.

    e. Keep the sheet of Saran Wrap, which provides a permanent record of the locations of the positive plaques. The colored spots on the original filters fade quite rapidly.

17. Allow the bacteriophage particles to elute from the agar plug for several hours at 4°C. Measure the titer of the bacteriophages in the eluate and then replate them so as to obtain approximately 3000 plaques per 90-mm plate. Rescreen the plaques as described above (from Step 4 onward) and repeat the process of screening and plating until a homogeneous population of immunopositive recombinant bacteriophages is obtained.

Screening Expression Libraries Constructed in Plasmid Vectors

In expression libraries constructed in most plasmid expression vectors, cDNAs are embedded in the region of the *lacZ* gene coding for the α-complementation fragment of β-galactosidase. If the cDNA sequences are inserted in the correct orientation and reading frame, an inducible fusion protein is produced that consists of β-galactosidase sequences linked to polypeptide sequences encoded by the cDNA. Recombinants expressing target polypeptides that display antigenic epitopes can be detected by screening the library with specific antibody probes.

In the screening experiment described in this protocol, a cDNA library constructed in a plasmid expression vector is plated on agar medium and then replicated onto filters, which are transferred to plates containing IPTG. After 2–4 hours of induction, the colonies are lysed with chloroform and then exposed to antibodies directed against the target protein. Three alternative methods to detect bound antibodies are described: radiochemical screening, chromogenic screening, and chemiluminescent screening. The advantages and disadvantages of these methods are discussed in MC3, pp. 14.1–14.3.

MATERIALS

CAUTION: Please see Appendix 4 for appropriate handling of materials marked with <!>.

Reagents and Solutions

Please see Appendix 1 for components of stock solutions, buffers, and reagents. Dilute stock solutions to the appropriate concentrations.

Blocking buffer
  10 mM Tris-HCl (pH 8.0)
  150 mM NaCl
  0.05% (v/v) Tween-20
  20% (v/v) fetal bovine serum
  Alternative blocking buffers that can be used if the backgrounds are unacceptably high are 5% (w/v) nonfat dried milk or TNT buffer containing 1% (w/v) gelatin and 3% (w/v) bovine serum albumin. Blocking buffers can be stored at 4°C and reused several times. In this case, sodium azide <!> should be added to a final concentration of 0.05% (w/v) to inhibit the growth of microorganisms.

Chloroform <!>

Lysis buffer
  100 mM Tris-Cl (pH 7.8)
  150 mM NaCl
  5 mM MgCl$_2$
  1.5% (w/v) bovine serum albumin
  1 μg/ml pancreatic DNase I
  40 μg/ml lysozyme
  Add DNase I and lysozyme to this buffer just before use in Step 11.

Reagents for detection of antigen-antibody complexes
1. Chromogenic screening with alkaline phosphatase–conjugated antibodies
   Alkaline phosphatase–conjugated anti-IgG antibodies
   AP buffer
     100 mM Tris-Cl (pH 9.5)
     100 mM NaCl
     5 mM MgCl$_2$
   5-Bromo-4-chloro-3-indolyl phosphate (BCIP) (50 mg/ml in 100% dimethylformamide) <!>
     Store in the dark.

Nitro blue tetrazolium (50 mg/ml in 70% dimethylformamide) <!>
  Store in the dark.

2. Chromogenic screening with horseradish peroxidase–conjugated antibodies
  4-Chloro-1-naphthol (60 mg dissolved in 20 ml of ice-cold methanol)
  H_2O_2 (30%) <!>
  HRP-conjugated anti-IgG antibodies
  NaCl (5 M)
  Tris-Cl (1 M, pH 7.5)
  Tris-Cl (10 mM, pH 7.5) containing 150 mM NaCl

3. Materials for chemiluminescent screening
  AP- or HRP-conjugated anti-IgG antibodies
  Chemiluminescent substrates available from, e.g., Tropix

4. ^{125}I-labeled protein A or ^{125}I-labeled anti-IgG antibodies (available commercially)
  Secondary antibody can be labeled as described in MC3, p. 14.5. These reagents are required only if radiochemical screening is used.

SM

TNT buffer
  10 mM Tris-Cl (pH 8.0)
  150 mM NaCl
  0.05% (v/v) Tween-20
  Approximately 1 liter of TNT is required to process ten filters.

Washing buffers
  TNT buffer containing 0.1% (w/v) bovine serum albumin
  TNT buffer containing 0.1% (w/v) bovine serum albumin and 0.1% (v/v) Nonidet P-40

Vectors and Hosts

Plasmid expression library, constructed in an appropriate expression vector (e.g., plasmids of the pUC, pUR, and pEX series; pBlueScript)
  Plasmid expression libraries are available commercially or can be constructed as described in Chapter 11.

Media and Antibiotics

LB or SOB agar plates, containing the appropriate antibiotic
LB or SOB agar plates, containing 1 mM IPTG

Radioactive Compounds

^{125}I-labeled protein A or ^{125}I-labeled anti-IgG antibodies <!>
Radioiodinated secondary antibody <!>
  These reagents are required only if radiochemical screening is used. ^{125}I-labeled protein A and ^{125}I-labeled anti-IgG are available commercially. Secondary antibody can be radioiodinated as described in MC3, p. 14.5.

Additional Items

Air incubators (37°C, 30°C, and 42°C), required only for vectors (e.g., pEX) equipped with the λ_{pR} promoter
Blunt-ended forceps (e.g., Millipore), for handling filters
Large glass petri dishes, crystallization dishes, or Pyrex baking dishes
Materials for autradiography/phosphorimaging (see MC3, Appendix A9)
Needle (18 gauge) and syringe loaded with India ink
Nitrocellulose filters (82 or 138 mm), free of Triton X-100, e.g., Millipore HATF
Primary antibody, directed against the target protein
  For advice on purifying primary and secondary antibodies, please see MC3, pp. 14.50–14.51.
Saran Wrap or Mylar sheets

Step 19 of this protocol requires the reagents listed in Protocol 14.1.
Sterile bent glass rod
Whatman 3MM filters (90 or 140 mm)

Additional Information

Antibodies	MC3, pp. A9.25–A9.34
Autoradiography/phosphorimaging	MC3, pp. A9.9–A9.15
BCIP	MC3, pp. A9.41–A9.42
Chemiluminescence	MC3, pp. A9.16–A9.20
Enzyme-conjugated antibodies	MC3, p. A9.34
Horseradish peroxidase	MC3, p. A9.35
Methods of detecting labeled antibodies	MC3, pp. A8.54–A8.55
Protein A	MC3, p. A9.46
Radiolabeling of antibodies	MC3, p. A9.30
Troubleshooting	MC3, p. 14.13
Validation of clones isolated by immunological screening	MC3, p. 14.12

METHOD

1. Use sterile blunt-ended forceps (e.g., Millipore) to place a sterile nitrocellulose filter, numbered side down, on an LB (or SOB) plate. Peel the filter from the agar, invert it, and replace it, numbered side up.

2. Apply the bacteria to the filter in a small volume of liquid (<0.5 ml containing up to 20,000 bacteria for a 138-mm filter; < 0.2 ml containing up to 10,000 bacteria for an 82-mm filter). Spread the liquid over the surface of the filter with a sterile bent glass rod. Leave a border 2–3 mm wide at the edge of the filter free of bacteria. Store the plates at room temperature until all of the liquid has been absorbed.

3. Invert the plates and incubate them until very small (0.1-mm diameter) colonies appear (8–10 hours).

   Grow colonies containing expression vectors carrying the *lac* promoter at 37°C. Grow colonies containing expression vectors carrying the bacteriophage λ_{pR} promoter at 30°C to prevent the expression of fusion proteins.

4. Wet a numbered, sterile nitrocellulose filter by touching it, numbered side up, to the surface of a fresh agar plate containing the appropriate antibiotic. Leave the filter in contact with the surface of the agar. The numbers on the set of replica filters should correspond to those on the master filters.

5. Use sterile blunt-ended forceps to remove the master filter gently from one of the agar plates (Step 3) and place it, colony side up, on a stack of sterile Whatman 3MM papers.

6. Carefully place the second, correspondingly numbered, wetted filter numbered side down on top of the master filter, being careful not to move the filters once contact has been made. Place a circle of 3MM paper on top of the filter sandwich. Place the bottom of an empty petri dish on top of the 3MM paper and exert hand pressure on the sandwich.

7. Use an 18-gauge needle to make a characteristic set of registration holes in the filters while they are sandwiched together. Gently peel the filters apart. Transfer the replica filter to the agar plate used for wetting (Step 4). Transfer the master filter, colony side up, to a fresh agar plate containing the appropriate antibiotic.

   If required, several replicas can be made from a single master filter. However, if the master filter is to be used to make more than two replicas, reincubate it for a few hours to allow the colonies to regenerate. Generally, it is best to make only two replicas from a single master to avoid problems caused by smearing of the colonies.

8. Repeat Steps 4–7 until all the master filters have been replicated.

9. Induce the expression of a gene cloned into an expression vector carrying the *lac* promoter as described below.

   a. Incubate the plates (masters and replicas) at 37°C until colonies 1–2 mm in diameter have appeared. Colonies on the master plates generally reach the desired size more rapidly.

   b. Allow the master plates to cool to room temperature on the laboratory bench, wrap them in plastic film, and store them at 4°C until the results of the immunological screens are available.

   c. Transfer the replica filters numbered side up to fresh agar plates, prewarmed to 37°C, containing IPTG at a concentration of 1 mM. Incubate the IPTG-containing plates for an additional 2–4 hours.

   d. To induce synthesis in expression vectors that carry the bacteriophage λ_{pR} promoter (e.g., the pEX vectors), transfer the filters to a series of prewarmed agar plates and incubate them for 2–4 hours at 42°C.

10. Use blunt-ended forceps to remove the nitrocellulose filters from the plates and place them on damp paper towels *in a chemical fume hood*. Cover the filters with an inverted plastic box. Place an open glass petri dish containing chloroform under the box with the filters. Expose the bacterial colonies on the filters to chloroform vapor for 15 minutes.

11. Transfer small groups of filters to petri dishes containing lysis buffer (6 ml per 82-mm filter; 12 ml per 138-mm filter). When all of the filters have been submerged, stack the petri dishes on a rotary platform and agitate the lysis buffer by gentle rotation of the platform. Lysis of the bacterial colonies takes 12–16 hours at room temperature.

12. Transfer the filters to petri dishes or glass trays containing TNT buffer. Incubate the filters for 30 minutes at room temperature.

13. Repeat Step 12 using fresh TNT buffer.

14. Transfer the filters, one by one, to a glass tray containing TNT buffer. Use Kimwipes to remove the residue of the colonies from the surfaces of the filters.

    IMPORTANT: Do not allow the filters to dry out during any of the subsequent steps. Antibodies bound nonspecifically and reversibly to wet filters become permanently attached if the filters dry out. It is also essential that the filters do not stick to one another when they are immersed in the various buffers and antibody solutions. This problem can be minimized by dividing the filters into small batches (e.g., five filters per batch) and using a separate large petri dish for each batch. The petri dishes can be stacked on top of one another on a slowly rotating platform shaker.

15. When all of the filters have been rinsed, transfer them one at a time to a fresh batch of TNT buffer. After transfer, agitate the buffer and filters gently for an additional 30 minutes at room temperature.

16. Use blunt-ended forceps to transfer the filters individually to glass trays or petri dishes containing blocking buffer (7.5 ml for each 82-mm filter; 15 ml for each 138-mm filter). When all of the filters have been submerged, agitate the buffer slowly on a rotary platform for 30 minutes at room temperature.

17. Use blunt-ended forceps to transfer the filters to fresh glass trays or petri dishes containing the primary antibody diluted in blocking buffer (7.5 ml for each 82-mm filter; 15 ml for each 138-mm filter). Use the greatest dilution of antibody that gives acceptable background yet still allows detection of 50–100 pg of denatured antigen. When all of the filters have been submerged, agitate the solution gently on a rotary platform for 2–4 hours at room temperature.

18. Wash the filters for 10 minutes in each of the buffers below in the order given. Transfer the filters individually from one buffer to the next. Use 7.5 ml of each buffer for each 82-mm filter and 15 ml for each 138-mm filter.

> TNT buffer containing 0.1% bovine serum albumin
> TNT buffer containing 0.1% bovine serum albumin and 0.1% Nonidet P-40
> TNT buffer containing 0.1% bovine serum albumin

19. Detect the antigen-antibody complexes with the chosen radiochemical, chromogenic, or chemiluminescent reagent. For radiochemical screening, use approximately 1 µCi of ^{125}I-labeled protein A or ^{125}I-labeled immunoglobulin per filter. Radiolabeled protein A is available commercially (sp. act. 2–100 µCi/µg). Radioiodinated secondary antibody is available commercially or can be prepared (MC3, p. A9.30). For chromogenic screening, antibodies coupled to horseradish peroxidase (HRP) or alkaline phosphatase (AP) that react with species-specific determinants on primary antibodies are available from commercial sources and should be used at the recommended dilution in accordance with manufacturer instructions. Typically, 5 µl of conjugated antiserum is used for each 82-mm filter in 7.5 ml of blocking buffer (without sodium azide). Chemiluminescence is the most sensitive method for detecting immunopositive colonies. Chemiluminescent detection is quick, produces a permanent record of the screened filter (an X-ray film or phosphorimage), and is sensitive (1–10 pg of antigen in a colony can be detected).

Radiochemical screening

a. Dilute radiolabeled ligands in blocking buffer (7.5 ml for each 82-mm filter; 15 ml for each 138-mm filter).

b. Incubate the filters for 1 hour at room temperature and then wash them several times in TNT buffer before establishing autoradiographs.

Continue at Step 20.

Chromogenic screening with AP-conjugated antibodies

a. Gently agitate the filters for 1.5–2 hours at room temperature in the solution of AP-conjugated antibody.

b. Wash the filters as described in Step 18.

c. Prepare stock solutions of BCIP (50 mg/ml in 100% dimethylformamide) and NBT (50 mg/ml in 70% dimethylformamide). These solutions are stable when stored in the dark.

d. Prepare the BCIP/NBT developing solution just before use as follows:

   i. Add 33 µl of the NBT solution to 5 ml of AP buffer and mix well.

   ii. Add 16.5 µl of the stock solution of BCIP. Mix again. Protect the solution from strong light and use it within 1 hour.

e. Blot the washed filters on paper towels.

f. Incubate the filters (Step d) in the BCIP/NBT developing solution (7.5 ml for each 82-mm filter; 15 ml for each 138-mm filter) for several hours at room temperature.

g. Rinse the filters briefly in two changes of H_2O. An intense purple color will develop at the site of antigen-antibody complexes.

Continue at Step 20.

Chromogenic screening with HRP-conjugated antibodies

a. Gently agitate the filters for 1.5–2 hours at room temperature in the solution of HRP-conjugated antibody.

b. Wash the filters as described in Step 18.

c. To prepare developing solution, dissolve 60 mg of 4-chloro-1-naphthol in 20 ml of ice-cold methanol. Just before use, mix the solution with 100 ml of 10 mM Tris-Cl (pH 7.5), 150 mM NaCl containing 60 µl of 30% H_2O_2.

d. Blot the washed filters on paper towels and wash them briefly in 10 mM Tris-Cl (pH 7.5) containing 150 mM NaCl.

e. Incubate the filters for 15–20 minutes at room temperature in the 4-chloro-1-naphthol developing solution (10 ml for each 82-mm filter; 25 ml for each 138-mm filter).

f. Wash the filters in two changes of H_2O. An intense purple color will develop at the site of antigen-antibody complexes.

Continue at Step 20.

Chemiluminescent screening

a. Gently agitate the filters for 1.5–2 hours at room temperature in the solution of AP- or HRP-conjugated antibody.

b. Wash the filters as described in Step 18.

c. Prepare the chemiluminescent substrates (MC3, pp. A9.16–A9.20) according to manufacturer instructions.

d. Incubate the washed filters in the chemiluminescent substrates for 1–5 minutes (again, consult the manufacturer's recommendations for optimal exposure times).

e. Drain the excess solution from the filters and immediately wrap the filters in Saran Wrap. Do not allow the filters to dry out.

f. Establish an autoradiogram. Typically, the initial exposure is for 1 minute. This interval provides enough information to establish the proper exposure time.

Continue at Step 20.

20. Identify the locations of positive colonies or, if made, compare the duplicate filters, searching for coincident signals. For screens involving radiolabeled or chemiluminescent probes, compare the resulting autoradiograms with the agar plates on a light box. For screens involving chromogenic reagents that leave a visible positive residue on the filter, perform the following steps.

a. Lay a sheet of Saran Wrap or Mylar film over the filters.

b. On the surface of the Saran Wrap, mark the locations of the holes in the filters and the locations of antigen-positive clones with different colored waterproof markers. Label the Saran Wrap to identify the plates from which the filters were derived.

c. Place the sheet of Saran Wrap on a light box and align the plates containing the original bacterial colonies on top of it.

d. Identify the areas containing the positive colonies and transfer a segment of each putative colony to 1 ml of LB medium containing the appropriate antibiotic. Incubate the cultures for 12–16 hours at the appropriate temperature.

e. Keep the sheet of Saran Wrap, which provides a permanent record of the locations of the positive colonies. The colored spots on the original filters fade quite rapidly.

21. Repeat the process of plating and screening until a homogeneous population of immunopositive colonies is obtained.

Removal of Cross-reactive Antibodies from Antiserum: Pseudoscreening

Polyclonal antisera often contain immunoglobulins that cross-react with bacterial and vector-encoded proteins. Several methods can be used to remove these antibodies, which otherwise can interfere with detection of recombinants encoding target proteins:

- incubating the antiserum with nitrocellulose filters carrying the debris of plaques formed by nonrecombinant bacteriophages (this protocol),
- incubating the antiserum with a lysate of *E. coli* host cells before screening (Protocol 14.4), and
- affinity chromatography, using a resin to which *E. coli* proteins have been covalently attached (Protocol 14.5).

It may be necessary to use more than one of these methods to reduce the level of cross-reactivity to an acceptable level.

MATERIALS

Reagents and Solutions

Please see Appendix 1 for components of stock solutions, buffers, and reagents. Dilute stock solutions to the appropriate concentrations.

Blocking buffer
10 mM Tris-HCl (pH 8.0)
150 mM NaCl
0.05% (v/v) Tween-20
blocking buffer
1% (w/v) gelatin
or 3% (w/v) bovine serum albumin
or 5% (w/v) nonfat dried milk

Blocking buffer can be stored at 4°C and reused several times. Sodium azide <!> should be added to a final concentration of 0.05% (w/v) to inhibit the growth of microorganisms.

Vectors and Hosts

Nonrecombinant bacteriophage λ and *E. coli* host
Use the same vector/host combination that was used to generate the cDNA library of interest.

Media and Antibiotics

LB agar plates, 2 days old
LB top agarose

Additional Items

Antibody preparation to be used for screening
Nitrocellulose filters (82 or 138 mm), free of Triton X-100, e.g., Millipore HATF
Step 1 of this protocol requires the reagents listed in Protocol 2.1.
Step 2 of this protocol requires the reagents listed in Protocol 14.1.

Adapted from Chapter 14, Protocol 3, p. 14.23 of MC3.

Additional Information

Alternative protocol (Adsorbing Antibodies with Lysates of Bacteriophage-infected Cells)	MC3, p. 14.25
Antibodies	MC3, pp. A9.25–A9.34

METHOD

1. On ten LB agar plates, plate out nonrecombinant bacteriophage λ so as to obtain semiconfluent lysis of the bacterial lawn (see Protocol 2.1).

2. Prepare imprints of the lysed lawns on nitrocellulose filters as described in Protocol 14.1, Steps 5–12, omitting the treatment with IPTG.

    It is not necessary to key the filters as described in Protocol 14.1, Step 8.

3. Dilute the preparation of antibody that is to be used for screening 1:10 with blocking buffer.

4. Incubate each of the filters with 2–3 ml of diluted antibody on a shaking platform for 6 hours at room temperature.

5. Recover the treated antibody and store it at 4°C in the presence of 0.05% (w/v) sodium azide until used for immunological screening.

    The method can easily be adapted to remove anti-*E. coli* antibodies by pseudoscreening of bacterial colonies. On nitrocellulose filters, establish bacterial colonies that carry the empty expression vector (see Protocol 14.2). Follow the cell lysis and washing procedures described in Protocol 14.2, Steps 10–16. Finally, treat the antiserum as described in Steps 3–5 above.

Removal of Cross-reactive Antibodies from Antiserum: Incubation with *E. coli* Lysate

This protocol describes how antibodies that react with bacterial-encoded proteins may be removed from polyclonal antisera by incubation with a bacterial lysate.

MATERIALS

Reagents and Solutions

Please see Appendix 1 for components of stock solutions, buffers, and reagents. Dilute stock solutions to the appropriate concentrations.

Blocking buffer
- 10 mM Tris-Cl (pH 8.0)
- 150 mM NaCl
- 0.05% (v/v) Tween-20
- blocking buffer
  - 1% (w/v) gelatin
  - *or* 3% (w/v) bovine serum albumin
  - *or* 5% (w/v) nonfat dried milk

Blocking buffer can be stored at 4°C and reused several times. Sodium azide <!> should be added to a final concentration of 0.05% (w/v) to inhibit the growth of microorganisms.

Cell resuspension buffer
- 50 mM Tris-Cl (pH 8.0)
- 10 mM EDTA (pH 8.0)

Vectors and Hosts

E. coli strain Y1090*hsd*R

Media and Antibiotics

LB medium (100 ml)

Centrifuges/Rotors/Tubes

Sorvall SLC-1500 rotor (4°C)

Additional Items

Antibody preparation to be used for screening
Probe sonicator for disruption of bacterial cells
Rotating wheel

Additional Information

Alternative protocol (Adsorbing Antibodies with Lysates of
   Bacteriophage-infected Cells)   MC3, p. 14.25
Antibodies   MC3, pp. A9.25–A9.34

Adapted from Chapter 14, Protocol 4, p. 14.26 of MC3.

METHOD

1. Grow a 100-ml culture of *E. coli* strain Y1090*hsd*R to saturation in LB medium.

2. Harvest the cells by centrifugation at 5000*g* (5500 rpm in a Sorvall SLC-1500 rotor) for 10 minutes at 4°C.

3. Resuspend the cells in 3 ml of cell resuspension buffer. Freeze and thaw the suspension several times and then sonicate it at full power for six periods of 20 seconds, each at 0°C.

    Make sure that the temperature of the cell suspension remains < 4°C.

4. Centrifuge the extract at maximum speed for 10 minutes at 4°C in a microfuge. Transfer the supernatant to a fresh tube. Store the lysate at –20°C.

5. Just before using the lysate, dilute the preparation of antibody that is to be used for screening 1:10 with blocking buffer.

6. Add 0.5 ml of lysate for every milliliter of antibody preparation to be processed. Incubate the mixture for 4 hours at room temperature on a slowly rotating wheel. The treated antibody may be stored at 4°C in the presence of 0.05% (w/v) sodium azide until used for immunological screening.

Removal of Cross-reactive Antibodies from Antiserum: Affinity Chromatography

This protocol describes a method for removing antibodies that react with bacterially encoded proteins by passing a crude preparation of immunoglobulins through a column containing immobilized bacterial proteins.

MATERIALS

CAUTION: Please see Appendix 4 for appropriate handling of materials marked with <!>.

Reagents and Solutions

Please see Appendix 1 for components of stock solutions, buffers, and reagents. Dilute stock solutions to the appropriate concentrations.

    Cell lysis buffer (sterile)
        0.1 M sodium borate (pH 8.0)
        1 M NaCl
        Approximately 100 ml of cell lysis buffer is required per liter of bacterial culture.
    NaOH (1 N) <!>
    TBS containing 0.2% (w/v) sodium azide <!>
    Tris-buffered saline (TBS)
    Triton X-100

Vectors and Hosts

    E. coli strain in which the expression library was propagated

Media and Antibiotics

    LB medium (1 liter)

Enzymes and Buffers

    Lysozyme, molecular biology grade
    Pancreatic DNase I

Centrifuges/Rotors/Tubes

    Sorvall SLC-1500 and SS-34 rotors (4°C)

Additional Items

    Antibody preparation that is to be used for screening
        This protocol works best when using an IgG fraction, prepared by fractionating total anti-
        serum on protein A–Sepharose.
    Chromatography column (5-ml plastic syringe plugged with glass wool or Bio-Rad
      Poly-Prep column)
    Cyanogen bromide–activated Sepharose 4B (GE Healthcare), or Affi-Gel 10 (Bio-Rad)

Additional Information

    Antibodies   MC3, pp. A9.25–A9.34

METHOD

1. Grow a 1-liter culture of the appropriate strain of *E. coli* (e.g., Y1090*hsd*R, XL1-Blue, or DH1) to stationary phase.

2. Recover the bacteria by centrifugation at 4000*g* (5000 rpm in a Sorvall SLC-1500 rotor) for 20 minutes at 4°C.

3. Pour off the medium and stand the centrifuge tubes in an inverted position to allow the last traces of medium to drain away.

4. Resuspend the pellet in 100 ml of cell lysis buffer.

5. Add 200 mg of lysozyme and incubate the bacterial suspension for 20 minutes at room temperature.

6. Add 1 mg of pancreatic DNase I and 200 µl of Triton X-100.

7. Incubate the bacterial suspension for 1 hour at 4°C or until the turbidity clears and the viscosity decreases.

8. Centrifuge the bacterial lysate at 8000*g* (8200 rpm in a Sorvall SS-34 rotor) for 20 minutes at 4°C. Carefully decant the supernatant into a fresh flask.

9. Adjust the pH of the supernatant to 9.0 with 1 N NaOH.

10. Determine the concentration of protein in the lysate using the Lowry, Bradford, or other method of measurement.

11. Chill the extract to 0°C and then bind the bacterial proteins to cyanogen bromide–activated Sepharose 4B or to Affi-Gel 10 according to the manufacturer's instructions.

12. Before use, equilibrate the Sepharose 4B or Affi-Gel 10 resin containing conjugated *E. coli* proteins in TBS containing 0.2% (w/v) sodium azide.

13. Use 1 ml of settled volume of resin coupled to *E. coli* antigen for each milligram of IgG protein to be purified by affinity chromatography. Mix the IgG and the coupled resin and incubate for 12–18 hours at room temperature on a rotating wheel.

14. Load the slurry into a chromatography column. Recover the antibody by washing the column with TBS. Collect fractions (0.2 column volume each) until the OD_{280} drops to zero. Pool the fractions containing antibody and store the pool at –20°C until it is used for immunological screening.

Identifying DNA-binding Proteins in Bacteriophage λ Expression Libraries

This protocol describes how to identify cloned cDNAs encoding proteins that bind to specific DNA sequences. The methods used are very similar to those used for immunological screening of expression libraries except that the nitrocellulose filters carrying immobilized proteins are screened with ^{32}P-labeled double-stranded DNA rather than with antibodies.

MATERIALS

CAUTION: Please see Appendix 4 for appropriate handling of materials marked with <!>.

Reagents and Solutions

Please see Appendix 1 for components of stock solutions, buffers, and reagents. Dilute stock solutions to the appropriate concentrations.

ATP (10 mM) in 25 mM Tris (pH 8.0)

ATP (50 mM) in 25 mM Tris (pH 8.0)

1x Binding buffer containing 1 mM DTT (see Steps 11–13 to calculate the volume of solution required)

1x Binding buffer containing 1 mM DTT and 0.25% (w/v) nonfat dry milk (~75 ml is needed per 82-mm filter or ~180 ml per 138-mm filter)

1x Binding buffer containing 5% (w/v) nonfat dry milk (~10 ml is needed per 82-mm filter or ~25 ml per 138-mm filter)

1x Binding buffer containing 0.25% (w/v) nonfat dry milk (~10 ml is needed per 82-mm filter or ~25 ml per 138-mm filter)

10x Binding buffer

    250 mM HEPES (pH 7.9)
    30 mM MgCl$_2$
    40 mM KCl

Denaturation buffer, freshly made

    Dilute 10x binding buffer with 5 volumes of distilled H$_2$O. The resulting solution (2x binding buffer) is then added to the appropriate amount of solid guanidine HCl <!> to make a 6 M solution. When the guanidine HCl has completely dissolved, adjust the concentration of the binding buffer to 1x with distilled H$_2$O and add DTT to a final concentration of 1 mM. About 15 ml of full-strength (6 M) denaturation buffer is needed for each 82-mm filter or 25 ml for each 138-mm filter.

Dithiothreitol (1 M)

EDTA (0.5 M, pH 8.0)

Ethanol

Phenol:chloroform (1:1, v/v) <!>

Screening buffer

    1x binding buffer
    0.25% (w/v) nonfat dry milk
    1 mM DTT
    10 µg/ml denatured salmon or herring sperm DNA

SDS (20% w/v)

Sodium acetate (3 M, pH 5.2)

TE (pH 8.0)

     Adapted from Chapter 14, Protocol 6, p. 14.31 of MC3.

Vectors and Hosts

Bacteriophage λ expression library (see Protocol 14.1)

Enzymes and Buffers

Bacteriophage T4 DNA ligase
Bacteriophage T4 polynucleotide kinase
10x Kinase/ligase buffer
    600 mM Tris-Cl (pH 7.6)
    100 mM MgCl$_2$

Nucleic Acids/Oligonucleotides

Synthetic oligonucleotides
    Single-stranded oligonucleotides of complementary sequence, 20–25 nucleotides in length, purified by gel electrophoresis and Sep-Pak chromatography (Protocol 10.1) should be dissolved separately in TE at a concentration of 0.2 mg/ml. When the two oligonucleotides are mixed under conditions favoring hybridization, they anneal to form a monomeric version of a binding site for the target protein, previously identified, for example, by gel retention assays or Southwestern blotting. At least one complementary pair of "mutant" oligonucleotides is also required in the second round of screening. The central region of a "mutant" oligonucleotide duplex forms a defective version of the binding site that is unable to bind the target protein. Sequential screening of an expression library with the wild-type and mutant duplexes eliminates most of the false positives.
    Both pairs of oligonucleotides should be designed with protruding cohesive termini so that duplexes can be ligated into highly efficient catenated probes containing multiple binding sites for the target protein.

Radioactive Compounds

[γ-^{32}P]ATP (10 mCi/ml, 5000 Ci/mmole) <!>

Gels/Loading Buffers

Nondenaturing polyacrylamide gel <!>

Additional Items

Adhesive luminescent labels, e.g., Glogos (Stratagene)
Baking dishes (Pyrex)
Blunt-ended forceps, e.g., Millipore
Crystallizing dish
Materials for autoradiography or phosphorimaging (see Appendix 9)
Nitrocellulose filters, free of Triton X-100, e.g., Millipore HATF
Spun column of Sephadex G-75 in TE (pH 7.6)
Steps 8–10 of this protocol require the reagents listed in Protocol 14.1.
Thermal cycler, programmed as described in Step 2
Water bath or heating/cooling block (16°C)
Whatman 3MM paper

Additional Information

Screening expression libraries with oligonucleotide probes   MC3, pp. 14.31–14.32

METHOD

1. Set up two separate phosphorylation reactions, each containing one of the synthetic oligonucleotides to be annealed:

oligonucleotide	200 ng
10x kinase/ligase buffer	2.5 µl
100 mM dithiothreitol	2.5 µl
[γ-^{32}P]ATP	100 µCi
H$_2$O	to 23 µl
bacteriophage T4 polynucleotide kinase (8–10 units/µl)	2 µl

   Incubate the reactions for 1 hour at 37°C.

2. Mix the two phosphorylation reactions together. Anneal the oligonucleotides by incubating the mixture in the following sequence, which is most conveniently carried out in a thermal cycler:

   2 minutes at 85°C
   15 minutes at 65°C
   15 minutes at 37°C
   15 minutes at 22°C
   15 minutes at 4°C

3. Add 4 µl of bacteriophage T4 DNA ligase (1 Weiss unit/µl) and 1 µl of 50 mM ATP. Incubate the mixture for 12 hours at 16°C.

4. Add 0.5 M EDTA (pH 8.0) to a final concentration of 5 mM.

5. Separate the labeled oligonucleotides from unused [γ-^{32}P]ATP, single-stranded oligonucleotides, and unligated double-stranded oligonucleotides by spun-column chromatography through a Sephadex G-75 column.

6. Estimate the specific activity of the final probe.

   The specific activity should be ≥2 x 10^6 cpm/pmole.

7. Analyze the size of the radiolabeled DNA by nondenaturing polyacrylamide gel electrophoresis and autoradiography. If all has gone well in the annealing and radiolabeling steps, the concatenated DNA should form a ladder of polymers of the original duplex oligonucleotide.

8. Prepare agar plates containing plaques of the bacteriophage λ expression library and numbered nitrocellulose filters exactly as described in Protocol 14.1, Steps 1–8.

9. Use blunt-ended forceps (e.g., Millipore) to remove the numbered nitrocellulose filters from the lawn of plaques and place them on Whatman 3MM paper with the side exposed to the plaques facing upward. Allow the filters to dry for 15 minutes at room temperature.

10. Lay a second (numbered) filter impregnated with IPTG on each agar plate (please see Protocol 14.1, Step 4). Use an 18-gauge needle to make holes in each filter in the same locations as the holes used to key the first filter to the lawn. Incubate the plates for 2 additional hours at 37°C and then remove the filters. Allow them to dry at room temperature as described in the preceding step.

    IMPORTANT: Perform all subsequent steps at 4°C. The first set of filters is probed directly without denaturation with guanidine HCl (i.e., omitting Steps 11–13), whereas the second set is processed as described in Steps 11–14.

11. Place the second set of numbered filters in a 12 x 8-inch baking dish containing denaturation solution at 4°C. Agitate the filters gently on a platform shaker for 5 minutes at 4°C. Decant the denaturation solution and replace it with fresh solution. Agitate the filters for an additional 5 minutes at 4°C.

12. Decant the second batch of denaturation solution into a graduated cylinder. Dilute the solution with an equal volume of 1x binding buffer containing dithiothreitol. Pour this solution into a clean glass dish and transfer the filters to the solution one at a time, making sure that each filter becomes thoroughly exposed to the diluted denaturation solution containing dithiothreitol.

13. Repeat the process described in Step 12 four more times, diluting the denaturation solution by a factor of 2 each time. The concentrations of guanidine HCl in the solutions are therefore 3 M (Step 11), 1.5 M, 0.75 M, 0.375 M, 0.187 M, and 0.094 M. Finally, wash the filters twice in 1x binding buffer containing dithiothreitol.

14. Place both sets of numbered filters (i.e., denatured and nondenatured) in 1x binding buffer containing 5% nonfat dried milk. Agitate the filters gently for 30 minutes at 4°C.

15. Rinse the filters in 1x binding buffer containing 0.25% nonfat dried milk.

16. In a crystallizing dish, add the ^{32}P-labeled concatenated DNA probe from Step 5 to screening buffer to make a hybridization solution (~10 ml for each 82-mm filter; ~25 ml for each 138-mm filter).

17. Transfer the filters to the radiolabeled probe solution in the crystallizing dish. Incubate the filters with gentle agitation on a rotating platform for 2–12 hours at 4°C.

18. Wash the filters for 5 minutes at 4°C in a large volume (25 ml for each 82-mm filter; 60 ml for each 138-mm filter) of binding buffer containing 1 mM dithiothreitol and 0.25% nonfat dried milk.

19. Repeat Step 18 twice more.

20. Decant the final wash buffer. Arrange the damp filters on a sheet of Saran Wrap. Cover the filters with another sheet of Saran Wrap. Apply adhesive labels marked with radioactive ink or chemiluminescent markers to several asymmetric locations on the Saran Wrap.

    Cover the radioactive labels with Scotch tape to prevent contamination with the radioactive ink of the film holder or intensifying screen.

21. Establish an autoradiograph.

22. Pick positive plaques and rescreen them with specific and nonspecific probes (see MC3, pp. 14.31–14.32).

Preparation of Lysates Containing Fusion Proteins Encoded by Bacteriophage λ Lysogens: Lysis of Bacterial Colonies

In this protocol, a bacterial lysogen is constructed from a recombinant bacteriophage λ encoding a fusion protein of interest. The resulting lysogenic colonies are induced to synthesize the fusion protein, which is then isolated in preparation for functional and biochemical analyses.

MATERIALS

CAUTION: Please see Appendix 4 for appropriate handling of materials marked with <!>.

Reagents and Solutions

Please see Appendix 1 for components of stock solutions, buffers, and reagents. Dilute stock solutions to the appropriate concentrations.

IPTG (1 M) (~40 μl is required to induce each lysogen)

Lysogen extraction buffer

    50 mM Tris-Cl (pH 7.5)

    1 mM EDTA (pH 8.0)

    5 mM dithiothreitol

    50 μg/ml phenylmethylsulfonyl fluoride (PMSF) <!>

    Add PMSF to the lysogen extraction buffer just before it is used in Steps 13 and 18. About 5 ml of lysogen extraction buffer is required for each induced lysogen.

NaCl (5 M)

Vectors and Hosts

E. coli strains Y1090*hsd*R and Y1089

    These strains, which are maintained on LB agar plates containing ampicillin, carry mutations in the *lon* gene, which encodes an ATP-dependent protease. Fusion proteins expressed in these strains are often more stable than those expressed in other strains. Y1089 carries the *hflA* mutation, which dramatically increases the frequency of lysogenization. In addition, because Y1089 lacks a suppressor tRNA gene, the amber mutation in the bacteriophage λ *S* (lysis) gene of many expression vectors is not suppressed. The inability to lyse the bacteria after lysogenetic induction allows large amounts of recombinant fusion proteins to accumulate to high levels.

Recombinant cDNA propagated in bacteriophage λ expression vector, e.g., λgt11, λgt18-23, λZAP, or λZipLox

    This protocol is optimized for a bacteriophage λgt11 recombinant that expresses a fusion protein. However, the protocol can be easily adapted for use with recombinants propagated in many other bacteriophage λ expression vectors that can form lysogens.

Media and Antibiotics

LB agar plates containing 50 μg/ml ampicillin

LB medium

LB medium containing 50 μg/ml ampicillin

LB medium containing 10 mM MgSO$_4$

LB medium containing 10 mM MgSO$_4$, 0.2% (w/v) maltose, and 50 μg/ml ampicillin

Enzymes and Buffers

Lysozyme (10 mg/ml) in 10 mM Tris (pH 8.0) (freshly prepared for use in Step 15)

Adapted from Chapter 14, Protocol 7, p. 14.37 of MC3.

Additional Items

Air incubators (32°C and 42°C)

Liquid nitrogen <!>

Millipore filters

> Circular (138 mm) Millipore, type VS filters, 0.025-mm pore size, are used to dialyze the lysates. As many as 20 lysates can be dialyzed per filter.

Rotary shaking incubator (32°C)

Shaking water bath (44°C)

Step 1 of this protocol requires the reagents listed in Protocols 2.2 and 2.3.

Toothpicks

Additional Information

Lysogeny	MC3, pp. 2.9–2.11 and 2.15–2.18
Plasmid and bacteriophage λ expression vectors	MC3, pp. 14.47–14.49

METHOD

1. Make a plate stock of each of the recombinant bacteriophage(s) of interest using the methods described in Protocols 2.2 and 2.3. The titer of the stocks, measured on *E. coli* strain Y1090*hsd*R, should be >10^{10} pfu/ml.

2. Grow a 2-ml culture of *E. coli* strain Y1089 to saturation in LB medium containing 10 mM MgSO$_4$, 0.2% maltose, and ampicillin (50 μg/ml).

3. Dilute 50 μl of the saturated culture with 2 ml of LB medium containing 10 mM MgSO$_4$. Transfer four 100-μl aliquots of the diluted culture to fresh culture tubes.

4. To each of three of the tubes, add 1×10^7, 5×10^7, and 2×10^8 plaque-forming units of the bacteriophage stock. The fourth tube should receive no bacteriophage. Incubate all four tubes for 20 minutes at 37°C to allow virus attachment.

5. Dilute 10 μl of each of the four cultures with 10 ml of LB medium. Immediately plate aliquots (100 μl) of each of the four diluted cultures onto LB agar plates containing 50 μg/ml ampicillin. Incubate the plates for 18–24 hours at 32°C.

6. Use sterile toothpicks to transfer a series of individual colonies onto two LB plates that contain 50 μg/ml ampicillin. Incubate one plate at 32°C and the other plate at 42°C for 12–16 hours. Clones that give rise to colonies at 32°C but not at 42°C are lysogenic for recombinant bacteriophage λgt11. Usually, 10–70% of the colonies tested are lysogens.

7. Inoculate 2 ml of LB medium containing 50 μg/ml ampicillin with individual bacteriophage λgt11 lysogens. Grow the cultures for 12–16 hours at 32°C with vigorous agitation (300 cycles/minute in a rotary shaker).

8. Add 50 μl of each culture to 4 ml of prewarmed (32°C) LB medium containing 50 μg/ml ampicillin. Continue incubation at 32°C with vigorous agitation.

9. Grow the cultures until the OD$_{600}$ = 0.45 (~3 hours of incubation).

   IMPORTANT: The OD$_{600}$ of the cultures should not exceed 0.5 (~2×10^8 bacteria/ml) before induction of the lysogens.

10. Transfer the cultures to a shaking water bath equilibrated to 44°C. Incubate the cultures for 15 minutes at 44°C.

11. Add IPTG to each culture to a final concentration of 10 mM and then incubate the cultures for 1 hour at 37°C with vigorous agitation.

12. Transfer 1.5-ml aliquots of each of the induced lysogenic cultures to two microfuge tubes. Immediately centrifuge the tubes in a microfuge at maximum speed for 30 seconds at 4°C.

13. Remove the medium by aspiration and then rapidly resuspend the bacterial pellets by vortexing in 100 µl of lysogen extraction buffer.

14. Close the caps of the tubes and place the tubes in liquid nitrogen.

15. After 2 minutes, recover the tubes from the liquid nitrogen. Hold the tubes in one hand to warm them until the lysates just thaw and then immediately add to each tube 20 µl of 10 mg/ml lysozyme. Store the tubes for 15 minutes in an ice bath.

16. Add 250 µl of 5 M NaCl to each tube. Mix the contents by flicking the side of each tube with a finger. Incubate the tubes for 30 minutes at 4°C on a rotating wheel.

17. Centrifuge the tubes in a microfuge at maximum speed for 30 minutes at 4°C.

18. Float a Millipore filter (Type VS, 0.025-µm pore size) on the surface of a petri dish (150 mm) filled with lysogen extraction buffer at 4°C.

19. Transfer the supernatants from the centrifuge tubes to the upper surface of the filter. Up to 20 different samples can be applied to the same filter.

20. After 1–2 hours at 4°C, transfer the dialyzed samples to fresh microfuge tubes, which can be stored at –70°C until needed.

21. Analyze the cell lysates directly for the presence of DNA-binding proteins, for example, in methylation protection experiments, by DNase I footprinting, or in gel electrophoresis DNA-binding assays (electrophoretic mobility-shift assays) (please see Chapter 17).

    Fusion proteins can be purified using commercially available affinity chromatography kits (e.g., ProtoSorb lacZ immunoaffinity adsorbent [Promega]) or as described in MC3, Chapter 15.

Preparation of Lysates Containing Fusion Proteins Encoded by Bacteriophage λ: Lytic Infection on Agar Plates

A lytic infection by a recombinant bacteriophage is established in soft agarose. Following induced expression, the recombinant fusion protein is recovered from the infected cells.

MATERIALS

CAUTION: Please see Appendix 4 for appropriate handling of materials marked with <!>.

Reagents and Solutions

Please see Appendix 1 for components of stock solutions, buffers, and reagents. Dilute stock solutions to the appropriate concentrations.

Control overlay solution (sterile)
10 mM $MgSO_4$
0.5x LB medium

IPTG (1 M) (~40 µl is required to induce each lysogen)

IPTG overlay solution <!>
Control overlay solution containing 5 mM IPTG.
12 ml of IPTG overlay solution is required for each 150-mm agar plate.

$MgSO_4$ (1 M)

NaCl (5 M)

SM

Vectors and Hosts

Bacteriophage λgt11 recombinant
Prepare a stock of recombinant bacteriophage λ of known titer by soaking an individual plaque in approximately 100 µl of SM for at least 2 hours at room temperature or by preparing a plate lysate (see MC3, Protocol 2.3). This protocol is optimized for a bacteriophage λgt11 recombinant that expresses a fusion protein. However, the protocol can be easily adapted for use with recombinants propagated in other bacteriophage λ expression vectors.

E. coli strain Y1090hsdR
This strain is maintained on LB agar plates containing ampicillin.

Media and Antibiotics

LB agar plates (150 mm), freshly poured
LB agar plates containing 50 µg/ml ampicillin
LB agar plates containing 50 µg/ml ampicillin and 10 mM $MgSO_4$
LB top agarose containing 10 mM $MgSO_4$ and 50 µg/ml ampicillin

Enzymes and Buffers

Lysozyme (10 mg/ml) in 10 mM Tris (pH 8.0), freshly prepared for use in Step 15

Centrifuges/Rotors/Tubes

Sorvall SS-34 rotor (room temperature)

Additional Items

Air incubators (42°C)

Additional Information

Lysogeny	MC3, pp. 2.9–2.11 and 2.15–2.18
Plasmid and bacteriophage λ expression vectors	MC3, pp. 14.47–14.49

METHOD

1. Inoculate 50 ml of LB medium containing 50 µg/ml ampicillin with a single colony of *E. coli* Y1090*hsd*R. Grow the culture overnight at 37°C with moderate agitation (250 cycles/minute in a rotary shaker).

2. Transfer the culture to a centrifuge tube and centrifuge the cells at 4000*g* (5800 rpm in a Sorvall SS-34 rotor) for 10 minutes at room temperature.

3. Discard the supernatant and resuspend the cell pellet in 20 ml of 10 mM $MgSO_4$. Measure the OD_{600} of a 1/100 dilution of the resuspended cells and prepare a plating stock by diluting the resuspended cells to a final concentration of 2.0 OD_{600}/ml with 10 mM $MgSO_4$.

4. Transfer three 0.2-ml aliquots of the plating stock of *E. coli* Y1090*hsd*R to fresh tubes. To two of the tubes, add 2×10^5 to 5×10^6 pfu of the recombinant bacteriophage λgt11 stock. The third tube serves as an uninfected control. Incubate the tubes for 20 minutes at 37°C to allow the virus to attach to the cells.

5. Add 7.5 ml of molten LB top agarose containing 10 mM $MgSO_4$ and 50 µg/ml ampicillin to one of the tubes. Mix the contents and immediately pour the top agarose onto a 150-mm LB agar plate.

6. Repeat Step 5 with each of the remaining tubes.

7. Incubate the agar plates for 4 hours at 42°C.

8. Remove the plates from the incubator. Add 6 ml of control overlay solution to one of the infected plates. Add 12 ml of IPTG overlay solution to the two remaining plates.

9. Return the plates to the 42°C incubator for 3–5 hours.

10. Remove the plates from the incubator and transfer the overlay solutions into individual sterile tubes.

11. Detect the fusion protein in the overlay solution by immunoblotting or by DNA-binding assays if the original phage was isolated as described in Protocol 14.6.

12. Purify the β-galactosidase fusion protein from the overlay solution by affinity chromatography using commercially available kits (e.g., Promega ProtoSorb) or as described in Protocol 15.1. Dialyze the overlay solution before purification of the protein to remove IPTG.

Preparation of Lysates Containing Fusion Proteins Encoded by Bacteriophage λ Lysogens: Lytic Infection in Liquid Medium

This rapid method is used to screen bacteriophage λgt11 clones for the production of immunodetectable fusion proteins. After optimizing the conditions of infection and induction, the method can be used to produce preparative amounts of a fusion protein.

MATERIALS

CAUTION: Please see Appendix 4 for appropriate handling of materials marked with <!>.

Reagents and Solutions

Please see Appendix 1 for components of stock solutions, buffers, and reagents. Dilute stock solutions to the appropriate concentrations.

Cell lysis buffer
> 50 mM Tris-Cl (pH 6.8)
> 100 mM dithiothreitol
> 2% (w/v) SDS
> 0.1% (w/v) bromophenol blue
> 10% (v/v) glycerol

IPTG (1 M) (~70 µl is required for each plaque analyzed)

$MgSO_4$ (1 M)

Phenylmethylsulfonyl fluoride (PMSF) (100 mM) <!>

SM

Vectors and Hosts

Bacteriophage λgt11 recombinant
> Prepare a stock of recombinant bacteriophage λ of known titer by soaking a single plaque in approximately 100 µl of SM for at least 2 hours or by preparing a plate lysate (see MC3, Protocol 2.3). This protocol is optimized for a bacteriophage λgt11 recombinant that expresses a fusion protein. However, the protocol can be easily adapted for use with recombinants propagated in other bacteriophage λ expression vectors.

E. coli strain Y1090*hsd*R
> This strain is maintained on LB agar plates containing ampicillin.

Media and Antibiotics

LB medium containing 50 µg/ml ampicillin

Enzymes and Buffers

Lysozyme (10 mg/ml) in 10 mM Tris (pH 8.0) (freshly prepared for use in Step 15)

Gels/Loading Buffers

SDS-polyacrylamide gel <!> (see MC3, pp. A8.40–A8.51)

Centrifuges/Rotors/Tubes

Sorvall SS-34 rotor (room temperature)

Additional Items

Boiling-water bath
Liquid nitrogen <!>
Rotary shaking incubator

Additional Information

Lysogeny	MC3, pp. 2.9–2.11 and 2.15–2.18
Plasmid and bacteriophage λ expression vectors	MC3, pp. 14.47–14.49
SDS-polyacrylamide gels	MC3, pp. A8.40–A8.51

METHOD

1. Inoculate 50 ml of LB medium containing 50 µg/ml ampicillin with a single colony of *E. coli* Y1090*hsd*R. Grow the culture overnight at 37°C with moderate agitation (250 cycles/minute in a rotary shaker).

2. Transfer the culture to a centrifuge tube and centrifuge the cells at 4000*g* (5800 rpm in a Sorvall SS-34 rotor) for 10 minutes at room temperature.

3. Discard the supernatant and resuspend the cell pellet in 25 ml of 10 mM $MgSO_4$. Store the bacterial suspension on ice until required.

4. In sterile 15-ml tubes, mix 8 ml of LB containing 50 µg/ml ampicillin, 400 µl of bacterial suspension, and 100 µl of phage lysate.

5. Place the tubes in a 37°C shaking water bath for 2 hours.

6. Transfer a 1-ml aliquot of each culture of infected cells to a sterile microfuge tube. Store the tightly capped tubes in liquid nitrogen. Add 70 µl of 1 M IPTG to the remainder of the infected cultures and continue the incubation at 37°C.

7. At hourly intervals thereafter, withdraw 1-ml aliquots of each culture of infected cells to microfuge tubes. Store the tightly capped tubes in liquid nitrogen. Collect aliquots in this fashion for a period of 4 hours.

8. Incubate the remainder of the infected cultures for an additional 12 hours at 37°C. Remove a final 1-ml sample from each culture. Place the tightly capped tubes in liquid nitrogen for 30 minutes.

9. Thaw all the samples and collect the infected bacteria by centrifugation at maximum speed for 1 minute at room temperature in a microfuge. Decant and discard the bacterial medium.

10. Add 100 µl of cell lysis buffer to each tube and rapidly resuspend the cell pellets by vigorous vortexing.

11. Place the samples in a boiling-water bath for 3 minutes. Transfer the samples to room temperature, add 1 µl of 100 mM PMSF, and mix the contents of the tubes by vortexing.

12. Analyze the samples directly by SDS-polyacrylamide gel electrophoresis and immunoblotting. Just before electrophoresis, spin the samples at maximum speed for 1 minute in a microfuge. Load a 25-µl aliquot of the supernatant from this spin onto the SDS-polyacrylamide gel.

CHAPTER 15

Expression of Cloned Genes in *Escherichia coli*

BACKGROUND INFORMATION

Background information found in *Molecular Cloning: A Laboratory Manual*, 3rd edition (hereafter MC3) unless otherwise indicated.

Choosing an expression system	MC3, pp. 15.2–15.4
Expression of cloned genes	MC3, pp. 15.55–15.59
Fusion proteins	MC3, pp. 15.4–15.8

Expression of Cloned Genes in *E. coli* Using IPTG-inducible Promoters

This protocol describes how to (1) clone sequences encoding open reading frames in plasmids carrying IPTG-inducible promoters, (2) optimize expression of target proteins in transformants carrying these recombinants, and (3) scale up production of foreign proteins.

MATERIALS

CAUTION: Please see Appendix 4 for appropriate handling of materials marked with <!>.

Reagents and Solutions

Please see Appendix 1 for components of stock solutions, buffers, and reagents. Dilute stock solutions to the appropriate concentrations.

Coomassie Brilliant Blue or silver stain (see Appendix 3, pp. 756–763)

IPTG (isopropyl-β-D-thiogalactosidase) (1 M) <!>

Vectors and Hosts

E. coli strain suitable for transformation and carrying either the *lacI^q* or *lacI^{q1}* allele

Some IPTG-inducible expression vectors carry the *lacI^q* allele on the expression plasmid (e.g., pMAL and pGEX). These plasmids can be used for expression of proteins in most laboratory strains of *E. coli*.

IPTG-inducible expression vector

Examples include pGEM-3Z (Promega), pGEX (GE Healthcare), and pMAL (New England BioLabs).

Positive-control plasmid (e.g., an IPTG-inducible recombinant known to express a LacZ fusion protein of defined size)

Media and Antibiotics

LB agar plates containing 50 µg/ml ampicillin

LB medium containing 50 µg/ml ampicillin

Nucleic Acids/Oligonucleotides

Coding sequence of interest

Gels/Loading Buffers

10% Polyacrylamide gel containing SDS <!> (see Appendix 3, pp. 761–763)

1x SDS gel-loading buffer

50 mM Tris-Cl (pH 6.8)

2% (w/v) SDS, electrophoresis grade

0.1% (w/v) bromophenol blue

10% glycerol

Add dithiothreitol from a 1 M stock to a final concentration of 100 mM just before the buffer is used in Step 10.

Centrifuges/Rotors/Tubes

Sorvall SLC-1500 rotor (4°C)

Adapted from Chapter 15, Protocol 1, p. 15.14 of MC3.

Additional Items

Boiling-water bath
Shaking incubators
Step 1 of this protocol requires the reagents listed in Protocol 8.7.
Step 2 of this protocol requires the reagents listed in Protocol 1.17 or 1.19.
Step 3 of this protocol requires the reagents listed in Protocols 1.23–1.26.
Step 4 of this protocol requires the reagents listed in Protocol 12.3.

Additional Information

SDS-polyacrylamide gel electrophoresis	Appendix 3, pp. 756–763
Troubleshooting protein expression from an inducible promoter	MC3, pp. 15.18–15.19

METHOD

1. Modify by PCR (Protocol 8.7), or isolate by restriction enzyme digestion, a fragment of coding DNA carrying 5′- and 3′-restriction enzyme sites compatible with sites in an IPTG-inducible expression vector.

    Constructs generated by PCR should be sequenced to ensure that no spurious mutations were introduced during the amplification reactions.

2. Ligate the DNA fragment coding for the polypeptide of interest into the expression vector (Protocol 1.17 or 1.19).

3. Transform an *E. coli* strain containing the *lacI^q* allele with the recombinant plasmid. If the plasmid vector itself carries the *lacI* gene, any appropriate strain of *E. coli* can be used. Plate aliquots of the transformation reaction on LB agar containing 50 μg/ml ampicillin. Incubate the cultures overnight at 37°C.

4. Screen transformants by colony hybridization and/or restriction enzyme analysis, oligonucleotide hybridization, or direct DNA-sequence analysis (please see Protocol 12.3) of plasmid minipreparations.

5. Inoculate 1-ml cultures (LB medium containing 50 μg/ml ampicillin) with one or two colonies containing the empty expression vector, the positive-control plasmid, and the recombinant expression plasmid. Incubate the cultures overnight at the appropriate temperature (20–37°C).

    Growth at lower temperature sometimes helps to suppress formation of insoluble inclusion bodies. See Dealing with Insoluble Proteins in MC3, pp. 15.9–15.11.

6. Inoculate 5 ml of LB medium containing 50 μg/ml ampicillin with 50 μl of each overnight culture. Incubate the cultures for >2 hours at 20–37°C in a shaking incubator until cells reach mid-log growth (A_{550} of 0.5–1.0).

    It is important to monitor the number of bacteria inoculated into the growth medium, the length of time cells are grown before induction, and the density to which cells are grown after induction.

7. Transfer 1 ml of each uninduced culture (zero-time aliquot) to microfuge tubes. Immediately process the zero-time aliquots as described in Steps 9 and 10.

8. Induce the remainder of each culture by adding IPTG to a final concentration of 1 mM and continue incubation at 20–37°C with aeration.

    The concentration of IPTG and the temperature of incubation (both before and after induction) can drastically affect the yield and solubility of the expressed polypeptide. If the yield of the native protein becomes an issue, a range of temperatures (15–42°C) and of concentrations of IPTG (0.01–5.0 mM) should be explored.

9. At various time points during the induction period (e.g., 1, 2, 4, and 6 hours), transfer 1 ml of each culture to a microfuge tube, measure the A_{550} in a spectrophotometer, and centrifuge the tubes at maximum speed for 1 minute at room temperature. Remove the supernatants by aspiration.

10. Resuspend each pellet in 100 μl of 1x SDS gel-loading buffer (containing 100 mM DTT) and heat the samples to 100°C for 3 minutes. Centrifuge the tubes at maximum speed for 1 minute at room temperature in a microfuge and store them on ice until all of the samples are collected and ready to load on a gel.

11. Warm the samples to room temperature and load 0.15 OD_{550} units (of original culture) or 40 μg of each suspension on a 10% SDS-polyacrylamide gel.

12. Run the gel at 8–15 V/cm until the bromophenol blue reaches the bottom of the resolving gel.

13. Stain the gel with Coomassie Brilliant Blue or silver, or perform an immunoblot to visualize the induced protein (see Appendix 3, pp. 756–763).

14. For large-scale expression and purification of the target protein, inoculate 50 ml of LB containing 50 μg/ml ampicillin in a 250-ml flask with a colony of *E. coli* containing the recombinant construct. Incubate the culture overnight at 20–37°C.

15. Inoculate 450–500 ml of LB containing 50 μg/ml ampicillin in a 2-liter flask with 5–50 ml of overnight culture of *E. coli*. Incubate with shaking at 20–37°C until the culture has reached the mid-log phase of growth (A_{550} = 0.5–1.0).

16. Induce expression of the target protein based on the optimal values of IPTG concentration, incubation time, and incubation temperature determined in the previous section.

17. After the induced cells have grown for the proper length of time, harvest the cells by centrifugation at 5000g (5500 rpm in a Sorvall SLC-1500 rotor) for 15 minutes at 4°C and proceed with a purification protocol:

    • Protocol 15.5 if the expressed protein is a fusion with glutathione *S*-transferase

    • Protocol 15.6 if the expressed protein is a fusion with maltose-binding protein

    • Protocol 15.7 if the expressed protein contains a polyhistidine tag

Expression of Cloned Genes in *E. coli* Using the Bacteriophage T7 Promoter

This protocol describes how to (1) clone sequences encoding open reading frames in plasmids carrying bacteriophage T7 promoters, (2) optimize expression of target proteins in transformants carrying these recombinants, and (3) scale up production of foreign proteins.

Reagents and Solutions

Please see Appendix 1 for components of stock solutions, buffers, and reagents. Dilute stock solutions to the appropriate concentrations.

Coomassie Brilliant Blue stain or silver stain (see Appendix 3, pp. 756–763)

IPTG (isopropyl-β-D-thiogalactosidase) (1 M) <!>

Vectors and Hosts

E. coli strains HMS174 (DE3) or BL21 (DE3)

In these strains, the gene encoding bacteriophage T7 RNA polymerase is carried on an integrated copy of bacteriophage λ DE3 (a.k.a. "Origami") (see MC3, p. A3.6, Table A3-2).

pET vector

For the many variations of this series of plasmids, see the Novagen pET System manual, available at www.emdbiosciences.com/.

Positive-control plasmid (e.g., encoding a foreign protein of known size whose expression is controlled by the bacteriophage T7 promoter)

Media and Antibiotics

NZCYM agar plates containing 50 µg/ml ampicillin

NZCYM medium containing 50 µg/ml ampicillin

Nucleic Acids/Oligonucleotides

Coding sequence of interest

Gels/Loading Buffers

10% Polyacrylamide gel containing SDS <!> (see Appendix 3, pp. 761–763)

1x SDS gel-loading buffer

50 mM Tris-Cl (pH 6.8)

2% (w/v) SDS, electrophoresis grade

0.1% (w/v) bromophenol blue

10% glycerol

Add dithiothreitol from a 1 M stock to a final concentration of 100 mM just before the buffer is used in Step 10.

Centrifuges/Rotors/Tubes

Sorvall SLC-1500 rotor (4°C)

Additional Items

Boiling-water bath
Shaking incubators
Step 1 of this protocol requires the reagents listed in Protocol 8.7.
Step 2 of this protocol requires the reagents listed in Protocol 1.17 or 1.19.
Step 3 of this protocol requires the reagents listed in Protocols 1.23–1.26.
Step 4 of this protocol requires the reagents listed in Protocol 12.3.

Additional Information

Expressing foreign proteins in the pET system	MC3, pp. 15.20–15.21
Regulating T7 promoter activity	MC3, p. 15.24
SDS-polyacrylamide gel electrophoresis	Appendix 3, pp. 756–763
Troubleshooting protein expression from	
an inducible promoter	MC3, pp. 15.18–15.19

METHOD

1. Modify by PCR (Protocol 8.7), or isolate by restriction enzyme digestion, a fragment of DNA carrying 5′- and 3′-restriction enzyme sites compatible with sites in a bacteriophage T7 promoter expression plasmid (e.g., pET vectors).

   Constructs generated by PCR should be sequenced to ensure that no spurious mutations were introduced during the amplification process.

2. Ligate the DNA fragment containing the cDNA/gene of interest into the expression vector (Protocol 1.17 or 1.19).

3. Transform *E. coli* strain BL21 (DE3) or HMS174 (DE3) with aliquots of the ligation reaction. Select for ampicillin-resistant transformants by plating aliquots of the transformation reaction on NZCYM agar plates containing 50 µg/ml ampicillin. Incubate the plates overnight at 37°C.

4. Screen the transformants by colony hybridization and/or restriction enzyme analysis, oligonucleotide hybridization, or direct DNA-sequence analysis (please see Protocol 12.3) of plasmid minipreparations.

5. Inoculate 1-ml cultures (NZCYM medium containing 50 µg/ml ampicillin) with a transformed colony containing positive-control vectors, negative-control vectors, or the recombinant vector. Incubate the cultures overnight at 37°C to obtain a saturated culture.

   Growth at lower temperature sometimes helps to suppress formation of insoluble inclusion bodies. See Dealing with Insoluble Proteins in MC3, pp. 15.9–15.11.

6. Inoculate 5 ml of NZCYM medium containing 50 µg/ml ampicillin in a 50-ml flask with 50 µl of a saturated culture. Incubate the cultures for 2 hours at 37°C.

   It is important to monitor the number of bacteria inoculated into the growth medium, the length of time cells are grown before induction, and the density to which cells are grown after induction.

7. Transfer 1 ml of each culture (zero-time aliquot) to a microfuge tube. Immediately process the zero-time aliquots as described in Steps 9 and 10.

8. Induce the remainder of each culture by adding IPTG to a final concentration of 1.0 mM and continue incubation at 20–37°C with aeration.

   The concentration of IPTG and the temperature of incubation (both before and after induction) can drastically affect the yield and solubility of the expressed polypeptide. If the yield of native protein becomes an issue, a range of temperatures (15–42°C) and of concentrations of IPTG (0.01–5.0 mM) should be explored.

9. At 0.5, 1, 2, and 3 hours after induction, transfer 1 ml of each culture to a microfuge tube, measure the A_{550} in a spectrophotometer, and centrifuge the tubes at maximum speed for 1 minute at room temperature in a microfuge. Remove the supernatants by aspiration.

10. Resuspend each pellet in 100 µl of 1x SDS gel-loading buffer and heat the samples to 100°C for 3 minutes. Centrifuge the tubes at maximum speed for 1 minute at room temperature in a microfuge and store them on ice until all of the samples are collected and ready to load on a gel.

11. Warm the samples to room temperature and load 0.15 OD_{550} units (of original culture) or 40 µg of each suspension on a 10% SDS-polyacrylamide gel.

12. Run the gel at 8–15 V/cm until the bromophenol blue reaches the bottom of the resolving gel.

13. Stain the gel with Coomassie Brilliant Blue or silver, or perform an immunoblot to visualize the induced protein.

14. For large-scale expression and purification of the target protein, inoculate 50 ml of NZCYM containing 50 µg/ml ampicillin in a 250-ml flask with individual colonies of *E. coli* containing the recombinant and control plasmids. Incubate the cultures overnight at 20–37°C.

15. Inoculate 450–500 ml of NZCYM containing 50 µg/ml ampicillin in a 2-liter flask with 5–50 ml of overnight culture of *E. coli* containing the recombinant plasmid. Incubate the culture with shaking at 20–37°C until the culture has reached the mid-log phase of growth ($A_{550} = 0.5$–1.0).

16. Induce expression of the target protein based on the optimal values of IPTG concentration, incubation time, and incubation temperature determined in the previous section.

17. After the induced cells have grown for the proper length of time, harvest the cells by centrifugation at 5000g (5500 rpm in a Sorvall SLC-1500 rotor) for 15 minutes at 4°C and proceed with a purification protocol:

    • Protocol 15.5 if the expressed protein is a fusion with glutathione *S*-transferase

    • Protocol 15.6 if the expressed protein is a fusion with maltose-binding protein

    • Protocol 15.7 if the expressed protein contains a polyhistidine tag

Expression of Cloned Genes in *E. coli* Using the Bacteriophage λ_{pL} Promoter

This protocol describes how to (1) clone sequences encoding open reading frames in plasmids carrying bacteriophage λ_{pL} promoters, (2) optimize expression of target proteins in transformants carrying these recombinants, and (3) scale up production of foreign proteins.

MATERIALS

CAUTION: Please see Appendix 4 for appropriate handling of materials marked with <!>.

Reagents and Solutions

Please see Appendix 1 for components of stock solutions, buffers, and reagents. Dilute stock solutions to the appropriate concentrations.

Coomassie Brilliant Blue stain or silver stain (see Appendix 3, pp. 756–763)

IPTG (isopropyl-β-D-thiogalactosidase) (1 M) <!>

Vectors and Hosts

E. coli harboring either the cIts857 allele (e.g., strain M5219, in which the cIts857 protein can be inactivated by heat shock) or the wild-type allele of the bacteriophage λ cI gene (e.g., strain G1724 [see MC3, p. A3.6, Table A3-2])

*p*L Expression vector (e.g., pAL-781, pAS1, pHUB, pPLc, or pTrxFus)

Positive control (e.g., a *p*L expression vector encoding a fusion protein of defined size)

Media and Antibiotics

LB agar

LB medium

LB medium, heated to 65°C (optional; see Step 13)

L-tryptophan (10 mg/ml)

M9 minimal medium, supplemented with 0.5% (w/v) glucose, 0.2% (w/v) casamino acids, and antibiotics (if needed)

This medium is required when using tryptophan-inducible cI-gene-based *p*L expression vectors.

Nucleic Acids/Oligonucleotides

Coding sequence of interest

Gels/Loading Buffers

10% Polyacrylamide gel containing SDS <!> (see Appendix 3, pp. 761–763)

1x SDS gel-loading buffer

50 mM Tris-Cl (pH 6.8)

2% (w/v) SDS, electrophoresis grade

0.1% (w/v) bromophenol blue

10% glycerol

Add dithiothreitol from a 1 M stock to a final concentration of 100 mM just before the buffer is used in Step 7.

Centrifuges/Rotors/Tubes

Sorvall SLC-1500 rotor (4°C)

Adapted from Chapter 15, Protocol 3, p. 15.25 of MC3.

> **Additional Items**
>
> Boiling-water bath
> Shaking incubators (30°C and 42°C)
> Step 1 of this protocol requires the reagents listed in Protocol 8.7.
> Step 2 of this protocol requires the reagents listed in Protocol 1.17 or 1.19.
> Step 3 of this protocol requires the reagents listed in Protocols 1.23–1.26.
> Step 4 of this protocol requires the reagents listed in Protocol 12.3.
>
> **Additional Information**
>
> | | |
> |---|---|
> | SDS-polyacrylamide gel electrophoresis | Appendix 3, pp. 756–763 |
> | Troubleshooting protein expression from an inducible promoter | MC3, pp. 15.18–15.19 |
> | Tryptophan-inducible expression systems | MC3, p. 15.26 |

METHOD

1. Modify by PCR (Protocol 8.7), or isolate by restriction enzyme digestion, a fragment of DNA carrying 5′- and 3′-restriction enzyme sites compatible with sites in a $_{pL}$ expression vector.

   Constructs generated by PCR should be sequenced to ensure that no spurious mutations were introduced during the amplification reactions.

2. Ligate the DNA fragment containing the cDNA/gene of interest into the expression vector (Protocol 1.17 or 1.19).

3. Transform an *E. coli* strain containing the *cI*ts857 allele or wild-type *cI* gene with aliquots of the ligation reaction. Plate aliquots of the transformation reaction on LB medium containing the appropriate selective antibiotic (usually ampicillin at 50 µg/ml) and incubate the cultures overnight at 30°C (strains harboring the *cI*ts857 allele) or at 37°C (strains harboring a tryptophan-inducible wild-type *cI* gene).

   As a negative control, transform additional cells of the same strain of *E. coli* with the empty expression vectors.

4. Screen the transformants by colony hybridization and/or restriction enzyme analysis, oligonucleotide hybridization, or direct DNA-sequence analysis of plasmid minipreparations (please see Protocol 12.3).

5. Determine the optimum conditions for the induction of target protein expression, which is driven by the down-regulation of the *cI* repressor protein, either by an increase in temperature or by the presence of tryptophan.

   It is important to monitor the number of bacteria inoculated into the growth medium, the length of time cells are grown before induction, and the density to which cells are grown after induction.

   When using a temperature-inducible system

   a. Inoculate 1-ml cultures of LB medium containing the appropriate antibiotics with one or two colonies of *E. coli* (carrying the *cI*ts857 allele) containing the empty expression vector, and one or two colonies containing the recombinant expression vector. Incubate the cultures overnight at 30°C.

      Include as a positive control an expression vector encoding a protein of known size.

   b. Inoculate 10 ml of LB medium containing antibiotic in a 50-ml flask with 50 µl of an overnight culture. Grow the culture to the mid-log phase of growth ($A_{550} = 0.5–1.0$) at 30°C.

   c. Transfer 1 ml of each uninduced culture (zero-time aliquot) to microfuge tubes. Immediately process the zero-time aliquots as described in Steps 6 and 7.

d. Induce the remainder of each culture by shifting the incubation temperature to 40°C. Proceed to Step 6.

When using a tryptophan-inducible system

a. Inoculate 1-ml cultures of supplemented M9 medium with one or two colonies of *E. coli* (carrying the cI wild-type allele) containing the empty expression vector and one or two colonies containing the recombinant expression vector. Incubate the cultures overnight at 37°C.

Include as a positive control an expression vector encoding a protein of known size.

b. Inoculate 10 ml of supplemented M9 medium in a 50-ml flask with 50 μl of the overnight cultures. Grow the cultures to the mid-log phase of growth (A_{550} = 0.5–1.0) at 37°C.

c. Transfer 1 ml of each uninduced culture (zero-time aliquot) to microfuge tubes. Immediately process the zero-time aliquots as described in Steps 6 and 7.

d. Induce the remainder of each culture by adding tryptophan to a final concentration of 100 μg/ml and continue incubation at 37°C.

6. At various time points during the induction period (e.g., 1, 2, 4, and 6 hours), transfer 1 ml of each culture to a microfuge tube, measure the A_{550} in a spectrophotometer, and centrifuge the tubes at maximum speed for 1 minute at room temperature in a microfuge. Remove the supernatants by aspiration.

7. Resuspend each pellet in 100 μl of 1x SDS gel-loading buffer and heat the samples to 100°C for 3 minutes. Centrifuge the tubes at maximum speed for 1 minute at room temperature in a microfuge and store them on ice until all of the samples are collected and ready to load on a gel.

8. Warm the samples to room temperature and load 0.15 OD_{550} units (of original culture) or 40 μg of each suspension on a 10% SDS-polyacrylamide gel.

9. Run the gel at 8–15 V/cm until the bromophenol blue reaches the bottom of the resolving gel.

10. Stain the gel with Coomassie Brilliant Blue or silver, or perform an immunoblot to visualize the induced protein.

11. For large-scale expression and purification of the target protein, inoculate 50 ml of LB containing antibiotic or supplemented M9 medium in a 250-ml flask with a colony of *E. coli* containing the recombinant construct. Incubate the cultures overnight at 30°C or at 37°C, respectively.

12. Inoculate 450 ml of LB plus antibiotic or supplemented M9 medium in a 2-liter flask with 50-ml overnight cultures of *E. coli* containing the recombinant plasmids. Incubate the cultures with agitation at 30°C or 37°C, respectively, until the cultures have reached the mid-log phase of growth (A_{550} = 0.5–1.0). Induce the culture according to either Step 13 or 14, as appropriate.

13. Induce the *E. coli* culture carrying the cIts857 allele by the addition of 500 ml of LB medium heated to 65°C. Incubate the culture at 40°C for the optimum time period determined in the previous section.

14. Induce the *E. coli* culture carrying the cI allele by the addition of tryptophan to 100 μg/ml. Incubate the culture at 37°C for the optimum time period determined in the previous section.

15. After the induced cells have grown for the proper length of time, harvest the cells by centrifugation at 5000*g* (5500 rpm in a Sorvall SLC-1500 rotor) for 15 minutes at 4°C and proceed with a purification protocol:

- Protocol 15.5 if the expressed protein is a fusion with glutathione *S*-transferase

- Protocol 15.6 if the expressed protein is a fusion with maltose-binding protein

- Protocol 15.7 if the expressed protein contains a polyhistidine tag

Expression of Secreted Foreign Proteins Using the Alkaline Phosphatase Promoter (*phoA*) and Signal Sequence

Secretion of a foreign protein into the periplasmic space of *E. coli* is accomplished by fusing the coding sequence downstream from a segment of DNA encoding the *phoA* signal peptide. The signal peptide is removed as the foreign protein is exported from the cytosol, in a cleavage reaction catalyzed by signal peptidase.

MATERIALS

CAUTION: Please see Appendix 4 for appropriate handling of materials marked with <!>.

Reagents and Solutions

Please see Appendix 1 for components of stock solutions, buffers, and reagents. Dilute stock solutions to the appropriate concentrations.
   Coomassie Brilliant Blue stain or silver stain (see Appendix 3, pp. 756–763)

Vectors and Hosts

   E. coli strain
      Essentially any strain of *E. coli* can be used to express genes cloned in *phoA* vectors. However, as with all bacterial expression systems, it may be necessary to try more than one *E. coli* strain grown under varying conditions of temperature.
   pTA1529 (Oka et al. 1985. *Proc. Natl. Acad. Sci.* **82:** 7212) or pBAce (available from C.C. Wang, Department of Pharmaceutical Chemistry, University of California, San Francisco)
   Positive control (e.g., a *phoA* expression vector encoding a fusion protein of defined size)

Media and Antibiotics

   Induction medium
      1x MOPS salts (for recipe, see below)
      0.2% (w/v) glucose
      0.2% casamino acids
      10 µg/ml adenine
      0.5 µg/ml thiamine
      0.1 M neutral phosphate buffer (for recipe, see below)
      Combine ingredients to a final volume of 1 liter in H_2O. Sterilize by filtration and store at 4°C.
   LB agar plates containing 50 µg/ml ampicillin
   LB medium containing 50 µg/ml ampicillin
   Micronutrients (used in 10x MOPS buffer)
      37 mg $(NH_4)_6Mo_7O_{24} \cdot 4H_2O$
      84 mg $CoCl_2$
      158 mg $MnCl_2 \cdot 4H_2O$
      247 mg H_3BO_3
      25 mg $CuSO_4$
      18 mg $ZnSO_4$
      Dissolve in a final volume of 10 ml H_2O. Sterilize by filtration and store at 4°C.
   10x MOPS salts (used in induction medium)
      400 mM 3-(*N*-morpholino)propane sulfonic acid (pH 7.4) (MOPS) <!>
      40 mM tricine (pH 7.4)
      0.1 mM $FeSO_4 \cdot 7H_2O$

95 mM NH$_4$Cl
2.8 mM K$_2$SO$_4$
5 μM CaCl$_2$·2H$_2$O
5.3 mM MgCl$_2$·6H$_2$O
0.5 M NaCl
Dissolve in a final volume of 1 liter of H$_2$O. Sterilize by filtration and store at 4°C. Add 10 ml of micronutrients per liter before use.

Neutral phosphate buffer (1 M, used in induction medium)
An equimolar mix of 1 M Na$_2$HPO$_4$ and 1 M NaH$_2$PO$_4$

Nucleic Acids/Oligonucleotides

Coding sequence of interest

Gels/Loading Buffers

10% Polyacrylamide gel containing SDS <!> (see Appendix 3, pp. 761–763)
1x SDS gel-loading buffer
50 mM Tris-Cl (pH 6.8)
2% (w/v) SDS, electrophoresis grade
0.1% (w/v) bromophenol blue
10% glycerol
Add dithiothreitol from a 1 M stock to a final concentration of 100 mM just before the buffer is used in Step 8.

Centrifuges/Rotors/Tubes

Sorvall SLC-1500 and SS-34 rotors (4°C)

Additional Items

Boiling-water bath
Shaking incubators
Step 1 of this protocol requires the reagents listed in Protocol 8.7.
Step 2 of this protocol requires the reagents listed in Protocol 1.17 or 1.19.
Step 3 of this protocol requires the reagents listed in Protocols 1.23–1.26.
Step 4 of this protocol requires the reagents listed in Protocol 12.3.

Additional Information

Additional protocol (Subcellular Localization of PhoA Fusion
  Proteins)   MC3, p. 15.35
SDS-polyacrylamide gel electrophoresis   Appendix 3, pp. 756–763

METHOD

1. Modify by PCR (Protocol 8.7), or isolate by restriction enzyme digestion, a fragment of DNA carrying 5'- and 3'-restriction enzyme sites compatible with the polycloning sites in pTA1529 or pBAce.

   Constructs generated by PCR should be sequenced to ensure that no spurious mutations were introduced during amplification.

2. Ligate the DNA fragment containing the cDNA/gene of interest into the expression vector (Protocol 1.17 or 1.19).

3. Transform an appropriate *E. coli* strain with aliquots of the ligation reaction. Plate aliquots of the transformation reaction on LB agar containing 50 μg/ml ampicillin and incubate the cultures overnight at 37°C.

   As a negative control, transform additional cells of the same strain of *E. coli* with the empty expression vector.

4. Screen the transformants by colony hybridization and/or restriction enzyme analysis, oligonucleotide hybridization, or direct DNA-sequence analysis (please see Protocol 12.3) of plasmid minipreparations.

5. Inoculate 1-ml cultures of LB medium containing 50 μg/ml ampicillin with one to two colonies containing the empty expression vector and one to two colonies containing the recombinant and control plasmids. Incubate the cultures overnight at 37°C.

   Include as a positive control an expression vector encoding a protein of known size.

6. Inoculate 5 ml of induction medium containing 50 μg/ml ampicillin in a 50-ml flask with 50 μl of an overnight culture. Incubate the cultures with shaking at 20–37°C.

   It is important to monitor the number of bacteria inoculated into the growth medium, the length of time cells are grown before induction, and the density to which cells are grown after induction.

7. At various time points after inoculation (e.g., 0, 6, 12, 18, and 24 hours), transfer 1 ml of each culture to a microfuge tube, measure the A_{550} in a spectrophotometer, and centrifuge the tubes at maximum speed for 1 minute at room temperature in a microfuge. Remove the supernatants by aspiration.

8. Resuspend each pellet in 100 μl of 1× SDS gel-loading buffer and heat the samples to 100°C for 3 minutes. Centrifuge the tubes at maximum speed for 1 minute at room temperature in a microfuge and store them on ice until all of the samples are collected and ready to load on a gel.

9. Warm the samples to room temperature and load 0.15 OD_{550} units (of original culture) or 40 μg of each suspension on a 10% SDS-polyacrylamide gel.

10. Run the gel at 8–15 V/cm until the bromophenol blue reaches the bottom of the resolving gel.

11. Stain the gel with Coomassie Brilliant Blue or silver, or perform an immunoblot to visualize the induced protein.

12. For large-scale expression and purification of the target protein, inoculate 25 ml of LB medium containing 50 μg/ml ampicillin in a 125-ml flask with a colony of *E. coli* containing the recombinant *phoA* construct. Incubate the culture overnight with agitation at 37°C.

13. Collect the cells by centrifugation at 5000*g* (6500 rpm in a Sorvall SS-34 rotor) for 15 minutes. Resuspend the cell pellet in 25 ml of induction medium and collect the cells again by centrifugation.

14. Resuspend the washed cells in 2.5 ml of fresh induction medium and inoculate the cells into 500 ml of induction medium in a 2-liter flask. Incubate the large-scale culture at the optimum temperature and for the optimum time determined in the previous section.

15. After the cells have grown for the proper length of time, harvest the cells by centrifugation at 4000*g* (5000 rpm in a Sorvall SLC-1500 rotor) for 15 minutes at 4°C.

Purification of Fusion Proteins by Affinity Chromatography on Glutathione Agarose

Recombinant proteins, expressed in pGEX vectors, are fused to glutathione *S*-transferase (GST) and can be purified to near homogeneity by affinity chromatography on glutathione-agarose beads. Bound GST-fusion proteins are readily displaced from the column by elution with buffers containing free glutathione.

MATERIALS

CAUTION: Please see Appendix 4 for appropriate handling of materials marked with <!>.

Reagents and Solutions

Please see Appendix 1 for components of stock solutions, buffers, and reagents. Dilute stock solutions to the appropriate concentrations.

Coomassie Brilliant Blue stain or silver stain (see Appendix 3, pp. 756–763)

Dithiothreitol (DTT) (1 M)

Glutathione elution buffer

> 10 mM reduced glutathione
> 50 mM Tris-HCl (pH 8.0)

Phosphate-buffered saline (PBS)

Triton X-100 (0.2% v/v)

Vectors and Hosts

E. coli cells expressing a GST-fusion protein from a pGEX vector

Enzymes and Buffers

DNase I (5 mg/ml in 20 mM Tris-Cl [pH 7.8])

Lysozyme

Pancreatic RNase (5 mg/ml in H_2O)

Protease appropriate for cleavage of fusion protein (e.g., thrombin, enterokinase, or Factor Xa [optional; see Step 14])

Gels/Loading Buffers

10% Polyacrylamide gel containing SDS <!> (see Appendix 3, pp. 761–763)

Centrifuges/Rotors/Tubes

Sorvall SS-34 rotor (4°C)

Additional Items

Boiling-water bath

Glutathione-agarose beads (available from Sigma-Aldrich and other commercial suppliers)

> About 1 ml of packed beads is required for each liter of *E. coli* culture (see MC3, p. 15.36).

Hypodermic needles (18, 22, and 25 gauge)

Rocking platform

Shaking incubators

Step 1 of this protocol requires the reagents listed in Protocol 8.7.

Step 2 of this protocol requires the reagents listed in Protocol 1.17 or 1.19.
Step 3 of this protocol requires the reagents listed in Protocols 1.23–1.26.
Step 4 of this protocol requires the reagents listed in Protocol 12.3.

Additional Information

Cleavage of fusion proteins	MC3, pp. 15.6–15.8
Purification of fusion proteins	MC3, pp. 15.4–15.6
SDS-polyacrylamide gel electrophoresis	Appendix 3, pp. 756–763

METHOD

1. Gently invert the container of the glutathione-agarose beads to mix the slurry.

2. Transfer an aliquot of the slurry to a 15-ml polypropylene tube (~0.5–1.0 ml of the slurry will be needed for each 100 ml of the original bacterial culture).

3. Centrifuge the tube at 500g (2100 rpm in a Sorvall SS-34 rotor) for 5 minutes at 4°C. Carefully remove and discard the supernatant.

4. Add 10 bed volumes of cold PBS to the beads and mix the slurry by inverting the tube several times. Centrifuge the tube at 500g (2100 rpm in a Sorvall SS-34 rotor) for 5 minutes at 4°C. Carefully remove and discard the supernatant.

5. Add 1 ml of cold PBS per milliliter of resin to make a 50% slurry. Mix the slurry by inverting the tube several times. Keep the suspension on ice until the cell extract has been prepared.

6. Resuspend the cell pellet (e.g., from Protocol 15.1, Step 17) in 4 ml of PBS per 100 ml of cell culture.

7. Add lysozyme to a final concentration of 1 mg/ml and incubate the cell suspension on ice for 30 minutes.

8. Add 10 ml of 0.2% Triton X-100. Use a syringe to inject the solution forcibly into the viscous cell lysate. Shake the tube vigorously several times to mix the solution of detergent and cell lysate. Add DNase and RNase to the tube, each to a final concentration of 5 μg/ml, and continue the incubation with rocking for 10 minutes at 4°C. Remove the insoluble debris by centrifugation at 3000g (5000 rpm in a Sorvall SS-34 rotor) for 30 minutes at 4°C. Collect the supernatant (cell lysate) in a fresh tube. Add dithiothreitol to a final concentration of 1 mM.

   It may be necessary to pass the supernatant through a 0.45-μm filter to prevent clogging of the resin during purification of the GST-fusion protein.

9. Combine the cell lysate with an appropriate amount of the 50% slurry of glutathione-agarose beads in PBS. Use 0.3–1.0 ml of slurry for each 100 ml of bacterial culture used to make the protein extract. Shake the mixture gently for 30 minutes at room temperature.

10. Centrifuge the mixture at 500g (2100 rpm in a Sorvall SS-34 rotor) for 5 minutes at 4°C. Carefully remove the supernatant. Save a small amount of the supernatant to analyze by SDS-polyacrylamide gel electrophoresis.

11. Wash unbound proteins from the resin by adding 10 bed volumes of PBS to the pellet and mix by inverting the tube several times.

12. Centrifuge at 500g (2100 rpm in a Sorvall SS-34 rotor) for 5 minutes at 4°C. Carefully remove the supernatant. Save a small amount of the supernatant to analyze by SDS-polyacrylamide gel electrophoresis.

13. Repeat Steps 11 and 12 twice more.

14. Bound GST-fusion protein may be eluted from the beads using glutathione elution buffer. Alternatively, GST-fusion proteins may be cleaved while still bound to the gel with an appropriate protease such as thrombin, enterokinase, or Factor Xa, liberating the protein of interest from the GST moiety.

Elution of the fusion protein using glutathione

a. Elute the bound protein from the beads by adding 1 bed volume of glutathione elution buffer to the pellet. Incubate the tube with gentle agitation for 10 minutes at room temperature.

b. Centrifuge the tube as in Step 12. Transfer the supernatant (which contains the eluted fusion protein) to a fresh tube.

c. Repeat Steps a and b twice more, pooling all three supernatants.

> Depending on the particular fusion protein being purified, a significant amount of protein may remain bound to the gel following the elution steps. The volume of elution buffer and the elution times may vary among fusion proteins. Additional elutions may be required. Monitor the eluates for GST protein by SDS-polyacrylamide gel electrophoresis.

Proteolytic cleavage of the target protein from the bound GST moiety

a. Add a protease such as thrombin, enterokinase, or Factor Xa (as appropriate for the cleavage site within the fusion protein) to the beads. Use 50 units of the appropriate protease in 1 ml of PBS for each milliliter of resin volume. Mix the solution by inverting the tube several times and incubate the mixture with shaking for 2–16 hours at room temperature. The exact time should be determined empirically using a series of small-scale trial reactions.

b. Centrifuge the tube at 500g (2100 rpm in a Sorvall SS-34 rotor) for 5 minutes at 4°C. Carefully transfer the supernatant to a fresh tube.

15. Analyze the protein profile of each step (cell extract, washes, and elution) on a 10% SDS-polyacrylamide gel.

Purification of Maltose-binding Fusion Proteins by Affinity Chromatography on Amylose Resin

Foreign proteins fused to maltose-binding protein can be readily purified to near homogeneity by affinity chromatography on resins containing cross-linked polysaccharides such as amylose.

MATERIALS

CAUTION: Please see Appendix 4 for appropriate handling of materials marked with <!>.

Reagents and Solutions

Please see Appendix 1 for components of stock solutions, buffers, and reagents. Dilute stock solutions to the appropriate concentrations.

Cell lysis buffer
30 mM Tris-Cl (pH 7.1)
0.1 mM EDTA (pH 8.0)
20% (w/v) sucrose
Cell wash buffer
10 mM Tris-Cl (pH 7.1)
30 mM NaCl
Column elution buffer
10 mM Tris-Cl (pH 7.5)
10 mM maltose
Column wash buffer
10 mM Tris-Cl (pH 7.1)
1 M NaCl
Dithiothreitol (1 M)
$MgCl_2$ (0.1 mM)
PMSF (phenylmethylsulfonyl fluoride) (17.4 mg/ml in isopropanol at –20°C) <!>
An alternative to PMSF, Pefabloc SC (Roche Applied Science), is nontoxic and stable in buffered aqueous solutions.
Tris-Cl (1 M, pH 7.1)

Vectors and Hosts

E. coli cells expressing a maltose-binding protein (MBP) fusion protein from a vector such as pMAL-c2 or pMAL-p2

Enzymes and Buffers

DNase I (5 mg/ml in 20 mM Tris-Cl [pH 7.8])
Lysozyme
Pancreatic RNase (5 mg/ml in H_2O)
Protease for cleavage of fusion protein (e.g., thrombin, enterokinase, or Factor Xa [optional; see Step 7])

Gels/Loading Buffers

10% Polyacrylamide gel containing SDS <!> (see Appendix 3, pp. 761–763)

Centrifuges/Rotors/Tubes

Beckman Ti 60 rotor (4°C)
Sorvall SS-34 and SLC-1500 rotors (4°C)

Additional Items

Amylose-agarose beads for affinity chromatography (New England BioLabs and other
    commercial suppliers)
    Capacity of the beads is approximately 3 mg of maltose-binding protein/ml.
Nalgene filters, disposable, 100-ml capacity, 0.45-µm pore size
Rocking platform
Sonicator, equipped with microtip (optional; see Step 3)
Step 1 of this protocol requires the reagents listed in Protocol 8.7.
Step 2 of this protocol requires the reagents listed in Protocol 1.17 or 1.19.
Step 3 of this protocol requires the reagents listed in Protocols 1.23–1.26.
Step 4 of this protocol requires the reagents listed in Protocol 12.3.

Additional Information

Cleavage of fusion proteins	MC3, pp. 15.6–15.8
Maltose-binding proteins	MC3, p. 15.40
Proteins secreted into periplasmic space of *E. coli*	MC3, p. 15.43
Purification of fusion proteins	MC3, pp. 15.4–15.6
SDS-polyacrylamide gel electrophoresis	Appendix 3, pp. 756–763

METHOD

1. Pour a 3 x 6-cm column of amylose agarose. Equilibrate in 10 mM Tris-Cl (pH 7.1) at 4°C.

2. Resuspend the cell pellet in 1/10 original culture volume (typically 50 ml) of ice-cold cell wash buffer. Collect the cells by centrifugation at 5000*g* (5500 rpm in a Sorvall SLC-1500 rotor) for 15 minutes at 4°C and again resuspend the pellet in 1/10 original culture volume (typically 50 ml) of ice-cold cell wash buffer.

3. Prepare the cell lysate.

   If an MBP vector without a signal sequence was used (pMAL-c2)

   a. Lyse cells by sonication with a microtip sonicator, using three bursts of 10 seconds each. Use a power setting of approximately 30 W and keep the cells cold (0°C) during sonication.

   b. Add PMSF to a final concentration of 1 mM and clarify the solution by centrifugation at 87,000*g* (35,000 rpm in a Beckman Ti 60 rotor) for 30 minutes at 4°C.
      Proceed to purify the fusion protein from the supernatant by amylose-agarose chromatography as described in Step 4.

   If an Mbp vector with a signal sequence was used (pMAL-p2)

   a. Collect the cells by centrifugation at 5000*g* (5500 rpm in a Sorvall SLC-1500 rotor) for 15 minutes at 4°C and resuspend the cell pellet in 1/20 original culture volume (typically 25 ml) of ice-cold cell wash buffer.

   b. Add PMSF to 1 mM. Stir the cell suspension for 15 minutes at room temperature. Collect the cells by centrifugation at 17,200*g* (12,000 rpm in a Sorvall SS-34 rotor) for 10 minutes at 4°C.

c. Spread the cell pellet around the sides of the centrifuge tube. Add ice-cold 0.1 mM MgCl$_2$ solution (100 ml/liter original culture volume) and stir the suspension for 10 minutes at 4°C.

d. Centrifuge the shocked cells at 17,200g (12,000 rpm in a Sorvall SS-34 rotor) for 10 minutes at 4°C. Add 10 mM Tris-Cl (pH 7.1) until the pH of the supernatant is 7.1.

e. Filter the supernatant through a 100-ml disposable Nalgene filter (0.45-µm nitrocellulose membrane) and dialyze the filtered solution against 100 volumes 10 mM Tris-Cl (pH 7.1) at 4°C.

> Proceed to purify the fusion protein from the supernatant by amylose-agarose chromatography as described in Step 4.

4. Pour the supernatant from Step 3 over the column. Rinse the column with 100 ml of 10 mM Tris-Cl (pH 7.1). Pass 100 ml of column wash buffer through the column.

5. Elute the bound fusion protein with 50 ml of column elution buffer and collect 1-ml fractions.

6. Analyze aliquots of the collected fractions by SDS-polyacrylamide gel electrophoresis to determine the location of the fusion protein in the series of elution fractions. Pool the fractions containing the fusion protein and store them at −70°C.

7. Use the appropriate protease to cleave the fusion protein from the MBP moiety.

a. Add thrombin, enterokinase, or Factor Xa (as appropriate for the cleavage site within the fusion protein) to the beads. Use 50 units of the appropriate protease in 1 ml of PBS for each milliliter of resin volume. Mix the solution by inverting the tube several times and incubate the mixture with shaking for 2–16 hours at room temperature.

b. Centrifuge the tube at 500g (2100 rpm in a Sorvall SS-34 rotor) for 5 minutes at 4°C. Carefully transfer the supernatant to a fresh tube.

> The protein of interest can be separated from the protease by conventional chromatography or by SDS-polyacrylamide gel electrophoresis.

Purification of Histidine-tagged Proteins by Immobilized Ni^{++} Absorption Chromatography

Recombinant proteins engineered to have a polyhistidine tail at either the carboxyl or amino terminus can easily be purified in one step by affinity chromatography on a resin carrying chelated nickel ions. Chromatography can be carried out in column or batch formats. After unbound proteins are washed away, the target protein is eluted using acid or imidazole, which is preferred because it typically preserves the antigenic and functional features of the protein.

MATERIALS

CAUTION: Please see Appendix 4 for appropriate handling of materials marked with <!>.

Reagents and Solutions

Please see Appendix 1 for components of stock solutions, buffers, and reagents. Dilute stock solutions to the appropriate concentrations.

Binding buffer (pH 7.8))
20 mM sodium phosphate (pH 6.0) (see Appendix 1, p. 695)
500 mM NaCl

Imidazole elution buffer (pH 6.0)
20 mM sodium phosphate (pH 6.0) (see Appendix 1, p. 695)
500 mM NaCl
Create a series of four elution buffers containing imidazole at concentrations of 10, 50, 100, and 150 mM by adding the appropriate amount of 3 M imidazole to the wash buffer.

PMSF (phenylmethylsulfonyl fluoride) (17.4 mg/ml in isopropanol at –20°C) <!>
An alternative to PMSF, Pefabloc SC (Roche Applied Science), is nontoxic and stable in buffered aqueous solutions.

Triton X-100 (10% v/v)

Wash buffer (pH 6.0)
20 mM sodium phosphate (pH 6.0) (see Appendix 1, p. 695)
500 mM NaCl

Vectors and Hosts

E. coli cells expressing a histidine-tagged protein from one of the pET series of vectors, e.g., pET-14–pET-16 or pET-19–pET-44 (Novagen)

Enzymes and Buffers

DNase I (5 mg/ml in 20 mM Tris-Cl [pH 7.8])
Lysozyme
Pancreatic RNase (5 mg/ml in H$_2$O)

Gels/Loading Buffers

10% Polyacrylamide gel containing SDS <!> (see Appendix 3, pp. 761–763)

Centrifuges/Rotors/Tubes

Sorvall SS-34 rotor (4°C)

Additional Items

Chromatography column, glass or polypropylene

Metal-chelate affinity support, e.g., ProBond (Invitrogen), which is iminodiacetic acid
(IDA) coupled to a highly cross-linked 6% agarose resin
   The resin is supplied charged with Ni^{++} and is available both as an individual item and in kit
   form. ProBond can bind ~5 mg of recombinant protein per 1 ml of resin. Similar resins are
   available from many other suppliers.
Nalgene filters, disposable, 100-ml capacity, 0.45-μm pore size (optional; see Step 8)
Rocking platform (4°C)

Additional Information

Additional protocol (Regeneration of NTA-Ni^{2+}-Agarose)	MC3, p. 15.48
Alternative protocol (Elution of Polyhistidine-tagged Proteins from Metal Affinity Columns Using Decreasing pH)	MC3, p. 15.47
Ni^{++} absorption chromatography	MC3, pp. 15.44–15.45
Purification of fusion proteins	MC3, pp. 15.4–15.6
SDS-polyacrylamide gel electrophoresis	Appendix 3, pp. 756–763

METHOD

1. Gently invert the bottle of Ni^{++}-charged chromatography resin to mix the slurry and transfer 2 ml to a small polypropylene or glass column. Allow the resin to pack under gravity flow.

2. Wash the resin with 3 column volumes of sterile H_2O.

3. Equilibrate the resin with 3 column volumes of binding buffer (pH 7.8). The column is now ready for use in Step 9.

4. Resuspend the bacterial pellet in 4 ml of binding buffer (pH 7.8) per 100 ml of cell culture.

5. Add lysozyme to a final concentration of 1 mg/ml and incubate the cell suspension on ice for 30 minutes.

   IMPORTANT: Protease inhibitors such as PMSF or Prefabloc SC may be added, but do not add EDTA or other chelators, which will remove the Ni^{++} from the affinity resin, destroying its ability to bind histidine.

6. Incubate the mixture on a rocking platform for 10 minutes at 4°C.

7. Add Triton X-100, DNase, and RNase to the tube to final concentrations of 1%, 5 μg/ml, and 5 μg/ml, respectively, and continue the incubation with rocking for another 10 minutes at 4°C.

8. Remove the insoluble debris by centrifugation at 3000g (5000 rpm in a Sorvall SS-34 rotor) for 30 minutes at 4°C. Collect the supernatant (cell lysate) in a fresh tube.

   It may be necessary to pass the supernatant through a 0.45-μm filter to prevent clogging of the resin during purification of the GST-fusion protein.

9. Allow the binding buffer above the resin to drain to the top of the column.

10. Immediately load the cell lysate (Step 8) onto the column. Adjust the flow rate to 10 column volumes per hour.

    Ni^{++} affinity resins will typically bind approximately 8–12 mg of protein per milliliter of resin. The amount of polyhistidine-tagged protein produced in *E. coli* will vary depending on the target protein, but it is typically in the range of 1–10 mg of protein per 100 ml of cell culture.

11. Wash the column with 6 column volumes of binding buffer (pH 7.8).

12. Wash the column with 4 volumes of wash buffer (pH 6.0). Continue washing the column until the A_{280} of the flowthrough is <0.01.

13. Elute the bound protein with 6 volumes of 10 mM imidazole elution buffer. Collect 1-ml fractions from the column and monitor the A_{280} of each fraction.

14. Repeat Step 13 using imidazole elution buffers containing increasing concentrations of imidazole (i.e., 50, 100, and 150 mM).

    Alternatively, elute the protein using a continuous gradient of increasing imidazole concentration from 10 to 100 mM. Most His-tagged proteins will elute between 50 and 100 mM imidazole.

15. Assay the fractions of interest for the presence of the polyhistidine-tagged protein by analyzing 20-μl aliquots by electrophoresis through a 10% SDS-polyacrylamide gel.

    Many commercial suppliers sell antisera specific for His(6) sequences. These antisera can be used in western blotting to identify proteins carrying His tags.

Purification of Expressed Proteins from Inclusion Bodies

The expression of foreign proteins at high levels in *E. coli* often results in the formation of inclusion bodies composed of insoluble aggregates of the expressed protein. The inclusion bodies are recovered from bacterial lysates by centrifugation and are washed with Triton X-100 and EDTA to remove as much bacterial protein as possible from the aggregated foreign protein.

To obtain soluble protein, the washed inclusion bodies are dissolved in denaturing agents and the released protein is then refolded by gradual removal of the denaturing reagents by dilution or dialysis. Whereas the procedure given here has been used to solubilize inclusion bodies, each protein may require a slightly different procedure, which must be determined empirically.

MATERIALS

CAUTION: Please see Appendix 4 for appropriate handling of materials marked with <!>.

Reagents and Solutions

Please see Appendix 1 for components of stock solutions, buffers, and reagents. Dilute stock solutions to the appropriate concentrations.

Cell lysis buffer I
> 50 mM Tris-Cl (pH 8.0)
> 1 mM EDTA (pH 8.0)
> 100 mM NaCl

Cell lysis buffer II
> 50 mM Tris-Cl (pH 8.0)
> 10 mM EDTA (pH 8.0)
> 100 mM NaCl
> 0.5% (v/v) Triton X-100

Deoxycholic acid, protein grade

HCl (12 M) (concentrated HCl) <!>

Inclusion body solubilization buffer I (prepare just before use)
> 50 mM Tris-Cl (pH 8.0)
> 1 mM EDTA (pH 8.0)
> 100 mM NaCl
> 8 M urea
> 0.1 M PMSF or Pefabloc SC

Inclusion body solubilization buffer II
> 50 mM KH_2PO_4 (pH 10.7)
> 1 mM EDTA (pH 8.0)
> 50 mM NaCl

KOH (10 N) <!>

PMSF (phenylmethylsulfonyl fluoride) (17.4 mg/ml in isopropanol at –20°C)
> An alternative to PMSF, Pefabloc SC (Roche Applied Science), is nontoxic and stable in buffered aqueous solutions.

Tris-Cl (0.1 M, pH 8.5) with urea
> For use in Method 2 of Step 7. Prepare 0.1 M Tris-Cl (pH 8.5) with increasing concentrations of urea (e.g., 0.5, 1, 2, and 5 M). Because urea decomposes in aqueous solutions, make the solution fresh from solid urea and use immediately.

Vectors and Hosts

E. coli cells (1-liter culture) expressing a protein of interest as an inclusion body

Enzymes and Buffers

DNase I (1 mg/ml in 20 mM Tris-Cl [pH 7.8])
Lysozyme

Gels/Loading Buffers

10% Polyacrylamide gel containing SDS <!> (see Appendix 3, pp. 761–763)
1x SDS gel-loading buffer
 50 mM Tris-Cl (pH 6.8)
 2% (w/v) SDS, electrophoresis grade
 0.1% (w/v) bromophenol blue
 10% glycerol
 Add dithiothreitol from a 1 M stock to a final concentration of 100 mM just before the buffer
 is used in Step 13.
2x SDS gel-loading buffer
 100 mM Tris-Cl (pH 6.8)
 4% (w/v) SDS, electrophoresis grade
 0.2% (w/v) bromophenol blue
 20% glycerol
 Add dithiothreitol from a 1 M stock to a final concentration of 100 mM just before the buffer
 is used in Steps 7 and 14.

Centrifuges/Rotors/Tubes

Sorvall SLC-1500 rotor (4°C)

Additional Items

Glass rod (polished)
pH paper

Additional Information

Additional protocol (Refolding Solubilized Proteins Recovered from Inclusion Bodies)	MC3, p. 15.53
Chaotropic reagents	MC3, p. 15.60
Purification of fusion proteins	MC3, pp. 15.4–15.7

METHOD

1. Centrifuge 1 liter of the cell culture of *E. coli* expressing the protein of interest at 5000*g* (5500 rpm in a Sorvall SLC-1500 rotor) for 15 minutes at 4°C in preweighed centrifuge bottles.
    IMPORTANT: Perform Steps 2–4 at 4°C.

2. Remove the supernatant and determine the weight of the *E. coli* pellet. For each gram (wet weight) of *E. coli*, add 3 ml of cell lysis buffer I. Resuspend the pellet by gentle vortexing or by stirring with a polished glass rod.

3. For each gram of *E. coli*, add 4 µl of 100 mM PMSF or Pefabloc SC and then 80 µl of 10 mg/ml lysozyme. Stir the suspension for 20 minutes.

4. Stirring continuously, add 4 mg of deoxycholic acid per gram of *E. coli*.

5. Store the suspension at 37°C and stir it occasionally with a glass rod. When the lysate becomes viscous, add 20 µl of 1 mg/ml DNase I per gram of *E. coli*.

6. Store the lysate at room temperature until it is no longer viscous (~30 minutes).

7. Purify and wash the inclusion bodies using one of the following two methods.

Method 1: Recover inclusion bodies using Triton X-100

a. Centrifuge the cell lysate at maximum speed for 15 minutes at 4°C in a microfuge.

b. Decant the supernatant. Resuspend the pellet in 9 volumes of cell lysis buffer II at 4°C.

c. Store the suspension for 5 minutes at room temperature.

d. Centrifuge the tube at maximum speed for 15 minutes at 4°C in a microfuge.

e. Decant the supernatant and set it aside for the next step. Resuspend the pellet in 100 µl of H_2O.

f. Remove 10-µl samples of the supernatant and of the resuspended pellet. Mix each sample with 10 µl of 2x SDS gel-loading buffer and analyze the samples by SDS-poly-acrylamide gel electrophoresis to determine which fraction contains the protein of interest.

g. If necessary, proceed with Step 8 to solubilize the inclusion bodies.

Method 2: Recover inclusion bodies using urea

a. Centrifuge the cell lysate at maximum speed for 15 minutes at 4°C in a microfuge.
   IMPORTANT: Perform Steps b, d, and f at 4°C.
   The following steps involve washing and solubilization of inclusion bodies with buffers containing different concentrations of urea.

b. Decant the supernatant. Resuspend the pellet in 1 ml of H_2O per gram of *E. coli*. Transfer 100-µl aliquots to four microfuge tubes and store the remainder of the suspension at 4°C.

c. Centrifuge the 100-µl aliquots at maximum speed for 15 minutes at 4°C in a microfuge.

d. Discard the supernatants. Resuspend each pellet in 100 µl of 0.1 M Tris-Cl (pH 8.5) containing a different concentration of urea (e.g., 0.5, 1, 2, and 5 M).

e. Centrifuge the tubes at maximum speed for 15 minutes at 4°C in a microfuge.

f. Decant the supernatants and set them aside for the next step. Resuspend each pellet in 100 µl of H_2O.

g. Remove 10-µl samples of each supernatant and each resuspended pellet. Mix each sample and resuspended pellet with 10 µl of 2x SDS gel-loading buffer and analyze by SDS-polyacrylamide gel electrophoresis to determine which concentration of urea yields the best recovery of the inclusion bodies.

h. Use the appropriate concentration of urea, determined in Step g, to wash the remaining pellet (from Step b) as described in this method.

i. If necessary, proceed with Step 8 to solubilize the inclusion bodies.

8. Centrifuge the appropriate resuspended pellets from Step 7 at maximum speed for 15 minutes at 4°C in a microfuge and suspend them in 100 µl of inclusion-body solubilization buffer I containing 0.1 mM PMSF or Pefabloc SC (freshly added).

9. Store the solution for 1 hour at room temperature.

10. Add this solution to 9 volumes of inclusion-body solubilization buffer II and incubate the mixture for 30 minutes at room temperature. Check that the pH is maintained at 10.7 by spotting small aliquots onto pH paper. If necessary, readjust the pH to 10.7 with 10 N KOH.

11. Adjust the pH of the solution to 8.0 with 12 M HCl and store the adjusted solution for at least 30 minutes at room temperature.

12. Centrifuge the solution at maximum speed for 15 minutes at room temperature in a microfuge.

13. Decant the supernatant and set it aside for the next step. Resuspend the pellet in 100 μl of 1x SDS gel-loading buffer.

14. Remove 10-μl samples of the supernatant and resuspended pellet. Mix the supernatant sample with 10 μl of 2x SDS gel-loading buffer. Analyze both samples by SDS-polyacrylamide gel electrophoresis to determine the degree of solubilization.

CHAPTER 16

Introducing Cloned Genes into Cultured Mammalian Cells

BACKGROUND INFORMATION

Background information found in *Molecular Cloning: A Laboratory Manual*, 3rd edition (hereafter MC3) unless otherwise indicated.

Commercial kits and reagents for transfection	MC3, p. 16.5
Review of transfection methods	MC3, pp. 16.2–16.6

INTRODUCTION

The last ten years have brought great advances in methods to introduce cloned DNA into cultured eukaryotic cells: A wide range of cell types can now be transfected efficiently by the traditional biochemical method of calcium phosphate coprecipitation, a variety of improved liposomal agents have become available, and physical methods such as electroporation and biolistic particle delivery have been used successfully with many cell lines that are refractory to transfection by other means. However, no single method works flawlessly with all types of cultured cells in all circumstances. The protocols in this chapter should therefore be viewed as a starting point for systematic optimization of transfection mediated by chemical and physical agents. Once a positive signal has been obtained from a transfected plasmid carrying a standard reporter gene, optimal conditions for transfection can be established by systematic variation of parameters such as initial cell density, amount and purity of the DNA, media and serum, and time and intensity of exposure to the transfection conditions. Table 16-1 summarizes the characteristics and uses of seven popular transfection methods used in Chapter 16.

TABLE 16-1 Transfection Methods

Method	Expression Transient	Stable	Cell Toxicity	Cell Types	Comments
Lipid-mediated Protocol 1	yes	yes	varies	adherent cells, primary cell lines, suspension cultures	Cationic lipids are used to create artificial membrane vesicles (liposomes) that bind DNA molecules. The resulting stable cationic complexes adhere to and fuse with the negatively charged cell membrane.
Calcium-phosphate-mediated Protocols 2 and 3	yes	yes	no	adherent cells (CHO, 293); suspension cultures	Calcium phosphate forms an insoluble coprecipitate with DNA, which attaches to the cell surface and is absorbed by endocytosis.
DEAE-dextran-mediated ing Protocol 4	yes	no	yes	BSC-1, CV-1, and COS	Positively charged DEAE-dextran binds to negatively charged phosphate groups of DNA, form- aggregates that bind to the negatively charged plasma membrane. Uptake into the cell is believed to be mediated by endocytosis, which is potentiated by osmotic shock.
Electroporation Protocol 5	yes	yes	no	many	Application of brief high-voltage electric pulses to a variety of mammalian and plant cells leads to the formation of nanometer-sized pores in the plasma membrane. DNA is taken directly into the cell cytoplasm either through these pores or as a consequence of the redistribution of membrane components that accompanies the closure of the pores. Electroporation can be extremely efficient and may be used for both transient and stable transfection.
Biolistics Protocol 6	yes	yes	no	primary cell lines; tissues, organs, plant cells	Small particles of tungsten or gold are used to bind DNA, in preparation for delivery into cells, tissues, or organelles by a particle accelerator system. This process has been variously called the microprojectile bombardment method, the gene gun method, and the particle acceleration method. Biolistics is used chiefly to transform cell types that are impossible or very difficult to transform by other methods.
Polybrene Protocol 7	yes	yes	varies	CHO and keratinocyte	The polycation Polybrene allows the efficient and stable introduction of low-molecular-weight DNAs (e.g., plasmid DNAs) into cell lines that are relatively resistant to transfection by other methods. The uptake of DNA is enhanced by osmotic shock and dimethylsulfoxide (DMSO), which may permeabilize the cell membrane.

DNA Transfection Mediated by Lipofection

A great many lipofecting agents are available commercially, but none are suitable for all circumstances. A good starting point is to purchase an optimization kit containing a series of lipids or combinations of lipids, e.g., PerFect Lipid Transfection Kit (Invitrogen), Transfection Optimization Kit (Invitrogen) or Tfx Reagents Transfection Trio (Promega). This protocol describes the use of two of the most commonly used lipofecting agents: the monocationic lipid DOTMA (Lipofectin, Invitrogen) and the polycationic lipid DOGS (Transfectam, Promega). Because a large number of variables affect the efficiency of lipofection with these and other agents, the following protocol should be used as a platform for systematic optimization to obtain the maximal ratio of signal to background and to minimize variability between replicate assays (see Lipofection, MC3, pp. 16.50–16.51).

MATERIALS

Reagents and Solutions

Please see Appendix 1 for components of stock solutions, buffers, and reagents. Dilute stock solutions to the appropriate concentrations.

Lipofection reagent (see Table 16-2)

Lipofectin, usually supplied mixed with a helper lipid such as DOPE

Polystyrene tubes must be used with Lipofectin because the lipid binds to polypropylene (e.g., standard microfuge tubes).

or

Transfectam, which may be substituted for Lipofectin in this protocol

Polyamines such as Transfectam do not require the use of polypropylene tubes.

NaCl (5 M)

Sodium citrate (20 mM, pH 5.5) containing 150 mM NaCl (optional)

Use this buffer as the diluent for the plasmid DNA if Transfectam is the lipofection reagent.

Vectors and Hosts

Plasmid DNA (1 µg/ml in H_2O [Lipofectin] or sodium citrate buffer [Transfectam])

If performing lipofection for the first time or if using an unfamiliar cell line, obtain an expression plasmid encoding *E. coli* β-galactosidase or green fluorescent protein. These can be purchased from several commercial manufacturers (pCMV-SPORT-β-gal [Invitrogen] or pEGFP-F [CLONTECH]). Closed circular plasmid DNAs should be purified by column chromatography as described in Chapter 1.

Cells, Tissues/Culture Media

Exponentially growing cultures of mammalian cells

This protocol is designed for cells grown in 60-mm culture dishes. If multiwell plates, flasks, or dishes of a different diameter are used, the number of cells and the volumes of culture media and reagents should be scaled appropriately.

Cell culture growth medium, complete, serum-free, and (optional) selective

Additional Items

Plasticware (polystyrene tubes must be used with Lipofectin)

Step 9 of this protocol may require the reagents in Protocol 17.7.

Additional Information

Additional protocol (Histochemical Staining of Cell Monolayers
    for β-Galactosidase)   MC3, p. 16.13
    Lipofection   MC3, pp. 16.50–16.51

METHOD

1. Twenty-four hours before lipofection, harvest exponentially growing mammalian cells by trypsinization and replate them on 60-mm tissue-culture dishes at a density of 10^5 cells/dish (or at 5×10^4 cells/35-mm dish). Add 5 ml (or 3 ml for 35-mm dish) of growth medium and incubate the cultures for 20–24 hours at 37°C in a humidified incubator with an atmosphere of 5–7% CO_2.

   The cells should be 75% confluent at the time of lipofection.

2. For each 60-mm dish of cultured cells to be transfected, dilute 1–10 μg of plasmid DNA into 100 μl of sterile deionized H_2O (if using Lipofectin) or 20 mM sodium citrate containing 150 mM NaCl (pH 5.5) (if using Transfectam) in a polystyrene or polypropylene test tube. In a separate tube, dilute 2–50 μl of the lipid solution to a final volume of 100 μl with sterile deionized H_2O or 300 mM NaCl.

   IMPORTANT: When transfecting with Lipofectin, use polystyrene test tubes; do not use polypropylene tubes, because the cationic lipid DOTMA can bind nonspecifically to polypropylene. For other cationic lipids, use the tubes recommended by the manufacturer.

3. Incubate the tubes for 10 minutes at room temperature.

4. Add the lipid solution to the DNA and mix the solution by pipetting up and down several times. Incubate the mixture for 10 minutes at room temperature.

5. While the DNA-lipid solution is incubating, wash the cells to be transfected three times with serum-free medium. After the third rinse, add 0.5 ml of serum-free medium to each 60-mm dish and return the washed cells to a 37°C humidified incubator with an atmosphere of 5–7% CO_2.

   It is very important to rinse the cells free of serum before the addition of the lipid-DNA liposomes.

6. After the DNA-lipid solution has incubated for 10 minutes, add 900 μl of serum-free medium to each tube. Mix the solution by pipetting up and down several times. Incubate the tubes for 10 minutes at room temperature.

7. Transfer each tube of DNA-lipid-medium solution to a 60-mm dish of cells. Incubate the cells for 1–24 hours at 37°C in a humidified incubator with an atmosphere of 5–7% CO_2.

8. After the cells have been exposed to the DNA for the appropriate time, wash them three times with serum-free medium. Feed the cells with complete medium and return them to the incubator.

9. If the objective is stable transformation of the cells, proceed to Step 10. Examine the cells 24–96 hours after lipofection using one of the following assays.

   - If a plasmid DNA expressing *E. coli* β-galactosidase was used, follow the steps outlined in Protocol 17.7 to measure enzyme activity in cell lysates.

   - If a green fluorescence protein expression vector was used, examine the cells with a microscope under 450–490-nm illumination.

   - For other gene products, newly synthesized protein may be analyzed by radioimmunoassay, immunoblotting, immunoprecipitation following in vivo metabolic labeling, or assays of enzymatic activity in cell extracts.

To isolate stable transfectants

10. After the cells have incubated for 48–72 hours in complete medium, trypsinize the cells and replate them in the appropriate selective medium. Change this medium every 2–4 days for 2–3 weeks to remove the debris of dead cells and to allow colonies of resistant cells to grow. Thereafter, individual colonies may be cloned and propagated for assay.

TABLE 16-2 Some Lipids Used in Lipofection

Abbreviation	IUPAC Name	Type	Product Name	Cell Lines Commonly Used for Transfection
DOTMA	*N*-[1-(2,3-dioleoyloxy)propyl]-*N,N,N*-trimethylammonium chloride	monocationic	Lipofectin	AS52 H187 mouse L cells NIH-3T3 HeLa
DOTAP	*N*-[1-(2,3-dioleoyloxy)propyl]-*N,N,N*-trimethylammonium methyl sulfate	monocationic	DOTAP	HeLa
DMRIE	1,2-dimyristyloxypropyl-3-dimethyl-hydroxyethylammonium bromide	monocationic	DMRIE-C	Jurkat CHO-K1 COS-7 BHK-21
DDAB	dimethyl dioctadecylammonium bromide	monocationic	LipofectACE	COS-7 CHO-K1 BHK-21 mouse L cells
Amidine	*N-t*-butyl-*N'*-tetradecyl-3-tetradecyl-aminopropionamide	monocationic	CLONfectin	A-431 HEK293 BHK-21 HeLa L6 CV-1
DC-Cholesterol	3β[*N*-(*N',N'*-dimethylaminoethane) carbamoyl]-cholesterol	monocationic	DC-Cholesterol	
DOSPER	1,3-dioleoyloxy-2-(6-carboxyspermyl) propylamide	dicationic	Tfx	CHO HeLa NIH-3T3
DOGS	spermine-5-carboxy-glycine dioctadecyl-amide	polycationic	Transfectam	293 HeLa HepG2 HC11 NIH-3T3
DOSPA	2,3-dioleoyloxy-*N*-[2(sperminecarboxy-amido)ethyl]-*N,N*-dimethyl-1-propan-aminium trifluoroacetate	polycationic	LipofectAMINE	HT-29 BHK-21 keratinocytes MDCK NIH-3T3
TM-TPS	*N,N',N'',N'''*-tetramethyl-*N, N',N'',N'''*-tetrapalmitylspermine	polycationic	CellFECTIN	CHO-K1 COS-7 BHK-21 Jurkat

Calcium-phosphate-mediated Transfection of Eukaryotic Cells with Plasmid DNAs

Calcium phosphate and DNA form an insoluble precipitate that attaches to cell surfaces and is taken into the cells by endocytosis. This protocol describes an optimized method for calcium-phosphate-mediated transfection of Chinese hamster ovary cells and the 293 line of human embryonic kidney cells. However, the protocol can readily be adapted for use with other stable lines of cultured cells, both adherent and nonadherent.

MATERIALS

CAUTION: Please see Appendix 4 for appropriate handling of materials marked with <!>.

Reagents and Solutions

Please see Appendix 1 for components of stock solutions, buffers, and reagents. Dilute stock solutions to the appropriate concentrations.

$CaCl_2$ (2.5 M)

Chloroquine (100 mM) (optional; see Step 5)
> Dissolve 52 mg of chloroquine diphosphate in 1 ml of deionized H_2O. Sterilize the solution by passage through a 0.22-μm filter. Store the solution in foil-wrapped tubes at –20°C.

Giemsa stain (10% w/v), freshly prepared in PBS or H_2O and filtered through Whatman No. 1 filter paper before use

Glycerol (15% v/v) in 1x HEPES-buffered saline (optional; see Step 5)
> Add autoclaved glycerol to sterile HEPES-buffered saline solution just before use.

2x HEPES-buffered saline
> 42 mM HEPES
> 14 mM Na_2HPO_4
> 274 mM NaCl
> 10 mM KCl
> 12 mM dextrose
> Adjust the pH of the solution to 7.10. Sterilize the solution by filtration and store it in 50–100-ml aliquots. Before use, check that the pH of the solution has not drifted.

Methanol <!>

Phosphate-buffered saline (PBS)

Sodium butyrate (500 mM) (optional; see Step 5)
> In a chemical fume hood, bring an aliquot of stock butyric acid solution to a pH of 7.0 with 10 N NaOH. Sterilize the solution by passing it through a 0.22-μm filter; store the solution in 1-ml aliquots at –20°C.

TE (pH 7.6)

Vectors and Hosts

Plasmid DNA (25 μg/ml in 0.1x TE [pH 7.6])
> Closed circular plasmid DNAs used for calcium-phosphate-mediated transfection of eukaryotic cells should be purified by column chromatography or centrifugation to equilibrium in ethidium bromide–CsCl gradients, as described in Chapter 1. If the amount of plasmid DNA is limiting, add carrier DNA, e.g., salmon sperm DNA or mammalian cell DNA, to adjust the final concentration to 25 μg/ml. Sterilize the carrier DNA before use by ethanol precipitation or extraction with chloroform. If performing transfecting for the first time or if using an unfamiliar cell line, obtain an expression plasmid encoding *E. coli* β-galactosidase or green fluorescent protein (pCMV-SPORT-β-gal [Invitrogen] or pEGFP-F [CLONTECH]).

Cells, Tissues/Culture Media

Cell culture growth medium, complete, serum-free, and (optional) selective
Exponentially growing cultures of mammalian cells

> This protocol is designed for cells grown in 60-mm culture dishes. If multiwell plates, flasks or dishes of a different diameter are used, the number of cells and the volumes of culture media and reagents should be scaled appropriately.

Additional Items

Plasticware
Step 6 of this protocol requires the reagents listed in MC3, Protocols 6.10 and 7.8.

Additional Information

Alternative protocol (High-efficiency Calcium-phosphate-mediated Transfection of Eukaryotic Cells with Plasmid DNAs)	MC3, pp. 16.19–16.20
Chloroquine diphosphate	MC3, p. 16.53
Selective agents for stable transformation	MC3, pp. 16.48–16.49
Transfection of mammalian cells with calcium phosphate–DNA coprecipitates	MC3, pp. 16.52–16.53

METHOD

1. Twenty-four hours before transfection, harvest exponentially growing cells by trypsinization and replate them at a density of 1×10^5 to 4×10^5 cells/cm^2 in 60-mm tissue-culture dishes or 12-well plates in the appropriate complete medium. Incubate the cultures for 20–24 hours at 37°C in a humidified incubator with an atmosphere of 5–7% CO_2. Change the medium 1 hour before transfection.

   > It is important to use exponentially growing cells.

2. Prepare the calcium phosphate–DNA coprecipitate as follows: Combine 100 µl of 2.5 M CaCl$_2$ with 25 µg of plasmid DNA in a sterile 5-ml plastic tube and, if necessary, bring the final volume to 1 ml with 0.1x TE (pH 7.6). Mix 1 volume of this 2x calcium-DNA solution with an equal volume of 2x HEPES-buffered saline at room temperature. Quickly tap the side of the tube to mix the ingredients and allow the solution to stand for 1 minute.

3. Immediately transfer the calcium phosphate–DNA suspension into the medium above the cell monolayer. Use 0.1 ml of suspension for each 1 ml of medium in a well or 60-mm dish. Rock the plate gently to mix the medium, which will become yellow-orange and turbid. Perform this step as quickly as possible because the efficiency of transfection declines rapidly once the DNA precipitate is formed. If the cells will be treated with transfection facilitators such as chloroquine, glycerol, and/or sodium butyrate, proceed directly to Step 5.

4. Transfected cells that will not be treated with transfection facilitators should be incubated at 37°C in a humidified incubator with an atmosphere of 5–7% CO_2. After 2–6 hours of incubation, remove the medium and DNA precipitate by aspiration. Add 5 ml of warmed (37°C) complete growth medium and return the cells to the incubator for 1–6 days. Proceed to Step 6 to assay for transient expression of the transfected DNA or proceed directly to Step 7 if the objective is stable transformation of the cells.

5. The uptake of DNA can be increased by treatment of the cells with chloroquine in the presence of the calcium phosphate–DNA coprecipitate or exposure to glycerol and sodium butyrate following removal of the coprecipitate solution from the medium.

Treatment of cells with chloroquine

a. Dilute 100 mM chloroquine diphosphate 1:1000 directly into the medium either before or after the addition of the calcium phosphate–DNA coprecipitate to the cells.

> The concentration of chloroquine added to the growth medium and the time of treatment are limited by the sensitivity of the cells to the toxic effect of the drug. The optimal concentration of chloroquine for the cell type used should be determined empirically.

b. Incubate the cells for 3–5 hours at 37°C in a humidified incubator with an atmosphere of 5–7% CO_2.

c. After the treatment with DNA and chloroquine, remove the medium, wash the cells with phosphate-buffered saline, and add 5 ml of warmed complete growth medium. Return the cells to the incubator for 1–6 days. Proceed to Step 6 to assay for transient expression of the transfected DNA or proceed directly to Step 7 if the objective is stable transformation of the cells.

Treatment of cells with glycerol

a. After cells have been exposed for 2–6 hours to the calcium phosphate–DNA coprecipitate in growth medium (± chloroquine), remove the medium by aspiration and wash the monolayer once with phosphate-buffered saline.

> This procedure may be used following treatment with chloroquine. Because cells vary widely in their sensitivity to the toxic effects of glycerol, each cell type must be tested in advance to determine the optimum time (30 seconds to 3 minutes) of treatment.

b. Add 1.5 ml of 15% glycerol in 1x HEPES-buffered saline to each monolayer and incubate the cells for the predetermined optimum length of time at 37°C.

c. Remove the glycerol by aspiration and wash the monolayers once with phosphate-buffered saline.

d. Add 5 ml of warmed complete growth medium and incubate the cells for 1–6 days. Proceed to Step 6 to assay for transient expression of the transfected DNA or proceed directly to Step 7 if the objective is stable transformation of the cells.

Treatment of cells with sodium butyrate

a. Following the glycerol shock, dilute 500 mM sodium butyrate directly into the growth medium (Step d, Treatment of cells with glycerol). Different concentrations of sodium butyrate are used depending on the cell type. For example:

CV-1	10 mM
NIH-3T3	7 mM
HeLa	5 mM
CHO	2 mM

The correct amount for other cell lines that may be transfected should be determined empirically.

b. Incubate the cells for 1–6 days. Proceed to Step 6 to assay for transient expression of the transfected DNA or proceed directly to Step 7 if the objective is stable transformation of the cells.

6. To assay the transfected cells for transient expression of the introduced DNA, harvest the cells 1–6 days after transfection. Analyze RNA or DNA using hybridization. Analyze newly synthesized protein by radioimmunoassay, immunoblotting, immunoprecipitation following in vivo metabolic labeling, or assays of enzymatic activity in cell extracts.

To isolate stable transfectants

7. a. Incubate the cells for 24–48 hours in nonselective medium to allow time for expression of the transferred gene(s).

   b. Either trypsinize and replate the cells in the appropriate selective medium or add the selective medium directly to the cells without further manipulation.

   c. Change the selective medium with care every 2–4 days for 2–3 weeks to remove the debris of dead cells and to allow colonies of resistant cells to grow.

   d. Clone individual colonies and propagate for appropriate assay.

   e. Obtain a permanent record of the numbers of colonies by fixing the remaining cells with ice-cold methanol for 15 minutes followed by staining with 10% Giemsa for 15 minutes at room temperature before rinsing in tap water.

Calcium-phosphate-mediated Transfection of Cells with High-molecular-weight Genomic DNA

This protocol is used chiefly to generate stable lines of cells carrying chromosomally integrated copies of the transfected DNA.

MATERIALS

Reagents and Solutions

Please see Appendix 1 for components of stock solutions, buffers, and reagents. Dilute stock solutions to the appropriate concentrations.

$CaCl_2$ (2 M), sterile

Giemsa stain (10% w/v), freshly prepared in PBS or H_2O and filtered through Whatman No. 1 filter paper just before use

Glycerol (15% v/v) in 1x HEPES-buffered saline (optional; see Step 10)

> Add autoclaved glycerol to sterile HEPES-buffered saline solution just before use.

1x HEPES-buffered saline, sterile

> 21 mM HEPES
>
> 0.7 mM Na_2HPO_4
>
> 137 mM NaCl
>
> 5 mM KCl
>
> 6 mM dextrose
>
> Adjust the pH of the solution to 7.10. Sterilize the solution by filtration and store it in 50–100-ml aliquots. Before use, check that of the solution has not drifted.

Isopropanol

NaCl (3 M), sterile

Vectors and Hosts

Genomic DNA (100 µg/ml in 0.1x TE [pH 7.6])

> Prepare high-molecular-weight genomic DNA from mammalian cells, as described in Protocol 6.3. The genomic DNA must be sheared to a size range of 45–60 kb before using it in transfection experiments. The appropriate conditions for shearing are best determined in preliminary experiments as follows: Pass aliquots of DNA through a 22-gauge needle for different numbers of times (e.g. three, five, six, seven, and eight times). Examine the DNA by electrophoresis through a 0.7% agarose gel followed by staining with ethidium bromide or SYBR Gold. For markers, use monomeric and dimeric forms of linear bacteriophage λ DNA.

Plasmid carrying selectable marker (optional; see notes to Steps 3 and 12, and see MC3, pp. 16.48–16.49)

Cells, Tissues/Culture Media

Cell culture growth medium, complete, serum-free, and (optional) selective

Exponentially growing cultures of mammalian cells

> This protocol is designed for cells grown in 60-mm culture dishes. If multiwell plates, flasks or dishes of a different diameter are used, the number of cells and the volumes of culture media and reagents should be scaled appropriately.

Additional Items

Polyethylene tubes (12 ml)

Shepherd's crook

Step 6 of this protocol requires the reagents listed in MC3, Protocols 6.10 and 7.8.

Additional Information

Alternative protocol (Calcium-phosphate-mediated Transfection of Adherent Cells)	MC3, p. 16.25
Alternative protocol (Calcium-phosphate-mediated Transfection of Cells Growing in Suspension)	MC3, p. 16.26
Alternative protocol (High-efficiency Calcium-phosphate-mediated Transfection of Eukaryotic Cells with Plasmid DNAs)	MC3, pp. 16.19–16.20
Chloroquine diphosphate	MC3, p. 16.53
Cotransformation	MC3, p. 16.47
Selective agents for stable transformation	MC3, pp. 16.48–16.49
Transfection of mammalian cells with calcium phosphate–DNA coprecipitates	MC3, pp. 16.52–16.53

METHOD

1. On day 1 of the experiment, plate exponentially growing cells (e.g., CHO cells) at a density of 5×10^5 cells per 90-mm culture dish in appropriate growth medium containing serum. Incubate the cultures for approximately 16 hours at 37°C in a humidified incubator with an atmosphere of 5% CO_2.

2. On day 2, shear an appropriate amount of high-molecular-weight DNA into fragments ranging in size from 45 to 60 kb, by passing it through a 22-gauge needle for the predetermined number of times.

   Cells should be transfected with 20–25 µg of genomic DNA per 90-mm dish.

3. Precipitate the sheared DNA by adding 0.1 volume of 3 M NaCl and 1 volume of isopropanol. Collect the DNA on a Shepherd's crook. Drain the precipitate briefly against the side of the tube and transfer it to a second tube containing HEPES-buffered saline (1 ml per 12–15 µg of DNA). Redissolve the DNA by gentle rotation for 2 hours at 37°C. Make sure that all of the DNA has dissolved before proceeding.

   When cotransfecting with a selectable marker, add a sterile solution of the appropriate plasmid to the genomic DNA, to a final concentration of 0.5 µg/ml.

4. Transfer 3-ml aliquots of sheared genomic DNA into 12-ml polyethylene tubes (one aliquot per two dishes to be transfected).

5. To form the calcium phosphate–DNA coprecipitate, gently vortex an aliquot of sheared genomic DNA and add 120 µl of 2 M $CaCl_2$ in a dropwise fashion. Incubate the tube for 15–20 minutes at room temperature.

   The solution should turn hazy, but it should not form visible clumps of precipitate.

6. Aspirate the medium from two dishes of cells (from Step 1) and gently add 1.5 ml of the calcium phosphate–DNA coprecipitate to each dish. Carefully rotate the dishes to swirl the medium and spread the precipitate over the monolayer of cells. Incubate the cells for 20 minutes at room temperature, rotating the dishes once during the incubation.

7. Gently add 10 ml of warmed (37°C) growth medium to each dish and incubate for 6 hours at 37°C in a humidified incubator with an atmosphere of 5% CO_2.

8. Repeat Steps 5–7 until all of the dishes of cells contain the calcium phosphate–DNA precipitate.

9. After 6 hours of incubation, examine each dish under a light microscope. A "peppery" precipitate should be seen adhering to the cells. The precipitate should be neither too fine nor clumpy.

10. In most cases, treatment with glycerol at this step will enhance the transfection frequency. To shock the cells with glycerol:

   a. Aspirate the medium containing the calcium phosphate–DNA coprecipitate.

   b. To each dish of cells, add 3 ml of 15% glycerol in 1x HEPES-buffered saline that has been warmed to 37°C. Incubate for *no longer than 3 minutes* at room temperature.

   > It is important that the glycerol in the HEPES-buffered saline *not* be left in contact with the cells for too long. The optimum time period usually spans a narrow range and varies from one cell line to another and from one laboratory to the next. For these reasons, treat only a few dishes at a time and take into account the length of time to aspirate the glycerol in the HEPES-buffered saline. Do not to exceed the optimum incubation period. Seconds can count!

   c. Aspirate the glycerol in the HEPES-buffered saline and rapidly wash the dishes twice with 10 ml of warmed growth medium.

   d. Add 10 ml of warmed growth medium and incubate the cultures for 12–15 hours at 37°C in a humidified incubator with an atmosphere of 5% CO_2.

11. Replace the medium with 10 ml of fresh growth medium. Continue the incubation overnight at 37°C in a humidified incubator with an atmosphere of 5% CO_2.

12. Microscopic examination of cells at this point (day 4) should reveal a normal morphology. Cells can be trypsinized and replated in selective medium on day 4. Continue the incubation for 2–3 weeks to allow growth of complemented and/or resistant colonies. Change the medium every 2–3 days.

   > The length of the selection period, the cell density of replating, and the selection conditions all depend on the gene being complemented or selected. Optimum cell density at Step 12 usually varies between 2.5×10^5 and 1×10^6 cells per 90-mm dish.

13. Thereafter, clone individual colonies and propagate them for the appropriate assay.

Transfection Mediated by DEAE-Dextran: High-efficiency Method

Transfection mediated by DEAE-dextran differs from calcium phosphate coprecipitation in three important ways. First, it is used to obtain a burst of transient expression of cloned genes and not for stable transformation of cells. Second, it works very efficiently with lines of simian cells such as BSC-1, COS, and CV-1 but is unsatisfactory with many other lines of cells. Third, smaller amounts of DNA are used for transfection with DEAE-dextran than with calcium phosphate coprecipitation. Maximal transfection efficiency of 10^5 simian cells is achieved with 0.1–1.0 μg of supercoiled plasmid DNA; larger amounts of DNA can be inhibitory. Finally, in contrast to transfection by calcium phosphate, where high concentrations of DNA are required to promote the formation of a coprecipitate, carrier DNA is rarely used with DEAE-dextran transfection.

Many variants of the technique have been described; all of them seek to minimize the cytotoxic effects of exposing cells to a high-molecular-weight, positively charged polymer. In this protocol, cells are exposed briefly to a high concentration of DEAE-dextran and then to chloroquine diphosphate—a facilitator of transfection.

MATERIALS

Reagents and Solutions

Please see Appendix 1 for components of stock solutions, buffers, and reagents. Dilute stock solutions to the appropriate concentrations.

Chloroquine (100 mM)

Dissolve 52 mg of chloroquine diphosphate in 1 ml of deionized H_2O. Sterilize the solution by passage through a 0.22-μm filter. Store the solution in foil-wrapped tubes at –20°C.

DEAE-dextran (50 mg/ml)

Dissolve DEAE-dextran (M_r = 500,000 [Pfizer]) in 2 ml of distilled H_2O. Sterilize the solution by autoclaving for 20 minutes at 15 psi (1.05 kg/cm²) on liquid cycle. Several manufacturers sell kits for DEAE-dextran-mediated transfection (e.g., Promega's kit, ProFection Mammalian Transfection).

Phosphate-buffered saline (PBS)

Tris-buffered saline with dextrose (TBS-D solution)

Immediately before use, add sterile dextrose solution (20% w/v, in H_2O) to TBS to a final concentration of 0.1%.

Cells, Tissues/Culture Media

Cell culture growth medium, with and without serum

Exponentially growing line of cultured simian cells (e.g., BSC-1, COS, or CV-1)

This protocol is designed for cells growing in 60-mm-diameter plastic tissue-culture dishes, but could easily be adapted for use with monolayers of cells grown in flasks or multiwell plates.

Nucleic Acids/Oligonucleotides

Plasmid DNA

Closed circular plasmid DNAs used for DEAE-dextran-mediated transfection of eukaryotic cells should be purified by column chromatography or centrifugation to equilibrium in ethidium bromide–CsCl gradients, as described in Chapter 1.

Additional Items

Step 8 of this protocol requires the reagents listed in MC3, Protocols 6.1 and 7.8.

Additional Information

Alternative protocol (Transfection Mediated by
DEAE-Dextran: Increased Cell Viability)   MC3, p. 16.32
Chloroquine diphosphate   MC3, p. 16.53

METHOD

1. Twenty-four hours before transfection, harvest exponentially growing cells by trypsinization and transfer them to 60-mm tissue-culture dishes at a density of 10^5 cells/dish (or 35-mm dishes at a density of 5×10^4 cells/dish). Add 5 ml (or 3 ml for a 35-mm dish) of complete growth medium and incubate the cultures for 20–24 hours at 37°C in a humidified incubator with an atmosphere of 5–7% CO_2.

   The cells should be 75% confluent at the time of transfection.

2. Prepare the DNA/DEAE-dextran/TBS-D solution by mixing 0.1–4 µg of supercoiled or circular plasmid DNA into 1 mg/ml DEAE-dextran in TBS-D.

   0.25 ml of the solution is required for each 60-mm dish; 0.15 ml is required for each 35-mm dish.

3. Remove the medium from the cell culture dishes by aspiration and wash the monolayers twice with warmed (37°C) PBS and once with warmed TBS-D.

4. Add the DNA/DEAE-dextran/TBS-D solution (250 µl per 60-mm dish; 150 µl per 35-mm dish). Rock the dishes gently to spread the solution evenly across the monolayer of cells. Return the cultures to the incubator for 30–90 minutes (the time will depend on the sensitivity of each batch of cells to the DNA/DEAE-dextran/TBS-D solution). At 15–20-minute intervals, remove the dishes from the incubator, swirl them gently, and check the appearance of the cells under the microscope. If the cells are still firmly attached to the substratum, continue the incubation. Stop the incubation when the cells begin to shrink and round up.

5. Remove the DNA/DEAE-dextran/TBS-D solution by aspiration. Gently wash the monolayers once with warmed TBS-D and then once with warmed PBS, taking care not to dislodge the transfected cells.

6. Add 5 ml (per 60-mm dish) or 3 ml (per 35-mm dish) of warmed medium supplemented with serum and chloroquine (100 µM final concentration) and incubate the cultures for 3–5 hours at 37°C in a humidified incubator with an atmosphere of 5–7% CO_2.

7. Remove the medium by aspiration and wash the monolayers three times with serum-free medium. Add to the cells 5 ml (per 60-mm dish) or 3 ml (per 35-mm dish) of medium supplemented with serum and incubate the cultures for 36–60 hours at 37°C in a humidified incubator with an atmosphere of 5–7% CO_2 before assaying for transient expression of the transfected DNA.

   The time of incubation should be optimized for the particular conditions.

8. To assay the transfected cells for transient expression of the introduced DNA, harvest the cells 36–60 hours after transfection. Analyze RNA or DNA using hybridization. Analyze newly synthesized protein by radioimmunoassay, immunoblotting, immunoprecipitation following in vivo metabolic labeling, or assays of enzymatic activity in cell extracts.

DNA Transfection by Electroporation

Pulsed electrical fields can be used to introduce DNA into a wide variety of animal cells. Electroporation works well with cell lines that are refractive to other techniques, such as calcium phosphate–DNA coprecipitation. But, as with other transfection methods, the optimal conditions for electroporating DNA into untested cell lines must be determined experimentally.

MATERIALS

CAUTION: Please see Appendix 4 for appropriate handling of materials marked with <!>.

Reagents and Solutions

Please see Appendix 1 for components of stock solutions, buffers, and reagents. Dilute stock solutions to the appropriate concentrations.

> Giemsa stain (10% w/v) (optional; see Step 11) freshly prepared in PBS or H_2O
> and filtered through Whatman No. 1 filter paper just before use
> Methanol <!>
> Phosphate-buffered saline (PBS)
> Sodium butyrate (500 mM) (optional; see Step 8)
>> For details, see Protocol 16.2, Step 5.

Cells, Tissues/Culture Media

> Cell culture medium, with and without serum
> Exponentially growing line of cultured cells
>> This protocol is designed for cells growing in 35-mm-diameter plastic tissue-culture dishes, but could easily be adapted for use with cells grown in multiwell plates.

Nucleic Acids/Oligonucleotides

> Carrier DNA (10 mg/ml), e.g., sonicated salmon sperm DNA (optional; see Step 6)
> Linearized or circular plasmid DNA (1 µg/µl in deionized H_2O)
>> Plasmid DNAs used for electroporation of eukaryotic cells should be purified by column chromatography or centrifugation to equilibrium in ethidium bromide–CsCl gradients, as described in Chapter 1.

Centrifuges/Rotors/Tubes

> Sorvall H-1000B rotor (4°C)

Additional Items

> Electroporation device and cuvettes, e.g., Gene Pulser II (Bio-Rad)
> Hemocytometer *or* other form of cell-counting device
> Rubber policeman
> Step 10 of this protocol may require the reagents listed in Protocol 17.7.

Additional Information

Electroporation: Mechanism and optimal conditions	MC3, pp. 16.54–16.57
Factors affecting the efficiency of transfection by electroporation	MC3, pp. 16.33–16.34
Sodium butyrate	MC3, pp. 16.17–16.18

METHOD

1. Harvest the cells to be transfected from cultures in the mid- to late-logarithmic phase of growth. Use either a rubber policeman or trypsin to release adherent cells. Centrifuge at 500g (1500 rpm in a Sorvall H-1000B rotor) for 5 minutes at 4°C.

2. Resuspend the cell pellet in 0.5x volume of the original growth medium and measure the cell number using a hemocytometer.

3. Collect the cells by centrifugation as described in Step 1 and resuspend them in growth medium or phosphate-buffered saline at room temperature at a concentration of 2.5 x 10^6 to 2.5 x 10^7 cells/ml.

4. Transfer 400-μl aliquots of the cell suspension (10^6 to 10^7 cells) into as many labeled electroporation cuvettes as needed. Place the loaded cuvettes on ice.

5. Set the parameters on the electroporation device. A typical capacitance value is 1050 μF. Voltages range from 200 to 350 V, depending on the cell line, but generally average 260 V. Use an infinite internal resistance value. Discharge a blank cuvette containing phosphate-buffered saline at least twice before beginning electroporation of cells.

6. Add 10–30 μg of plasmid DNA in a volume of up to 40 μl to each cuvette containing cells. (Some investigators add carrier DNA [e.g., salmon sperm DNA] to bring the total amount of DNA to 120 μg.) Gently mix the cells and DNA by pipetting the solution up and down. Proceed to Step 7 without delay.

    IMPORTANT: Do not introduce air bubbles into the suspension during the mixing step.

7. Immediately transfer the cuvette to the electroporator and discharge the device. After 1–2 minutes, remove the cuvette, place it on ice, and proceed immediately to the next step.

8. Transfer the electroporated cells to a 35-mm culture dish using a micropipettor equipped with a sterile tip. Rinse out the cuvette with a fresh aliquot of growth medium and add the washings to the culture dish. Transfer the dish to a humidified incubator at 37°C with an atmosphere of 5–7% CO_2.

    To incorporate a sodium butyrate shock (please see Protocol 15.2, Step 5), rinse the cuvette with growth medium containing the optimal amount of sodium butyrate (determined in preliminary control experiments) and combine the rinse with the electroporated cells in the culture dish. Transfer the dish to an incubator. After 24 hours of incubation, remove the medium containing sodium butyrate and replace it with normal growth medium.

9. Repeat Steps 6–8 until all of the DNA cell samples in cuvettes are shocked. Record the actual pulse time for each cuvette to facilitate comparisons between experiments.

10. If the objective is stable transformation of the cells, proceed directly to Step 11. For transient expression, examine the cells 24–96 hours after electroporation using one of the following assays:

    • If a plasmid DNA expressing *E. coli* β-galactosidase was used, follow the steps outlined in Protocol 17.7 to measure enzyme activity in cell lysates.

    • If a green fluorescence protein expression vector was used, examine the cells with a microscope under 450–490-nm illumination.

    • For other gene products, analyze the newly synthesized protein by radioimmunoassay, immunoblotting, immunoprecipitation following in vivo metabolic labeling, or assays of appropriate enzymatic activity in cell extracts.

To isolate stable transfectants

11. After incubation for 48–72 hours in complete medium, trypsinize the cells and replate them in the appropriate selective medium. The selective medium should be changed every 2–4 days for 2–3 weeks to remove the debris of dead cells and to allow colonies of resistant cells to grow. Thereafter, clone individual colonies and propagate for the appropriate assay.

    A permanent record of the numbers of colonies may be obtained by fixing the remaining cells with ice-cold methanol for 15 minutes, followed by staining with 10% Giemsa (freshly prepared) for 15 minutes at room temperature before rinsing in tap water and air drying.

DNA Transfection by Biolistics

In biolistics, DNA is mixed with small metal particles that are then fired into the host cell at very high speeds. The idea for the biolistic gun came from a collaboration between Edward Wolf and Nelson Allen of the Cornell University Nanofabrication Facility and two plant scientists, John Sanford and Theodore Klein. Their first biolistic guns were versions of tools used to sink nails into concrete, modified to fire tungsten particles. Today's guns use helium propellant and far more sophisticated delivery mechanisms and allow cloned DNAs to be introduced with high efficiency into organisms with rigid walls, such as bacteria, plants, and yeasts, and into slices of fresh mammalian tissue and cultures of adherent cells.

MATERIALS

Reagents and Solutions

Please see Appendix 1 for components of stock solutions, buffers, and reagents. Dilute stock solutions to the appropriate concentrations.

$CaCl_2$ (2.5 M)

Ethanol

> Use a fresh bottle of absolute ethanol that has not been previously opened. Because ethanol is hygroscopic, it absorbs small amounts of moisture from air, which reduces the efficiency of biolistic transfection.

Ethanol (70%)

Glycerol (50% v/v, in H_2O)

Spermidine (0.1 M)

> Dissolve an appropriate amount of spermidine (free-base form) in deionized H_2O and sterilize it by filtration through an 0.22-μm nitrocellulose filter. The solution may be stored in small aliquots for up to 1 month at –20°C.

Cells, Tissues/Culture Media

Cells or tissue to be transfected

> Adherent cells should be bombarded at 20–80% confluency. Plant cells grown in suspension should be collected by sterile filtration onto Whatman No. 1 filter papers (7-cm diameter) using a Buchner funnel and placed on sterile filter papers soaked with culture medium of high osmolarity.
>
> Freshly dissected mammalian tissue should be sectioned at approximately 400 μm and maintained in culture dishes.
>
> Bacteria and yeast in the mid- to late-logarithmic phase of growth should be collected by centrifugation, resuspended in a small volume of culture medium of high osmolarity, and plated (1 × 10^8 to 2 × 10^9 cells) on a thin layer of agar atop a piece of filter paper in a petri dish before being shot.

Media appropriate for the species and type of cells to be bombarded

Nucleic Acids/Oligonucleotides

Linearized or circular plasmid DNA (1 μg/μl in deionized H_2O)

> When performing an experiment for the first time with a gene gun or when using an unfamiliar tissue or cell type, obtain an expression plasmid encoding *E. coli* β-galactosidase, green fluorescent protein, β-glucuronidase (for plant cells), or neomycin resistance. These plasmids, which can be used to optimize the system, can be purchased from commercial manufacturers. Plasmid DNAs used for biolistic bombardment of cells should be purified by centrifugation to equilibrium in ethidium bromide–CsCl gradients, as described in Chapter 1.

   Adapted from Chapter 16, Protocol 6, p. 16.37 of MC3.

Additional Items

Biolistics gun
> Biolistic guns sold by Bio-Rad include a bombardment chamber with separate connections for vacuum and helium lines.

Helium gas
> Use a high-pressure tank (2400–2600 psi) of helium gas (>99.999% pure), safely anchored to the bench.

Lens paper
Step 8 of this protocol may require the reagents listed in Protocol 17.7.

Tungsten or gold particles
> DNA is delivered to cells on tungsten or gold particles that vary in diameter from 0.6 to 5 μm (Bio-Rad and Sylvania). The optimum pellet diameter for a given cell or tissue type must be determined empirically.

Additional Information

Additional protocol (Histochemical Staining of Cell Monolayers or Tissue for β-Glucuronidase)	MC3, p. 16.42
Biolistics: Mechanism	MC3, pp. 16.37–16.38

METHOD

1. Prepare tungsten or gold particles.

   a. Weigh 60 mg of gold or tungsten particles into a 1.5-ml microfuge tube.

   b. Add 1 ml of 70% ethanol to the particles and vortex the tube continuously for 5 minutes at room temperature. Store the tube on the benchtop for 15 minutes.

   c. Collect the particles by centrifugation at maximum speed for 5 seconds in a microfuge.

   d. Gently remove the supernatant. Resuspend the metal particles in 1 ml of sterile H_2O and vortex the suspension for 1 minute. Store the tube on the benchtop for 1 minute.

   e. Collect the metal particles by centrifugation at maximum speed for 5 seconds in a microfuge.

   f. Repeat the H_2O wash (Steps d and e) three more times.

   g. Remove the supernatant after the fourth H_2O wash. Resuspend the particles in 1 ml of sterile 50% glycerol.
   > The washed particles are assumed to have a concentration of 60 mg/ml and may be stored at room temperature for a maximum of 2 weeks.

2. For every six dishes of cells or slices of tissue to be shot, prepare an aliquot of DNA-coated particles as follows:

   a. While continuously vortexing the stock solution of microcarrier particles, remove a 50-μl aliquot (~3 mg).

   b. Transfer the aliquot to a fresh microfuge tube and, while vortexing, add the following to the tube:

plasmid DNA (~2.5 μg)	2.5 μl
2.5 M $CaCl_2$	50 μl
0.1 M spermidine	20 μl

   After all ingredients are added, continue vortexing the tube for an additional 3 minutes.
   > It is very important that the microfuge tube be continuously vortexed during this procedure to ensure uniform coating of the particles with plasmid DNA.

c. Stand the tube on the bench for 1 minute to allow the particles to settle and then collect them by centrifugation at maximum speed for 2 seconds in a microfuge.

d. Remove the supernatant and carefully layer 140 µl of 70% ethanol over the pelleted particles. Remove the 70% ethanol and add 140 µl of absolute ethanol, again without disturbing the particles. Remove the supernatant and replace with 50 µl of ethanol.

e. Resuspend the particle pellet by tapping the side of the tube, followed by gentle vortexing for 2–3 seconds.

3. Place a macrocarrier in the metal holder of the gene gun apparatus using the seating device supplied by the manufacturer. Wash the sheet twice with 6-µl aliquots of ethanol. Between washes, blot the sheet dry with lens paper.

4. Vortex the pellet sample from Step 2e for 1 minute. While vortexing, withdraw 6 µl of the pellet slurry (~500 µg of particles) and, as quickly as possible, spread the aliquot around the central 1 cm of the macrocarrier.

5. Repeat Steps 3 and 4 until the desired number of loaded macrocarriers has been prepared. Allow the ethanol solution containing the DNA-coated particles to dry on the macrocarrier.

6. Load a macrocarrier into the gene gun and, following manufacturer directions, shoot a plate of cells or tissue slice.

7. After the vacuum has returned to atmospheric pressure, remove the wounded cells or tissue and place in appropriate culture conditions. Remove the ruptured macrocarrier and repeat Steps 6 and 7 until all plates are shot.

8. If the objective is stable transformation of the cells, proceed directly to Step 9. For transient expression, examine the cells 24–96 hours after shooting, using one of the following assays.

   - If a plasmid DNA expressing *E. coli* β-galactosidase was used, follow the steps outlined in Protocol 17.7 to measure enzyme activity in cell lysates. Alternatively, perform a histochemical staining assay as detailed in the additional protocol (Histochemical Staining of Cell Monolayers for β-Galactosidase) in MC3, p. 16.13.

   - If a green fluorescence protein expression vector was used, examine the cells with a microscope under 450–490-nm illumination.

   - If a plasmid DNA expressing β-glucuronidase was used, assay for β-glucuronidase activity.

   - For other gene products, analyze the newly synthesized protein by radioimmunoassay, immunoblotting, immunoprecipitation following in vivo metabolic labeling, or assays of appropriate enzymatic activity in cell extracts.

9. To isolate stable transfectants, after the cells have incubated for 48–72 hours in complete medium, transfer the bombarded cells to selective medium. The concentration of selective agent and the culture conditions will vary depending on the cell type.

DNA Transfection Using Polybrene

Polybrene is a polycation that, in the presence of dimethylsulfoxide (DMSO), is used to achieve stable transformation of cultured mammalian cells by plasmid DNA. Variables that influence the efficiency of transformation include the relative and absolute concentrations of the two facilitators, the amount of DNA, its molecular weight, and the length of time that the cells are left in contact with the Polybrene/DMSO mixture. When using Polybrene/DMSO for the first time on a cell line, the optimal combination of these variables should be established in preliminary experiments using plasmids carrying easily scored markers.

MATERIALS

CAUTION: Please see Appendix 4 for appropriate handling of materials marked with <!>.

Reagents and Solutions

Please see Appendix 1 for components of stock solutions, buffers, and reagents. Dilute stock solutions to the appropriate concentrations.

Dimethylsulfoside (DMSO) (30% v/v)

> Dilute HPLC-grade DMSO in the culture medium containing serum just before use in Step 3.

Methanol <!>

Polybrene (10 mg/ml)

> Dissolve Polybrene (Sigma-Aldrich) at a concentration of 10 mg/ml in H_2O and sterilize the solution by filtration through a 0.22-μm filter. Store the solution in small aliquots at –20°C.

Sodium butyrate (500 mM) (optional; see Step 5)

> In a chemical fume hood, bring an aliquot of stock butyric acid solution to a pH of 7.0 with 10 N NaOH. Sterilize the solution by passage through a 0.22-μm filter and store in 1-ml aliquots at –20°C.

Cells, Tissues/Culture Media

Exponentially growing cell cultures

> This protocol is designed for CHO cells grown on 90-mm culture dishes but can easily be adapted for use with plates, dishes, and wells of other sizes and other types of cells.

Media, e.g., MEM-α, with and without 10% fetal bovine serum and/or selective agents

Nucleic Acids/Oligonucleotides

Linearized or circular plasmid DNA (1 μg/μl in deionized H_2O)

> When performing an experiment for the first time with a gene gun or when using an unfamiliar tissue or cell type, obtain an expression plasmid encoding *E. coli* β-galactosidase, green fluorescent protein, β-glucuronidase (for plant cells), or neomycin resistance. These plasmids, which can be used to optimize the system, can be purchased from commercial manufacturers. Plasmid DNAs used for biolistic bombardment of cells should be purified by centrifugation to equilibrium in ethidium bromide–CsCl gradients, as described in Chapter 1.

Additional Items

Step 6 of this protocol may require the reagents listed in Protocol 17.7.

Additional Information

Polybrene: Background information	MC3, p. 16.43
Selective agents for stable transformation	MC3, pp. 16.48–16.49

METHOD

1. Harvest exponentially growing cells by trypsinization and replate them at a density of 5×10^5 cells per 90-mm tissue-culture dish in 10 ml of MEM-α containing 10% fetal calf serum. Incubate the cultures for 18–20 hours at 37°C in a humidified incubator with an atmosphere of 5–7% CO_2.

2. Replace the medium with 3 ml of warmed (37°C) medium containing serum, DNA (5 ng to 40 µg; no carrier DNA), and 30 µg of Polybrene. Mix the DNA with the medium before adding the 10 mg/ml Polybrene. Return the cells to the incubator for 6–16 hours. Gently rock the dishes every 90 minutes during the early stages of this incubation to ensure even exposure of the cells to the DNA-Polybrene mixture.

3. Remove the medium containing the DNA and Polybrene by aspiration. Add 5 ml of 30% DMSO in serum-containing medium. Gently swirl the DMSO medium around the dish to ensure even exposure of the cells to the solvent and place the dishes in the incubator.

4. After 4 minutes of incubation, remove the dishes from the incubator and immediately aspirate the DMSO solution. Wash the cells once or twice with warmed (37°C) serum-free medium and add 10 ml of complete medium containing 10% fetal calf serum. If a sodium butyrate boost is to be included, proceed to Step 5. If not, incubate the cultures for 48 hours at 37°C in a humidified incubator with an atmosphere of 5–7% CO_2. Then proceed directly to either Step 6 (to assay for transient expression) or Step 7 (to establish stable transformants).

5. (*Optional*) To facilitate the transfection of cells treated with DMSO and Polybrene:

    a. Add 500 mM sodium butyrate directly to the growth medium to a final concentration of 2.5–10 mM.

    b. Incubate the cells for 20–24 hours at 37°C in a humidified incubator with an atmosphere of 5–7% CO_2.

    c. Remove the medium containing sodium butyrate and replace it with butyrate-free medium containing 10% fetal bovine serum. Return the cells to the incubator.

6. If the objective is stable transformation of the cells, proceed directly to Step 7. For transient expression, examine the cells 1–2 days after transfection using one of the following assays:

    - If a plasmid DNA expressing *E. coli* β-galactosidase was used, follow the steps outlined in Protocol 17.7 to measure enzyme activity in cell lysates. Alternatively, perform a histochemical staining assay as detailed in the additional protocol (Histochemical Staining of Cell Monolayers for β-Galactosidase) in MC3, p. 16.13.

    - If a green fluorescence protein expression vector was used, examine the cells with a microscope under 450–490-nm illumination.

    - For other gene products, analyze the newly synthesized protein by radioimmunoassay, immunoblotting, immunoprecipitation following in vivo metabolic labeling, or assays of enzymatic activity in cell extracts.

To isolate stable transfectants

7. After the cells have incubated for 48 hours in nonselective medium (to allow expression of the transferred gene[s] to occur [Step 4]), either trypsinize or replate the cells in the appropriate selective medium or add the selective medium directly to the cells without further manipulation. Change this medium every 2–4 days for 2–3 weeks to remove the debris of dead cells and to allow colonies of resistant cells to grow.

8. Thereafter, clone and propagate individual colonies of transformed cells for specific experimental purposes.

CHAPTER 17

Analysis of Gene Expression in Mammalian Cells

BACKGROUND INFORMATION

Background information found in *Molecular Cloning: A Laboratory Manual*, 3rd edition (hereafter MC3) unless otherwise indicated.

Footprinting DNA	MC3, pp. 17.75–17.78
Gel retardation assays	MC3, pp. 17.78–17.80
Reporter genes	MC3, pp. 17.30–17.32

Mapping Protein-binding Sites on DNA by DNase I Footprinting

In this protocol, DNase I is used to fragment a radiolabeled target DNA in the presence and absence of a nuclear extract. A "footprint" is generated when a protein binds to the target and protects a specific segment of DNA from the nucleolytic activity of DNase I. By comparing the electrophoretic mobility of the DNase I cleavage products to those of a sequence ladder derived from the same DNA fragment, the position(s) of the DNA sequences recognized by DNA-binding proteins can be determined.

MATERIALS

CAUTION: Please see Appendix 4 for appropriate handling of materials marked with <!>.

Reagents and Solutions

Please see Appendix 1 for components of stock solutions, buffers, and reagents. Dilute stock solutions to the appropriate concentrations.

Bradford reagent for quantifying protein concentration

Cell homogenization buffer
- 10 mM HEPES-KOH (pH 7.9) <!>
- 1.5 mM $MgCl_2$
- 10 mM KCl
- 0.5 mM dithiothreitol
- 0.5 mM phenylmethylsulfonyl fluoride (PMSF) <!>

An alternative to PMSF, Pefabloc SC (Roche Applied Science), is nontoxic and stable in buffered aqueous solutions.

Cell homogenization buffer containing 0.05% (v/v) Nonidet P-40

Cell resuspension buffer
- 40 mM HEPES-KOH (pH 7.9)
- 0.4 M KCl
- 1 mM dithiothreitol
- 10% (v/v) glycerol
- 0.1 mM PMSF <!>
- 0.1% (w/v) aprotinin (a protease inhibitor)

An alternative to PMSF, Pefabloc SC (Roche Applied Science), is nontoxic and stable in buffered aqueous solutions.

Cell rinse buffer
- 40 mM Tris-Cl (pH 7.4)
- 1 mM EDTA
- 0.15 M NaCl

Ethanol

Ethanol (70%)

Ficoll 400 (20% [w/v] in sterile H_2O)

$MgCl_2$-$CaCl_2$ solution, sterile
- 10 mM $MgCl_2$
- 5 mM $CaCl_2$

NaCl (5 M)

Phenol:chloroform (1:1, v/v) <!>

Phosphate-buffered saline (PBS) lacking calcium and magnesium salts

Polyvinyl alcohol (10% [w/v] in H_2O)

Stop mix
- 20 mM EDTA (pH 8.0)
- 1% (w/v) SDS

Adapted from Chapter 17, Protocol 1, p. 17.4 of MC3.

0.2 M NaCl

125 µg/ml yeast tRNA

Tissue homogenization buffer

10 mM HEPES-KOH (pH 7.9)

25 mM KCl

0.15 mM spermine

0.5 mM spermidine

1 mM EDTA (pH 8.0)

2 M sucrose

10% (v/v) glycerol

The buffer should be used ice cold. Add protease inhibitors (e.g., 0.5 mM PMSF, Pefabloc, 1 µg/ml leupeptin, 1 µg/ml pepstatin, or others as needed) just before use (see MC3, p. A5.1).

Tissue resuspension buffer

5 mM HEPES-KOH (pH 7.9)

1.5 mM MgCl$_2$

0.5 mM dithiothreitol

0.5 mM PMSF *or* Pefabloc

26% (v/v) glycerol

Trypan Blue dye (0.4% [w/v] in PBS lacking calcium and magnesium salts)

Cells, Tissues/Culture Media

Fresh tissue, cultured cells, or protein extracted from them

Enzymes and Buffers

DNase I (1 mg/ml in 10 mM Tris-Cl, pH 8.0)

Nucleic Acids/Oligonucleotides

Poly(dl-dC) (1 mg/ml in H$_2$O)

The copolymer is used to suppress nonspecific binding of proteins to the radiolabeled DNA fragment. The optimum concentration (usually between 0 and 100 µg/ml) of the copolymer should be determined empirically and in comparison to other competitors such as sheared genomic DNA (*E. coli*, salmon sperm, poly[dA-dT], etc.).

Radioactive Compounds

^{32}P end-labeled DNA (200–500 bp in length, sp. act. ≥2.5 x 10^7 cpm/µg [i.e., ≥5000 cpm/fmole]) <!>

For end-labeling methods, see Protocols 9.13–9.16, 10.2, 10.7, 9.10, 9.11, or 8.1. Ideally, the binding site for the target protein should be located at least 30 bp from the labeled end of the DNA. Sites that are closer to the termini may not be recognized by the DNA-binding protein of DNase I.

Gels/Loading Buffers

Denaturing 6% or 8% polyacrylamide DNA-sequencing gel, cast and run in 1x TBE (see Protocol 12.8) <!>

Formamide dye mix

10 ml formamide

10 mg xylene cyanol FF

10 mg bromophenol blue

Radiolabeled DNA size standards, e.g., the products of a dideoxysequencing reaction of a DNA whose 5′ terminus is identical to the radiolabeled end of the test DNA

Centrifuges/Rotors/Tubes

Beckman SW 28 rotor or equivalent (4°C)

Sorvall H-1000B rotor or equivalent (4°C)

Additional Items

Boiling-water bath

Dounce homogenizer with type-B pestle

Gel dryer

Liquid nitrogen <!>

Materials for autoradiography or phosphorimaging

Rubber policemen

Steps 7–10 of this protocol require the reagents listed in Protocols 12.8, 12.11, and 12.12.

Additional Information

Alternative protocol (Mapping Protein-binding Sites on DNA by Hydroxyl Radical Footprinting)	MC3, p. 17.12
Poly(dI-dC)	MC3, p. 17.14
Troubleshooting and optimization of DNase I footprinting	MC3, p. 17.11

METHOD

1. Prepare nuclear extracts using one of the following three methods. Alternatively, fractions derived from purification of cellular proteins can be used directly in Step 2.

Preparation of nuclear extracts from tissue

a. Dissect and mince 10–15 g of tissue. Adjust the volume of minced tissue to 30 ml with ice-cold tissue homogenization buffer. Homogenize in a tight-fitting Dounce homogenizer until >80–90% of the cells are broken as determined by microscopy.

b. To monitor lysis, mix 10 µl of the cell suspension with an equal volume of 0.4% Trypan Blue dye and examine the solution under a microscope equipped with a 20x objective. Lysed cells take up the dye and stain blue, whereas intact cells exclude dye and remain translucent. Continue to homogenize the tissue until >80–90% of the cells are broken.

c. Dilute the homogenate to 85 ml with ice-cold tissue homogenization buffer. Layer 27-ml aliquots over 10-ml cushions of ice-cold tissue homogenization buffer in ultraclear or polyallomer swinging-bucket centrifuge tubes. Centrifuge the tubes at 103,900*g* (24,000 rpm in a Beckman SW 28 rotor) for 40 minutes at 4°C.

d. Decant the supernatant and allow the tubes to drain in an inverted position for 1–2 minutes. Place the tubes on ice.

   (*Optional*) Use a razor blade to cut off the top two thirds of the tube and place the bottom one-third containing the nuclei on ice.

e. Resuspend the pellet of nuclei in 2 ml of ice-cold tissue resuspension buffer. Accurately measure the volume of the resuspended nuclei and add ice-cold 5 M NaCl to a final concentration of 300 mM. Mix the suspension gently. Incubate the suspension for 30 minutes on ice.

f. Recover the nuclei by centrifugation at 103,900*g* (24,000 rpm in a Beckman SW 28 rotor) for 20 minutes at 4°C. Carefully transfer the supernatant to a fresh tube. Divide the supernatant into aliquots of 100–200 µl. Reserve an aliquot for protein concentration determination. Snap-freeze the remainder of the aliquots in liquid nitrogen and store them in liquid nitrogen.

g. Determine the protein concentration of the supernatant by the Bradford method.

Preparation of nuclear extracts from cultured mammalian cells

a. Harvest 0.5×10^8 to 1×10^8 cells from their culture flasks, plates, or wells. Collect the cells by centrifugation at 250*g* (1100 rpm in a Sorvall H-1000B rotor) for 10 minutes at room temperature. Rinse the cells several times with PBS without calcium and magnesium salts.

b. Resuspend the cell pellet in 5 volumes of ice-cold cell homogenization buffer. Incubate the cells for 10 minutes on ice and then collect them by centrifugation as before.

c. Resuspend the cell pellet in 3 volumes of ice-cold cell homogenization buffer containing 0.05% (v/v) Nonidet P-40 and homogenize the cells with 20 strokes of a tight-fitting Dounce homogenizer. The body of the homogenizer should be buried in ice during the homogenization process, during which the swollen cells lyse and release intact nuclei.

d. Collect the nuclei by centrifugation at 250*g* (1100 rpm in a Sorvall H-1000B rotor) for 10 minutes at 4°C. Remove the supernatant and resuspend the pellet of nuclei in 1 ml of cell resuspension buffer. Accurately measure the volume of the resuspended nuclei and add 5 M NaCl to a final concentration of 300 mM. Mix the suspension gently and incubate for 30 minutes on ice.

e. Recover the nuclei by centrifugation at 103,900*g* (24,000 rpm in a Beckman SW 28 rotor) for 20 minutes at 4°C. Carefully transfer the supernatant to a chilled, fresh tube. Divide the supernatant into aliquots of 100–200 μl. Reserve an aliquot for protein concentration determination. Quick-freeze the aliquots in liquid nitrogen and store them in liquid nitrogen. Determine the protein concentration of the supernatant by the Bradford method.

Preparation of nuclear extracts from small numbers of cultured mammalian cells

a. Rinse the cells with several changes of cell rinse buffer. Add 1 ml of the cell rinse buffer to each dish and scrape the cells into the buffer using a rubber policeman.

> This procedure is suitable for cells transfected with plasmids expressing cDNAs encoding transcription factors.

b. Transfer the cell suspension to a 1.5-ml microfuge tube and pellet the cells by centrifuging at maximum speed for 2 minutes at room temperature in a microfuge.

c. Resuspend the cell pellet in 300 μl of cell resuspension buffer per 150-mm dish of original cells. Subject the resuspended cells to three cycles of freezing and thawing.

d. Remove the cellular debris by centrifuging the tubes at maximum speed for 5 minutes at 4°C in a microfuge. Store the supernatant (i.e., the cell lysate) in small aliquots at –70°C.

2. To an appropriate number of 1.5-ml microfuge tubes, add:

nuclear extract or protein fraction (5–10 μg)	1–23 μl
^{32}P end-labeled DNA	1–10 fmoles
1 mg/ml poly(dI-dC)	1 μl
H$_2$O	to 25 μl
Optional additions:	
20% Ficoll 400	12 μl
or	
10% polyvinyl alcohol	10 μl

Centrifuge the tubes for 5 seconds at 4°C in a microfuge to deposit the reaction mixtures at the bottom of the tubes. Incubate the reaction mixtures for 10–30 minutes on ice.

> For each DNA fragment or fraction to be assayed, set up two control reactions: one control without the nuclear extract, the other without addition of DNase 1 in Step 3.

3. Add 50 μl of MgCl$_2$-CaCl$_2$ solution at room temperature and mix gently. Incubate the reactions for 1 minute at room temperature. Add 1–8 μl of diluted DNase I solution to the microfuge tubes, mix gently, and incubate the reactions for 1 minute at room temperature.

4. Stop the reactions by adding 75 μl of stop mix. Vortex briefly and extract the reactions with an equal volume of phenol:chloroform.

5. Transfer the aqueous phases to fresh microfuge tubes and precipitate the nucleic acids with 2.5 volumes of ethanol. Chill the ethanolic solution for 15 minutes at –70°C and collect precipitates by centrifugation at maximum speed for 10 minutes at 4°C in a microfuge. Rinse the pellets with 1 ml of 70% ethanol, centrifuge again, and air dry to remove the last traces of ethanol.

6. Solubilize the DNA pellets in 5–10 μl of formamide dye mix by vigorous vortexing. Denature the DNA solutions by boiling for 3–5 minutes.

7. Set up a denaturing 6% or 8% polyacrylamide sequencing gel and run the gel for at least 30 minutes before loading the DNA samples.

8. Load the DNA samples in the following order:
   sequence ladder
   control DNA digested with DNase I in the absence of nuclear extract
   target DNA from reactions digested with DNase I in the presence of nuclear extract
   target DNA incubated with nuclear extract and no DNase I

9. Run the gel at sufficient constant power to maintain a temperature of 45–50°C.

   The time required to achieve optimal resolution of the sequence of interest must be determined empirically.

10. After electrophoresis is complete, pry the glass plates apart and transfer the gel to a piece of thick blotting paper. Dry the gel under vacuum for approximately 1 hour and expose it to X-ray film without an intensifying screen for 12–16 hours at –20°C. Alternatively, subject the dried gel to phosphorimage analysis for 1–3 hours.

Gel Retardation Assay for DNA-binding Proteins

This protocol exploits differences in electrophoretic mobility through a nondenaturing polyacrylamide gel between a rapidly migrating target DNA and a more slowly migrating DNA–protein complex.

MATERIALS

CAUTION: Please see Appendix 4 for appropriate handling of materials marked with <!>.

Reagents and Solutions

Please see Appendix 1 for components of stock solutions, buffers, and reagents. Dilute stock solutions to the appropriate concentrations.

Bradford reagent for quantifying protein concentration

Ficoll 400 (20% [w/v] in sterile H_2O)

Polyvinyl alcohol (10% [w/v] in H_2O)

Cells, Tissues/Culture Media

Nuclear extract or protein fractions prepared from target cells and control cells

Prepare the nuclear extract by one of the methods outlined in Protocol 17.1.

Nucleic Acids/Oligonucleotides

Poly(dI-dC) (1 mg/ml in H_2O)

The copolymer is used to suppress nonspecific binding of proteins to the radiolabeled DNA fragment. The optimum concentration (usually between 0 and 100 µg/ml) of the copolymer should be determined empirically.

Radioactive Compounds

^{32}P-labeled control DNA <!>

^{32}P-labeled target DNA (>20 bp in length, sp. act. ≥2.5 x 10^7 cpm/µg

[i.e., [≥5000 cpm/fmole]) <!>

Labeling the DNA fragment can be accomplished by phosphorylation (Protocols 9.10 and 10.2), end filling (Protocol 9.10), or PCR (Protocol 8.1).

Gels/Loading Buffers

Neutral (nondenaturing) 4–7% polyacrylamide gel, approximately 1.5 mm thick, cast and run in 0.5x TBE (see Protocol 12.8) <!>

Sucrose dye solution

0.25% (w/v) bromophenol blue

0.25% (w/v) xylene cyanol

40% (w/v) sucrose

Additional Items

Gel dryer

Materials for autoradiography or phosphorimaging

Additional Information

Additional protocol (Competition Assays)	MC3, p. 17.17
Additional protocol (Supershift Assays)	MC3, p. 17.17
Alternative protocol (Mapping Protein-binding Sites on DNA by Hydroxyl Radical Footprinting)	MC3, p. 17.12
Gel retardation assays	MC3, pp. 17.78–17.80
Poly(dI-dC)	MC3, p. 17.14
Troubleshooting and optimization	MC3, p. 17.16

METHOD

1. To a sterile 1.5-ml microfuge tube, add:

^{32}P-labeled target DNA	1 ng (1–10 fmoles)
1 mg/ml poly(dI-dC)	1 μl
nuclear extract (5–10 μg)	≤10 μl
or	
protein fraction	≤10 μl
20% Ficoll 400	5 μl
or	
10% polyvinyl alcohol	4 μl
H$_2$O	to 20 μl

   Include control reactions with every experiment. Positive-control reactions contain a nuclear extract (or protein fraction) and a radiolabeled DNA fragment carrying a sequence recognized by a DNA-binding protein that is abundant in the extract and has high affinity for the DNA sequence. Examples are a DNA fragment containing an Sp1, C/EBP, or NF-1 site and mammalian cell nuclear extract, or a *lacI* recognition site and extract derived from a *lacI^q* strain of *E. coli*. The negative-control reactions contain the radiolabeled target DNA fragment, but no nuclear extract.

2. Centrifuge the reaction tubes for several seconds in a microfuge to deposit the reaction mixtures at the bottom of the tubes. Incubate the reactions for 10–30 minutes on ice.

3. Add 3 μl of sucrose dye solution to each tube. Load the samples into the slots of a neutral 4–7% polyacrylamide gel.

4. Run the gel in either 0.5x Tris-glycine buffer or 0.5x TBE buffer at 200–250 V and 20 mA for ≥2 hours.

   Depending on the lability of the binding protein(s) and the affinity of the binding reaction(s), it may be necessary to run the gel at 4°C.

5. After electrophoresis is complete, pry the gel plates apart, transfer the gel to a piece of sturdy blotting paper, and dry the gel for approximately 1 hour on a gel dryer.

6. Expose the dried gel to X-ray film for ≥1 hour at –20°C to visualize radiolabeled DNA fragments. Less abundant DNA–protein complexes can be detected after 1–3 hours on a phosphorimager.

Mapping DNase I Hypersensitive Sites

In this protocol, nuclei isolated from mammalian cells are incubated with varying amounts of DNase I. Genomic DNA is then isolated from the nuclei and digested with a restriction enzyme, analyzed by gel electrophoresis, and probed by Southern hybridization. If the probe corresponds to the 5′ end of the gene, intact restriction fragments arising from that region will be detected in DNA isolated from control nuclei not treated with DNase I. If DNase I hypersensitive sites exist in one or more of the fragments recognized by the probe, shorter DNAs will be detected on the Southern blot.

MATERIALS

CAUTION: Please see Appendix 4 for appropriate handling of materials marked with <!>.

Reagents and Solutions

Please see Appendix 1 for components of stock solutions, buffers, and reagents. Dilute stock solutions to the appropriate concentrations.

    Bradford reagent for quantifying protein concentration

    Buffer A (4°C)

        50 mM Tris-Cl (pH 7.9)

        100 mM NaCl

        3 mM $MgCl_2$

        1 mM dithiothreitol

        0.2 mM phenylmethylsulfonyl fluoride (PMSF) <!> or Pefabloc

        Add PMSF and dithiothreitol just before using the buffer.

    EDTA (0.5 M, pH 8.0)

    Ethanol

    Lysis buffer (4°C)

        50 mM Tris-Cl (pH 7.9)

        100 mM KCl

        5 mM $MgCl_2$

        0.05% (v/v) saponin

        50% (v/v) glycerol

        200 mM β-mercaptoethanol <!>

        Add β-mercaptoethanol just before using the buffer.

    Phenol:chloroform <!>

    Phosphate-buffered saline (PBS) without calcium and magnesium salts, ice cold

    SDS buffer

        20 mM Tris-Cl (pH 7.9)

        100 mM NaCl

        70 mM EDTA (pH 8.0)

        2% (w/v) SDS

    TE (pH 7.9)

    Trypan Blue dye (0.4% w/v)

Cells, Tissues/Culture Media

Approximately 10^8 mammalian cells are required per DNase I hypersensitivity experiment.

Enzymes and Buffers

DNase I, RNase-free (10 units/ml) in DNase I dilution buffer
10 mM HEPES-KOH (pH 7.9) <!>
30 mM $CaCl_2$
30 mM $MgCl_2$
50% (v/v) glycerol
Store the diluted enzyme at –20°C.
Proteinase K (0.2 mg/ml) in 50 mM Tris-Cl (pH 7.9) containing 100 mM NaCl
Store the diluted enzyme at –20°C.
RNase A, DNase-free (0.5 mg/ml), in TE (pH 8.0)

Centrifuges/Rotors/Tubes

Sorvall H-1000B rotor (4°C)

Additional Items

Gel dryer
Materials for autoradiography or phosphorimaging
Shaking water baths (37°C, 50°C, and 55°C)
Step 18 of this protocol requires the reagents listed in Protocols 6.8 and 6.10.

Additional Information

Controls for DNase I hypersensitivity mapping	MC3, p. 17.22
Hybridization probes for DNase I hypersensitivity mapping	MC3, p. 17.21

METHOD

1. Harvest approximately 10^8 cells from spinner cultures, flasks, or dishes and wash the cells twice with 25-ml aliquots of ice-cold PBS without calcium and magnesium salts.

    Alternatively, if starting with fresh tissue, isolate the nuclei as described in Protocol 17.1, Step 1.

2. Resuspend the cell pellet from the final wash in 1.5 ml of ice-cold lysis buffer. Incubate the cells for 10 minutes on ice to allow cell lysis to occur.

3. Mix a 10-µl aliquot of cell lysate with an equal volume of 0.4% Trypan Blue dye and examine the solution under a microscope equipped with a 20x objective. Lysed cells and nuclei take up the dye and appear blue, whereas unlysed cells are impermeable to the dye and remain translucent. Continue the incubation on ice until >80% of cells are lysed.

4. Recover nuclei from the lysed cells by centrifugation of the suspension at 1300*g* (2500 rpm in a Sorvall H-1000B rotor) for 15 minutes at 4°C.

5. Carefully remove the supernatant and resuspend the pellet of nuclei in 1.5 ml of ice-cold buffer A. Collect the nuclei by centrifugation as described in Step 4.

6. Resuspend the nuclear pellet in 4 ml of ice-cold buffer A.

7. Set up a series of dilutions of the standard DNase I solution (1/40, 1/80, 1/160, 1/320, 1/640, 1/1280, and 1/2560 in DNase dilution buffer). Store the dilutions on ice.

8. Label a series of tubes 1–9 and add 180 µl of resuspended nuclei from Step 6 to each tube.

9. To Tube 1, add 20 µl of DNase I dilution buffer containing no DNase I and store the tube on ice until Step 12, below. To Tube 2, add 20 µl of DNase I dilution buffer containing no DNase and incubate as described in Step 11, below. Tubes 1 and 2 are controls.

10. To Tubes 3–9, add 20 μl of each of the progressive dilutions, i.e., to Tube 3, add 20 μl of the 1/2560 dilution; to Tube 4, add 20 μl of the 1/1280 dilution; etc.

11. Incubate Tubes 2–9 for 20 minutes at 37°C.

12. Terminate the reactions by adding three individual aliquots of 16.6 μl of 0.5 M EDTA to each tube, with vortexing between additions. When all of the tubes have been treated, add 12 μl of RNase solution to each tube. Incubate the reactions for 30 minutes at 37°C to allow digestion of nuclear RNA.

13. Digest nuclear proteins by adding 40 μl of proteinase K solution to each tube. Mix the solution gently by pipetting up and down. Add 100 μl of SDS buffer to each tube and mix once more. Incubate the tubes for 16 hours at 50°C with rotation or rocking.

14. Add an additional aliquot of 100 μl of proteinase K solution and continue the digestion for an additional 2–3 hours at 50°C.

15. Extract the digestion mixtures three times with phenol:chloroform. Be gentle. Precipitate the DNA with the addition of 3 volumes of ice-cold ethanol, incubate for 30 minutes on ice, and collect the DNA precipitates by centrifugation at 1200g (2400 rpm in a Sorvall H-1000B rotor). Decant the supernatant and drain the last dregs of ethanol from the tubes on a paper towel.

16. Add 200 μl of TE to each tube and allow the DNA to redissolve with rocking or rotation overnight at 55°C.

    IMPORTANT: An extended incubation is required for complete solubilization and recovery of the DNase-treated DNA.

17. Determine the A_{260} of the resuspended DNA and estimate the concentration.

18. Digest the DNA with a restriction enzyme(s), followed by Southern blotting and hybridization as described in Protocols 6.8 and 6.10. Load 15–30 μg of restricted genomic DNA per lane on the agarose gel.

    It is crucial to use high-specific-activity radioactive probes when mapping hypersensitive sites. The specific activity of the probe should be >5 x 10^8 cpm/μg. Single-stranded DNA probes derived from bacteriophage M13 templates (Protocol 3.4) are ideal for high-resolution hypersensitivity mapping experiments.

Transcriptional Run-on Assays

In this protocol, nuclei isolated from cells expressing the gene of interest are incubated with radiolabeled UTP, which is incorporated into nascent RNA transcripts by RNA polymerase molecules that were actively transcribing at the time the cells were harvested. Because very little de novo initiation of RNA synthesis occurs in isolated nuclei, transcription of the target gene can be measured by hybridizing the radiolabeled RNA to an excess of the target gene immobilized on a nitrocellulose or nylon membrane. The fraction of the RNA that hybridizes to the immobilized DNA reflects the contribution of the target gene to the total transcriptional activity of the cell. All test tubes and solutions must be prepared RNase-free.

MATERIALS

CAUTION: Please see Appendix 4 for appropriate handling of materials marked with <!>.

Reagents and Solutions

Please see Appendix 1 for components of stock solutions, buffers, and reagents. Dilute stock solutions to the appropriate concentrations.

Chloroform <!>

Chloroform:isoamyl alcohol (24:1, v/v) <!>

DNA denaturing solution

    2 M NaCl

    0.1 M NaOH <!>

Ethanol

Ethanol (70%)

Glycerol storage buffer

    50 mM Tris-Cl (pH 8.3)

    5 mM $MgCl_2$

    0.1 mM EDTA (pH 8.0)

    40% (v/v) glycerol

    Store at 4°C.

HSB buffer

    10 mM Tris-Cl (pH 7.4)

    50 mM $MgCl_2$

    2 mM $CaCl_2$

    0.5 M NaCl

    Store at room temperature.

Labeling buffer

    20 mM Tris-Cl (pH 8.0 at 4°C)

    140 mM KCl

    10 mM $MgCl_2$

    1 mM $MoCl_2$

    20% (v/v) glycerol

    Store the solution at 4°C. Just before use, add the following five ingredients to final concentrations of

        14 mM β-mercaptoethanol <!>

        1 mM each of ATP, GTP, and CTP

        10 mM phosphocreatine

        100 µg/ml phosphocreatine kinase

        0.1 µM [α-^{32}P]UTP (sp. act. 500–5000 Ci/mmole) <!>

LiCl (5 M)

Lysis buffer

    10 mM Tris-Cl (pH 8.4 at 4°C)

    1.5 mM $MgCl_2$

    0.14 M NaCl

    Store the buffer at 4°C.

NaCl (2 M)

NaOH (0.1 M) <!>

Nonidet P-40 (5% v/v)

Nuclei wash buffer

    20 mM Tris-Cl (pH 8.0 at 4°C)

    140 mM KCl

    10 mM $MgCl_2$

    1 mM $MoCl_2$

    20% (v/v) glycerol

    Store at 4°C. Just before use, add β-mercaptoethanol to a final concentrated of 14 mM.

Phenol <!>, water saturated and equilibrated to pH 7.2

Phosphate-buffered saline (PBS)

Prehybridization/hybridization buffer <!>

    50% (v/v) formamide

    6x SSC

    5 mM sodium pyrophosphate

    2x Denhardt's solution

    0.5% (w/v) SDS

    10 µg/ml poly(A)

    100 µg/ml salmon sperm DNA (see Protocol 6.10)

2x Reaction buffer

    10 mM Tris-Cl (pH 8.0 at 4°C)

    5 mM $MgCl_2$

    0.3 M KCl

    Store at 4°C. Just before use, add the following to 1 ml of 2x reaction buffer.

        5 µl dithiothreitol (1 M)

        100 µl 100-mM ATP

        10 µl 100-mM CTP

        10 µl 100-mm GTP

SDS (0.5% w/v)

6x SSC

Stop buffer

    50 mM Tris-Cl (pH 7.5)

    20 mM EDTA (pH 8.0)

    0.8% (w/v) SDS

Tissue homogenization buffer

    10 mM HEPES-KOH (pH 7.6)

    25 mM KCl

    0.15 mM spermine

    0.5 mM spermidine

    1 mM EDTA (pH 8.0)

    2 M sucrose

    10% (v/v) glycerol

    Store at 4°C. Just before use, add dithiothreitol to a final concentration of 1 mM and protease inhibitors (e.g., PMSF or Pefabloc to 0.5 mM; leupeptin *and/or* pepstatin to 1 µg/ml).

Trypan Blue dye (0.4% w/v) in PBS lacking calcium and magnesium salts

Vectors and Hosts

Nonrecombinant plasmid vector

Recombinant plasmid containing the gene of interest

Cells, Tissues/Culture Media

10^7–10^8 Cultured mammalian cells *or* 10–15 g of fresh tissue

Enzymes and Buffers

DNase I, RNase-free (2 mg/ml), in 0.0025 N HCl, 50% (v/v) glycerol
> Store the diluted enzyme at –20°C.

Proteinase K (optional; see Step 4)
> Dissolve the enzyme at a concentration of 1 mg/ml in 50 mM Tris-Cl (pH 7.9) containing 100 mM NaCl. Store the enzyme at –20°C.

RNasin or equivalent RNase inhibitor (see MC3, p. 7.83)
Restriction enzymes and 10x buffers

Radioactive Compounds

[α-^{32}P]UTP (sp. act. 500–5000 Ci/mmole) <!>
> Use 100 μCi for each sample of nuclei.

Centrifuges/Rotors/Tubes

Beckman SW 28 rotor (4°C)
Sorvall H-1000B and SS-34 rotors (4°C)

Additional Items

Boiling-water bath
Dounce homogenizer with type-B pestle
Gel dryer
Materials for autoradiography or phosphorimaging
Materials for TCA precipitation of radiolabeled RNA (see Appendix 3, p. 745)
Razor blades
Rubber policeman
Shaking water bath (30°C)
Slot-blotting apparatus
Step 11 of this protocol requires the reagents and equipment listed in Protocol 7.9.
Step 13 of this protocol requires the reagents listed in Protocol 6.8.
Steps 14–16 of this protocol require the reagents listed in Protocols 6.8 and 6.10.
Water baths (4°C and 65°C)

Additional Information

Controls and troubleshooting   MC3, p. 17.29

METHOD

1. Isolate nuclei from either cultured cells or fresh tissue.

   ### Isolation of nuclei from cultured cells

   a. Using a rubber policeman, scrape the cells from culture dishes and wash them twice with ice-cold PBS. Resuspend 1 x 10^7 to 1 x 10^8 cells in 1 ml of ice-cold lysis buffer in a 17 x 100-mm polypropylene tube. Add 2–4 μl of 5% Nonidet P-40 and incubate the suspension for 10 minutes on ice.

   b. Mix 10 μl of the suspension with an equal volume of 0.4% Trypan Blue dye and examine the solution under a microscope equipped with a 20x objective. Lysed cells take up the dye and appear blue, whereas unlysed cells are impermeable to the dye and remain translucent. Continue adding 2-μl aliquots of 5% Nonidet P-40 and check cell lysis until >80% of the cells are lysed.

c. Recover the nuclei by centrifugation at 1300*g* (2500 rpm in a Sorvall H-1000B rotor) for 1 minute in a benchtop centrifuge. Remove and discard the supernatant. Wash the pellet of nuclei twice in 1 ml of ice-cold nuclei wash buffer. Proceed to Step 2.

Isolation of nuclei from tissue

a. Dissect and mince 10–15 g of tissue. Adjust the volume of minced tissue to 30 ml with ice-cold tissue homogenization buffer and homogenize in a tight-fitting Dounce homogenizer.

b. To monitor lysis, mix 10 μl of the cell suspension with an equal volume of 0.4% Trypan Blue dye and examine the solution under a microscope equipped with a 20x objective. Lysed cells take up the dye and stain blue, whereas intact cells exclude dye and remain translucent. Continue to homogenize the tissue until >80–90% of the cells are broken.

c. Dilute the homogenate to 85 ml with ice-cold tissue homogenization buffer. Layer 27-ml aliquots over 10-ml cushions of ice-cold homogenization buffer in ultraclear or polyallomer swinging-bucket centrifuge tubes. Centrifuge the tubes at 103,900*g* (24,000 rpm in a Beckman SW 28 rotor) for 40 minutes at 4°C.

d. Decant the supernatant and allow the tubes to drain in an inverted position for 1–2 minutes. Place the tubes on ice. Resuspend each pellet in 2 ml of glycerol storage buffer by pipetting the mixture up and down.

e. Mix 10 μl of resuspended nuclei with 990 μl of 0.5% SDS. Measure the OD_{260} in a UV spectrophotometer and dilute resuspended nuclei with glycerol storage buffer to a final concentration of 50 OD_{260}/ml. Divide the preparation of nuclei into 200-μl aliquots in 1.5-ml microfuge tubes, snap-freeze the aliquots in liquid nitrogen, and store them at –70°C. Proceed to Step 2.

2. Radiolabel the nascent RNA transcripts in the isolated nuclei.

Radiolabeling of the transcripts in nuclei isolated from cultured cells

a. Remove as much supernatant as possible from the last wash (Step 1c) and resuspend the nuclei in 50–100 μl of nuclei labeling buffer. Incubate the nuclei for 15–20 minutes at 30°C in a shaking water bath.

b. Pellet the nuclei by centrifugation at 800*g* (1960 rpm in a Sorvall H-1000B rotor) for 5 minutes in a benchtop centrifuge and carefully discard the supernatant as radioactive waste. Proceed to Step 3.

Radiolabeling of the transcripts in nuclei isolated from tissue

a. Transfer an appropriate number of aliquots of the nuclear preparation from –70°C to an ice bucket. When the aliquots have thawed, add 400 units of RNasin to each tube. Add 200 μl of 2x reaction buffer supplemented with nucleotides and dithiothreitol to each tube of nuclei. Add 100 μCi of [α-^{32}P]UTP.

b. Incubate the nuclei for 20 minutes in a 30°C shaking water bath. Recover the nuclei by centrifugation at 2000 rpm for 1–2 minutes. Carefully discard the supernatant as radioactive waste. Proceed to Step 3.

3. Resuspend the nuclear pellet in 1 ml of ice-cold HSB buffer and add 10 μl of DNase solution. Pipette the nuclei up and down until they are resuspended and the viscosity of the solution is reduced (1–5 minutes). Add 2 ml of stop buffer.

4. (*Optional*) To increase the yield of radiolabeled RNA, add proteinase K after the DNase step. Following the addition of 2 ml of stop buffer, add proteinase K to a final concentration of 100 µg/ml. Incubate the solution for 30 minutes at 42°C. Proceed to Step 5.

5. Add 3 ml of phenol and incubate the mixture for 15 minutes at 65°C with vortexing every 5 minutes. Add 3 ml of chloroform:isoamyl alcohol, vortex, and separate the organic and aqueous phases by centrifugation at 1900*g* (3000 rpm in a Sorvall H-1000B rotor). Extract the aqueous layer again with 3 ml of chloroform, centrifuge as before, and transfer the aqueous layer to a fresh tube.

6. Add 0.3 ml of 5 M LiCl and 2.5 volumes of ethanol; mix well. Collect precipitated nucleic acids by centrifugation at 12,000*g* (10,000 rpm in a Sorvall SS-34 rotor) for 10 minutes.

7. Resuspend the pellet in 0.4 ml of H_2O and transfer it to a 1.5-ml microfuge tube. Add 40 µl of 5 M LiCl and 2.5 volumes of ethanol. Centrifuge the solution for 10 minutes at maximum speed in a microfuge.

8. Resuspend the pellet in 100 µl of H_2O and measure the cpm/µl in a liquid scintillation counter.

9. Linearize 10 µg of recombinant plasmid DNA containing the cDNA or gene of interest and 10 µg of empty plasmid vector using a restriction enzyme whose sites of cleavage are present in the vector sequences.

10. Recover the cleaved DNAs using standard ethanol precipitation and resuspend each of the pellets separately in 20 µl of DNA denaturation solution. Boil the resuspended DNAs for 2 minutes and then add 180 µl of 6x SSC to each tube.

11. Cut a piece of nylon or nitrocellulose membrane to the appropriate size for use in a dot- or slot-blotting apparatus. Wet the membrane in H_2O and then soak it for 5–10 minutes in 6x SSC. Clamp the wet membrane in the blotting apparatus, attach a vacuum line, and apply suction to the device (please see Protocol 7.9).

12. Filter the denatured DNAs through separate slots and wash each filter with 200 µl of 6x SSC.

13. Dismantle the device, air dry the membrane, and fix the DNA to the membrane by baking or by exposure to UV light.

14. Place the membrane in prehybridization solution and incubate it for at least 16 hours at an appropriate temperature (e.g., 42°C for solvents containing 50% formamide).

15. Add the radiolabeled probe (2×10^6 to 4×10^6 cpm/ml of ^{32}P-labeled RNA from Step 8) directly to the prehybridization solution and incubate the filter for an additional 72 hours.

16. Wash the membrane at high stringency and expose it to X-ray film or a phosphorimager plate. Typical exposure times are 24–72 hours for X-ray film and 4–24 hours for a phosphorimager plate.

Measurement of Chloramphenicol Acetyltransferase in Extracts of Mammalian Cells

In this protocol, extracts prepared from cells transfected with a chloramphenicol acetyltransferase (CAT) reporter plasmid are incubated with radiolabeled chloramphenicol. The acetylated products generated by the action of CAT are separated from the unmodified drug by thin-layer chromatography and quantitated by scraping the spots from the thin-layer plates and counting them by scintillation spectroscopy.

MATERIALS

CAUTION: Please see Appendix 4 for appropriate handling of materials marked with <!>.

Reagents and Solutions

Please see Appendix 1 for components of stock solutions, buffers, and reagents. Dilute stock solutions to the appropriate concentrations.

Bradford reagent for quantifying protein concentration

CAT lysis buffer (optional; see Step 4)

> 0.1 M Tris-Cl (pH 7.8)
>
> 0.5% (v/v) Triton X-100
>
> It is important to use a highly purified preparation of Triton X-100; lower-quality preparations can inhibit CAT activity. If problems arise, substitute Nonidet P-40 (0.125% v/v) for Triton X-100 in the lysis buffer.
>
> The same lysis buffer can also be used to prepare cell extracts for measurements of β-galactosidase activity (see Protocol 17.7).

CAT reaction mixture 1

> 50 μl 1 M Tris-Cl (pH 7.8)
>
> 10 μl [^{14}C]chloramphenicol (60 mCi/mmole), diluted in H$_2$O to 0.1 mCi/ml <!>
>
> 20 μl acetyl-CoA (freshly prepared at a concentration of 3.5 mg/ml in H$_2$O)
>
> Prepare 80 μl of CAT reaction mixture 1 per 50 μl of cell lysate to be assayed.

Ethyl acetate <!>

> Use to separate acetylated and nonacetylated forms of chloramphenicol.

Phosphate-buffered saline (PBS) without calcium and magnesium salts

Thin-layer chromatography (TLC) solvent

> 190 ml chloroform <!>
>
> 10 ml methanol <!>
>
> Prepare 200 ml per standard TLC tank. Two TLC plates can be developed per tank.

Tris-Cl (1 M, pH 7.8)

Cells, Tissues/Culture Media

Cultured mammalian cells transfected with pCAT vector (e.g., of the pCAT3 series) carrying the DNA of interest

> The cells (growing in 90-mm dishes) should be transfected with a pCAT reporter construct and a plasmid containing a reporter gene (e.g., pCMV-SPORT-β-gal) suitable for normalizing the results of the CAT assay.

Radioactive Compounds

[^{14}C]chloramphenicol (60 mCi/ml) <!>

Additional Items

Dry-ice/ethanol bath (optional; see Step 4) <!>

Hair dryer

> Liquid scintillation spectroscope
> Luminescent adhesive labels (used in Step 13 to orient the autoradiograph with the TLC plate)
> Marking pen with ethanol-insoluble ink
> Materials for autoradiography or phosphorimaging
> Rotary vacuum evaporator (Savant SpeedVac or equivalent)
> Rubber policeman
> Soft-lead pencil
> Thin-layer chromatography plates
> Thin-layer chromatography tank (27.5 x 27.5 x 7.5 cm)
> Water bath (65°C)
>
> **Additional Information**
>
> Alternative protocol (Measurement of CAT by Diffusion of Reaction Products into Scintillation Fluid)   MC3, p. 17.41
> Alternative protocol (Measurement of CAT by Extraction with Organic Solvents)   MC3, p. 17.40
> Chloramphenicol acetyltransferase   MC3, pp. 17.94–17.95

METHOD

1. Use gentle aspiration to remove the medium from transfected monolayers of cells growing in 90-mm tissue-culture dishes. Wash the monolayers three times with 5 ml of PBS without calcium and magnesium salts.

2. Stand the dishes at an angle for 2–3 minutes to allow the last traces of PBS to drain to one side. Remove the last traces of PBS by aspiration. Add 1 ml of PBS to each plate and use a rubber policeman to scrape the cells into microfuge tubes. Store the tubes in ice until all of the plates have been processed.

3. Recover the cells by centrifugation at maximum speed for 10 seconds at room temperature in a microfuge. Gently resuspend the cell pellets in 1 ml of ice-cold PBS and again recover the cells by centrifugation. Remove the last traces of PBS from the cell pellets and from the walls of the tubes. Store the cell pellets at –20°C for future analysis or prepare cell extracts by either of the methods in Step 4.

4. Lyse the cells either by repeated cycles of freezing and thawing or by incubating the cells in detergent-containing buffers. The latter is a quicker and easier method of cell lysis that permits CAT, β-galactosidase, and other marker gene assays to be adapted to a 96-well microtiter plate format (Protocol 17.6).

   Lysis of cells by repeated freezing and thawing

   a. Resuspend the cell pellet from one 90-mm dish in 100 µl of 0.25 M Tris-Cl (pH 7.8). Vortex the suspension vigorously to break up clumps of cells.

   b. Disrupt the cells by three cycles of freezing in a dry-ice/ethanol bath and thawing at 37°C. Make sure that the tubes have been marked with ethanol-insoluble ink.

   c. Centrifuge the suspension of disrupted cells at maximum speed for 5 minutes at 4°C in a microfuge. Transfer the supernatant to a fresh microfuge tube. Set aside 50 µl of this supernatant for the CAT assay and store the remainder of the extract at –20°C.

Lysis of cells using detergent-containing buffers

a. To lyse cells with detergent, resuspend the cell pellets from Step 3 in 500 µl of CAT lysis buffer. Incubate the mixture for 15 minutes at 37°C.

> Use 100 µl of this lysis buffer per cell pellet for extracts prepared from cells grown in 35-mm dishes.

b. Remove the cellular debris by centrifuging the tubes at maximum speed for 10 minutes in a microfuge. Recover the supernatant. Assay CAT activity using one of the methods described in this protocol. Snap-freeze the remainder of the cleared lysates in liquid nitrogen and store them at −70°C.

5. Incubate a 50-µl aliquot of the cell extract for 10 minutes at 65°C to inactivate endogenous deacetylases. If the extract is cloudy or opaque at this stage, remove the particulate material by centrifugation at maximum speed for 2 minutes at 4°C in a microfuge.

6. Mix each of the samples to be assayed with 80 µl of CAT reaction mixture 1 and incubate the reactions at 37°C. The length of the incubation depends on the concentration of CAT in the cell extract, which in turn depends on the strength of the promoter and the cell type under investigation. In most cases, incubation for 30 minutes to 2 hours is sufficient.

7. Add 1 ml of ethyl acetate to each sample and mix the solutions thoroughly by vortexing for three periods of 10 seconds. Centrifuge the mixtures at maximum speed for 5 minutes at room temperature in a microfuge.

8. Use a pipette to transfer exactly 900 µl of the upper phase to a fresh tube, carefully avoiding the lower phase and the interface. Discard the tube containing the lower phase in the radioactive waste.

9. Evaporate the ethyl acetate under vacuum by placing the tubes in a rotary evaporator (e.g., Savant SpeedVac) for approximately 1 hour.

10. Add 25 µl of ethyl acetate to each tube and dissolve the reaction products by gentle vortexing.

11. Apply 10–15 µl of the dissolved reaction products to the origin of a 25-mm silica gel TLC plate. The origin on the plate can be marked with a soft-lead pencil. Apply 5 µl at a time and evaporate the sample to dryness with a hair dryer after each application.

12. Prepare a TLC tank containing 200 ml of TLC solvent. Place the TLC plate in the tank, close the chamber, and allow the solvent front to move approximately 75% of the distance to the top of the plate.

13. Remove the TLC plate from the tank and allow it to dry at room temperature. Place adhesive luminescent labels on the TLC plate to align the plate with the film and then expose the plate to X-ray film. Alternatively, enclose the plate in a phosphorimaging cassette. Store the cassette at room temperature for an appropriate period of time.

> Do not cover the TLC plate with Saran Wrap, because this coverage will block the relatively weak radiation emitted by the ^{14}C isotope.

14. Develop the X-ray film and align it with the plate. Alternatively, expose the chromatogram to the imager plate of a phosphorimager device or subject the plate to scanning.

15. To quantitate CAT activity, cut the radioactive spots from the TLC plate and measure the amount of radioactivity they contain in a liquid scintillation counter. Use another aliquot of the cell extract (from Step 3 above) to determine the concentration of protein in the extract, using a rapid colorimetric assay such as the Bradford assay. Reduce the concentration of Triton X-100 to ≤0.1% by dilution before determining the concentration of protein to prevent interference with the assay. Express the CAT activity as pmoles of acetylated product formed per unit time per milligram of cell extract protein.

Assay for Luciferase in Extracts of Mammalian Cells

In this protocol, cells transfected with a luciferase reporter plasmid are lysed in a detergent-containing buffer. Luciferase in the extract catalyzes an oxidation reaction in which D-luciferin is converted to oxyluciferin, with production of light at 556 nm that can be quantified in a luminometer.

MATERIALS

Reagents and Solutions

Please see Appendix 1 for components of stock solutions, buffers, and reagents. Dilute stock solutions to the appropriate concentrations.

Bradford reagent for quantifying protein concentration

Luciferase assay buffer

> 15 mM potassium phosphate (pH 7.8)
> 25 mM glycylglycine
> 15 mM MgSO$_4$
> 4 mM EGTA
> 2 mM ATP
>
> Just before use, add dithiothreitol from a 1 M stock to a final concentration of 1 mM. Approximately 400 μl of luciferase enzyme assay buffer is needed per luminometer assay tube.

Luciferase cell lysis buffer

> 25 mM glycylglycine (pH 8.0)
> 15 mM MgSO$_4$
> 4 mM EGTA
> Triton X-100 (1% v/v)
>
> Just before use, add dithiothreitol from a 1 M stock to a final concentration of 1 mM. Approximately 1 ml of luciferase lysis buffer is needed per 90-mm plate of cells.

Luciferin solution

> 25 mM glycylglycine
> 15 mM MgSO$_4$
> 4 mM EGTA (pH 8.0)
> 0.2 mM luciferin
>
> Just before use, add dithiothreitol from a 1 M stock to a final concentration of 1 mM. Approximately 200 μl of luciferin solution is needed per luminometer assay tube. The addition of acetyl CoA to a final concentration of 1–2 μM just before the luciferin solution is used in Step 5 will result in a more extended and sustained emission of light.
>
> Luciferin is a generic name for substrates that generate light in reactions catalyzed by a luciferase enzyme—of which there are many. The enzyme most commonly used in molecule biology is derived from the firefly and acts on a particular luciferin substrate. Luciferases from other sources uses chemically different substrates. Make sure that the substrate used in the assay matches the enzyme expressed from the reporter plasmid.

Phosphate-buffered saline (PBS) lacking calcium and magnesium salts

Cells, Tissues/Culture Media

Cultured mammalian cells transfected with a luciferase reporter recombinant carrying a DNA sequence of interest

> The cells (growing in 90-mm dishes) should be transfected with a luciferase reporter recombinant (constructed, for example, in a vector of the pGL3 series [see MC3, p. 17.43]) and a plasmid containing a reporter gene (e.g., pCMV-SPORT-β-gal) suitable for normalizing the results of the luciferase assay.

Additional Items

Luminometer (e.g., BD Monolight™, model 3010C) and luminometer tubes
Rubber policeman

Additional Information

Alternative protocol (Assay for Luciferase in Cells Growing in 96-well Plates)	MC3, p. 17.47
Alternative protocol (Using a Scintillation Counter to Measure Luciferase)	MC3, p. 17.46
Luciferase	MC3, p. 17.96
Optimizing the measurement of luciferase enzyme activity	MC3, p. 17.45

METHOD

1. Between 24 and 72 hours after transfection, wash the cells three times at room temperature with PBS without calcium and magnesium salts. Add and remove the PBS gently, because some mammalian cells (e.g., human embryonic kidney 293 cells) can easily be displaced from the dish by vigorous pipetting.

2. Add 1 ml of ice-cold luciferase cell lysis buffer per 100-mm dish of transfected cells. Swirl the buffer gently and scrape the lysed cells from the dish using a rubber policeman. Transfer the cell lysate to a 1.5-ml microfuge tube.

3. Centrifuge the cell lysate at maximum speed for 5 minutes at 4°C in a microfuge. Carefully transfer the supernatant to a fresh 1.5-ml microfuge tube.

4. Determine the concentration of protein in the lysate using a rapid colorimetric assay, such as the Bradford assay. Reduce the concentration of Triton X-100 to ≤0.1% by dilution before determining the concentration of protein to prevent interference with the assay.

    The cell lysates may be stored at this stage at –70°C. Luciferase is unstable when stored at 4°C or –20°C in standard lysis buffer. However, enzyme activity is stable in lysis buffer containing 15% (v/v) glycerol and 1% (w/v) bovine serum albumin.

5. Tap the side of the tube containing the lysate to gently mix the contents. Add 5–200-μl aliquots of cell lysate to individual luminometer tubes containing 360 μl of luciferase assay buffer at room temperature. Place a tube in the luminometer.

6. To start the assay, inject 200 μl of luciferin solution into the luminometer tube and measure the light output for a period of 2–60 seconds at room temperature.

    The optimal time of light collection must be determined empirically.

7. Measure the relative light units generated in each tube and determine the linear range of the assay. Use the amount of cell lysate protein that produces a response in the middle of the linear range in subsequent assays. This amount will vary depending on the strength of the promoter being studied and, to a lesser extent, on the efficiency of transfection in individual experiments. Express luciferase activity as relative light units/mg of protein in the cell lysate.

Assay for β-galactosidase in Extracts of Mammalian Cells

The assay for β-galactosidase relies on the ability of the enzyme to catalyze the hydrolysis of ONPG (o-nitrophenyl-β-D-galactopyranoside) to free o-nitrophenol, which absorbs light at 420 nm. In this protocol, extracts of cells transfected with a β-galactosidase reporter plasmid are incubated with ONPG. When the substrate is in excess, the OD_{420} of the assay solution increases with time and is proportional to the enzyme concentration.

MATERIALS

CAUTION: Please see Appendix 4 for appropriate handling of materials marked with <!>.

Reagents and Solutions

Please see Appendix 1 for components of stock solutions, buffers, and reagents. Dilute stock solutions to the appropriate concentrations.

100x Mg^{++} solution
  0.1 M $MgCl_2$
  4.5 M β-mercaptoethanol <!>, added just before buffer is used in Step 2
Na_2CO_3 (1 M)
1x ONPG (4 mg/ml in 0.1 M sodium phosphate [pH 7.5])
Sodium phosphate (0.1 M, pH 7.5)
  Mix 41 ml of 0.2 M Na_2HPO_4, 9 ml of 0.2 M NaH_2PO_4, and 50 ml of H_2O.
Tris-Cl (1 M, pH 7.8)

Cells, Tissues/Culture Media

Cultured mammalian cells transfected with the β-galactosidase reporter recombinant
  carrying a DNA sequence of interest
    The cells (growing in 90-mm dishes) should be transfected with a reporter recombinant (constructed, e.g., in a vector of the pβ-gal series [see MC3, p. 17.49]).

Enzymes and Buffers

E. coli β-galactosidase (e.g., Sigma-Aldrich)

Additional Items

Spectrophotometer and cuvettes (420 nm)
Water bath (50°C) (optional; see Step 1)

Additional Information

β-galactosidase	MC3, p. 17.97–17.99
β-galactosidase substrates	MC3, p. 17.50
β-galactosidase units of activity	MC3, p. 17.51
ONPG	MC3, p. 17.50
Phosphate buffers	p. 695, this volume

METHOD

1. Prepare cell extracts from the transfected cells as described in Protocol 17.5, Steps 1–4. Set aside approximately 30 µl of the extract for the β-galactosidase assay. The exact amount of extract required will depend on the strength of the promoter driving the expression of the β-galactosidase gene, the efficiency of transfection, and the incubation time of the assay. If a heat treatment is to be used to inactivate endogenous β-galactosidases, incubate the cell lysates for 45–60 minutes at 50°C before assay. Luciferase activity is also inactivated by preheating; assay luciferase and β-galactosidase activities in separate aliquots of cell lysate if a preheating step has been used.

2. For each sample of transfected cell lysate to be assayed, mix:

100x Mg^{++} solution	3 µl
1x ONPG	66 µl
cell extract	30 µl
0.1 M sodium phosphate (pH 7.5)	201 µl

   It is essential to include positive and negative controls. These assays check for the presence of endogenous inhibitors and β-galactosidase, respectively. All of the controls should contain 30 µl of cell extract from mock-transfected cells. In addition, the positive controls should include 1 µl of a commercial preparation of *E. coli* β-galactosidase (50 units/ml). The commercial enzyme preparation should be dissolved at a concentration of 3000 units/ml in 0.1 M sodium phosphate (pH 7.5). Just before use, transfer 1 µl of the stock solution of β-galactosidase into 60 µl of 0.1 M sodium phosphate (pH 7.5) to make a working stock of the enzyme containing 50 units/ml. One unit of *E. coli* β-galactosidase is defined as the amount of enzyme that will hydrolyze 1 µmole of ONPG substrate in 1 minute at 37°C.

3. Incubate the reactions for 30 minutes at 37°C or until a faint yellow color has developed. In most cell types, the background of endogenous β-galactosidase activity is very low, allowing incubation times as long as 4–6 hours to be used.

4. Stop the reactions by adding 500 µl of 1 M Na_2CO_3 to each tube. Read the optical density of the solutions at a wavelength of 420 nm in a spectrophotometer.

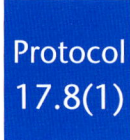

Tetracycline as a Regulator of Inducible Gene Expression in Mammalian Cells
Stage 1: Stable Transfection of Fibroblasts with pTet-tTAk

The following protocol uses an autoregulatory system in which the transcriptional *trans*-activator tTA drives its own expression and that of a target gene. The first stage of the protocol describes how to generate stable lines of NIH-3T3 cells that express either tTA alone or tTA and the tetracycline-regulated target gene.

MATERIALS

Reagents and Solutions

Please see Appendix 1 for components of stock solutions, buffers, and reagents. Dilute stock solutions to the appropriate concentrations.

$CaCl_2$ (2 M), sterilized by filtration

Glycerol (15% v/v) in HEPES-buffered saline

HEPES-buffered saline

Phosphate-buffered saline (PBS)

Vectors and Bacterial Strains

For descriptions of the plasmids used in this protocol, please see MC3, pp. 17.52–17.59. Plasmids used in this protocol should be purified by CsCl–ethidium bromide equilibrium centrifugation (Protocol 1.10) or column chromatography (Protocol 1.9).

pSV2-His (see MC3, p. 17.61)

> Vectors carrying selectable markers, e.g., *neo*^{r} or *hyg*^{r}, can be substituted for pSV2-His in this protocol, with the appropriate selection media.

pTet-tTAk (Invitrogen)

Cells, Tissues/Culture Media

Cultures of NIH-3T3 cells, growing in 60- or 90-mm plates

> This protocol can be easily adapted for use with smaller cultures and other cell lines.

DMEM complete containing 0.5 µg/ml tetracycline-HCl

> To make a stock solution of tetracycline-HCl, dissolve 10 mg of the antibiotic in 1 ml 70% ethanol. Store the solution at –70°C. Tissue-culture media containing tetracycline may be stored, protected from light, at 4°C for periods of up to 1 month.

Dulbecco's modified Eagle's complete medium (DMEM)

> 100 units/ml penicillin
> 100 µg/ml streptomycin
> 2 mM glutamine
> 10% bovine calf serum

Selection medium containing L-histidinol and 0.5 µg/ml tetracycline-HCl

> histidine-free DMEM (Irvine Scientific)
> 100 units/ml penicillin
> 100 µg/ml streptomycin
> 2 mM glutamine
> 10% donor bovine calf serum
> 125, 250, or 500 µM L-histidinol
> 0.5 µg/ml tetracycline-HCl
> Stock solutions of L-histidinol (125 µM in H_2O) should be stored at –20°C.

Trypsin-EDTA

Adapted from Chapter 17, Protocol 8 Stage 1, p. 17.60 of MC3.

Enzymes and Buffers

Appropriate restriction enzymes and buffers

Additional Items

Plastic cloning rings
Autoclave the rings in an upright position with the bottom embedded in a thin layer of vacuum grease.

Additional Information

Alternative protocol (Tetracycline-regulated Induction of Gene Expression in Transiently Transfected Cells Using the Autoregulatory tTA System)	MC3, p. 17.70
Chloroquine diphosphate	MC3, pp. 16.52–16.53
Tetracycline	MC3, pp. 17.52–17.53
Tetracycline as regulator of gene expression in mammalian cells	MC3, pp. 17.53–17.59
Troubleshooting	MC3, p. 17.59

METHOD

1. Culture adherent cells in complete DMEM. The day before transfection, transfer the cells into complete DMEM containing 0.5 µg/ml tetracycline-HCl (tetracycline). Apply enough cells per 9-cm dish so that on the day of the transfection, the cells will be 33% confluent.

   IMPORTANT: From this point on, maintain cells in the presence of 0.5 µg/ml tetracycline-HCl at 37°C, in an atmosphere of 5% CO_2, unless otherwise stated.

2. Linearize the plasmids at an appropriate restriction endonuclease site and adjust the DNA concentration of each plasmid to ≥0.5 mg/ml. Mix 10–20 µg of pTet-tTAk plasmid and 1–2 µg of pSV2-His (~10:1 molar ratio of Tet plasmid to the selectable marker plasmid) with 500 µl of HEPES-buffered saline in a clear 4-ml polystyrene tube.

   Prepare a control for mock transfection containing the HEPES-buffered saline and no DNA. All of the mock-transfected cells should die in histidine-free DMEM containing L-histidinol.

3. Add 32.5 µl of 2 M $CaCl_2$ to the DNA mixture. Immediately mix the solution by gentle vortexing. Store the solution at room temperature, mixing it from time to time. A cloudy precipitate should form over the course of 15–30 minutes.

4. Aspirate all of the medium from the dishes of cells prepared in Step 1.

5. Mix the $CaCl_2$-DNA precipitate a few times by pipetting with a pasteur pipette. Apply the mixture dropwise and distribute it evenly over the cell monolayers.

6. Incubate the cells in an atmosphere of 5% CO_2 for 30 minutes at 37°C, rocking the plate after 15 minutes to ensure even coverage of the DNA precipitate.

7. To each dish of cells, add 10 ml of complete DMEM containing tetracycline. Incubate the cells in an atmosphere of 5% CO_2 for 4–5 hours at 37°C.

8. Gently aspirate the medium from the cells. Avoid disruption of the precipitate that has settled onto the cells.

9. Subject the cells to a glycerol shock by adding 2.5 ml of 15% glycerol in HEPES-buffered saline warmed to 37°C. Store the cells for 2.5 minutes at room temperature. Add the glycerol dropwise to the culture.

10. Aspirate the glycerol solution after *exactly* 2.5 minutes. Work quickly, because glycerol can be very toxic to cells.

11. Immediately, gently, and quickly wash the cells by adding 10 ml of complete DMEM containing tetracycline. Immediately remove the medium by aspiration and repeat the wash.

    IMPORTANT: Because the cells tend to detach easily from the plate after glycerol shock, add all of the medium to a single spot on the plate.

12. Add 10 ml of complete DMEM containing tetracycline to the cells and incubate the cultures overnight at 37°C.

13. Approximately 16–24 hours after transfection, aspirate the medium and replace it with 10 ml of complete DMEM containing tetracycline. Incubate the cultures for a total of 48 hours at 37°C after the transfection (i.e., the sum of the incubation times in Steps 12 and 13).

14. Forty-eight hours after transfection, passage several dilutions of the cells into DMEM selection medium containing 125 μM L-histidinol and 0.5 μg/ml tetracycline-HCl. Cell densities should range from approximately 1×10^6 to 3×10^4 cells per 10-cm plate. Include several plates containing approximately 1×10^5 cells.

15. After incubating the cultures in DMEM selection medium for 4 days, feed them with an additional 3–4 ml of DMEM selection medium containing 125 μM L-histidinol and 0.5 μg/ml tetracycline-HCl.

16. When colonies have formed (typically after ~10–12 days of selection), replace the medium with DMEM selection medium containing 250 μM L-histidinol and 0.5 μg/ml tetracycline-HCl.

17. When colonies are well established (at day 12–15 of selection), delineate their borders by drawing a circle on the bottom of the culture dish around each colony. Aspirate the medium from the plate and place a sterile plastic cloning ring on the plate to surround an individual clone. Repeat this process for each colony to be picked. Choose cells from plates on which individual colonies are well spaced and can be easily distinguished.

    IMPORTANT: After stable transfection with pTet-tTAk, it is imperative that the cells be maintained in medium containing 0.5 μg/ml tetracycline to prevent any toxic effects of tTA expression and subsequent selection against clones expressing high levels of tTA.

18. Quickly wash the clones with approximately 100 μl of PBS. To release the cells, add 2 drops of 1x trypsin-EDTA (~100 μl) and incubate for 30–60 seconds. Loosen the cells by pipetting up and down with a pasteur pipette. Transfer each colony to one well of a 24-well tissue-culture plate that contains 1 ml of selection medium containing 250 μM L-histidinol and tetracycline.

19. When the cells in the wells have grown to 80% confluency, transfer them into 6-cm tissue-culture dishes in DMEM selection medium containing 500 μM L-histidinol and tetracycline.

20. Expand the cells (typically use a 1:5 to 1:10 dilution of the cells) in DMEM selection medium containing 500 μM L-histidinol and tetracycline.

21. When the cell monolayers are again approximately 80% confluent, recover a portion of each clone of cells and store in aliquots in liquid nitrogen. Passage the remainder of the cells until they have expanded sufficiently to allow testing for inducible expression of protein.

22. If cells were cotransfected with both tTA and the target plasmids, directly analyze the products of the target genes, as described in Protocol 17.8 Stage 3. If cells were transfected with only the pTet-tTAk plasmid, prepare the cells to be tested for inducible expression as follows:

    a. The night before induction, plate the cells in DMEM selection medium containing 500 μM L-histidinol and 0.5 μg/ml tetracycline at an appropriate density so that they will be subconfluent at the time of harvest.

b. The next day, wash the cells three times with PBS, swirling the plates gently each time.

c. After the third wash, immediately add selection medium containing 500 µM L-histidinol, but lacking tetracycline. Culture the cells in the presence or absence of tetracycline for 6–48 hours.

23. Test the cells for inducible expression of tTA by northern analysis or immunoblotting. Cell lines expressing tTA may then be transfected with the target plasmid(s) as described in Protocol 17.8 Stage 2.

> It is essential to include controls that are maintained in selection medium containing 500 µM L-histidinol and 0.5 µg/ml tetracycline.

Tetracycline as a Regulator of Inducible Gene Expression in Mammalian Cells

Protocol 17.8(2)

Stage 2: Stable Transfection of Inducible tTA-expressing NIH-3T3 Cells with Tetracycline-regulated Target Genes

Stage 2 of this protocol involves transfecting target genes into cell lines already expressing inducible tTA. In this example, the target genes are transfected on a plasmid that carries puromycin resistance as a selectable marker.

MATERIALS

CAUTION: Please see Appendix 4 for appropriate handling of materials marked with <!>.

Reagents and Solutions

Please see Appendix 1 for components of stock solutions, buffers, and reagents. Dilute stock solutions to the appropriate concentrations.
> Calf serum (10%)
> HEPES-buffered saline
> Phosphate-buffered saline (PBS)
> Trypsin-EDTA

Vectors and Hosts

Plasmids used in this protocol should be purified by CsCl–ethidium bromide equilibrium centrifugation (Protocol 1.10) or column chromatography (Protocol 1.9).
> pPGKPuro
>> pPGKPuro ontains the puromycin resistance gene encoding puromycin N-acetyltransferase cloned into pBlueScript SK+ (Stratagene). Transcription of the N-acetyltransferase gene in transfected mammalian cells is controlled by upstream (phosphoglycerate kinase [PGK] promoter) and downstream (PGK polyadenylation sequences) elements. pPGKPuro is available from a gift from P.W. Laird, University of Southern California. Vectors carrying other selectable markers, e.g., neo^r or hyg^r, can be substituted for pPGKPuro in this protocol, in combination with the appropriate selective conditions.
> pTet-Splice (Invitrogen) carrying a reporter gene, e.g., pUHC13-3 (optional)
> pTet-Splice carrying the target gene open reading frame(s)

Cells, Tissues/Culture Media

> Selection medium, which is histidine-free DMEM (Irving Scientific) containing 500 μM
> L-histidinol (stock solutions of L-histidinol [125 μM in H_2O] should be stored at –20°C)
>> 100 units/ml penicillin
>> 100 μg/ml streptomycin
>> 2 mM glutamine
>> 10% donor bovine calf serum
>> 3 μg/ml puromycin (when appropriate)
>> 0.5 mg/ml tetracycline-HCl
>> Tissue-culture media containing tetracycline may be stored, protected from the light, at 4°C for periods of up to 1 month.
> Stable cell lines that inducibly express autoregulatory tTA (Protocol 17.8 Stage 1)

Enzymes and Buffers

> Appropriate restriction enzymes and buffers

---

648   Adapted from Chapter 17, Protocol 8 Stage 2, p. 17.65 of MC3.

Additional Items

Liquid nitrogen <!> and materials for freezing and storing clones of cells
Plastic cloning rings
> Autoclave the rings in an upright position with the bottom embedded in a thin layer of vacuum grease.

Step 3 of this protocol requires the reagents listed in Protocol 17.8 Stage 1.

Additional Information

Alternative protocol (Tetracycline-regulated Induction of Gene Expression in Transiently Transfected Cells Using the Autoregulatory tTA System	MC3, p. 17.70
Tetracycline	MC3, pp. 17.52–17.53
Tetracycline as regulator of gene expression in mammalian cells	MC3, pp. 17.53–17.59
Troubleshooting	MC3, p. 17.59

METHOD

1. Culture stable cell lines that inducibly express autoregulatory tTA (isolated in Protocol 17.8 Stage 1) in complete selection medium containing 500 μM L-histidinol and 0.5 μg/ml tetracycline-HCl. The day before transfection, passage the cells into 10-cm tissue-culture dishes containing complete selection medium. Transfer enough cells per dish so that on the day of the transfection, the cell monolayers will be 33% confluent.

   > IMPORTANT: From this point on, maintain the cells in the presence of 0.5 μg/ml tetracycline-HCl at 37°C, in an atmosphere of 5% CO_2 unless otherwise stated.

2. Linearize the plasmids to be used for transfection and adjust the DNA concentration of each to ≥0.5 mg/ml. Mix 10–20 μg of each target gene plasmid(s) and 1–2 μg of pPGKPuro (a 10:1 molar ratio of each tetracycline plasmid to selectable marker plasmid) with 500 μl of HEPES-buffered saline in a clear 4-ml polystyrene tube.

   > Prepare a control for mock transfection containing HEPES-buffered saline and no DNA. All of the mock-transfected cells should die when incubated in a medium containing puromycin. Please see the note to Step 4.

3. Carry out Steps 3–13 from Protocol 17.8 Stage 1.

   > IMPORTANT: Be sure to substitute the selection medium containing 500 μM L-histidinol and 0.5 μg/ml tetracycline-HCl in this transfection, whenever Stage 1 calls for complete DMEM containing tetracycline.

4. Forty-eight hours after transfection, passage the cells into DMEM selection medium containing 500 μM L-histidinol, 3 μg/ml puromycin, and 0.5 μg/ml tetracycline at several dilutions ranging from approximately 1×10^6 to 3×10^4 cells per 10-cm plate. Include several plates containing approximately 1×10^5 cells.

   > The lowest concentration of puromycin that kills all untransfected cells within a few days should be determined empirically before transfection and varies with the cell type. A concentration of 3 μg/ml puromycin is sufficient for selection of transfected NIH-3T3 cells. Most types of cells are killed efficiently in concentrations of puromycin ranging from 0.1 to 10 μg/ml.

5. After incubating cultures in DMEM selection medium for 4 days, feed them with an additional 3–4 ml of DMEM selection medium containing 500 μM L-histidinol, 3 μg/ml puromycin, and 0.5 μg/ml tetracycline.

6. When colonies are well established (at day 12–14 of selection), delineate their borders by drawing a circle on the bottom of the culture dish around each colony. Aspirate the medium from the plate and place a sterile plastic cloning ring on the plate to surround an individual

clone. Repeat the procedure with each colony that is to be picked. Choose cells from plates on which individual colonies are well spaced and can easily be distinguished.

7. Quickly wash the clones with approximately 100 µl of PBS. To release the cells, add 2 drops of 1x trypsin-EDTA (~100 µl) and incubate for 30–60 seconds. Loosen the cells by pipetting up and down with a pasteur pipette. Transfer each colony to one well of a 24-well tissue-culture plate that contains 1 ml of DMEM selection medium containing 500 µM L-histidinol, 3 µg/ml puromycin, and 0.5 µg/ml tetracycline.

   Perform all subsequent passaging of cells by a standard procedure, such as (i) a quick PBS wash, and (ii) a 1–3-minute incubation with trypsin-EDTA (2 ml per confluent 10-cm plate) using 10% calf serum (3 ml) to dilute/stop the activity of the trypsin.

8. When cells in the wells have grown to 80% confluency, transfer them into 6-cm dishes that contain DMEM selection medium containing 500 µM L-histidinol, 3 µg/ml puromycin, and 0.5 µg/ml tetracycline.

   NIH-3T3 cells become 80% confluent in approximately 4–7 days; however, the time required to reach 80% confluency varies from cell line to cell line and even from clone to clone.

9. Expand the cells (typically use a 1:5 to 1:10 dilution of the cells) in DMEM selection medium containing 500 µM L-histidinol, 3 µg/ml puromycin, and 0.5 µg/ml tetracycline.

10. When cell monolayers are again approximately 80% confluent, recover a portion of each clone of the cells and store them in aliquots in liquid nitrogen. Passage the remainder of the cells until they have expanded sufficiently to allow testing for inducible expression of the target gene product(s) as described in Protocol 17.8 Stage 3.

   When the frozen cells are later used, they should be revived and grown in selection medium containing 500 µM L-histidinol, 3 µg/ml puromycin, and 0.5 µg/ml tetracycline-HCl.

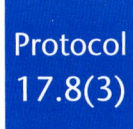

Tetracycline as a Regulator of Inducible Gene Expression in Mammalian Cells
Stage 3: Analysis of Protein Expression in Transfected Cells

Stably transfected cells, generated in the first two stages of this protocol, are induced for expression of the target gene. After harvesting and lysis, the lysates are analyzed by SDS-PAGE and immunoblotting.

MATERIALS

CAUTION: Please see Appendix 4 for appropriate handling of materials marked with <!>.

Reagents and Solutions

Please see Appendix 1 for components of stock solutions, buffers, and reagents. Dilute stock solutions to the appropriate concentrations.
  Phosphate-buffered saline (PBS)

Cells, Tissues/Culture Media

  Selection medium, which is histidine-free DMEM (Irving Scientific) containing 500 μM
    L-histidinol (stock solutions of L-histidinol [125 μM in H$_2$O] should be stored at –20°C)
      100 units/ml penicillin
      100 μg/ml streptomycin
      2 mM glutamine
      10% donor bovine calf serum
      3 μg/ml puromycin
      0.5 mg/ml tetracycline-HCl (when appropriate)
      Tissue-culture media containing tetracycline may be stored, protected from the light, at 4°C
      for periods of up to 1 month.
  Stable cell lines that inducibly express autoregulatory tTA (Protocol 17.8 Stage 1) and
    contain the plasmids harboring the DNAs of interest

Enzymes and Buffers

  Appropriate restriction enzymes and buffers

Gels/Loading Buffers

  Tris-glycine SDS-polyacrylamide gel
    Cast the gel with a concentration of acrylamide appropriate for the molecular mass of the target protein (see pp. 756–763).

Additional Items

  Boiling-water bath
  Liquid nitrogen <!> and materials for freezing and storing clones of cells
  Polyvinylidene difluoride (PVDF) membranes for western blotting
  Step 13 of this protocol requires the reagents and equipment for SDS-gel electrophoresis
    and immunoblotting (see pp. 756–763).

Additional Information

  Alternative protocol (Tetracycline-regulated Induction
    of Gene Expression in Transiently Transfected Cells
    Using the Autoregulatory tTA System   MC3, p. 17.70
  Tetracycline   MC3, pp. 17.52–17.53

Tetracycline as regulator of gene expression in mammalian cells	MC3, pp. 17.53–17.59
Troubleshooting	MC3, p. 17.59

METHOD

1. The night before induction, plate the cells in DMEM selection medium containing 500 μM L-histidinol, 3 μg/ml puromycin, and 0.5 μg/ml tetracycline at an appropriate density so that they will be subconfluent at the time of harvest.

2. The next day, wash the cells three times with PBS, swirling the plates gently each time.

3. After the third wash, immediately add DMEM selection medium containing 500 μM L-histidinol and 3 μg/ml puromycin, but lacking tetracycline. Culture the cells in the presence or absence of tetracycline for 6–48 hours.

    It is essential to include controls that are maintained in selection medium containing 500 μM L-histidinol, 3 μg/ml puromycin, and 0.5 μg/ml tetracycline.

4. After the cells have grown for the appropriate length of time, harvest them quickly and place them in a tube in an ice bucket.

5. For each clone and control, transfer 0.5×10^6 cells to a microfuge tube and centrifuge all of the tubes at 3000 rpm (low to moderate speed) for 5 minutes at 4°C.

6. Wash the cell pellets by adding 1 ml of ice-cold PBS. Pellet the cells as in Step 5 and gently aspirate the supernatant without disturbing the cell pellet.

7. Keep the cell pellets on ice and loosen them by gently and quickly running the tubes over the open holes of a microfuge rack before freezing the cell pellets at –70°C.

8. Resuspend each cell pellet in 30 μl of protein sample buffer by gently pipetting the cells and then vortexing the tubes.

9. Boil the cells in protein sample buffer for 10 minutes.

10. Recover the cell debris by centrifugation at maximum speed for 2 minutes in a microfuge.

11. Load 10 μl of cell lysate per lane of a Tris-glycine SDS-polyacrylamide gel.

12. Run the gel for the appropriate length of time.

13. Electrotransfer the proteins from the SDS-polyacrylamide gel to a PVDF membrane and probe the membrane with the appropriate antibodies.

Ecdysone as a Regulator of Inducible Gene Expression in Mammalian Cells

This protocol is adapted from the information supplied by Invitrogen as part of their edysone-inducible mammalian expression system.

MATERIALS

Reagents and Solutions

Please see Appendix 1 for components of stock solutions, buffers, and reagents. Dilute stock solutions to the appropriate concentrations.

Kits containing materials for ecdysone-inducible expression systems (Stratagene and Invitrogen)

Neomycin, a broad spectrum antibiotic that interferes with protein synthesis

Ponasterone A or muristerone A (Sigma-Aldrich), members of the ecdysteroid family

Zeocin (InvivoGen), a glycopeptide antibiotic of the bleomycin family

Media and Antibiotics

Medium suitable for growth of the cells to be transfected, variously supplemented with Neomycin Ponasterone A or Muristerone A and Zeocin

Vectors and Hosts

pVgRXR (Invitrogen)

Cells, Tissues/Culture Media

Cultured mammalian cells

Additional Items

Step 4 of this protocol requires the reagents listed in Protocol 17.6.

Step 6 of this protocol requires the reagents listed in Protocol 6.8, 7.8, or Appendix 3.

Steps 1, 2, and 6 of this protocol require the reagents listed in the appropriate transfection protocol in Chapter 16.

Additional Information

Ecdysone as a regulator of transcription   MC3, pp. 17.71–17.72

METHOD

1. Stably transfect cells with pVgRXR using the preferred method of transfection for the cells under study (for a selection of transfection protocols, please see Chapter 16).

   Perform a control mock transfection with no DNA added to the cells. All of the mock-transfected cells should die in the presence of Zeocin.

2. To choose clones capable of selectively inducing gene expression in the presence of ecdysone or one of its analogs, transiently transfect Zeocin-resistant colonies obtained in Step 1 with an ecdysone-inducible expression plasmid carrying a luciferase reporter gene (again, for a selection of transfection protocols, please see Chapter 16).

3. Twenty-four to ninety-six hours after transfection, induce the expression of luciferase by replacing the medium with fresh medium containing Zeocin, neomycin, and 5 μM ponasterone A. Incubate the cells for 20 hours at 37°C in a humidified incubator with an atmosphere of 5–7% CO_2.

> It is essential to include control cultures that are not exposed to ponasterone A. Incubate these control cultures in medium containing only Zeocin and neomycin.

4. Assay for luciferase activity according to Protocol 17.6.

5. Use clones that exhibit the desired level of ecdysone-induced luciferase activity. Expand the cell culture in medium containing Zeocin and neomycin.

6. Stably transfect cells from Step 5 with an ecdysone-inducible plasmid harboring the gene of interest and a hygromycin resistance marker using the preferred method of transfection for the cells under study (for a selection of transfection protocols, please see Chapter 16).

7. Expand the chosen colonies of cells that are resistant to both antibiotics, hygromycin and neomycin.

8. Analyze expression of the target gene by immunoblotting, northern hybridization, or other appropriate assay.

Protein Interaction Technologies

BACKGROUND INFORMATION

Background information found in *Molecular Cloning: A Laboratory Manual*, 3rd edition (hereafter MC3) unless otherwise indicated.

Two-hybrid and Other Two-component Systems
Stage 1: Characterization of a Bait-LexA Fusion Protein

This protocol describes how to generate a plasmid construct (pBAIT) that expresses a target protein fused to the bacterial LexA protein. pBAIT is cotransformed into yeast with a *lexAop-lacZ* reporter plasmid carrying the bacterial *lacZ* gene under the control of the *lexA* operator. The recipient yeast strain contains a chromosomally integrated *leu2* reporter gene, also under the control of the *lexA* operator. pBAIT is analyzed in transformants to establish whether the bait protein is expressed as a stable nuclear protein of the correct size that does not independently activate transcription of either of the *lexA* operator-reporter genes to a significant extent.

MATERIALS

CAUTION: Please see Appendix 4 for appropriate handling of materials marked with <!>.

Vectors and Hosts

S. cerevisiae strains for selection and propagation of vectors (see MC3, Table 18-6)
Vectors carrying *LexA*, activation domain fusion sequences, and *LacZ* reporter plasmids (see http://www.fccc.edu/research/labs/golemis/com_sources1.html or MC3, Tables 18-1, 18-2, and 18-5)

Media and Antibodies

Please see Appendix 2 for components of yeast media. Use Table 18-1 to prepare the necessary selective media.
Complete minimal (CM) selective medium (see Table 18-1)
Monoclonal antibody to LexA (CLONTECH), polyclonal antibody to LexA (Invitrogen), or specific antibody to the fusion domain of the target protein (if available)
Yeast selective X-gal medium
   i. Prepare the base medium in 900 ml of H_2O, according to Table 18-1. Autoclave the base medium and cool it to 55°C.
   ii. In a separate bottle, autoclave 7 g of dibasic sodium phosphate and 3 g of monobasic sodium phosphate in 100 ml of distilled H_2O.
   iii. Mix the two solutions together and add 0.8 ml of X-gal (100 mg/ml in *N,N*-dimethylformamide <!>). Pour the plates.

TABLE 18-1 Yeast Complete Minimal (CM) Selective Media for Two-hybrid Analysis

Ingredients	CM(Glu) -U-H Medium	CM(Glu) -U Agar	CM(Glu) -U-H Agar	CM(Glu) -U-H-L Agar	CM(Glu, X-gal)-U Agar[a]	CM(Gal) -U-H-L Agar	CM(Gal, X-gal)-U-H-L Agar[a]
YNB[b]	6.7 g	6.7 g	6.7 g	6.7 g	6.7 g	6.7 g	6.7 g
Glucose	20 g	20 g	20 g	20 g	20 g		
Galactose						20 g	20 g
Dropout mix	2 g	2 g	2 g	2 g	2 g	2 g	2 g
Leucine	15 ml	15 ml	15 ml		15 ml		
Tryptophan	10 ml	10 ml	10 ml	10 ml	10 ml	10 ml	10 ml
Histidine		5 ml			5 ml		
Agar		20 g	20 g	20 g	20 g	20 g	20 g

[a]For media containing X-gal, prepare the appropriate base medium in 900 ml of H_2O and use it in the recipe for yeast selective X-gal medium.
[b]Use the form of yeast nitrogen base (YNB) that contains ammonium sulfate.
To prepare liquid or agar medium, mix the ingredients in a final volume of 1 liter of H_2O. Autoclave for 20 minutes. Cool the agar medium to 55°C before pouring plates: Leucine stock = 4 mg/ml; tryptophan stock = 4 mg/ml; histidine stock = 4 mg/ml.

Adapted from Chapter 18, Protocol 1 Stage 1, p. 18.17 of MC3.

YPD medium

20 g peptone

10 g yeast extract

20 g glucose

20 g agar (if for plates)

Add 1 liter of distilled H_2O and autoclave the medium for 20 minutes. Before pouring plates, cool the medium to 55°C.

Nucleic Acids/Oligonucleotides

Target DNA encoding the protein of interest (bait)

Gels/Loading Buffers

2x SDS gel-loading buffer

100 mM Tris-Cl (pH 6.8)

200 mM dithiothreitol

4% SDS (w/v), electrophoresis grade

0.2% bromophenol blue

20% glycerol

Store the 2x buffer without dithiothreitol at room temperature. Add dithiothreitol from a 1 M stock just before use.

SDS-polyacrylamide gel <!>

Please see Appendix 3 for the preparation of SDS gels used in the separation of proteins.

Additional Items

Dry-ice/ethanol bath <!>

Heating block, thermal cycler, or water bath (100°C)

Shaking or orbital incubators (30°C)

Step 1 of this protocol requires the reagents for subcloning listed in Protocol 1.17.

Step 2 of this protocol requires the reagents for transformation of yeast (see Spector et al. 1988. *Cells: A Laboratory Manual*, Chapter 21, Cold Spring Harbor Laboratory Press, Cold Spring Harbor, New York).

Step 17 of this protocol requires the reagents for immunoblotting in Appendix 3.

Toothpicks (flat edged), sterilized by autoclaving

Additional Information

Alternative protocol (Assay of β-Galactosidase Activity by Chloroform Overlay)	MC3, p. 18.28
Review of two-hybrid and other two-component systems	MC3, pp. 18.6–18.16
Troubleshooting and modifications	MC3, p. 18.27

METHOD

1. Clone the target DNA encoding the protein to be used as bait into the polylinker of a LexA fusion vector (e.g., pMW101 or pMW103) to synthesize an in-frame fusion to LexA. Ensure that a translational stop sequence is present at the carboxyl terminus of the desired bait sequence. The resulting plasmid is referred to as pBait.

2. Set up a series of transformations of the EGY48 *lexAop-LEU2* selection strain of yeast using the following combinations of LexA fusion and *lexAop-lacZ* reporter plasmids:

   a. pBait + pMW112 (test for activation)

   b. pSH17-4 + pMW112 (positive control for activation)

    c. pRFHM1 + pMW112 (negative control for activation)

    d. pBait + pJK101 (test for repression/DNA binding)

    e. pRFHM1 + pJK101 (positive control for repression)

    f. pJK101 alone (negative control for repression)

3. Plate each transformation mixture on selective dropout plates: CM(Glu)-Ura-His (for plasmid combinations **a–e**) or CM(Glu)-Ura (for plasmid combination **f**), as appropriate. Incubate the plates for 2–3 days at 30°C to select for transformed yeast colonies that contain the plasmids.

4. Make a master plate of transformants, from which specific colonies can be assayed for the phenotype of activation of *lacZ* and *LEU2* reporters as described in Steps 5–9.

    Steps 5–9 are used to test the bait-LexA fusion protein for transcriptional activity and to demonstrate that the fusion of the bait does not affect LexA DNA-binding activity. Several independent colonies are assayed for each combination of plasmids transformed in Step 2.

5. From each transformation **a–f** (from Step 2), use sterile, standard flat-edged toothpicks to pick approximately eight colonies. Touch a clean toothpick to the colony to pick up cells and restreak them as a 1-cm-long streak in a grid on a fresh CM(Glu)-Ura-His or CM(Glu)-Ura plate. As many as 60–80 streaks can generally be grown on a single plate. Incubate the plates overnight at 30°C.

6. On the second day, restreak from the two master plates to each of the following:

    Transformations a–f: Streak onto CM(Glu, X-gal)-Ura and onto CM(Gal, X-gal)-Ura

    Transformations a–c: Streak onto CM(Glu)-Ura-His-Leu and onto CM(Gal)-Ura-His-Leu

    Only uracil is dropped out (histidine is present) from the X-gal plates, to allow side-by-side comparison of the JK101 plasmid-only transformation (**f**) with **d** and **e**. Lack of selection for the LexA-fused plasmid does not notably affect transcriptional activation over the period of this assay.

7. Incubate the plates for up to 4 days at 30°C.

8. Assay for repression and activation activities:

    a. For repression, observe the X-gal phenotype at approximately 12–24 hours after streaking.

    b. For activation, observe the X-gal phenotype between 18 and 72 hours after streaking.

    c. Observe the Leu2 phenotype between 48 and 96 hours. The expected results for a well-behaved bait are summarized below.

      • Optimally, at 12–24 hours after streaking to CM(Gal, X-gal)-Ura, the **d** + **e** transformants should be discernibly lighter in color than **f**.

      • At 48 hours after streaking to CM(Glu, X-gal)-Ura, **d** transformants should be bright blue, **c** should be white, and **a** should be white or very light blue.

      • At 48 hours after streaking, **b** transformants should be as well grown on CM(Glu)-Ura-His-Leu or CM(Gal)-Ura-His-Leu as on a CM(Glu)-Ura-His master plate, whereas **a** and **c** should show no growth.

      • Ideally, **a** transformants will still display no apparent growth at 96 hours after streaking.

9. Select the appropriate candidate colonies, based on the results of the repression and activation assays.

10. On the master plate, mark the colonies that are to be assayed for protein expression. Use the colony that has been shown to express bait appropriately as the founder to grow a culture for transformation of a library (in Protocol 18.1 Stage 2).

11. Analyze at least two primary transformants for each novel bait construct. Include two transformants as positive controls for protein expression (e.g., pRFHM1).

a. Use a sterile toothpick to pick colonies from the CM(Glu)-Ura-His master plate into CM(Glu)-Ura-His liquid medium.

b. Grow the cultures overnight on a roller drum or other shaker at 30°C.

c. In the morning, dilute the saturated cultures into fresh tubes containing 3–5 ml of CM(Glu)-Ura-His, with a starting density of OD_{600} of approximately 0.15. Incubate the cultures for 4–6 hours at 30°C until the optical density has doubled approximately twice (OD_{600} ~0.45–0.7).

12. Transfer 1.5 ml of each culture to a microfuge tube and centrifuge the cells at maximum speed for 3–5 minutes in a microfuge. The volume of the visible pellet should be 2–5 µl. Carefully decant or aspirate the supernatant.

13. Add 50 µl of 2x SDS gel-loading buffer to the tube and vortex the tube rapidly to resuspend the pellet. Immediately place the tube either on dry ice or in a dry-ice/ethanol bath.

14. Transfer the samples from the dry ice or –70°C directly to 100°C and boil them for 5 minutes.

15. Chill the samples on ice and centrifuge them at maximum speed for 5–30 seconds in a microfuge to pellet large cell debris. Load 20–50 µl into each lane of a SDS-polyacrylamide gel.

16. Run the gel and analyze the products to determine whether bait protein of the expected size is expressed at reasonable levels.

17. To anticipate and forestall potential problems, analyze the lysates of yeast containing LexA-fused baits by immunoblotting as described in Appendix 3.

Two-hybrid and Other Two-component Systems
Stage 2: Selecting an Interactor

In Stage 2 of this protocol, a mammalian cDNA library constructed in an expression plasmid such as pJG4-5 is transformed into yeast strains containing pBAIT and the *lexAop-lacZ* reporter plasmid. pJG4-5 expresses the cloned cDNAs from a cassette containing a transcriptional activation domain and other moieties under the control of the yeast GAL1 promoter. The yeast transformants are plated under both noninducing and inducing conditions on appropriate dropout media. Plasmids encoding putative target proteins that (1) allow growth in the absence of leucine and (2) cause expression of *lacZ* are selected for further analysis.

MATERIALS

CAUTION: Please see Appendix 4 for appropriate handling of materials marked with <!>.

Reagents and Solutions

Please see Appendix 1 for components of stock solutions, buffers, and reagents. Dilute stock solutions to the appropriate concentrations.
Dimethylsulfoxide (DMSO) <!>
Ethanol (optional; see Step 9)
TE (pH 7.5) containing 0.1 M lithium acetate
TE (pH 7.5) containing 40% (w/v) PEG 4000 and 0.1 M lithium acetate
TE (pH 7.5), sterile

Vectors and Hosts

Library to be screened for interaction
 LacZ reporter plasmids and interaction trap libraries are available commercially (see http://www. fccc.edu/research/labs/golemis/lib_sources1.html).
S. cerevisiae candidate strains expressing the bait and *lexAop-lacZ* reporter (from Protocol 18.1 Stage 1)

Media and Antibiotics

Please see Appendix 2 for components of yeast media.
Complete minimal (CM) selective medium (see Table 18-1)
Sterile glycerol solution for freezing transformants
 65% (v/v) glycerol, sterile
 0.1 M $MgSO_4$
 25 mM Tris-HCl (pH 8.0)
Yeast selective X-gal medium
 i. Prepare the base medium in 900 ml of H_2O, according to Table 18-1. Autoclave the base medium and cool it to 55°C.
 ii. In a separate bottle, autoclave 7 g of dibasic sodium phosphate and 3 g of monobasic sodium phosphate in 100 ml of distilled H_2O.
 iii. Mix the two autoclaved solutions together and add 0.8 ml of X-gal (100 mg/ml in *N,N*-dimethylformamide <!>). Pour the plates.

Nucleic Acids/Oligonucleotides

Sheared salmon sperm DNA, for use as carrier
 This DNA, which should be of very high quality, can be prepared as described in Protocol 6.10 or purchased commercially. Poor-quality carrier DNA can reduce transformation frequencies by one or two orders of magnitude.

Centrifuges/Rotors/Tubes

   Sorvall SLC-1500 and H-1000B rotors (room temperature)

Additional Items

   Culture plates (24 x 24 cm) for selective media
   Glass beads (0.45-mm diameter), sterile (Sigma-Aldrich) (optional; see Step 9)
   Heating block, thermal cycler, or water bath (42°C)
   Microtiter plates (96 well) (optional; see Step 21)
   Multichannel pipettor or inoculating manifold/frogger (e.g., DanKar) (optional; see
      Step 21)
   Shaking or orbital incubators (30°C)

Additional Information

Review of two-hybrid and other two-component systems	MC3, pp. 18.6–18.16
Troubleshooting	MC3, pp. 18.35 and 18.37

METHOD

1. Select a yeast colony, expressing the bait and *lexAop-lacZ* reporter, that was shown in Protocol 18.1 Stage 1 to be optimal in the initial control experiments. From this colony, grow a 20-ml culture in CM(Glu)-Ura-His liquid dropout medium overnight at 30°C on a roller drum.

   > IMPORTANT: The bait and *lexAop-lacZ* reporter plasmids should have been transformed into the yeast 7–10 days before retransformation with the library. It is important to maintain sterile conditions throughout.

2. Dilute the 20-ml overnight culture into 300 ml of CM(Glu)-Ura-His liquid dropout medium such that the diluted culture has an OD_{600} of approximately 0.10–0.15. Incubate the culture at 30°C on an orbital shaker until the culture has gone through 1.5 doublings, to reach an OD_{600} of approximately 0.50.

3. Transfer the culture to sterile 250-ml centrifuge bottles and centrifuge at 1000–1500*g* (2500–3000 rpm in a Sorvall SLC-1500 rotor) for 5 minutes at room temperature. Remove the supernatant and add 30 ml of sterile H_2O to the pellet. Resuspend the pellet by gently rapping the bottle against a countertop. Transfer the slurry to a sterile 50-ml Falcon tube.

4. Centrifuge the cells at 1000–1500*g* (2500–3000 rpm in a Sorvall SLC-1500 rotor) for 5 minutes. Pour off the H_2O and resuspend the yeast cells in 1.5 ml of TE (pH 7.5) containing 0.1 M lithium acetate.

5. Add 1 μg of library DNA and 50 μg of freshly denatured carrier DNA to each of 30 sterile 1.5-ml microfuge tubes. Immediately add 50 μl of the yeast suspension (from Step 4) to each of the 30 tubes.

   > A good transformation should yield approximately 10^5 transformants/μg of library DNA. For optimal efficiency, set up several small-scale transformation reactions rather than a single large one and use small volumes of DNA (<10 μl/transformation tube).

6. To each tube of the cell suspensions, add 300 μl of sterile TE (pH 7.5) containing 40% PEG 4000 and 0.1 M lithium acetate. Mix by gently inverting the tubes a number of times (do not vortex). Incubate the tubes for 30–60 minutes at 30°C.

7. To each tube, add 40 μl of DMSO and mix the suspensions by inversion. Place the tubes in a heating block for 10 minutes at 42°C.

8. Plate the transformation mixtures as follows:

For 28 of the tubes to be used solely to generate transformants

a. Pipette the contents of each tube onto 24 x 24-cm CM(Glu)-Ura-His-Trp dropout plates.

b. Spread the cells evenly and incubate the plates at 30°C until colonies appear.

> Each 24 x 24-cm plate requires 250–300 ml of agar medium, which should be allowed to dry for 1–2 days before use.

For the two remaining tubes to be used to assess transformation efficiency

a. Pipette 350 μl from each tube to 24 x 24-cm CM(Glu)-Ura-His-Trp dropout plates.

b. Spread the cells evenly and incubate the plates at 30°C until colonies appear.

c. Pipette the remaining 40 μl from each tube to make a series (at least three) of 1:10 dilutions in sterile TE (pH 7.5) or H_2O.

d. Plate 100 μl of each dilution on 100-mm CM(Glu)-Ura-His-Trp dropout plates and incubate the plates at 30°C until colonies appear.

> Before proceeding, read the material on harvesting and pooling of primary transformants in MC3, p. 18.33.

9. Harvest the library using one of the following methods.

> In Step 9, the first technique (by agitation) is faster to perform and allows induction of the library and screening on selective plates to be performed on the same day. It also minimizes the time the plates are open. About one third of the yeast slurry will be left on the plates; however, normally no more than 2% of the collected slurry is used, so it is important to ensure that colonies are washed from the plates with approximately equal efficiency. The second technique (by scraping) in Step 9 is more economical with regard to reagents and may be easier to use on plates from which molds and contaminants have been excised.

To harvest the library by agitation

a. Pour 10 ml of sterile H_2O and approximately 30 sterile glass beads on each of five 24 x 24-cm plates containing transformants.

b. Stack the five plates on top of one another and, holding the plates tightly, shake the stack until all of the colonies are resuspended (1–2 minutes).

c. Use a sterile pipette to collect 5 ml of yeast slurry from each plate (tilt the plate). Pool the slurry into sterile 50-ml conical tubes.

d. Proceed to the next five plates and repeat Steps a–c. Continue harvesting yeast cells from all 30 plates, resulting in a total volume of 150 ml of liquid contained in three 50-ml tubes.

To harvest the library by scraping

a. Wearing gloves, place the 30 24 x 24-cm plates containing transformants at 4°C to harden the agar (generally, 2 hours to overnight is acceptable).

b. Sterilize a glass microscope slide by immersing it in ethanol and flaming; use this slide to scrape yeast cells gently from the transformation plates into 50-ml conical tubes. Reflame or use a fresh slide at intervals (every 5–10 plates).

10. If necessary, fill each conical tube containing yeast to the 40–45-ml mark with sterile TE (pH 7.5) or H_2O. Vortex or invert the tubes to suspend cells.

11. Centrifuge the tubes in a benchtop centrifuge at 1000–1500*g* (2200–2700 rpm using a Sorvall H-1000B rotor) for 5 minutes at room temperature and discard the supernatant.

12. Repeat Steps 10 and 11.

13. Resuspend the packed cell pellet in 1 volume of sterile yeast glycerol solution. Combine the contents of the different Falcon tubes and mix thoroughly.

14. Transfer 1-ml aliquots into a series of sterile microfuge tubes and freeze at –70°C (cells remain stable for at least 1 year).

    If proceeding directly to plating on selective medium (which requires 5 hours to complete), leave one aliquot unfrozen and perform the next sequence of steps. Assume that the viability of the unfrozen culture is 100%.

    Before proceeding, read the material on screening for interacting proteins and troubleshooting in MC3, p. 18.33.

15. Thaw one aliquot of library-transformed yeast (from Step 14) and dilute 1:10 with CM(Gal-Raff)-Ura-His-Trp dropout medium. Incubate the yeast with shaking for 4 hours at 30°C to induce transcription from the GAL1 promoter on the library.

    Aliquots of the transformed library are plated on -Leu selective media to test for the inability to promote *LEU2* transcription.

16. Plate 10^6 cells (or 50 µl of a culture at $OD_{600} = 1.0$) on each of an appropriate number of 100-mm CM(Gal/Raff)-Ura-His-Trp-Leu dropout plates.

    IMPORTANT: The value of 10^6 cells per plate is the highest plating density that generally can be effectively used. Plating at higher densities (e.g., 3×10^6) can result in cross-feeding between yeast cells, resulting in high background growth.

17. Incubate the plates for 5 days at 30°C.

18. Observe the plates for growth and mark colonies as they appear.

19. At day 5, create a master plate (CM[Glu]-Ura-His-Trp) on which colonies are grouped by day of appearance.

20. Incubate the plates at 30°C until patches/colonies form.

21. Assay for transcription activation.

    Steps 21 and 22 test for galactose-inducible transcriptional activation of both the *lexAop-LEU2* and *lexAop-lacZ* reporters. Simultaneous activation of both reporters in a galactose-specific manner generally indicates that the transcriptional phenotype is due to expression of library-encoded proteins, rather than mutation of the yeast host. A master plate containing glucose and leucine is used as a source for test colonies.

    To assay by direct streaking

    a. On each of the four following plates, use a flat-edged toothpick to replicate the grid from the master plate. Use the same toothpick to streak an individual colony across the four plates. Try to get a thick streak of yeast on plates containing X-gal and a thin streak of yeast on plates lacking leucine.

CM(Glu/X-gal)-Ura-His-Trp	1 plate
CM(Gal/Raff/X-gal)-Ura-His-Trp	1 plate
CM(Glu)-Ura-His-Trp-Leu	1 plate
CM(Gal/Raf)-Ura-His-Trp-Leu	1 plate

    b. Incubate the plates for 3–4 days at 30°C.

    To assay using a manifold/frogger

    a. Deliver 25–30 µl of sterile H_2O into 48 wells of a 96-well microtiter plate.

    b. Use a frogger to transfer patches simultaneously from a pregridded master plate to the wells of a microtiter plate. Agitate the plate gently.

c. Use the frogger to transfer yeast from the microtiter plate to each of the following plates. This approach allows approximately equal quantities of cells to be transferred to each plate.

CM(Glu)-Ura-His-Trp	2 plates
CM(Gal)-Ura-His-Trp	1 plate
CM(Glu)-Ura-His-Trp-Leu	1 plate
CM(Gal/Raff)-Ura-His-Trp-Leu	1 plate

d. Incubate the plates for 3–4 days at 30°C. After 1 day of incubation, use one CM(Glu)-Ura-His-Trp plate and one CM(Gal)-Ura-His-Trp plate to assay for activation of the *lacZ* reporter, using the chloroform overlay assay. Please see the panel in the alternative protocol, Assay of β-Galactosidase Activity by Chloroform Overlay, in MC3, p. 18.28.

e. Continue to monitor growth: Differential growth on leucine will usually be apparent between 48 and 72 hours on the -Leu plates. The second CM(Glu)-Ura-His-Trp plate can be taken as a fresh master plate.

22. Interpret the results. Colonies and the library plasmids that they contain are designated as first-round positives if

- X-Gal analysis indicates blue color following culture on CM(Gal)-Ura-His-Trp plates.

- X-Gal analysis indicates white, or only faintly blue, following culture on CM(Glu)-Ura-His-Trp plates.

- Colonies grow well on CM(Gal/Raff)-Ura-His-Trp-Leu plates.

- Colonies grow poorly or not at all on CM(Glu)-Ura-His-Trp-Leu plates.

Plasmids encoding putative target proteins that (1) allow growth of transformants in the absence of leucine and (2) cause expression of *lacZ* are recovered by passage through *E. coli*, characterized by restriction analysis, and retransformed into "virgin" strains of *lexAop-LEU2/lexAop-lacZ/pBAIT* yeast to confirm the specificity of the interaction of the putative target with pBAIT.

MATERIALS

CAUTION: Please see Appendix 4 for appropriate handling of materials marked with <!>.

Reagents and Solutions

Please see Appendix 1 for components of stock solutions, buffers, and reagents. Dilute stock solutions to the appropriate concentrations.

Ammonium acetate (7.5 M)
Chloroform <!>
Ethanol
Ethanol (70%)
Isopropanol
Phenol, water saturated and equilibrated to pH 8.0 <!>
SDS (10% w/v)
STES lysis solution
    100 mM NaCl
    10 mM Tris-HCl (pH 8.0)
    1 mM EDTA
    0.1% (w/v) SDS
TE (pH 8.0)
Yeast lysis solution
    Zymolyase 100T dissolved at 2–5 mg/ml in rescue buffer
    or
    β-glucuronidase (100,000 units/ml) (Sigma-Aldrich) diluted 1:50 in yeast rescue buffer
Yeast rescue buffer (freshly prepared)
    50 mM Tris-HCl (pH 7.5)
    10 mM EDTA
    0.3% (v/v) β-mercaptoethanol <!>

Vectors and Hosts

E. coli strain DH5α or strain KC8 (CLONTECH), competent for electrotransformation (see Protocol 1.26)
Nonspecific bait plasmid
    If the pBait initially used in the screen was able to activate transcription, we strongly recommend using a nonspecific bait that can weakly activate transcription as a control. In general, baits that activate transcription poorly are difficult to distinguish from the background of false positives.
pBait (from Protocol 18.1 Stage 1)
Plasmids pMW112, pRFHM1
Yeast colonies with the appropriate phenotype growing on a CM(Glu)-Ura-His-Trp master plate (from Protocol 18.1 Stage 2, Step 21)
Yeast strain EGY48

Media

Please see Appendix 2 for components of yeast media.

  Complete minimal (CM) selective medium (see Table 18-1)

  Minimal (-trp) medium for bacteria

  1. Prepare the following solutions (sterile):
     i. 20% (w/v) MgSO$_4$
     ii. 4 mg/ml uracil
     iii. 4 mg/ml histidine
     iv. 4 mg/ml leucine
     v. 20% (w/v) glucose
     vi. 50 mg/ml ampicillin
     vii. 1% (w/v) thiamine hydrochloride
  2. Autoclave the following two solutions separately:
     viii. 15 g agar in 800 ml distilled H$_2$O
     ix. 10.5 g K$_2$HPO$_4$
         4.5 g KH$_2$HPO$_4$
         1 g (NH$_4$)$_2$SO$_4$
         0.5 g sodium citrate
         160 ml distilled H$_2$O

  Cool Solutions viii and ix to 50°C; mix them together. Quickly add 1 ml of solution i, 10 ml of solution ii, 10 ml of solution iii, 10 ml of solution iv, 10 ml of solution v, 1 ml of solution vi, and 0.5 ml of solution vii. Mix the final solution well and pour the plates immediately.

  Sterile glycerol solution for freezing transformants
      65% (v/v) glycerol, sterile
      0.1 M MgSO$_4$
      25 mM Tris-Cl (pH 8.0)

  Yeast selective X-gal medium
      i. Prepare the base medium in 900 ml of H$_2$O, according to Table 18-2. Autoclave the base medium and cool it to 55°C.
      ii. In a separate bottle, autoclave 7 g of dibasic sodium phosphate and 3 g of monobasic sodium phosphate in 100 ml of distilled H$_2$O.
      iii. Mix the two solutions together and add 0.8 ml of X-gal (in *N,N*-dimethylformamide <!>). Pour the plates.

Enzymes and Buffers

  Restriction enzymes *Eco*RI, *Xho*I, *Hae*III, and 10x buffers
  Zymolyase or β-glucoronidase (see Reagents and Solutions, above)

Centrifuges/Rotors/Tubes

  Sorvall H-1000B MPC and H-6000A MPC rotors, for centrifuging microtiter plates
  Sorvall SLC-1500 rotor (room temperature)

Additional Items

  Glass beads (0.45-mm diameter), sterile (Sigma-Aldrich)
  Microtiter plate (24 or 96 well)
  Repeating pipettor (optional; see Step 1)
  Shaking incubator (30°C)
  Step 2 of this protocol requires the reagents listed in Protocol 1.26.
  Step 10 of this protocol requires the reagents listed in Protocol 4.13.
  Steps 3 and 4 of this protocol require the reagents listed in Protocol 1.1.

Additional Information

Alternative protocol (Rapid Screen for Interaction Trap Positives)	MC3, pp. 18.46–18.47
Review of two-hybrid and other two-component systems	MC3, pp. 18.6–18.16
Subsequence characterization of positives	MC3, p. 18.45
Troubleshooting	MC3, pp. 18.35 and 18.37

TABLE 18-2 Yeast Complete Minimal (CM) Selective Media for Two-hybrid Analysis

Ingredients	CM (Glu) -T Medium	CM (Glu) -U-H Agar	CM (Glu) -U-H-T Agar	CM (Glu) -U-H-T-L Agar	CM (Glu, X-gal) -U-H-T Agar[a]	CM(Gal, Raff, X-gal) -U-H-T Agar[a]	CM (Gal,Raff) -U-H-T-L Agar
YNB[b]	6.7 g	6.7 g	6.7 g	6.7 g	6.7 g	6.7 g	6.7 g
Glucose	20 g	20 g	20 g	20 g	20 g		
Galactose						20 g	20 g
Raffinose						10 g	10 g
Dropout mix	2 g	2 g	2 g	2 g	2 g	2 g	2 g
Leucine	15 ml	15 ml	15 ml		15 ml	15 ml	
Tryptophan		10 ml					
Histidine	5 ml						
Uracil	5 ml						
Agar		20 g	20 g	20 g	20 g	20 g	20 g

[a]For media containing X-gal, prepare the appropriate base medium in 900 ml of H$_2$O and use it in the recipe for yeast selective X-gal medium.

[b]Use the form of yeast nitrogen base (YNB) that contains ammonium sulfate.

To prepare liquid or agar medium, mix the ingredients together in a final volume of 1 liter of H$_2$O. Autoclave for 20 minutes. Cool the medium to 50°C before pouring plates: Leucine stock = 4 mg/ml; tryptophan stock = 4 mg/ml; histidine stock = 4 mg/ml; uracil stock = 4 mg/ml.

METHOD

1. Prepare cell lysates from positive colonies.

   Two approaches are given for the isolation of discrete library plasmids. For the isolation of a small number of colonies (24–36 colonies), cells are lysed in SDS. For the isolation of a large number of colonies, cells are lysed in Zymolyase.

For isolation of a small number of colonies

   a. Starting from the CM(Glu)-Ura-His-Trp master plate (Protocol 18.1 Stage 2, Step 21), pick colonies that display the appropriate phenotype on selective plates into 5 ml of -Trp glucose medium. Grow the cultures overnight at 30°C.

   b. Centrifuge 1 ml of each culture at maximum speed for 1 minute at room temperature in a microfuge. Resuspend the pellets in 200 μl of STES lysis solution and add 100 μl of 0.45-mm-diameter sterile glass beads. Vortex the tubes vigorously for 1 minute.

   c. Add 200 μl of equilibrated phenol to each tube and vortex the tubes vigorously for another minute.

   d. Centrifuge the emulsions at maximum speed for 2 minutes at room temperature in a microfuge and transfer each aqueous phase to a fresh microfuge tube.

   e. Add 200 μl of equilibrated phenol and 100 μl of chloroform to each aqueous phase; vortex for 30 seconds. Centrifuge the emulsions at maximum speed for 2 minutes at room temperature in a microfuge and transfer each aqueous phase to a fresh tube.

   f. Add two volumes (400 μl) of ethanol to each aqueous phase, mix by inversion, and chill the tubes for 20 minutes at –20°C. Recover the nucleic acid by centrifugation at maximum speed for 15 minutes at 4°C in a microfuge.

   g. Pour off the supernatants. Wash the pellets with ice-cold 70% ethanol and dry the pellets briefly under vacuum. Resuspend each pellet in 5–10 μl of TE (pH 8.0). Proceed to Step 2.

For isolation of a large number of colonies

a. Transfer 2 ml of 2x CM(Glu)-Trp medium into each well of a 24-well microtiter plate. Use a toothpick to pick a putative positive colony from a master plate into each well (Protocol 18.1 Stage 2, Step 21). Grow the cultures overnight at 30°C with shaking.

b. Centrifuge the plate(s) in a centrifuge with microplate holders at 1500g (3000 rpm in a Sorvall H-1000B MPC rotor) for 5 minutes at 4°C. Shake off the supernatant with a snap of the wrist and return the plate to an upright position. Swirl or lightly vortex the plate to resuspend each cell pellet in the remaining liquid. Add 1 ml of H_2O to each well and swirl the plate gently.

c. Centrifuge as in Step b, shake off the supernatant, resuspend the cells in residual supernatant, and add 1 ml of rescue buffer.

d. Centrifuge as in Step b, shake off the supernatant, and resuspend the cells in the small volume of liquid remaining in the plate. To each well, add 25 µl of lysis solution. Swirl or vortex the plate. Incubate the plate (with the cover on) on a rotary shaker for approximately 1 hour at 37°C.

e. Add 25 µl of 10% SDS to each well. Disperse the precipitates completely by swirling the plates. Allow the plates to rest on the bench for 1 minute at room temperature. After 1 minute, the wells should contain a clear, somewhat viscous solution.

f. To each well, add 100 µl of 7.5 M ammonium acetate. Swirl the plates gently and store them for 15 minutes at –70°C or at –20°C until the lysates are frozen.

g. Remove the plates from the freezer. Once they begin to thaw, centrifuge the plates at 3000g (3800 rpm in a Sorvall H-6000A MPC rotor) for 15 minutes at 4°C. Transfer 100–150 µl of the resulting clear supernatants to fresh 24-well plates.

> IMPORTANT: In general, some contamination of the supernatants with pelleted material cannot be avoided. However, it is better to sacrifice yield to maintain purity.

h. To each well, add approximately 0.7 volume of isopropanol. Mix the solutions by swirling and allow the nucleic acids to precipitate for 2 minutes at room temperature. Centrifuge as in Step g. Shake off the supernatants with a snap of the wrist.

i. To each well, add 1 ml of cold 70% ethanol. Swirl the plates and then centrifuge them at 3000g (3800 rpm in a Sorvall H-6000A MPC rotor) for 5 minutes at 4°C. Shake off the supernatant with a snap of the wrist, invert the plates, and blot them well onto paper towels. Allow the plates to air dry.

j. To each well, add 100 µl of TE (pH 8.0). Swirl the plates and allow them to rest on the bench for several minutes until the pellets appear fully dissolved. Transfer the preparations to microfuge tubes or the wells of a 96-well plate for storage at –20°C. Proceed to Step 2.

> 1–5 µl of each of the resulting preparations can be used to transform competent *E. coli* by electroporation. If insufficient numbers of colonies are obtained, reprecipitate the DNAs and dissolve them in 20 µl rather than 100 µl of TE, to concentrate the DNA stock.

2. Introduce 1–5 µl of each preparation of plasmid DNA into competent *E. coli* strain DH5α or strain KC8 (*pyrF leuB600 trpC hisB463*) by electroporation (see Protocol 1.26). Plate the bacteria on LB agar containing 50 µg/ml of ampicillin and incubate the plates overnight at 37°C.

> If the bait is cloned in one of the specialized LexA-fusion plasmids that carries an ampicillin resistance marker, a proportion of the *E. coli* DH5α transformants will not contain the library plasmid. One option to resolve this problem is to analyze multiple transformants from each DNA preparation. Alternatively, it is possible to passage plasmids through a strain of *E. coli* possessing a *trpC* mutation and select for a library plasmid by the ability of the yeast *TRP1* gene to complement the *E. coli trpC* mutation.

3. If plasmid DNA was transformed into DH5α, proceed to Step 4. If plasmid DNA was transformed into KC8,

a. Use restreaking or replica plating to transfer colonies from LB/ampicillin plates to minimal (*-trp*) medium for bacteria. Incubate the bacteria overnight at 37°C.

b. Prepare miniprep DNA from an isolated colony and use the DNA to transform DH5α cells as described in Step 2.

4. Prepare miniprep DNA from DH5α cells carrying the library plasmid.

> **IMPORTANT:** It is a good idea to prepare DNA from two or three separate bacterial colonies generated from each original positive interactor. In contrast to bacteria, yeast can tolerate multiple 2μ plasmids with identical selective markers. If a single yeast cell contains two or three distinct library plasmids, and only one of which encodes an interacting protein, the relevant cDNA clone can be lost at the stage of plasmid isolation. In general, this is not a major problem; on average, perhaps 10% of yeast cells will contain two or more library plasmids.

5. Confirm by restriction endonuclease digestion that the duplicate samples prepared for each positive contain identical inserts and/or determine whether a small number of cDNAs have been isolated repeatedly.

> **IMPORTANT:** Some investigators sequence DNAs at this stage. A lot of money can be wasted on sequencing nonspecific interactors. It is strongly recommended that the transformation into yeast and specificity tests, as described below, be completed before sequencing.

6. Transform yeast strain EGY48 with the following sets of plasmids and select colonies on CM(Glu)-Ura-His plates.

> The final test of the specificity of interacting proteins is the retransformation of library plasmids from *E. coli* into "virgin" *lexAop-LEU2/lexAop-lacZ*/pBait-containing strains. The aim is to verify that interaction-dependent phenotypes are still observed and are specific to the starting pBait. This test will eliminate false positives, library-encoded cDNAs that interact with the LexA DNA-binding domain, and library-encoded proteins that are "sticky" and interact with multiple pBaits in a promiscuous manner.

a. pMW112 + pBait

b. pMW112 + pRFHM1

c. pMW112 + a nonspecific bait

> If the pBait initially used in the screen was able to activate transcription, even slightly, we strongly recommend that a nonspecific control bait that can also weakly activate transcription be included as a control. In general, baits that activate transcription poorly are difficult to distinguish from the background of false positives. Some false positives interact generically with weakly activating LexA-fused proteins.

7. After 2–3 days, transformed yeast from Step 6 should be available. Use electroporation to introduce library plasmids prepared from *E. coli* KC8 or DH5α into individual transformants **a–c**. Plate each transformation mixture on CM(Glu)-Ura-His-Trp dropout plates and incubate the plates at 30°C until colonies grow (2–3 days).

> **IMPORTANT:** As a negative control, also electroporate examples of transformants **a–c** with the pJG4-5 library vector.

8. Create a CM(Glu)-Ura-His-Trp master dropout plate for each library plasmid being tested. It is generally helpful to streak transformants **a–c** for each library plasmid in close proximity on the plate to facilitate detection of nonspecific interactions.

> **IMPORTANT:** Each plate should also contain the **a–c** series transformed with the pJG4-5 negative control.

9. Test for β-galactosidase activity and for leucine auxotrophy, exactly as described in Protocol 18.1 Stage 2, Step 21.

10. Analyze the results of these specificity tests and sequence the positive isolates (please see Protocol 4.13).

Detection of Protein–Protein Interactions Using Far Western with GST Fusion

Far western analysis was developed to screen cDNA expression libraries for clones that can interact with a ^{32}P-labeled target protein fused to GST. The target-GST fusion protein is synthesized in bacteria, purified by affinity chromatography on glutathione-agarose beads, and labeled in an in vitro reaction catalyzed by a commercially available protein kinase. The fusion protein is then digested with a protease to remove the GST moiety, and the labeled target is used to probe an expression library and/or membranes containing putative interacting proteins that have been separated by SDS-PAGE.

MATERIALS

CAUTION: Please see Appendix 4 for appropriate handling of materials marked with <!>.

Reagents and Solutions

Please see Appendix 1 for components of stock solutions, buffers, and reagents. Dilute stock solutions to the appropriate concentrations.

GST basic buffer
>   20 mM HEPES (pH 7.5)
>   50 mM KCl
>   10 mM MgCl$_2$
>   1 mM dithiothreitol, added just before the buffer is used
>   0.1% Nonidet P-40

GST blocking buffer
>   5% nonfat dry milk in GST basic buffer

GST fusion protein attached to glutathione-agarose beads
>   The screening method described in this protocol is for proteins containing a protein kinase A phosphorylation site in the fusion protein. However, the protocol can be easily adapted for other methods of screening.

GST interaction buffer
>   1% nonfat dry milk in GST basic buffer

GST wash buffer 1
>   Phosphate-buffered saline (PBS) containing 0.2% (v/v) Triton X-100

GST wash buffer 2
>   PBS containing 0.2% (v/v) Triton X-100 and 100 mM KCl

Enzymes and Buffers

2x PK buffer
>   100 mM KPO$_4$
>   20 mM MgCl$_4$
>   10 mM NaF <!>
>   9 mm dithiothreitol, added just before the buffer is used

Protease (optional; see Step 3)

Protein kinase A
>   Prepare fresh at each use according to manufacturer instructions.

Radioactive Compounds

[γ-^{32}P]ATP (6000 Ci/mmole) <!>

Additional Items

Materials for autoradiography or phosphorimaging
Nitrocellulose or PVDF membranes for western blotting (MC3, pp. A8.52–A8.53)
Sephadex G-50 spun column, equilibrated in 1x PK buffer (optional; see Step 4)
Step 3 of this protocol requires the reagents for the cleavage of the fusion protein from its carrier or from the glutathione-agarose beads as listed in Protocol 15.5.
Step 5 of this protocol requires either an SDS-polyacrylamide gel containing the proteins to be probed or plates containing a cDNA expression library (Protocol 14.2) and reagents for immunoblotting (Appendix 3).

A negative control of GST alone or a nonspecific protein should be loaded onto the SDS-poly-acrylamide gel with the target proteins.

Additional Information

Additional protocol (Refolding of Membrane-bound Proteins)	MC3, p. 18.53
Alternative protocol (Detection of Protein–protein Interactions with Anti-GST Antibodies)	MC3, p. 18.54
Review of far-western method applied to GST fusions	MC3, pp. 18.48–18.50
Troubleshooting	MC3, p. 18.53

METHOD

1. Prepare the following reaction mixture in a microfuge tube:

[γ-^{32}P]ATP (6000 Ci/mmole)	5 µl
protein kinase A	1 unit/µl
GST fusion protein on glutathione-agarose beads	1–3 µg
2x PK buffer	12.5 µl
H$_2$O	to 25 µl

Incubate the reaction mixture for 30 minutes at 37°C.

The fusion protein may be cleaved with a protease before the labeling reaction (see MC3, pp. 15.6–15.8). In this case, substitute the cleaved protein for the protein bound to glutathione-agarose beads in the labeling reaction, incubate for 30 minutes at 37°C, and proceed to Step 4.

If the GST moiety is retained on the fusion protein, the experiment must be replicated with GST alone (bound to glutathione-agarose beads) as a negative control. In the case of a library screen where this would be impractical, it is important to test positive plaques for their ability to bind GST alone. This control is usually performed after purification and before clone characterization.

2. After the labeling reaction is complete, wash the beads by adding 200 µl of 1x PK buffer to the tube and centrifuge the tube at maximum speed for 1 minute in a microfuge. Discard the supernatant containing the free radiolabeled nucleotide in an appropriate manner. Repeat the wash one more time.

3. Either cleave the labeled protein with a protease or elute the labeled GST fusion protein from the beads with 20 mM reduced glutathione in 50 mM Tris (pH 8.0) (please see Protocol 15.5). Store the radiolabeled protein probe in an ice bucket and use it on the same day that it is made.

4. If the labeled protein was cleaved from the GST moiety before the labeling reaction (please see the note to Step 1), load the radiolabeled protein onto a Sephadex G-50 spun column equilibrated with 1x PK buffer to remove the free radiolabeled nucleotide. Once the probe protein is separated from the free nucleotide, it is ready for use. Store the radiolabeled protein probe in an ice bucket and use it on the same day that it is made.

Probing the membranes

5. Prepare the membrane to be probed by transferring proteins to the membrane according to standard western blotting techniques.

6. Cover the membrane completely with GST basic buffer and wash it for 10 minutes at 4°C with gentle agitation.

7. Discard the GST basic buffer. Cover the membrane completely with GST blocking buffer and incubate it with gentle agitation for 4 hours to overnight at 4°C.

8. Prepare a solution of labeled protein by adding 1–3 μg of the stock probe (from either Step 3 or 4) to enough GST interaction buffer to make a final concentration of 1–5 nM. Transfer the membrane to a dish containing the diluted probe. Make sure that the probe solution contacts the entire surface of the membrane evenly. Incubate the membrane for 4–5 hours at 4°C with gentle agitation.

9. Discard the radioactive probe solution in an appropriate manner. Cover the membrane completely with GST wash buffer 1 and incubate the membrane for 10 minutes at 4°C with gentle agitation. Repeat the wash three more times.

10. Cover the membrane completely with GST wash buffer 2 and wash the membrane for 10 minutes at 4°C with gentle agitation. Repeat this once.

11. Wrap the membrane carefully in plastic wrap and expose it to X-ray film.

Detection of Protein–Protein Interactions Using the GST Protein Pulldown Technique

GST pulldown experiments are used to identify novel interactions between a probe protein and unknown targets and to confirm suspected interactions between a probe and a known protein. In both cases, the probe protein is a GST fusion, which is expressed in bacteria and purified by affinity chromatography on glutathione beads. Target proteins are usually lysates of cells, which may be labeled with [^{35}S]methionine or unlabeled, depending on the method used to assay interaction between the target and the probe. The cell lysate and the GST fusion protein probe are incubated together with glutathione-agarose beads, which are then collected and washed. Complexes recovered from the beads are resolved by SDS-PAGE and processed for further analysis by western blotting, autoradiography, or staining.

MATERIALS

CAUTION: Please see Appendix 4 for appropriate handling of materials marked with <!>.

Reagents and Solutions

Please see Appendix 1 for components of stock solutions, buffers, and reagents. Dilute stock solutions to the appropriate concentrations.

Lysis buffer
> 20 mM Tris-Cl (pH 8.0)
> 200 mM NaCl
> 1 mM EDTA (pH 8.0)
> 0.5% (v/v) Nonidet P-40
> Just before use, add protease inhibitors to the following final concentrations: 2 µg/µl aprotinin, 1 µg/µl leupeptin, 0.7 µg/ml pepstatin, and 25 µg/ml phenylmethylsulfonyl fluoride (PMSF) <!>.

Reduced glutathione (20 mM) in 50 mM Tris-Cl (pH 8.0) (optional; see Step 8)

Cells, Tissues/Culture Media

Cell lysate containing proteins that have been labeled in vivo with ^{35}S-methionine
> It is possible to use unlabeled cell lysates depending on the goals of the experiment and the desired detection method (see MC3, pp. 18.55–18.56).

Probes

GST fusion protein carrying the "bait" or probe sequence (see Protocol 15.1)
GST protein

Gels/Loading Buffers

2x SDS-PAGE gel-loading buffer
> 100 mM Tris-Cl (pH 6.8)
> 4% (w/v) SDS (electrophoresis grade)
> 0.2% (w/v) bromophenol blue
> 20% (v/v) glycerol
> 200 mM dithiothreitol or β-mercaptoethanol added just before use <!>

SDS-polyacrylamide gel <!> (see Appendix 3)

Additional Items

Boiling-water bath
End-over-end sample rotator
Glutathione-agarose beads (prepared as a 50% slurry in lysis buffer)

Step 11 of this protocol requires the reagents for immunoblotting and staining proteins separated by SDS-polyacrylamide gel electrophoresis (Appendix 3).

Additional Information

Review of GST pulldown	MC3, pp. 18.55–18.57
Troubleshooting	MC3, p. 18.59

METHOD

1. Incubate the cell lysate with 50 μl of a 50% slurry of glutathione-agarose beads and 25 μg of GST for 2 hours at 4°C with end-over-end mixing. The amount of lysate needed to detect an interaction is highly variable. Start with a volume of lysate equivalent to 1×10^6 to 1×10^7 cells.

   Because the aim of the experiment is to compare GST with a GST fusion protein, it is necessary to prepare enough precleared lysate for each reaction. Efficient mixing of reagents is the key to success. This is best achieved if the reaction is performed in a reasonable volume: 500–1000 μl is a good starting point.

2. Centrifuge the mixture at maximum speed for 2 minutes at 4°C in a microfuge.

3. Transfer the supernatant (i.e., the precleared cell lysate) to a fresh microfuge tube.

4. Set up two microfuge tubes containing equal amounts of precleared cell lysate and 50 μl of glutathione-agarose beads. To one tube, add approximately 10 μg of GST protein; to the other tube, add approximately 10 μg of the GST fusion probe protein. The amount of probe and control protein added should be equimolar in the two reactions (i.e., the final molar concentration of GST should be the same as the GST fusion probe protein). Incubate the tubes for 2 hours at 4°C with end-over-end mixing.

   IMPORTANT: If the bound proteins will be removed from the beads by boiling (Step 10), it is important to include a control tube containing only glutathione-agarose beads and cell lysate. This allows the detection of proteins that bind nonspecifically to the beads.

5. Centrifuge the samples at maximum speed for 2 minutes in a microfuge.

6. Save the supernatants at 4°C in fresh microfuge tubes. These samples will be analyzed by SDS-polyacrylamide gel electrophoresis analysis in Step 10.

7. Wash the beads with 1 ml of ice-cold GST lysis buffer. Centrifuge the tubes at maximum speed for 1 minute in a microfuge. Discard the supernatants. Repeat the washes three times.

8. (*Optional*) Elute the GST fusion protein and any proteins bound to it by adding 50 μl of 20 mM reduced glutathione in 50 mM Tris-Cl (pH 8.0) to the beads. Centrifuge the tubes at maximum speed for 2 minutes in a microfuge.

9. Mix the beads (from Step 7) or the eluted proteins (from Step 8) with an equal volume of 2x SDS-PAGE gel-loading buffer.

10. Boil the samples for 4 minutes and analyze them by SDS-polyacrylamide gel electrophoresis.

11. The method of detecting proteins associated with the GST fusion protein will depend on whether the cell lysate was radiolabeled and on the goal of the experiment.

    - If the goal is to detect all of the ^{35}S-labeled proteins associated with the fusion protein, dry the gel on a gel dryer and expose it to X-ray film to produce an autoradiograph.

    - If the goal is to detect specific associated proteins, transfer the proteins from the SDS-polyacrylamide gel to a membrane and perform immunoblotting.

    - If the goal is to determine the sizes and abundance of proteins associated with the fusion protein from a nonradioactive lysate, stain the gel with Coomassie Blue or silver nitrate.

Identification of Associated Proteins by Coimmunoprecipitation

Coimmunoprecipitation is most commonly used to test whether two proteins of interest are associated in vivo, but it can also be used to identify novel interacting partners of a target protein. In both cases, the cells, which may have been labeled with [^{35}S]methionine, are harvested and lysed under conditions that preserve protein–protein interactions. The target protein is specifically immunoprecipitated from the cell extracts and the immunoprecipitates are fractionated by SDS-PAGE. Coimmunoprecipitated proteins are detected by autoradiography and/or by western blotting with an antibody directed against that protein. The identity of interacting proteins may be established or confirmed by Edman degradation of tryptic peptides. This protocol was originally used to identify pVHL-associated proteins, but is easily adaptable for analysis of other proteins.

MATERIALS

CAUTION: Please see Appendix 4 for appropriate handling of materials marked with <!>.

Reagents and Solutions

Please see Appendix 1 for components of stock solutions, buffers, and reagents. Dilute stock solutions to the appropriate concentrations.

Acetonitrile (50%) <!>

Coomassie Blue R-250 staining solution (see p. 712)

EBC lysis buffer

    50 mM Tris-Cl (pH 8.0)

    120 mM NaCl

    0.5% (v/v) Nonidet P-40

    0.2 mM sodium orthovanadate

    100 mM NaF <!>

    Just before use, add protease inhibitors to the following final concentrations: 10 µg/ml aprotinin, 5 µg/ml leupeptin, and 50 µg/ml PMSF <!>.

NETN

    20 mM Tris-Cl (pH 8.0)

    1 mM EDTA (pH 8.0)

    100 mM NaCl

    0.5% (v/v) Nonidet P-40

NETN containing 900 mM NaCl

Phosphate-buffered saline (PBS)

Trifluoroacetic acid (TFA) (0.1% w/v), sequencing grade <!>

Cells, Tissues/Culture Media

Appropriate cell lines growing in culture (6 x 10^7 to 8 x 10^7 cells are required for this protocol)

Enzymes and Buffers

Trypsin digestion buffer

    0.02% (v/v) Tween-20

    200 mM ammonium bicarbonate (pH 8.9)

Trypsin (sequencing grade [Roche Applied Science]), 250 µg/ml in 200 mM ammonium bicarbonate (pH 8.9)

Antibodies

Control antibody
Polyclonal or monoclonal antibodies capable of precipitating the target protein

Radioactive Compounds

$[\gamma\text{-}^{32}P]ATP$ (6000 Ci/mmole) <!>

Gels/Loading Buffers

Discontinuous SDS-PAGE gradient gel <!>
> The separating gel (7.5–15% at pH 8.8) should be 20–30 cm long, the stacking gel (5% at pH 6.8) should be 10 cm long, and the loading well should be 1–2 cm deep. Pour the stacking gel without a multiwell comb and construct the wells by inserting spacers vertically between the glass plates so that they form wells large enough to take the samples.

1x SDS-PAGE gel-loading buffer
> 50 mM Tris-Cl (pH 6.8)
> 2% (w/v) SDS (electrophoresis grade)
> 0.1% (w/v) bromophenol blue
> 10% (v/v) glycerol
> 100 mM dithiothreitol or β-mercaptoethanol added just before use <!>

Additional Items

Boiling-water bath
Protein A–Sepharose (1:1 slurry in NETN buffer)
Razor blades
Rocking platform

Additional Information

Detailed description of coimmunoprecipitation   MC3, pp. 18.60–18.66
Steps 10 and 11 of this protocol require the reagents and equipment for mapping, purification, and sequencing of peptides. See Spector et al. 1998. *Cells: A Laboratory Manual*, Chapters 62 and 63. Cold Spring Harbor Laboratory Press, Cold Spring Harbor, New York.

METHOD

1. Wash 30 10-cm plates of the appropriate cells (a total of ~6×10^7 cells) in PBS. Scrape each plate of cells into 1 ml of ice-cold EBC lysis buffer.

2. Transfer each milliliter of cell suspension into a microfuge tube and centrifuge the tubes at maximum speed for 15 minutes at 4°C in a microfuge.

3. Pool the supernatants (~30 ml) and add 30 µg of the appropriate antibody. Rock the immunoprecipitate for 1 hour at 4°C.

4. Add 0.9 ml of the protein A–Sepharose slurry. Rock the immunoprecipitate for another 30 minutes at 4°C.

5. Wash the protein A–Sepharose mixture in NETN containing 900 mM NaCl. Repeat this wash five more times. Finally, wash the mixture once in NETN.

6. Remove the liquid portion of the mixture by aspiration. Add 800 µl of 1x SDS gel-loading buffer to the beads and boil them for 4 minutes.

7. Load the sample into the large well of the discontinuous SDS-PAGE gradient gel and run the gel at 10 mA constant current overnight.

8. Visualize the protein bands by staining with Coomassie Blue (see p. 761).

9. Using a razor blade, excise the band of interest from the gel, place it in a microfuge tube, and wash it twice for 3 minutes each in 1 ml of 50% acetonitrile.

10. Digest the protein with trypsin while it is still in the gel and electroelute the peptides.

    a. Remove the gel slice to a clean surface and allow it to partially dry.

    b. Add 5 μl of trypsin digestion buffer and 2 μl of trypsin solution.

    c. After the gel absorbs the trypsin solution, add 5-μl aliquots of trypsin buffer until the gel slice regains its original size.

    d. Place the gel slice in a microfuge tube, immerse it in trypsin digestion buffer, and incubate it for 4 hours at 30°C. Stop the reaction by addition of 1.5 μl of 0.1% trifluoroacetic acid.

11. Fractionate the peptides by narrow-bore high-performance liquid chromatography. Subject the collected peptides to automated Edman degradation sequencing.

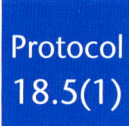

Probing Protein Interactions Using GFP and Fluorescence Resonance Energy Transfer

Stage 1: Labeling Proteins with Fluorescent Dyes

The first stage of this protocol describes how components (usually Fab fragments of antibodies) that are to be introduced into cells are labeled with fluorescent sulfoindocyanine dyes. The example used here involves labeling a monoclonal antibody directed against PKCα. However, the protocol can easily be adapted for use with other probes.

MATERIALS

CAUTION: Please see Appendix 4 for appropriate handling of materials marked with <!>.

Reagents and Solutions

Please see Appendix 1 for components of stock solutions, buffers, and reagents. Dilute stock solutions to the appropriate concentrations.

Bicine (0.1 M, pH 7.5)

Bicine (1 M, pH 9.0–9.5)

Bradford reagent for quantifying protein concentration

N,N-dimethylformamide (DMF) <!>

> Because H_2O competes for free dye in the labeling reaction, the DMF must be kept dry by the addition of approximately 1/3 volume of hygroscopic resin to the storage vessel, for example, AG 501-X8 mixed bed resin (Bio-Rad), which can be placed permanently in the stock bottle of DMF.

Phosphate-buffered saline (PBS)

Sodium phosphate (20 mM) with 10 mM EDTA (pH 7.0)

TE (pH 7.0)

Enzymes and Buffers

Papain immobilized to agarose beads (50% slurry in digestion buffer [Sigma-Aldrich])

Antibodies

Purified monoclonal or polyclonal antibody, or protein to be labeled

> The labeling reaction must be performed in a buffer free of amine groups. Antibodies purchased from commercial sources are usually supplied in buffers rich in amine groups. However, a number of suppliers will provide the antibody in PBS, if requested.

Labeled Compounds

Cy3 and Cy5 multifunctional sulfoindocyanine succinimide esters (GE Healthcare)

> These dyes are supplied as lyophilized samples that must be maintained in a desiccated environment at all times.

Gels/Loading Buffers

2x SDS-PAGE gel-loading buffer

190 mM Tris-Cl (pH 6.8)

4% (w/v) SDS (electrophoresis grade)

0.2% (w/v) bromophenol blue

20% (v/v) glycerol

200 mM dithiothreitol or β-mercaptoethanol added just before use <!>

SDS-polyacrylamide gel <!> (see Appendix 3)

Additional Items

    Boiling-water bath
    Centricon concentrators YM-10, Y-30, and YM-100 (Millipore)
    Gel-filtration/size exclusion chromatography columns (prepacked Econo-Pac disposable
    [Bio-Rad])
    Protein A–Sepharose column (Econo-Pac, 2 ml [Bio-Rad])

Additional Information

Detailed description of FLIM-FRET	MC3, pp. 18.69–18.79
GFP	MC3, pp. 17.85–17.89

METHOD

1. Concentrate the antibodies to 15–20 mg/ml in 20 mM sodium phosphate, 10 mM EDTA (pH 7.0), by centrifugation using a disposable concentrator such as a Centricon YM-100.

2. To 250 µl of the concentrated antibody preparation (~4–5 mg), add 500 µl of Ab digestion buffer and 500 µl of a 50% slurry (in digestion buffer) of papain immobilized on agarose beads. Allow the digestion to proceed for 6–10 hours at 37°C with shaking (300–400 rpm in a shaking incubator).

   IMPORTANT: Excessive digestion (i.e., in excess of 16 hours) will result in the cleavage of the Fab fragments into smaller polypeptides. It is therefore advisable to perform a pilot experiment where small aliquots of the digestion (~5 µl) are withdrawn at 1-hour intervals for analysis by reducing SDS-polyacrylamide gel electrophoresis. Under reducing conditions, the heavy and light chains of IgG migrate as 50- and 25-kD polypeptides. The papain-digested antibody migrates at approximately 25 kD.

3. Purify the Fab fragments from the Fc portion and nondigested antibodies by passage through a protein A–Sepharose column. Apply the digestion mixture to the equilibrated column and immediately begin collecting 1-ml fractions of the flowthrough.

4. To identify which fraction contains the purified Fab fragments, measure the OD_{280} of each fraction. Collect and pool three or four fractions containing the highest concentration of Fab.

5. Concentrate the pooled fractions by centrifugation through a Centricon YM-10 (10-kD cut-off [Millipore]) to a volume of <0.5 ml.

6. Apply the sample to a low-molecular-mass (6-kD exclusion) gel-filtration chromatography column equilibrated in PBS. Elute the Fab fragments with PBS (according to manufacturer instructions).

7. Following elution, concentrate the Fab fragments through a Centricon concentrator and determine the protein concentration using the Bradford reagent. Add 1/10 volume of 1 M bicine (pH 9.0).

8. Resuspend the contents of a single vial of lyophilized dye in 20 µl of dry *N,N*-dimethylformamide, to give a Cy solution with a concentration of approximately 10 mM.

9. To prepare the dye, dilute 1 µl of the dye mixture 1:10,000 in PBS and calculate the dye concentration from the visible peak absorption.

   The extinction coefficients for Cy3 and Cy5 are 150,000 and 250,000 mM^{-1} at 554 and 650 nm, respectively.

10. To label the proteins, suspend them in a buffer that does not contain free amino groups. If the protein solution is not in an appropriate buffer, perform Steps 6 and 7.

11. React the protein sample (antibody, Fab fragment, or protein) with a 10- to 40-fold molar excess of Cy3 for 30 minutes at room temperature (prepared as described in Steps 8 and 9). Add the dye very slowly while stirring the solution with a pipette tip (carefully avoiding direct dye contact with the microfuge tube).

    To avoid protein denaturation by DMF, the volume of Cy3/DMF added must not exceed 10% of the total volume.

12. Terminate the labeling reaction by the addition of a free-amine-containing buffer such as Tris to a final concentration of 10 mM.

13. Remove the excess unreacted dye by exchanging the buffer using gel-filtration column chromatography (Bio-Rad) and eluting the protein into PBS.

    a. Equilibrate the column with 3 bed volumes of PBS or the buffer of choice (~30 ml).

    b. To maximize separation, load the labeling reaction mixture directly onto the resin, taking care to apply it in as small a volume as possible.

    c. Wash the column with 2.5 ml of buffer (~3.3 ml is void) and discard.

    d. Add an additional 2 ml of buffer and collect the visible protein fraction, discarding the remainder. The labeled product will run at the leading front and should be visible. Free dye migrates more slowly through the column.

14. Concentrate the labeled protein once again using a Centricon concentrator (of the appropriate molecular-mass cutoff).

15. Analyze the labeled product by electrophoresis through an SDS-acrylamide gel to verify successful covalent labeling of the protein. Following electrophoresis, examine the gel directly on a UV transilluminator (302 nm) to visualize the labeled product.

16. Following the labeling reaction, calculate the labeling ratio using the following formula:

    $$A_\lambda \times M/[(A_{280} - f \times A_\lambda) \times \varepsilon_\lambda]$$

    where A_λ is the absorption of the dye at its absorption maximum at wavelength λ, A_{280} is the absorption of the protein at 280 nm, M is the molecular mass of the protein in kD, and ε_λ is the molar extinction coefficient of the dye at wavelength λ in mM^{-1} cm^{-1}. The equation also corrects for absorption of the dye at 280 nm. This factor f is the ratio between absorption of the dye at 280 nm and its maximal visible absorption at wavelength λ. For example, the labeling ratio formula for Cy3-labeled antibody becomes

    $$A_{554} \times 170/[(A_{280} - 0.05 \times A_{554}) \times 150]$$

    To verify that dye coupling has not damaged the biological function of the Fab, whole antibody, or protein, compare the specific activity of the labeled product with its unlabeled counterpart.

Probing Protein Interactions Using GFP and Fluorescence Resonance Energy Transfer

Stage 2: Preparation for FLIM-FRET Analysis

In preparation for FLIM-FRET analysis, the appropriate donor and acceptor components must be introduced into live or fixed cells. The method of introduction depends on the nature of the components and the state of the cells. For example, plasmid DNAs encoding a protein of interest fused to a variant of GFP may be introduced into live cells by transfection or microinjection, whereas labeled antibodies are delivered by microinjection. For studies on fixed cells, plasmid DNA is introduced by transfection or microinjection, and the cells are subsequently fixed in paraformaldehyde before staining with labeled protein or antibody. This protocol describes the introduction of plasmid DNA into live cells by transfection.

MATERIALS

Reagents and Solutions

Please see Appendix 1 for components of stock solutions, buffers, and reagents. Dilute stock solutions to the appropriate concentrations.

> Gelatin solution (0.1%) (optional; see Step 1)
> Phosphate-buffered saline (PBS)
> Poly-L-lysine (optional; see Step 1; available commercially from Sigma-Aldrich as a 0.1% solution in H_2O; dilute 1:10 in H_2O before use)

Vectors and Hosts

> Recombinant plasmid DNA, e.g., GFP fusion vector carrying sequences encoding the protein of interest

Media and Antibiotics

> CO_2-independent imaging medium (Invitrogen)

Cells, Tissues/Culture Media

> Cell line to be transfected

Antibodies

> Purified monoclonal or polyclonal antibody, or protein to be labeled
>> The labeling reaction must be performed in a buffer free of amine groups. Antibodies purchased from commercial sources are usually supplied in buffers rich in amine groups. However, a number of suppliers will provide the antibody in PBS, if requested.

Labeled Compounds

> Cy3 and Cy5 multifunctional sulfoindocyanine succinimide esters (GE Healthcare)
>> These dyes are supplied as lyophilized samples that must be maintained in a desiccated environment at all times.

Additional Items

> Glass-bottomed tissue-culture dishes (35 mm [MatTek Corporation])
> *or*
> Glass or coverglass chamber slides (Labtek; Nunc)

Tissue-culture plates (6 or 12 well) (Nunc)

Step 2 of this protocol requires the reagents for transfection (see Chapter 16 protocols).

Additional Information

Alternative protocol (Microinjection of Live Cells)	MC3, pp. 18.88–18.89
Alternative protocol (Preparation of Fixed Cells for FLIM-FRET Analysis)	MC3, pp. 18.87–18.88
Detailed description of FLIM-FRET	MC3, pp. 18.69–18.79
GFP	MC3, pp. 17.85–17.89

METHOD

1. Seed the cells onto an appropriate surface for microscopy.

   For live cell preparations

   Seed the cells onto glass-bottomed dishes or coverglass chamber slides.

   For cell preparations to be fixed after transfection

   Seed the cells onto glass coverslips placed in 6- or 12-well tissue-culture dishes.

   > For suspension cells, immobilization can be facilitated with the use of media such as gelatin (0.1% [w/v] in PBS; autoclaved) or poly-L-lysine (0.01% [w/v] in H_2O). In either case, coat the surface by covering the wells, coverslips, or coverglass slides for approximately 30 minutes with the chosen medium and then aspirate the excess. Cells can then be directly seeded onto the coated surface.

2. Transfect the cells with the plasmid(s) encoding the GFP-tagged protein(s) of interest, using any one of the transfection methods described in Chapter 16.

3. Incubate the transfected cells under the appropriate conditions for 16–24 hours to allow the cells to express the protein of interest.

4. Identify cells that are expressing the protein of interest.

5. (*Optional*) If appropriate for the experimental design, introduce another probe (e.g., labeled protein) using one of the alternative protocols following Step 6: microinjection (for live cells) or fixation and staining (for fixed cells).

6. Immediately before performing an experiment at the microscope, replace the culture medium with CO_2-independent imaging medium (commercial medium or Dulbecco's modified Eagle's medium).

Probing Protein Interactions Using GFP and Fluorescence Resonance Energy Transfer

Stage 3: FLIM-FRET Measurements

This stage of the protocol presents a basic plan for capturing a series of images by fluorescent lifetime imaging microscopy (FLIM). The protocol specifically describes data acquisition for a particular variant of GFP (EGFP) or Oregon Green as a donor fluorophore, but it can be adapted for image acquisition of other chromophore systems as described in Table 18-3.

TABLE 18-3 Filter Specifications for FLIM-FRET Using Selected Donor and Acceptor Pairs

FRET Pair	Donor Filter Set			Acceptor Filter Set		
Donor and Acceptor	Excitation	Dichroic	Emission	Excitation	Dichroic	Emission
ECFP and EYFP	laser line	455DRLP	480DF30	HQ500/40	Q525LP	HQ555/50
	457.9 nm	(Omega)	(Omega)	(Chroma)	(Chroma)	(Chroma)
EGFP and DsRed	laser line	Q495LP	HQ510/20	HQ545/30	HQ565LP	HQ610/75
	488 nm	(Chroma)	(Chroma)	(Chroma)	(Chroma)	(Chroma)
Cy3 and Cy5	laser line	Q565LP	HQ610/75	HQ620/60	HQ660LP	HQ700/75
	514.5 nm	(Chroma)	(Chroma)	(Chroma)	(Chroma)	(Chroma)
EGFP and Cy3	laser line	Q495LP	HQ510/20	HQ545/30	HQ565LP	HQ610/75
	488 nm	(Chroma)	(Chroma)	(Chroma)	(Chroma)	(Chroma)

MATERIALS

CAUTION: Please see Appendix 4 for appropriate handling of materials marked with <!>.

Cells and Tissues

Live or fixed cells expressing the protein of interest, prepared as described in Protocol 18.5 Stage 2

Additional Items

Amplifier (ENI [Electronic Navigation Industries] 403LA or IntraAction PA-4)
Argon laser (Innova 70C [Coherent]) <!>
Broadband dielectric mirrors (two) (Newport Corporation)
Circulating water bath (Boekel Scientific)
Detectors
    MCP: Hamamatsu Photonics C5825
    CCD: Photometrics Quantix with Kodak KAF 1400 chip
Filters
    GFP, OG (Q495LP, HQ510/20) (Chroma)
    Cy3 (HQ545/30, Q565LP, HQ610/75) (Chroma)
Flexure mirror mounts (two), lens holders (two), posts (five), and postholders (five) (Newport)
Inverted microscope (Zeiss 135 TV)
IPLab spectrum (Signal Analytics)
Iris diaphragm (Edmund Optics)
Lenses (12- and 7.6-cm focal length and 2.5- and 3.8-cm focal length, respectively) (Newport)
Mercury arc lamp (100 W) (Zeiss; HBO AttoArc)
Multimode fiber (step-indexed, 1-mm core) (Newport)

Adapted from Chapter 18, Protocol 5 Stage 3, p. 18.90 of MC3.

683

Oil objective (100x/1.4 NA) (Zeiss)
Optical breadboard (2 x 1 m) (TMC)
Power meter and head (Ophir)
Power PC (Apple) equipped with PCI-GPIB card (National Instuments)
Shutter (high speed) (Uniblitz VS25 [Vincent Associates])
Shutter driver (Vincent Associates)
Standing wave acousto-optic modulator (AOM) (80 mHz [IntraAction])
Two-frequency synthesizers (IFR 2023)
Variable density filter wheel (Laser Components)

Additional Information

GFP	MC3, pp. 17.85–17.89
Theoretical and experimental details of FLIM-FRET	MC3, pp. 18.69–18.80

METHOD

1. Select the 488-nm excitation wavelength on the argon laser by adjusting the wavelength selector prism. Optimize the output power by fine-tuning the position of the high-reflector mirror using the control knobs at the back of the laser.

2. Set the frequency synthesizer to drive the AOM to approximately 40 MHz at a resonance frequency (for the experiments described herein, a driving frequency of 40.112 MHz is used). This gives rise to intensity oscillations in the laser light beam at twice the driving frequency (80.224 MHz).

3. Optimize the diffraction in the AOM and thereby the modulation depth by adjusting the angle of incidence and monitoring the intensity of the undiffracted zero-order beam with a power meter. The optimal angle of diffraction (corresponding to maximal diffraction) gives rise to a minimum in the output power of the zero-order beam.

4. Turn on the MCP and the CCD. Set the bias of the photocathode voltage to –2 V and adjust the gain to match the full dynamic range on the CCD. This is dependent on the fluorescence intensity of the sample and must be determined empirically. Ideally, keep the gain as low as possible to reduce noise. Typically, the gain is set at 1 for the Hamamatsu C5825, with the gain of the Photometrics Quantix CCD set at 3. Set the readout of the CCD to 2 x 2 binning.

5. Set the master frequency synthesizer driving the MCP to a value exactly double (for the above example: 80.224 MHz) of that driving the AOM.

6. Choose the most suitable objective for the experiment. For this example, a Zeiss Fluar 100x/1.4 NA oil objective is used.

7. To obtain a zero lifetime reference image, record a cycle of 16 phase-dependent images, each separated by 22.5° from a strong scatterer (e.g., a small piece of aluminum foil placed on the imaging surface of a coverslip or glass-bottom dish).

   a. Exchange the fluorescence filter set for a half-silvered mirror and reduce the intensity of the incident beam to a minimum with a variable density filter wheel. Adjust the focus on the foil surface.

      IMPORTANT: When setting up the foil image, take extreme care not to look directly into the microscope occular until the incident laser source has been reduced to a minimum.

   b. Take a single image of the aluminum foil with an exposure time of approximately 100 msec. This image is taken to select an ROI and to gauge the likely exposure time required. Because the phase of the master frequency synthesizer may not be at a maximum, and to avoid saturation of the detector, select an exposure time that generates approximately 1000 counts.

c. Record a cycle of 16 phase images, each separated by 22.5°. Determine the phase setting at which maximum intensity is reached in the image series. Reset this phase to zero degrees on the master frequency synthesizer.

d. Record another cycle of 16 phase images from the foil to save as a zero lifetime reference.

While performing lifetime imaging, we recommend that phase stability be monitored over time. In our setup, the phase is stable within 0.3° during a period of 1 hour. For our setup, a reference foil sequence is recorded and saved each hour.

8. Restore the incident excitation source to a maximum (using the variable density filter wheel); the system is now ready for the acquisition of cell imaging data.

9. Acquire a donor image using the GFP filter set (dichroic: Q495LP, emission: HQ510/20). Select an exposure time such that 75% of the dynamic range of the CCD (3000 counts for a 12-bit CCD) is filled. Select a region of interest in the image on which to measure fluorescence lifetimes.

10. Create a donor fluorescence lifetime map.

a. Take two contiguous series of 16 phase-dependent images (45° phase-stepped), one forward and one reverse cycle to correct for photobleaching. Optimize the set exposure time for each phase image as in Step 9.

b. Take an additional image in the absence of sample illumination. This background offset image is subtracted from all phase images in the series. Each of the pairs of equivalent phase images of the forward and reverse cycles is then summed to first order to correct for bleaching of the donor.

c. From these data and the zero lifetime reference (foil), calculate a donor fluorescence lifetime map as described in the discussion on Image Processing in MC3, p. 18.75.

11. Record an image of the acceptor and then photobleach the acceptor by changing to the 100-W mercury arc lamp (Zeiss; HBO AttoArc) as a source of illumination. Move the Cy3 filter set (excitation: HQ545/30, dichroic: Q565LP, emission: HQ610/75 chroma) into the detection path. Take an image of the acceptor by optimizing the exposure time to occupy the full dynamic range of the CCD. Close the detector port and illuminate the acceptor until there is no discernible Cy3 fluorescence.

IMPORTANT: When performing acceptor photobleaching experiments, it is crucial to establish that there is no fluorescence from the photoproduct of the photobleached acceptor in the donor channel.

Analysis of Interacting Proteins with SPR Spectroscopy Using BIAcore
Stage 1: Preparation of the Capture Surface and Test Binding

The introduction of commercial instruments, such as BIAcore, that are capable of measuring surface plasmon resonance (SPR), has simplified the study of macromolecular interactions by providing a format that may be used to measure molecular interactions in real time with small analytical amounts of material. This protocol describes the sequence of steps for analyzing a typical antibody-antigen system in a BIAcore experiment. The antibody (antithyroid stimulating hormone [anti-TSH]) and its ligand (TSH) are available commercially and can be used to optimize the parameters of the system. A complete description of the setup and use of the instrument may be found at www.biacore.com/. Stage 1 describes how to (1) prepare a sensor chip containing immobilized rabbit antimouse C domain, (2) capture mouse anti-TSH on the prepared surface, and (3) test for the binding of TSH to its captured antibody.

MATERIALS

CAUTION: Please see Appendix 4 for appropriate handling of materials marked with <!>.

Reagents and Solutions

Please see Appendix 1 for components of stock solutions, buffers, and reagents. Dilute stock solutions to the appropriate concentrations.

EDC (0.2 M N-ethyl-N'-(dimethylaminopropyl)-carbodiimide in H_2O) <!>
Ethanolamine (1 M ethanolamine hydrochloride, adjusted to pH 8.5 with NaOH) <!>
HCl (20 mM) <!>
HEPES-buffered saline buffer
    10 mM HEPES (pH 7.4)
    150 mM NaCl
    3 mM EDTA (pH 8.0)
    0.005% (v/v) Tween-20
    This buffer is available from BIAcore.
NHS (0.05 M N-hydroxysuccinimide in H_2O)
Note: EDC, ethanolamine, and NHS are available as part of BIAcore's amine-coupling kit.

Antibodies

The Molecular Interactions Research Group (MIRG) within the Association of Biomolecular Resource Facilities (ABRF) (www.abrf.org) recommends that protein samples be prepared for BIAcore analysis by buffer exchange dialysis (three times in 500 volumes) against HEPES-buffered saline, centrifuged at maximum speed in a refrigerated microfuge, and finally passed through a 0.22-μm filter. The resulting protein concentration is determined by absorbance at 280 nm using the calculated extinction coefficient for the particular protein or peptide.

Mouse anti-TSH monoclonal antibody (1 mg/ml in HEPES-buffered saline) (Alexon-Trend)
Rabbit antimouse Fc domain (RAMc; 30 μg/ml in 10 mM sodium acetate [pH 5.0] [BIAcore])

Additional Items

BIAcore instrument with control and evaluation software (see www.biacore.com)
Sensor chip CM5 (BIAcore)
TSH (20 μM in HEPES-buffered saline) (Alexon-Trend)

Additional Information

Theoretical and experimental details of surface
    plasmon resonance

MC3, pp. 18.96–18.103

Adapted from Chapter 18, Protocol 6 Stage 1, p. 18.104 of MC3.

METHOD

1. Dock Sensor Chip CM5 in the BIAcore instrument.

   The procedure for primary amine coupling of RAMc to the sensor chip CM5 surface (Steps 1–14) takes approximately 45 minutes. The commands are accessed through the pulldown menus or by the icons of the toolbar of the BIAcore control software.

2. Prime using filtered and degassed HEPES-buffered saline.

3. Place tubes containing 100 μl each of NHS, EDC, ethanolamine, RAM Fc, and 20 mM HCl in appropriate positions in the autosampler rack in the BIAcore.

4. Place an empty tube in the autosampler rack in the BIAcore.

5. Start the instrument at a flow of 5 μl/minute over one flow cell.

6. Transfer 75 μl of NHS to the empty tube.

7. Transfer 75 μl of EDC to the same tube.

8. Mix the contents of the tube containing the NHS and EDC.

9. Inject 35 μl of the NHS/EDC mixture to activate the surface.

10. Inject 35 μl of the RAM Fc to couple the antibody to the activated surface.

11. Inject 35 μl of ethanolamine to deactivate excess reactive groups.

12. Quickinject 10 μl of 20 mM HCl followed by Extraclean to remove noncovalently bound material.

13. Determine the level of RAMc bound by placing a baseline report point before the start of the RAMc injection and a second report point 2 minutes after the end of the 20 mM HCl injection.

14. Stop the flow, close the command queue window, and save the report file.

15. Start the instrument at a flow of 10 μl/minute over the flow cell with the coupled RAMc.

    The procedure for testing the binding of the anti-TSH to the RAMc surface (Steps 15–20) takes approximately 20 minutes. The commands are accessed through the pulldown menus or by the icons of the toolbar of the BIAcore control software.

16. Inject 10 μl of 2 μg/ml anti-TSH (this requires a total of 40 μl of anti-TSH solution).

    10 μl of the anti-TSH should result in an increase of approximately 250 RUs. If this level of binding to the RAMc surface is not obtained, regenerate (see Step 17) and repeat the injection of anti-TSH using different volumes until the necessary injection volume is determined.

17. Quickinject 10 μl of 20 mM HCl followed by Extraclean to regenerate the RAMc surface.

18. To test the reproducibility of the binding to the RAMc surface, repeat the injection using the volume of anti-TSH required to give 250 RUs bound to anti-TSH.

19. Quickinject 10 μl of 20 mM HCl followed by Extraclean to regenerate the RAMc surface.

20. Stop the flow, close the command queue window, and save the report file.

21. Start the instrument at a flow of 10 μl/minute over the flow cell with the coupled RAMc.

    The procedure for testing the binding of TSH to the captured anti-TSH surface (Steps 21–25) takes approximately 20 minutes. The commands are accessed through the pulldown menus or by the icons of the toolbar of the BIAcore control software.

22. Inject 10 μl of 2 μg/ml anti-TSH (or the volume determined to give 250 RUs bound anti-TSH as determined in Steps 15–20).

23. Inject 25 μl of 200 nM TSH, specifying a 120-second dissociation time (this requires a total of 65 μl of TSH solution). The 120-second dissociation time allows the rate at which the antibody-antigen complex dissociates to be gauged.

24. Quickinject 10 μl of 20 mM HCl followed by Extraclean to regenerate the RAMc surface.

25. Stop the flow, close the command queue window, and save the report file.

Analysis of Interacting Proteins with SPR Spectroscopy Using BIAcore
Stage 2: Kinetic Analysis of the Antibody-Antigen Interaction

Stage 2 presents a series of method blocks or analysis programs that allow the determination of the equilibrium constants for the interaction between ligand pairs.

MATERIALS

Reagents and Solutions

Please see Appendix 1 for components of stock solutions, buffers, and reagents. Dilute stock solutions to the appropriate concentrations.
 HEPES-buffered saline buffer
  10 mM HEPES (pH 7.4)
  150 mM NaCl
  3 mM EDTA (pH 8.0)
  0.005% (v/v) Tween-20
  This buffer is available from BIAcore.

Antibodies

 Mouse anti-TSH monoclonal antibody (1 mg/ml in HEPES-buffered saline) (Alexon-Trend)

Additional Items

 BIAcore instrument with control and evaluation software (see www.biacore.com)
 Sensor chip CM5, coupled to RAMc, as described in Protocol 18.6 Stage 1
 TSH (20 μM in HEPES-buffered saline) (Alexon-Trend)
  Prepare serial dilutions of TSH for injection:
   i. Prepare serial dilutions of 200, 100, 50, 25, 10, and 5 nM TSH, each in 280 μl of the running buffer used in the experiment.
   ii. Dispense 70 μl of each concentration into four plastic vials and cap each vial.
   iii. Place the vials in the appropriate position in the BIAcore rack, as directed by the sample loop parameters.

Additional Items

Analysis of data	MC3, pp. 18.112–18.113
Theoretical and experimental details of surface plasmon resonance	MC3, pp. 18.96–18.103

METHOD

1. For an example of a sequence of commands used to determine the equilibrium constants of an interaction by the analysis program, please see the description that begins on p. 18.108 in MC3.

2. The sequence of commands consists of three method blocks or sections: the MAIN, the DEFINE APROG, and the DEFINE SAMPLE. There are four DEFINE APROG blocks, one for each of the four APROGs referred to in the MAIN (bind1, bind2, bind3, and bind4). When the BIAcore software edits a method, the method commands will be written in uppercase letters and the user-defined parameters will be in lowercase letters (additional details may be found at www.biacore.com).

Appendices

Preparation of Reagents and Buffers Used in Molecular Cloning

BACKGROUND INFORMATION

Background information found in *Molecular Cloning: A Laboratory Manual*, 3rd edition (hereafter MC3) unless otherwise indicated.

BUFFERS

Tris Buffers

Biological reactions work well only within a narrow concentration range of hydrogen ions. Paradoxically, however, many of these reactions themselves generate or consume protons. Buffers are substances that undergo reversible protonation within a particular pH range and therefore maintain the concentration of hydrogen ions within acceptable limits. Perfect buffers are, like the Holy Grail, always beyond reach. An ideal biological buffer should

- have a pK_a between pH 6.0 and 8.0
- be inert to a wide variety of chemicals and enzymes
- be highly polar, so that it is both exquisitely soluble in aqueous solutions and also unlikely to diffuse across biological membranes and thereby affect intracellular pH
- be nontoxic
- be cheap
- not be susceptible to salt or temperature effects
- not absorb visible or ultraviolet light

None of the buffers used in molecular biology fulfill all of these criteria. Very few weak acids are known that have dissociation constants between 10^{-7} and 10^{-9}. Among inorganic salts, only borates, bicarbonates, phosphates, and ammonium salts lie within this range. However, they are all incompatible in one way or another with physiological media.

In 1946, George Gomori suggested that organic polyamines could be used to control pH in the range 6.5–9.7. One of the three compounds he investigated was Tris(2-amino-2-hydroxymethyl-1,3-propanediol), which had been first described in 1897 by Piloty and Ruff. Tris turned out to be an extremely satisfactory buffer for many biochemical purposes and today is the standard buffer used for most enzymatic reactions in molecular cloning.

TABLE A1-1 Preparation of Tris Buffers of Various Desired pH Values

Desired pH (25°C)	Volume of 0.1 N HCl (ml)
7.10	45.7
7.20	44.7
7.30	43.4
7.40	42.0
7.50	40.3
7.60	38.5
7.70	36.6
7.80	34.5
7.90	32.0
8.00	29.2
8.10	26.2
8.20	22.9
8.30	19.9
8.40	17.2
8.50	14.7
8.60	12.4
8.70	10.3
8.80	8.5
8.90	7.0

Tris buffers (0.05 M) of the desired pH can be made by mixing 50 ml of 0.1 M Tris base with the indicated volume of 0.1 N HCl and then adjusting the volume of the mixture to 100 ml with water. For preparation of stock solutions of 1 M Tris, see p. A1.7.

TRIS BUFFERS

One of Tris' first commercial successes, which received wide attention, was the reduction of mortality during handling and hauling of fish. In the 1940s, live fish were carried to market in tanks of seawater. Unfortunately, many of the fish died because of the decline in pH resulting from an accumulation of CO_2. This problem was only partially alleviated by including anesthetics in the water that minimized the fishes' metabolic activities. What these anesthetics did to the people who ate the fish is not recorded. Tris certainly reduced the mortality rate of the fish by stabilizing the pH of the seawater and may also have kept the fish eaters more alert. Tris also turned out to be an extremely satisfactory buffer for many biochemical purposes and today is the standard buffer used for most enzymatic reactions in molecular cloning.

Tris [Tris(hydroxymethyl)aminomethane] has a very high buffering capacity, is highly soluble in water, and is inert in a wide variety of enzymatic reactions. However, Tris also has a number of deficiencies:

- *The pK_a of Tris is pH 8.0 (at 20°C),* which means that its buffering capacity is very low at pHs below 7.5 and above 9.0.

- *Temperature has a significant effect on the dissociation of Tris.* The pH of Tris solutions decreases approximately 0.03 pH units for each 1°C increase in temperature. For example, a 0.05 M solution has pH values of 9.5, 8.9, and 8.6 at 5°C, 25°C, and 37°C, respectively. By convention, the pH of Tris solutions given in the scientific literature refers to the pH measured at 25°C. When preparing stock solutions of Tris, it is best to bring the pH into the desired range and then allow the solution to cool to 25°C before making final adjustments to the pH.

- *Tris reacts with many types of pH electrodes* that contain linen-fiber junctions, apparently because Tris reacts with the linen fiber. This effect is manifested in large liquid-junction potentials, electromotive force (emf) drift, and long equilibration times. Electrodes with linen-fiber junctions, therefore, cannot accurately measure the pH of Tris solutions. Use only those electrodes with ceramic or glass junctions that are warranted by the manufacturer to be suitable for Tris.

- *Concentration has a significant effect on the dissociation of Tris.* For example, the pHs of solutions containing 10 and 100 mM Tris will differ by 0.1 of a pH unit, with the more concentrated solution having the higher pH.

- *Tris is toxic to many types of mammalian cells.*

- *Tris, a primary amine, cannot be used with fixatives* such as glutaraldehyde and formaldehyde. Tris also reacts with glyoxal. Phosphate or MOPS buffer is generally used in place of Tris with these reagents.

Good Buffers

Tris is a poor buffer at pH values below 7.5. In the mid-1960s, Norman Good and his colleagues responded to the need for better buffers in this range by developing a series of *N*-substituted aminosulfonic acids that behave as strong zwitterions at biologically relevant pH values. Without these buffers, several techniques central to molecular cloning either would not exist at all or would work at greatly reduced efficiency. These techniques include high-efficiency transfection of mammalian cells (HEPES, Tricine, and BES), gel electrophoresis of RNA (MOPS), and high-efficiency transformation of bacteria (MES).

TABLE A1-2 Properties of Good Buffers

Acronym	Chemical Name	FW	pK_a	Useful Range (in pH units)
MES	2-(*N*-morpholino)ethanesulfonic acid	195.2	6.1	5.5–6.7
Bis-Tris	*bis*(2-hydroxyethyl)imino*tris*(hydroxymethyl)methane	209.2	6.5	5.8–7.2
ADA	*N*-(2-acetamido)-2-iminodiacetic acid	190.2	6.6	6.0–7.2
ACES	2-[(2-amino-2-oxoethyl)amino]ethanesulfonic acid	182.2	6.8	6.1–7.5
PIPES	piperazine-*N,N'-bis*(2-ethanesulfonic acid)	302.4	6.8	6.1–7.5
MOPSO	3-(*N*-morpholino)-2-hydroxypropanesulfonic acid	225.3	6.9	6.2–7.6
Bis-Tris Propane	1,3-*bis*[*tris*(hydroxymethyl)methylamino]propane	282.3	6.8[a]	6.3–9.5
BES	*N,N-bis*(2-hydroxyethyl)-2-aminoethanesulfonic acid	213.2	7.1	6.4–7.8
MOPS	3-(*N*-morpholino)propanesulfonic acid	209.3	7.2	6.5–7.9
HEPES	*N*-(2-hydroxyethyl)piperazine-*N'*-(2-ethanesulfonic acid)	238.3	7.5	6.8–8.2
TES	*N-tris*(hydroxymethyl)methyl-2-aminoethanesulfonic acid	229.2	7.4	6.8–8.2
DIPSO	3-[*N,N-bis*(2-hydroxyethyl)amino]-2-hydroxypropanesulfonic acid	243.3	7.6	7.0–8.2
TAPSO	3-[*N-tris*(hydroxymethyl)methylamino]-2-hydroxypropanesulfonic acid	259.3	7.6	7.0–8.2
TRIZMA	*tris*(hydroxymethyl)aminomethane	121.1	8.1	7.0–9.1
HEPPSO	*N*-(2-hydroxyethyl)piperazine-*N'*-(2-hydroxypropanesulfonic acid)	268.3	7.8	7.1–8.5
POPSO	piperazine-*N,N'-bis*(2-hydroxypropanesulfonic acid)	362.4	7.8	7.2–8.5
EPPS	*N*-(2-hydroxyethyl)piperazine-*N'*-(3-propanesulfonic acid)	252.3	8.0	7.3–8.7
TEA	triethanolamine	149.2	7.8	7.3–8.3
Tricine	*N-tris*(hydroxymethyl)methylglycine	179.2	8.1	7.4–8.8
Bicine	*N,N-bis*(2-hydroxyethyl)glycine	163.2	8.3	7.6–9.0
TAPS	*N-tris*(hydroxymethyl)methyl-3-aminopropanesulfonic acid	243.3	8.4	7.7–9.1
AMPSO	3-[(1,1-dimethyl-2-hydroxyethyl)amino]-2-hydroxypropane-sulfonic acid	227.3	9.0	8.3–9.7
CHES	2-(*N*-cyclohexylamino)ethanesulfonic acid	207.3	9.3	8.6–10.0
CAPSO	3-(cyclohexylamino)-2-hydroxy-1-propanesulfonic acid	237.3	9.6	8.9–10.3
AMP	2-amino-2-methyl-1-propanol	89.1	9.7	9.0–10.5
CAPS	3-(cyclohexylamino)-1-propanesulfonic acid	221.3	10.4	9.7–11.1

Data are compiled from various sources, including *Biochemical and Reagents for Life Science Research* 1994 (Sigma-Aldrich) and references therein.

[a]pK_a = 9.0 for the second dissociation stage.

Phosphate Buffers (Gomori Buffers)

The most commonly used phosphate buffers are named after their inventor: George Gomori, a histochemist who worked at the University of Chicago in the 1940s and 1950s. They consist of a mixture of monobasic dihydrogen phosphate and dibasic monohydrogen phosphate. By varying the amount of each salt, a range of buffers can be prepared that buffer well between pH 5.8 and 8.0 (please see Tables A1-3A and A1-3B). Phosphates have a very high buffering capacity and are highly soluble in water. However, they have a number of potential disadvantages:

- Phosphates inhibit many enzymatic reactions and procedures that are the foundation of molecular cloning, including cleavage of DNA by many restriction enzymes, ligation of DNA, and bacterial transformation.

- Because phosphates precipitate in ethanol, it is not possible to precipitate DNA and RNA from buffers that contain significant quantities of phosphate ions.

- Phosphates sequester divalent cations such as Ca^{++} and Mg^{++}.

TABLE A1-3A Preparation of 0.1 M Potassium Phosphate Buffer at 25°C

pH	Volume of 1 M K_2HPO_4 (ml)	Volume of 1 M KH_2PO_4 (ml)
5.8	8.5	91.5
6.0	13.2	86.8
6.2	19.2	80.8
6.4	27.8	72.2
6.6	38.1	61.9
6.8	49.7	50.3
7.0	61.5	38.5
7.2	71.7	28.3
7.4	80.2	19.8
7.6	86.6	13.4
7.8	90.8	9.2
8.0	94.0	6.0

Data from Green A.A. 1933. *J. Am. Chem. Soc.* **55:** 2331–2336.

Dilute the combined 1 M stock solutions to 1 liter with distilled H_2O. pH is calculated according to the Henderson-Hasselbalch equation:

$$pH = pK' + \log \left\{ \frac{(proton\ acceptor)}{proton\ donor} \right\}$$

where $pK' = 6.86$ at 25°C.

TABLE A1-3B Preparation of 0.1 M Sodium Phosphate Buffer at 25°C

pH	Volume of 1 M Na_2HPO_4 (ml)	Volume of 1 M NaH_2PO_4 (ml)
5.8	7.9	92.1
6.0	12.0	88.0
6.2	17.8	82.2
6.4	25.5	74.5
6.6	35.2	64.8
6.8	46.3	53.7
7.0	57.7	42.3
7.2	68.4	31.6
7.4	77.4	22.6
7.6	84.5	15.5
7.8	89.6	10.4
8.0	93.2	6.8

Data from ISCO. 1982. *ISCOTABLES: A handbook of data for biological and physical scientists*, 8th edition. ISCO, Inc., Lincoln, Nebraska.

Dilute the combined 1 M stock solutions to 1 liter with distilled H_2O. pH is calculated according to the Henderson-Hasselbalch equation:

$$pH = pK' + \log \left\{ \frac{(proton\ acceptor)}{proton\ donor} \right\}$$

where $pK' = 6.86$ at 25°C.

ACIDS AND BASES

TABLE A1-4 Concentrations of Acids and Bases: Common Commercial Strengths

Substance	Formula	M.W.	Moles/Liter[a]	Grams/Liter	% by Weight	Specific Gravity	ml/Liter to Prepare 1 M Solution
Acetic acid, glacial	CH_3COOH	60.05	17.4	1045	99.5	1.05	57.5
Acetic acid		60.05	6.27	376	36	1.045	159.5
Formic acid	HCOOH	46.02	23.4	1080	90	1.20	42.7
Hydrochloric acid	HCl	36.5	11.6	424	36	1.18	86.2
			2.9	105	10	1.05	344.8
Nitric acid	HNO_3	63.02	15.99	1008	71	1.42	62.5
			14.9	938	67	1.40	67.1
			13.3	837	61	1.37	75.2
Perchloric acid	$HClO_4$	100.5	11.65	1172	70	1.67	85.8
			9.2	923	60	1.54	108.7
Phosphoric acid	H_3PO_4	80.0	18.1	1445	85	1.70	55.2
Sulfuric acid	H_2SO_4	98.1	18.0	1766	96	1.84	55.6
Ammonium hydroxide	NH_4OH	35.0	14.8	251	28	0.898	67.6
Potassium hydroxide	KOH	56.1	13.5	757	50	1.52	74.1
			1.94	109	10	1.09	515.5
Sodium hydroxide	NaOH	40.0	19.1	763	50	1.53	52.4
			2.75	111	10	1.11	363.6

[a]With some acids and bases, stock solutions of different molarity/normality are in common use. These are often abbreviated "conc" for concentrated stocks and "dil" for dilute stocks.

TABLE A1-5 Approximate pH Values for Various Concentrations of Stock Solutions

Substance	1 N	0.1 N	0.01 N	0.001 N
Acetic acid	2.4	2.9	3.4	3.9
Hydrochloric acid	0.10	1.07	2.02	3.01
Sulfuric acid	0.3	1.2	2.1	
Citric acid		2.1	2.6	
Ammonium hydroxide	11.8	11.3	10.8	10.3
Sodium hydroxide	14.05	13.07	12.12	11.13
Sodium bicarbonate		8.4		
Sodium carbonate		11.5	11.0	

PREPARATION OF BUFFERS AND STOCK SOLUTIONS FOR USE IN MOLECULAR BIOLOGY

CAUTION: Please see Appendix 4 for appropriate handling of materials marked with <!>.

pH Buffers

Phosphate-buffered Saline (PBS)
137 mM NaCl
2.7 mM KCl
10 mM Na_2HPO_4
2 mM KH_2PO_4

Dissolve 8 g of NaCl, 0.2 g of KCl, 1.44 g of Na_2HPO_4, and 0.24 g of KH_2PO_4 in 800 ml of distilled H_2O. Adjust the pH to 7.4 with HCl. Add H_2O to 1 liter. Dispense the solution into aliquots and sterilize them by autoclaving for 20 minutes at 15 psi (1.05 kg/cm²) on liquid cycle or by filter sterilization. Store the buffer at room temperature.

> This recipe for PBS lacks divalent cations. If necessary, PBS may be supplemented with 1 mM $CaCl_2$ and 0.5 mM $MgCl_2$.

10x Tris EDTA (TE)
pH 7.4
100 mM Tris-Cl (pH 7.4)
10 mM EDTA (pH 8.0)

pH 7.6
100 mM Tris-Cl (pH 7.6)
10 mM EDTA (pH 8.0)

pH 8.0
100 mM Tris-Cl (pH 8.0)
10 mM EDTA (pH 8.0)

Sterilize solutions by autoclaving for 20 minutes at 15 psi (1.05 kg/cm²) on liquid cycle. Store the buffer at room temperature.

Tris-Cl (1 M)
Dissolve 121.1 g of Tris base in 800 ml of H_2O. Adjust the pH to the desired value by adding concentrated HCl <!>.

pH	HCl
7.4	70 ml
7.6	60 ml
8.0	42 ml

Allow the solution to cool to room temperature before making final adjustments to the pH. Adjust the volume of the solution to 1 liter with H_2O. Dispense into aliquots and sterilize by autoclaving.

If the 1 M solution has a yellow color, discard it and obtain Tris of better quality. The pH of Tris solutions is temperature dependent and decreases approximately 0.03 pH units for each 1°C increase in temperature. For example, a 0.05 M solution has pH values of 9.5, 8.9, and 8.6 at 5°C, 25°C, and 37°C, respectively.

Tris Magnesium Buffer (TM)
50 mM Tris-Cl (pH 7.8)
10 mM $MgSO_4$

Tris-buffered Saline (TBS)
Dissolve 8 g of NaCl, 0.2 g of KCl, and 3 g of Tris base in 800 ml of distilled H_2O. Add 0.015 g of phenol red and adjust the pH to 7.4 with HCl. Add distilled H_2O to 1 liter. Dispense the solution into aliquots and sterilize them by autoclaving for 20 minutes at 15 psi (1.05 kg/cm²) on liquid cycle. Store the buffer at room temperature.

Enzyme Stocks and Buffers

Enzyme Stocks

Lysozyme (10 mg/ml)

Dissolve solid lysozyme at a concentration of 10 mg/ml in 10 mM Tris-Cl (pH 8.0) immediately before use. Check that the pH of the Tris solution is 8.0 before dissolving the protein. Lysozyme will not work efficiently if the pH of the solution is less than 8.0.

Lyticase (67 mg/ml)

Dissolve at 67 mg/ml (900 units/ml) in 0.01 M sodium phosphate containing 50% glycerol just before use.

Pancreatic DNase I (1 mg/ml)

Dissolve 2 mg of crude pancreatic DNase I in 1 ml of

10 mM Tris-Cl (pH 7.5)
150 mM NaCl
1 mM MgCl$_2$

When the DNase I is dissolved, add 1 ml of glycerol to the solution and mix by gently inverting the closed tube several times. Take care to avoid creating bubbles and foam. Store the solution in aliquots of –20°C.

Pancreatic RNase (1 mg/ml)

Dissolve 2 mg of crude pancreatic RNase I in 2 ml of TE (pH 7.6).

Proteinase K (20 mg/ml)

Purchase as a lyophilized powder and dissolve at a concentration of 20 mg/ml in sterile 50 mM Tris (pH 8.0), 1.5 mM calcium acetate. Divide the stock solution into small aliquots and store at –20°C. Each aliquot can be thawed and refrozen several times but should then be discarded. Unlike much cruder preparations of proteolytic enzymes (e.g., pronase), proteinase K need not be self-digested before use. (For futher information please see MC3, p. A4.50.)

Trypsin

Prepare bovine trypsin (Sequencer grade, Roche Applied Science) at a concentration of 250 μg/ml in 200 mM ammonium bicarbonate (pH 8.9). Store the solution in aliquots at –20°C.

Zymolyase 5000 (2 mg/ml)

Purchase from Kirin Breweries. Dissolve at 2 mg/ml in 0.01 M sodium phosphate containing 50% glycerol just before use.

Enzyme Dilution Buffers

DNase I Dilution Buffer

10 mM Tris-Cl (pH 7.5)
150 mM NaCl
1 mM MgCl$_2$

Polymerase Dilution Buffer

50 mM Tris-Cl (pH 8.1)
1 mM dithiothreitol
0.1 mM EDTA (pH 8.0)
0.5 mg/ml bovine serum albumin
5% (v/v) glycerol

Prepare solution fresh for each use.

Sequenase Dilution Buffer

10 mM Tris-Cl (pH 7.5)
5 mM dithiothreitol
0.5 mg/ml bovine serum albumin

Store the solution at –20°C.

Taq **Dilution Buffer**
  25 mM Tris (pH 8.8)
  0.01 mM EDTA (pH 8.0)
  0.15% (v/v) Tween-20
  0.15% (v/v) Nonidet P-40

Enzyme Reaction Buffers

IMPORTANT: Wherever possible, use the 10x reaction buffer supplied by the manufacturer of the enzyme. The recipes given here are good standbys.

10x Amplification Buffer
  500 mM KCl
  100 mM Tris-Cl (pH 8.3 at room temperature)
  15 mM $MgCl_2$

Autoclave the 10x buffer for 10 minutes at 15 psi (1.05 kg/cm²) on liquid cycle. Divide the sterile buffer into aliquots and store them at –20°C.

10x Bacteriophage T4 DNA Ligase Buffer
  200 mM Tris-Cl (pH 7.6)
  50 mM $MgCl_2$
  50 mM dithiothreitol
  0.5 mg/ml bovine serum albumin (Fraction V; Sigma-Aldrich) (*optional*)

Divide the buffer in small aliquots and store at –20°C. Add ATP when setting up the reaction to an appropriate concentration (e.g., 1 mM).

10x Bacteriophage T4 DNA Polymerase Buffer
  330 mM Tris-acetate (pH 8.0)
  660 mM potassium acetate
  100 mM magnesium acetate
  5 mM dithiothreitol
  1 mg/ml bovine serum albumin (Fraction V; Sigma-Aldrich)

Divide the 10x stock into small aliquots and store frozen at –20°C.

10x Bacteriophage T4 Polynucleotide Kinase Buffer
  700 mM Tris-Cl (pH 7.6)
  100 mM $MgCl_2$
  50 mM dithiothreitol

Divide the 10x stock into small aliquots and store frozen at –20°C.

5x BAL 31 Buffer
  3 M NaCl
  60 mM $CaCl_2$
  60 mM $MgCl_2$
  100 mM Tris-Cl (pH 8.0)
  1 mM EDTA (pH 8.0)

10x Dephosphorylation Buffer (for use with calf intestinal alkaline phosphase [CIP])
  100 mM Tris-Cl (pH 8.3)
  10 mM $MgCl_2$
  10 mM $ZnCl_2$

10x Dephosphorylation Buffer (for use with shrimp alkaline phosphatase [SAP])
  200 mM Tris-Cl (pH 8.8)
  100 mM $MgCl_2$
  10 mM $ZnCl_2$

1x *Eco*RI Methylase Buffer

50 mM NaCl
50 mM Tris-Cl (pH 8.0)
10 mM EDTA
80 μM *S*-adenosylmethionine

Store the buffer in small aliquots at −20°C.

10x Exonuclease III Buffer

660 mM Tris-Cl (pH 8.0)
66 mM MgCl$_2$
100 mM β-mercaptoethanol <!>

Add β-mercaptoethanol just before use.

10x Klenow Buffer

0.4 M potassium phosphate (pH 7.5)
66 mM MgCl$_2$
10 mM β-mercaptoethanol <!>

10x Linker Kinase Buffer

600 mM Tris-Cl (pH 7.6)
100 mM MgCl$_2$
100 mM dithiothreitol
2 mg/ml bovine serum albumin

Prepare fresh just before use.

Nuclease S1 Digestion Buffer

0.28 M NaCl
0.05 M sodium acetate (pH 4.5)
4.5 mM ZnSO$_4$·7H$_2$O

Store aliquots of nuclease S1 buffer at −20°C and add nuclease S1 to a concentration of 500 units/ml just before use.

10x Proteinase K Buffer

100 mM Tris-Cl (pH 8.0)
50 mM EDTA (pH 8.0)
500 mM NaCl

10x Reverse Transcriptase Buffer

500 mM Tris-Cl (pH 8.3)
750 mm KCl
30 mM MgCl$_2$

RNase H Buffer

20 mM Tris-Cl (pH 7.6)
20 mM KCl
0.1 mM EDTA (pH 8.0)
0.1 mM dithiothreitol

Prepare fresh just before use.

5x Terminal Transferase Buffer

Most manufacturers supply a 5x reaction buffer, which typically contains

500 mM potassium cacodylate (pH 7.2) <!>
10 mM CoCl$_2$·6H$_2$O
1 mM dithiothreitol

5x terminal transferase (or tailing) buffer may be prepared according to the following method:

1. Equilibrate 5 g of Chelex 100 (Bio-Rad) with 10 ml of 3 M potassium acetate at room temperature.
2. After five minutes, remove excess liquid by vacuum suction. Wash the Chelex three times with 10 ml of deionized H_2O.
3. Prepare a 1 M solution of potassium cacodylate. Equilibrate the cacodylate solution with the treated Chelex resin.
4. Recover the cacodylate solution by passing it through a Buchner funnel fitted with Whatman No. 1 filter paper.
5. To the recovered cacodylate, add (in order) H_2O, dithiothreitol, and cobalt chloride to make the final concentrations of 500 mM potassium cacodylate, 1 mM dithiothreitol, and 20 mM $CoCl_2$.

Store the buffer in aliquots at –20°C.

10x Universal KGB (Restriction Endonuclease) Buffer
1 M potassium acetate
250 mM Tris-acetate (pH 7.6)
100 mM magnesium acetate tetrahydrate
5 mM β-mercaptoethanol <!>
0.1 mg/ml bovine serum albumin

Store the 10x buffer in aliquots at –20°C.

Hybridization Buffers

Alkaline Transfer Buffer (for alkaline transfer of DNA to nylon membranes)
0.4 N NaOH <!>
1 M NaCl

Church Buffer
1% (w/v) bovine serum albumin
1 mM EDTA
0.5 M phosphate buffer*
7% (w/v) SDS

*0.5 M phosphate buffer is 134 g of $Na_2HPO_4 \cdot 7H_2O$, 4 ml of concentrated (85%) H_3PO_4 <!>, H_2O to 1 liter.

Denaturation Solution (use to denature double-stranded DNA before transfer from gels to membranes)
1.5 M NaCl
0.5 M NaOH <!>

HCl (2.5 N)
Add 25 ml of concentrated HCl <!> (11.6 N) to 91 ml of sterile H_2O. Store the diluted solution at room temperature.

Hybridization Buffer with formamide (for RNA)
40 mM PIPES (pH 6.8)
1 mM EDTA (pH 8.0)
0.4 M NaCl
80% (v/v) deionized formamide <!>

Use the disodium salt of PIPES to prepare the buffer and adjust the pH to 6.4 with 1 N HCl.

Hybridization Buffer without formamide (for RNA)
40 mM PIPES (pH 6.4)
0.1 mM EDTA (pH 8.0)
0.4 M NaCl

Use the disodium salt of PIPES to prepare the buffer and adjust the pH to 6.4 with 1 N HCl.

Neutralization Buffer I (for neutral transfer of denatured DNA to uncharged membranes)
1 M Tris-Cl (pH 7.4)
1.5 M NaCl

Neutralization Buffer II (for alkaline transfer of DNA to nylon membranes)
0.5 M Tris-Cl (pH 7.2)
1 M NaCl

Neutralizing Solution (for transfer of DNA to nylon membranes)
0.5 M Tris-Cl (pH 7.4)
1.5 M NaCl

Prehybridization Solution (for dot, slot, and northern hybridization)
0.5 M sodium phosphate (pH 7.2)*
7% (w/v) SDS
1 mM EDTA (pH 7.0)

*0.5 M phosphate buffer is 134 g of $Na_2HPO_4 \cdot 7H_2O$, 4 ml of concentrated (85%) H_3PO_4 <!> H_2O to 1 liter.

Prehybridization and Hybridization Solutions

Prehybridization/Hybridization Solution (for plaque/colony lifts)
50% (v/v) formamide (*optional*) <!>
6x SSC (or 6x SSPE)
0.05x BLOTTO (see p. A1.13)

As an alternative to the above solution, use Church Buffer (please see recipe in MC3, p. A1.12). For advice on which hybridization solution to use, please see the panel Prehybridization and Hybridization Solutions in MC3, p. 1.141; for advice on the use of formamide, please see the information panel Formamide and Its Uses in Molecular Cloning in MC3, p. 6.59.

Prehybridization/Hybridization Solution (for hybridization in aqueous buffer)
6x SSC (or 6x SSPE)
5x Denhardt's reagent (see p. A1.13)
0.5% (w/v) SDS
1 µg/ml poly(A)
100 µg/ml salmon sperm DNA

Prehybridization/Hybridization Solution (for hybridization in formamide buffers)
6x SSC (or 6x SSPE)
5x Denhardt's reagent (see p. A1.13)
0.5% (w/v) SDS
1 µg/ml poly(A)
100 µg/ml salmon sperm DNA
50% (v/v) formamide <!>

After thorough mixing, filter the solution through a 0.45-µm disposable cellulose acetate membrane (Schleicher & Schuell Uniflow syringe membrane or its equivalent). To decrease background when hybridizing under conditions of reduced stringency (e.g., 20–30% formamide), it is important to use formamide that is as pure as possible.

Prehybridization/Hybridization Solution (for hybridization in phosphate-SDS buffer)
0.5 M phosphate buffer (pH 7.2)*
1 mM EDTA (pH 8.0)
7% (w/v) SDS
1% (w/v) bovine serum albumin

Use an electrophoresis grade of bovine serum albumin. No blocking agents or hybridization rate enhancers are required with this particular prehybridization/hybridization solution.

*0.5 M phosphate buffer is 134 g of $Na_2HPO_4 \cdot 7H_2O$, 4 ml of concentrated (85%) H_3PO_4 <!> H_2O to 1 liter.

20x SSC

Dissolve 175.3 g of NaCl and 88.2 g of sodium citrate in 800 ml of H_2O. Adjust the pH to 7.0 with a few drops of a 14 N solution of HCl <!>. Adjust the volume to 1 liter with H_2O. Dispense into aliquots. Sterilize by autoclaving. The final concentrations of the ingredients in 20x SSC are 3.0 M NaCl and 0.3 M sodium citrate.

20x SSPE

Dissolve 175.3 g of NaCl, 27.6 g of $NaH_2PO_4 \cdot H_2O$, and 7.4 g of EDTA in 800 ml of H_2O. Adjust the pH to 7.4 with NaOH <!> (~6.5 ml of a 10 N solution). Adjust the volume to 1 liter with H_2O. Dispense into aliquots. Sterilize by autoclaving. The final concentrations of the ingredients in 20x SSPE are 3.0 M NaCl, 0.2 M NaH_2PO_4, and 0.02 M EDTA.

Blocking Agents

Blocking agents prevent ligands from sticking to surfaces. They are used in molecular cloning to stop non-specific binding of probes in Southern, northern, and western blotting. If left to their own devices, these probes would bind tightly and nonspecifically to the supporting nitrocellulose or nylon membrane. Without blocking agents, it would be impossible to detect anything but the strongest target macromolecules.

No one knows for sure what causes nonspecific binding of probes. Hydrophobic patches, lignin impurities, excessively high concentrations of probe, overbaking or underbaking of nitrocellulose filters, and homopolymeric sequences in nucleic acid probes have all been blamed from time to time, together with a host of less likely culprits. Whatever the cause, the solution is generally simple: Treat the filters with a blocking solution containing a cocktail of substances that will compete with the probe for nonspecific binding sites on the solid support. Blocking agents work by brute force. They are used in high concentrations and generally consist of a cocktail of high-molecular-weight polymers (heparin, polyvinylpyrrolidine, nucleic acids), proteins (bovine serum albumin, nonfat dried milk), and detergents (SDS or Nonidet P-40). The following recommendations apply only to nylon and nitrocellulose filters. Charged nylon filters should be treated as described by the individual manufacturer.

Blocking Agents Used for Nucleic Acid Hybridization. Two blocking agents in common use in nucleic acid hybridization are Denhardt's reagent and BLOTTO (bovine lacto transfer technique optimizer). Usually, the filters carrying the immobilized target molecules are incubated with the blocking agents for 1 or 2 hours before the probe is added. In most cases, background hybridization is completely suppressed when filters are incubated with a blocking agent consisting of 6x SSC or SSPE containing 5x Denhardt's reagent, 1.0% SDS, and 100 mg/ml denatured, sheared salmon sperm DNA. This mixture should be used whenever the ratio of signal to noise is expected to be low, for example, when performing northern analysis of low-abundance RNAs or Southern analysis of single-copy sequences of mammalian DNA. However, in most other circumstances (Grunstein-Hogness hybridization, Benton-Davis hybridization, Southern hybridization of abundant DNA sequences, etc.), a less expensive alternative is 6x SSC or SSPE containing 0.25–0.5% nonfat dried milk (BLOTTO).

Blocking agents are usually included in both prehybridization and hybridization solutions when nitrocellulose filters are used. However, when the target nucleic acid is immobilized on nylon membranes, the blocking agents are often omitted from the hybridization solution. This is because high concentrations of protein are believed to interfere with the annealing of the probe to its target. Quenching of the hybridiza-

---

DENHARDT'S REAGENT

Denhardt's reagent is used for

- northern hybridization
- single-copy Southern hybridization
- hybridizations involving DNA immobilized on nylon membranes

Denhardt's reagent is usually made up as a 50x stock solution, which is filtered and stored at –20°C. The stock solution is diluted tenfold into prehybridization buffer (usually 6x SSC or 6x SSPE containing 1.0% SDS and 100 µg/ml denatured salmon sperm DNA). 50x Denhardt's reagent contains in H_2O:

1% (w/v) Ficoll 400
1% (w/v) polyvinylpyrrolidone
1% (w/v) bovine serum albumin (Fraction V; Sigma-Aldrich)

BLOTTO <!>

BLOTTO is used for

- Grunstein-Hogness hybridization
- Benton-Davis hybridization
- all Southern hybridizations other than single-copy dot blots and slot blots

1x BLOTTO is 5% (w/v) nonfat dried milk dissolved in H_2O containing 0.02% sodium azide <!>. 1x BLOTTO is stored at 4°C and is diluted 10- to 25-fold into prehybridization buffer before use. BLOTTO should not be used in combination with high concentrations of SDS, which will cause the milk proteins to precipitate. If background hybridization is a problem, Nonidet P-40 may be added to a final concentration of 1% (v/v).

BLOTTO may contain high levels of RNase and should be treated with diethylpyrocarbonate or heated overnight to 72°C when used in northern hybridizations and when RNA is used as a probe. BLOTTO is not as effective as Denhardt's solution when the target DNA is immobilized on nylon filters.

---

tion signal by blocking agents is particularly noticeable when oligonucleotides are used as probes. This problem can often be solved by performing the hybridization step in a solution containing high concentrations of SDS (6–7%), sodium phosphate (0.4 M), bovine serum albumin (1%), and EDTA (0.02 M).

Heparin is sometimes used instead of Denhardt's solution or BLOTTO when hybridization is performed in the presence of the accelerator, dextran sulfate. It is used at a concentration of 500 mg/ml in hybridization solutions containing dextran sulfate. In hybridization solutions that do not contain dextran sulfate, heparin is used at a concentration of 50 mg/ml. Heparin (Sigma-Aldrich porcine grade II or its equivalent) is dissolved at a concentration of 50 mg/ml in 4x SSPE or SSC and stored at 4°C.

Blocking Agents Used for Western Blotting. The best and least expensive blocking reagent is nonfat dried milk. It is easy to use and is compatible with all of the common immunological detection systems. The only time nonfat dried milk should not be used is when western blots are probed for proteins that may be present in milk.

One of the following recipes may be used to prepare blocking buffer. A blocking solution for western blots is phosphate-buffered saline containing 5% (w/v) nonfat dried milk, 0.01% Antifoam, and 0.02% sodium azide.

Blocking Buffer (TNT buffer containing a blocking agent)
10 mM Tris-Cl (pH 8.0)
150 mM NaCl
0.05% (v/v) Tween-20
blocking agent (1% [w/v] gelatin, 3% [w/v] bovine serum albumin, or
    5% [w/v] nonfat dried milk)

Opinion about which of these blocking agents is best varies from laboratory to laboratory. We recommend performing preliminary experiments to determine which agent works best. Blocking buffer can be stored at 4°C and reused several times. Sodium azide <!> should be added to a final concentration of 0.05% (w/v) to inhibit the growth of microorganisms.

Extraction/Lysis Buffers and Solutions

Alkaline Lysis Solution I (plasmid preparation)
50 mM glucose
25 mM Tris-Cl (pH 8.0)
10 mM EDTA (pH 8.0)

Prepare solution I from standard stocks in batches of approximately 100 ml, autoclave for 15 minutes at 15 psi (1.05 kg/cm²) on liquid cycle, and store at 4°C.

Alkaline Lysis Solution II (plasmid preparation)
0.2 N NaOH (freshly diluted from a 10 N stock) <!>
1% (w/v) SDS

Prepare solution II fresh and use at room temperature.

Alkaline Lysis Solution III (plasmid preparation)

5 M potassium acetate	60.0 ml
glacial acetic acid <!>	11.5 ml
H_2O	28.5 ml

The resulting solution is 3 M with respect to potassium and 5 M with respect to acetate. Store the solution at 4°C and transfer it to an ice bucket just before use.

STET

10 mM Tris-Cl (pH 8.0)
0.1 M NaCl
1 mM EDTA (pH 8.0)
5% (v/v) Triton X-100

Make sure that the pH of STET is 8.0 after all ingredients are added. There is no need to sterilize STET before use.

Electrophoresis and Gel-loading Buffers

Commonly Used Electrophoresis Buffers

Buffer	Working Solution	Stock Solution/Liter
TAE	1x 40 mM Tris-acetate 1 mM EDTA	50x 242 g of Tris base 57.1 ml of glacial acetic acid <!> 100 ml of 0.5 M EDTA (pH 8.0)
TBE[a]	0.5x 45 mM Tris-borate 1 mM EDTA	5x 54 g of Tris base 27.5 g of boric acid 20 ml of 0.5 M EDTA (pH 8.0)
TPE	1x 90 mM Tris-phosphate 2 mM EDTA	10x 108 g of Tris base 15.5 ml of phosphoric acid <!> (85%, 1.679 g/ml) 40 ml of 0.5 M EDTA (pH 8.0)
Tris-glycine[b]	1x 25 mM Tris-Cl 250 mM glycine 0.1% SDS	5x 15.1 g of Tris base 94 g of glycine (electrophoresis grade) 50 ml of 10% SDS (electrophoresis grade)

[a]TBE is usually made and stored as a 5x or 10x stock solution. The pH of the concentrated stock buffer should be approximately 8.3. Dilute the concentrated stock buffer just before use and make the gel solution and the electrophoresis buffer from the same concentrated stock solution. Some investigators prefer to use more concentrated stock solutions of TBE (10x as opposed to 5x). However, 5x stock solution is more stable because the solutes do not precipitate during storage. Passing the 5x or 10x buffer stocks through a 0.22-μm filter can prevent or delay formation of precipitates.

[b]Use Tris-glycine buffers for SDS-polyacrylamide gels (see Appendix 3).

Specialized Electrophoresis Buffers

10x Alkaline Agarose Gel Electrophoresis Buffer

500 mM NaOH <!>
10 mM EDTA

Add 50 ml of 10 N NaOH and 20 ml of 0.5 M EDTA (pH 8.0) to 800 ml of H_2O and then adjust the final volume to 1 liter. Dilute the 10x alkaline agarose gel electrophoresis buffer with H_2O to generate a 1x working solution immediately before use. Use the same stock of 10x alkaline agarose gel electrophoresis buffer to prepare the alkaline agarose gel and the 1x working solution of alkaline electrophoresis buffer.

10x BPTE Electrophoresis Buffer
100 mM PIPES
300 mM Bis-Tris
10 mM EDTA

The final pH of the 10x buffer is approximately 6.5. The 10x buffer can be made by adding 3 g of PIPES (free acid), 6 g of Bis-Tris (free base), and 2 ml of 0.5 M EDTA to 90 ml of distilled H_2O and then treating the solution with diethylpyrocarbonate <!> (final concentration 0.1%; for more details, please see the information panel Diethylpyrocarbonate in MC3, p. 7.84).

10x MOPS Electrophoresis Buffer
0.2 M MOPS (pH 7.0) <!>
20 mM sodium acetate
10 mM EDTA (pH 8.0)

Dissolve 41.8 g of MOPS in 700 ml of sterile DEPC-treated <!> H_2O. Adjust the pH to 7.0 with 2 N NaOH. Add 20 ml of DEPC-treated 1 M sodium acetate and 20 ml of DEPC-treated 0.5 M EDTA (pH 8.0). Adjust the volume of the solution to 1 liter with DEPC-treated H_2O. Sterilize the solution by passing it through a 0.45-μm Millipore filter and store it at room temperature protected from light. The buffer yellows with age if it is exposed to light or is autoclaved. Straw-colored buffer works well, but darker buffer does not.

MOPS (3[*N*-MORPHOLINO])PROPANESULFONIC ACID

FW	pK_α (20°C)	ΔpK_α/°C	Molar strength of saturated solution at 0°C
209.3	7.15	−0.013	3.1

MOPS is one of the buffers developed in Robert Good's laboratories in the 1970s to facilitate isolation of chloroplasts and other plant organelles. In molecular cloning, MOPS is a component of buffers used for the electrophoresis of RNA through agarose gels.

TAFE Gel Electrophoresis Buffer
20 mM Tris-acetate (pH 8.2)
0.5 mM EDTA

Use acetic acid to adjust the pH of the Tris solution to 8.2 and use the free acid of EDTA, not the sodium salt. Concentrated solutions of TAFE buffer can also be purchased (e.g., Beckman Coulter).

IMPORTANT: The TAFE gel electrophoresis buffer must be cooled to 14°C before use.

Gel-loading Buffers

6x Alkaline Gel-loading Buffer
300 mM NaOH <!>
6 mM EDTA
18% (w/v) Ficoll (Type 400; Pfizer)
0.15% (w/v) bromocresol green
0.25% (w/v) xylene cyanol

Bromophenol Blue Solution (0.4%, w/v)
Dissolve 4 mg of solid bromophenol blue in 1 ml of sterile H_2O. Store the solution at room temperature.

TABLE A1-6 6x Gel-loading Buffers

Buffer Type	6x Buffer	Storage Temperature
I	0.25% (w/v) bromophenol blue 0.25% (w/v) xylene cyanol FF 40% (w/v) sucrose in H_2O	4°C
II	0.25% (w/v) bromophenol blue 0.25% (w/v) xylene cyanol FF 15% (w/v) Ficoll (Type 400; Pfizer) in H_2O	room temperature
III	0.25% (w/v) bromophenol blue 0.25% (w/v) xylene cyanol FF 30% (v/v) glycerol in H_2O	4°C
IV	0.25% (w/v) bromophenol blue 40% (w/v) sucrose in H_2O	4°C

Bromophenol Blue Sucrose Solution
0.25% (w/v) bromophenol blue
40% (w/v) sucrose

Cresol Red Solution (10 mM)
Dissolve 4 mg of the sodium salt of cresol red (Sigma-Aldrich) in 1 ml of sterile H_2O. Store the solution at room temperature.

10x Formaldehyde Gel-loading Buffer
50% (v/v) glycerol (diluted in DEPC-treated <!> H_2O)
10 mM EDTA (pH 8.0)
0.25% (w/v) bromophenol blue
0.25% (w/v) xylene cyanol FF

Formamide-loading Buffer
80% (w/v) deionized formamide <!>
10 mM EDTA (pH 8.0)
1 mg/ml xylene cyanol FF
1 mg/ml bromophenol blue

Purchase a distilled deionized preparation of formamide and store in small aliquots under nitrogen at –20°C. Alternatively, deionize reagent-grade formamide as described in Appendix 3.

RNA Gel-loading Buffer
95% (v/v) deionized formamide <!>
0.025% (w/v) bromophenol blue
0.025% (w/v) xylene cyanol FF
5 mM EDTA (pH 8.0)
0.025% (w/v) SDS

2x SDS Gel-loading Buffer
100 mM Tris-Cl (pH 6.8)
4% (w/v) SDS (electrophoresis grade)
0.2% (w/v) bromophenol blue
20% (v/v) glycerol
200 mM dithiothreitol or β-mercaptoethanol <!>

1x and 2x SDS gel-loading buffer lacking thiol reagents can be stored at room temperature. Add the thiol reagents from 1 M (dithiothreitol) or 14 M (β-mercaptoethanol) stocks just before the buffer is used.

2.5x SDS-EDTA Dye Mix
0.4% (v/v) SDS
30 mM EDTA
0.25% bromophenol blue
0.25% xylene cyanol FF
20% (w/v) sucrose

Special Buffers and Solutions

Elution Buffer (Qiagen)
50 mM Tris-Cl (pH 8.1–8.2)
1.4 M NaCl
15% (v/v) ethanol

KOH/Methanol Solution
This solution is for cleaning the glass plates used to cast sequencing gels. It is prepared by dissolving 5 g of KOH <!> pellets in 100 ml of methanol <!>. Store the solution at room temperature in a tightly capped glass bottle.

λ Annealing Buffer
100 mM Tris-Cl (pH 7.6)
10 mM MgCl$_2$

LB Freezing Buffer
36 mM K$_2$HPO$_4$ (anhydrous)
13.2 mM KH$_2$PO$_4$
1.7 mM sodium citrate
0.4 mM MgSO$_4$·7H$_2$O
6.8 mM ammonium sulfate
4.4% (v/v) glycerol
in LB broth

LB freezing buffer is best made by dissolving the salts in 100 ml of LB to the specified concentrations. Measure 95.6 ml of the resulting solution into a fresh container and add 4.4 ml of glycerol. Mix the solution well and then sterilize by passing it through a 0.45-μm disposable Nalgene filter. Store the sterile freezing medium at a controlled room temperature (15–25°C).

MgCl$_2$–CaCl$_2$ Solution
80 mM MgCl$_2$
20 mM CaCl$_2$

P3 Buffer (Qiagen)
3 M potassium acetate (pH 5.5)

PEG–MgCl$_2$ Solution
40% (w/v) polyethylene glycol (PEG 8000)
30 mM MgCl$_2$

Dissolve 40 g of PEG 8000 in a final volume of 100 ml of 30 mM MgCl$_2$. Sterilize the solution by passing it through a 0.22-μm filter and store it at room temperature.

QBT Buffer (Qiagen)
750 mM NaCl <!>
50 mM MOPS (pH 7.0) <!>
15% (v/v) isopropanol
0.15% (v/v) Triton X-100

Radioactive Ink <!>
Radioactive ink is made by mixing a small amount of ^{32}P with waterproof black drawing ink. We find it convenient to make the ink in three grades: very hot (>2000 cps on a handheld minimonitor), hot

(>500 cps on a handheld minimonitor), and cool (>50 cps on a handheld minimonitor). Use a fiber-tip pen to apply ink of the desired activity to the pieces of tape. Attach radioactive-warning tape to the pen and store it in an appropriate place.

Sephacryl Equilibration Buffer
50 mM Tris-Cl (pH 8.0)
5 mM EDTA
0.5 M NaCl

SM and SM Plus Gelatin
Per liter:

NaCl	5.8 g
$MgSO_4 \cdot 7H_2O$	2 g
1 M Tris-Cl (pH 7.5)	50 ml
2% (w/v) gelatin solution	5 ml
H_2O	to 1 liter

Sterilize the buffer by autoclaving for 20 minutes at 15 psi (1.05 kg/cm^2) on liquid cycle. After the solution has cooled, dispense 50-ml aliquots into sterile containers. SM may be stored indefinitely at room temperature. Discard each aliquot after use to minimize the chance of contamination.

Sorbitol Buffer
1 M sorbitol
0.1 M EDTA (pH 7.5)

STE
10 mM Tris-Cl (pH 8.0)
0.1 M NaCl
1 mM EDTA (pH 8.0)

Sterilize by autoclaving for 15 minutes at 15 psi (1.05 kg/cm^2) on liquid cycle. Store the sterile solution at 4°C.

10x TEN Buffer
0.1 M Tris-Cl (pH 8.0)
0.01 M EDTA (pH 8.0)
1 M NaCl

TES
10 mM Tris-Cl (pH 7.5)
1 mM EDTA (pH 7.5)
0.1% (w/v) SDS

Tris-Sucrose
50 mM Tris-Cl (pH 8.0)
10% (w/v) sucrose

Sterilize the solution by passing it through a 0.22-μm filter and store it at room temperature. Solutions containing sucrose should not be autoclaved because the sugar tends to carbonize at high temperatures.

Triton/SDS Solution
10 mM Tris-Cl (pH 8.0)
2% (v/v) Triton X-100
1% (w/v) SDS
100 mM NaCl
1 mM EDTA (pH 8.0)

Sterilize the solution by passing it through a 0.22-μm filter and store it at room temperature.

Wash Buffer (Qiagen)
50 mM MOPS-KOH <!> (pH 7.5–7.6)
0.75 M NaCl
15% (v/v) ethanol

When making this buffer, adjust the pH of a MOPS/NaCl solution before adding the ethanol.

Yeast Resuspension Buffer
50 mM Tris-Cl (pH 7.4)
20 mM EDTA (pH 7.5)

PREPARATION OF ORGANIC REAGENTS

CAUTION: Please see Appendix 4 for appropriate handling of materials marked with <!>.

Phenol

Most batches of commercial liquefied phenol <!> are clear and colorless and can be used in molecular cloning without redistillation. Occasionally, batches of liquefied phenol are pink or yellow, and these should be rejected and returned to the manufacturer. Crystalline phenol is not recommended because it must be redistilled at 160°C to remove oxidation products, such as quinones, that cause the breakdown of phosphodiester bonds or cause cross-linking of RNA and DNA.

Equilibration of Phenol

Before use, phenol must be equilibrated to a pH of >7.8 because the DNA partitions into the organic phase at acid pH. Wear gloves, full face protection, and a lab coat when performing this procedure.

1. Store liquefied phenol at –20°C. As needed, remove the phenol from the freezer, allow it to warm to room temperature, and then melt it at 68°C. Add hydroxyquinoline to a final concentration of 0.1%. This compound is an antioxidant, a partial inhibitor of RNase, and a weak chelator of metal ions. In addition, its yellow color provides a convenient way to identify the organic phase.

2. To the melted phenol, add an equal volume of buffer (usually 0.5 M Tris-Cl [pH 8.0] at room temperature). Stir the mixture on a magnetic stirrer for 15 minutes. Turn off the stirrer, and when the two phases have separated, aspirate as much as possible of the upper (aqueous) phase using a glass pipette attached to a vacuum line equipped with appropriate traps (please see Appendix 3, Figure A3-2).

3. Add an equal volume of 0. 1 M Tris-Cl (pH 8.0) to the phenol. Stir the mixture on a magnetic stirrer for 15 minutes. Turn off the stirrer and remove the upper aqueous phase as described in Step 2. Repeat the extractions until the pH of the phenolic phase is >7.8 (as measured with pH paper).

4. After the phenol is equilibrated and the final aqueous phase has been removed, add 0.1 volume of 0.1 M Tris-Cl (pH 8.0) containing 0.2% β-mercaptoethanol <!>. The phenol solution may be stored in this form under 100 mM Tris-Cl (pH 8.0) in a light-tight bottle at 4°C for periods of up to 1 month.

Phenol:Chloroform:Isoamyl Alcohol (25:24:1)

A mixture consisting of equal parts of equilibrated phenol and chloroform:isoamyl alcohol <!> (24:1) is frequently used to remove proteins from preparations of nucleic acids. The chloroform denatures proteins and facilitates the separation of the aqueous and organic phases, and the isoamyl alcohol reduces foaming during extraction. Neither chloroform nor isoamyl alcohol requires treatment before use. The phenol:chloroform:isoamyl alcohol mixture may be stored under 100 mM Tris-Cl (pH 8.0) in a light-tight bottle at 4°C for periods of up to 1 month.

Deionization of Formamide

Many batches of reagent-grade formamide <!> are sufficiently pure to be used without additional treatment. However, if any yellow color is present, deionize the formamide by stirring on a magnetic stirrer with Dowex XG8 mixed-bed resin for 1 hour and filtering it twice through Whatman No. 1 paper. Store deionized formamide in small aliquots under nitrogen at –70°C.

Deionization of Glyoxal

Commercial stock solutions of glyoxal (40% or 6 M) contain various hydrated forms of glyoxal, as well as oxidation products such as glyoxylic acid, formic acid, and other compounds that can degrade RNA. These contaminants must be removed by treatment with a mixed-bed resin such as Bio-Rad AG-510-X8 until the indicator dye in the resin is exhausted. To deionize the glyoxal,

1. Immediately before use, mix the glyoxal with an equal volume mixed-bed ion-exchange resin (Bio-Rad AG-510-X8). Alternatively, pass the glyoxal through a small column of mixed-bed resin and proceed to Step 3.

2. Separate the deionized material from the resin by filtration (e.g., through a Uniflow Plus filter, Schleicher & Schuell).

3. Monitor the pH of the glyoxal by mixing 200 μl of glyoxal with 2 μl of a 10-mg/ml solution of bromocresol green in H_2O and observing the change in color. Bromocresol green is yellow at pH <4.8 and blue-green at pH >5.2.

4. Repeat the deionization process (Steps 1 and 2) until the pH of the glyoxal is >5.5.

Deionized glyoxal can be stored indefinitely at –20°C under nitrogen in tightly sealed microfuge tubes. Use each aliquot only once and then discard.

CHEMICAL STOCK SOLUTIONS

CAUTION: Please see Appendix 4 for appropriate handling of materials marked with <!>.

Acrylamide Solution (45% w/v)

acrylamide (DNA-sequencing grade) <!>	434 g
N,N′-methylenebisacrylamide <!>	16 g
H_2O	to 600 ml

Heat the solution to 37°C to dissolve the chemicals. Adjust the volume to 1 liter with distilled H_2O. Filter the solution through a nitrocellulose filter (e.g., Nalgene Nunc, 0.45-μm pore size) and store the filtered solution in dark bottles at room temperature.

Actinomycin D (5 mg/ml)

Dissolve actinomycin D <!> in methanol <!> at a concentration of 5 mg/ml. Store the stock solution at –20°C in the dark. Please see the information panel Actinomycin D in MC3, p. 7.88.

Adenosine Diphosphate (ADP) (1 mM)

Dissolve solid adenosine diphosphate in sterile 25 mM Tris-Cl (pH 8.0). Store small aliquots (~20 μl) of the solution at –20°C.

Ammonium Acetate (10 M)

To prepare a 1-liter solution, dissolve 770 g of ammonium acetate in 800 ml of H_2O. Adjust volume to 1 liter with H_2O. Sterilize by filtration. Alternatively, to prepare a 100-ml solution, dissolve 77 g of ammonium acetate in 70 ml of H_2O at room temperature. Adjust the volume to 100 ml with H_2O. Sterilize the solution by passing it through a 0.22-μm filter. Store the solution in tightly sealed bottles at 4°C or at room temperature. Ammonium acetate decomposes in hot H_2O and solutions containing it should not be autoclaved.

Ammonium Persulfate (10% w/v)

ammonium persulfate <!>	1 g
H_2O	to 10 ml

Dissolve 1 g of ammonium persulfate in 10 ml of H_2O and store at 4°C. Ammonium persulfate decays slowly in solution, so replace the stock solution every 2–3 weeks. Ammonium persulfate is used as a catalyst for the copolymerization of acrylamide and bisacrylamide gels. The polymerization reaction is driven by free radicals generated by an oxido-reduction reaction in which a diamine (e.g., TEMED) is used as the adjunct catalyst.

ATP (10 mM)

Dissolve an appropriate amount of solid ATP in 25 mM Tris-Cl (pH 8.0). Store the ATP solution in small aliquots at –20°C.

Calcium Chloride (CaCl$_2$, 2.5 M)

Dissolve 11 g of CaCl$_2$·6H$_2$O in a final volume of 20 ml of distilled H$_2$O. Sterilize the solution by passing it through a 0.22-µm filter. Store in 1-ml aliquots at 4°C.

Coomassie Staining Solution

Dissolve 0.25 g of Coomassie Brilliant Blue R-250 in 90 ml of methanol:H$_2$O <!> (1:1, v/v) and 10 ml of glacial acetic acid <!>. Filter the solution through a Whatman No. 1 filter to remove any particulate matter. Store at room temperature. Please see the entry on Coomassie Staining in Appendix 3.

Deoxyribonucleoside Triphosphates (dNTPs)

Dissolve each dNTP in H$_2$O at an approximate concentration of 100 mM. Use 0.05 M Tris base and a micropipette to adjust the pH of each of the solutions to 7.0 (use pH paper to check the pH). Dilute an aliquot of the neutralized dNTP appropriately and read the optical density at the wavelengths given in the table below. Calculate the actual concentration of each dNTP. Dilute the solutions with H$_2$O to a final concentration of 50 mM dNTP. Store each separately at –70°C in small aliquots.

Base	Wavelength (nm)	Extinction Coefficient (E)(M^{-1} cm^{-1})
A	259	1.54×10^4
G	253	1.37×10^4
C	271	9.10×10^3
T	267	9.60×10^3

For a cuvette with a path length of 1 cm, absorbance = EM. One hundred micromolar stock solutions of each dNTP are commercially available (Pfizer).

For polymerase chain reactions (PCRs), adjust the dNTP solution to pH 8.0 with 2 N NaOH. Commercially available solutions of PCR-grade dNTPs require no adjustment.

Dimethylsulfoxide (DMSO)

Purchase a high grade of DMSO <!> (HPLC grade or better). Divide the contents of a fresh bottle into 1-ml aliquots in sterile tubes. Close the tubes tightly and store at –20°C. Use each aliquot only once and then discard.

Dithiothreitol (DTT, 1 M)

Dissolve 3.09 g of DTT in 20 ml of 0.01 M sodium acetate (pH 5.2) and sterilize by filtration. Dispense into 1-ml aliquots and store at –20°C. Under these conditions, DTT is stable to oxidation by air.

EDTA (0.5 M, pH 8.0)

Add 186.1 g of disodium EDTA·2H$_2$O to 800 ml of H$_2$O. Stir vigorously on a magnetic stirrer. Adjust the pH to 8.0 with NaOH (~20 g of NaOH pellets <!>). Dispense into aliquots and sterilize by autoclaving. The disodium salt of EDTA will not go into solution until the pH of the solution is adjusted to approximately 8.0 by the addition of NaOH.

EGTA (0.5 M, pH 8.0)

EGTA is ethylene glycol bis(β-aminoethyl ether) *N,N,N',N'*-tetraacetic acid. A solution of EGTA is made up essentially as described for EDTA above and sterilized by either autoclaving or filtering. Store the sterile solution at room temperature.

Ethidium Bromide (10 mg/ml)

Add 1 g of ethidium bromide <!> to 100 ml of H$_2$O. Stir on a magnetic stirrer for several hours to ensure that the dye has dissolved. Wrap the container in aluminum foil or transfer the solution to a dark bottle and store at room temperature.

Gelatin (2% w/v)

Add 2 g of gelatin to a total volume of 100 ml of H$_2$O and autoclave the solution for 15 minutes at 15 psi (1.05 kg/cm^2) on liquid cycle.

Glycerol (10% v/v)

Dilute 1 volume of molecular-biology-grade glycerol in 9 volumes of sterile, pure H$_2$O. Sterilize the solution by passing it through a prerinsed 0.22-µm filter. Store in 200-ml aliquots at 4°C.

IPTG (20% [w/v], 0.8 M)

IPTG is isopropylthio-β-D-galactoside. Make a 20% solution of IPTG by dissolving 2 g of IPTG in 8 ml of distilled H_2O. Adjust the volume of the solution to 10 ml with H_2O and sterilize by passing it through a 0.22-μm disposable filter. Dispense the solution into 1-ml aliquots and store them at –20°C.

KCl (4 M)

Dissolve an appropriate amount of solid KCl in H_2O, autoclave for 20 minutes on liquid cycle, and store at room temperature. Ideally, this solution should be divided into small (~100 μl) aliquots in sterile tubes and each aliquot thereafter used one time.

Lithium Chloride (LiCl, 5 M)

Dissolve 21.2 g of LiCl in a final volume of 90 ml of H_2O. Adjust the volume of the solution to 100 ml with H_2O. Sterilize the solution by passing it through a 0.22-μm filter or by autoclaving for 15 minutes at 15 psi (1.05 kg/cm²) on liquid cycle. Store the solution at 4°C.

Maltose (20% w/v)

Dissolve 20 g of maltose in a final volume of 100 ml of H_2O and sterilize by passing it through a 0.22-μm filter. Store the sterile solution at room temperature.

$MgCl_2·6H_2O$ (1 M)

Dissolve 203.3 g of $MgCl_2·6H_2O$ in 800 ml of H_2O. Adjust the volume to 1 liter with H_2O. Dispense into aliquots and sterilize by autoclaving. $MgCl_2$ is extremely hygroscopic. Buy small bottles (e.g., 100 g) and do not store opened bottles for long periods of time.

$MgSO_4$ (1 M)

Dissolve 12 g of $MgSO_4$ in a final volume of 100 ml of H_2O. Sterilize by autoclaving or filter sterilization. Store at room temperature.

NaCl (sodium chloride, 5 M)

Dissolve 292 g of NaCl in 800 ml of H_2O. Adjust the volume to 1 liter with H_2O. Dispense into aliquots and sterilize by autoclaving. Store the NaCl solution at room temperature.

NaOH (10 N)

The preparation of 10 N NaOH <!> involves a highly exothermic reaction, which can cause breakage of glass containers. Prepare this solution with extreme care in plastic beakers. To 800 ml of H_2O, slowly add 400 g of NaOH pellets <!>, stirring continuously. As an added precaution, place the beaker on ice. When the pellets have dissolved completely, adjust the volume to 1 liter with H_2O. Store the solution in a plastic container at room temperature. Sterilization is not necessary.

PEG 8000

Working concentrations of PEG <!> range from 13% to 40% (w/v). Prepare the appropriate concentration by dissolving PEG 8000 in sterile H_2O, warming if necessary. Sterilize the solution by passing it through a 0.22-μm filter. Store the solution at room temperature.

Polyethylene glycol (PEG) is a straight-chain polymer of a simple repeating unit $H(OCH_2CH_2)_nOH$. PEG is available in a range of molecular weights whose names reflect the number (n) of repeating units in each molecule. In PEG 400, for example, $n = 8$–9, whereas in PEG 4000, n ranges from 68 to 84. PEG induces macromolecular crowding of solutes in aqueous solution and has a range of uses in molecular cloning, including:

- *Precipitation of DNA molecules according to their size.* The concentration of PEG required for precipitation is in inverse proportion to the size of the DNA fragments.

- *Precipitation and purification of bacteriophage particles.*

- *Increasing the efficiency of reassociation of complementary chains* of nucleic acids during hybridization, blunt-end ligation of DNA molecules, and end-labeling of DNA with bacteriophage T4 polynucleotide kinase (please see the information panel Condensing and Crowding Reagents in MC3, p. 1.152).

- *Fusion of cultured cells with bacterial protoplasts.*

Potassium Acetate (5 M)

5 M potassium acetate	60 ml
glacial acetic acid <!>	11.5 ml
H_2O	28.5 ml

The resulting solution is 3 M with respect to potassium and 5 M with respect to acetate. Store the buffer at room temperature.

SDS (20% w/v)

Also called sodium lauryl sulfate. Dissolve 200 g of electrophoresis-grade SDS <!> in 900 ml of H_2O. Heat to 68°C and stir with a magnetic stirrer to assist dissolution. If necessary, adjust the pH to 7.2 by adding a few drops of concentrated HCl <!>. Adjust the volume to 1 liter with H_2O. Store at room temperature. Sterilization is not necessary. Do not autoclave.

Silver Stain. *Please see staining section (Appendix 3).*

Sodium Acetate (3 M, pH 5.2 and 7.0)

Dissolve 408.3 g of sodium acetate·$3H_2O$ in 800 ml of H_2O. Adjust the pH to 5.2 with glacial acetic acid <!> or adjust the pH to 7.0 with dilute acetic acid. Adjust the volume to 1 liter with H_2O. Dispense into aliquots and sterilize by autoclaving.

Spermidine (1 M)

Dissolve 1.45 g of spermidine (free-base form) in 10 ml of deionized H_2O and sterilize by passing it through a 0.22-μm filter. Store the solution in small aliquots at –20°C. Make a fresh stock solution of this reagent every month.

SYBR Gold Staining Solution

SYBR Gold <!> (Invitrogen) is supplied as a stock solution of unknown concentration in dimethylsulfoxide. Agarose gels are stained in a working solution of SYBR Gold, which is a 1:10,000 dilution of SYBR Gold nucleic acid stain in electrophoresis buffer. Prepare working stocks of SYBR Gold daily and store in the dark at regulated room temperature. For a discussion of staining agarose gels, please see MC3, p. 5.14.

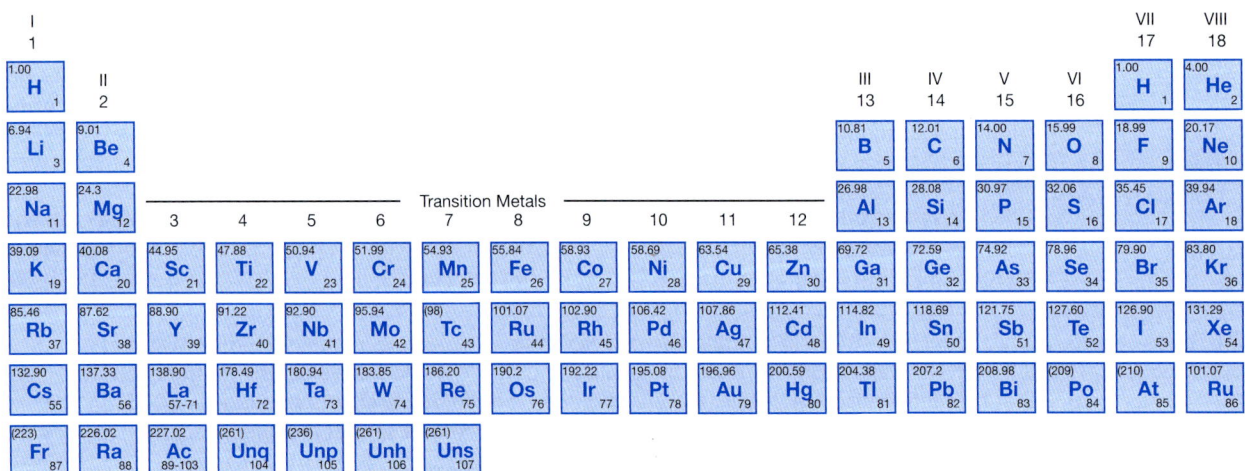

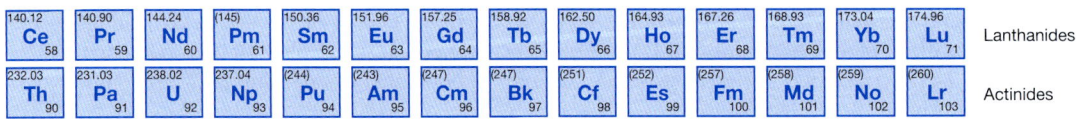

FIGURE A1-1 Periodic Table

Numbers in parentheses are the mass numbers of the most stable isotope of that element.

Trichloroacetic Acid (TCA; 100% solution)

To a previously unopened bottle containing 500 g of TCA <!>, add 227 ml of H_2O. The resulting solution will contain 100% (w/v) TCA.

X-gal Solution (2% w/v)

X-gal is 5-bromo-4-chloro-3-indolyl-β-D-galactoside. Make a stock solution by dissolving X-gal in dimethylformamide <!> at a concentration of 20 mg/ml solution. Use a glass or polypropylene tube. Wrap the tube containing the solution in aluminum foil to prevent damage by light and store at –20°C. It is not necessary to sterilize X-gal solutions by filtration. Please see the information panel X-gal in MC3, p. 1.149.

REAGENTS AND BUFFERS INDEX

Media

BACKGROUND INFORMATION

Background information found in *Molecular Cloning: A Laboratory Manual,* 3rd edition (hereafter MC3) unless otherwise indicated.

LIQUID MEDIA FOR GROWTH OF *E. COLI*

IMPORTANT: Use distilled deionized H_2O in all recipes. Unless otherwise stated, sterile media can be stored at room temperature.

GYT Medium

10% (v/v) glycerol
0.125% (w/v) yeast extract
0.25% (w/v) tryptone

Sterilize the medium by passing it through a prerinsed 0.22-μm filter. Store in 2.5-ml aliquots at 4°C.

LB Medium (Luria-Bertani Medium)

Per liter:

To 950 ml of deionized H_2O, add:

tryptone	10 g
yeast extract	5 g
NaCl	10 g

Shake until the solutes have dissolved. Adjust the pH to 7.0 with 5 N NaOH (~0.2 ml). Adjust the volume of the solution to 1 liter with deionized H_2O. Sterilize by autoclaving for 20 minutes at 15 psi (1.05 kg/cm²) on liquid cycle.

M9 Minimal Medium

Per liter:

To 750 ml of sterile H_2O (cooled to 50°C or less), add:

5x M9 salts*	200 ml
1 M $MgSO_4$	2 ml
20% solution of the appropriate carbon source (e.g., 20% glucose)	20 ml
1 M $CaCl_2$	0.1 ml
sterile deionized H_2O	to 980 ml

If necessary, supplement the M9 medium with stock solutions of the appropriate amino acids and vitamins.

*5x M9 salts is made by dissolving the following salts in deionized H_2O to a final volume of 1 liter:

$Na_2HPO_4 \cdot 7H_2O$	64 g
KH_2PO_4 (anhydrous)	15 g
NaCl	2.5 g
NH_4Cl	5.0 g

Divide the salt solution into 200-ml aliquots and sterilize by autoclaving for 15 minutes at 15 psi (1.05 kg/cm²) on liquid cycle.

Prepare the $MgSO_4$ and $CaCl_2$ solutions separately, sterilize by autoclaving, and add the solutions after diluting the 5x M9 salts to 980 ml with sterile H_2O. Sterilize the glucose by passing it through a 0.22-μm filter before it is added to the diluted M9 salts.

When using *E. coli* strains that carry a deletion of the proline biosynthetic operon [Δ(*lac-proAB*)] in the bacterial chromosome and the complementing *proAB* genes on the F' plasmid, supplement the M9 minimal medium with the following:

0.4% (w/v) glucose (dextrose)
5 mM $MgSO_4 \cdot 7H_2O$
0.01% thiamine

NZCYM Medium

Per liter:

To 950 ml of deionized H_2O, add:

NZ amine	10 g
NaCl	5 g
yeast extract	5 g
casamino acids	1 g
$MgSO_4 \cdot 7H_2O$	2 g

Shake until the solutes have dissolved. Adjust the pH to 7.0 with 5 N NaOH (~0.2 ml). Adjust the volume of the solution to 1 liter with deionized H_2O. Sterilize by autoclaving for 20 minutes at 15 psi (1.05 kg/cm²) on liquid cycle.

> NZ amine: Casein hydrolysate enzymatic (ICN Biochemicals). NZCYM, NZYM, and NZM are also available as dehydrated media from BD Biosciences.

NZM Medium

NZM medium is identical to NZYM medium, except that yeast extract is omitted.

NZYM Medium

NZYM medium is identical to NZCYM medium, except that casamino acids are omitted.

SOB Medium

Per liter:

To 950 ml of deionized H_2O, add:

tryptone	20 g
yeast extract	5 g
NaCl	0.5 g

Shake until the solutes have dissolved. Add 10 ml of a 250 mM solution of KCl. (This solution is made by dissolving 1.86 g of KCl in 100 ml of deionized H_2O.) Adjust the pH of the medium to 7.0 with 5 N NaOH <!> (~0.2 ml). Adjust the volume of the solution to 1 liter with deionized H_2O. Sterilize by autoclaving for 20 minutes at 15 psi (1.05 kg/cm²) on liquid cycle. Just before use, add 5 ml of a sterile solution of 2 M $MgCl_2$. (This solution is made by dissolving 19 g of $MgCl_2$ in 90 ml of deionized H_2O. Adjust the volume of the solution to 100 ml with deionized H_2O and sterilize by autoclaving for 20 minutes at 15 psi [1.05 kg/cm²] on liquid cycle.)

SOC Medium

SOC medium is identical to SOB medium, except that it contains 20 mM glucose. After the SOB medium has been autoclaved, allow it to cool to 60°C or less. Add 20 ml of a sterile 1 M solution of glucose. (This solution is made by dissolving 18 g of glucose in 90 ml of deionized H_2O. After the sugar has dissolved, adjust the volume of the solution to 100 ml with deionized H_2O and sterilize by passing it through a 0.22-μm filter.)

Terrific Broth (also known as TB)

Per liter:

To 900 ml of deionized H_2O, add:

tryptone	12 g
yeast extract	24 g
glycerol	4 ml

Shake until the solutes have dissolved and then sterilize by autoclaving for 20 minutes at 15 psi (1.05 kg/cm²) on liquid cycle. Allow the solution to cool to 60°C or less and then add 100 ml of a sterile solution of 0.17 M KH_2PO_4, 0.72 M K_2HPO_4. (This solution is made by dissolving 2.31 g of KH_2PO_4 and 12.54 g of K_2HPO_4 in 90 ml of deionized H_2O. After the salts have dissolved, adjust the volume of the solution to 100 ml with deionized H_2O and sterilize by autoclaving for 20 minutes at 15 psi [1.05 kg/cm²] on liquid cycle.)

2x YT Medium

Per liter:

To 900 ml of deionized H_2O, add:

tryptone	16 g
yeast extract	10 g
NaCl	5 g

Shake until the solutes have dissolved. Adjust the pH to 7.0 with 5 N NaOH <!>. Adjust the volume of the solution to 1 liter with deionized H_2O. Sterilize by autoclaving for 20 minutes at 15 psi (1.05 kg/cm²) on liquid cycle.

MEDIA CONTAINING AGAR OR AGAROSE

IMPORTANT: Use distilled deionized H_2O in all recipes.

Prepare liquid media according to the recipes given above. Just before autoclaving, add one of the following:

Bacto Agar (for plates)	15 g/liter
Bacto Agar (for top agar)	7 g/liter
agarose (for plates)	15 g/liter
agarose (for top agarose)	7 g/liter

Sterilize by autoclaving for 20 minutes at 15 psi (1.05 kg/cm^2) on liquid cycle. When the medium is removed from the autoclave, swirl it gently to distribute the melted agar or agarose evenly throughout the solution. *Be careful!* The fluid may be superheated and may boil over when swirled. Allow the medium to cool to 50°C–60°C before adding thermolabile substances (e.g., antibiotics). To avoid producing air bubbles, mix the medium by swirling. Plates can then be poured directly from the flask; allow approximately 30–35 ml of medium per 90-mm plate. To remove bubbles from medium in the plate, flame the surface of the medium with a Bunsen burner before the agar or agarose hardens. Set up a color code (e.g., two red stripes for LB-ampicillin plates; one black stripe for LB plates, etc.) and mark the edges of the plates with the appropriate colored markers.

When the medium has hardened completely, invert the plates and store them at 4°C until needed. The plates should be removed from storage 1–2 hours before they are used. If the plates are fresh, they will "sweat" when incubated at 37°C. When this condensation drops on the agar/agarose surface, it allows bacterial colonies or bacteriophage plaques to spread and increases the chances of cross-contamination. This problem can be avoided by wiping off the condensation from the lids of the plates and then incubating the plates for several hours at 37°C in an inverted position before they are used. Alternatively, remove the liquid by shaking the lid with a single, quick motion.

STORAGE MEDIA

IMPORTANT: Use distilled deionized H_2O in all recipes.

Liquid Cultures

Bacteria growing on plates, or in liquid culture, can be stored in aliquots of LB medium containing 30% (v/v) sterile glycerol. Aliquots of 1 ml of LB with glycerol should be prepared and vortexed to ensure that the glycerol is completely dispersed. Alternatively, bacterial strains may be stored in LB freezing buffer:

LB freezing buffer
36 mM K_2HPO_4 (anhydrous)
13.2 mM KH_2PO_4
1.7 mM sodium citrate
0.4 mM $MgSO_4 \cdot 7H_2O$
6.8 mM ammonium sulfate
4.4% (v/v) glycerol
in LB

LB freezing buffer is best made by dissolving the salts in 100 ml of LB to the specified concentrations. Measure 95.6 ml of the resulting solution into a fresh container and add 4.4 ml of glycerol. Mix the solution well and then sterilize by passing it through a 0.45-μm disposable Nalgene filter. For more information on storage of bacterial cultures, please see Appendix 3.

Stab Cultures

Prepare stab cultures in glass vials (2–3 ml) with screw-on caps fitted with rubber gaskets. Add molten LB agar until the vials are two-thirds full. Autoclave the partially filled vials (with their caps loosely screwed on) for 20 minutes at 15 psi (1.05 kg/cm^2) on liquid cycle. Remove the vials from the autoclave, let them cool to room temperature, and then tighten the caps. Store the vials at room temperature until needed.

ANTIBIOTICS

TABLE A2-1 Commonly Used Antibiotic Solutions

	Stock Solution[a]		Working Concentration	
	Concentration	Storage	Stringent Plasmids	Relaxed Plasmids
Ampicillin	50 mg/ml in H_2O	–20°C	20 µg/ml	50 µg/ml
Carbenicillin	50 mg/ml in H_2O	–20°C	20 µg/ml	60 µg/ml
Chloramphenicol	34 mg/ml in ethanol	–20°C	25 µg/ml	170 µg/ml
Kanamycin	10 mg/ml in H_2O	–20°C	10 µg/ml	50 µg/ml
Streptomycin	10 mg/ml in H_2O	–20°C	10 µg/ml	50 µg/ml
Tetracycline[b]	5 mg/ml in ethanol	–20°C	10 µg/ml	50 µg/ml

Magnesium ions are antagonists of tetracycline. Use media without magnesium salts (e.g., LB medium) for selection of bacteria resistant to tetracycline.

[a]Sterilize stock solutions of antibiotics dissolved in H_2O by filtration through a 0.22-µm filter.

[b]Antibiotics dissolved in ethanol need not be sterilized. Store solutions in light-tight containers.

TABLE A2-2 Antibiotic Modes of Action

Antibiotic	Molecular Weight	Mode of Action	Additional Information
Actinomycin C₁ (actinomycin D)	1255.4	Inhibits synthesis of RNA by binding to double-stranded DNA.	
Amphotericin	924.1	Broad-spectrum antifungal agent from *Streptomyces*.	
Ampicillin	349.4	Inhibits cell-wall synthesis by interfering with peptidoglycan cross-linking.	Please see the information panel Ampicillin and Carbenicillin (MC3, p. 1.148).
Bleomycin	n.a.[a]	Inhibits DNA synthesis; cleaves single-stranded DNA.	
Carbenicillin (disodium salt)	422.4	Inhibits bacterial wall synthesis.	
Chloramphenicol	323.1	Inhibits translation by blocking peptidyl transferase on the 50S ribosomal subunit; at higher concentrations can inhibit eukaryotic DNA synthesis.	Please see the information panel Chloramphenical (MC3, pp. 1.143–1.144).
Geneticin (G418 geneticin disulfate)	692.7	Aminoglycoside is toxic to a broad range of cell types (bacterial, higher plant, yeast, mammaliam, protozoans, helminths); used in selection of eukaryotic cells transformed with neomycin resistance genes.	
Gentamycin	692.7	Inhibits protein synthesis by binding to L6 protein of the 50S ribosomal subunit.	
Hygromycin B	527.5	Inhibits protein synthesis.	
Kanamycin monosulfate	582.6	Broad-spectrum antibiotic; binds to 70S ribosomal subunit and inhibits growth of gram-positive and -negative bacteria and mycoplasmas.	Please see the information panel Kanamycins (MC3, p. 1.145).
Methotrexate	454.45	Folic acid analog; powerful inhibitor of the enzyme dihydrofolate reductase.	
Mitomycin C	334.33	Inhibits DNA synthesis; antibacterial to gram-positive, -negative, and acid-fast bacilli.	
Neomycin B sulfate	908.9	Binds to 30S ribosomal subunit and inhibits bacterial protein synthesis.	
Novobiocin sodium salt	634.62	Bacteriostatic antibiotic; inhibits growth of gram-positive bacteria.	
Penicillin G sodium salt	356.4	Inhibits peptidoglycan synthesis in bacterial cell walls.	
Puromycin dihydrochloride	544.4	Inhibits protein synthesis by acting as an analog of aminoacyl-tRNA (causes premature chain termination).	
Rifampicin	823.0	Strongly inhibits prokaryotic RNA polymerases and mammalian RNA polymerase to a lesser degree.	
Streptomycin sulfate	1457.4	Inhibits protein synthesis; binds to 30S ribosomal subunit.	
Tetracycline hydrochloride	480.9	Inhibits bacterial protein synthesis; blocks ribosomal binding of aminoacyl-tRNA.	Please see the information panel Tetracycline (MC3, p. 1.147).

[a]Not available

SOLUTIONS FOR WORKING WITH BACTERIOPHAGE λ

IMPORTANT: Use distilled deionized H_2O in all recipes.

Maltose

Maltose, an inducer of the gene (*lamB*) that encodes the bacteriophage λ receptor, is often added to the medium during growth of bacteria that are to be used for plating bacteriophage λ. Add 1 ml of a sterile 20% maltose solution for every 100 ml of medium. For a further discussion of the use of maltose, please see the Materials list in MC3, p. 2.25. Make up a sterile 20% stock solution of maltose as follows:

maltose	20 g
H_2O	to 100 ml

Sterilize the solution by passing it through a 0.22-μm filter. Store the sterile solution at room temperature.

SM

This buffer is used for storage and dilution of bacteriophage λ stocks.
Per liter:

NaCl	5.8 g
$MgSO_4\cdot7H_2O$	2 g
1 M Tris-Cl (pH 7.5)	50 ml
2% gelatin solution	5 ml
H_2O	to 1 liter

Sterilize the buffer by autoclaving for 20 minutes at 15 psi (1.05 kg/cm²) on liquid cycle. After the solution has cooled, dispense 50-ml aliquots into sterile containers. SM may be stored indefinitely at room temperature.

A 2% gelatin solution is made by adding 2 g of gelatin to a total volume of 100 ml of H_2O and autoclaving the solution for 15 minutes at 15 psi (1.05 kg/cm²) on liquid cycle.

TM

Per liter:

1 M Tris-Cl (pH 7.5)	50 ml
$MgSO_4\cdot7H_2O$	2 g
H_2O	to 1 liter

Sterilize the buffer by autoclaving for 20 minutes at 15 psi (1.05 kg/cm²) on liquid cycle. After the solution has cooled, dispense 50-ml aliquots into sterile containers. TM may be stored indefinitely at room temperature.

MEDIA FOR THE PROPAGATION AND SELECTION OF YEAST*

CAUTION: Please see Appendix 4 for appropriate handling of materials marked with <!>.

IMPORTANT: Use distilled deionized H_2O in all recipes. Unless otherwise stated, media and solutions are sterilized by autoclaving at 15 psi (1.05 kg/cm²) for 15–20 minutes.

Complete Minimal (CM) or Synthetic Complete (SC) and Dropout Media

To test the growth requirements of strains, it is useful to have media in which each of the commonly encountered auxotrophies is supplemented except the one of interest (dropout media). Dry growth supplements are stored premixed. CM (or SC) is a medium in which the dropout mix contains all possible supplements (i.e., nothing is "dropped out").

yeast nitrogen base without amino acids*	6.7 g
glucose	20 g
Bacto Agar	20 g
dropout mix	2 g
H_2O	to 1000 ml

*Yeast nitrogen base without amino acids (YNB) is sold either with or without ammonium sulfate. This recipe is for YNB with ammonium sulfate. If the batch of YNB lacks ammonium sulfate, add 5 g of ammonium sulfate and use only 1.7 g of YNB.

Dropout Mix

Combine the appropriate ingredients, minus the relevant supplements, and mix in a sealed container. Turn the container end over end for at least 15 minutes; add a few clean marbles to help mix the solids.

Adenine	0.5 g
Alanine	2.0 g
Arginine	2.0 g
Asparagine	2.0 g
Aspartic acid	2.0 g
Cysteine	2.0 g
Glutamine	2.0 g
Glutamic acid	2.0 g
Glycine	2.0 g
Histidine	2.0 g
Inositol	2.0 g
Isoleucine	2.0 g
Leucine	10.0 g
Lysine	2.0 g
Methionine	2.0 g
para-Aminobenzoic acid	0.2 g
Phenylalanine	2.0 g
Proline	2.0 g
Serine	2.0 g
Threonine	2.0 g
Tryptophan	2.0 g
Tyrosine	2.0 g
Uracil	2.0 g
Valine	2.0 g

Reprinted from Adams et al. 1998. *Methods in yeast genetics: A laboratory course manual.* Cold Spring Harbor Laboratory Press, Cold Spring Harbor, New York.

TABLE A2-3 Components of Supplemented Minimal Media

Constituent	Stock Concentration (g/100 ml)	Volume for 1 Liter of Stock of Medium (ml)	Final Concentration in Medium (mg/liter)	Volume of Stock to Spread on Plate (ml)
Adenine sulfate	0.2[a]	10	20	0.2
Uracil	0.2[a]	10	20	0.2
L-Tryptophan	1	2	20	0.1
L-Histidine HCl	1	2	20	0.1
L-Arginine LiCl	1	2	20	0.1
L-Methionine	1	2	20	0.1
L-Tyrosine	0.2	15	30	0.2
L-Leucine	1	10	100	0.1
L-Isoleucine	1	3	30	0.1
L-Lysine HCl	1	3	30	0.1
L-Phenylalanine	1[a]	5	50	0.1
L-Glutamic acid	1[a]	10	100	0.2
L-Aspartic acid	1[a,b]	10	100	0.2
L-Valine	3	5	150	0.1
L-Threonine	4[a,b]	5	200	0.1
L-Serine	8	5	400	0.1

[a]Store at room temperature.
[b]Add after autoclaving the medium.

Supplemented Minimal Medium (SMM)

SMM is SD (please see below) to which various growth supplements have been added. These solutions can then be stored for extended periods. Some should be stored at room temperature to prevent precipitation, whereas the other solutions may be refrigerated. Wherever applicable, HCl salts of amino acids are preferred.

Prepare the medium by adding the appropriate volumes of the stock solutions to the ingredients of SD medium and then adjusting the total volume to 1 liter with distilled H_2O. Add threonine and aspartic acid solutions separately to the medium after it is autoclaved.

Alternatively, it is often more convenient to prepare the medium by spreading a small quantity of the supplement(s) on the surface of an SD plate. Allow the solution(s) to thoroughly onto the plate before inoculating it with yeast strains.

Table A2-3 provides the concentrations of the stock solutions, the volume of stock solution necessary for mixing 1 liter of medium, the volume of stock solution to spread on SD plates, and the final concentration of each constituent in SMM.

Synthetic Dextrose Minimal Medium (SD)

SD is a synthetic minimal medium containing salts, trace elements, vitamins, a nitrogen source (yeast nitrogen base without amino acids), and glucose.

yeast nitrogen base without amino acids*	6.7 g
glucose	20 g
Bacto Agar	20 g
H_2O	1000 ml

*Please see note to recipe for CM on p. 723.

X-Gal Indicator Plates for Yeast

Because 5-bromo-4-chloro-3-indolyl-β-D-galactoside (X-gal <!>) does not work for yeast at the normal acidic pH of SD medium, a medium at neutral pH medium is used. This choice is clearly a trade-off because many yeast strains will not grow well at neutral pH. For each liter of X-gal indicator plates, prepare the following solutions:

Solution I

10x phosphate-buffered stock solution	100 ml
1000x mineral stock solution (see recipe below)	1 ml
dropout mix (see p. 723)	2 g

Adjust the volume to 450 ml with distilled H_2O if the medium is to contain glucose or to 400 ml if it is to contain galactose.

10x Phosphate-buffered Stock Solution

KH_2PO_4 (1 M)	136.1 g
$(NH_4)_2SO_4$ (0.15 M)	19.8 g
KOH (0.75 N) <!>	42.1 g
H_2O	1000 ml

Adjust the pH to 7.0 and autoclave.

1000x Mineral Stock Solution

$FeCl_3$ (2 mM)	32 mg
$MgSO_4 \cdot 7H_2O$ (0.8 M)	19.72 g
H_2O	100 ml

Autoclave and store. This solution will develop a fine yellow precipitate that should be resuspended before use.

Solution II

Mix in a 2-liter flask:

Bacto Agar	20 g
H_2O	500 ml

- Autoclave solutions I and II separately.

- After cooling to <65°C, add the following components to solution I:
  glucose or other sugar to a final concentration of 2%

X-gal (20 mg/ml dissolved in dimethylformamide <!>)	2 ml
100x vitamin stock solution	10 ml

- Include any other heat-sensitive supplements at this point.

- Mix solutions I and II together and pour approximately 30 ml/plate.

100x Vitamin Stock Solution

thiamine (0.04 mg/ml)	4 mg
biotin (2 µg/ml)	0.2 mg
pyridoxine (0.04 mg/ml)	4 mg
inositol (0.2 mg/ml)	20 mg
pantothenic acid (0.04 mg/ml)	4 mg
H_2O	100 ml

Sterilize by passing the solution through a 0.22-µm filter.

X-Gal Plates for Lysed Yeast Cells on Filters

These plates are used for checking β-galactosidase activity in cells that have been lysed and are immobilized on Whatman 3MM filters.

Bacto Agar	20 g
1 M Na_2HPO_4	57.7 ml
1 M NaH_2PO_4	42.3 ml
$MgSO_4$	0.25 g
H_2O	900 ml

After autoclaving, add 6 ml of X-gal solution (20 mg/ml in dimethylformamide).

YPD (YEPD) Medium

YPD is a complex medium for routine growth of yeast.

yeast extract	10 g
peptone	20 g
glucose	20 g
H_2O	to 1000 ml

To prepare plates, add 20 g of Bacto Agar (2%) before autoclaving.

MEDIA INDEX

Commonly Used Techniques in Molecular Cloning

Adapted from Appendix 8, p. A8.1 of MC3.

BACKGROUND INFORMATION

Background information found in *Molecular Cloning: A Laboratory Manual,* 3rd edition (hereafter MC3) unless otherwise indicated.

PREPARATION OF GLASSWARE AND PLASTICWARE

CAUTION: Please see Appendix 4 for appropriate handling of materials marked with <!>.

All glassware should be sterilized by autoclaving or baking. Some, but not all, plasticware can be auto-claved, depending on the type of plastic. Many items of sterilized plasticware are commercially available. All of the procedures commonly used in molecular cloning should be carried out in sterile glassware or plasticware; there is no significant loss of material by adsorption onto the surfaces of these containers. However, for certain procedures (e.g., handling very small quantities of single-stranded DNA or sequencing by the Maxam-Gilbert technique), it is best to use glassware or plasticware that has been coated with a thin film of silicone. A simple procedure for siliconizing small items such as pipettes, tubes, and beakers is given below. To siliconize large items such as glass plates, please refer to the note at the end of the protocol.

Siliconizing Glassware, Plasticware, and Glass Wool

The following method was supplied by Brian Seed.

1. Place the items to be siliconized inside a large, glass desiccator.

2. Add 1 ml of dichlorodimethylsilane <!> to a small beaker inside the desiccator.

3. Attach the desiccator, through a trap, to a vacuum pump. Turn on the vacuum and continue to apply suction until the dichlorodimethylsilane begins to boil. Immediately clamp the connection between the vacuum pump and the desiccator. Switch off the vacuum pump. The desiccator should maintain a vacuum.

   It is essential to turn off the vacuum pump as soon as the dichlorodimethylsilane begins to boil. Otherwise, the volatile agent will be sucked into the pump and cause irreparable damage to the vacuum seals.

4. When the dichlorodimethylsilane has evaporated (1–2 hours), open the desiccator in a chemical fume hood. After the fumes of dichlorodimethylsilane have dispersed, remove the glassware or plasticware. Bake glassware and glass wool for 2 hours at 180°C before use. Rinse plasticware extensively with H_2O before use; do not autoclave.

Notes

- Large items of glassware can be siliconized by soaking or rinsing them in a 5% solution of dichlorodimethylsilane in chloroform or heptane. Commercial preparations for siliconizing are also available (e.g., Sigmacoat).

- As the organic solvent evaporates, the dichlorodimethylsilane is deposited on the glassware, which must be rinsed numerous times with H_2O or baked for 2 hours at 180°C before use.

Preparation of RNase-free Glassware

Guidelines for the treatment of glassware for use with RNA are given in the information panel How to Win the Battle with RNase in MC3, p. 7.82.

PREPARATION OF DIALYSIS TUBING

The separation of molecules across a semipermeable membrane is driven by the concentration differential between the solutions on either side of the membrane and is constrained by the size (molecular weight) of the molecules relative to the size of the pores within the membrane. The pore size determines the molecular-weight cutoff, defined as the molecular weight at which 90% of the solute will be retained by the membrane. The exact permeability of a solute is dependent not only on the size of the molecule, but also on the shape of the molecule, its degree of hydration, and its charge. Each of these parameters may be influenced by the nature of the solvent, its pH, and its ionic strength. As a consequence, the molecular-weight cutoff should be used as a guide and not an absolute predictor of performance with every type of solute and solvent. Dialysis membranes are available in an enormous range of pore sizes (from 100 daltons to 2000 kD). For dialysis of most plasmid DNAs and many proteins, a molecular-weight cutoff of 12,000 to 14,000 is suitable.

1. Cut the tubing into pieces of convenient length (10–20 cm).

2. Boil the tubing for 10 minutes in a large volume of 2% (w/v) sodium bicarbonate and 1 mM EDTA (pH 8.0).

3. Rinse the tubing thoroughly in distilled H_2O.

4. Boil the tubing for 10 minutes in 1 mM EDTA (pH 8.0).

5. Allow the tubing to cool and then store it at 4°C. Make sure that the tubing is always submerged.

   IMPORTANT: From this point onward, wear gloves when handling the tubing.

6. Before use, wash the tubing inside and out with distilled H_2O.

Note

- Instead of boiling for 10 minutes in 1 mM EDTA (pH 8.0) (Step 4), the tubing may be autoclaved at 20 psi (1.40 kg/cm²) for 10 minutes on liquid cycle in a loosely capped jar filled with H_2O.

STORAGE OF BACTERIAL CULTURES

Stab Cultures

To store a bacterial culture in solid medium, pick a single, well-isolated colony with a sterile inoculating needle and stab the needle several times through the agar to the bottom of a stab vial (for the preparation of stab vials, please see p. 720). Replace and tighten the cap; label both the vial and the cap. Store the vial in the dark at room temperature.

Cultures Containing Glycerol

Storage of Bacterial Cultures Growing in Liquid Media

1. To 1.5 ml of bacterial culture, add 0.5 ml of sterile 60% glycerol (sterilized by autoclaving for 20 minutes at 15 psi [1.05 kg/cm²] on liquid cycle).

2. Vortex the culture to ensure that the glycerol is evenly dispersed.

3. Transfer the culture to a labeled storage tube equipped with a screw cap and an air-tight gasket.

4. Freeze the culture in ethanol–dry ice or liquid nitrogen and then transfer the tube to –70°C for long-term storage.

5. To recover the bacteria, scrape the frozen surface of the culture with a sterile inoculating loop. Immediately streak the bacteria that adhere to the needle onto the surface of an LB agar plate containing the appropriate antibiotic. Return the frozen culture to storage at –70°C. Incubate the plate overnight at 37°C.

Storage of Bacterial Cultures Growing on Agar Plates

1. Scrape the bacteria growing on the surface of an agar plate into 2 ml of LB medium in a sterile tube. Add an equal volume of LB medium containing 30% sterile glycerol.

2. Vortex the mixture to ensure that the glycerol is completely dispersed.

3. Dispense aliquots of the glycerinated culture into sterile tubes equipped with screw caps and air-tight gaskets. Freeze the cultures as described above.

   This method is useful for storing copies of cDNA libraries established in plasmid vectors.

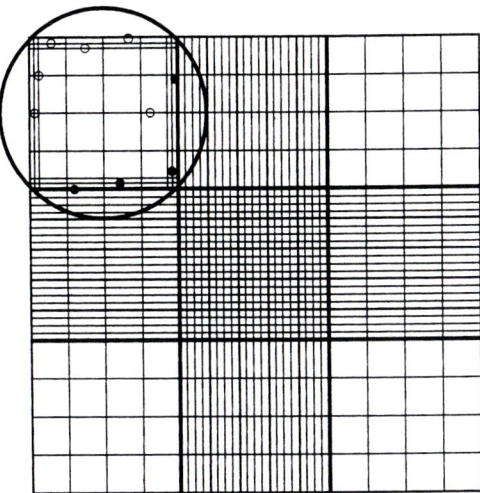

FIGURE A3-1 Standard Hemocytometer Chamber
The circle indicates the approximate area covered at 100x microscope magnification (10x ocular and 10x objective). Count the cells on top and left touching the middle line (*open circles*). Do not count the cells touching the middle line at bottom and right (*closed circles*). Count the four corner squares and the middle square in both chambers (only one chamber is represented here).

ESTIMATION OF CELL NUMBER*

The number of mammalian cells in a defined volume of medium can be measured using a hemocytometer. Automated methods using cell-counting devices such as those produced by Coulter are desirable when large numbers of individual samples are to be counted. A method to estimate the number of live cells in a population by staining with a vital dye is also provided here.

Hemocytometry Counting

A hemocytometer contains two chambers, each of which when filled and coverslipped contains a total volume of 9 μl. Each chamber is ruled into nine major squares and each square is 1 x 1 mm with a depth of 0.1 mm. Thus, when coverslipped, the volume of each square is 0.1 mm^3 or 0.1 μl. Additional subdivisions of the major nine squares are not necessary for counting and can be ignored. A representation of the marking on a hemocytometer is shown in Figure A3-1.

1. Trypsinize the cells (please see MC3, p. 17.63, Step 18) and resuspend them in growth medium.

2. Use pasteur pipettes to remove two independent samples from the cell suspension to be counted. Deliver each sample of cell suspension into one side of the coverslipped hemocytometer by capillary action.

    Fluid should just fill the chamber and not overflow into the troughs outside the counting face. Load the first sample into one chamber and the second sample into the second chamber.

3. Count the total number of cells in five of the nine large squares in each of two sides of the hemocytometer for a total of ten squares.

    The microscope field using a 10x objective and a 10x ocular should encompass the majority of one of the nine squares of the chamber and is a convenient magnification to use for counting. Cells that overlap the

---

EXAMPLE

One ml of a 10-ml suspension of cells is diluted with 4 ml of medium. The diluted suspension is then sampled with a Pasteur pipette twice. The first sample is delivered to one chamber of the hemocytometer. The second sample is delivered to the second side. Five squares are counted from each side of the hemocytometer.

Number of cells/square: 45, 37, 52, 40, 60, 48, 54, 70, 58, 60
Total count: 524
Dilution factor: (1 + 4)/1 = 5

Cells/ml (in original): $524 \times 10^3 \times 5 = 2.62 \times 10^6$/ml cells

---

*Adapted from Spector et al. 1998. *Cells: A laboratory manual.* Cold Spring Harbor Laboratory Press, Cold Spring Harbor, New York.

border on two sides of the square should be included in the cell count and not counted on the other two sides. If the initial dilution results in more than 50–100 cells/square, make an additional dilution to improve counting accuracy and speed the process of determining cell numbers.

4. Add the number of cells in a total of ten chambers (five from one side and five from the other) to give the number of cells in 1×10^{-3} ml (1×10^{-4} ml/square $\times$ 10 squares = a volume of 10^{-3} ml). Multiply the total number of cells by 1000 to give the number of cells/ml in the sample counted.

   If dilutions from the original cell suspension have been made, this factor must also be incorporated.

5. Immediately after use, clean the hemocytometer and coverslip by rinsing in distilled H_2O followed by 70% ethanol. Dry with lens paper.

   IMPORTANT: Do not allow the cell suspension to dry on the hemocytometer.

Notes

Errors that may result from using hemocytometer counts are the result of the following:

- *Variable sampling from the original cell suspension.* The cell suspension must be agitated; do not allow the cells to settle to the bottom of the container.
- *Inadequate or excessive filling of the hemocytometer chamber.* The volume in the chambers counted is based on the coverslip resting on the sides of the hemocytometer. Overflow increases the volume counted.
- *Cell clumping.* Large clumps of cells may be too large to enter the chamber through capillary action and will be excluded from the cell count. Small clumps that are able to enter the chamber are difficult to count with accuracy. It is important to have a monodisperse suspension of cells for accurate counting. The cells must be thoroughly mixed to achieve uniformity.

Viability Staining

Various manipulations of cells, including passaging, freezing, and dissociation from primary tissue, can result in cell death. Exclusion of the dye, Trypan Blue, can be used to determine the number of surviving cells in a population. Normal healthy cells are able to exclude the dye, but Trypan Blue diffuses into cells in which membrane integrity has been lost. The dye exclusion method is an approximate estimate of cell viability and often does not distinguish within a 10–20% difference. Additionally, cells that exclude dye are not necessarily capable of attachment and prolonged survival or proliferation.

1. Trypsinize the cells (please see MC3, p. 17.63, Step 8) and aseptically dilute 0.5 ml of cells into PBS to a concentration from 2×10^5 to 4×10^5 cells/ml.

2. Aseptically transfer 0.5 ml of the diluted cell suspension in PBS to a fresh tube and add 0.5 ml of a solution of Trypan Blue (0.4% w/v).

3. Allow the cells to remain in the dye solution for no less than 3 minutes and no longer than 10 minutes. Use a pasteur pipette to sample the cells in dye and deliver them to a hemocytometer by capillary action.

4. Count a total of at least 500 cells, keeping a separate count of blue cells. Determine the frequency of those that are blue, i.e., have not excluded the dye.

5. Determine the percent of viability from the number of cells that have not excluded the dye.

EXAMPLE

A monolayer culture is trypsinized and resuspended in 5 ml of medium; 0.5 ml of cells is mixed with 4.5 ml of PBS, and 0.5 ml of the suspended cells in PBS is transferred to a small tube and mixed with 0.5 ml of Trypan Blue solution. In the sample transferred to the hemocytometer, 540 cells are counted; 62 of the cells fail to exclude the dye and are blue. The percent of viability equals 88.5%.

540 – 62 = the number of cells that excluded the dye

540 = the total number of cells counted

$$\frac{540 - 62 \times 100}{540} = 88.5\% \text{ viability}$$

PURIFICATION OF NUCLEIC ACIDS

CAUTION: Please see Appendix 4 for appropriate handling of materials marked with <!>.

Perhaps the most basic of all procedures in molecular cloning is the purification of nucleic acids. The key step, the removal of proteins, can often be carried out simply by extracting aqueous solutions of nucleic acids with phenol:chloroform <!> and chloroform <!>. Such extractions are used whenever it is necessary to inactivate and remove enzymes that are used in one step of a cloning operation before proceeding to the next. However, additional measures are required when nucleic acids are purified from complex mixtures of molecules such as cell lysates. In these cases, it is usual to remove most of the protein by digestion with proteolytic enzymes such as pronase or proteinase K (please see MC3, p. A4.50, Table A4-8), which are active against a broad spectrum of native proteins, before extracting with organic solvents.

Extraction with Phenol:Chloroform

The standard way to remove proteins from nucleic acid solutions is to extract first with phenol:chloroform (optionally containing hydroxyquiniline at 0.1%) and then with chloroform. This procedure takes advantage of the fact that deproteinization is more efficient when two different organic solvents are used instead of one. Furthermore, although phenol denatures proteins efficiently, it does not completely inhibit RNase activity, and it is a solvent for RNA molecules that contain long tracts of poly(A). Both of these problems can be circumvented by using a mixture of phenol:chloroform:isoamyl alcohol (25:24:1). The subsequent extraction with chloroform removes any lingering traces of phenol from the nucleic acid preparation. Extraction with ether, which was widely used for this purpose for many years, is no longer required or recommended for routine purification of DNA.

1. Transfer the sample to a polypropylene tube and add an equal volume of phenol:chloroform.

    The nucleic acid will tend to partition into the organic phase if the phenol has not been adequately equilibrated to a pH of 7.8–8.0.

2. Mix the contents of the tube until an emulsion forms.

3. Centrifuge the mixture at 80% of the maximum speed that the tubes can bear for 1 minute at room temperature. If the organic and aqueous phases are not well separated, centrifuge again for a longer time.

    Normally, the aqueous phase forms the upper phase. However, if the aqueous phase is dense because of salt (>0.5 M) or sucrose (>10%), it will form the lower phase. The organic phase is easily identifiable because of the yellow color contributed by the 8-hydroxyquinoline that is added to phenol during equilibration.

4. Use a pipette to transfer the aqueous phase to a fresh tube. For small volumes (<200 µl), use an automatic pipettor fitted with a disposable tip. Discard the interface and organic phase.

    To achieve the best recovery, the organic phase and interface may be "back-extracted" as follows: After the first aqueous phase has been transferred as described above, add an equal volume of TE (pH 7.8) to the organic phase and interface. Mix well. Separate the phases by centrifugation as in Step 3. Combine this second aqueous phase with the first and proceed to Step 5.

5. Repeat Steps 1–4 until no protein is visible at the interface of the organic and aqueous phases.

6. Add an equal volume of chloroform and repeat Steps 2–4.

7. Recover the nucleic acid by standard precipitation with ethanol.

    Occasionally, ether <!> is used to remove traces of chloroform from preparations of high-molecular-weight DNA (please see the Notes below).

Notes

The organic and aqueous phases may be mixed by vortexing when isolating small DNA molecules (<10 kb) or by gentle shaking when isolating DNA molecules of moderate size (10–30 kb). When isolating large DNA molecules (>30 kb), the following precautions must be taken to avoid shearing (please see also MC3, Chapter 6).

- Mix the organic and aqueous phases by rotating the tube slowly (20 rpm) on a wheel.

- Use large-bore pipettes to transfer the DNA from one tube to another.

Drop Dialysis

Low-molecular-weight contaminants that may inhibit restriction digestion or DNA sequencing can be removed from DNA in solution by drop dialysis.

1. Spot a drop (~50 μl) of DNA in the center of a Millipore Series V membrane (0.025 μm), floating shiny side up on 10 ml of sterile H_2O in a 10-cm-diameter petri dish.

2. Dialyze the DNA for 10 minutes.

3. Remove the drop to a clean microfuge tube and use aliquots of the dialyzed DNA for restriction enzyme digestion and/or DNA sequencing.

CONCENTRATING NUCLEIC ACIDS

CAUTION: Please see Appendix 4 for appropriate handling of materials marked with <!>.

Ethanol Precipitation

Precipitation with ethanol is the standard method to recover nucleic acids from aqueous solutions. It is rapid, virtually foolproof, and efficient: Subnanogram amounts of DNA and RNA can be quantitatively precipitated with ethanol, collected by centrifugation, and redissolved within minutes.

Ethanol depletes the hydration shell from nucleic acids and exposes negatively charged phosphate groups. Counterions such as Na^+ bind to the charged groups and reduce the repulsive forces between the polynucleotide chains to the point at which a precipitate can form. Ethanol precipitation can therefore only occur if cations are available in sufficient quantity to neutralize the charge on the exposed phosphate residues. The most commonly used cations are shown in Table A3-1 and are described below.

- *Ammonium acetate* is frequently used to reduce the coprecipitation of unwanted contaminants (e.g., dNTPs or oligosaccharides) with nucleic acids. For example, two sequential precipitations of DNA in the presence of 2 M ammonium acetate result in the removal of >99% of dNTPs from preparations of DNA. Ammonium acetate is also the best choice when nucleic acids are precipitated after digestion of agarose gels with agarase. The use of this cation reduces the possibility of coprecipitation of oligosaccharide digestion products. However, ammonium acetate should not be used when the precipitated nucleic acid is used as a substrate for phosphorylation by bacteriophage T4 polynucleotide kinase, which is inhibited by ammonium ions.

- *Lithium chloride* is frequently used when high concentrations of ethanol are required for precipitation (e.g., when precipitating RNA). LiCl is very soluble in ethanolic solutions and is not coprecipitated with the nucleic acid. Small RNAs (tRNAs and 5S RNAs) are soluble in solutions of high ionic strength (without ethanol), whereas large RNAs are not. Because of this difference in solubility, precipitation in high concentrations of LiCl (0.8 M) can be used to purify large RNAs.

- *Sodium chloride* (0.2 M) should be used if the DNA sample contains SDS. The detergent remains soluble in 70% ethanol.

- *Sodium acetate* (0.3 M, pH 5.2) is used for most routine precipitations of DNA and RNA.

Until recently, ethanol precipitation was routinely performed at low temperature (e.g., in a dry-ice/methanol bath). This is now known to be unnecessary. At 0°C in the absence of carrier, DNA concentrations as low as 20 ng/ml will form a precipitate that can be quantitatively recovered by centrifugation in a microfuge. However, when lower concentrations of DNA or very small fragments (<100 nucleotides in length) are processed, more extensive centrifugation may be necessary to cause the pellet of nucleic acid to adhere tightly to the centrifuge tube. Centrifugation at 100,000g for 20–30 minutes allows the recovery of picogram quantities of nucleic acid in the absence of carrier.

TABLE A3-1 Salt Solutions

Salt	Stock Solution (M)	Final Concentration (M)
Ammonium acetate	10.0	2.0–2.5
Lithium chloride	8.0	0.8
Sodium chloride	5.0	0.2
Sodium acetate	3.0 (pH 5.2)	0.3

When dealing with small amounts of DNA, it is prudent to save the ethanolic supernatant from each step until all of the DNA has been recovered. This precaution is especially important after precipitates of DNA have been washed with 70% ethanol, a treatment that often loosens the precipitates from the wall of the tube.

Dissolving DNA Precipitates

Until recently, DNA precipitates recovered after ethanol precipitation were dried under vacuum before being redissolved. This practice has now been abandoned because (1) desiccated pellets of DNA dissolve slowly and inefficiently and (2) small fragments of double-stranded DNA (<400 bp) become denatured on drying, probably as a result of loss of the stabilizing shell of bound water molecules.

These days, the best practice is to remove ethanol from the nucleic acid pellet and from the sides of the tube by gentle aspiration and then to store the open tube on the bench for about 15 minutes to allow most of the residual ethanol to evaporate. The still-damp pellet of nucleic acid can then be dissolved rapidly and completely in the appropriate buffer. If necessary, the open tube containing the redissolved DNA can be incubated for 2–3 minutes at 45°C in a heating block to allow any traces of ethanol to evaporate.

The precipitated DNA is not all found at the bottom of the tube after centrifugation in an angle-head rotor. In the case of microfuge tubes, for example, at least 40% of the precipitated DNA is plastered on the wall of the tube. To maximize recovery of DNA, use a pipette tip to roll a bead of solvent several times over the appropriate segment of the wall. If the sample of DNA is radioactive, check that no detectable radioactivity remains in the tube after the dissolved DNA has been removed.

Carriers

Carriers (or coprecipitants) are inert substances that are used to improve the recovery of small quantities of nucleic acids during ethanol precipitation. Insoluble in ethanolic solutions, carriers form a precipitate that traps the target nucleic acids. During centrifugation, carriers generate a visible pellet that facilitates handling of the target nucleic acids. This may be their major virtue: As discussed above, ethanol precipitation—even of small amounts of nucleic acids in dilute solution—is remarkably efficient. Carriers do little, other than provide visual clues to the location of the target nucleic acid. Three substances are commonly used as carriers: yeast tRNA, glycogen, and linear polyacrylamide. Their advantages and disadvantages are listed in Table A3-2.

Standard Ethanol Precipitation of DNA in Microfuge Tubes

1. Estimate the volume of the DNA solution.

2. Adjust the concentration of monovalent cations either by dilution with TE (pH 8.0) if the DNA solution contains a high concentration of salts or by addition of one of the salt solutions shown in Table A3-1.

   If the volume of the final solution is 400 μl or less, perform precipitation in a single microfuge tube. Larger volumes can be divided among several microfuge tubes or the DNA can be precipitated and centrifuged in tubes that will fit in a medium-speed centrifuge or ultracentrifuge.

3. Mix the solution well. Add exactly 2 volumes of ice-cold ethanol and again mix the solution well. Store the ethanolic solution on ice to allow the precipitate of DNA to form.

   Usually 15–30 minutes is sufficient, but when the size of the DNA is small (<100 nucleotides) or when it is present in small amounts (<0.1 μg/ml), extend the period of storage to at least 1 hour and add MgCl$_2$ to a final concentration of 0.01 M.

   DNA can be stored indefinitely in ethanolic solutions in sealed tubes at 0°C or at –20°C.

TABLE A3-2 Carriers

Carrier	Working Concentration (μg/ml)	Advantages/Disadvantages
Yeast tRNA	10–20	Yeast tRNA is inexpensive, but it has the disadvantage that it cannot be used for precipitating nucleic acids that will be used as substrates in reactions catalyzed by polynucleotide kinase or terminal transferase. The termini of yeast RNA are excellent substrates for these enzymes and would compete with the termini contributed by the target nucleic acid.
Glycogen	50	Glycogen is usually used as a carrier when nucleic acids are precipitated with 0.5 M ammonium acetate and isopropanol. Glycogen is not a nucleic acid and therefore does not compete with the target nucleic acids in subsequent enzymatic reactions. However, it can interfere with interactions between DNA and proteins.
Linear polyacrylamide	10–20	Linear polyacrylamide is an efficient neutral carrier for precipitating picogram amounts of nucleic acids with ethanol and proteins with acetone.

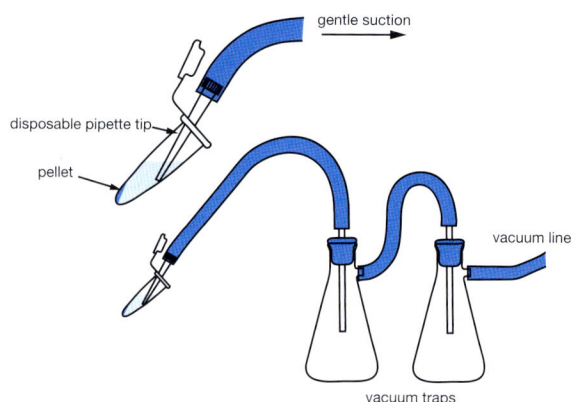

FIGURE A3-2 Aspiration of Supernatants

Hold the open microfuge tube at an angle, with the pellet on the upper side. Use a disposable pipette tip attached to a vacuum line to withdraw fluid from the tube. Insert the tip just beneath the miniscus on the lower side of the tube. Move the tip toward the base of the tube as the fluid is withdrawn. Use a gentle suction to avoid drawing the pellet into the pipette tip. Keep the end of the tip away from the pellet. Finally, vacuum the walls of the tube to remove any adherent drops of fluid.

4. Recover the DNA by centrifugation at 0°C.

> For most purposes, centrifugation at maximum speed for 10 minutes in a microfuge is sufficient. However, as discussed above, when low concentrations of DNA (<20 ng/ml) or very small fragments are being processed, more extensive centrifugation may be required.

5. Carefully remove the supernatant with an automatic micropipettor or a disposable pipette tip attached to a vacuum line (please see Figure A3-2). Take care not to disturb the pellet of nucleic acid (which may be invisible). Use the pipette tip to remove any drops of fluid that adhere to the walls of the tube.

> It is best to save the supernatant from valuable DNA samples until recovery of the precipitated DNA has been verified.

6. Fill the tube half way with 70% ethanol and recentrifuge at maximum speed for 2 minutes at 4°C in a microfuge.

7. Repeat Step 5.

8. Store the open tube on the bench at room temperature until the last traces of fluid have evaporated.

> It was once common practice to dry pellets of nucleic acid in a lyophilizer. This step is not only unnecessary, but also undesirable, because it causes denaturation of small (<400-nucleotide) fragments of DNA and greatly reduces the recovery of larger fragments of DNA.

9. Dissolve the DNA pellet (which is often invisible) in the desired volume of buffer (usually TE [pH between 7.6 and 8.0]). Rinse the walls of the tube well with the buffer.

Notes

- After centrifugation in a microfuge, not all of the DNA is deposited on the bottom of the microfuge tube. Up to 40% of the DNA is smeared on the wall of the tube. To recover all of the DNA, it is necessary to work a bead of fluid backward and forward over the appropriate quadrant of wall. This step can easily be done by pushing the bead of fluid over the surface with a disposable pipette tip attached to an automatic micropipettor.

- One volume of isopropanol <!> may be used in place of 2 volumes of ethanol to precipitate DNA. Precipitation with isopropanol has the advantage that the volume of liquid to be centrifuged is smaller. However, isopropanol is less volatile than ethanol and is therefore more difficult to remove; moreover, solutes such as sucrose or sodium chloride are more easily coprecipitated with DNA when isopropanol is used. In general, precipitation with ethanol is preferable, unless it is necessary to keep the volume of fluid to a minimum.

- In general, DNA precipitated from solution by ethanol can be redissolved easily in buffers of low ionic strength, such as TE (pH 8.0). Occasionally, difficulties arise when buffers containing $MgCl_2$ or >0.1 M NaCl are added directly to the DNA pellet. It is therefore preferable to dissolve the DNA in a small volume of low-ionic-strength buffer and to adjust the composition of the buffer later. If the sample does not dissolve easily in a small volume, add a larger volume of buffer and repeat the precipitation with ethanol. The second precipitation may help eliminate additional salts or other components that may be preventing dissolution of the DNA.

Precipitation of RNA with Ethanol

RNA is efficiently precipitated with 2.5–3.0 volumes of ethanol from solutions containing 0.8 M LiCl, 5 M ammonium acetate, or 0.3 M sodium acetate. The choice among these salts is determined by the way in which the RNA will be used later. Since the potassium salt of dodecyl sulfate is extremely insoluble, avoid potassium acetate if the precipitated RNA is to be dissolved in buffers that contain SDS, for example, buffers that are used for chromatography on oligo(dT)-cellulose. For the same reason, avoid potassium acetate if the RNA is already dissolved in a buffer containing SDS. Avoid LiCl when the RNA is to be used for cell-free translation or reverse transcription. LiCl ions inhibit initiation of protein synthesis in most cell-free systems and suppress the activity of RNA-dependent DNA polymerase.

Note

- Solutions used for precipitation of RNA must be free of RNase (please see MC3, p. 7.82).

Precipitation of Large RNAs with Lithium Chloride

Whereas small RNAs (tRNAs and 5S RNAs) are soluble in solutions of high ionic strength, large RNAs (e.g., rRNAs and mRNAs) are insoluble and can be removed by centrifugation.

1. Measure the volume of the sample and add 0.2 volume of RNase-free 8 M LiCl. Mix the solution well and store it on ice for at least 2 hours.

2. Centrifuge the solution at 15,000g for 20 minutes at 0°C. Discard the supernatant and dissolve the precipitated high-molecular-weight RNA in 0.2 volume of H_2O.

3. Repeat Steps 1 and 2.

4. Recover the high-molecular-weight RNA from the resuspended pellet by precipitation with 2 volumes of ethanol.

Concentrating and Desalting Nucleic Acids with Microconcentrators

Ultrafiltration is an alternative to ethanol precipitation for the concentration and desalting of nucleic acid solutions. It requires no phase change and is particularly useful for dealing with very low concentrations of nucleic acids. The Microcon cartridge, supplied by Millipore, is a centrifugal ultrafiltration device that can desalt and concentrate nucleic acid samples efficiently. The protocol presented below and the accompanying notes have been adapted from those provided on the Millipore Web site (www.millipore.com). Complete directions may be found on this Web site.

1. Select a Microcon unit with a nucleotide cutoff equal to or smaller than the molecular size of the nucleic acid of interest (please see Table A3-3).

TABLE A3-3 Nucleotide Cutoffs for Microcon Concentrators

Microcon Model	Color Code	Nucleotide Cutoff[a]		Maximum Recommended g Force	Spin Time in Minutes	
		ss	ds		4°C	25°C
3	yellow	10	10	14,000	185	95
10	green	30	20	14,000	50	35
30	clear	60	50	14,000	15	8
50	rose	125	100	14,000	10	6
100	blue	300	125	500	25	15

Note that ultrafiltration alone does not change buffer composition. The salt concentration in a sample concentrated by spinning in a Microcon will be the same as that in the original sample. For desalting, the concentrated sample is diluted with H_2O or buffer to its original volume and spun again (called discontinuous diafiltration). This removes the salt by the concentration factor of the ultrafiltration. For example, if a 500-μl sample containing 100 mM salt is concentrated to 25 μl (20x concentration factor), 95% of the total salt in the sample will be removed. The salt concentration in the sample will remain at 100 mM. Rediluting the sample to 500 μl in H_2O will bring the salt concentration to 5 mM. Concentrating to 25 μl once more will remove 99% of the original total salt. The concentrated sample will now be in 0.25 mM salt. For more complete salt removal, an additional redilution and spinning cycle will remove 99.9% of the initial salt content.

[a]ss indicates single stranded and ds indicates double stranded.

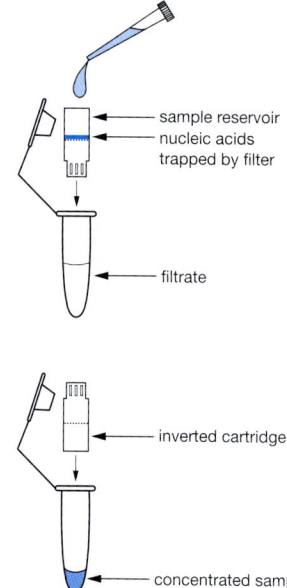

FIGURE A3-3 Concentration/Desalting of Nucleic Acid Solutions Using Micron Ultracentrifugation

2. Insert the Microcon cartridge into one of the two vials supplied, as shown in Figure A3-3.

3. To concentrate (without affecting salt concentration), pipette up to 500 μl of sample (DNA or RNA) into the reservoir. Centrifuge for the recommended time, not exceeding the *g* force shown in Table A3-3.

4. To exchange salt, add the proper amount of appropriate diluent to bring the concentrated sample to 500 μl. Centrifuge for the recommended time, not exceeding the *g* force shown in Table A3-3. To achieve a lower salt concentration, repeat the entire step as necessary. Please see the footnote below to Table A3-3.

    IMPORTANT: Do not overfill the filtrate vial.

5. Remove the reservoir from the vial and invert the reservoir into a new vial (save the filtrate until the sample has been analyzed).

6. Centrifuge at 500–1000*g* for 2 minutes in a microfuge to recover nucleic acid in the vial.

7. Remove reservoir. Cap the vial to store the sample.

Concentrating Nucleic Acids by Extraction with Butanol

During extraction of aqueous solutions with solvents such as secondary butyl alcohol (isobutanol) or *n*-butyl alcohol (*n*-butanol <!>), some of the water molecules are partitioned into the organic phase. By performing several cycles of extraction, the volume of a nucleic acid solution can be reduced significantly. This method of concentration is used to reduce the volume of dilute solutions to the point at which the nucleic acid can be recovered easily by precipitation with ethanol.

1. Measure the volume of the nucleic acid solution and add an equal volume of isobutanol. Mix the solution well by vortexing.

    Addition of too much isobutanol can result in removal of all of the H_2O and precipitation of the nucleic acid. If this happens, add H_2O to the organic phase until an aqueous phase (which should contain the nucleic acid) reappears.

2. Centrifuge the solution at maximum speed for 20 seconds at room temperature in a microfuge or at 1600*g* for 1 minute in a benchtop centrifuge. Use an automatic micropipettor to remove and discard the upper (isobutanol) phase.

3. Repeat Steps 1 and 2 until the desired volume of aqueous phase is achieved.

    Because isobutanol extraction does not remove salt, the salt concentration increases in proportion to the reduction in the volume of the solution. The nucleic acid can be transferred to the desired buffer by spun-column chromatography or by precipitation with ethanol.

QUANTITATION OF NUCLEIC ACIDS

CAUTION: Please see Appendix 4 for appropriate handling of materials marked with <!>.

Two types of methods are widely used to measure the amount of nucleic acid in a preparation. If the sample is pure (i.e., without significant amounts of contaminants such as proteins, phenol, agarose, or other nucleic acids), spectrophotometric measurement of the amount of UV irradiation absorbed by the bases is simple and accurate. If the amount of DNA or RNA is very small or if the sample contains significant quantities of impurities, the amount of nucleic acid can be estimated from the intensity of fluorescence emitted by ethidium bromide or Hoechst 33258. A summary of the methods commonly used to measure the concentrations of DNA in solution is listed in Table A3-4. A more detailed discussion of the methods follows the table.

TABLE A3-4 Measuring Nucleic Acid Concentrations

Method	Instrument	Comments
Absorbance at 260 nm	spectrophotometer	Useful only for highly purified preparations of nucleic acid, because it detects any compound that absorbs significantly at 260 nm, which includes, for example, DNA, RNA, EDTA, and phenol. The ratio of absorbance at 260 and 280 nm is often used as a test for contamination of a preparation of DNA and RNA with protein. Despite its popularity, this test is of questionable worth. Nucleic acids absorb so strongly at 260 nm that only a significant level of protein contamination will cause a significant change in the ratio of absorbance at the two wavelengths (please see the panel on Absorption Spectroscopy of Nucleic Acids on the following page).
		The specific absorption coefficients of both DNA and RNA are affected by the ionic strength and the pH of the solution. Accurate measurements of concentration can be made only when the pH is carefully controlled and the ionic strength of the solution is low.
		It is difficult to measure the absorbance of small volumes of solution and the method is reliable only over a fairly narrow range of concentrations (5 to 90 µg/ml).
Emission at 458 nm in the presence of Hoechst 33258	fluorometer	Hoechst 33258 is one of a class of *bis*-benzimidazole fluorescent dyes that bind nonintercalatively and with high specificity to double-stranded DNA. After binding, the fluorescent yield increases from 0.01 to 0.6; Hoechst 33258 is therefore a useful fluorochrome for fluorometric detection and quantitation of double-stranded DNA. Hoechst 33258 interacts preferentially with A/T-rich regions of the DNA helix, with the $\log_{10}$ of the intensity of fluorescence increasing in proportion to the A+T content of the DNA. The fluorescent yield of Hoechst 33258 is approximately threefold lower with single-stranded DNA.
		Fluorometry assays with Hoechst 33258 do not work at extremes of pH and are affected by both detergents and salts. Assays are therefore usually carried out in 0.2 M NaCl, 10 mM EDTA at pH 7.4. The concentration of DNA in the unknown sample is estimated from a standard curve constructed using a set of reference DNAs (10–250 ng/ml) whose base composition is the same as the unknown sample. The intensity of emission is nearly linear over a 1000-fold range of DNA concentrations.
		The DNAs must be of high molecular weight because Hoechst 33258 does not bind efficiently to small fragments of DNA. All DNAs and solutions must be free of ethidium bromide, which quenches the fluorescence of Hoechst 33258. However, because Hoechst 33258 has little affinity for proteins or rRNA, measurements can be performed using cell lysates or purified preparations of DNA.
Dipstick (a kit from Invitrogen)		This method is effective only for solutions containing low concentrations of DNA and RNA (<10 µg/ml), and is both expensive and relatively slow (30–40 minutes).
Ethidium bromide spot test	UV transilluminator	A fast and sensitive method that utilizes the UV-induced fluorescence emitted by intercalated ethidium bromide molecules. The DNA preparations under test are spotted onto an agarose plate containing 0.5 µg/ml ethidium bromide. A series of DNAs of known concentration are used as standards. Because the amount of fluorescence is proportional to the total mass of DNA, the quantity of DNA can be estimated by comparing the light emitted at 590 nm by the test preparations and the standards. The results of the assay can be recorded on film. In a similar, older test developed in the early 1970s, DNA samples and standards are spotted onto a sheet of Saran Wrap, mixed with a dilute solution of ethidium bromide, and photographed.
		The chief problem with the method is that it is sensitive to interference by RNA.

Spectrophotometry of DNA or RNA

For quantitating the amount of DNA or RNA, readings are taken at wavelengths of 260 and 280 nm. The reading at 260 nm allows calculation of the concentration of nucleic acid in the sample. An OD of 1 corresponds to approximately 50 µg/ml for double-stranded DNA, 40 µg/ml for single-stranded DNA and RNA, and 33 µg/ml for single-stranded oligonucleotides. The ratio between the readings at 260 and 280 nm (OD_{260}:OD_{280}) provides an estimate of the purity of the nucleic acid. Pure preparations of DNA and RNA have OD_{260}:OD_{280} values of 1.8 and 2.0, respectively. If there is significant contamination with protein or phenol, the OD_{260}:OD_{280} will be less than the values given above, and accurate quantitation of the amount of nucleic acid will not be possible.

Because it is rapid, simple, and nondestructive, absorption spectroscopy has long been the method of choice to measure the amount of DNA and RNA in concentrated pure solutions. However, absorption spectroscopy is comparatively insensitive and, with most laboratory spectrophotometers, nucleic acid concentrations of at least 1 µg/ml are required to obtain reliable estimates of A_{260}. In addition, absorption spectroscopy cannot readily distinguish between DNA and RNA, and it cannot be used with crude preparations of nucleic acids. Because of these limitations, a number of alternative methods have been devised to measure the concentration of DNA and RNA (please see Table A3-4).

ABSORPTION SPECTROSCOPY OF NUCLEIC ACIDS

Purines and pyrimidines in nucleic acids absorb UV light. As described by the Beer-Lambert Law, the amount of energy absorbed at a particular wavelength is a function of the concentration of the absorbing material:

$$I = I_o \, 10^{-\varepsilon dc}$$

where I = intensity of transmitted light
I_o = intensity of incident light
ε = molar extinction coefficient (also known as the molar absorption coefficient)
d = optical path length (in cm)
c = concentration of absorbing material (mole/liter)

ε is numerically equal to the absorbance of a 1 M solution in a 1-cm light path and is therefore expressed in M^{-1} cm^{-1}. Absorbance data are collected using a UV spectrometer and are generally reported as the absorbance A (log I/I_o). When $D = 1$ cm, A is called the optical density or OD at a particular wavelength, λ.

$$OD_\lambda = \varepsilon \, c$$

Because the absorption spectra of DNA and RNA are maximal at 260 nm, absorbance data for nucleic acids are almost always expressed in A_{260} or OD_{260} units. For double-stranded DNA, one A_{260} or OD_{260} unit corresponds to a concentration of 50 µg/ml. The Beer-Lambert law is valid at least to an OD = 2 and the concentration of a solution of nucleic acid is therefore easily calculated by simple interpolation. For example, a solution whose $OD_{260} = 0.66$ contains 33 µg/ml of double-stranded DNA. For nucleic acids, ε decreases as the ring systems of adjacent purines and pyrimidines become stacked in a polynucleotide chain. The value of ε therefore decreases in the following series:

free base
↓
small oligonucleotides
↓
single-stranded nucleic acids
↓
double-stranded nucleic acids

This means that single-stranded nucleic acids have a higher absorbance at 260 nm than double-stranded nucleic acids. Thus, the molar extinction coefficient of double-stranded DNA at 260 nm is 6.6, whereas the molar extinction coefficient of single-stranded DNA and RNA is approximately 7.4. Note, however, that the extinction coefficients of both DNA and RNA are affected by the ionic strength and the pH of the solution. Accurate measurements of concentration can be made only when the pH is carefully controlled and the ionic strength of the solution is low.

(Continued on facing page.)

The extinction coefficients of nucleic acids are the sum of the extinction coefficients of each of their constituent nucleotides. For large molecules, where it is both impractical and unnecessary to sum the coefficients of all the nucleotides, an average extinction coefficient is used. For double-stranded DNA, the average extinction coefficient is 50 $(\mu g/ml)^{-1}$ cm^{-1}; for single-stranded DNA or RNA, the average coefficient is 38 $(\mu g/ml)^{-1}$ cm^{-1}. These values mean that

$$1 \ OD_{260} \ unit =$$

50 $\mu g/ml$ double-stranded DNA

or

38 $\mu g/ml$ single-stranded DNA or RNA

For small molecules such as oligonucleotides, it is best to calculate an accurate extinction coefficient from the base composition. Because the concentrations of oligonucleotides are commonly reported as mmole/liter, a millimolar extinction coefficient (E) is conventionally used in the Beer-Lambert equation

$$E = A \ (15.3) + G \ (11.9) + C \ (7.4) + T \ (9.3)$$

where A, G, C, and T are the number of times each nucleotide is represented in the sequence of the oligonucleotide. The numbers in parentheses are the molar extinction coefficients for each deoxynucleotide at pH 7.0.

OD$_{260}$:OD$_{280}$ Ratios

Although it is possible to estimate the concentration of solutions of nucleic acids and oligonucleotides by measuring their absorption at a single wavelength (260 nm), this is not good practice. The absorbance of the sample should be measured at several wavelengths because the ratio of absorbance at 260 nm to the absorbance at other wavelengths is a good indicator of the purity of the preparation. Significant absorption at 230 nm indicates contamination by phenolate ions, thiocyanates, and other organic compounds, whereas absorption at higher wavelengths (330 nm and higher) is usually caused by light scattering and indicates the presence of particulate matter. Absorption at 280 nm indicates the presence of protein, because aromatic amino acids absorb strongly at 280 nm.

For many years, the ratio of the absorbance at 260 and 280 nm (OD_{260}:OD_{280}) has been used as a measure of purity of isolated nucleic acids. This method dates from Warburg and Christian who showed that the ratio is a good indicator of contamination of protein preparations by nucleic acids. The reverse is not true! Because the extinction coefficients of nucleic acids at 260 and 280 nm are so much greater than that of proteins, significant contamination with protein will not greatly change the OD_{260}:OD_{280} ratio of a nucleic acid solution (please see Table A3-5). Nucleic acids absorb so strongly at 260 nm that only a significant level of protein contamination will cause a significant change in the ratio of absorbance at the two wavelengths.

TABLE A3-5 Absorbance of Nucleic Acids and Proteins

% Protein	% Nucleic Acid	OD$_{260}$:OD$_{280}$	% Protein	% Nucleic Acid	OD$_{260}$:OD$_{280}$
100	0	0.57	45	55	1.89
95	5	1.06	40	60	1.91
90	10	1.32	35	65	1.93
85	15	1.48	30	70	1.94
80	20	1.59	25	75	1.95
75	25	1.67	20	80	1.97
70	30	1.73	15	85	1.98
65	35	1.78	10	90	1.98
60	40	1.81	5	95	1.99
55	45	1.84	0	100	2.00
50	50	1.87			

Using the predicted values in this table, Glasel derived an empirical equation to describe %N for a range of OD_{260}:OD_{280} ratios: %$N = F([11.16R - 6.32],[2.16 - R])$, where $R = OD_{260}$:OD_{280}. Note that estimates of purity of nucleic acids based on OD_{260}:OD_{280} ratios are accurate only when the preparations are free of phenol. Water saturated with phenol absorbs with a characteristic peak at 270 nm and an OD_{260}:OD_{280} ratio of 2. Nucleic acid preparations free of phenol should have OD_{260}:OD_{270} ratios of approximately 1:2.

Fluorometric Quantitation of DNA Using Hoechst 33258

Measuring the concentration of DNA using fluorometry is simple and more sensitive than spectrophotometry, and allows the detection of nanogram quantities of DNA. The assay can only be used to measure the concentration of DNAs whose sizes exceed approximately 1 kb, because Hoechst 33258 binds poorly to smaller DNA fragments. In this assay, DNA preparations of known and unknown concentrations are incubated with Hoechst 33258 fluorochrome. Absorption values for the unknown sample are compared with those observed for the known series, and the concentration of the unknown sample is estimated by interpolation.

HOECHST 33258

Hoechst 33258 is one of a class of *bis*-benzimidazole fluorescent dyes that bind nonintercalatively and with high specificity into the minor groove of double-stranded DNA. After binding, the fluorescent yield increases from 0.01 to 0.6, and Hoechst 33258 can therefore be used for fluorometric detection and quantification of double-stranded DNA in solution. Hoechst 33258 is preferred to ethidium bromide for this purpose because of its greater ability to differentiate double-stranded DNA from RNA and single-stranded DNA.

Like many other nonintercalative dyes, Hoechst 33258 binds preferentially to A/T-rich regions of the DNA helix, with the $\log_{10}$ of the intensity of fluorescence increasing in proportion to the A+T content of the DNA. The fluorescent yield of Hoechst 33258 is approximately threefold lower with single-stranded DNA.

Facts and Figures

- Hoechst 33258 in free solution has an excitation maximum at approximately 356 nm and an emission maximum at 492 nm. However, when bound to DNA, Hoechst 33258 absorbs maximally at 365 nm and emits maximally at 458 nm.

- Fluorometry assays with Hoechst 33258 do not work at extremes of pH and are affected by both detergents and salts. Assays are therefore usually carried out under standard conditions (0.2 M NaCl, 10 mM EDTA at pH 7.4). However, two different salt concentrations are required to distinguish double-stranded from single-stranded DNA and RNA. The concentration of DNA in the unknown sample is estimated from a standard curve constructed using a set of reference DNAs (10–250 ng/ml) whose base composition is the same as the unknown sample. The DNAs must be of high molecular weight because Hoechst 33258 does not bind efficiently to small fragments of DNA. Measurements should be performed rapidly to minimize photobleaching and shifts in fluorescence emission because of changes in temperature. Either a fixed wavelength fluorometer (e.g., Hoefer) or a scanning fluorescence spectrometer (e.g., PerkinElmer) can be used.

- The concentration of Hoechst 33258 ($M_r = 533.9$) in the reaction should be kept low (5×10^{-7} to 2.5×10^{-6} M), because quenching of fluorescence occurs when the ratio of dye to DNA is high. However, two concentrations of dye are sometimes used to extend the dynamic range of the assay.

- All DNAs and solutions should be free of ethidium bromide, which quenches the fluorescence of Hoechst 33258. However, because Hoechst 33258 has little affinity for proteins or rRNA, measurements can be performed using cell lysates or purified preparations of DNA.

- Unlike ethidium bromide, Hoechst dyes are cell-permeant.

---

1. Turn on the fluorometer 1 hour before the assay is performed to allow the machine to warm up and stabilize.

    When bound to high-molecular-weight double-stranded DNA, Hoechst 33258 dye absorbs maximally at 365 nm and emits maximally at 458 nm.

2. Prepare an appropriate amount of diluted Hoechst 33258 dye solution by combining 50 µl of concentrated Hoechst 33258 dye solution per 100 ml of fluorometry buffer (please see Appendix 1). Each tube in the DNA assay requires 3 ml of diluted Hoechst 33258 dye solution. Transfer 3 ml of diluted dye solution to an appropriate number of clean glass tubes. Include six extra tubes for a blank and the standard curve.

    The concentrated Hoechst 33258 dye solution is prepared in H_2O at 0.2 mg/ml and can be stored at room temperature in a foil-wrapped test tube.

3. Prepare a standard curve by adding 100, 200, 300, 400, and 500 ng of DNA from the reference stock solution to individual tubes. Mix and read the absorbance on the prewarmed fluorometer of each tube immediately after addition of the DNA.

   The reference stock solution of DNA is prepared in TE to a concentration of 100 μg/ml. Because the binding of Hoechst 33258 dye to DNA is influenced by the base composition, the DNA used to construct the standard curve should be from the same species as the test sample.

4. Add 0.1 (i.e., 1 μl of a 1:10 dilution), 1.0, and 10 μl of the preparation of DNA, whose concentration is being determined, to individual tubes containing diluted dye solution. Immediately read the fluorescence.

5. Construct a standard curve plotting fluorescence on the ordinate and weight of reference DNA (in nanograms) on the abscissa. Estimate the concentration of DNA in the unknown sample by interpolation.

   If the reading for the unknown DNA solution falls outside that of the standard curve, read the fluorescence of a larger sample or make an appropriate dilution of the sample and repeat the assay.

Notes

- Binding of Hoechst 33258 is adversely influenced by pH extremes, the presence of detergents near or above their critical micelle concentrations, and salt concentrations above 3 M. If these conditions or reagents are used to prepare the DNA and improbable results are obtained in the fluorometry assay, precipitate an aliquot of the DNA with ethanol, rinse the pellet of nucleic acid in 70% ethanol, dissolve the dried pellet in TE, and repeat the assay.

- If the preparation of test DNA is highly viscous, sampling with standard yellow tips may be so inaccurate that the dilutions of unknown DNA will not track with the standard curve. In this case, the best solution is to withdraw two samples (10–20 μl) with an automatic pipettor equipped with a cutoff yellow tip. Each sample is then diluted with approximately 0.5 ml of TE (pH 8.0) and vortexed vigorously for 1–2 minutes. Different amounts of the diluted samples can then be transferred to the individual tubes containing diluted dye solution. The results obtained from the two sets of samples should be consistent.

- Use scissors or a dog nail clipper (e.g., Fisher-Scientific) to generate cutoff yellow tips. Alternatively, the tips can be cut with a sharp razor blade. Sterilize the cutoff tips before use, either by autoclaving or by immersion in 70% alcohol for 2 minutes followed by air drying. Presterilized, purpose-made wide-bore tips can be purchased from a number of commercial companies (e.g., Bio-Rad).

Quantitation of Double-stranded DNA Using Ethidium Bromide

Sometimes there is not enough DNA (<250 ng/ml) to assay spectrophotometrically, or the DNA may be heavily contaminated with other substances that absorb UV irradiation and therefore impede accurate analysis. A rapid way to estimate the amount of DNA in such samples is to use the UV-induced fluorescence emitted by ethidium bromide <!> molecules intercalated into the DNA. Because the amount of fluorescence is proportional to the total mass of DNA, the quantity of DNA in the sample can be estimated by comparing the fluorescent yield of the sample with that of a series of standards. As little as 1–5 ng of DNA can be detected by this method. For more information on ethidium bromide, please see MC3, pp. 5.14 and A9.3.

Saran Wrap Method Using Ethidium Bromide or SYBR Gold

1. Stretch a sheet of Saran Wrap over a UV transilluminator or over a sheet of black paper.

2. Spot 1–5 μl of the DNA sample onto the Saran Wrap.

3. Spot equal volumes of a series of DNA concentration standards (0.1, 2.5, 5, 10, and 20 μg/ml) in an ordered array on the Saran Wrap.

   The standard DNA solutions should contain a single species of DNA approximately the same size as the expected size of the unknown DNA. The DNA standards are stable for many months when stored at –20°C.

4. Add to each spot an equal volume of TE (pH 7.6) containing 2 μg/ml ethidium bromide or an equal volume of a 1:5000 dilution of dimethylsulfoxide <!> (DMSO)/SYBR Gold <!> stock. Mix by pipetting up and down with a micropipette.

5. Photograph the spots using short-wavelength UV illumination for ethidium bromide or 300-nm transillumination for SYBR Gold (please see MC3, p. 5.15). Estimate the concentration of DNA by comparing the intensity of fluorescence in the sample with that of the standard solutions.

Agarose Plate Method

Contaminants that may be present in the DNA sample can either contribute to or quench the fluorescence. To avoid these problems, the DNA samples and standards can be spotted onto the surface of a 1% agarose slab gel containing ethidium bromide (0.5 µg/ml). Allow the gel to stand at room temperature for a few hours so that small contaminating molecules have a chance to diffuse away. Photograph the gel as described in MC3, p. 5.16.

Minigel Method

Electrophoresis through minigels (please see MC3, p. 5.13) provides a rapid and convenient way to measure the quantity of DNA and to analyze its physical state at the same time. This is the method of choice if there is a possibility that the samples may contain significant quantities of RNA.

1. Mix 2 µl of the DNA sample with 0.4 µl of gel-loading buffer IV (bromophenol blue only; please see Table A1-6 in Appendix 1) and load the solution into a slot in a 0.8% agarose minigel containing ethidium bromide (0.5 µg/ml).

   SYBR Gold is too expensive to use routinely in this technique.

2. Mix 2 µl of each of a series of standard DNA solutions (0, 2.5, 5, 10, 20, 30, 40, and 50 µg/ml) with 0.4 µl of gel-loading buffer IV. Load the samples into the wells of the gel.

   The standard DNA solutions should contain a single species of DNA approximately the same size as the expected size of the unknown DNA. The DNA standards remain stable for many months when stored at –20°C.

3. Perform electrophoresis until the bromophenol blue has migrated about 1–2 cm.

4. Destain the gel by immersing it for 5 minutes in electrophoresis buffer containing 0.01 M MgCl$_2$.

5. Photograph the gel using short-wavelength UV irradiation (please see MC3, p. 5.16). Compare the intensity of fluorescence of the unknown DNA with that of the DNA standards and estimate the quantity of DNA in the sample.

MEASUREMENT OF RADIOACTIVITY IN NUCLEIC ACIDS

CAUTION: Please see Appendix 4 for appropriate handling of materials marked with <!>.

Radioactive isotopes <!> are used as tracers to monitor the progress of many reactions used to synthesize DNA and RNA. To calculate the efficiency of such reactions, it is necessary to accurately measure the proportion of the radioactive precursor incorporated into the desired product. This goal can be achieved by two methods: (1) differential precipitation of the nucleic acid products with trichloroacetic acid <!> (TCA) and (2) differential adsorption of the products onto positively charged surfaces (e.g., DE81 paper).

Precipitation of Nucleic Acids with Trichloroacetic Acid

1. Use a soft-lead pencil to label the appropriate number of Whatman GF/C glass fiber filters (2.4-cm diameter). Impale each of the filters on a pin stuck into a polystyrene support.

2. Spot an accurately known volume (up to 5 µl) of each sample to be assayed on the center of each of two labeled filters.

   One of the filters is used to measure the total amount of radioactivity in the reaction (i.e., acid-soluble and acid-precipitable radioactivity). The other filter is used to measure only the acid-precipitable radioactivity. Under the conditions described, DNA and RNA molecules >50 nucleotides long will be precipitated on the surface of the filter.

3. Store the filters at room temperature until all of the fluid has evaporated. This process can be accelerated by using a heat lamp, although this is not usually necessary.

4. Use blunt-end forceps (e.g., Millipore) to transfer one of each pair of filters to a beaker containing 200–300 ml of ice-cold 5% TCA and 20 mM sodium pyrophosphate. Swirl the filters in the acid solution for 2 minutes and then transfer them to a fresh beaker containing the same volume of the ice-cold 5% TCA/20 mM sodium pyrophosphate mixture. Repeat the washing twice more.

During washing, the unincorporated nucleotide precursors are eluted from the filters and the radioactive nucleic acids are fixed to them.

Commercially available, vacuum-driven filtration manifolds that hold up to 24 filters may also be used to wash the filters.

5. Transfer the washed filters to a beaker containing 70% ethanol and allow them to remain there briefly. Dry the filters either at room temperature or under a heat lamp.

6. Insert each of the filters (washed and unwashed) into a scintillation vial. Measure the amount of radioactivity on each filter.

    ^{32}P can be detected on dry filters by Cerenkov counting (in the ^3H channel of a liquid scintillation counter). The efficiency with which Cerenkov radiation can be measured varies from instrument to instrument and also depends on the geometry of the scintillation vials and the amount of H_2O remaining in the filters. With dry filters, the efficiency of Cerenkov counting is about 25% (one radioactive decay in four can be detected). Alternatively, ^{32}P can be measured with 100% efficiency by adding a few milliliters of toluene-based scintillation fluid to the dried filters and counting in the ^{32}P channel of the liquid scintillation counter.

    To measure other isotopes (^3H, ^{14}C, ^{35}S, ^{33}P, etc.), it is essential to use toluene-based scintillation fluid and the appropriate channel of a liquid scintillation counter. The efficiency of counting these isotopes varies from counter to counter and should be determined for each instrument.

7. Compare the amount of radioactivity on the unwashed filter with the amount on the washed filter and then calculate the proportion of the precursor that has been incorporated as described in the panel below.

Adsorption to DE81 Filters

DE81 filters are positively charged and strongly adsorb and retain nucleic acids, including oligonucleotides that are too small to be precipitated efficiently with TCA. Unincorporated nucleotides stick less tightly to the filters and can be removed by washing the filter extensively in sodium phosphate. The procedure is essentially identical to that described for precipitation of nucleic acids by TCA, except that the DE81 filters are washed in 0.5 M Na_2HPO_4 (pH 7.0) instead of TCA/sodium pyrophosphate.

CALCULATION OF THE SPECIFIC ACTIVITY OF A RADIOLABELED PROBE

To calculate the specific activity of a radiolabeled probe in dpm/µg, use the following equation:

$$\text{specific activity} = \frac{L\,(2.2 \times 10^9)\,(PI)}{m + [(1.3 \times 10^3)\,(PI)\,(L/S)]}$$

where

- L = input radioactive label (µCi)
- PI = proportion of the precursor that has been incorporated (cpm in washed filter/cpm in unwashed filter; please see above)
- m = mass of template DNA (ng)
- S = specific activity of input label (µCi/nmole)

The numerator of this equation is the product of three terms: the total dpm in the reaction [$(L)(2.2 \times 10^6$ dpm/µCi)], the proportion of these dpm that were incorporated (PI), and a factor to convert the final value for specific activity from dpm/ng to dpm/µg (10^3).

The denominator represents the total mass of DNA (in nanograms) at the end of the reaction, equal to the starting mass (m) plus the mass (in nanograms) synthesized during the reaction. The latter is calculated from the number of nanomoles of dNMP incorporated [$(PI)(L/S)$] multiplied by four times the average molecular mass of the four dNMPs (4×325 ng/nmole = 1.3×10^3 ng/nmole).

EXAMPLE

In a random priming reaction in which 50% of the radioactivity has been incorporated into TCA-precipitable material, from a starting reaction containing 25 ng of template DNA and 50 µCi of radiolabeled dNTP with a specific activity of 3000 Ci/mmole: L = 50 µCi, PI = 0.5, m = 25 ng, and S = 3000 Ci/mmole.

$$\text{specific activity of the probe} = \frac{50\,(2.2 \times 10^9)(0.5)}{25 + [(1.3 \times 10^3)(0.5)(50/3000)]}$$

$$= 1.5 \times 10^9 \text{ dpm/µg}$$

DECONTAMINATION OF SOLUTIONS CONTAINING ETHIDIUM BROMIDE

CAUTION: Please see Appendix 4 for appropriate handling of materials marked with <!>.

Removing Ethidium Bromide from DNA

The reaction between ethidium bromide <!> and DNA is reversible, but the dissociation of the complex is very slow and is measured in days rather than minutes or hours. For practical purposes, dissociation is achieved by passing the complex through a small column packed with a cation-exchange resin such as Dowex AG 50W-X8 or by extracting with organic solvents such as isopropanol or *n*-butanol <!>. The former method has been shown to result in the removal of ethidium bromide to a binding ratio below that detectable by fluorescence, a molar ratio of dye:nucleic acid of 1:4000.

Disposing of Ethidium Bromide

Ethidium bromide itself is not highly mutagenic, but it is metabolized by microsomal enzymes to compound(s) that are moderately mutagenic in yeast and *Salmonella typhimurium*. A number of methods have been described to decontaminate solutions and surfaces that contain or have been exposed to ethidium bromide. The concentration of ethidium bromide in solution may be reduced to <0.5 μg/ml with activated charcoal, which can then be incinerated. Alternatively, ethidium bromide can be degraded by treatment with sodium nitrite <!> and hypophosphorous acid <!>.

Decontamination of Concentrated Solutions of Ethidium Bromide (solutions containing >0.5 mg/ml)

Method 1

This method of Lunn and Sansone reduces the mutagenic activity of ethidium bromide in the *Salmonella/* microsome assay by approximately 200-fold.

1. Add sufficient H_2O to reduce the concentration of ethidium bromide to <0.5 mg/ml.

2. To the resulting solution, add 0.2 volume of fresh 5% hypophosphorous acid and 0.12 volume of fresh 0.5 M sodium nitrite. Mix carefully.

    IMPORTANT: Make sure that the pH of the solution is <3.0.

    Hypophosphorous acid is usually supplied as a 50% solution, which is corrosive and must be handled with care. Freshly dilute the acid immediately before use.

    Sodium nitrite solution (0.5 M) should be freshly prepared by dissolving 34.5 g of sodium nitrite in H_2O to a final volume of 500 ml.

3. After incubation for 24 hours at room temperature, add a large excess of 1 M sodium bicarbonate. The solution may now be discarded.

Method 2

This method of Quillardet and Hofnung reduces the mutagenic activity of ethidium bromide in the *Salmonella*/microsome assay by approximately 3000-fold. However, there are reports of mutagenic activity in occasional batches of "blanks" treated with the decontaminating solutions.

1. Add sufficient H_2O to reduce the concentration of ethidium bromide to <0.5 mg/ml.

2. Add 1 volume of 0.5 M $KMnO_4$ <!>. Mix carefully and add 1 volume of 2.5 N HCl. Mix carefully and allow the solution to stand at room temperature for several hours.

3. Add 1 volume of 2.5 N NaOH <!>. Mix carefully and then discard the solution.

Decontamination of Dilute Solutions of Ethidium Bromide (e.g., electrophoresis buffer containing 0.5 μg/ml ethidium bromide)

Method 1

The following method is from Lunn and Sansone.

1. Add 2.9 g of Amberlite XAD-16 (Sigma-Aldrich) for each 100 ml of solution. Amberlite XAD-16 is a nonionic, polymeric absorbent.

2. Store the solution for 12 hours at room temperature, shaking it intermittently.

3. Filter the solution through a Whatman No. 1 filter and discard the filtrate.

4. Seal the filter and Amberlite resin in a plastic bag; dispose of the bag in a hazardous waste container.

Method 2

The following method is from Bensaude.

1. Add 100 mg of powdered activated charcoal for each 100 ml of solution.

2. Store the solution for 1 hour at room temperature, shaking it intermittently.

3. Filter the solution through a Whatman No. 1 filter and discard the filtrate.

4. Seal the filter and activated charcoal in a plastic bag; dispose of the bag in a hazardous waste container.

Notes

- Treatment of dilute solutions of ethidium bromide with hypochlorite (bleach) is not recommended as a method of decontamination. Such treatment reduces the mutagenic activity of ethidium bromide in the *Salmonella*/microsome assay by approximately 1000-fold, but it converts the dye into a compound that is mutagenic in the absence of microsomes.

- Ethidium bromide decomposes at 262°C and is unlikely to be hazardous after incineration under standard conditions.

- Slurries of Amberlite XAD-16 or activated charcoal can be used to decontaminate surfaces that become contaminated by ethidium bromide.

Commercial Decontamination Kits

Several commercial companies sell devices to extract ethidium bromide from solutions with a minimum of fuss and bother, e.g., EtBr GreenBag (Q•BIOgene).

GEL-FILTRATION CHROMATOGRAPHY

CAUTION: Please see Appendix 4 for appropriate handling of materials marked with <!>.

This technique, which uses gel filtration to separate high-molecular-weight DNA from smaller molecules, is used most often to separate unincorporated labeled dNTPs from DNA that has been radiolabeled. However, it is also used at several stages during the synthesis of double-stranded cDNA, during addition of linkers to blunt-ended DNA, to remove oligonucleotide primers from polymerase chain reaction (PCR), and, in general, whenever it is necessary to change the composition of the buffer in which DNA is dissolved.

Two methods are available: conventional column chromatography, which is used when it is necessary to collect fractions that contain components of different sizes, and centrifugation through gel matrices packed in disposable syringes, which is a rapid method used to free DNA from smaller molecules. The two most commonly used gel matrices are Sephadex and Bio-Gel, both of which are available in several porosities. Sephadex G-50 and Bio-Gel P-60 are ideal for purifying DNA larger than 80 nucleotides in length. Smaller molecules are retained in the pores of the gel, whereas the larger DNA is excluded and passes directly through the column. Bio-Gel P-2 can be used to separate oligonucleotides from phosphate ions or dNTPs. Bio-Gel is supplied in the form of a gel and need only be equilibrated in running buffer before use. Sephadex is supplied as a powder that must be hydrated before use.

Preparation of Sephadex

1. Slowly add Sephadex of the desired grade to distilled sterile H_2O in a 500-ml beaker or bottle (10 g of Sephadex G-50 granules yields 160 ml of slurry). Wash the swollen resin with distilled sterile H_2O several times to remove soluble dextran, which can create problems by precipitating during ethanol precipitation.

2. Equilibrate the resin in TE (pH 7.6), autoclave at 10 psi (0.70 kg/cm²) for 15 minutes, and store at room temperature.

Column Chromatography

1. Prepare Sephadex or Bio-Gel columns in disposable 5-ml borosilicate glass pipettes or pasteur pipettes plugged with a small amount of sterile glass wool. Use a long, narrow pipette (e.g., a disposable 1-ml plastic pipette) to push the wool to the bottom of the glass or pasteur pipette.

2. Use a pasteur pipette to fill the column with a slurry of the Sephadex or Bio-Gel, taking care to avoid producing bubbles. There is no need to close the bottom of the column. Continue to add gel until it packs to a level 1 cm below the top of the column. Wash the gel with several volumes of 1x TEN buffer (pH 8.0) (please see p. 709).

3. Apply the DNA sample (in a volume of 200 µl or less) to the top of the gel. Wash out the sample tube with ~100 µl of 1x TEN buffer and load the washing on the column as soon as the DNA sample has entered the gel. When the washing has entered the gel, immediately fill the column with 1x TEN buffer.

   > WARNING: Columns used to separate radiolabeled DNA from radioactive precursors should be run behind Lucite screens to protect against exposure to radioactivity.

4. Immediately start to collect fractions (~200 µl) in microfuge tubes.

   > If the DNA is labeled with ^{32}P <!>, measure the radioactivity in each of the tubes by using a handheld minimonitor or by Cerenkov counting in a liquid scintillation counter. Add more 1x TEN buffer to the top of the gel as required from time to time.

   > The DNA will be excluded from the gel and will be found in the void volume (usually 30% of the total column volume). The leading peak of radioactivity therefore consists of nucleotides incorporated into DNA, and the trailing peak consists of unincorporated [^{32}P]dNTPs.

5. Pool the radioactive fractions in the leading peak and store at –20°C.

Notes

Instead of collecting individual fractions, it is possible with practice to follow the progress of the incorporated and unincorporated [^{32}P]dNTPs down the column using a handheld minimonitor.

- Collect the leading peak into a sterile polypropylene tube as it elutes from the column.
- Clamp the bottom of the column and disconnect the buffer reservoir.
- Discard the column in the radioactive waste.

Spun-column Chromatography

This method is used to separate DNA, which passes through the gel-filtration matrix, from lower-molecular-weight substances that are retained on the column. Spun-column chromatography is particularly useful when separating labeled DNA from radioactive precursors. However, it is also used extensively for other purposes, for example, to remove unwanted nucleotide primers or double-stranded linkers, to change the buffer in which small amounts of DNA are dissolved, or to free crude preparations of minipreparations of plasmid or bacteriophage DNA from inhibitors that prevent cleavage by restriction enzymes. Several samples of DNA can be handled simultaneously. In this respect, spun-column chromatography is much superior to conventional column chromatography.

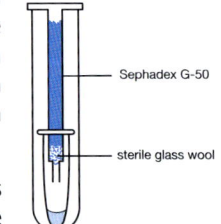

1. Plug the bottom of a 1-ml disposable syringe with a small amount of sterile glass wool. This is best accomplished by using the barrel of the syringe to tamp the glass wool in place.

2. Fill the syringe with Sephadex G-50 or Bio-Gel P-60, equilibrated in 1x TEN buffer (pH 8.0) (please see p. 709). Start the buffer flowing by tapping the side of the syringe barrel several times. Continue to add more resin until the syringe is completely full.

3. Insert the syringe into a 15-ml disposable plastic tube. Centrifuge at 1600g for 4 minutes at room temperature in a swinging-bucket rotor in a benchtop centrifuge. Do not be alarmed by the appearance of the column. The resin packs down and becomes partially dehydrated during centrifugation. Continue to add more resin and recentrifuge until the volume of the packed column is approximately 0.9 ml and remains unchanged after centrifugation.

4. Add 0.1 ml of 1x TEN buffer to the columns and recentrifuge as in Step 3.

5. Repeat Step 4 twice more.

> Spun columns may be stored at this stage if desired. Several spun columns can be prepared simultaneously and stored at 4°C for periods of 1 month or longer before being used. Fill the syringes with 1x TEN buffer and wrap Parafilm around them to prevent evaporation. Store the columns upright at 4°C. Spun columns stored in this way should be washed once with sterile 1x TEN buffer as described in Step 4 just before they are used.

6. Apply the DNA sample to the column in a total volume of 0.1 ml (use 1x TEN buffer to make up the volume). Place the spun column in a fresh disposable tube containing a decapped microfuge tube (please see figure above).

7. Centrifuge again as in Step 3, collecting the effluent from the bottom of the syringe (~100 µl) into the decapped microfuge tube.

8. Remove the syringe, which will contain unincorporated radiolabeled dNTPs or other small components. Using forceps, carefully recover the decapped microfuge tube, which contains the eluted DNA, and transfer its contents to a capped, labeled microfuge tube.

> A rough estimate of the proportion of radioactivity that has been incorporated into nucleic acid may be obtained by holding the syringe and the eluted DNA to a handheld minimonitor.

9. If the syringe is radioactive <!>, carefully discard it in the radioactive waste container. Store the eluted DNA at –20°C until needed.

Note

- Not all resins are suitable for spun-column centrifugation: DEAE-Sephacel forms an impermeable lump during centrifugation, and the larger grades of Sephadex cannot be used because the beads are crushed by centrifugation. If a coarser-sieving resin is required, use Sepharose CL-4B.

SEPARATION OF SINGLE-STRANDED AND DOUBLE-STRANDED DNAS BY HYDROXYAPATITE CHROMATOGRAPHY

Nucleic acids bind to hydroxyapatite by virtue of interactions between the phosphate groups of the polynucleotide backbone and calcium residues in the matrix. Bound nucleic acids can be eluted in phosphate buffers. This step is usually carried out at 60°C, although there is no good reason to do so since the adsorption and elution profiles of nucleic acids are indistinguishable between 25°C and 60°C. The affinity of nucleic acids is determined by the number of phosphate groups that are available to bind to the matrix. Both single- and double-stranded nucleic acids bind to hydroxyapatite in 0.05 M sodium phosphate (pH 6.8). Double-stranded molecules, with their well-ordered and evenly spaced sets of phosphate residues, make many regular contacts with the matrix and therefore require high concentrations of phosphate (0.4 M) for elution. Single-stranded molecules are more disordered and a smaller proportion of their phosphate residues are available for contact with the matrix. Hence, single-stranded DNA is eluted in lower concentrations of phosphate (~0.12 M). Partial duplexes and DNA-RNA hybrid molecules elute at intermediate concentrations.

Nucleic acids are often eluted in such large volumes that they need to be concentrated before they can be used. Ethanol precipitation must be avoided until the phosphate ions have been removed from the solution. This is best achieved by concentrating the eluate by extraction with isobutanol and then removing the salt by chromatography through Sephadex G-50 columns.

Batches of hydroxyapatite vary slightly in their characteristics, and it is therefore important to perform preliminary experiments to determine the optimal phosphate concentrations for elution of single- and double-stranded nucleic acids. This can be accomplished by setting up two hydroxyapatite columns as described below. One of the columns is loaded with a small amount (~10^5 cpm) of ^{32}P-labeled DNA that has been denatured by boiling for 10 minutes in TE (pH 7.6). The other column receives an equal amount (~10^5 cpm) of ^{32}P-labeled native DNA. Each of the columns is washed with a series of buffers containing increasing concentrations of sodium phosphate (0.01, 0.12, 0.16, 0.20, 0.24, 0.28, 0.32, 0.36, and 0.40 M). The amount of radioactivity eluting at each phosphate concentration is then measured in a liquid scintillation counter (either by Cerenkov counting or in a water-miscible fluor). Usually, single-stranded DNA elutes in 0.14–0.16 M sodium phosphate (pH 6.8), whereas double-stranded DNA is not removed from the column until the phosphate concentration exceeds 0.36 M. In the protocol that follows, SS buffer contains the phosphate concentration that is optimal for elution of single-stranded DNA; DS buffer contains the concentration that is optimal for elution of double-stranded DNA.

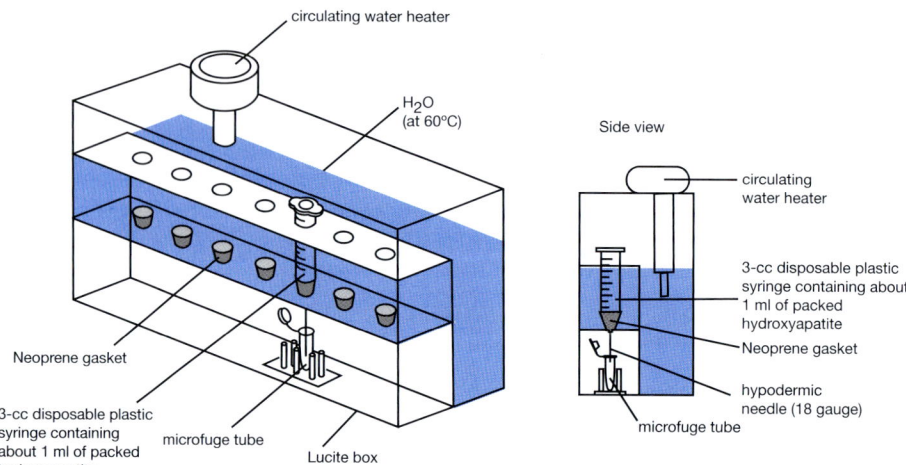

FIGURE A3-4 Apparatus for Hydroxyapatite Chromatography

1. Determine the concentrations of sodium phosphate that are optimal for elution of single-stranded DNA (SS buffer) and double-stranded DNA (DS buffer) as described above.

2. Prepare SS and DS buffers by diluting 2 M sodium phosphate (pH 6.8).

   2 M sodium phosphate (pH 6.8) is made by mixing equal volumes of 2 M NaH_2PO_4 and 2 M Na_2HPO_4.

3. Suspend the hydroxyapatite powder (Bio-Gel HTP) in 0.01 M sodium phosphate (pH 6.8). Approximately 0.5 ml of packed Bio-Gel HTP is required for each column.

   Bio-Gel HTP has a capacity of 100–200 µg of native DNA/ml of bed volume.

4. Prepare the hydroxyapatite columns in disposable 3-cc plastic syringes as follows:

   a. Remove the barrel from the syringe.

   b. Use the wide end of a pasteur pipette to push a Whatman GF/C filter to the bottom of the syringe. The filter should completely cover the bottom of the syringe.

   c. Attach an 18-gauge hypodermic needle to the syringe.

   d. Insert the syringe through a Neoprene gasket in the apparatus shown in Figure A3-4.

   e. Use a pasteur pipette to add enough of the slurry of hydroxyapatite to the syringe to form a column whose packed volume is 0.5–1.0 ml. Wash the column with several volumes of 0.01 M sodium phosphate (pH 6.8). The column will not be harmed if it runs dry; simply rewet before use.

   f. Seal the bottom of the column by placing a small Neoprene stopper on the end of the hypodermic needle.

5. Load the sample containing the nucleic acid onto the column.

   The concentration of phosphate in the sample should be less than 0.08 M.

6. Remove the Neoprene stopper and allow the sample to flow through the column.

   There is usually no need to collect and save the loading buffer that elutes from the column.

7. Wash the column with 3 ml of 0.01 M sodium phosphate.

8. Seal the bottom of the column with a Neoprene stopper and add 1 column volume of SS buffer preheated to 60°C.

9. After 5 minutes, remove the Neoprene stopper and collect the eluate in microfuge tubes. No more than 0.5 ml should be collected in any one microfuge tube. Repeat Steps 8 and 9 twice more.

10. Seal the bottom of the column with a Neoprene stopper and add 1 column volume of DS buffer preheated to 60°C.

11. After 5 minutes, remove the Neoprene stopper and collect the eluate in microfuge tubes. No more than 0.5 ml should be collected in any one microfuge tube. Repeat Steps 10 and 11 twice more.

12. Allow the eluates to cool to room temperature. DNA can then be extracted as follows:

    a. Add an equal volume of isobutanol to each of the tubes containing the desired nucleic acids.

    b. Mix the two phases by vortexing and centrifuge the mixture at maximum speed for 20 seconds at room temperature in a microfuge.

    c. Discard the upper (organic) phase.

    d. Repeat the extraction with isobutanol until the volume of the aqueous phase is 100–125 µl.

    e. Remove salts from the DNA by chromatography on, or centrifugation through, a small column of Sephadex G-50 equilibrated in TE (pH 8.0).

    f. Recover the DNA by precipitation with 2 volumes of ethanol at 0°C.

Note

- In molecular cloning, nucleic acids fractionated by hydroxyapatite are usually radiolabeled, and the tubes containing the desired fractions can be easily identified by Cerenkov counting in a liquid scintillation counter.

FRAGMENTATION OF DNA

The fragmentation of DNA is often a necessary step preceding library construction or subcloning for DNA sequencing. Fragmentation is typically achieved by physical or enzymatic methods; the most commonly used of these are described in Table A3-6. Although each approach is reasonably successful for generating a range of fragments from a large contiguous segment of DNA, each has its particular limitations. Because they are independent of sequence composition, physical methods for shearing DNA typically result in more uniform and random disruption of the target DNA than enzymatic methods. In particular, methods involving hydrodynamic shearing resulting from physical stress induced by sonication or nebulization produce collections of appropriately random fragments. The variety in lengths of these fragments is quite large, however, and their use usually requires a subsequent size selection step to narrow the range of fragments that are acceptable for cloning or sequencing.

AUTOMATED SHEARING

During recent years, a method for hydrodynamic shearing, initially based on the use of high-performance liquid chromatography (HPLC) and called the "point-sink" flow system, has become increasingly refined and, finally, automated. In the point-sink system, an HPLC pump is used to apply pressure to the DNA sample, thereby forcing it through tubing of very small diameter. In the automated process known as HydroShear, a sample of DNA is repeatedly passed through a small hole until the sample is fragmented to products of a certain size. The final size distribution is determined by both the flow rate of the sample and the size of the opening, parameters that are controlled and monitored by the automated system. At any given setting, DNA fragments larger than a certain length are broken, whereas shorter fragments are unaffected by passage through the opening. The resulting sheared products therefore have a narrow size distribution: Typically 90% of the sheared DNA falls within a twofold size range of the target length. It is reasonable to expect that libraries constructed from these DNA fragments are likely to be of higher quality than those made using one of the "old-fashioned" ways. They will certainly contain clones of more uniform size and possibly may be more comprehensive in their coverage of the genome. However, libraries constructed from sonicated or hydrodynamically sheared DNA, although imperfect, are certainly workable. Perfectionists will feel that the machine is necessary; pragmatists will find it merely desirable.

TABLE A3-6 Hydrodynamic Shearing Methods Used to Fragment DNA

Method	Pros and Cons
Sonication	Requires relatively large amounts of DNA (10–100 µg); fragments of DNA distributed over a broad range of sizes; only a small fraction of the fragments are of a length suitable for cloning and sequencing; requires ligation of DNA before sonication and end-repair afterward; DNA may be damaged by hydroxyl radicals generated during cavitation.
Nebulization	Easy and quick; requires only small amounts of DNA (0.5–5 µg) and large volumes of DNA solution; no preference for A+T-rich regions; size of fragments easily controlled by altering the pressure of the gas blowing through the nebulizer; fragments of DNA distributed over a narrow range of sizes (700–1330 bp); requires ligation of DNA before nebulization and end-repair afterward.
Circulation through an HPLC pump	Requires expensive apparatus, ligation of DNA before sonication, and 1–100 µg of DNA; fragments of DNA distributed over a narrow range of sizes that can be adjusted by changing the flow rate; end-repair of fragments before cloning not necessary.
Passage through the orifice of a 28-gauge hypodermic needle	Cheap, easy, and requires only small amounts of DNA; however, the fragments are a little larger (1.5–2.0 kb) than required for shotgun sequencing; requires ligation of DNA before cleavage and end-repair afterward.

Sonication

DNA samples are subjected to hydrodynamic shearing by exposure to brief pulses of sonication. DNA sonicated for excessive periods of time is extremely difficult to clone, perhaps because of damage caused by free radicals generated by cavitation. Most sonicators will not shear DNA to a size smaller than 300–500 bp, and it is tempting to continue sonication until the entire population of DNA fragments has been reduced to this size. However, the yield of subclones is usually greater if sonication is stopped when the fragments of target DNA first reach a size of approximately 700 bp.

Calibration of the Sonicator

1. Transfer 10 µg of bacteriophage λ DNA (or some other large DNA of defined molecular weight) to a microfuge tube. Add TE (pH 7.6) to a final volume of 150 µl. Distribute 25-µl aliquots of the DNA solution into five microfuge tubes. Store the remaining DNA in an ice bath.

2. Fill the cup horn of the sonicator with a mixture of ice and H_2O. Clamp the five microfuge tubes containing the bacteriophage λ DNA just above the probe.

   If the temperature of the sample rises during sonication, the speed and vigor of fragmentation will increase. It is therefore important to mix the ice and H_2O after each burst of sonication and to add fresh ice when necessary.

3. Sonicate at maximum output and continuous power for bursts of 10 seconds. After each burst, remove one of the microfuge tubes from the sonicator and store it on ice.

4. After sonication is completed, analyze the size of the DNA fragments in each stored sample by electrophoresis through a 1.4% agarose gel. Use suitable standards for molecular-weight markers (please see MC3, p. A6.14).

5. Stain and photograph the gel and then estimate the amount of sonication required to produce a reasonable yield of fragments of the desired size (500 bp to 2 kb).

   The times of sonication given in this method are for a cup-horn sonicator with a nominal peak output energy of 475 W. Because the actual output of different sonicators varies widely, it is necessary to calibrate each instrument. A probe-type sonicator can be used with a microtip if the volume of the DNA sample is increased to approximately 250 µl to accommodate the probe. After sonication is completed, concentrate the DNA by precipitation with 2 volumes of ethanol and dissolve in 25 µl of TE (pH 7.6).

Sonication of Target DNA

6. Sonicate the chosen DNA sample for the length of time estimated to produce a reasonable yield of fragments of the desired size (500 bp to 1 kb). Confirm that the fragmentation has gone according to plan by analyzing an aliquot of the sample (~200 ng) by electrophoresis through a 1.4% agarose gel as described above.

Nebulization

Nebulization is performed by collecting the fine mist created by forcing DNA in solution through a small hole in the nebulizer unit (Figure A3.5). The size of the fragments is determined chiefly by the speed at which the DNA solution passes through the hole, the viscosity of the solution, and the temperature.

Modification of the Nebulizer

1. Modify a nebulizer model number CA-209 (CIS-US Inc., Bedford, Massachusetts) by sealing the mouthpiece hole in the top cover with a QS-T plastic cap (Isolab). Connect a length of Nalgene tubing to the smaller hole. This tubing will be connected to a source of nitrogen gas.

Calibration of the Nebulizer

2. Prepare a sample containing 25 μg of bacteriophage λ DNA or other large DNA of defined size in 500 μl of TE (pH 7.6) containing 25% glycerol. Store the DNA solution in an ice bath for 5 minutes.

3. Place the DNA in the cup of a nebulizer connected to a nitrogen gas source and place the nebulizer in an ice-water bath.

4. Nebulize the DNA sample at 10 psi (69 KPa) for 90 seconds.

   The nebulizer usually leaks a little.

5. After nebulization is completed, collect the DNA by placing the entire unit in a rotor bucket of a benchtop centrifuge fitted with pieces of Styrofoam to cushion. Centrifuge the nebulizer at 2500 rpm for 30 seconds. Analyze the size of the DNA fragments by electrophoresis through a 1.4% agarose gel with appropriate standards, e.g., a 100-bp DNA ladder (New England BioLabs).

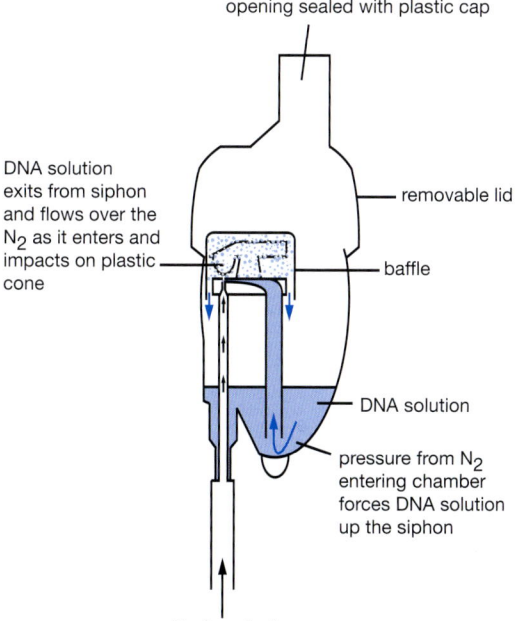

opening sealed with plastic cap

DNA solution exits from siphon and flows over the N₂ as it enters and impacts on plastic cone

removable lid

baffle

DNA solution

pressure from N₂ entering chamber forces DNA solution up the siphon

N₂ from tank

FIGURE A3-5 Nebulizer for Random Fragmentation of DNA

A DNA solution containing glycerol, for viscosity, is placed in the nebulizer. The nebulizer is attached to a nitrogen tank. Pressure from nitrogen entering the chamber siphons the DNA solution from the bottom of the chamber to the top. The solution exits the siphon and impacts on a small plastic cone suspended near the top of the chamber, thus shearing the DNA.

6. If the fragments are too large (>1.0 kb), repeat the procedure increasing either the pressure of the nitrogen gas (to 14 psi or 96.5 KPa) or the length of nebulization (2 minutes). Repeat the procedure until conditions are found that produce a reasonable yield of fragments of the desired size (500 bp to 1 kb).

Nebulization and Recovery of the Target DNA

7. Nebulize the target DNA to produce a reasonable yield of fragments of the desired size (500 bp to 1 kb).

8. Confirm that the fragmentation has gone according to plan by analyzing an aliquot of the sample (~200 ng) by electrophoresis through a 1.4% agarose gel as described above.

9. Distribute the remainder of the sample into two microfuge tubes and recover the DNA by precipitation with 3 volumes of ethanol in the presence of 2.5 M ammonium acetate.

10. Centrifuge the DNA sample at maximum speed for 5 minutes in a microfuge, wash the pellet with 0.5 ml of 70% ethanol at room temperature, and centrifuge again. Remove the ethanol and allow the DNA to air dry for a few minutes. Dissolve the pellet of DNA in 25 μl of TE (pH 7.6).

CENTRIFUGATION

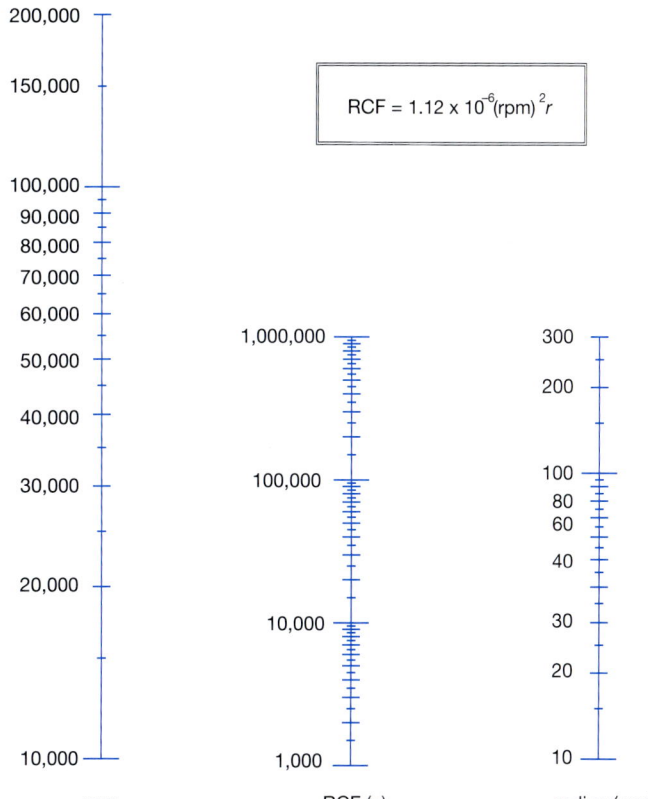

FIGURE A3-6 Nomogram for Conversion of Rotor Speed (rpm) and Relative Centrifugal Force (RCF)

The symbol r represents the radial distance from the center of the rotor at the point at which RCF is required. This is generally equivalent to r_{max} of the rotor. For the example marked, rotor speed = 80,000 rpm and r = 20 mm. RCF can be read as approximately 145,000g from the middle scale. Using the equation given above, RCF can be calculated as 142,080g. (Figure kindly provided by Siân Curtis.)

$$RCF = 1.12 \times 10^{-6}(rpm)^2 r$$

TABLE A3-7 Commonly Used Rotors

	r_{max}	g_{max}	Max rpm
Sorvall Rotors			
GS-3 Superspeed	151.3	13,700	9,000
SLC-1500 Superspeed	145.6	27,400	13,000
HB-6	146.3	27,617	13,000
AH-629 Ultraspeed	161.0	151,000	29,000
SS-34 Superspeed	107.0	20,500	50,200
Beckman Rotors			
SW 28.1	171.3	28,000	150,000
VTi 50	86.6	50,000	242,000
SW 40 Ti	158.8	40,000	285,000
SW 41 Ti	153.1	41,000	288,000
SW 50.1	107.3	50,000	300,000
Type 70 Ti	39.5	70,000	504,000

SDS-POLYACRYLAMIDE GEL ELECTROPHORESIS OF PROTEINS

Almost all analytical electrophoreses of proteins are carried out in polyacrylamide gels under conditions that ensure dissociation of the proteins into their individual polypeptide subunits and that minimize aggregation. Most commonly, the strongly anionic detergent SDS is used in combination with a reducing agent and heat to dissociate the proteins before they are loaded onto the gel. The denatured polypeptides bind SDS and become negatively charged. Because the amount of SDS bound is almost always proportional to the molecular weight of the polypeptide and is independent of its sequence, SDS-polypeptide complexes migrate through polyacryl-amide gels in accordance with the size of the polypeptide. At saturation, about 1.4 g of detergent is bound per gram of polypeptide. By using markers of known molecular weight, it is possible to estimate the molecular weight of the polypeptide chain(s). Modifications to the polypeptide backbone, such as N- or O-linked glyco-sylation and phosphorylation, however, have a significant impact on the apparent molecular weight. Thus, the apparent molecular weight of glycosylated proteins is not a true reflection of the mass of the polypeptide chain.

In most cases, SDS-polyacrylamide gel electrophoresis is carried out with a discontinuous buffer system in which the buffer in the reservoirs is of a pH and ionic strength different from that of the buffer used to cast the gel. The SDS-polypeptide complexes in the sample that is applied to the gel are swept along by a moving boundary created when an electric current is passed between the electrodes. After migrating through a stacking gel of high porosity, the complexes are deposited in a very thin zone (or stack) on the surface of the resolving gel. The ability of discontinuous buffer systems to concentrate all of the complex-es in the sample into a very small volume greatly increases the resolution of SDS-polyacrylamide gels.

The discontinuous buffer system that is most widely used was originally devised by Ornstein and Davis. The sample and the stacking gel contain Tris-Cl (pH 6.8), the upper and lower buffer reservoirs contain Tris-glycine (pH 8.3), and the resolving gel contains Tris-Cl (pH 8.8). All components of the system contain 0.1% SDS. The chloride ions in the sample and stacking gel form the leading edge of the moving bound-ary, and the trailing edge is composed of glycine molecules. Between the leading and trailing edges of the moving boundary is a zone of lower conductivity and steeper voltage gradient, which sweeps the polypep-tides from the sample and deposits them on the surface of the resolving gel. There, the higher pH of the resolving gel favors the ionization of glycine, and the resulting glycine ions migrate through the stacked polypeptides and travel through the resolving gel immediately behind the chloride ions. Freed from the moving boundary, the SDS-polypeptide complexes move through the resolving gel in a zone of uniform voltage and pH and are separated according to size by sieving.

Polyacrylamide gels are composed of chains of polymerized acrylamide that are cross-linked by a bifunctional agent such as *N,N'*-methylenebisacrylamide (please see Figure A3-7).

The effective range of separation of SDS-polyacrylamide gels depends on the concentration of polyacryl-amide used to cast the gel and on the amount of cross-linking. Polymerization of acrylamide in the absence of cross-linking agents generates viscous solutions that are of no practical use. Cross-links formed from bisacrylamide add rigidity and tensile strength to the gel and form pores through which the SDS-polypep-tide complexes must pass. The size of these pores decreases as the bisacrylamide:acrylamide ratio increases, reaching a minimum when the ratio is approximately 1:20. Most SDS-polyacrylamide gels are cast with a molar ratio of bisacrylamide:acrylamide of 1:29, which has been empirically shown to be capable of resolv-ing polypeptides that differ in size by as little as 3%.

The sieving properties of the gel are determined by the size of the pores, which is a function of the absolute concentrations of acrylamide and bisacrylamide used to cast the gel. Table A3-8 shows the linear range of sep-aration of proteins obtained with gels cast with concentrations of acrylamide that range from 5% to 15%.

Reagents

- *Acrylamide and* **N,N'**-*methylenebisacrylamide.* Several manufacturers sell electrophoresis-grade acryl-amide that is free of contaminating metal ions. A stock solution containing 29% (w/v) acrylamide and 1% (w/v) *N,N'*-methylenebisacrylamide should be prepared in deionized warm H_2O (to assist the disso-lution of the bisacrylamide). Acrylamide and bisacrylamide are slowly converted during storage to acrylic acid and bisacrylic acid. This deamination reaction is catalyzed by light and alkali. Check that the pH of the solution is 7.0 or less and store the solution in dark bottles at room temperature. Fresh solu-tions should be prepared every few months. Note that prepackaged, premixed stock solutions are com-mercially available (e.g., Invitrogen). These gel systems provide all of the components except ammoni-um persulfate and are certainly convenient but perhaps a bit expensive.

- *Sodium dodecyl sulfate (SDS).* Several manufacturers sell special grades of SDS that are sufficiently pure for electrophoresis. Although any one of these will give reproducible results, they are not interchange-able. The pattern of migration of polypeptides may change quite drastically when one manufacturer's SDS is substituted for another's. We recommend exclusive use of one brand of SDS. A 10% (w/v) stock

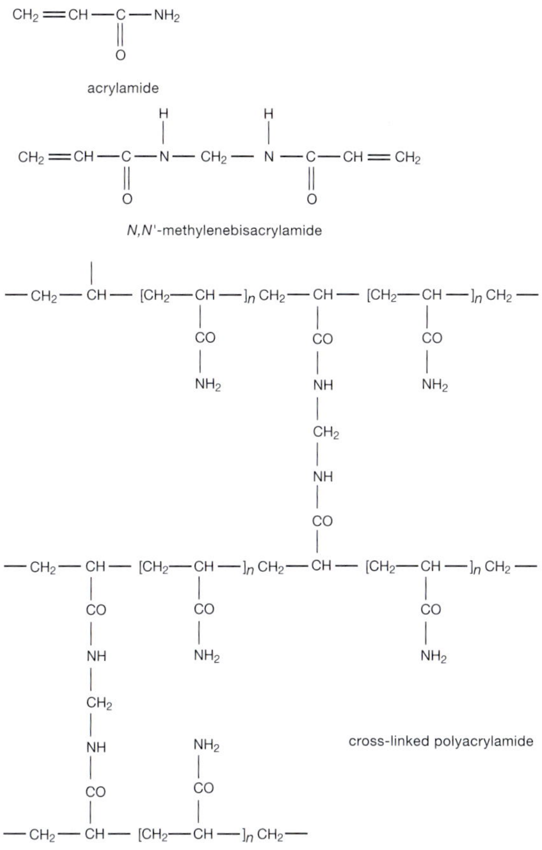

acrylamide

N,N'-methylenebisacrylamide

cross-linked polyacrylamide

FIGURE A3-7 Chemical Structure of Polyacrylamide

Monomers of acrylamide are polymerized into long chains in a reaction initiated by free radicals. In the presence of N,N'-methylenebisacrylamide, these chains become cross-linked to form a gel. The porosity of the resulting gel is determined by the length of chains and degree of cross-linking that occurs during the polymerization reaction.

solution should be prepared in deionized H_2O and stored at room temperature. If proteins are to be eluted from the gel for sequencing, electrophoresis-grade SDS should be further purified.

- *Tris buffers for the preparation of resolving and stacking gels.* It is essential that these buffers be prepared with Tris base. After the Tris base has been dissolved in deionized H_2O, adjust the pH of the solution with HCl as described on p. 697. If Tris-Cl or Trizma is used to prepare buffers, the concentration of salt will be too high and polypeptides will migrate anomalously through the gel, yielding extremely diffuse bands.

- *TEMED (N,N,N',N'-tetramethylethylenediamine).* TEMED accelerates the polymerization of acrylamide and bisacrylamide by catalyzing the formation of free radicals from ammonium persulfate. Use the electrophoresis grade sold by several manufacturers. Because TEMED works only as a free base, polymerization is inhibited at low pH.

- *Ammonium persulfate.* Ammonium persulfate provides the free radicals that drive polymerization of acrylamide and bisacrylamide. A small amount of a 10% (w/v) stock solution should be prepared in deionized H_2O and stored at 4°C. Ammonium persulfate decomposes slowly, and fresh solutions should be prepared weekly.

- *Tris-glycine electrophoresis buffer.* This buffer contains 25 mM Tris base, 250 mM glycine (electrophoresis grade) (pH 8.3), 0.1% SDS.

TABLE A3-8 Effective Range of Separation of SDS-Polyacrylamide Gels

Acrylamide Concentration (%)	Linear Range of Separation (kD)
15	10–43
12	12–60
10	20–80
7.5	36–94
5.0	57–212

Molar ratio of bisacrylamide:acrylamide is 1:29.

MATERIALS

CAUTION: Please see Appendix 4 for appropriate handling of materials marked with <!>.

Reagents and Solutions

Please see Appendix 1 for components of stock solutions, buffers, and reagents. Dilute stock solutions to the appropriate concentrations.

Acrylamide solutions <!>

> Please see Table A3-9 for resolving gel recipes and Table A3-10 for stacking gel recipes.

Protein markers

> Standard molecular-weight markers are commercially available (e.g., Invitrogen and Promega).

Protein samples

> Samples to be resolved, for example, can be purified protein or cell lysates.

1x SDS gel-loading buffer

> 50 mM Tris-Cl (pH 6.8)
> 100 mM dithiothreitol
> 2% (w/v) SDS (electrophoresis grade)
> 0.1% bromophenol blue
> 10% (v/v) glycerol
> Store 1x SDS gel-loading buffer lacking dithiothreitol at room temperature. Add dithiothreitol from a 1 M stock just before the buffer is used.

1x Tris-glycine electrophoresis buffer

> 25 mM Tris
> 250 mM glycine (electrophoresis grade) (pH 8.3)
> 0.1% (w/v) SDS
> Prepare a 5x stock of electrophoresis buffer by dissolving 15.1 g of Tris base and 94 g of glycine in 900 ml of deionized H_2O. Add 50 ml of a 10% (w/v) stock solution of electrophoresis-grade SDS and adjust the volume to 1000 ml with H_2O.

Additional Items

Power supply

> A device capable of supplying up to 500 V and 200 mA is needed.

Vertical electrophoresis apparatus

> The use of discontinuous buffer systems requires SDS-polyacrylamide electrophoresis to be performed in vertical gels. Although the basic design of the electrophoresis tanks and plates has changed little since Studier introduced the system, many small improvements have been incorporated into the apparatuses. Standard vertical as well as minigel vertical electrophoresis systems for separation and blotting of proteins are now sold by many manufacturers (e.g., Invitrogen). Which of these systems to purchase is a matter of personal choice, but it is sensible for a laboratory to use only one type of apparatus. This type of standardization makes it easier to compare the results obtained by different investigators and allows parts of broken apparatuses to be scavenged and reused.

METHOD

Pouring SDS-polyacrylamide Gels

1. Assemble the glass plates according to manufacturer instructions.

2. Determine the volume of the gel mold (this information is usually provided by the manufacturer). In an Erlenmeyer flask or disposable plastic tube, prepare the appropriate volume of solution containing the desired concentration of acrylamide for the resolving gel, using the values given in Table A3-9. Mix the components in the order shown. Polymerization will begin as soon as the TEMED has been added. Without delay, swirl the mixture rapidly and proceed to the next step.

    > The concentration of ammonium persulfate that we recommend, which is higher than that used by some investigators, eliminates the need to rid the acrylamide solution of dissolved oxygen (which retards polymerization) by degassing.

TABLE A3-9 Solutions for Preparing Resolving Gels for Tris-glycine SDS-Polyacrylamide Gel Electrophoresis

Components/Gel Volume	Volume (ml) of Components Required to Cast Gels of Indicated Volumes and Concentrations							
	5	10	15	20	25	30	40	50
6% gel								
H₂O	2.6	5.3	7.9	10.6	13.2	15.9	21.2	26.5
30% acrylamide mix <!>	1.0	2.0	3.0	4.0	5.0	6.0	8.0	10.0
1.5 M Tris (pH 8.8)	1.3	2.5	3.8	5.0	6.3	7.5	10.0	12.5
10% SDS	0.05	0.1	0.15	0.2	0.25	0.3	0.4	0.5
10% ammonium persulfate <!>	0.05	0.1	0.15	0.2	0.25	0.3	0.4	0.5
TEMED <!>	0.004	0.008	0.012	0.016	0.02	0.024	0.032	0.04
8% gel								
H₂O	2.3	4.6	6.9	9.3	11.5	13.9	18.5	23.2
30% acrylamide mix <!>	1.3	2.7	4.0	5.3	6.7	8.0	10.7	13.3
1.5 M Tris (pH 8.8)	1.3	2.5	3.8	5.0	6.3	7.5	10.0	12.5
10% SDS	0.05	0.1	0.15	0.2	0.25	0.3	0.4	0.5
10% ammonium persulfate <!>	0.05	0.1	0.15	0.2	0.25	0.3	0.4	0.5
TEMED <!>	0.003	0.006	0.009	0.012	0.015	0.018	0.024	0.03
10% gel								
H₂O	1.9	4.0	5.9	7.9	9.9	11.9	15.9	19.8
30% acrylamide mix <!>	1.7	3.3	5.0	6.7	8.3	10.0	13.3	16.7
1.5 M Tris (pH 8.8)	1.3	2.5	3.8	5.0	6.3	7.5	10.0	12.5
10% SDS	0.05	0.1	0.15	0.2	0.25	0.3	0.4	0.5
10% ammonium persulfate <!>	0.05	0.1	0.15	0.2	0.25	0.3	0.4	0.5
TEMED <!>	0.002	0.004	0.006	0.008	0.01	0.012	0.016	0.02
12% gel								
H₂O	1.6	3.3	4.9	6.6	8.2	9.9	13.2	16.5
30% acrylamide mix <!>	2.0	4.0	6.0	8.0	10.0	12.0	16.0	20.0
1.5 M Tris (pH 8.8)	1.3	2.5	3.8	5.0	6.3	7.5	10.0	12.5
10% SDS	0.05	0.1	0.15	0.2	0.25	0.3	0.4	0.5
10% ammonium persulfate <!>	0.05	0.1	0.15	0.2	0.25	0.3	0.4	0.5
TEMED <!>	0.002	0.004	0.006	0.008	0.01	0.012	0.016	0.02
15% gel								
H₂O	1.1	2.3	3.4	4.6	5.7	6.9	9.2	11.5
30% acrylamide mix <!>	2.5	5.0	7.5	10.0	12.5	15.0	20.0	25.0
1.5 M Tris (pH 8.8)	1.3	2.5	3.8	5.0	6.3	7.5	10.0	12.5
10% SDS	0.05	0.1	0.15	0.2	0.25	0.3	0.4	0.5
10% ammonium persulfate <!>	0.05	0.1	0.15	0.2	0.25	0.3	0.4	0.5
TEMED <!>	0.002	0.004	0.006	0.008	0.01	0.012	0.016	0.02

Modified from Harlow and Lane. 1988. *Antibodies: A laboratory manual.* Cold Spring Harbor Laboratory, Cold Spring Harbor, New York.

3. Pour the acrylamide solution into the gap between the glass plates. Leave sufficient space for the stacking gel (the length of the teeth of the comb plus 1 cm). Use a pasteur pipette to carefully overlay the acrylamide solution with 0.1% SDS (for gels containing ~8% acrylamide) or isobutanol (for gels containing ~10% acrylamide). Place the gel in a vertical position at room temperature.

   The overlay prevents oxygen from diffusing into the gel and inhibiting polymerization. Isobutanol dissolves the plastic of some minigel apparatuses.

4. After polymerization is complete (30 minutes), pour off the overlay and wash the top of the gel several times with deionized H₂O to remove any unpolymerized acrylamide. Drain as much fluid as possible from the top of the gel and remove any remaining H₂O with the edge of a paper towel.

5. Prepare the stacking gel as follows: In a disposable plastic tube, prepare the appropriate volume of solution containing the desired concentration of acrylamide, using the values given in Table A3-10. Mix the

TABLE A3-10 Solutions for Preparing 5% Stacking Gels for Tris-glycine SDS-polyacrylamide Gel Electrophoresis

Components/Gel Volume	Volume (ml) of Components Required to Cast Gels of Indicated Volumes							
	1	2	3	4	5	6	8	10
H_2O	0.68	1.4	2.1	2.7	3.4	4.1	5.5	6.8
30% acrylamide mix <!>	0.17	0.33	0.5	0.67	0.83	1.0	1.3	1.7
1.0 M Tris (pH 6.8)	0.13	0.25	0.38	0.5	0.63	0.75	1.0	1.25
10% SDS	0.01	0.02	0.03	0.04	0.05	0.06	0.08	0.1
10% ammonium persulfate <!>	0.01	0.02	0.03	0.04	0.05	0.06	0.08	0.1
TEMED <!>	0.001	0.002	0.003	0.004	0.005	0.006	0.008	0.01

Modified from Harlow and Lane. 1988. *Antibodies: A laboratory manual.* Cold Spring Harbor Laboratory, Cold Spring Harbor, New York.

components in the order shown. Polymerization will begin as soon as the TEMED has been added. Without delay, swirl the mixture rapidly and proceed to the next step.

> The concentration of ammonium persulfate is higher than that used by some investigators. This eliminates the need to rid the acrylamide solution of dissolved oxygen (which retards polymerization) by degassing.

6. Pour the stacking gel solution directly onto the surface of the polymerized resolving gel. Immediately insert a clean Teflon comb into the stacking gel solution, being careful to avoid trapping air bubbles. Add more stacking gel solution to fill the spaces of the comb completely. Place the gel in a vertical position at room temperature.

> Teflon combs should be cleaned with H_2O and dried with ethanol just before use.

Preparation of Samples and Running the Gel

7. While the stacking gel is polymerizing, prepare the samples in the appropriate volume of 1x SDS gel-loading buffer and heat them to 100°C for 3 minutes to denature the proteins.

> Be sure to denature a sample containing marker proteins of known molecular weights. Mixtures of appropriately sized polypeptides are available from commercial sources.

> Extremely hydrophobic proteins, such as those containing multiple transmembrane domains, may precipitate or multimerize when boiled for 3 minutes at 100°C. To avoid these pitfalls, heat the samples for 1 hour at a lower temperature (45–55°C) to effect denaturation.

8. After polymerization is complete (30 minutes), remove the Teflon comb carefully. Use a squirt bottle to wash the wells immediately with deionized H_2O to remove any unpolymerized acrylamide. If necessary, straighten the teeth of the stacking gel with a blunt hypodermic needle attached to a syringe. Mount the gel in the electrophoresis apparatus. Add Tris-glycine electrophoresis buffer to the top and bottom reservoirs. Remove any bubbles that become trapped at the bottom of the gel between the glass plates. This is best done with a bent hypodermic needle attached to a syringe.

> IMPORTANT: Do not prerun the gel before loading the samples, because this will destroy the discontinuity of the buffer systems.

9. Load up to 15 µl of each of the samples in a predetermined order into the bottom of the wells. This is best done with a Hamilton microliter syringe or a micropipettor equipped with gel-loading tips that is washed with buffer from the bottom reservoir after each sample is loaded. Load an equal volume of 1x SDS gel-loading buffer into any wells that are unused.

10. Attach the electrophoresis apparatus to an electric power supply (the positive electrode should be connected to the bottom buffer reservoir). Apply a voltage of 8 V/cm to the gel. After the dye front has moved into the resolving gel, increase the voltage to 15 V/cm and run the gel until the bromophenol blue reaches the bottom of the resolving gel (~4 hours). Then turn off the power supply.

11. Remove the glass plates from the electrophoresis apparatus and place them on a paper towel. Use an extra gel spacer to carefully pry the plates apart. Mark the orientation of the gel by cutting a corner from the bottom of the gel that is closest to the leftmost well (slot 1).

> IMPORTANT: Do not cut the corner from gels that are to be used for immunoblotting.

12. At this stage, the gel can be fixed, stained with Coomassie Brilliant Blue or silver salts, fluorographed or autoradiographed, or used to establish an immunoblot, all as described on the following pages.

STAINING SDS-POLYACRYLAMIDE GELS

Unlabeled proteins separated by polyacrylamide gel electrophoresis typically are detected by staining, either with Coomassie Brilliant Blue or with silver salts. In a relatively rapid and straightforward reaction, Coomassie Brilliant Blue binds nonspecifically to proteins but not to the gel, thereby allowing visualization of the proteins as discreet blue bands within the translucent matrix of the gel. Silver staining, although somewhat more difficult to perform, is significantly more sensitive. The use of silver staining allows detection of proteins resolved by gel electrophoresis at concentrations nearly 100-fold lower than those detected by Coomassie Brilliant Blue staining. The identification of proteins by silver staining is based on the differential reduction of silver ions, in a reaction similar to that used in photographic processes. Reagents for staining with Coomassie Brilliant Blue as well as kits are commercially available. Kits for silver staining are commercially available from Pierce and Bio-Rad.

Staining SDS-Polyacrylamide Gels with Coomassie Brilliant Blue

Coomassie Brilliant Blue is an aminotriarylmethane dye that forms strong but not covalent complexes with proteins, most probably by a combination of van der Waals forces and electrostatic interactions with NH_3^+ groups. Coomassie Brilliant Blue is used to stain proteins after electrophoresis through polyacrylamide gels. The uptake of dye is approximately proportional to the amount of protein, following the Beer-Lambert law.

Polypeptides separated by SDS-polyacrylamide gels can be simultaneously fixed with methanol:glacial acetic acid and stained with Coomassie Brilliant Blue R-250, a triphenylmethane textile dye also known as Acid Blue 83. The gel is immersed for several hours in a concentrated methanol:acetic acid solution of the dye, and excess dye is then allowed to diffuse from the gel during a prolonged period of destaining.

MATERIALS

CAUTION: Please see Appendix 4 for appropriate handling of materials marked with <!>.

Coomassie Brilliant Blue R-250
Methanol:acetic acid solution <!>
  Combine 900 ml of methanol:H_2O (500 ml of methanol and 400 ml of H_2O) and 100 ml of glacial acetic acid.

METHOD

1. Separate proteins by electrophoresis through an SDS-polyacrylamide gel as described beginning on p. 756.

2. Prepare the staining solution by dissolving 0.25 g of Coomassie Brilliant Blue R-250 per 100 ml of methanol:acetic acid solution. Filter the solution through a Whatman No. 1 filter to remove any particulate matter.

3. Immerse the gel in at least 5 volumes of staining solution and place on a slowly rotating platform for a minimum of 4 hours at room temperature.

4. Remove the stain and save it for future use. Destain the gel by soaking it in the methanol:acetic acid solution without the dye on a slowly rocking platform for 4–8 hours, changing the destaining solution three or four times.

   The more thoroughly the gel is destained, the smaller the amount of protein detected by staining with Coomassie Brilliant Blue. Destaining for 24 hours usually allows as little as 0.1 μg of protein to be detected in a single band.

   A more rapid rate of destaining can be achieved by the following methods:

   - Destaining in 30% methanol, 10% acetic acid. If destaining is prolonged, there will be some loss in the intensity of staining of protein bands.

   - Destaining in the normal destaining buffer at higher temperatures (45°C).

   - Including a few grams of an anion exchange resin or a piece of sponge in the normal destaining buffer. These absorb the stain as it leaches from the gel.

   - Destaining electrophoretically in apparatuses that are sold commercially for this purpose.

5. After destaining, store the gels in H$_2$O in a sealed plastic bag.

> Gels may be stored indefinitely without any diminution in the intensity of staining; however, fixed polyacrylamide gels stored in H$_2$O will swell and may distort during storage. To avoid this problem, store fixed gels in H$_2$O containing 20% glycerol. Stained gels should not be stored in destaining buffer, because the stained protein bands will fade.

6. To make a permanent record, either photograph the stained gel or dry the gel as described on p. 763.

Staining SDS-Polyacrylamide Gels with Silver Salts

A number of methods have been developed to stain polypeptides with silver salts after separation by SDS-polyacrylamide gel electrophoresis. In every case, the process relies on differential reduction of silver ions that are bound to the side chains of amino acids. These methods fall into two major classes: those that use ammoniacal silver solutions and those that use silver nitrate. Although both types of staining are approximately 100- to 1000-fold more sensitive than staining with Coomassie Brilliant Blue R-250 and are capable of detecting as little as 0.1–1.0 ng of polypeptide in a single band, silver nitrate solutions are easier to prepare and, in contrast to ammoniacal silver salts, do not generate potentially explosive by-products. The method given below is a modification of the staining procedure originally devised by Sammons et al., which has since undergone several improvements. For further information, please see the discussion on silver staining in MC3, pp. A9.5–A9.7.

MATERIALS

CAUTION: Please see Appendix 4 for appropriate handling of materials marked with <!>.

Acetic acid (1%) <!>
Developing solution
> Prepare fresh for each use an aqueous solution of 2.5% sodium carbonate, 0.02% formaldehyde. <!>

Ethanol (30%)
Fixing solution
> Ethanol:glacial acetic acid:H$_2$O (30:10:60) <!>

Photographic reducing solution (optional; see Step 10)
> *Prepare solution A:* Dissolve 37 g of NaCl and 37 g of CuSO$_4$ in 850 ml of deionized H$_2$O. Add concentrated NH$_4$OH <!> until a deep blue precipitate forms and then dissolves. Adjust the volume to 1 liter with H$_2$O.
>
> *Prepare solution B:* Dissolve 436 g of sodium thiosulfate in 900 ml of deionized H$_2$O. Adjust the volume to 1 liter with H$_2$O.
>
> Mix equal volumes of solutions A and B, dilute the mixture with 3 volumes of H$_2$O, and use the diluted mixture immediately.

Silver nitrate solution
> Prepare fresh for each use a 0.1% solution of AgNO$_3$ <!> diluted from a 20% stock and stored in a tightly closed, brown glass bottle at room temperature.

METHOD

IMPORTANT: Wear gloves and handle the gel gently because pressure and fingerprints produce staining artifacts. In addition, it is essential to use clean glassware and deionized H$_2$O because contaminants greatly reduce the sensitivity of silver staining.

1. Separate proteins by electrophoresis through an SDS-polyacrylamide gel as described beginning on p. 756.

2. Fix the proteins by incubating the gel for 4–12 hours at room temperature with gentle shaking in at least 5 gel volumes of fixing solution.

3. Discard the fixing solution and add at least 5 gel volumes of 30% ethanol. Incubate the gel for 30 minutes at room temperature with gentle shaking.

4. Repeat Step 3.

5. Discard the ethanol and add 10 gel volumes of deionized H_2O. Incubate the gel for 10 minutes at room temperature with gentle shaking.

6. Repeat Step 5 twice.

    The gel will swell slightly during rehydration.

7. Discard the last of the H_2O washes and, wearing gloves, add 5 gel volumes of silver nitrate solution. Incubate the gel for 30 minutes at room temperature with gentle shaking.

8. Discard the silver nitrate solution and wash both sides of the gel (20 seconds each) under a stream of deionized H_2O.

    Allowing the surface of the gel to dry out will result in staining artifacts.

9. Add 5 gel volumes of fresh developing solution. Incubate the gel at room temperature with gentle agitation. Watch the gel carefully. Stained bands of protein should appear within a few minutes. Continue incubation until the desired contrast is obtained.

    Prolonged incubation leads to a high background of silver staining within the body of the gel.

10. Quench the reaction by washing the gel in 1% acetic acid for a few minutes. Then wash the gel several times with deionized H_2O (10 minutes per wash).

    A shiny gray film of silver sometimes forms on the surface of the gel. This can be removed by washing the gel for 2–3 seconds in a 1:4 dilution of photographic reducing solution. Rinse the treated gel extensively in deionized H_2O.

11. Preserve the gel by drying as described on the following pages.

DRYING SDS-POLYACRYLAMIDE GELS

SDS-polyacrylamide gels containing proteins radiolabeled with 35S-labeled amino acids must be dried before autoradiographic images can be obtained. The major problems encountered when a gel is dried are (1) shrinkage and distortion and (2) cracking of the gel. The first of these problems can be minimized if the gel is attached to a piece of Whatman 3MM paper before it is dehydrated. (For nonradioactive gels, note that the preference may be to dry the gel between acetate sheets to allow transillumination and easy visualization of the dried gel.) However, there is no guaranteed solution to the second problem, which becomes more pronounced with thicker gels containing more polyacrylamide. Cracking generally occurs when the gel is removed from the drying apparatus before it is completely dehydrated. It is therefore essential to keep the drying apparatus in good condition, to use a reliable vacuum line that has few fluctuations in pressure, and to use the thinnest gel possible to achieve the desired purpose. An excellent alternative method is soaking the gel in 3% glycerol, followed by air drying using a simple apparatus (available from GE Healthcare or Owl Separation Systems).

MATERIALS

CAUTION: Please see Appendix 4 for appropriate handling of materials marked with <!>.

Fixing solution
Glacial acetic acid:methanol:H_2O (10:20:70 [v/v/v]) <!>
Gel dryer
Gel dryers are available from a number of commercial sources (e.g., Invitrogen and Promega). It is best to purchase the dryer from the manufacturer of the SDS-polyacrylamide gel electrophoresis tanks to ensure that the size of the dryer will be tailored to that of the gels and will accommodate several SDS-polyacrylamide gels simultaneously.
Methanol (20%) containing 3% glycerol <!> (optional; see Step 1)
Whatman 3MM paper

METHOD

1. Remove the gel from the electrophoresis apparatus and incubate it at room temperature in 5–10 volumes of fixing solution. The bromophenol blue will turn yellow as the acidic fixing solution diffuses into the gel. Continue fixation for 5 minutes after all of the blue color has disappeared and wash the gel briefly in deionized H_2O.

   > If cracking of polyacrylamide gels during drying is a constant problem, soak the fixed gel in 20% methanol, 3% glycerol overnight before proceeding to Step 2.

2. On a piece of Saran Wrap slightly larger than the gel, arrange the gel with its cut corner on the lower right-hand side.

3. Place a piece of dry Whatman 3MM paper on the damp gel. The paper should be large enough to create a border (1–2 cm) around the gel and small enough to fit on the gel dryer. Do not attempt to move the 3MM paper once contact has been made with the gel.

4. Arrange another piece of dry 3MM paper on the drying surface of the gel dryer. This piece should be large enough to accommodate all of the gels that are to be dried at the same time.

5. Place the sandwich of 3MM paper/gel/Saran Wrap on the piece of 3MM paper on the gel dryer. The Saran Wrap should be uppermost.

6. Close the lid of the gel dryer and apply suction so that the lid makes a tight seal around the gels. If the dryer is equipped with a heater, apply low heat (50–65°C) to speed up the drying process.

7. Dry the gel for the time recommended by the manufacturer (usually 2 hours for standard 0.75-mm gels). If heat was applied, turn off the heat for a few minutes before releasing the vacuum.

8. Remove the gel, which is now attached to a piece of 3MM paper, from the dryer.

9. Remove the piece of Saran Wrap and establish an autoradiograph as described in MC3, pp. A9.9–A9.15, or store the dehydrated gel as a record of the experiment.

IMMUNOBLOTTING

CAUTION: Please see Appendix 4 for appropriate handling of materials marked with <!>.

Immunoblotting is used to identify and measure the size of macromolecular antigens (usually proteins) that react with a specific antibody. The proteins are first separated by electrophoresis through SDS-polyacrylamide gels and then transferred electrophoretically from the gel to a solid support, such as a nitrocellulose, polyvinylidene difluoride (PVDF), or cationic nylon membrane. After the unreacted binding sites of the membrane are blocked to suppress nonspecific adsorption of antibodies, the immobilized proteins are reacted with a specific polyclonal or monoclonal antibody. Antigen-antibody complexes are finally located by radiographic, chromogenic, or chemiluminescent reactions.

Much mumbo jumbo has been written about ways to avoid the problems that commonly arise in immunoblotting. These problems include inefficient transfer of proteins, loss of antigenic sites, low sensitivity, high background, and nonquantitative detection methods. Although no magic incantation can eliminate all of these undesirable difficulties for every antigen, a small amount of experimentation is usually sufficient to cure all but the most obdurate technical problems.

Transfer of Proteins from Gels to Filters

Electrophoretic transfer of proteins from polyacrylamide gels to membranes is far more efficient and much quicker than capillary transfer (see Table A3-11). Transfer is carried out perpendicularly from the direction of travel of proteins through the separating gel, using electrodes and membranes that cover the entire area of the gel. Most commercial electrophoretic transfer devices use large electrodes made of graphite, platinum wire mesh, or stainless steel. In older devices, vertical electrodes were submerged in a tank of transfer buffer in a plastic cradle surrounding the gel and the membrane. The more modern devices use the efficient "semidry" method, in which Whatman 3MM paper saturated with transfer buffer is used as a reservoir. For transfer from SDS gels, the membrane is placed on the side of the gel facing the anode. The conditions used for transfer vary according to the design of the apparatus, and it is therefore best to follow manufacturer instructions at this stage.

TABLE A3-11 Buffers for Transfer of Proteins from Polyacrylamide Gels to Membranes

Type of Transfer	Buffer
Semidry	24 mM Tris base
	192 mM glycine
	20% methanol <!>
Immersion	48 mM Tris base
	39 mM glycine
	20% methanol <!>
	0.0375% SDS

Methanol minimizes swelling of the gel and increases the efficiency of binding of proteins to nitrocellulose membranes. The efficiency of transfer may be affected by the presence of SDS in the electrophoresis buffer, the pH of the transfer buffer, and whether the proteins were stained in the gel before transfer. To maximize transfer of protein to membranes, the concentration of SDS should be ≤ 0.1% and the pH should be ≥8.0. CAPS buffer should be used for transfer if the protein is to be sequenced on the membrane. Glycine interferes with this procedure.

Types of Membranes

Three types of membranes are used for immunoblotting: nitrocellulose, nylon, and polyvinylidene fluoride. Different proteins may bind with different efficiencies to these membranes, and particular antigenic epitope(s) may be better preserved in one case than another. It is therefore worthwhile wherever possible to test the efficiency with which the antigen of interest can be detected on various membranes, using several antibodies.

- *Nitrocellulose* (pore size 0.45 μm) remains a standard membrane used for immunoblotting, although membranes with a smaller pore size (0.22 or 0.1 μm) are recommended for immunoblotting of small proteins of M_r <14,000. The capacity of nitrocellulose to bind and retain proteins ranges from 80 to 250 μg/cm². depending on the protein. Proteins bind to nitrocellulose chiefly by hydrophobic interactions, although hydrogen bonding between amino acid side chains and the nitro group of the membrane may also be involved. In any event, partial dehydration of the proteins by methanol or salt in the transfer buffer ensures a longer-lasting bond between the protein and the membrane. Even so, proteins may be lost from the membrane during processing, particularly if buffers containing nonionic detergents are used. Many investigators therefore fix the proteins to nitrocellulose membranes to reduce loss during washing and incubation with antibody. However, it is important to check that the treatments used for fixation (glutaraldehyde, cross-linking, UV irradiation) do not destroy the antigenic epitope under study. These treatments can also increase the brittleness of nitrocellulose filters that are allowed to dry after transfer.

- *Nylon and positively charged nylon membranes* are tougher than nitrocellulose and bind proteins tightly by electrostatic interactions. Their capacity varies from one type of nylon to the next and from one protein to another but is usually in the range of 150 to 200 μg/cm². The advantage of nylon and charged nylon membranes over nitrocellulose is that they can be probed multiple times with different antibodies. However, nylon membranes have two potential disadvantages. First, as discussed below, no simple and sensitive procedure is available to stain proteins immobilized on nylon and charged nylon membranes. Second, because it is difficult to block all of the unoccupied sites on these membranes, antibodies tend to bind nonspecifically to the filter, resulting in a high background, especially when a highly sensitive detection method such as enhanced chemiluminescence (ECL) is used. In many cases, extended blocking in solutions containing 6% heat-treated casein and 1% polyvinylpyrrolidone is required to achieve satisfactory results.

- *Polyvinylidene fluoride* (PVDF) is mechanically strong and manifests a strong interfacial (hydrophobic) interaction with proteins. Before transfer, it is necessary to wet the hydrophobic surface of the membrane with methanol. The capacity of PVDF membranes is approximately equal to that of nylon membranes (~170 μg protein/cm²). Proteins bind approximately sixfold more tightly to PVDF membranes than to nitrocellulose and are retained more efficiently during the subsequent detection steps. Proteins immobilized on PVDF can be visualized with standard stains such as Amido Black, India ink, Ponceau-S, and Coomassie Brilliant Blue.

Staining of Proteins during Immunoblotting

Separation of proteins in gels and transfer to membranes can be confirmed by staining. This is a simple procedure, but it requires careful choice of a stain that is sufficiently sensitive and appropriate for the type of membrane. Staining can be performed at several stages in the immunoblotting procedure as outlined below.

- *Staining gels before transfer to membranes.* Proteins can be stained in polyacrylamide gels with conventional dyes such as Coomassie Brilliant Blue, destained, and then transferred electrophoretically to nitrocellulose or PVDF filters for immunoblotting. The chief advantage of this method is that proteins remain stained during immunodetection, thereby providing a set of internal markers. However, in some cases, staining of proteins in gels appears to reduce the efficiency of electroelution and/or to interfere with binding of antibody. (The use of prestained protein markers [Invitrogen] provides a set of internal markers during protein transfer without the need to stain the entire gel.)

- *Staining proteins after transfer to membranes.* The entire area of nitrocellulose and PVDF membranes can be stained with the removable but insensitive stain Ponceau-S. When more permanent stains are used (e.g., India ink, Amido Black, colloidal gold, or silver), it is usually necessary to cut a reference lane from the membrane.

Brief exposure to alkali enhances staining with both India ink and colloidal gold, perhaps by reducing loss of protein from the filter during washing. Under these conditions, it is easily possible to detect a band containing as little as a few nanograms of protein. There is no satisfactory method to stain proteins immobilized on nylon or cationic nylon membranes. The high density of charge on these membranes causes dye molecules to bind indiscriminately to the surface, producing high backgrounds that obscure all but the strongest protein bands.

Blocking Agents

Traditional blocking agents such as 0.5% low-fat dry milk or 5% bovine serum albumin are suitable for use with chromogenic detection systems based on horseradish peroxidase. However, these solutions are usually rich in residual alkaline phosphatase and should not be used in detection systems that use this enzyme. This is particularly true with chemiluminescent systems, where the sensitivity is determined not by the strength of the emitted signal but by the efficiency of suppression of background. The best blocking solution for alkaline-phosphatase-based systems contains 6% casein, 1% polyvinylpyrrolidone, 10 mM EDTA in phosphate-buffered saline. The blocking solution should be heated to 65°C for 1 hour to inactivate residual alkaline phosphatase and then stored at 4°C in the presence of 3 mM sodium azide.

Probing and Detection

The antibody that reacts with the epitope of interest can be either polyclonal or monoclonal. In either case, it is not radiolabeled or conjugated to an enzyme but is merely diluted into an appropriate buffer for formation of antibody-antigen complexes. In general, backgrounds in immunoblotting are unacceptably high unless the primary antibody can be diluted at least 1:1000 when enzymatic methods of detection are used and at least 1:5000 when chemiluminescent methods are used. After washing, the bound antibody is detected by a radiolabeled or enzyme-conjugated secondary reagent that recognizes common features of the primary antibody and carries a reporter enzyme or group. Secondary reagents include

- *Radioiodinated antibodies or* **staphylococcal** *protein A,* which were used in the first immunoblots and for a few years thereafter. However, radiolabeled secondary reagents have now been replaced by nonradioactive detection systems such as enhanced chemiluminscence that are less hazardous and more sensitive. They remain the most accurate method for semiquantitative immunoblotting.

- *Antibodies conjugated to enzymes,* such as horseradish peroxidase or alkaline phosphatase, for which a variety of chromogenic, fluorescent, and chemiluminescent substrates are available.

- *Antibodies coupled to biotin* that can then be detected by labeled or conjugated streptavidin.

Images of radiolabeled reagents are captured on X-ray film or phosphorimagers, whereas the results of chromogenic and fluorogenic reactions are best recorded by conventional photography. Table A3-12 shows the approximate sensitivity with which the best of these methods can detect a standard antigen using antibodies of high titer and specificity. For more information about these detection methods, please see MC3, Appendix 9.

TABLE A3-12 Chromogenic and Chemiluminescent Methods of Detection of Immobilized Antigens

Enzyme	Reagent	Sensitivity	Comments
Chromogenic			
Horseradish peroxidase	4-chloro-1 naphthol/ H_2O_2	1 ng	The purple color of oxidized products fades rapidly on exposure to light.
	diaminobenzidine <!>/ H_2O_2	250 pg	Potentially carcinogenic. The diamino-benzidine reaction generates a brown precipitate, which is enhanced by the addition of cobalt, silver, and nickel salts.
	3,3′,5,5′-tetramethyl-benzidine	100 pg	Deep purple precipitate.
Alkaline phosphatase	nitro blue tetrazolium/ 5-bromo-4-chloroindolyl phosphate	100 pg	Steel-blue precipitate.
Chemiluminescent			
Horseradish peroxidase	luminol/4-iodo-phenol/ H_2O_2	300 pg	Oxidized luminol emits blue light that is captured on X-ray film. Luminescence generated by intense bands appears within a few seconds, whereas faint bands need at least 30 minutes to develop.
Alkaline phosphatase	AMPPD 3-(4-methoxyspiro[1,2-dioxetane-3′2′-tricyclo-[3.3.1^{3,7}]decan]-4-yl)-phenylphosphate	1 pg	The enzymatically dephosphorylated product emits light. Because of its high turnover number, alkaline phosphatase rapidly generates a strong signal that provides an exquisitely sensitive method of immunodetection.

TECHNIQUES INDEX

APPENDIX 4

Cautions

GENERAL CAUTIONS

The following general cautions should always be observed.

- Become **completely familiar** with the properties of all substances used before beginning the procedure.

- **The absence of a warning** does not necessarily mean that the material is safe, since information may not always be complete or available.

- If **exposed** to toxic substances, contact the local safety office immediately for instructions.

- **Use proper disposal procedures** for all chemical, biological, and radioactive waste.

- **For specific guidelines on appropriate gloves**, consult the local safety office.

- **Handle concentrated acids and bases** with great care. Wear goggles and appropriate gloves, as well as a face shield if handling large quantities.

  Do not mix strong acids with organic solvents because they may react. Sulfuric and nitric acid especially may react highly exothermically and cause fires and explosions.

  Do not mix strong bases with halogenated solvent because they may form reactive carbenes that can lead to explosions.

  When preparing diluted solutions of acids from concentrated stocks, add acid to water ("If you do what you oughta, add acid to wata").

- **Never pipette** solutions using mouth suction. This method is not sterile and can be dangerous. Always use a pipette aid or bulb.

- **Keep halogenated and nonhalogenated** solvents separately (e.g., mixing chloroform and acetone can cause unexpected reactions in the presence of bases). Halogenated solvents are organic solvents such as chloroform, dichloromethane, trichlorotrifluoroethane, and dichloroethane. Some nonhalogenated solvents are pentane, heptane, ethanol, methanol, benzene, toluene, N,N-dimethylformamide (DMF), dimethylsulfoxide (DMSO), and acetonitrile.

- **Laser radiation**, visible or invisible, can cause severe damage to the eyes and skin. Take proper precautions to prevent exposure to direct and reflected beams. Always follow manufacturer safety guidelines and consult the local safety office. For more detailed information, see caution below.

- **Flash lamps**, because of their light intensity, can be harmful to the eyes and may explode on occasion. Wear appropriate eye protection and follow manufacturer guidelines.

- **Photographic fixatives and developers** contain harmful chemicals. Handle them with care and follow manufacturer directions.

- **Power supplies and electrophoresis equipment** pose serious fire hazard and electrical shock hazards if not used properly.

- **Microwave ovens and autoclaves** in the lab require certain precautions. If the screw top on the bottle is not loose enough, and there is not enough space for the steam to vent, the bottle can explode when the containers are removed from the microwave or autoclave. Always loosen bottle caps before microwaving or autoclaving.

- Use extreme caution when handling **cutting devices** such as microtome blades, scalpels, razor blades, or needles. Microtome blades are extremely sharp! If unfamiliar with their use, have an experienced person demonstrate proper procedures. For proper disposal, use a "sharps" disposal container in the lab. Discard used needles unshielded, with the syringe still attached. This method prevents injuries (and possible infections) while manipulating used needles because many accidents can occur while trying to replace the needle shield. Injuries may also be caused by broken pasteur pipettes, coverslips, or slides.

- **Ultrasonicators** are high-frequency sound waves (16–100 kHz) used for cell disruption and other purposes. This "ultrasound," conducted through air, does not pose a direct hazard to humans, but the associated high volumes of audible sound can cause a variety of effects, including headache, nausea, and tinnitus. Direct contact of the body with high-intensity ultrasound (not medical imaging equipment) should be avoided. Use appropriate ear protection and display signs on the door(s) of laboratories in which these units are used.

GENERAL PROPERTIES OF COMMON CHEMICALS

The hazardous materials list can be summarized in the following categories:

- Inorganic acids, such as hydrochloric, sulfuric, nitric, or phosphoric, are colorless liquids with stinging vapors. Avoid spills on skin or clothing. Dilute spills with large amounts of water. The concentrated forms of these acids can destroy paper, textiles, and skin, as well as cause serious injury to the eyes.

- Salts of heavy metals are usually colored, powdered solids that dissolve in water. Many of them are potent enzyme inhibitors and therefore toxic to humans and to the environment (e.g., fish and algae).

- Most organic solvents are flammable volatile liquids. Breathing their vapors can cause nausea or dizziness. Avoid skin contact.

- Other organic compounds, including organosulphur compounds such as mercaptoethanol or organic amines, have very unpleasant odors. Others are highly reactive and must be handled with appropriate care.

- If improperly handled, dyes and their solutions can stain not only the sample, but also skin and clothing. Some of them are also mutagenic (e.g., ethidium bromide), carcinogenic, and toxic.

- Nearly all names ending with "ase" (e.g., catalase, β-glucuronidase, or zymolyase) refer to enzymes. There are also other enzymes with nonsystematic names such as pepsin. Many of them are provided by manufacturers in preparations containing buffering substances, etc. Be aware of the individual properties of materials contained in these substances.

- Toxic compounds often used to manipulate cells (e.g., cycloheximide, actinomycin D, and rifampicin) can be dangerous and should be handled appropriately.

- Be aware that several of the compounds listed have not been thoroughly studied with respect to their toxicological properties. Handle each chemical with the appropriate respect. Although the toxic effects of a compound can be quantified (e.g., LD_{50} values), this is not possible for carcinogens or mutagens where one single exposure can have an effect. Also realize that dangers related to a given compound may also depend on its physical state (e.g., fine powder vs. large crystals, diethylether vs. glycerol, dry ice vs. carbon dioxide under pressure in a gas bomb). Anticipate under which circumstances during an experiment exposure is most likely to occur and how best to protect yourself and your environment.

HAZARDOUS MATERIALS

NOTE: In general, proprietary materials are not listed here. Kits and other commercial items as well as most anesthetics, dyes, fixatives, and stains are not included either. Anesthetics also require special care. Follow manufacturer safety guidelines that accompany these products.

Acetic acid (concentrated) must be handled with great care. It may be harmful by inhalation, ingestion, or skin absorption. Wear appropriate gloves and goggles. Use in a chemical fume hood.

Acetonitrile is very volatile and extremely flammable. It is an irritant and a chemical asphyxiant that can exert its effects by inhalation, ingestion, or skin absorption. Treat cases of severe exposure as cyanide poisoning. Wear appropriate gloves and safety glasses. Use only in a chemical fume hood. Keep away from heat, sparks, and open flame.

Acrylamide (unpolymerized) is a potent neurotoxin and is absorbed through the skin (effects are cumulative). Avoid breathing the dust. Wear appropriate gloves and a face mask when weighing powdered acrylamide and methylene-bisacrylamide. Use in a chemical fume hood. Polyacrylamide is considered to be nontoxic, but it should be handled with care because it might contain small quantities of unpolymerized acrylamide.

Actinomycin D is a teratogen and a carcinogen. It is highly toxic and may be fatal if inhaled, ingested, or absorbed through the skin. It may also cause irritation. Avoid breathing the dust. Wear appropriate gloves and safety glasses. Always use in a chemical fume hood. Solutions of actinomycin D are light sensitive.

S-**Adenosylmethionine** is toxic and may be harmful by inhalation, ingestion, or skin absorption. Wear appropriate gloves and safety glasses. Use in a chemical fume hood. Do not breathe the dust.

AgNO₃, *see* **Silver nitrate**

α-**Amanitin** is highly toxic and may be fatal by inhalation, ingestion, or skin absorption. Symptoms may be delayed for as long as 6–24 hours. Wear appropriate gloves and safety glasses. Always use in a chemical fume hood.

Aminobenzoic acid may be harmful by inhalation, ingestion, or skin absorption. Wear appropriate gloves and safety glasses.

Ammonium acetate, $H_3CCOONH_4$, may be harmful by inhalation, ingestion, or skin absorption. Wear appropriate gloves and safety glasses. Use in a chemical fume hood.

Ammonium chloride, NH_4Cl, may be harmful by inhalation, ingestion, or skin absorption. Wear appropriate gloves and safety glasses. Use in a chemical fume hood.

Ammonium formate, *see* **Formic acid**

Ammonium hydroxide, NH_4OH, is a solution of ammonia in water. It is caustic and should be handled with great care. As ammonia vapors escape from the solution, they are corrosive, toxic, and can be explosive. Use only with mechanical exhaust. Wear appropriate gloves. Use only in a chemical fume hood.

Ammonium molybdate, $(NH_4)_6Mo_7O_{24} \cdot 4H_2O$, (or its tetrahydrate) may be harmful by inhalation, ingestion, or skin absorption. Wear appropriate gloves and safety glasses. Use in a chemical fume hood.

Ammonium persulfate, $(NH_4)_2S_2O_8$, is extremely destructive to tissue of the mucous membranes and upper respiratory tract, eyes, and skin. Inhalation may be fatal. Wear appropriate gloves, safety glasses, and protective clothing. Always use in a chemical fume hood. Wash thoroughly after handling.

Ammonium sulfate, $(NH_4)_2SO_4$, may be harmful by inhalation, ingestion, or skin absorption. Wear appropriate gloves and safety glasses.

Ampicillin may be harmful by inhalation, ingestion, or skin absorption. Wear appropriate gloves and safety glasses. Use in a chemical fume hood.

Aprotinin may be harmful by inhalation, ingestion, or skin absorption. It may also cause allergic reactions. Exposure may cause gastrointestinal effects, muscle pain, blood pressure changes, or bronchospasm. Wear appropriate gloves and safety glasses. Do not breathe the dust. Use only in a chemical fume hood.

Arc lamps are potentially explosive. Follow manufacturer guidelines. When turning on arc lamps, make sure nearby computers are turned off to avoid damage from electromagnetic wave components. Computers may be restarted once the arc lamps are in operation.

Aspartic acid is a possible mutagen and poses a risk of irreversible effects. It may be harmful by inhalation, ingestion, or skin absorption. Wear appropriate gloves and safety glasses. Use in a chemical fume hood. Do not breathe the dust.

Bacterial strains (shipping of): The Department of Health, Education, and Welfare (HEW) has classified various bacteria into different categories with regard to shipping requirements (please see Sanderson and Zeigler, *Methods Enzymol. 204:* 248–264 [1991] or the instruction brochure by Alexander and Brandon (*Packaging and Shipping of Biological Materials at ATCC* [1986] available from the American Type Culture Collection [ATCC], Rockville, Maryland). Nonpathogenic strains of *Escherichia coli* (such as K-12) and *Bacillus subtilis* are in Class 1 and are considered to present no or minimal hazard under normal shipping conditions. However, *Salmonella, Haemophilus*, and certain strains of *Streptomyces* and *Pseudomonas* are in Class 2. Class 2 bacteria are "Agents of ordinary potential hazard: agents which produce disease of varying degrees of severity...but which are contained by ordinary laboratory techniques."

BCIG, *see* **5-Bromo-4-chloro-3-indolyl-β-D-galactopyranoside**

Biotin may be harmful by inhalation, ingestion, or skin absorption. Wear appropriate gloves and safety glasses. Use in a chemical fume hood.

Bisacrylamide is a potent neurotoxin and is absorbed through the skin (the effects are cumulative). Avoid breathing the dust. Wear appropriate gloves and a face mask when weighing powdered acrylamide and methylene-bisacrylamide.

Blood (human) and blood products and Epstein-Barr virus. Human blood, blood products, and tissues may contain occult infectious materials such as hepatitis B virus and HIV that may result in laboratory-acquired infections. Investigators working with EBV-transformed lymphoblast cell lines are also at risk of EBV infection. Any human blood, blood products, or tissues should be considered a biohazard and handled accordingly. Wear disposable appropriate gloves, use mechanical pipetting devices, work in a biological safety cabinet, protect against aerosol generation, and disinfect all waste materials before disposal. Autoclave contaminated plasticware before disposal; autoclave contaminated liquids or treat with bleach (10% [v/v] final concentration) for at least 30 minutes before disposal. Consult the local institutional safety officer for specific handling and disposal procedures.

Boric acid, H_3BO_3, may be harmful by inhalation, ingestion, or skin absorption. Wear appropriate gloves and goggles.

5-Bromo-4-chloro-3-indolyl-β-D-galactopyranoside (BCIG; X-gal) is toxic to the eyes and skin and may be harmful by inhalation, ingestion, or skin absorption. Wear appropriate gloves and safety goggles.

Bromophenol blue may be harmful by inhalation, ingestion, or skin absorption. Wear appropriate gloves and safety glasses. Use in a chemical fume hood.

n-**Butanol** is irritating to the mucous membranes, upper respiratory tract, skin, and especially the eyes. Avoid breathing the vapors. Wear appropriate gloves and safety glasses. Use in a chemical fume hood. *n*-Butanol is also highly flammable. Keep away from heat, sparks, and open flame.

Cacodylate contains arsenic, is highly toxic, and may be fatal if inhaled, ingested, or absorbed through the skin. It

is a possible carcinogen and may be mutagenic. Wear appropriate gloves and safety glasses. Use in a chemical fume hood.

Cacodylic acid is toxic and a possible carcinogen. It may be mutagenic and is harmful by inhalation, ingestion, or skin absorption. Wear appropriate gloves and safety glasses. Use only in a chemical fume hood. Do not breathe the dust.

Carbenicillin may cause sensitization by inhalation, ingestion, or skin absorption. Wear appropriate gloves and safety glasses.

Carbon dioxide, CO_2, in all forms may be fatal by inhalation, ingestion, or skin absorption. In high concentrations, it can paralyze the respiratory center and cause suffocation. Use only in well-ventilated areas. In the form of dry ice, contact with carbon dioxide can also cause frostbite. Do not place large quantities of dry ice in enclosed areas such as cold rooms. Wear appropriate gloves and safety goggles.

Cesium chloride, CsCl, may be harmful by inhalation, ingestion, or skin absorption. Wear appropriate gloves and safety glasses.

Cetylpyridinium bromide (CPB) causes severe irritation to the eyes, skin, and respiratory tract. Wear appropriate gloves and safety glasses. Use in a chemical fume hood.

Cetyltrimethylammonium bromide (CTAB) is toxic and an irritant and may be harmful by inhalation, ingestion, or skin absorption. Wear appropriate gloves and safety glasses. Avoid breathing the dust.

CH_3CH_2OH, *see* **Ethanol**

$C_6H_5CH_2SO_2F$, *see* **Phenylmethylsulfonyl fluoride**

$CHCl_3$, *see* **Chloroform**

$C_7H_7FO_2S$, *see* **Phenylmethylsulfonyl fluoride**

Chloramphenicol may be harmful by inhalation, ingestion, or skin absorption and is a carcinogen. Wear appropriate gloves and safety glasses. Use in a chemical fume hood.

Chloroform, $CHCl_3$, is irritating to the skin, eyes, mucous membranes, and respiratory tract. It is a carcinogen and may damage the liver and kidneys. It is also volatile. Avoid breathing the vapors. Wear appropriate gloves and safety glasses. Always use in a chemical fume hood.

Citric acid is an irritant and may be harmful by inhalation, ingestion, or skin absorption. It poses a risk of serious damage to the eyes. Wear appropriate gloves and safety goggles. Do not breathe the dust.

CO_2, *see* **Carbon dioxide**

Cobalt chloride, $CoCl_2$, may be harmful by inhalation, ingestion, or skin absorption. Wear appropriate gloves and safety glasses.

$CoCl_2$, *see* **Cobalt chloride**

Coomassie Brilliant Blue may be harmful by inhalation, ingestion, or skin absorption. Wear appropriate gloves and safety glasses.

Copper sulfate, $CuSO_4$, may be harmful by inhalation or ingestion. Wear appropriate gloves and safety glasses.

CPB, *see* **Cetylpyridinium bromide**

m-**Cresol** may be fatal if inhaled, ingested, or absorbed through the skin. It may also cause burns and is extremely destructive to the eyes, skin, mucus membranes, and upper respiratory tract. Wear appropriate gloves and safety glasses. Use in a chemical fume hood.

CsCl, *see* **Cesium chloride**

CTAB, *see* **Cetyltrimethylammonium bromide**

CuSO₄, *see* **Copper sulfate**

Cysteine is an irritant to the eyes, skin, and respiratory tract. It may be harmful by inhalation, ingestion, or skin absorption. Wear appropriate gloves and safety glasses. Do not breathe the dust.

DEAE, *see* **Diethylaminoethanol**

DEPC, *see* **Diethyl pyrocarbonate**

Dichloromethylsilane, *see* **Dichlorosilane**

Dichlorosilane is highly flammable and toxic and may be fatal if inhaled. It is harmful by inhalation, ingestion, or skin absorption. Wear appropriate gloves and safety goggles. Use only in a chemical fume hood. It reacts violently with water. Keep away from heat, sparks, and open flame. Take precautionary measures against static discharges.

Diethylamine, NH(C₂H₅)₂, is corrosive, toxic, and extremely flammable. It may be harmful by inhalation, ingestion, or skin absorption. Wear appropriate gloves and safety glasses. Use only in a chemical fume hood. Keep away from heat, sparks, and open flame.

Diethylaminoethanol (DEAE) may be harmful by inhalation, ingestion, or skin absorption. Wear appropriate gloves and safety glasses. Use in a chemical fume hood.

Diethyl ether, Et₂O or (C₂H₅)₂O, is extremely volatile and flammable. It is irritating to the eyes, mucous membranes, and skin. It is also a CNS depressant with anesthetic effects. It may be harmful by inhalation, ingestion, or skin absorption. Avoid breathing the vapors. Wear appropriate gloves and safety glasses. Always use in a chemical fume hood. Explosive peroxides can form during storage or on exposure to air or direct sunlight. Keep away from heat, sparks, and open flame.

Diethyl pyrocarbonate (DEPC) is a potent protein denaturant and a suspected carcinogen. Aim bottle away from you when opening it; internal pressure can lead to splattering. Wear appropriate gloves and lab coat. Use in a chemical fume hood.

Diethyl sulfate (DES), (C₂H₅)₂SO₄, is a mutagen and suspected carcinogen. It is also volatile. Avoid breathing the vapors. Wear appropriate gloves. Use in a chemical fume hood. Use screw-cap tubes for all DES-treated cultures and mechanical pipettors to manipulate DES solutions. Dispose of all DES-treated cultures in bleach.

N,N-Dimethylformamide (DMF), HCON(CH₃)₂, is irritating to the eyes, skin, and mucous membranes. It can exert its toxic effects through inhalation, ingestion, or skin absorption. Chronic inhalation can cause liver and kidney damage. Wear appropriate gloves and safety glasses. Use in a chemical fume hood.

Dimethylsulfate (DMS), (CH₃)₂SO₄, is extremely toxic and is a carcinogen. Avoid breathing the vapors. Wear appropriate gloves and safety glasses. Use only in a chemical fume hood. Dispose of solutions containing dimethylsulfate by pouring them slowly into a solution of sodium hydroxide or ammonium hydroxide and allowing them to sit overnight in the chemical fume hood. Contact the local safety office before reentering the lab to clean up a spill.

Dimethylsulfoxide (DMSO) may be harmful by inhalation or skin absorption. Wear appropriate gloves and safety glasses. Use in a chemical fume hood. DMSO is also

combustible. Store in a tightly closed container. Keep away from heat, sparks, and open flame.

Dinitrophenol (DNP) may be fatal by inhalation, ingestion, or skin absorption. Wear appropriate gloves and safety glasses. Use only in a chemical fume hood.

Diphenyloxazole (PPO) may be carcinogenic. It may be harmful by inhalation, ingestion, or skin absorption. Wear appropriate gloves and safety glasses. Consult the local institutional safety officer for specific handling and disposal procedures.

Disodium citrate, *see* **Citric acid**

Dithiothreitol (DTT) is a strong reducing agent that emits a foul odor. It may be harmful by inhalation, ingestion, or skin absorption. When working with the solid form or highly concentrated stocks, wear appropriate gloves and safety glasses. Use in a chemical fume hood.

DMF, *see* **N,N-Dimethylformamide**

DMS, *see* **Dimethylsulfate**

DMSO, *see* **Dimethylsulfoxide**

DNP, *see* **Dinitrophenol**

Dry ice, *see* **Carbon dioxide**

DTT, *see* **Dithiothreitol**

EDC, *see* **N-Ethyl-N′-(dimethylaminopropyl)-carbodiimide**

EMS, *see* **Ethyl methane sulfonate**

Ethanol (EtOH), CH₃CH₂OH, may be harmful by inhalation, ingestion, or skin absorption. Wear appropriate gloves and safety glasses.

Ethanolamine, HOCH₂CH₂NH₂, is toxic and harmful by inhalation, ingestion, or skin absorption. Handle with care and avoid any contact with the skin. Wear appropriate gloves and goggles. Use in a chemical fume hood. Ethanolamine is highly corrosive and reacts violently with acids.

Ether, *see* **Diethyl ether**

Ethidium bromide is a powerful mutagen and is toxic. Consult the local institutional safety officer for specific handling and disposal procedures. Avoid breathing the dust. Wear appropriate gloves when working with solutions that contain this dye.

Ethyl acetate may be fatal by ingestion and harmful by inhalation or skin absorption. Wear appropriate gloves and safety goggles. Do not breathe the dust. Use in a well-ventilated area.

N-Ethyl-N′-(dimethylaminopropyl)-carbodiimide (EDC) is irritating to the mucus membranes and upper respiratory tract. It may be harmful by inhalation, ingestion, or skin absorption. Wear appropriate gloves and safety glasses. Handle with care.

Ethyl methane sulfonate (EMS) is a volatile organic solvent that is a mutagen and carcinogen. It is harmful if inhaled, ingested, or absorbed through the skin. Discard supernatants and washes containing EMS in a beaker containing 50% sodium thiosulfate. Decontaminate all material that has come in contact with EMS by treatment in a large volume of 10% (w/v) sodium thiosulfate. Use extreme caution when handling. When using undiluted EMS, wear protective appropriate gloves and use in a chemical fume hood. Store EMS in the cold. DO NOT mouth pipette EMS. Pipettes used with undiluted EMS should not be too warm; chill them in the refrigerator before use to minimize the volatility of EMS. All glassware coming in contact with EMS should be immersed in a large beaker of 1 N NaOH or laboratory bleach before recycling or disposal.

EtOH, *see* **Ethanol**

FeCl₃, *see* **Ferric chloride**

Ferric chloride, FeCl₃, may be harmful by inhalation, ingestion, or skin absorption. Wear appropriate gloves and safety glasses. Use only in a chemical fume hood.

Formaldehyde, HCOH, is highly toxic and volatile. It is also a carcinogen. It is readily absorbed through the skin and is irritating or destructive to the skin, eyes, mucous membranes, and upper respiratory tract. Avoid breathing the vapors. Wear appropriate gloves and safety glasses. Always use in a chemical fume hood. Keep away from heat, sparks, and open flame.

Formamide is teratogenic. The vapor is irritating to the eyes, skin, mucous membranes, and upper respiratory tract. It may be harmful by inhalation, ingestion, or skin absorption. Wear appropriate gloves and safety glasses. Always use in a chemical fume hood when working with concentrated solutions of formamide. Keep working solutions covered as much as possible.

Formic acid, HCOOH, is highly toxic and extremely destructive to tissue of the mucous membranes, upper respiratory tract, eyes, and skin. It may be harmful by inhalation, ingestion, or skin absorption. Wear appropriate gloves and safety glasses (or face shield) and use in a chemical fume hood.

β-Galactosidase is an irritant and may cause allergic reactions. It may be harmful by inhalation, ingestion, or skin absorption. Wear appropriate gloves and safety glasses.

Giemsa may be fatal or cause blindness by ingestion and is toxic by inhalation and skin absorption. There is a possible risk of irreversible effects. Wear appropriate gloves and safety goggles. Use only in a chemical fume hood. Do not breathe the dust.

Glassware, pressurized, must be used with extreme caution. Handle glassware under vacuum, such as desiccators, vacuum traps, drying equipment, or a reactor for working under argon atmosphere, with appropriate caution. Always wear safety glasses.

Glass wool may be harmful by inhalation and may cause skin irritation. Wear appropriate gloves and mask.

Glutaraldehyde is toxic. It is readily absorbed through the skin and is irritating or destructive to the skin, eyes, mucous membranes, and upper respiratory tract. Wear appropriate gloves and safety glasses. Always use in a chemical fume hood.

Glycine may be harmful by inhalation, ingestion, or skin absorption. Wear gloves and safety glasses. Avoid breathing the dust.

Guanidine hydrochloride is irritating to the mucous membranes, upper respiratory tract, skin, and eyes. It may be harmful by inhalation, ingestion, or skin absorption. Wear appropriate gloves and safety glasses. Avoid breathing the dust.

Guanidine thiocyanate may be harmful by inhalation, ingestion, or skin absorption. Wear appropriate gloves and safety glasses.

Guanidinium hydrochloride, *see* **Guanidine hydrochloride**

Guanidinium isothiocyanate, *see* **Guanidine thiocyanate**

Guanidinium thiocyanate, *see* **Guanidine thiocyanate**

H₃BO₃, *see* **Boric acid**

H₃CCOONH₄, *see* **Ammonium acetate**

HCl, *see* **Hydrochloric acid**

HCOH, *see* **Formaldehyde**

H₃COH, *see* **Methanol**

HCON(CH₃)₂, *see* **Dimethylformamide**

HCOOH, *see* **Formic acid**

Heptane may be harmful by inhalation, ingestion, or skin absorption. Wear appropriate gloves and safety glasses. It is extremely flammable. Keep away from heat, sparks, and open flame.

HNO₃, *see* **Nitric acid**

H₂O₂, *see* **Hydrogen peroxide**

HOCH₂CH₂NH₂, *see* **Ethanolamine**

HOCH₂CH₂SH, *see* **β-Mercaptoethanol**

H₃PO₂, *see* **Hypophosphorous acid**

H₃PO₄, *see* **Phosphoric acid (concentrated)**

H₂S, *see* **Hydrogen sulfide**

H₂SO₄, *see* **Sulfuric acid**

Hydrazine, N₂H₄, is highly toxic and explosive in the anhydrous state. It may be harmful by inhalation, ingestion, or skin absorption. Avoid breathing the vapors. Wear appropriate gloves, goggles, and protective clothing. Use only in a chemical fume hood. Dispose of solutions containing hydrazine in accordance with MSDS recommendations. Keep away from heat, sparks, and open flame.

Hydrochloric acid, HCl, is volatile and may be fatal if inhaled, ingested, or absorbed through the skin. It is extremely destructive to mucous membranes, upper respiratory tract, eyes, and skin. Wear appropriate gloves and safety glasses. Use with great care in a chemical fume hood. Wear goggles when handling large quantities.

Hydrogen peroxide, H₂O₂, is corrosive, toxic, and extremely damaging to the skin. It may be harmful by inhalation, ingestion, and skin absorption. Wear appropriate gloves and safety glasses. Use only in a chemical fume hood.

Hydrogen sulfide, H₂S, is an extremely toxic gas that causes paralysis of the respiratory center. It is irritating and corrosive to tissues and may cause olfactory fatigue. Do not rely on odor to detect its presence. Take great care when handling it. Keep H₂S tanks in a chemical fume hood or in a room equipped with appropriate ventilation. Wear appropriate gloves and safety glasses. It is also very flammable. Keep away from heat, sparks, and open flame.

N-Hydroxysuccinimide is an irritant and may be harmful by inhalation, ingestion, or skin absorption. Wear appropriate gloves and safety glasses.

Hygromycin B is highly toxic and may be fatal if inhaled, ingested, or absorbed through the skin. Wear appropriate gloves and safety goggles. Use only in a chemical fume hood. Do not breathe the dust.

Hypophosphorous acid, H₃PO₂, is usually supplied as a 50% solution, which is corrosive and should be handled with care. It should be freshly diluted immediately before use. Wear appropriate gloves and safety glasses. Use in a chemical fume hood.

Inositol may be harmful by inhalation, ingestion, or skin absorption. Wear appropriate gloves and safety glasses.

IPTG, *see* **Isopropyl-β-D-thiogalactopyranoside**

Isoamyl alcohol may be harmful by inhalation, ingestion, or skin absorption and presents a risk of serious damage to the eyes. Wear appropriate gloves and safety goggles. Keep away from heat, sparks, and open flame.

Isobutanol, *see* **Isobutyl alcohol**

Isobutyl alcohol (Isobutanol) is extremely flammable and may be harmful by inhalation or ingestion. Wear appropriate gloves and safety glasses. Keep away from heat, sparks, and open flame.

Isopropanol is irritating and may be harmful by inhalation, ingestion, or skin absorption. Wear appropriate gloves and safety glasses. Do not breathe the vapor. Keep away from heat, sparks, and open flame.

Isopropyl-β-D-thiogalactopyranoside (IPTG) may be harmful by inhalation, ingestion, or skin absorption. Wear appropriate gloves and safety glasses.

Isotope ^{125}**I** accumulates in the thyroid and is a potential health hazard. Consult the local radiation safety office for further guidance in the appropriate use and disposal of radioactive materials. Wear appropriate gloves when handling radioactive substances. The ^{125}I$_2$ formed during oxidation of Na^{125}I is volatile. Work in an approved chemical fume hood with a charcoal filter when exposing the Na^{125}I to oxidizing reagents such as chloramine-T, IODO-GEN, or acids. Because the oxidation proceeds very rapidly and releases large amounts of volatile ^{125}I$_2$ when chloramine-T is used, it is important to be well prepared for each step of the reaction, so that the danger of contamination from volatile radiation can be minimized. Shield all forms of the isotope with lead. When handling the isotope, wear one or two pairs of appropriate gloves, depending on the amount of isotope being used and the difficulty of the manipulation required.

KCl, *see* **Potassium chloride**

K₃Fe(CN)₆, *see* **Potassium ferricyanide**

K₄Fe(CN)₆·3H₂O, *see* **Potassium ferrocyanide**

KH₂PO₄/K₂HPO₄/K₃PO₄, *see* **Potassium phosphate**

KMnO₄, *see* **Potassium permanganate**

KOH, *see* **Potassium hydroxide**

Laser radiation, both visible and invisible, can be seriously harmful to the eyes and skin and may generate airborne contaminants, depending on the class of laser used. High-power lasers produce permanent eye damage, can burn exposed skin, ignite flammable materials, and activate toxic chemicals that release hazardous by-products. Avoid eye or skin exposure to direct or scattered radiation. Do not stare at the laser and do not point the laser at someone else. Wear appropriate eye protection and use suitable shields that are designed to offer protection for the specific type of wavelength, mode of operation (continuous wave or pulsed), and power output (watts) of the laser being used. Avoid wearing jewelry or other objects that may reflect or scatter the beam. Some nonbeam hazards include electrocution, fire, and asphyxiation. Entry to the area in which the laser is being used must be controlled and posted with warning signs that indicate when the laser is in use. Always follow suggested safety guidelines

that accompany the equipment and contact your local safety office for further information.

LiCl, *see* **Lithium chloride**

Liquid nitrogen can cause severe damage due to extreme temperature. Handle frozen samples with extreme caution. Do not breathe the vapors. Seepage of liquid nitrogen into frozen vials can result in an exploding tube on its removal from liquid nitrogen. Use vials with O-rings when possible. Wear cryomitts and a face mask.

Lithium chloride, LiCl, is an irritant to the eyes, skin, mucous membranes, and upper respiratory tract. It may be harmful by inhalation, ingestion, or skin absorption. Wear appropriate gloves and safety goggles. Use in a chemical fume hood. Do not breathe the dust.

Lysozyme is caustic to mucus membranes. Wear appropriate gloves and safety glasses.

Magnesium chloride, MgCl₂, may be harmful by inhalation, ingestion, or skin absorption. Wear appropriate gloves and safety glasses. Use in a chemical fume hood.

Magnesium sulfate, MgSO₄, may be harmful by inhalation, ingestion, or skin absorption. Wear appropriate gloves and safety glasses. Use in a chemical fume hood.

Manganese chloride, MnCl₂, may be harmful by inhalation, ingestion, or skin absorption. Wear appropriate gloves and safety glasses. Use in a chemical fume hood.

MeOH or H₃COH, *see* **Methanol**

β-Mercaptoethanol (2-Mercaptoethanol), HOCH₂CH₂SH, may be fatal if inhaled or absorbed through the skin and is harmful if ingested. High concentrations are extremely destructive to the mucous membranes, upper respiratory tract, skin, and eyes. β-Mercaptoethanol has a very foul odor. Wear appropriate gloves and safety glasses. Always use in a chemical fume hood.

MES, *see* **2-(N-morpholino)ethanesulfonic acid**

Methanol, MeOH or H₃COH, is poisonous and can cause blindness. It may be harmful by inhalation, ingestion, or skin absorption. Adequate ventilation is necessary to limit exposure to vapors. Avoid inhaling these vapors. Wear appropriate gloves and goggles. Use only in a chemical fume hood.

Methotrexate (MTX) is a carcinogen and teratogen. It may be harmful by inhalation, ingestion, or skin absorption. Exposure may cause gastrointestinal effects, bone marrow suppression, and liver or kidney damage. It may also cause irritation. Avoid breathing the vapors. Wear appropriate gloves and safety glasses. Always use in a chemical fume hood.

N,N′-Methylenebisacrylamide is a poison and may affect the central nervous system. It may be harmful by inhalation, ingestion, or skin absorption. Wear appropriate gloves and safety glasses. Do not breathe the dust.

Methylene blue is irritating to the eyes and skin. It may be harmful by inhalation, ingestion, or skin absorption. Wear appropriate gloves and safety glasses.

Methylmercuric hydroxide is extremely toxic and may be harmful by inhalation, ingestion, or skin absorption. It is also volatile. Therefore, perform all manipulations of solutions containing concentrations of methylmercuric hydroxide in excess of 10^{-2} M in a chemical fume hood and wear appropriate gloves when handling such solutions. Treat all solid and liquid wastes as toxic materials and dispose of in accordance with MSDS recommendations.

MgCl₂, *see* **Magnesium chloride**

MgSO₄, *see* **Magnesium sulfate**

2-(N-morpholino)ethanesulfonic acid (MES) may be harmful by inhalation, ingestion, or skin absorption. Wear appropriate gloves and safety glasses.

3-(N-morpholino)propanesulfonic acid (MOPS) may be harmful by inhalation, ingestion, or skin absorption. It is irritating to mucous membranes and the upper respiratory tract. Wear appropriate gloves and safety glasses. Use in a chemical fume hood.

MnCl₂, *see* **Manganese chloride**

MOPS, *see* **3-(N-morpholino)propanesulfonic acid**

MTX, *see* **Methotrexate**

NaF, *see* **Sodium fluoride**

Na₂HPO₄, *see* **Sodium hydrogen phosphate**

NaN₃, *see* **Sodium azide**

NaNO₃, *see* **Sodium nitrate**

NaOH, *see* **Sodium hydroxide**

N₂H₄, *see* **Hydrazine**

NH₄Cl, *see* **Ammonium chloride**

(NH₄)₆Mo₇O₂₄·4H₂O, *see* **Ammonium molybdate**

NH₄OH, *see* **Ammonium hydroxide**

(NH₄)₂SO₄, *see* **Ammonium sulfate**

(NH₄)₂S₂O₈, *see* **Ammonium persulfate**

Nickel sulfate, **NiSO₄**, is a carcinogen and may cause heritable genetic damage. It is a skin irritant and may be harmful by inhalation, ingestion, or skin absorption. Wear appropriate gloves and safety glasses. Use in a chemical fume hood. Do not breathe the dust.

NiSO₄, *see* **Nickel sulfate**

Nitric acid, **HNO₃**, is volatile and must be handled with great care. It is toxic by inhalation, ingestion, and skin absorption. Wear appropriate gloves and safety goggles. Use in a chemical fume hood. Do not breathe the vapors. Keep away from heat, sparks, and open flame.

PEG, *see* **Polyethyleneglycol**

Perchloric acid may be fatal by inhalation, ingestion, or skin absorption. Wear appropriate gloves and safety glasses. Use only in a chemical fume hood.

Phenol is extremely toxic, highly corrosive, and can cause severe burns. It may be harmful by inhalation, ingestion, or skin absorption. Wear appropriate gloves, goggles, and protective clothing. Always use in a chemical fume hood. Rinse any areas of skin that come in contact with phenol with a large volume of water and wash with soap and water; do not use ethanol!

Phenylmethylsulfonyl fluoride (PMSF), **C₇H₇FO₂S** or **C₆H₅CH₂SO₂F**, is a highly toxic cholinesterase inhibitor. It is extremely destructive to the mucous membranes of the respiratory tract, eyes, and skin. It may be fatal by inhalation, ingestion, or skin absorption. Wear appropriate gloves and safety glasses. Always use in a chemical fume hood. In case of contact, immediately flush eyes or skin with copious amounts of water and discard contaminated clothing.

Phosphoric acid, **H₃PO₄**, is highly corrosive and may be harmful by inhalation, ingestion, or skin absorption. Wear appropriate gloves and safety glasses.

Piperidine is highly toxic and is corrosive to the eyes, skin, respiratory tract, and gastrointestinal tract. It reacts violently with acids and oxidizing agents and may be harmful by inhalation, ingestion, or skin absorption. Do not breathe the vapors. Keep away from heat, sparks, and open flame. Wear appropriate gloves and safety glasses. Use in a chemical fume hood.

PMSF, *see* **Phenylmethylsulfonyl fluoride**

Polyacrylamide is considered to be nontoxic, but it should be treated with care because it may contain small quantities of unpolymerized material (*see* **Acrylamide**).

Polyethyleneglycol (PEG) may be harmful by inhalation, ingestion, or skin absorption. Avoid inhalation of powder. Wear appropriate gloves and safety glasses.

Polyvinylpyrrolidone may be harmful by inhalation, ingestion, or skin absorption. Wear appropriate gloves and safety glasses. Use in a chemical fume hood.

Potassium cacodylate, *see* **Cacodylate**

Potassium chloride, **KCl**, may be harmful by inhalation, ingestion, or skin absorption. Wear appropriate gloves and safety glasses.

Potassium ferricyanide, **K₃Fe(CN)₆**, may be fatal by inhalation, ingestion, or skin absorption. Wear appropriate gloves and safety glasses. Always use with extreme care in a chemical fume hood. Keep away from strong acids.

Potassium ferrocyanide, **K₄Fe(CN)₆·3H₂O**, may be fatal by inhalation, ingestion, or skin absorption. Wear appropriate gloves and safety glasses. Always use with extreme care in a chemical fume hood. Keep away from strong acids.

Potassium hydroxide, **KOH** and **KOH/methanol**, can be highly toxic. It may be harmful by inhalation, ingestion, or skin absorption. Solutions are caustic and should be handled with great care. Wear appropriate gloves.

Potassium permanganate, **KMnO₄**, is an irritant and a strong oxidant. It may form explosive mixtures when mixed with organics. Use all solutions in a chemical fume hood. Do not mix with hydrochloric acid.

Potassium phosphate, **KH₂PO₄/K₂HPO₄/K₃PO₄**, may be harmful by inhalation, ingestion, or skin absorption. Wear appropriate gloves and safety glasses. Do not breathe the dust. K₂HPO₄·3H₂O is *dibasic* and KH₂PO₄ is *monobasic*.

PPO, *see* **Diphenyloxazole**

Probe DNA or **RNA**, *see* **Radioactive substances**

Proteinase K is an irritant and may be harmful by inhalation, ingestion, or skin absorption. Wear appropriate gloves and safety glasses.

Putrescine is flammable and corrosive and may be harmful by inhalation, ingestion, or skin absorption. Wear appropriate gloves and safety glasses. Keep away from heat, sparks, and open flame.

Radioactive substances. When planning an experiment that involves the use of radioactivity, consider the physicochemical properties of the isotope (half-life, emission type, and energy), the chemical form of the radioactivity, its radioactive concentration (specific activity), total amount, and its chemical concentration. Order and use only as much as needed. Always wear appropriate gloves, lab coat, and safety goggles when handling radioactive material. **X rays** and **gamma rays** are electromagnetic waves of very short wavelengths either generated by technical devices or emitted by radioactive materials. They might be emitted isotropically from the source or may be focused into a beam. Their potential dangers depend on the time period of exposure, the intensity experienced, and the wavelengths used. Be aware that appropriate shielding is usually made of lead or other similar material. The thickness of the shielding is determined by the energy(s) of the X rays or gamma rays. Consult the local safety office for further guidance in the appropriate use and disposal of radioactive materials. Always monitor thoroughly after using radioisotopes. Graphpad has a collection of free resources on its Web site, among them a calculator for radioactivity calculations. Please see http://www.graphpad.com/.

SDS, *see* **Sodium dodecyl sulfate**

Silane may be harmful by inhalation, ingestion, or skin absorption. It is extremely flammable. Keep away from heat, sparks, and open flame. The vapor is irritating to the eyes, skin, mucous membranes, and upper respiratory tract. Wear appropriate gloves and safety glasses. Always use in a chemical fume hood.

Silica is an irritant and may be harmful by inhalation, ingestion, or skin absorption. Wear appropriate gloves and safety glasses. Do not breathe the dust.

Silver nitrate, $AgNO_3$, is a strong oxidizing agent and should be handled with care. It may be harmful by inhalation, ingestion, or skin absorption. Avoid contact with skin. Wear appropriate gloves and safety glasses. It can cause explosions on contact with other materials.

Sodium acetate, *see* **Acetic acid**

Sodium azide, NaN_3, is highly poisonous. It blocks the cytochrome electron transport system and may be harmful by inhalation, ingestion, or skin absorption. Wear appropriate gloves and safety goggles and handle it with great care. Solutions containing sodium azide should be clearly marked.

Sodium cacodylate may be carcinogenic and contains arsenic. It is highly toxic and may be fatal by inhalation, ingestion, or skin absorption. It also may cause harm to an unborn child. Effects of contact or inhalation may be delayed. Do not breathe the dust. Wear appropriate gloves and safety goggles. Use only in a chemical fume hood. *See also* **Cacodylate.**

Sodium citrate, *see* **Citric acid**

Sodium dodecyl sulfate (SDS) is toxic, an irritant, and poses a risk of severe damage to the eyes. It may be harmful by inhalation, ingestion, or skin absorption. Wear appropriate gloves and safety goggles. Do not breathe the dust.

Sodium fluoride, NaF, is highly toxic and causes severe irritation. It may be fatal by inhalation, ingestion, or skin absorption. Wear appropriate gloves and safety glasses. Use only in a chemical fume hood.

Sodium hydrogen phosphate, Na_2HPO_4 (sodium phosphate, dibasic), may be harmful by inhalation, ingestion, or skin absorption. Wear appropriate gloves and safety glasses. Use in a chemical fume hood.

Sodium hydroxide, NaOH, and **solutions containing NaOH** are highly toxic and caustic and should be handled with great care. Wear appropriate gloves and a face mask. All concentrated bases should be handled in a similar manner.

Sodium nitrate, $NaNO_3$, may be harmful by inhalation, ingestion, or skin absorption. Wear appropriate gloves and safety glasses. Use in a chemical fume hood.

Sodium nitrite, $NaNO_2$, is irritating to the eyes, mucous membranes, upper repiratory tract, and skin. It may be harmful by inhalation, ingestion, or skin absorption. Wear appropriate gloves and safety glasses and always use in a chemical fume hood. Keep away from acids.

Sodium pyrophosphate is an irritant and may be harmful by inhalation, ingestion, or skin absorption. Wear appropriate gloves and safety glasses. Do not breathe the dust.

Sodium salicylate is an irritant and may be harmful by inhalation, ingestion, or skin absorption. Wear appropriate gloves and safety glasses. Do not breathe the dust.

Spermidine may be corrosive and harmful by inhalation, ingestion, or skin absorption. Wear appropriate gloves and safety glasses. Use in a chemical fume hood.

Streptomycin is toxic and a suspected carcinogen and mutagen. It may cause allergic reactions. It may be harmful by inhalation, ingestion, or skin absorption. Wear appropriate gloves and safety glasses.

Sulfuric acid, H_2SO_4, is highly toxic and extremely destructive to tissue of the mucous membranes and upper respiratory tract, eyes, and skin. It causes burns, and contact with other materials (e.g., paper) may cause fire. Wear appropriate gloves, safety glasses, and lab coat. Use in a chemical fume hood.

SYBR Green I/Gold is supplied by the manufacturer as a 10,000-fold concentrate in DMSO that transports chemicals across the skin and other tissues. Wear appropriate gloves and safety glasses and decontaminate according to Safety Office guidelines. *See* **DMSO.**

TCA, *see* **Trichloroacetic acid**

TEMED, *see* **N,N,N',N'-Tetramethylethylenediamine**

Tetracycline may be harmful by inhalation, ingestion, or skin absorption. Wear appropriate gloves and safety glasses and use in a chemical fume hood. Solutions of tetracycline are sensitive to light.

N,N,N',N'-Tetramethylethylenediamine (TEMED) is extremely destructive to tissues of the mucous membranes and upper respiratory tract, eyes, and skin. Inhalation may be fatal. Prolonged contact can cause severe irritation or burns. Wear appropriate gloves, safety glasses, and other protective clothing. Use only in a chemical fume hood. Wash thoroughly after handling. Flammable: Vapor may travel a considerable distance to source of ignition and flash back. Keep away from heat, sparks, and open flame.

TFA, *see* **Trifluoroacetic acid**

Thiourea may be carcinogenic and harmful by inhalation, ingestion, or skin absorption. Wear appropriate gloves and safety glasses. Use in a chemical fume hood.

Tissues (human), *see* **Blood (human)** and **blood products**

Trichloroacetic acid (TCA) is highly caustic. Wear appropriate gloves and safety goggles.

Trichlorotrifluoroethane may be harmful by inhalation, ingestion, or skin absorption. Wear appropriate gloves and safety goggles. Use in a chemical fume hood. Keep away from heat, sparks, and open flame.

Trifluoroacetic acid (TFA) (concentrated) may be harmful by inhalation, ingestion, or skin absorption. Concentrated acids must be handled with great care. Decomposition causes toxic fumes. Wear appropriate gloves and a face mask. Use in a chemical fume hood.

Tris may be harmful by inhalation, ingestion, or skin absorption. Wear appropriate gloves and safety glasses.

Triton X-100 causes severe eye irritation and burns. It may be harmful by inhalation, ingestion, or skin absorption. Wear appropriate gloves and safety goggles.

Trypan blue may be a carcinogen and harmful by inhalation, ingestion, or skin absorption. Do not breathe the dust. Wear appropriate gloves and safety glasses.

Tryptophan may be harmful by inhalation, ingestion, or skin absorption. Wear appropriate gloves and safety glasses.

Urea may be harmful by inhalation, ingestion, or skin absorption. Wear appropriate gloves and safety glasses.

UV light and/or **UV radiation** is dangerous and can damage the retina of the eyes. Never look at an unshielded UV-light source with naked eyes. Examples of UV-light sources that are common in the laboratory include handheld lamps and transilluminators. View only through a filter or safety glasses that absorb harmful wavelengths. UV radia-tion is also mutagenic and carcinogenic. To minimize exposure, make sure that the UV-light source is adequately shielded. Wear protective appropriate gloves when holding materials under the UV-light source.

Valine may be harmful by inhalation, ingestion, or skin absorption. Wear appropriate gloves and safety glasses.

X-gal may be toxic to the eyes and skin. Observe general cautions when handling the powder. Note that stock solutions of X-gal are prepared in DMF, an organic solvent. For details, see *N,N*-**dimethylformamide**. *See also* **5-Bromo-4-chloro-3-indolyl-β-D-galactopyranoside (BCIG).**

X rays, *see* **Radioactive substances**

Xylene is flammable and may be narcotic at high concentrations. It may be harmful by inhalation, ingestion, or skin absorption. Wear appropriate gloves and safety glasses. Use only in a chemical fume hood. Keep away from heat, sparks, and open flame.

Xylene cyanol, *see* **Xylene**

Zinc chloride, ZnCl$_2$, is corrosive and poses possible risk to an unborn child. It may be harmful by inhalation, ingestion, or skin absorption. Wear appropriate gloves and safety glasses. Do not breathe the dust.

ZnCl$_2$, *see* **Zinc chloride**

Zymolyase may be harmful by inhalation, ingestion, or skin absorption. Wear appropriate gloves and safety glasses.

Index